BEACHES OF THE WESTERN AUSTRALIAN COAST: EUCLA TO ROEBUCK BAY

A guide to their nature, characteristics, surf and safety

ANDREW D SHORT

Coastal Studies Unit
School of Geosciences
University of Sydney
Sydney NSW 2006

SYDNEY UNIVERSITY PRESS

Coastal Studies Unit and **Surf Life Saving Australia Ltd**
School of Geosciences F09
University of Sydney
Sydney NSW 2006

1 Notts Ave
Locked Bag. 2
Bondi Beach NSW 2026

Short, Andrew D
Beaches of the Western Australian Coast: Eucla to Roebuck Bay 0-9586504-3-8
A guide to their nature, characteristics, surf and safety Published 2005

Other books in this series by A D Short:

- *Beaches of the New South Wales Coast, 1993* 0 646 15055 3
- *Beaches of the Victorian Coast and Port Phillip Bay, 1996* 0 9586504 0 3
- *Beaches of the Queensland Coast: Cooktown to Coolangatta, 2000* 0 9586504 1 1
- *Beaches of the South Australian Coast and Kangaroo Island, 2001* 0 9586504 2-X

Forthcoming books:

Beaches of the Tasmania Coast and Islands (publication. 2006) 1-920898 12-3
Beaches of the Northern Australian Coast: The Kimberley, Northern Territory and Cape York 1-920898-16-6
Beaches of the New South Wales Coast (2nd edition) 1-920898-15-8

Published by:
Sydney University Press
University of Sydney
www.sydney.edu.au /sup

Copies of all books in this series may be purchased online from Sydney University Press at:

http://purl.library.usyd.edu.au/sup/marine

Western Australian beach database:
Inquiries about the Western Australian beach database should be directed to Surf Life Saving Australia at sls.com.au

Cover photographs: Wharton Beach, Mullaloo Beach
Cape Peron North, Point Maud-Coral Bay (A D Short)
Cover design: Jacqui Owen

Table of Contents

To Ben, Pip and Bonnie
who have also wandered these beaches

Preface

This book had it origins in 1990 when the then New South Wales Beach Safety Program was expanded into the Australia Beach Safety and Management Program, thereby encompassing the entire country including Western Australia.

This book is part of a series of books produced by the Australian Beach Safety and Management Program (ABSMP). One of the aims of the program is to develop a better understanding of the location, type, characteristics, nature, hazards and public risk along all of Australia's 10 685 mainland beaches. To this end the program has completed a database on the beaches of each state and the Northern Territory. It is also publishing books summarising the main characteristics of each beach on a state by state basis. This book is the fifth in the series with *Beaches of the New South Wales Coast* published in 1993, and *Beaches of the Victorian Coast and Port Phillip Bay* in 1996, the *Beaches of the Queensland Coast: Cooktown to Coolangatta* in 2000, the *Beaches of the South Australian Coast and Kangaroo Island* in 2001, while the *Beaches of Tasmania* and the *Beaches of Northern Australia: The Kimberley, North Territory and Cape York*, the sixth and seventh in the series are well underway.

The ABSMP was initiated in New South Wales in 1986 as the New South Wales Beach Safety Program in collaboration with Andrew May of Surf Life Saving New South Wales (SLSNSW), with Chris Hogan as Research Officer. In 1990 with support from the Australian Research Council and in cooperation with Surf Life Saving Australia Ltd, the project was expanded to encompass all states and the Northern Territory.

The Coastal Studies Unit commenced scientific investigations in Western Australian in 1980 with a major field program at Cable Beach (Broome) studying the impact of the large tides on the beach morphodynamics.

Studies of the Western Australian coast specifically for the present project commenced in March 1991 with a three month field trip from Eucla right around the coast to Cape Leveque. With subsequent trips in 1996, an aerial photography flight of the entire WA coast in 1997, and a vessel-based field trip around the Kimberley coast in 2001.

In compiling a book of this magnitude there will be errors and omissions, particularly with regard to the names of beaches, many of which have no official name, and many local factors. If you notice any errors or wish to comments on any aspects of the book please communicate them to the author at the Coastal Studies Unit, University of Sydney, Sydney NSW 2006, phone (02) 9351 3625, fax (02) 9351 3644, email: A.Short@geosci.usyd.edu.au or via Surf Life Saving Western Australia (08) 9244 1222 or Surf Life Saving Australia (02) 9597 5588. In this way we an update the beach database and ensure that future editions are more up to date and correct.

Andrew D Short
Narrabeen Beach, July 2005

Acknowledgements

This book is the result of numerous field trips to Western Australia by land, air and sea, and while written in Sydney many people from both states have contributed to the final publication.

The author's first venture to Western Australia was in an old FJ across the old Nullabor (Eyre) highway in 1968, when the main aim was to surf the Margaret River region. In 1980 I returned driving across the top as part of a Coastal Studies Unit field investigation into the impact of the high tide ranges on beach systems, which was conducted at Cable Beach in Broome using the old Bali Hai caravan park as our base. Three major field trips were undertaken specifically to investigate the beach systems for this project. The first was April-July 1991 when with my family and a caravan we commenced at Eucla and followed the coast to Cape Leveque. Next was in October-November 1996 when I decided to fill in some more difficult to access locations between Eucla and Israelite Bay, and then carried on to Steep Point. Again I was accompanied by my family and met up with Steve Kennedy, from Busselton, at Eucla. Steve knows the Bight coast and was a great guide and companion on this leg. In October 1997 I flew the entire WA coast in an Australian Aerial Patrol Par Navier piloted by Dean Franklin and Harry Mitchell. This was part of a memorable flight that also covered western South Australia and all the Northern Territory coast and resulted in digital images of all the Western Australian beaches. Finally in June-July 2001 with Graham Lloyd we inspected the entire Kimberley coast between Wyndham and Broome in the University vessel *CSU 3*. This trip was generously assisted by Paspaley Pearls and Broome Pearls, both of whom supplied fuel. The Kimberley beaches will however be reported on in a later book on the Northern Australia beaches. Following this trip I drove from Broome down to Margaret River filling in some gaps left from the previous trips.

At Surf Life Saving Australia (SLSA) where the project is based it received full support from the Greg Nance (CEO), while at Surf Life Saving Western Australia (SLSWA), first Cameron O'Beirne and then Grant Trew rendered valuable assistance. Also at SLSA the Project Research Officer Katherine McLeod has been an essential component of ABSMP since 1996 and manages the ABSMP database.

The project received an Australian Research Grant from 1990-1992, which enabled it to expand nationally, with additional project funding from a joint Australian Research Council Collaborative Research Grant (1996-1998) and a Strategic Partnership with Industry- Research and Training (SPIRT) Grant (1999-2002).

Cape Du Couedic and Cape Sorrell waverider data was kindly provided by the Bureau of Meteorology through Ray Rice at Lawson and Treloar Pty Ltd. Professor JL (Jack) Davies kindly provided his vertical aerial photograph collection of the entire Western Australian coast, which proved an extremely valuable resource, both in the office and in the field. A special thanks also to Peter Johnson who drafted all the beach maps and figures. All photographs are by the author.

At the University of Sydney thanks to Glen Harris who analysed most of the beach sediment samples; at University Printing Service to Jacqui Owen for the cover design and Josh Fry his assistance in the production, and Ross Coleman at Sydney University Press for assistance in publication and marketing.

Finally, as the entire beach database was complied and the book was written at my home office, I thank my wife Julia, and children Ben, Pip and Bonnie for putting up with its intrusion into our home life, as well as accompanying me to most parts of the Western Australian coast.

Abstract

This book is about the entire Western Australian coast between Eucla and Roebuck Bay and includes Rottnest Island. It begins with three chapters that provide a background to the physical nature and evolution of the Western Australian coast and its 2051 mainland beach systems. Chapter 1 covers the geological evolution of the coast and the role climate, wave, tides and wind in shaping the present coast and beaches. Chapter 2 presents in more detail the sixteen types of beach systems that occur along the Western Australian coast, while chapter 3 discusses they types of beach hazards along the coast and the role of Surf Lifesaving Western Australia in mitigating these hazards. Finally the long chapter 4 presents a description of every one of the 2051 mainland beaches (Eucla-Roebuck Bay), as well as 63 beaches on Rottnest Island. The description of each beach covers its name, location, physical characteristics, access and facilities, with specific comments on its surf zone character and physical hazards, as well as its suitability for swimming, surfing and fishing. Based on the physical characteristics each beach is rated in terms of the level of beach hazards from the least hazardous rated 1 (safest) to the most hazardous 10 (least safe). The book contains 512 figures which include 409 photographs, which illustrate all beach types, as well as beach maps and photographs of all beaches patrolled by surf lifesavers and many other popular beaches.

Keywords: beaches, surf zone, rip currents, beach hazards, beach safety, Western Australia

Australian Beach Safety and Management Program (ABSMP)

Awards

NSW Department of Sport, Recreation and Racing
Water Safety Award – Research 1989
Water Safety Award – Research 1991

Surf Life Saving Australia
Innovation Award 1993

International Life Saving
Commemorative Medal 1994

New Zealand Coastal Survey
In 1997 Surf Life Saving New Zealand adopted and modified the ABSMP in order to compile a similar database on New Zealand beaches.

Great Britain Beach Hazard Assessment
In 2002 the Royal National Lifeboat Institute adopted and modified the ABSMP in order to compile a similar database on the beaches of Great Britain.

Hawaiian Ocean Safety
In 2003 the Hawaiian Lifeguard Association adopted ABSMP as the basis for their Ocean Safety survey and hazard assessment of all Hawaiian beaches.

Handbook on Drowning 2004
This handbook was product of the World Congress on Drowning held in Amsterdam in 2002. The Congress and the Handbook endorse as the international standard the ABSMP approach to assessing beach hazards.

AUSTRALIAN BEACH BOOKS

Published by the Sydney University Press for the
Australian Beach Safety and Management Program
a joint project of
Coastal Studies Unit, University of Sydney and Surf Life Saving Australia

by

Andrew D Short
Coastal Studies Unit, University of Sydney

BEACHES OF THE NEW SOUTH WALES COAST
Publication: 1993 **ISBN:** 0 646 15055 3
358 pages, 167 original figures, including 18 photographs; glossary, general index, beach index, surf index.

BEACHES OF THE VICTORIAN COAST & PORT PHILLIP BAY
Publication: 1996 **ISBN:** 0 9586504 0 3
298 pages, 132 original figures, including 41 photographs; glossary, general index, beach index, surf index.

BEACHES OF THE QUEENSLAND COAST: COOKTOWN TO COOLANGATTA
Publication: 2000 **ISBN** 0 9586504 1 1
369 pages, 174 original figures, including 137 photographs, glossary, general index, beach index, surf index.

BEACHES OF THE SOUTH AUSTRALIAN COAST & KANGAROO ISLAND
Publication: 2001 **ISBN** 0 9586504 2-X
346 pages, 286 original figures, including 238 photographs, glossary, general index, beach index, surf index.

BEACHES OF THE WESTERN AUSTRALIAN COAST : EUCLA TO ROEBUCK BAY
Publication: 2005 **ISBN** 0-9586504-3-8
433 pages, 517 original figures, including 409 photographs, glossary, general index, beach index, surf index.

Order online from **Sydney University Press** at

http://www.sup.usyd.edu.au/marine

forthcoming titles:

BEACHES OF THE TASMANIAN COAST AND ISLANDS (publication 2006)

BEACHES OF THE NORTHERN AUSTRALIA: THE KIMBERLEY, NORTHERN TERRITORY & CAPE YORK (publication 2007)

BEACHES OF THE NEW SOUTH WALES COAST (revised, expanded and updated, publication 2007)

1 The Western Australian Coast

Western Australia is Australia's largest state and has Australia's longest coastline and greatest number of beaches. The state and its coastline are expansive in all respects, as is the range of climates, ocean and coastal processes active around the coast. The result is both a large number of mainland beaches (3426) and a wide range of types of beach systems, spread along the 10 194 km of coast. This book is about all 2051 beaches in Western Australia between Eucla and Roebuck Bay, together with another 63 on Rottnest Island. It leaves the coverage of the 1375 Kimberley coast beaches to a later book covering the beaches of northern Australia.

The beaches in this book lie between 18°S and 35°S, a latitudinal distance of 2000 km. They extend from the monsoonal tropics of the northwest, exposed to periodic tropical cyclones, along the western desert coast, to the cooler more humid south coast exposed to strong westerly winds and high waves. The beaches occupy 63% (3687 km) of the coast and average 1.8 km in length. However they vary considerably in length with a standard deviation of 4.5 km, while the longest continuous beach, down in the Bight, extends for 103 km.

The beaches are composed of a mixture of quartz sands derived from the hinterland, and carbonate sands swept in from the sea, with an average composition of 53% carbonate material. The proportion of quartz and carbonate varies regionally around the coast, with generally more quartz sand along the more humid southwest coast, and carbonate-rich beaches along much of the west and northwest coast with 20% of the beaches having more than 90% carbonate material. In other words more than half the sand on the Western Australian coast originates from marine organisms supplying their shells and debris to the shoreline to build the beaches and backing dune systems.

While this book is about the beaches, the coastal dune systems are entirely dependent on the beach for their sand and the two are intimately linked. Just as the state has extensive beach systems it also has extensive, and in places massive, coastal sand dunes including Australia's longest active dune (38 km) and highest active dune (125 m). In total coastal dunes back 57% of the coast and cover an area of 7731 km^2, with a total sand volume of 23 km^3.

This book divides the coast into eight regions, which are illustrated in Figure 1.1 and listed in Table 1.1, together with some of their shoreline attributes. Regions 1 to 7 and Rottnest Island are covered in this book.

This book is a product of the *Australian Beach Safety and Management Program*, a co-operative project of the University of Sydney's Coastal Studies Unit, Surf Life Saving Australia (SLSA) and Surf Life Saving Western Australia (SLSWA). It is part of the most comprehensive study ever undertaken of beaches on any part of the

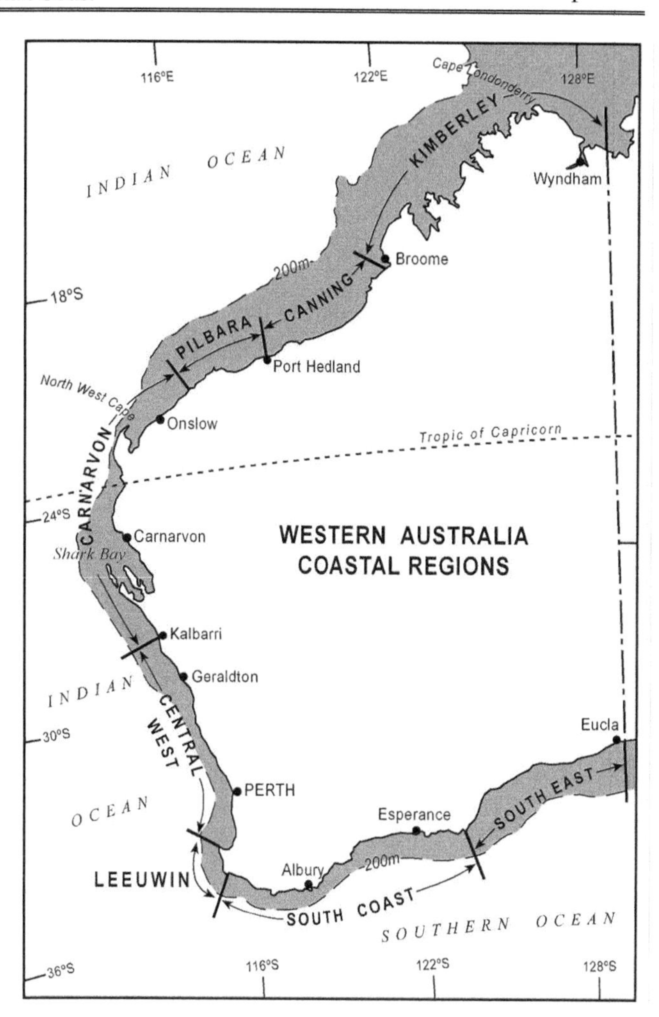

Figure 1.1 Western Australian coastal regions and continental shelf (shaded). This book covers the Southeast to Canning regions.

world's coast. In Western Australia it has investigated every beach on the mainland coast and Rottnest Island. The project has already published similar books on the beaches of the New South Wales (1993), Victorian (1996), Queensland (2000) and South Australian (2001) coasts, with books on the Tasmanian and Northern Australian coasts to follow. The program also maintains a database on every Australian mainland beach, together with those on 30 major inhabited islands. For information about this database contact SLSA.

Chapter 1 of this book begins by examining the nature of the Western Australian coast, including its geological evolution, climate and ocean processes. Chapter 2 details the types of beaches that occur around the coast, while Chapter 3 looks at beach usage, physical hazards and safety issues. The bulk of the book (Chapter 4) is devoted to a description of every beach on the coast (Eucla to Roebuck Bay) including Rottnest Island. Information is provided on each beach: its number, name, location, access, facilities, physical characteristics and wave conditions. Specific comments are made regarding each beach's suitability for swimming and surfing, together with a physical beach hazard rating from 1 (least hazardous) to 10 (most hazardous).

Table 1.1 Western Australia beach statistics

	WA coastal regions	No. of beaches	Beach length (km)	Rocky/ mangrove length (km)	Total (km)	%% Carbonate beach/dune sand
1	Southeast	52	469	168	637	45
2	South	566	773	379	1182	22
3	Leeuwin	112	68	43	111	83
4	Central West	403	747	58	805	67
5	Carnarvon	574	1035	1000	2035	60
6	Pilbara	188	132	333	465	60
7	Canning	156	463	156	619	57
	Subtotal[1]	**2051**	**3687**	**2137**	**5854**	**51**
8	Kimberley[2]	1375	713	3627	4340	54
	WA total	**3426**	**4398**	**5796**	**10 194**	**52**
	Rottnest Is[1]	63	19	44	36	87

[1] This book
[2] In Short (in prep) 'Beaches of the Northern Australian Coast'

Beach Systems

A *beach* is a wave-deposited accumulation of sediment, usually sand, but occasionally cobbles and boulders. It extends from the upper limit of wave swash, approximately 3 m above sea level, out across the surf zone or sand flats and seaward to the depth to which average waves can move sediment shoreward. Western Australia has some of the world's highest and lowest energy beaches. On the exposed, high wave energy sections of the south Western Australian coast (Fig. 1.2a), the beaches can extend to depths between 30 and 50 m and as much as several kilometres offshore. Along parts of the Central West coast, in Shark Bay and along the Pilbara and Canning coasts, wave height and energy decrease substantially and the beaches range from the more exposed, with wind waves averaging about 1 m (Fig. 1.2b), to essentially waveless, sandy tidal flats (Fig. 1.2c). The low energy beaches and sand flats usually terminate at low tide. As waves decrease, tides and tide range become increasingly important, particularly north of Exmouth where the tide gradually increases to reach 9 m at Broome.

Most Western Australians live within an hour of a beach and even those who live inland often travel to the coast for their holidays. Many are frequent beachgoers and have their favourite beach. The beaches most of us have been to, or go to, are usually close by our home or holiday area. They are usually at the end of a sealed road, with a car park and other facilities. Often, they are patrolled by lifesavers and/or professional lifeguards. These popular, developed beaches, however, represent only a minority of the state's beaches. In Western Australia there are 27 surf lifesaving clubs (SLSC), including nine beaches also patrolled by professional lifeguards (Table 1.2). Most of these are located in the Central West region, with 14 of the clubs in the Perth region. The patrolled beaches, however, represent only 1% of all Western Australian beaches. The vast majority are unpatrolled and many are extremely hazardous. Furthermore only 301 (9%) of the beaches have sealed road access, while 310 (9%) lie at the end of a gravel road. Off-road (4WD) vehicles are required to reach 947 beaches (28%), while the majority (1868, 54%) have no vehicle access and are only accessible on foot or by boat. Finally 50 beaches (1%) located at the base of steep cliffs are only accessible by boat.

To most of us the beach is that part of the dry sand we sit on or cross to reach the shoreline and the adjacent surf zone. It is an area that has a wide variety of uses and users (Table 1.3). This book will focus on the dry or subaerial beach plus the surf zone or area of wave breaking, typical of the beaches illustrated in Figure 1.2.

Evolution and Geology of the Coast

Western Australia contains the ancient core or cratons of the Australian continent. The large Yilgarn Block (657 000 km^3) in the south and the Pilbara Block (183 000 km^3) in the central west are both approximately 2500 million years in age and have been relatively stable since their formation. They form the core against, and to the east of which, the rest of the Australian continent has accreted. In Western Australia between the ancient cratons is a series of sedimentary basins, dating back up to 2000 million years, starting in the south with the Eucla and Bremer basins, then along the west coast the elongate Perth Basin followed by the southern and northern Carnarvon and Canning basins, and finally the northern Kimberley and smaller Bonaparte basins (Table 1.4). The cratons and surrounding basins, as well as Capricorn and King Leopold orogens, today comprise Western Australia's geological framework (Fig. 1.3).

a.	b.	c.

*Figure 1.2	Examples of the range of Western Australian beach systems. a) High energy Owingup on the South Coast; b) moderate energy Dide Bay on the Central West coast; and c) low energy Eagle Bluff in Shark Bay.*

*Table 1.2	Western Australian beaches - lifesaving facilities and access type*

	Eucla-Roebuck Bay	%	Kimberley	%	WA	WA %
Surf lifesaving clubs	25	1.2	1	-	26	0.7
Lifeguards (SLSWA)	9	0.5	0	-	9	-
Sealed road access	296	14	5	0.4	301	9
Gravel road access	283	14	27	2	310	9
Dirt road access (4WD)	857	42	90	6.5	947	28
Foot access only	570	28	1248	91	1818	53
No access	45	2	5	0.4	50	1
Total beaches	2051	100	1375	100	3426	100

Table 1.3 Types of Western Australian beach users and their activities

Type	User	Location
Passive	sightseer, tourist	road, car park, lookout
Passive-active	sunbakers, picnickers, beach sports	dry beach
Active	beachcombers, joggers	dry beach, swash zone
Active	fishers, swimmers	swash, inner surf zone
Active	surfers, water sports	breakers & surf zone
Active	skis, kayaks, windsurfers	breakers & beyond
Active	IRB, fishing boats	beyond breakers

*Table 1.4	Major geological provinces (**cratons** and basins) of the Western Australian coast*

	Province	Age Ma*	Coast location	Geology	Coastal morphology
1	Eucla Basin	400-0	Eucla-Israelite Bay	limestone	cliffs & coastal plains
2	**Yilgarn Craton**	3800-2500	Pt Malcolm-Cape Arid; Pt Hood-Pt D'Entrecasteaux	granites & gneiss	prominent headlands & beaches, estuaries
3	Bremer Basin	150-0	IsraeliteBay-Pt D'Entrecasteaux	siltstone, sandstone	headlands & beaches
4	Perth Basin	450-0	Augusta-Murchison R	sandstones	low coastal plain, long sandy beaches
5	Carnarvon Basin	450-0	Murchison R-Cape Preston	sandstone, limestone	beaches, calcarenite & sandstone cliffs
6	**Pilbara Craton**	3800-2500	Cape Preston-Port Hedland	granite-greenstone	bedrock, deltas, coastal plain
7	Canning Basin	500-0	Port Hedland-King Sound	limestone	low coastal pindan plain
8	Kimberley Basin	2000-400	Kimberley coast	sandstone, shales, basalts	bedrock control, deeply weathered
9	Bonaparte Basin	500-0	Cambridge Gulf	sedimentary	tidal flats & bedrock

* Ma = million years

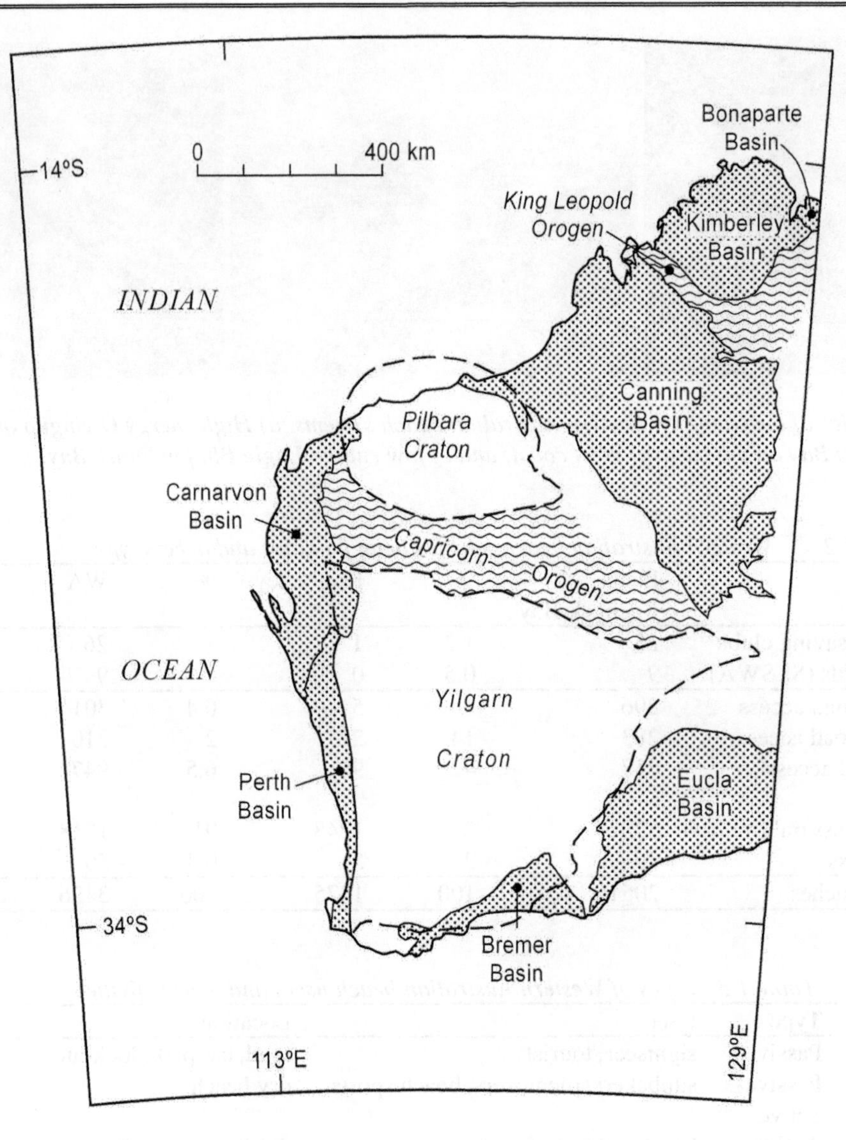

Figure 1.3 Major geological provinces of Western Australia

The cratons, basins and orogens impinge on the coast to varying degrees and in doing so influence the nature of the coastal geology, topography and Holocene geomorphology and the beach systems. In Western Australia the coast has gone through six major phases in its evolution, starting with the initiation of the ancient cratons 3800 Ma and continuing until the present.

i. Yilgarn and Pilbara Cratons (3800-2500 Ma)

The two ancient *Yilgarn* and *Pilbara* cratons contain some of the oldest remnants of the earth's original crust. They represent the accumulation of deep-seeded volcanic greenstone terrains, including granite-gneiss complexes, which began accumulating in association with the earth's then thin crust as early as 3800 Ma. The Pilbara was possibly cratonised by 3000 Ma and the Yilgarn by 2500 Ma.

ii. Mobile Cratons and Orogens (2500-900 Ma)

Following the formation of the Yilgarn and Pilbara cratons they underwent a long period of mobility resulting in the building of mountain chains between and adjacent to the cratons, a process known as orogens (or mountain building). The latter led to the formation of new cratons to the east including the *Northern Australian Craton* (1850-1600 Ma) and the *Gawler Craton* in South Australia (1450 Ma), thereby increasing the size of the core of the Australian continent. In Western Australia the Capricorn Orogen formed between the Pilbara and Yilgarn cratons between about 2000 and 1600 Ma, the Albany-Fraser Orogen bordered the south and eastern side of the Yilgarn craton between 2000 and 1800 Ma, and the King Leopold Orogen surrounded the Kimberley basin between 1900 and 400 Ma.

Today the great *Yilgarn Craton*, which dominates the southwest of the state, only outcrops at the coast as prominent granite and gneiss headlands between Point Malcolm, Cape Pasley and Cape Arid; further west between Hood Point and Point D'Entrecasteaux; and in the southwest corner between Cape Leeuwin and Cape Naturaliste. The *Pilbara Craton* provides granite and greenstones outcrops between Cape Preston and the more prominent Dampier Peninsula.

iii. Kimberley Basin (2000-400 Ma)

The *Kimberley Basin* is the oldest of Western Australia's 18 sedimentary basins with rocks dating from 2000-400 Ma. The core of the Kimberley, covering the largest area, is a thick central basin sequence of sandstone, shale and basalt that were formed between 1800-1650 Ma and have since been subject to only minor faulting and warping. This now uplifted central region forms a plateau, which has been heavily weathered and dissected. It is surrounded in the east and south by the King Leopold and Halls Creek Orogens, both remnants of mountain belts, which ceased tectonic activity by 1800 Ma. The mountains have since been eroded and the area was also glaciated between 700 and 600 Ma. By 400 Ma the Kimberley region had moved to the tropics and was covered by shallow seas, leaving coral reefs, which now form narrow limestone ranges, including Geikie Gorge. A more southern location and glacial activity prevailed by 250 Ma. With the breakup of Gondwanaland and northern movement of the continent the Kimberley has been in tropical regions for the past 100 Ma, with more humid conditions until about 30 Ma.

iv. Australian Precambrian Craton (900 Ma)

By 900 Ma all the Precambrian cratons finally welded together to form the *Australian Precambrian Craton*, which covers two-thirds of the present continent, extending from the Kimberley in the northwest, around to the Gawler Craton in the southeast. The eastern side of the continent did not yet exist. At this time the craton was part of the Gondwana supercontinent, attached to Antarctica together with India, Africa and South America.

v. Western Basins (500-0 Ma)

Three sedimentary basins now form much of the west coast, with a small basin on the northern border. The *Perth Basin* is a deep linear trough extending for 1,000 km from the South Coast to the Murchison River and averaging about 120 km in width. It is a polycyclic basin containing nearly 15 000 m of sediment laid down in shallow marginal seas and deposited between the Silurian and the Quaternary (450-0 Ma). The sediments include sandstones, shales and coal measures and are strongly faulted. Pleistocene coastal limestone dominates the shoreline exposures.

The *Carnarvon Basin* is an elongate basin extending for about 1000 km between the Murchison and the Dampier Archipelago of the Pilbara craton, essentially from Kalbarri to Karratha. It contains up to 7000 m of mainly Palaeozoic sediments deposited from 400-0 Ma. The sediments are a mix of fluvial, coastal and shallow marine deposits and include the red sandstones that dominate the Kalbarri coast. Elsewhere, Tertiary limestone dominates the coastal deposits.

The *Canning Basin* is the largest in Western Australia, occupying the coast between Port Hedland and King Sound, with much of the basin extending out onto the wide northwest continental shelf and covering an area of 415 000 km^2. It contains up to 10 000 m of shallow marine sediments ranging in age from the Ordovician to the Holocene (500-0 Ma). Today its onshore extension is largely blanketed by the longitudinal dunes of the Great Sandy Desert, while the low-lying Eighty Mile Beach and Roebuck Bay dominate much of the shoreline.
The *Bonaparte Basin* is a synclinal basin located east of the Kimberley and extending into the Northern Territory. At the coast it occupies the eastern side of Cambridge Gulf to the border and beyond. It contains a wide range of sediments up to 6000 m thick dating from the Cambrian to Quaternary (500-0 ma).

vi. Southern Basins

Two younger basins dominate much of the south coast of the state. The *Eucla Basin* extends from Fowlers Bay in South Australia west to Israelite Bay. The sediments range in age from Early Cretaceous to Holocene (120-0 Ma) and do not exceed 750 m in thickness. The lower sequences are sandstone and siltstone, while the upper carbonate sequences occur in two phases: the lower Wilson Bluff limestone deposited in shallow seas during the late Eocene (35 Ma) and the capping Toolina and Nullarbor Limestone deposited during the Early Miocene (20 Ma). Subsequent uplift of the Eucla Basin formed the flat Nullarbor Plain. Today the southern boundary of the plain is exposed as 60-90 m high cliffs and bluffs capped by Nullarbor Limestone. These form vertical cliffs for 210 km between Head of Bight and the SA-WA border at Wilson Bluff. In Western Australia the cliffs continue, at first as the Hampton Bluffs, which are fronted by the Roe Plain. They re-emerge at the coast forming the 180 km long Baxter Cliffs capped by Toolina Limestone. These then run inland as the Wyllie escarpment south of Point Culver, finally terminating inland from Point Malcolm, 550 km west of the border. The cliffs and bluffs have a total length (Head of Bight-Point Malcolm) of 760 km, the longest continuous escarpment in Australia.

The *Bremer Basin* extends along the south coast between Israelite Bay and Point D'Entrecasteaux, covering only 20 000 km^2 onshore, with a maximum onshore thickness of about 200 m. Sediments are Eocene in age (40 Ma) and were deposited in several broad topographic depressions ranging from valleys to shallow seaways. The sediments consist of a lower siltstone, sandstone and claystone sequence, overlain by an upper siltstone and sponglite sequence.

Formation of the Western Australia Coast

While the rocks that make up the core of Western Australia date back thousands of millions of years and most of the surrounding basins hundreds of millions of years, the coast as we know it today only began taking shape in the Jurassic (~160 Ma) when the great supercontinent Gondwanaland began to break up. The break-up began along the northwest margin, separating East Gondwanaland (Australia, Antarctica, India, Madagascar and Southern Africa) from West Gondwanaland (Africa, Arabia and South America). The eastward continuation of this spreading led to the separation of Argo Land from the North West Shelf defining the first part of the Western Australia margin some 160 Ma (Fig. 1.4).

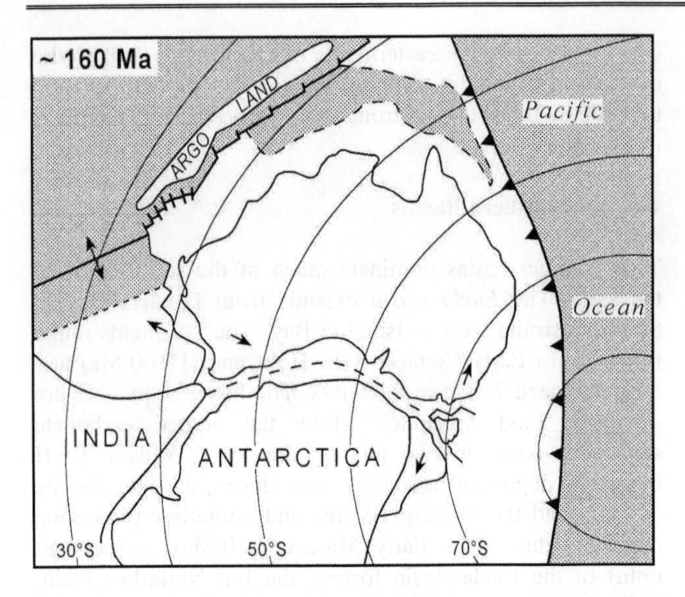

Figure 1.4 Australia during the Jurassic (~160 Ma) when the break-up of Gondwanaland and Australia's separation from Antarctica began in the northwest. Shaded areas represent seas.

The rifting initiated in the northwest shelf region then proceeded anticlockwise southward during the Cretaceous (140-65 Ma), causing rifting between India and Australia, and later eastward between Australia and Antarctica (120 Ma) (Fig. 1.5). By the Late Cretaceous, seafloor spreading took place in the east between the Lord Howe Rise and the mainland (75-55 Ma).

During the Late Cretaceous, the rifting across the southern Australian coast produced buckling of the plate and a series of basins (Eucla, Otway and Gippsland), which began filling with shallow marine sediments. The Otway and Gippsland basins are now submerged, while the Eucla Basin was uplifted during the late Tertiary.

Quaternary Climate and Sea Level

The Quaternary covers the past 2 Ma of earth's history and is divided into the Pleistocene (2 Ma-10 ka) and Holocene (10 ka to present). This period has been characterised by cyclical fluctuations in the earth's surface temperature, with cooler periods leading to the growth of massive ice sheets in North America and Europe, followed by their melting during warmer periods. These fluctuations are also accompanied by a fall in sea level as the ice builds up, and a rise as it melts, with maximum changes in sea level up to 120 m. During the Quaternary there have been as many as 20 major shifts in sea level, when sea level has fallen up to 120 m then risen to approximately its present level, and many more secondary shifts, when it has oscillated between these extremes.

In Australia the cooler periods result in the overall climate of the continent becoming cooler, drier and windier, with a minor accumulation of permanent ice on the slopes of Mount Kosciuszko. However during cooler periods as sea level falls the continental shelf is exposed, greatly increasing the size of the Australian continent, and changing the

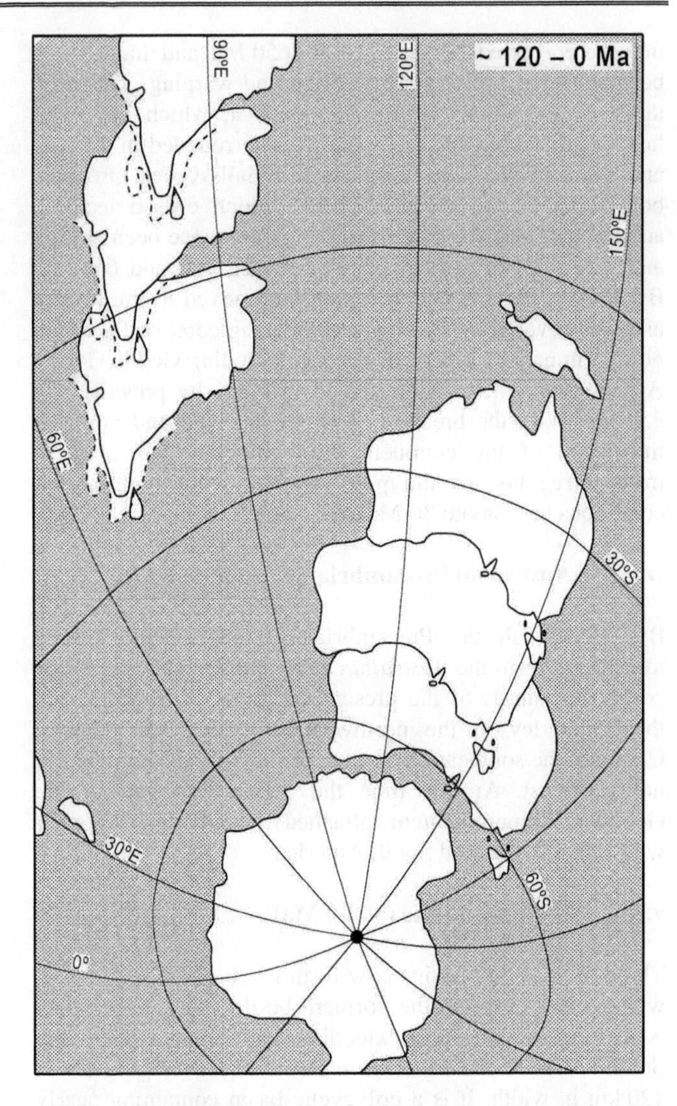

Figure 1.5 During the past 120 Ma the Australian plate, containing Australia, New Guinea and New Zealand, has been moving northward at a rate of several centimetres per year. This figure shows Australia and India as they have both moved northward away from the original core of Antarctica.

location of the shoreline. During warmer periods, as at present, the climate is warmer, more humid and less windy, and sea level rises to about its present position, flooding the continental shelf and shrinking the size of the continent. Figure 1.6 illustrates the fluctuating level of the sea in the Australian region during the late Quaternary, and the rapid rise in sea level to its present level during the Holocene.

Quaternary climate and sea level changes have had four major impacts around the entire Western Australian coast. *First*, they have periodically exposed and flooded the continental shelf, at the same time shifting the shoreline tens to hundreds of kilometres backwards and forwards across the shelf. *Second*, each time sea level has risen, accompanied by waves and winds, it has, particularly in the south, driven huge volume of sediment eroded from the shelf towards and onto the shore to form beaches and massive coast dunes. *Third*, during periods of low sea level, these deposits are stranded well above sea level, during which time most have been exposed to soil forming processes, which have led to the formation of dune calcarenite, a partially cemented dune soil. *Fourth*, when sea

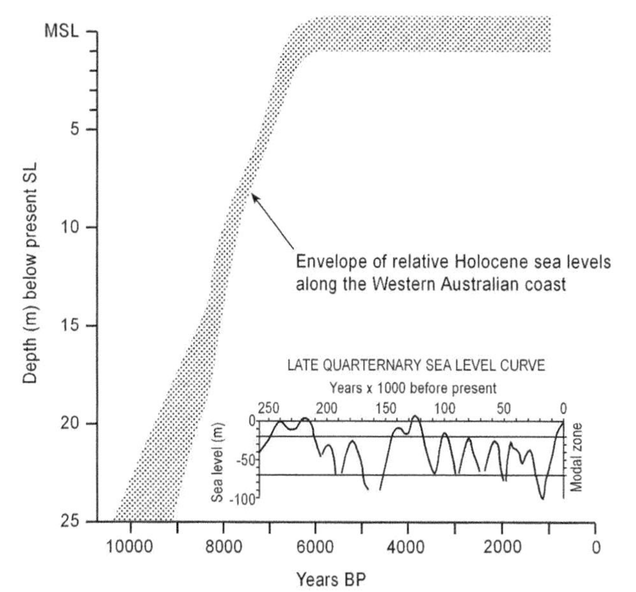

Figure 1.6 Two plots showing the Holocene rise and still-stand in sea level (upper) and the larger scale late Quaternary sea level fluctuations (lower). Along many parts of the Western Australian coast the Holocene sea level rose to 1-2 m above the present level about 6 ka, before falling to its present location.

level has returned each time to near its present level it has in some locations deposited multiple sequences of beaches and dunes, either in front of the older beaches or on top of these deposits, leading in places to the formation of 100-200 m thick dune sequences.

As a consequence of the above, much of the modern (Holocene) coast of Western Australia lies either over or in front of older Quaternary deposits. The best example of overlapping deposits is the Zuytdorp cliffs, which are formed from successive layers of Pleistocene calcarenite capped with Holocene dune sands. In most locations, and particularly along the Perth Basin, the modern beach lies in front of the older Pleistocene deposits, often separated by a lagoon or tidal flats.

Coastal Regions

There are seven mainland coastal regions covered in this book (Fig 1.1). The regions have been chosen to delineate major changes in the geology, nature and in some cases orientation of the coast. In addition there are accompanying changes in wave and tide conditions between some regions and the resulting types of beaches. The geology of each is briefly described below.

1. The *South East* region extends for 640 km between Eucla and Cape Pasley and includes all of the Eucla Basin and its associated limestone cliffs and bluffs, the latter fronted by the Roe Plain in the east and the Bilbunya barrier systems in the south. Much of the coast between Eucla and Twilight Cove is fronted by shallow carbonate-rich seafloor, with extensive seagrass meadows fringing the shore. The Roe Plain is a remnant of a similar seafloor deposited during a higher Pleistocene sea level, which reached as far as the former sea cliffs now known as the Hampton Bluffs. South of Twilight Cove along the Baxter Cliffs and Bilbunya beach wave energy increases as the seafloor deepens.

2. The *South Coast* region commences at Cape Pasley and has a shoreline length of 1180 km west to Cape Leeuwin. The coast is a combination of the prominent granite and gneiss outcrops of the ancient Yilgarn Craton, together with younger siltstones of the Bremer Basin, the latter more prominent between Israelite Bay and Point D'Entrecasteaux. The coast faces into the prevailing high southwest waves and winds with higher energy beaches and extensive coastal dunes.

3. The small 100 km long *Leeuwin* section between Cape Leeuwin and Cape Naturaliste is composed entirely of the granites of the Yilgarn Craton, largely overlapped by Quaternary calcarenite dunes. The granite and particularly the calcarenite also outcrops along the shore as numerous islets and reefs.

4. The *Central West* coastline extends from Cape Naturaliste for 800 km north to Kalbarri and includes all the low-lying Perth Basin and the southernmost portion of the Carnarvon Basin. The coast is largely dominated by Quaternary carbonate-rich barriers and backing dunes, paralleled by largely submerged Pleistocene calcarenite barriers, which form near continuous reefs in the nearshore.

5. The *Carnarvon* coast between Kalbarri and Cape Preston covers the Carnarvon Basin. It has a shoreline length of 2036 km, which includes highly crenulate Shark Bay. At the coast the basin sediments have been overlaid in the south by the calcarenite Zuytdorp Cliffs together with Bernier and Dorre islands. These are backed by the low-energy calcareous shores of Shark Bay, including the sandstone Peron Peninsula. North of the Gascoyne River higher energy beaches and dunes and Pleistocene calcarenite prevail, and then between Amherst Point and North West Cape the shoreline is fronted by the Ningaloo reef system. The Cape Range backs much of the reefs, as well as forming the western boundary of Exmouth Gulf. Generally low energy island and reef-protected shore, dominated by mangroves and calcarenite barriers, extends up to Cape Preston.

6. The 460 km long *Pilbara* coast between Cape Preston and Port Hedland has some outcrops of the ancient craton at Cape Preston and in particular around the Dampier Peninsula, with gneiss and granites dominating. In between and in sheltered embayment are generally low energy often mangrove-dominated shores.

7. The *Canning* coast covers much of the Canning Basin and extends for 620 km from Port Hedland to Roebuck Bay and abuts the Great Sandy Desert. At the coast low gradients and desert dunes back Eighty Mile Beach and Roebuck Bay, with some deeply weathered Cretaceous sandstones outcropping along the coast between Cape Bossut and Roebuck Bay.

Climate

Climate contributes to the formation and dynamics of beaches in two main areas. First, climate interacts with the geology and biology to provide the geo- and bio-chemistry to weather the land surface, which, together with the physical forces of rain, runoff, rivers and gravity, erode and transport sediments to the coast. Secondly, at the coast it is also the climate, particularly winds that interact with the ocean to generate waves and currents, that are essential to move and build this sediment into beaches and dunes.

On a global scale, beaches can be classified by their climate. The *polar beaches* of the Arctic Ocean and those surrounding Antarctica, including Australia's large Antarctic Territory, are all dark in colour and composed of coarse sand, cobbles and boulders. They receive only low waves and have little or no surf. The coarse beaches have steep swash zones and, because of the coarse sediment and often frozen surfaces, there are no dunes. All these characteristics are a product of the cold polar climate, which permits little chemical weathering, hence the dark unstable minerals in the sediments, while the dominance of cold physical weathering ensures a supply of coarse sediments.

Temperate middle latitude beaches typical of southern Australia have sediments composed predominantly of well-weathered quartz sand grains, with variable amounts of biotic fragments (shell, algae, coral, etc). In addition, wind and wave energy associated with the strong mid latitude westerly wind stream is relatively high. The waves produce energetic surf zones, while the winds can build massive coastal dune systems, as are typical of the exposed southern coasts of Western Australia.

Tropical beaches of northern Australia, including the Kimberley, reside in areas of lower winds of the equatorial low pressure system (the doldrums) during the summer, while exposed to the more moderate velocity easterly Trade winds during the winter. Consequently wave energy is low to moderate at best, particularly in north Western Australia where the Trades blow offshore. The areas of lower waves and winds tend to have steep high tide beaches fronted by tidal flats with little or no surf and few dunes. Their sediments are often well-weathered quartz sands derived from plentiful tropical rivers and bleached coral and algal fragments derived from fringing coral and algal reefs. Fringing reefs also dominate many of the tropical Kimberley beaches. The most exposed tropical beaches face into the Trade winds and receive moderately high waves and may be backed by transgressive dune systems. In north Western Australia where the Trades tend to blow offshore there are only a few such exposed beaches.

Western Australian Climate

The Western Australian coast extends from Cape Londonderry in the tropical Kimberley at 13.5°S, to West Cape Howe on the temperate south coast at 35°S, a latitudinal distance of 2600 km. From north to south the climate is dominated by three pressure systems – the summer equatorial low, the year-round subtropical high and in winter the mid-latitude cyclones. Their area of influence and seasonal shifts are illustrated in Figure 1.7.

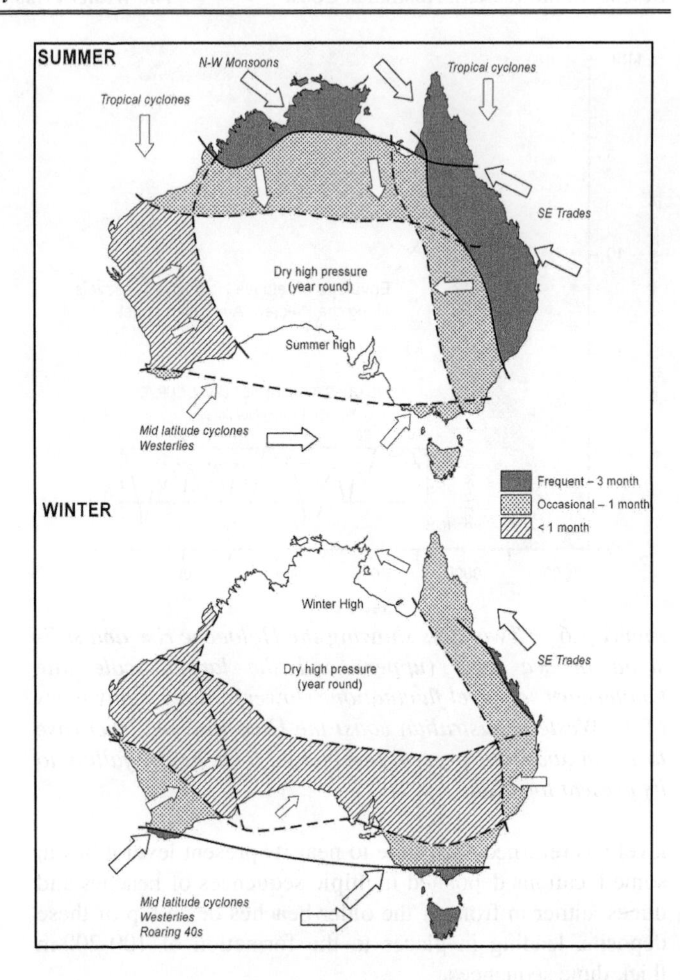

Figure 1.7 Australian summer and winter pressure systems and their area of influence. The Australian climate is dominated year round by the dry high pressure anticyclones that sit over much of Australia. In summer, the high is centred over 30°S and brings dry conditions to the southern parts of the continent, while permitting the northwest monsoons to penetrate across the north, bringing The Wet. In winter, the high moves north to dominate central and northern Australia with dry conditions prevailing, while mid latitude cyclones and their cold fronts bring cool weather and rain across southern Australia.

Equatorial Low

During summer the centre of the subtropical highs shifts south over Perth (35°S) and permits the equatorial low to form across northern Australia. On the southern side of the low intense heat lows form during the summer, with the Cloncurry Low in the east and the Pilbara Low in the west. The hot rising air is replaced by the humid tropical northwest winds from the eastern Indian Ocean, Timor and Arafura Seas. The light winds bring the summer northwest monsoons, The Wet: these usually penetrate deep into the Kimberley and as far south as Broome, bringing tropical summer afternoon thunderstorms and torrential rain. The summer rains can occasionally penetrate as far south as Shark Bay. The Pilbara heat low can at times extend south bringing hot northerly winds and heat wave conditions to Perth and the southwest.

Tropical Cyclones

Tropical cyclones are also associated with the summer monsoons, and usually form off the northwest coast when favourable conditions occur. These include the warm Indian Ocean waters, a spiralling convergence along the Intertropical Convergence Zone (ITCZ) and latitudes south of 5°S, where the Coriolis force can cause the hot, humid converging winds to spiral and rise. The northern Western Australia region receives on average 6 tropical cyclones a year, with most (80%) occurring between December and March, and February the peak month with an average of 1.4 cyclones per year. The number of actual cyclones varies considerably from a low of 0 in 1954 and 1 in 1950, to 17 in 1966 (Figure 1.8). The tropical cyclones bring strong winds, high seas, storm surges and torrential rains and can have a devastating impact where they make a landfall. Fortunately because of their relatively low frequency and variable trajectory and intensity, most coastal locations experience the effect of such cyclones for only a few days every few years. The main area of cyclone landfall and impact is between Exmouth and Port Hedland, centred on the small town of Onslow, the 'cyclone capital' of Australia.

Subtropical High

The subtropical high dominates the climate of Western Australia, its influence extending from the north to south coast. In winter the highs move north centred at 30°S and dominate central and northern Australia, resulting in easterly Trade winds, which blow offshore and maintain dry conditions across the north, The Dry. In the south the northerly shift permits the prevailing westerly wind stream and associated low pressure cyclones, and their cold fronts bring periodic winter frontal rain to the southwestern corner, as well as cool to occasional cold weather. In between the lows, the high can re-establish itself, bringing drier conditions. Winter conditions in the south are therefore cool to mild and wetter, with all coastal locations receiving a significant winter maximum in their rainfall.

During summer when the high is centred over Perth at 35°S the summer monsoons penetrate into the north, while the mid latitude cyclones cross south of the continent with only minor rainfall in the southwest corner. In between the two lows the high dominates year round maintaining dry conditions.

Sea Breeze Systems

Associated with high pressure conditions are the coastal sea breezes. Sea breeze is a result of the differential heating of the land and sea surface, particularly during summer, when the land can heat up considerably. In the absence of zonal westerly winds, the hot air over the land rises to be replaced by cooler air flowing in from the sea, hence the *sea breezes*. The breeze usually commences in mid to late morning after the land has heated and continues into the afternoon. At night the reverse may occur, with the land cooling to a lower temperature, particularly on calm clear nights, relative to the sea surface. If the land is cooler than the sea, then rising air over the sea may be replaced by cooler air

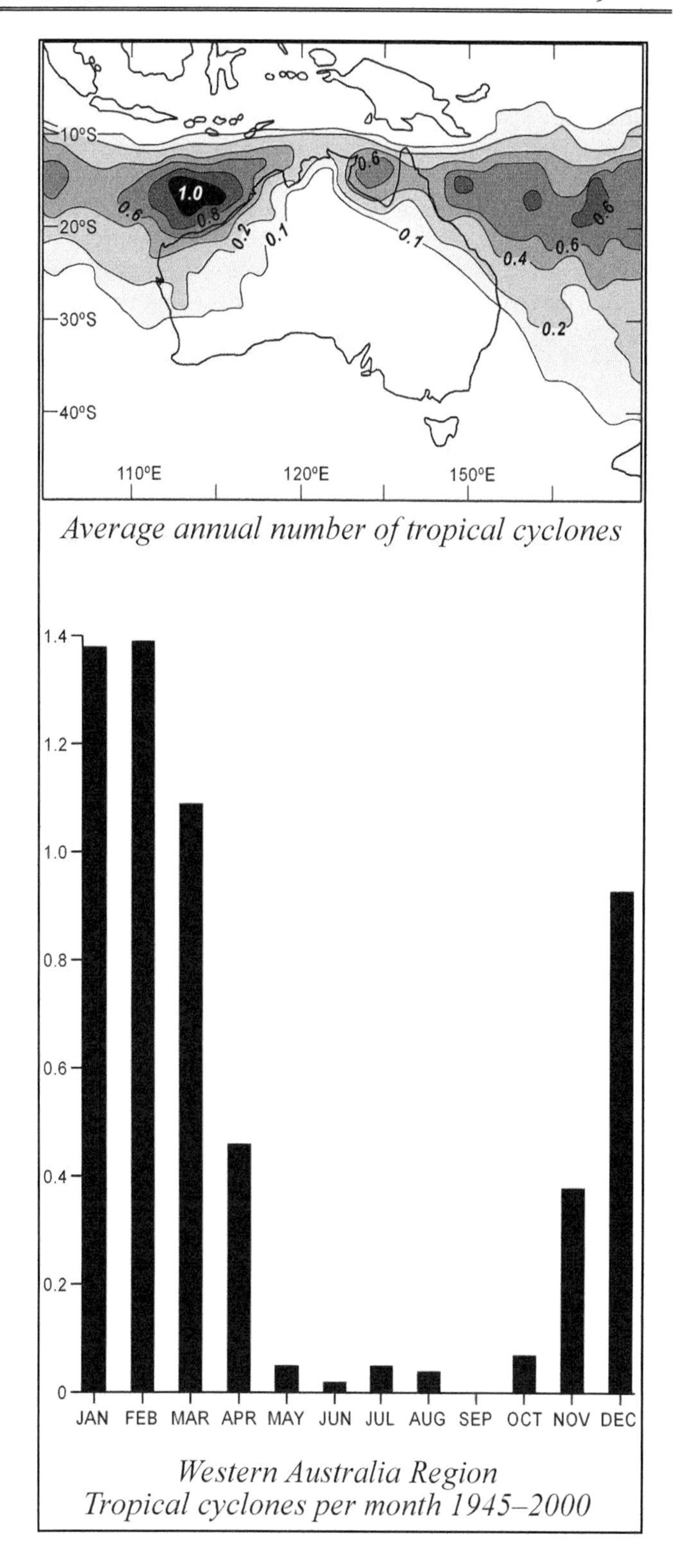

Figure 1.8 a) Average annual number of tropical cyclones; and b) Seasonal frequency of tropical cyclones in Western Australia.

from over the land, leading to an early morning land breeze. The land breezes are usually less frequent and lower velocity than the sea breezes. The Central West coast experiences some of the world's strongest sea breezes. The combination of hot interior conditions during summer and resulting strong temperature and pressure gradients induce a strong southerly sea breeze which is reinforced by the general zonal southwesterly flow of winds.

Mid Latitude Cyclones

Mid latitude cyclones are part of the subpolar low pressure system that extends around the entire Southern Hemisphere in a belt centred on 40° to 60°S latitude, the so called 'Roaring Forties and Raging Fifties'. The northern side of this system forms the southern boundary of the subtropical high pressure system. Like the high it shifts with the seasons, moving closer to the southern Australian continent in winter and further south in summer. The lows or cyclones that are embedded in this belt are continually moving from west to east. On average, one passes south of Australia every three to four days with between 80 and 100 passing south of the continent each year. These cyclones are responsible for both the spiralling westerly wind streams and the persistent southerly waves that arrive on average 350 days a year along the southern Australian coast. While they tend to pass south of the state during summer, during winter they shift northwards and penetrate into the southwest and occasionally as far north as Shark Bay, as indicated by the winter rainfall pattern (Fig. 1.9a).

Local winds and sea breezes play a major role in the climate and coastal processes around much of the Western Australian coast. Figure 1.10 illustrates the summer and winter 9 am and 3 pm wind roses. During summer winds tend to be lighter and more variable in the morning, whereas in the afternoon the onshore sea breeze dominates all locations, though its direction ranges from southeast, to south to west around the coast. The sea breeze is particularly strong along the west coast (see Perth, Carnarvon, Port Hedland, Broome) and contributes to the local wind waves climate, as well as locally wind driven currents. In winter winds tend to be offshore in the morning and light onshore in the afternoon, indicating a weaker sea breeze.

Figure 1.9a shows the annual *rainfall* pattern, with the humid northwest and southwest receiving up to 800 mm, and the large arid central coastal zone and centre with rainfall less than 400 mm. The seasonality is also highlighted and clearly shows the summer maxima in the north associated with the northwest monsoons and occasional cyclones, which bring 400 mm as far south as Broome, and lesser amounts as far south as Shark Bay. In winter the reverse applies with the humid cold fronts bringing up to 800 mm to the southwest corner, and penetrating as far north as Shark Bay. The coast between Shark Bay and Eighty Mile beach and the interior however remain relatively dry with annual rainfall less than 400 mm. The annual and seasonal temperature pattern is illustrated in Figure 1.9b. It highlights the latitudinal gradation in *temperature*, with annual means of 27°C in the north grading down to 18°C along the south coast. During summer high temperatures extend south with all of the state above a 20°C mean, while in winter cooler temperatures prevail, with only the northwest averaging over 20°C. There is a slight moderating in summer temperatures along the coast owing to the impact of the southerly sea breezes. Likewise in winter temperatures are slightly warmer along the coast adjacent to the warmer ocean waters, as the interior cools under the clear dry high pressure conditions.

Ocean Processes

Oceans occupy 71% of the world's surface. They therefore influence much of what happens on the remaining land surfaces. Nowhere is this more the case than at the coast and nowhere are coasts more dynamic than on sandy beach systems. The oceans are the immediate source of most of the energy that drives coastal systems. Approximately half the energy arriving at the world's coastlines is derived from waves; much of the rest arrives as tides, with the remainder contributed by other forms of ocean and regional currents. In addition to supplying physical energy to build and reshape the coast, the ocean also influences beaches through its temperature, salinity and the rich biosphere that it hosts.

There are eight types of ocean processes that impact the Western Australian coast, namely: wind and swell waves, tides, shelf waves, ocean currents, local wind driven currents, upwelling and downwelling, sea surface temperature and ocean biota (Table 1.5). Each of these and their impacts are discussed briefly below.

Ocean Waves

There are many forms of waves in the ocean ranging from small ripples to wind waves, swell, tidal waves, tsunamis and long waves, including standing and edge waves; the latter lesser known but very important for beaches. In this book, the term 'waves' refers to the wind waves and swell, while other forms of waves are referred to by their full name. The major waves and their impact on beaches are discussed in the following sections.

Wind Waves

Wind waves, or sea, are generated by wind blowing over the ocean. They are the waves that occur in what is called the area of wave generation; as such they are called '*sea*'. Five factors determine the size of wind waves:

- *Wind velocity* - wave height will increase exponentially as velocity increases;
- *Wind duration* - the longer the wind blows with a constant velocity and direction, the larger the waves will become until a fully arisen sea is reached; that is, the maximum size sea for a given velocity and duration;
- *Wind direction* will determine, together with the Coriolis force, the direction the waves will travel;
- *Fetch* - the area of sea or ocean surface is also important; the longer the stretch of water the wind can blow over, called the fetch, the larger the waves can become;
- *Water depth* is important, as shallow seas will cause wave friction and possibly breaking. This is not a problem in the deep ocean, which averages 4.2 km in depth, but is very relevant once waves start to cross the Western Australian continental shelf, which averages less than 100 m, and particularly as waves encounter the shallow rocks and reefs which dominate much of the coast resulting in smaller waves at the shore.

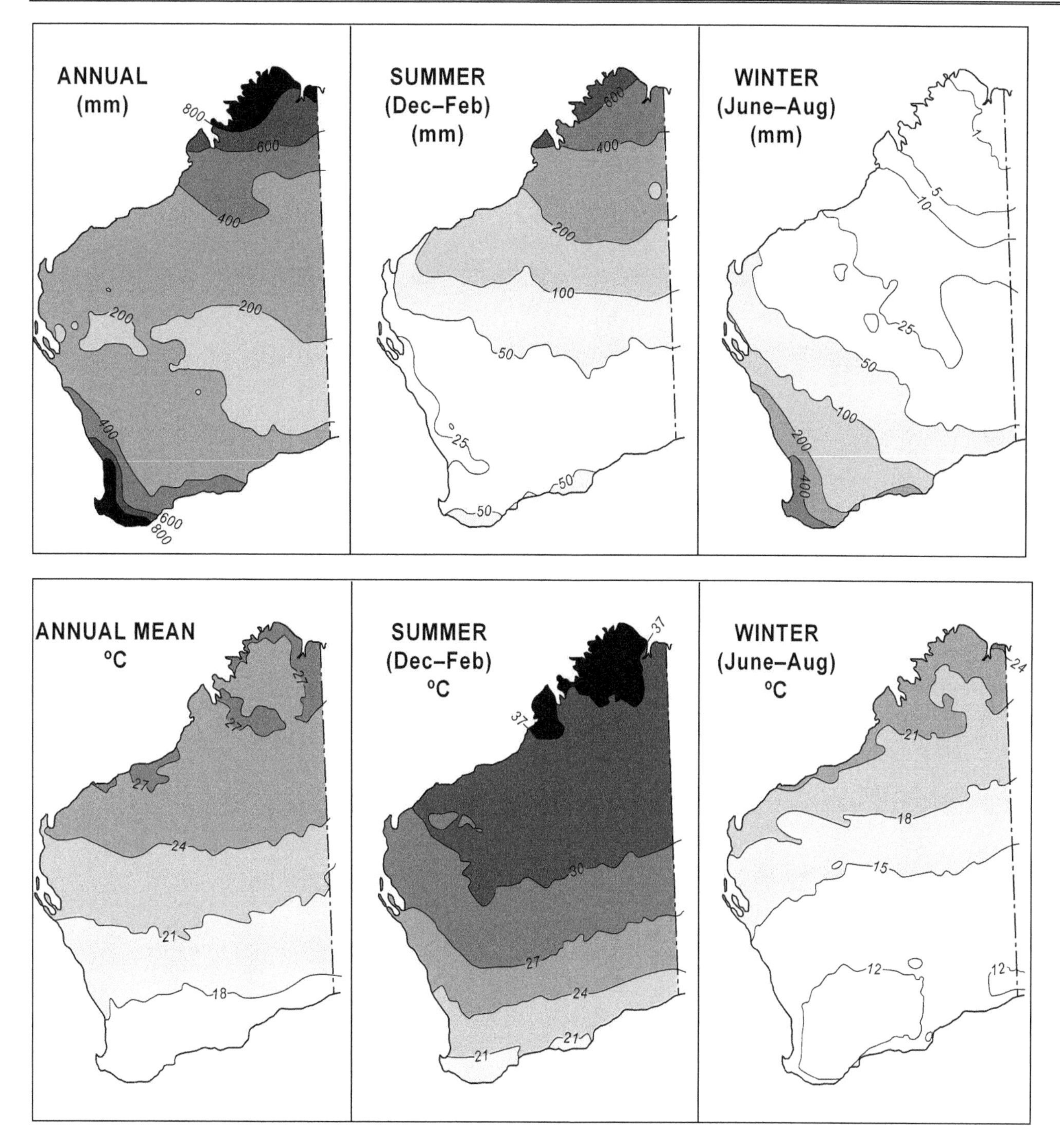

Figure 1.9 a) Western Australian seasonal rainfall pattern; b) Western Australian seasonal temperature pattern.

Table 1.5 Major ocean processes impacting the coast

Process	Area of coastal impact	Type of impact
waves – sea & swell	shallow coast & beach	wave currents, breaking waves, wave bores
tides	shoreline & inlets	sea level, currents
shelf waves	shoreline & inlets	sea level fluctuations
ocean currents	continental shelf	currents
local wind currents	nearshore & shelf	currents
upwelling & downwelling	nearshore & shelf	currents & temperature
storm surge	shallow coast & beach	rise in sea level
ocean temperature	entire coast	sea temperature
biota	entire coast	varies with environment & depth contributes carbonate sediments

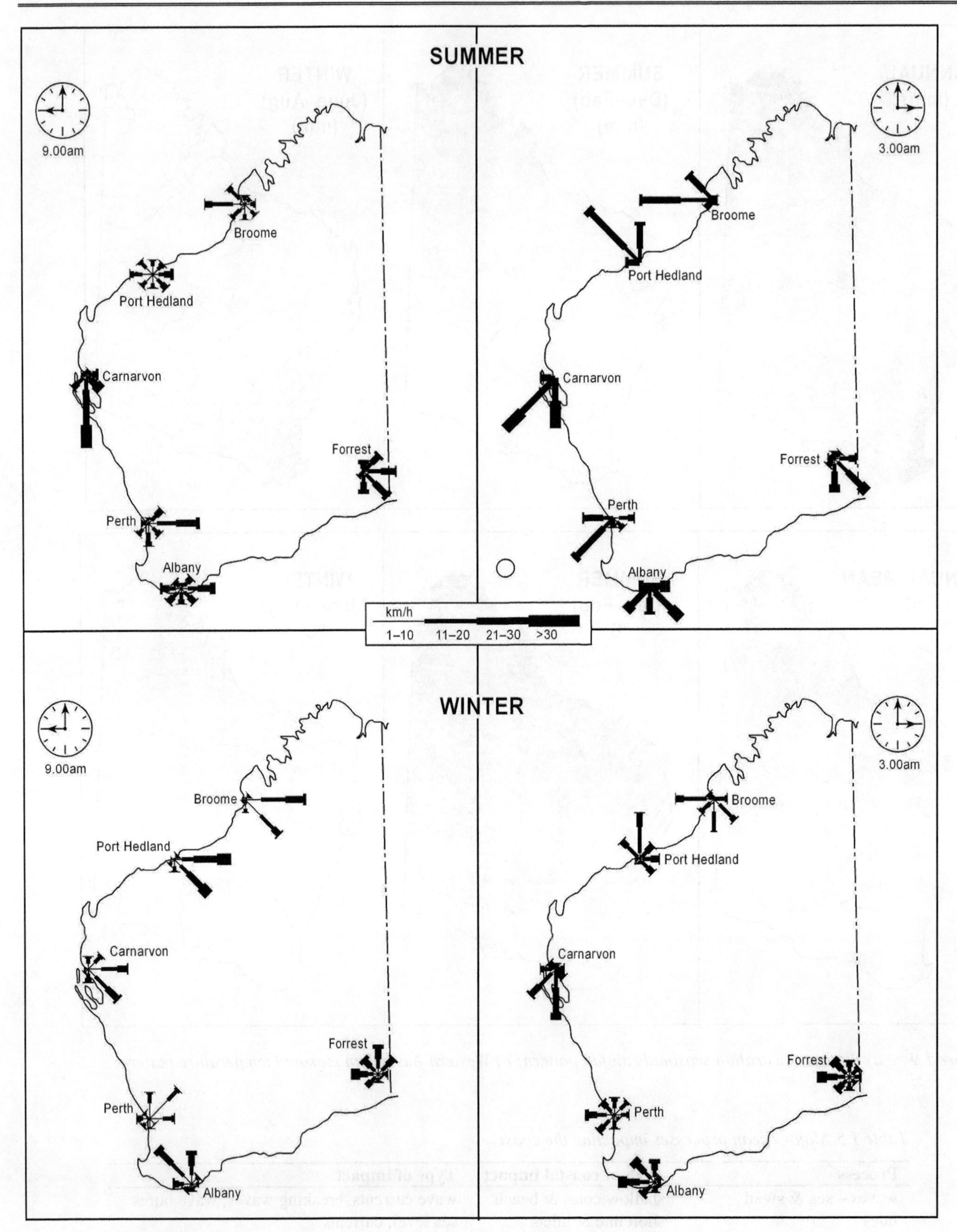

Figure 1.10 Western Australia summer and winter 9 am and 3 pm coastal wind roses.

The biggest seas occur in those parts of the world where strong winds of a constant direction and long duration blow over a long stretch of ocean. The part of the globe where these factors occur most frequently is in the southern oceans between 40° and 50°S, where the Roaring Forties and their westerly gales prevail. Satellite sensing of all the world's oceans found that the world's biggest waves, averaging 6 m and reaching up to 20 m, occurred most frequently around the Southern Ocean, centred on 40ºS and including a band across the south of Australia. Waves from this source are recorded at three Waverider buoys located at Perth (32ºS), Cape Du Couedic on Kangaroo Island (34ºS) and Cape Sorell on Tasmania's west coast (36ºS).

Swell

Wind waves become swell when they leave the area of wave generation, by either travelling out of the area where the wind is blowing or when the wind stops blowing. Wind waves and swell are also called free waves or progressive waves. This means that, once formed, they are free of their generating mechanism, the wind, and they can travel long distances without any additional forcing. They are also progressive, as they can move or progress unaided over great distances.

Once swell leaves the area of wave generation, the waves undergo a transformation that permits them to travel great distances with minimum loss of energy. Whereas in a sea the waves are highly variable in height, length and direction, in swell the waves decrease in height, increase in length and become more uniform and unidirectional. As the speed of a wave is proportional to its length, they also increase in speed (Fig. 1.11).

A quick and simple way to accurately calculate the speed of waves in deep water is to measure their period, that is, the time between two successive wave crests. The speed is equal to the wave period multiplied by 1.56 m. Therefore a 12 sec wave travels at 12 x 1.56 m/s, which equals 18.7 m/s or 67 km/hr. In contrast, a 5 sec wave in the gulfs travels at 28 km/hr, whereas a 14 sec wave travels at 79 km/hr. What this means is that the long ocean swell is travelling much faster than the short seas and that as sea and swell propagate across the ocean, the longest waves travel fastest and arrive first at the shore.

Swell also travels in what surfers call 'sets' or more correctly *wave groups*, that is, groups of higher waves followed by lower waves. These wave groups are a source of long, low waves (the length of the groups) that become very important in the surf zone, as discussed later.

Swell and seas will move across the ocean surface as *progressive waves*, through a process called orbital motion. This means the wave particles move in an orbital path as the wave crest and trough pass overhead. This is the reason the wave form moves, while the water, or a person or boat floating at sea, simply goes up and down, or more correctly around and around. However when waves enter water where the depth is less than 25% of their wave length (wave length equals wave period squared, multiplied by 1.56; for example, a 12 sec wave will be 12 x 12 sec x 1.56 = 225 m in length and a 5 second wave only 39 m long) they begin to transform into shallow water waves, a process that may ultimately end in wave breaking. Using the above calculations this will happen on the open coast when the long 12-14 sec swell reaches between 50 and 80 m depth, while in the northwest the short 3-5 sec seas will begin to feel bottom between 5 and 10 m.

Wave Shoaling

As waves move into shallow water and begin to interact with the seabed or 'feel bottom', four processes take place, affecting the wave speed, length, height, energy and ultimately the type of wave breaking (Fig. 1.12).

- *Wave speed* decreases with decreasing water depth.
- Variable water depth produces variable wave speed, causing the wave crest to travel at different speeds over variable seabed topography. At the coast this leads to *wave refraction*. This is a process that bends the wave crests, as that part of the wave moving faster in deeper water overtakes that part moving more slowly in shallower water.
- At the same time that the waves are refracting and slowing, they are interacting with the seabed, a process called *wave attenuation*. At the seafloor, some potential wave energy is transformed into kinetic energy as it interacts with the seabed, doing work such as moving sand. The loss of energy causes a decrease or attenuation in the overall wave energy and therefore lowers the height of the wave.
- Finally, as the water becomes increasingly shallow, the waves shoal, which causes them to slow further; and, decrease in length but increase in height, as the crest attempts to catch up to the trough. The speed and distance over which this takes place determine the type of *wave breaking*.

WAVE TYPES: sea and swell

Waves are generated by wind blowing over water surfaces. Large waves require very strong winds, blowing for many hours to days, over long stretches of deep ocean.

Sea waves occur in the area of wave generation, in close vicinity to the mid latitude cyclones, with sea breeze conditions around the coast and accompanying the northwest monsoon and Trade winds in the north.

Swells are sea waves that have travelled out of the area of wave generation. Swell dominates the southern Australian open coast, however much energy is lost along the west coast as they encounter shallow nearshore reefs.

Wave Breaking

Waves break basically because the wave trough reaches shallower water (such as the sand bar) ahead of the following crest. The trough therefore slows down, while the crest in deeper water is still travelling a little faster. Depending on the slope of the bar and the speed and distance over which this occurs, the crest will attempt to 'overtake' the trough by spilling or even plunging forward and thereby breaking (Fig. 1.13).

There are three basic types of breaking wave:

- *Surging breakers,* which occur when waves run up a steep slope without appearing to break. They transform from an unbroken wave to beach swash in the process of breaking. Such waves can be commonly observed on steeper beaches when waves are low, or after larger waves have broken offshore and reformed in the surf

OCEAN WAVE GENERATION, TRANSFORMATION AND BREAKING

Wave Type	**Breaking**	**Shoaling**	**Swell**	**Sea**
Environment	Shallow water – surf zone	Inner continental shelf	Deep water >> 100m	Deep water >>100m Long fetch = sea/ocean surface Wind velocity↑waves↑ Wind duration↑waves↑ Wind direction = wave direction
Distance travelled	~100 m	1 to 100 km	100s to 1000s km	100s to 1000s km
Time required	Seconds	Minutes	Hours to days	Hours to days
Wave profile			←SWELL	←SEA
Water depth	1.5 x wave height	<100m	>>100m	>>100m
Wave character	Wave breaks Wave bore Swash	Higher Shorter Steeper Same speed	Regular Lower Longer Flatter Faster	Variable height High Short Steep Slow
Example:				
Height (m)	2.5 to 3	2 to 2.5	2 to 3	3 to 5
Period (sec)	12	12	12	6 to 8
Length (m)	0 to 50	50 to 220	220	50 to 100
Speed (km/h)	0 to 15	15 to 60	66	33 to 45
Distance travelled (km/day)			1600	800 to 1100

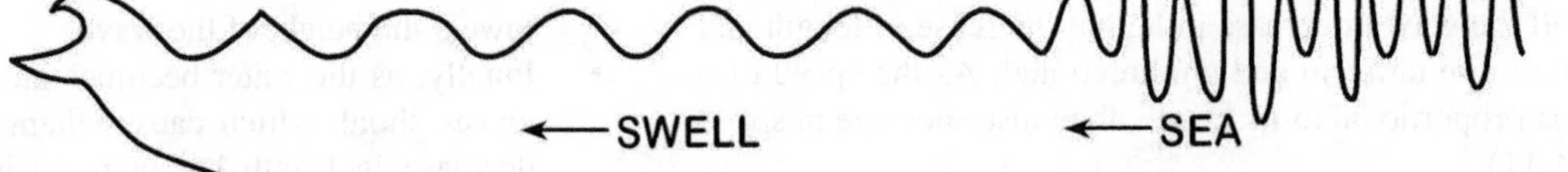

Figure 1.11 Waves begin life as 'sea' produced by winds blowing over the ocean or sea surface. If they leave the area of wave generation they transform into lower, longer, faster and more regular 'swell', which can travel for hundreds to thousands of kilometres. As all waves reach shallow water, they undergo a process called 'wave shoaling', which causes them to slow, shorten, steepen and finally break. This figure provides information on the characteristics of each type of wave. The Western Australian coast receives year round swell on all open southwest and south coasts, with shorter wind waves associated with strong sea breeze conditions, especially along the west coast, and the northern northwest summer monsoons.

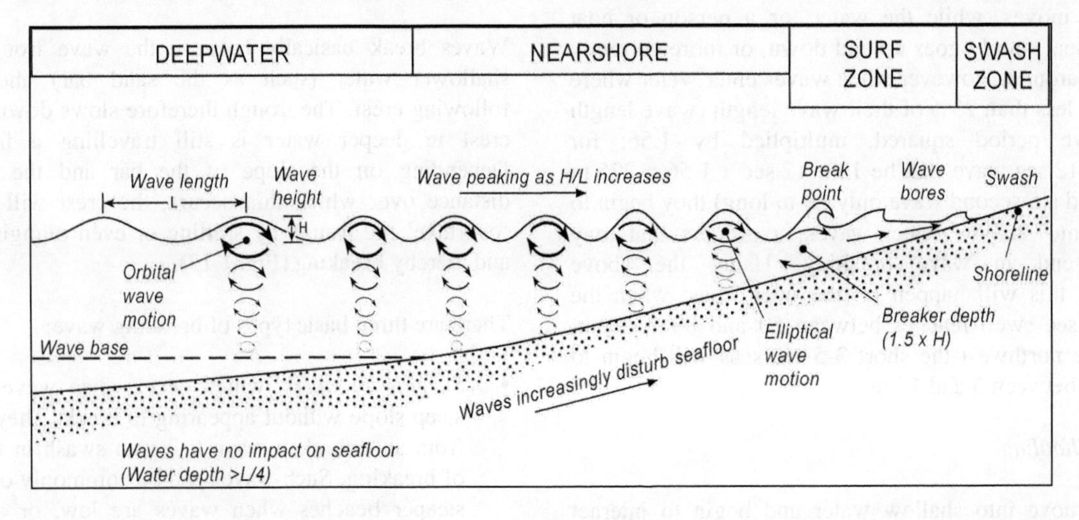

Figure. 1.12 As waves move into shallow coastal waters they shoal and in doing so slow, shorten, steepen and may increase in height. At the break point they break and move across the surf zone as wave bores (broken white water) and finally up the beach face as swash . Below the surface, the orbital wave currents are also interacting with the seabed, doing work by moving sand and ultimately building and forever changing the beach systems.

Figure 1.13 Long regular swell breaking at Sand Patch, South Coast.

zone. They then may reach the shore as a lower wave, which finally surges up the beach face as swash.

- *Plunging or dumping waves,* which surfers know as a tubing or curling wave, occur when shoaling takes place rapidly, such as when the waves hit a reef or a steep bar and/or are travelling fast. As the trough slows, the following crest continues racing ahead and as it runs into the stalling trough, its forward momentum causes it to both move upward, increasing in height, and throw forward, producing a curl or tube.

- *Spilling breakers* on the other hand occur when the seabed shoals gently and/or waves are moving slowly, resulting in the wave breaking over a wide zone. As the wave slows and steepens, only the top of the crest breaks and spills down the face of the wave. Whereas a plunging wave may rise and break in a distance of a few metres, spilling waves may break over tens or even hundreds of metres.

Broken Waves

As waves break they are transformed from a progressive wave to a mass of turbulent white water and foam called a *wave bore*. It is also called a *wave of translation*, as unlike the unbroken progressive wave, the water actually moves or translates shoreward. Surfers, assisted by gravity, surf on the steep part of the breaking wave. Once the wave is broken, boards, bodies and whatever can be propelled shoreward with the leading edge of the wave bore, while the turbulence in the wave bore is well known to anyone who has dived into or under the white water.

Surf Zone Processes

Ocean waves can originate thousands of kilometres from a beach and can travel as swell for days to reach their destination. However, on reaching the coast, they can undergo wave refraction, shoaling and breaking in a matter of minutes to seconds. Once broken and heading for shore, the wave has been transformed from a progressive wave containing potential energy to a wave bore or wave of translation with kinetic energy, which can do work in the surf zone.

There are four major forms of wave energy in the surf zone - broken waves, swash, surf zone currents and long waves (Table 1.6).

- *Broken waves* consist of wave bores and perhaps reformed or unbroken parts of waves. These move shoreward to finally reach the shoreline where the bore immediately collapses and becomes swash.
- *Swash* runs up and down the beach as uprush and stops, then end of the wave. Some soaks into the beach, while the remainder runs back down as backwash. Some of the backwash may reflect out to sea as a reflected wave, albeit a much smaller version of the original source.
- *Surf zone currents* are generated by broken and unbroken waves, wave bores and swash. They include orbital wave motions under unbroken or reformed waves; shoreward moving wave bores; the uprush and backwash of the swash across the beach face; the concentrated movement of the water along the beach as a longshore current; and where two converging longshore or rip feeder currents meet, their seaward moving rip currents.
- The third mechanism is a little more complex and relates to *long waves* produced by wave groups and, at times, other mechanisms. Long waves accompany sets of higher and lower waves. However, the long waves that accompany them are low (a few centimetres high), long (perhaps a few hundred metres) and invisible to the naked eye. As sets of higher and lower waves break, the accompanying long waves also go through a transformation. Like ocean waves they also become much shorter as they pile up in the surf zone, but, unlike ocean waves, they do not break, but instead increase in energy and height toward the shore. Their increase in energy is due to what is called 'red shifting', a shift in wave energy to the red or lower frequency part of the wave energy field. These waves become very important in the surf zone, as their dimensions ultimately determine the number and spacing of bars and rips.

As waves arrive and break every few seconds, the energy they release at the break point diminishes shoreward, as the wave bores decrease in height toward the beach. The energy released from these bores goes into driving the surf zone currents and into building the long waves. The long wave crest attains its maximum height at the shoreline. Here it is visible to the naked eye in what is called *wave set-up* and *set-down*. These are low frequency, long wave motions, with periods in the order of several times the breaking wave period (30 sec to a few minutes), that are manifest as a periodic rise (set-up) and fall (set-down) in the water level at the shoreline. If you sit and watch the swash, particularly during high waves, you will notice that every minute or two the water level and maximum swash level rises then rapidly falls.

Table 1.6 Wave motions in the surf zone

Wave form	Motion	Impact
Unbroken wave	orbital	stirs sea bed, builds ripples
Breaking wave	crest moves rapidly shoreward	wave collapses
Wave bore	all bore moves shoreward	shoreward moving turbulence
Swash	up-down beach face	controls beach face accretion & erosion
Surf zone currents	water flows shoreward, longshore and seaward (rips)	moves water and sediment in surf, builds bars, erodes troughs
Long waves	slow on-off shore	determines location of bars & rips

The height of wave set-up is a function of wave height and also increases with larger waves and on lower gradient beaches, to reach as much as one third to one half the height of the breaking waves. This means that if you have 1, 2, 5 and 10 m waves, the set-up could be as much as 0.3, 0.6, 1.5 and 3.0 m high, respectively. For this reason, wave set-up is a major hazard on low gradient, high energy Western Australian beaches, as occur along parts of the South East and South coasts.

Because the waves set up and set down in one place, the crest does not progress. They are therefore also referred to as a *standing wave*, one that stands in place with the crest simply moving up and down. These standing long waves are extremely important in the surf zone as they help determine the number and spacing of bars and rips.

Western Australian Wave Climate

While it is easy to see waves and to make an estimate of their height, period, length and direction, accurate measures of these statistics are more difficult. If we are, however, to properly design for and manage the coast we need to know just what types of waves are arriving at the shore. Traditionally, wave measurements have been made by observers on ships and at lighthouses visually estimating wave height, length and direction. This was the case in Western Australian lighthouses until they were automated in the 1970s. Since then the only wave measuring system has been operating off Perth.

The Datawell Waverider buoy is the present state-of-the-art on-site wave recording device. It operates using an accelerometer housed in a watertight buoy, about 50 cm in diameter. The buoy is chained to the sea floor, usually in about 80 m water depth. As the waves cause the buoy to rise and fall, the vertical displacement of the buoy is recorded by the accelerometer. This information is transmitted to a shore station and then by phone line to a central computer, where it is recorded. The first Australian Waverider buoy was installed off the Gold Coast in 1968. Today, Queensland and New South Wales have a network of Waverider buoys stretching from Weipa to Eden, the most extensive in the world.

There is just one permanent Waverider off Perth for the Western Australian coast. A number of buoys have been deployed temporarily at 40 sites, primarily near ports, between Esperance and Ningaloo. Data from these deployments is available from the Department of Infrastructure and Planning. In the Southern Ocean the Bureau of Meteorology maintains Waverider buoys off Cape Du Couedic in South Australia and Cape Sorell in Tasmania. The waves recorded at these sites are discussed later.

Wave climate refers to the seasonal variation in the source, size and direction of waves arriving at a location. Waves on the Western Australian coast originate from six possible sources - mid latitude cyclones and the large high pressure systems, and the more localised sea breeze systems in the south and west, and in the northwest from the offshore Trade winds and summer onshore westerly monsoonal winds, as well as infrequent tropical cyclones (Table 1.7)

Swell

Mid latitude cyclones produce year round swell that arrives right across the southern Australian coast (Fig. 1.14), with a slight summer minimum and late winter maximum. The swell is long (10-14 sec), moderate to high (2-3 m) and arrives predominantly from the west (20%) and southwest (60%), being more southerly in the north of the Bight region and more westerly along the South Coast. Swell height is also higher toward the south, decreasing toward the equator, as indicated in Figure 1.14.

Table 1.7 Wave sources and characteristics along the Western Australian coast

Source	Location	Direction	Season	Characteristics
Mid latitude cyclones	40-60°S	S-SW	year-round	moderate-high S-SW swell (see Fig. 1.14)
High pressures	centred on 30°S	westerly	Winter max in south	westerly seas on swell
Sea breeze	along entire coast	S-SW-south W-NW-north	summer max	short, low seas
Trade winds	central-northwest coast	E-SE	year round, winter max	short, seas to 3 m
Monsoonal winds	Kimberley region	W-NW	summer	short seas to 2 m
Tropical cyclones	northwest coast	W	summer	high seas to several metres

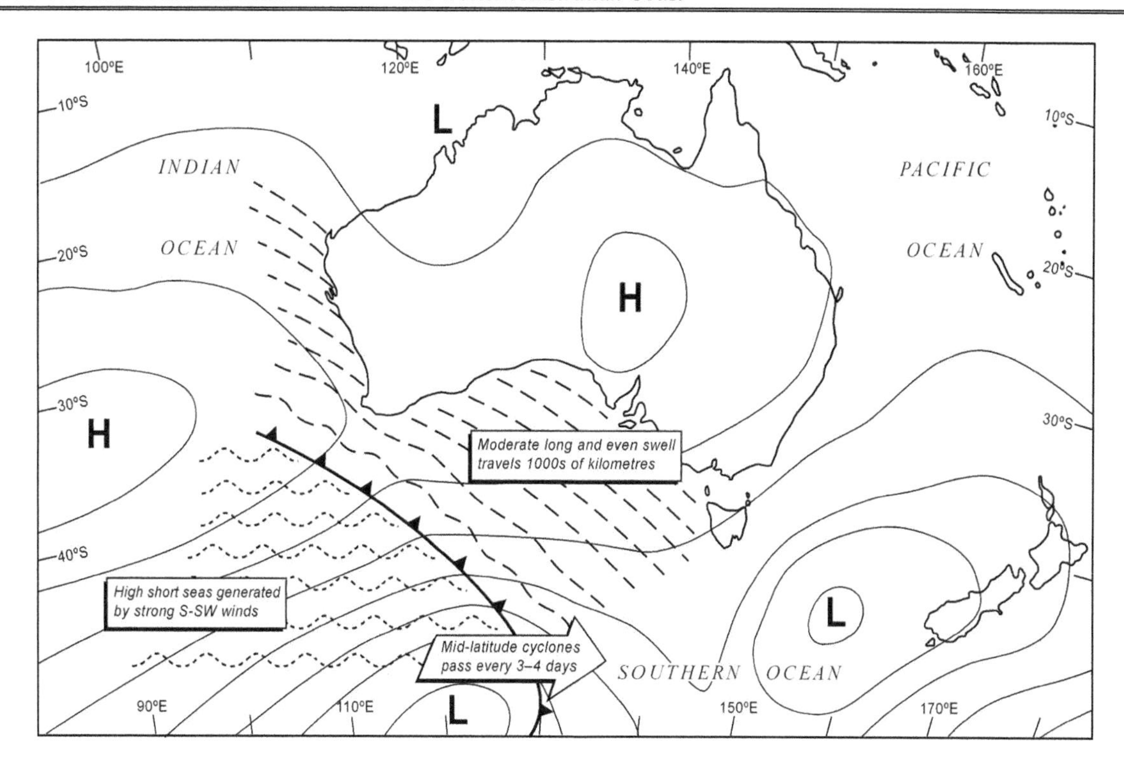

Figure 1.14 An example of a mid latitude cyclone and cold front passing south of Australia. As the cyclone traverses the Southern Ocean, its strong west to southwesterly winds, blowing over a long stretch of ocean, produce high seas. As the sea waves travel north of the cyclone, they become more regular, long crested high swell. In this summer example the cyclones are located well south of the continent and the swell can take one to two days to reach the south Western Australian coast. In winter the lows are closer to the coast and occasionally cross the coast, producing bigger seas and swell and accompanying strong westerly winds.

Figure 1.15 illustrates the global wave environment based on satellite altimetry. The figure clearly indicates that the world's highest average waves occur across the Southern Ocean, including south of Western Australia. In coastal waters the waves average 4-5 m for 10% of the year, 2-3 m for 50% and 1-2 m for 90%.

Figure 1.16 plots the monthly wave conditions for the three southern Australian wave recording sites. The table displays three variables: H_s the significant wave height which records the one-third highest waves; H_{max} the maximum wave height and records the 10% highest waves; and T_p the peak wave period, the time between each wave. Four characteristics are obvious. First, waves are high (>1.7 m) year round peaking during the winter months (>2.2 m) when the mid latitude cyclones move closer to Australia. Second, very high waves (3 to 5 m) occur year round. Third, wave period is long (>12 s) and also increases slightly during the winter months, due in part to the higher waves and also to the presence of shorter period summer sea breeze waves. Fourth, there is a progressive decrease in wave height between Cape Sorell at 42°S with a mean of 3 m, Cape Du Couedic at 36°S with 2.7 m and Perth at 32°S with 2.1 m. This decrease continues up the west coast.

In summary, the dominant source of swell waves affecting the southern half of the state, south of North West Cape, are the year-round mid latitude cyclones centred between 40 and 60°S. Waves arrive predominantly from the southwest and west. GEOSAT data (Fig. 1.13) indicates that the waves are rarely less than 0.5 m (<10%), that they average between 2 m and 3 m and reach 3 m 40% of the year, 4 m 15% and 5 m about 5% of the time. The waves are high year round, with a slight decrease in summer when the mid latitude cyclones are located further to the south and decrease slightly in intensity, with the lowest waves in January (1.5-2.5 m) and the highest in late winter (July-October, 2.5-3.5 m) and year-round average of 3 m at Cape Sorell, 2.7 m at Cape Du Couedic and 2.1 m at Perth. In total, the southern half of Western Australia, particularly the south coast, faces squarely into one of the world's highest energy deepwater wave environments. Waves reach the coastal zone as a moderate to high year-round swell with peak periods of 12-14 s.

Sea breeze flows onshore around the entire coast, particularly during the hotter summer months (Fig. 1.10). Their direction varies around the coast from more southerly in the southern half to more westerly in the northern half. In the southwest they are particularly strong. They average 25 km/hr and can reach over 50 km/hr generating short, steep southerly seas (Hb <1 m, T≈3-5s), usually from late morning to afternoon. These produce a distinctive summer wave climate with usually calmer conditions in the morning and strong sea breeze waves in the late morning to afternoon. In addition these winds are responsible for much of the coastal dune activity along the southwest coast.

In the northwest, two seasonal wind systems generate coastal seas. For most of the year the moderate to occasionally fresh easterly *Trade winds* blow predominantly offshore, producing low waves and calms along the coast. During summer the light to moderate *northwest*

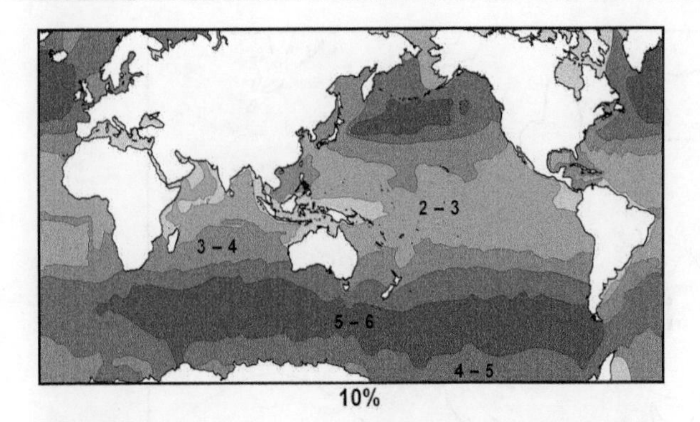

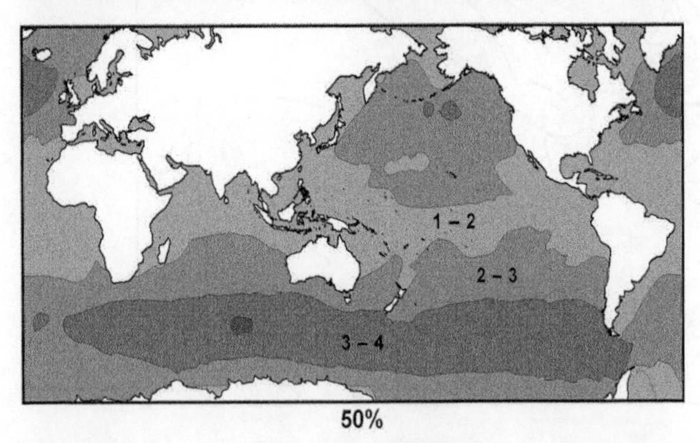

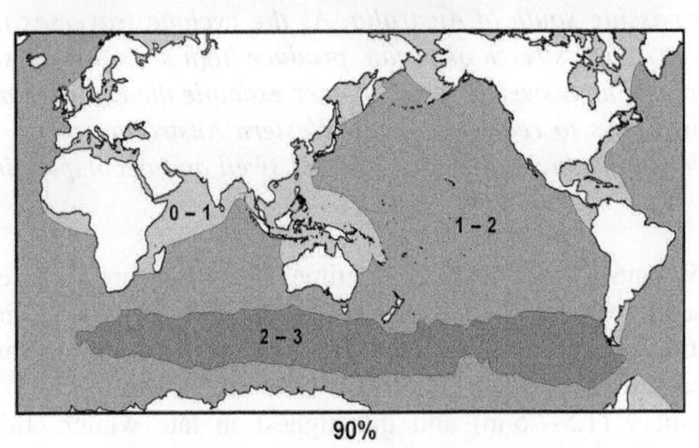

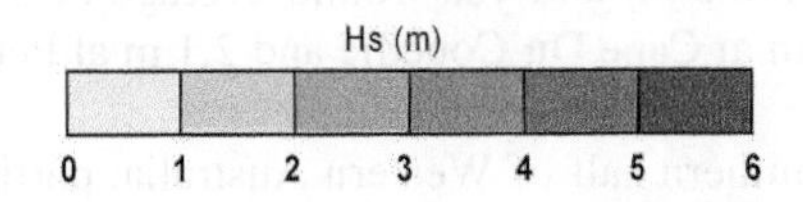

Figure 1.15 Average global wave heights. Note the dominance of high waves south of Australia year round, with deepwater waves south of Western Australia exceeding 5-6 m for 10%, 3-4 m for 5% and 2-3 m for 90% of the year (based on satellite data from Young and Holland, 1995).

monsoonal winds generate low short seas (Hb <1 m, T ≈ 3-5 sec) along all west-facing shores.

The highest seas and most damaging waves are generated by the periodic summer *tropical cyclones*, which also impact the northwest (Fig. 1.8). The cyclones can produce waves to several metres and up to 10 s in period. They however only occur infrequently at any one location.

Sea Waves

Four types of sea waves impact parts of the Western Australian coast. In the south while the mid latitude cyclones pass often well south of the continent (40-50°S), the coast is usually under the direct influence of *high pressure systems* and their westerly wind stream. These winds generate westerly sea conditions, which are superimposed on top of the swell. The occurrence and size of the seas is entirely dependent on local wind conditions. Under strong westerly winds they may reach a height of a few metres with periods of several seconds.

Wave Shoaling

To reach the shore deepwater, swell and seas must cross the continental shelf and nearshore zone, and in doing so considerable wave transformation can take place. Exposed beaches fronted by deeper shelf and nearshore zones and free of reef receive most of the deepwater energy, as along much of the South Coast. Four factors combine to lower waves along the coast.

- *Calcarenite reefs*: Submerged Pleistocene calcarenite reefs parallel much of the Central West and parts of the Leeuwin and Carnarvon coasts. The moderate to high deepwater waves either break on or are attenuated in crossing the reefs, resulting in lower waves at the shore. At the same time the sea breeze seas, generated between the reefs and the shore, become in places more important for beach processes than the attenuated ocean swell.
- *Coral reefs*: Along Ningaloo Reef and in the Kimberley, fringing coral reefs provide additional protection to much of the shore. The intertidal reefs cause heavy wave breaking with always low energy conditions in lee of the reefs.

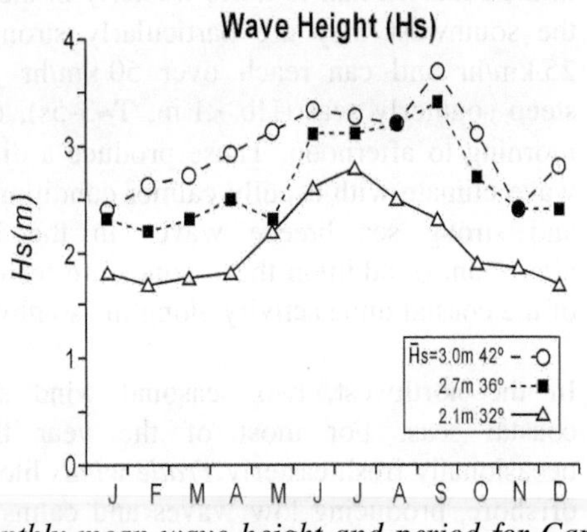

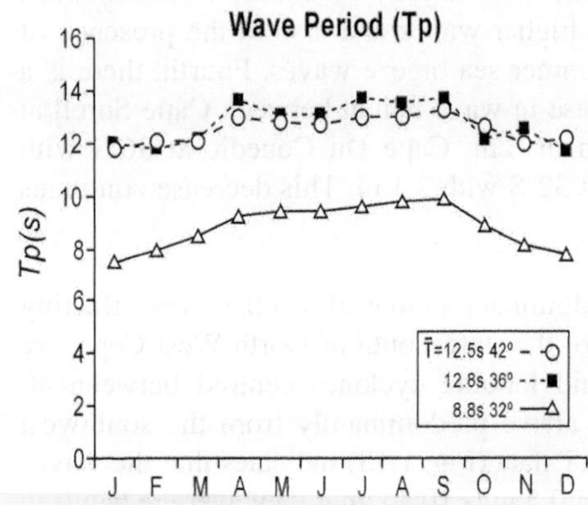

Figure 1.16 Monthly mean wave height and period for Cape Sorell (42°S), Cape Du Couedic (36°S) and Perth (32°S). Box indicates mean annual wave height and period.

- *Islands*: Numerous bedrock, calcarenite and reef islands and island groups extend around the coast, from the Recherche Archipelago in the south to numerous small islands along the west coast, many associated with the calcarenite reefs, and larger groups off Shark Bay and the northern Carnarvon, Pilbara and Kimberley coasts. All islands will block and intercept ocean waves and cause lower energy wave shadows in their lee.
- *Low nearshore gradients*: North of Exmouth Gulf a wide shallow continental shelf and low nearshore gradients dominate all the way to Cape Leveque. The low gradients result in greater wave attenuation and lower waves at the shore.

The breaker wave climate, as opposed to the deepwater wave climate, is therefore a function of the deepwater wave height, less the amount of wave energy (and height) lost as the waves cross the continental shelf and nearshore zone. This loss can range from near zero on deep, steep shelves, to 100% on wide and/or shallow shelves, particularly on sections of the west coast protected by calcarenite reefs. As a consequence of the calcarenite reefs and islands, and the variable coastal orientations along the Western Australian coast, breaker wave height, derived from deep ocean waves, ranges from a maximum average of 3 m to zero.

Freak waves, king waves, rogue waves and tidal waves

Freak waves do not exist.
All waves travel in wave groups or sets. A so-called 'freak', 'king' or 'rogue' wave is simply the largest wave or waves in a wave group. The fact that these waves can also break in deep water, due to over steepening, also adds to their demeanour.

Unusually high waves are more likely in a sea than a swell. For this reason they are more likely to be encountered by yachtsmen than surfers or rock fishermen.
Tidal waves arrive on the Western Australian coast twice each day. These are related to the predictable movement of the tides and not the damaging *tsunamis* with which they are commonly confused. Tidal waves are discussed in the next section.

How to estimate wave height from shore

Wave height is the vertical distance from the trough to the crest of a wave. It is easier to make a visual measurement when waves are relatively low and if there are surfers in the water to give you a reference scale. If a surfer is standing up and the wave is waist height, then it's about 1 m; if as high as the surfer, about 1.5 m; if a little higher, then it's about 2 m. For big waves, just estimate how many surfers you could 'stack' on top of each other to get a general estimate, i.e. two surfers, about 4 m, three, about 6 m and so on. Many surfers prefer to underestimate wave height by as much as 60% and a substantial number still estimate height in the old imperial measure of 'feet', using the USA or Hawaiian system. In Hawaii this system has been criticised as it attracts inexperienced surfers to what they think are lower wave conditions.

Actual wave height	Surfer's (under) estimate	% underestimated
0.5 m	1 ft (0.3 m)	60%
1.0 m	2 ft (0.6 m)	60%
1.5 m	3 ft (0.9 m)	60%
2.0 m	5 ft (1.5 m)	80%
2.5 m	6 ft (1.8 m)	70%
3.0 m	8 ft (2.4 m)	80%

Tides

Tides are the periodic rise and fall in the ocean surface, due to the gravitational force of the moon and the sun acting on a rotating earth. The amount of force is a function of the size of each and their distance from the earth. While the sun is much larger than the moon, the moon exerts 2.16 times the force of the sun because it is much closer to earth. Therefore, approximately two-thirds of the tidal force is due to the moon and are called the *lunar* tides. The other one-third is due to the sun and these are called *solar* tides.
Because the rotation and orbit of the earth and the orbit of the moon and sun are all rigidly fixed, the *lunar* tidal period, or time between successive high or low tides, is an exact 12.4 hours; while the *solar* period is 24.07 hours. As these periods are not in phase, they progressively go in and out of phase. When they are in phase, their combined force acts together to produce higher than average tides, called *spring tides*. Fourteen days later when they are 90° out of phase, they counteract each other to produce lower than average tides, called *neap tides*. The whole cycle takes 28 days and is called the lunar cycle, over a lunar month.

The actual tide is in fact a wave, more correctly called a *tidal wave,* and not to be confused with tsunamis. Tidal waves consist of a crest and trough, but are hundreds of kilometres in length. When the crest arrives it is called *high tide* and the trough *low tide*. Ideally, the tidal waves would like to travel around the globe. However the varying size, shape and depth of the oceans, plus the presence of islands, continents, continental shelves and small seas, complicate matters. The result is that the tide breaks down into a series of smaller tidal waves that rotate around an area of zero tide called an *amphidromic point*. The Coriolis force causes the tidal waves to rotate in a clockwise direction in the Southern Hemisphere, and anticlockwise in the Northern Hemisphere.

Tides in the deep ocean are zero at the amphidromic point and average less than 20 cm over much of the ocean. However three processes cause them to be amplified in shallow water and at the shore. The first is due to shoaling of the tidal waves across the relatively shallow (< 150 m) due to wave shoaling processes and increase in northwest, height (tide range) up to 1 to 9 m across much of the deep) continental shelf. Like breaking waves they are amplified and in some locations even break as a tidal bore. Secondly, when two tidal waves arriving from different directions converge, they may be amplified. Finally, in certain large embayments the tidal wave can be amplified by a process of wave resonance, which causes the tide to reach heights of several metres, as occurs in King Sound.

Tides are classified as being micro-tidal when their range is less than 2 m, meso-tidal between 2 and 4 m, macro-tidal between 4 and 8 m and mega-tidal when greater than 8 m. Western Australia has all four tidal types. Tides range from a low of 0.5 m in the south and southwest to up to 12 m in King Sound.

Western Australian Tides

The Western Australian coast receives its tides from two amphidromic systems, one located to the south with an amphidromic point near Tasmania which delivers low tides to all the southern coast about the same time. The second is located off the southwest tip of the state and delivers low tide along the southwest coast about the same time, then undergoes substantial amplification (2-12 m) and a progressive 2-5 hour delay in the time of the tide along the northwest and Kimberley coast, and further amplification (3-6 m) and a 10 hour delay by the time the tide reaches Wyndham (Fig. 1.16).

There is a prominent north versus south variation in the nature of the Western Australia tides (Table 1.8). In the south from the Bight right around to Kalbarri tides are low (<1 m), with a range of often less than 0.5 m, amongst the lowest ocean tides in the world. The southern tides also tend to arrive an hour either side of the arrival of the tide at Fremantle. North of Kalbarri there is a substantial delay in the tide as it enters and reaches the lower embayments within Shark Bay: while the height rises to 1.8 m at Monkey Mia, it is a low 0.5 m at the sheltered Hamelin Pool located deep inside the bay where it arrives 6.5 hours later. Between Carnarvon and North West Cape the tides increase to 1.5 m and are approximately 1 hour ahead of Fremantle, owing to the more westerly location of this section of coast.

Substantial amplification of the tides commences in Exmouth Gulf with the tide range increasing south into the gulf from 2.0 to 2.6 m. Then from Onslow to the east along the Pilbara and Canning coast the tide undergoes a progressive increase from 2.5 m at Onslow, to 4.4 m at Dampier, 6.8 m at Port Hedland and 9.4 m at Broome, and reaches a maximum in Australia of 12 m at King Sound (Derby), one of the highest tide ranges in the world. The timing of the tides along the northwest coast ranges from 1.5 hours later at Kalbarri, to 2-2.5 hours later between Exmouth and Broome. It then slows considerably as it moves round the Kimberley coast, arriving at the northern Cape Londonderry 4 hours later and at Wyndham 10 hours after Perth.

As a result, the Western Australian coast has a micro-tide range along the entire south and southwest coast, with a spring range that varies from between 0.6 and 1.4 m. Meso-tidal range commences along the Pilbara coast gradually increasing to macro-tidal around much of the Kimberley (Fig. 1.16 and Table 1.8).

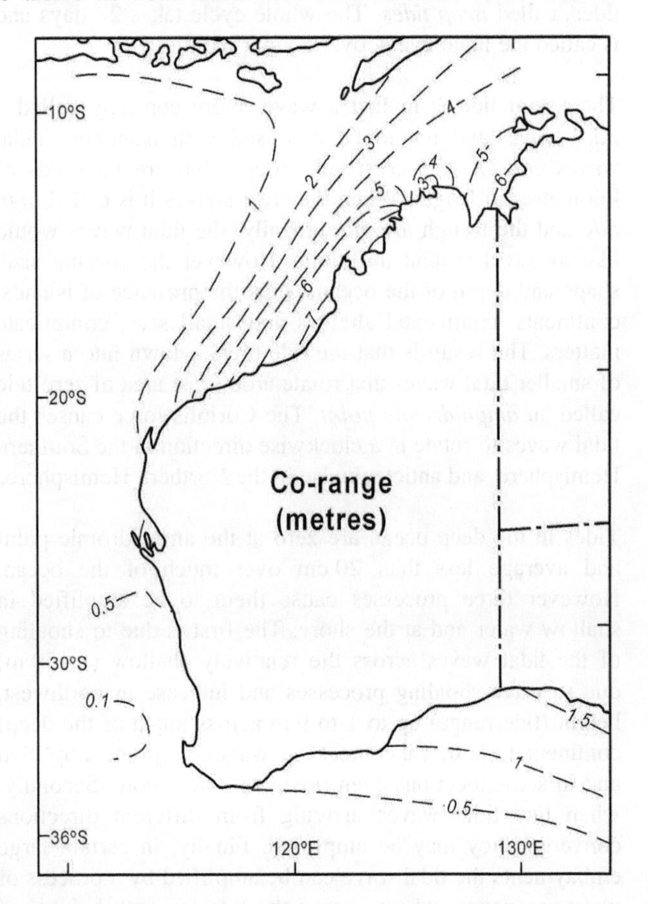

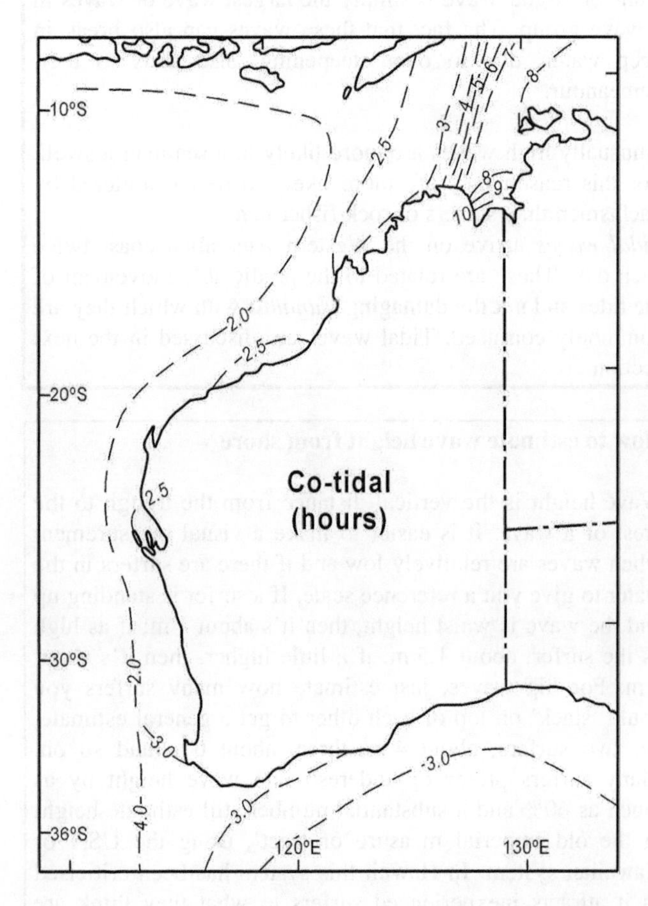

Figure 1.16 Co-range and co-tidal lines for Western Australia. Co-range locates areas of equal tide range, co-tidal locates areas with the same time of the tide.

Table 1.8 Tidal characteristics of Western Australian coast

Location	Mean spring high tide (m)	Mean spring low tide (m)	Mean spring tide range (m)	Relative time of arrival 0 hr=Perth - = before + = after
South East				
Eucla	1.3	0.3	1.0	0.0
South Coast				
Mary Ann Haven	1.1	0.4	0.7	-0.1
Albany	1.0	0.4	0.6	-0.5
Windy Harbour	0.8	0.4	0.4	-0.3
Flinders Bay (Augusta)	0.8	0.2	0.6	-0.5
Central West Coast				
Cape Hamelin	1.0	0.2	0.8	-0.5
Cowaramup	0.9	0.2	0.7	-1.4
Busselton	0.9	0.4	0.5	-0.1
Bunbury	0.8	0.3	0.5	-0.1
Bunker Bay	0.8	0.1	0.7	0.0
Fremantle	1.0	0.3	0.0	0.0
Rottnest Island	0.8	0.4	0.4	-0.1
Lancelin	0.8	0.2	0.6	0.0
Cervantes	0.9	0.2	0.7	0.0
Jurien	0.8	0.3	0.5	-0.1
Leeman	0.8	0.1	0.7	-0.6
Port Denison	0.9	0.2	0.7	-0.5
Geraldton	0.8	0.3	0.5	-0.5
Port Gregory	0.8	0.2	0.6	-1.1
Kalbarri	0.7	0.3	0.4	+1.4
Shark Bay				
Useless Loop	1.0	0.4	0.6	+0.9
Denham	1.2	0.4	0.8	
Monkey Mia	1.8	0.6	1.2	+2.1
Hamelin Pool	0.5	0.2	0.3	+6.5
Carnarvon Coast				
Carnarvon	1.5	0.6	0.9	+0.5
Coral Bay	1.4	0.2	1.2	-1.1
Norwegian Bay	1.4	0.2	1.2	-0.9
Tantabiddi	1.5	0.5	1.0	-0.6
Exmouth Gulf				
Point Murat	2.0	0.5	1.5	+0.1
Exmouth	2.3	0.5	1.8	+0.1
Learmonth	2.6	0.5	2.1	0.0
Pilbara Coast				
Onslow	2.5	0.6	1.9	+0.5
Fortescue Road	3.7	0.3	3.4	+0.1
Dampier	4.4	0.7	3.7	+0.5
Hauy Island	4.8	0.7	4.1	0.0
Depuch Island	5.7	0.7	5.0	+0.2
Port Walcott	5.5	0.8	4.7	+0.5
Port Hedland	6.8	0.9	5.9	+0.6
Kimberley Coast				
Broome	9.4	1.1	8.3	+0.5
Derby	11.2	1.1	10.1	+1.0
Port Warrender	7.0	0.6	6.4	+3.0
Cape Voltaire	6.5	0.9	5.4	+3.0
Hall Point	9.2	0.4	8.8	+3.5
Napier Broome Bay	2.6	0.2	2.4	+3.5
Lesueur Island	2.8	0.1	2.7	+5.5
Cape Domett	6.9	1.4	5.5	+9.0
Lacrosse Island	5.9	1.3	4.7	+9.0
Wyndham	7.7	1.2	6.5	+10.0
Pelican Island	6.9	1.6	5.3	+10.0

Note: **Spring tides** are also called 'king' tides and are highest around New Year and Christmas.

Spring or *king tides* are not responsible for beach erosion, unless they happen to coincide with large waves.

In areas of high tide range, tides can influence wave height, as high tide and deeper nearshore waters result in less wave shoaling and higher waves at the shore, while shallower water at low tide induces greater wave shoaling and lower waves at the shore.

Tidal Currents

Tidal currents are generally weak on the open coast, but strengthen considerably around the northwest coast as the tide range increases. Strong tidal currents are a feature across the North West Shelf and around the Kimberley coast. Some of the strongest currents occur at the entrance to King Sound and in Cambridge Gulf. In addition, all tidal creeks and lagoons are dominated by tidal flows.

Other Oceanographic Processes

Shelf Waves

Shelf waves are periodic oscillations in sea level that occur along the coast. Across the southern Australian coast there are two additional meteorologically driven sources of sea level oscillations: one a progressive wave, the other a standing wave. The progressive shelf waves move from west to east across the southern coast, with a period between waves of 5 to 20 days, that is, the time between crest and trough. They simultaneously reach the southern half of the west coast and move across the southern coast and on to eastern Australia. These waves raise sea level, for a few days at a time, by as much as 1.5 m, but more normally 50 cm or less.

The second standing wave type is more seasonal, with periods of several months. They affect the entire coast simultaneously, that is, they stand against the coast. These can also result in sea level oscillations, as much as 0.5 m. As tides on the open coast are normally of the order of a few decimetres, these waves can cause abnormally high or low tides, depending on whether the crest or trough is at the coast. They also make it difficult for the casual observer to determine the actual height of the tide.

Ocean Currents

Ocean currents refer to the wind driven movement of the upper 100 to 200 m of the ocean. The major wind systems blowing over the ocean surface drive currents that move in large ocean gyres, spanning millions of square kilometres. Two major currents affect the Western Australian coast. In the north, warm water pools in the eastern Indian Ocean and moves down the west coast as a warm, narrow current called the Leeuwin Current. This current brings warmer tropical waters south along the coast and is partly responsible for the world's most poleward coral reefs at Houtman Abrolhos. The current reaches as far south as Cape Leeuwin and periodically moves around the cape and east into the Great Australian Bight.

The dominant current south of Australia is the great Antarctic Circumpolar Current, which runs westerly at relatively high speed (1-3 knots), driven by the strong westerly wind stream. This current is continuous around the southern oceans, centred between 50 and 60°S, well south of the continent. Between the main current and the southern Australian coast, the currents, while still predominantly westerly, decrease in velocity owing to increasing distance from the main westerly wind stream, more variable wind direction and the impact of the relatively shallow continental shelf. Closer to the coast the currents become increasingly dependent on local winds, which still however are predominantly westerly

Tsunamis

Following the Boxing Day 2004 tsunami off Sumatra, there has been a tremendous increases public awareness of these waves and the potential damage and destruction they can do. Tsunamis are waves generated by a sudden impulse or impact in the sea. The source of the impulse is usually an earthquake that displaces the sea floor up or down. In the case of Sumatra a 1000 km long by 200 km wide area of seafloor was suddenly uplifted 10 m. They are also generated by large undersea and coastal landslides (usually triggered by an earthquake), volcanic explosions (Krakatoa near Java generated a massive tsunami in 1883) and rarely meteorite-comet impacts, which can generate mega-tsunamis. The usually single impulse generates several waves as the ocean surface gradually returns to normal, in much the same was as the waves generated by a rock thrown into a pond.

In the deep ocean tsunamis travel very fast (up to 800 km/hr), are spaced approximately 200 km apart and have a period of about 15 minutes. Tsunamis have a crest and a trough like all waves. The trough usually arrives first, resulting in a draw-down in the water and sea level, which unfortunately can attract people onto the exposed seafloor. However within 5-10 minute the first crest will arrive. While in the deep ocean tsunamis may only be a few decimetres high, because they are so long and contain so much water they build in height as they approach the shallow coastline also slowing down to 40-50 km/hr.

Once they hit the coast the wave not only increase in height, up to 10 m and more in Sumatra and Sri Lanka, but because of its great length just keeps on coming for several minutes rising the water level and flowing inland until its reaches ground higher than the tsunami. It then retreats for several minutes before the second and usually largest wave arrives. This is followed by a series of increasingly smaller waves.

Tsunamis occur relatively frequently, particularly in the Pacific Basin, and major tsunamis the size of the Aceh tsunami occur on average every one hundred years. The devastation and lost of life caused by the Aceh tsunami was a result of a of a large 8.9 scale undersea earthquake and associated uplift of the seafloor occurring close to a low-lying densely populated coast. The earthquake generated a large tsunami which arrived at the Sumatra coast within 10-15 minutes as a 10 m high wall of water that rapidly flooded the densely populated low lying coastline. The tsunami wreaked havoc further a field in Thailand, Sri Lanka. India and east Africa, The tsunami also reached the Western Australian several hours later and eventually entire Australian coast. However by the time of arrival it was much lower (a few centimetres to 1 m) and no damage was recorded in Australia. While Australia does receive numerous small tsunamis, which largely go unnoticed apart from oscillations in the tidal gauges, there is no evidence of major tsunami impact or damage over the past few thousand years.

Other Currents

There are several other forms of ocean currents driven by winds, density, tides, shelf waves and ocean waves. It is not uncommon to have several operating simultaneously. Each will have a measurable impact on the overall current structure and must be taken into account if one needs to know the finer detail of the coastal currents, their direction, velocity and temperature. Around the Western Australian coast, in addition to the ocean and tidal currents, the next most important are those associated with the shelf waves and wind driven upwelling and downwelling. The latter are associated with regional winds, including sea breezes. Along the southern coasts strong west through southwest winds push the surface waters to the left (north) causing a piling up of warmer ocean water at the coast which downwells to return seaward. When strong (and often hot) northerly to easterly winds prevail, the water is also pushed to the left (west) pushing the warmer surface away from the shore, particularly along the south coast, which is replaced at the shore by the upwelling of cooler, deeper bottom water. The combination of this cooler water and warm air can lead to condensation in the overlying warmer air and the formation of a sea fog or mist along the coast, not uncommon along the south coast and Bight region.

Sea Surface Temperature

The sea temperature along the Western Australian coast is a product of four main processes: firstly, the latitudinal location of the coast between 13.5 and 35°S determines the overhead position of the sun and the amount of solar radiation available to warm the ocean water; secondly, Leeuwin Current brings warmer waters south along the entire west coast; third, the generally shallow shelf-nearshore waters along much of the west and northwest coast, including the shallow Shark Bay and Exmouth Gulf, permit greater summer warming; and fourth, the wind driven impact of upwelling and downwelling.

On a seasonal basis, the summer water temperature ranges from the low 30s around the Kimberley coast, to the mid 20s along the southwest coast, to the high teens (18-19°C) in the South East rising to 22°C in the northern Bight. In winter, the northern waters drop slightly to the mid to high 20s; the Central West coast cools to the low 20s, whole the South Coast and Bight drop to the mid teens (13-15°C).

Salinity

All oceans and seas contain dissolved salts derived from the erosion of land surfaces over hundreds of millions of years. Chlorine and sodium dominate and, together with several other minerals, account for the dissolved 'salt'. The salts are well mixed and globally average 35 parts per thousand, increasing slightly into the dry subtropics and decreasing slightly in the wetter latitudes. Along the open coast salinity maintains this average. However in Shark Bay, evaporation combined with limited circulation and essentially no runoff permits a build up in salinity. Within the bay salinity increases southward into each of the embayments reaching $70^{o}/_{oo}$ twice that of seawater in Hamelin Pool. The high salinity permits stromatolites to grow in the shallow waters of the Hamelin Pool Marine Nature Reserve.

Biological Processes

Biological processes are extremely important along the entire Western Australian coast and continental shelf. The climate ranges from hot humid grading to arid in the North West, to temperate humid in the South West grading to arid in the South East. Water temperatures range from warm in the north to cool in the south. As a consequence a wide range of dune, shoreline and subtidal habitats exists around the coast. These can be broadly divided into coastal dunes, supratidal (samphire) flats, intertidal, subtidal and shelf.

Coastal Dune Vegetation

Coastal dunes occur around the entire Western Australian coast, with small pockets of dunes even along the Kimberley coast. All dunes are vegetated by a predictable succession of plants beginning with herbs and grasses on the foredunes (Fig.1.17), grading landward into a combination of sedgelands and shrublands. Only in the humid South West do Tuart forests (*Eucalyptus gomphocephala*) form the climax dune vegetation. While the structure of the dune vegetation is similar around the coast the species vary from south to north as the climate changes. The following lists some of the major plant communities and species found on dunes around the coast.

Figure 1.17 A low vegetated foredune near Eucla.

South and Central West

- Beach Spinifex community on incipient foredune (*S. hirsutus, S. longifolia, Cakile maritima, Arctotheca populifolia*).
- Seaward side of foredune – low shrubs (*Acacia cyclops, Acacia rostellifera, Olearia axillaris*).
- Crest to leeward side of foredune closer and larger shrubs.
- Stable dune community – shrub community (*Acacia rostellifera, Olearia axillaris*).
- Tuart forest in humid regions (*Eucalyptus gomphocephala*).

Pilbara

- Beach spinifex community on the active incipient foredune (*S. longifolia, Ipomoea brasiliensis, Sahola kali*).
- Grass-dominated community on seaward faces and crest of foredune (*Whiteocloa airoides*).
- Diverse wattle community on the stable dunes (*Acacia sp., Whiteocloa airoides, Euphorbia sp.*).
- Soft spinifex community on leeward slopes and hind dunes (*Triodia pungens, Rhynchosia minima, Evolvulus alsinoides*).
-

Kimberley

- Incipient foredune – coastal grasses (*S. longifolius, Ipomoea brasiliensis, Sahola kali, Fimbristylis cymosa, Fimbristylis sericea, Cyperus bulbosus*).
- Foredune – low shrubs (*Acacia bivenosa, Lysiphyllum cunninghamii, Canavalia rosea*).
- Hind dune and hollows – dense shrub community of diverse plants, some *Pandanus spiralis*.
- Grades into pindan or rocky vegetation.

Salt Marsh-Samphire Vegetation

Samphire vegetation in association with algae grows along lower energy sections of coast in the lower swales between beach ridges and in the saline back barrier depressions and dry lagoons. The samphire vegetation is usually low (<1 m) and scrubby and forms a boundary between the shoreline and the landward terrestrial vegetation. In the South West there is a succession from high water landward of the following three communities:

1. The low succulent *Salicornia australis* extends close to high water, with *Arthrocnemum arbuscula* slightly landward.
2. These are backed by the *Arthrocnemum halocnemoides* community, a low woody, succulent glasswort shrub, in the zone rarely reached by high tide. It grows in association with *Saliconia australis, Suaeda australis, Scirpus nodosus* and *Samolus junceus* and the perennial grass *Sporobolus virginicus*.
3. The sedge zone of tussocky grasses occupies the most landward position, and is dominated by *Gahnia trifida, Juncus maritima* and *Scirpus nodosus*.

In the Pilbara and Kimberley regions tides are higher, waves lower and the climate hotter resulting in wide inter to supratidal zones and greater climate stress. Four communities of plants can occur in favourable locations.

1. A more seaward community of succulent samphires *Suaeda arbusculoides* at the seaward fringe, sometimes separated from *Halosarcia halocnemoides* by mud or sand flats covered with dense mats of blue-green algae.
2. A mixed herbaceous and grasses community in the mid to upper marsh level containing *Limonium salicorniaceum*, water couch grass (*Sporobolus virginicus)* and rice grass (*Xerochloa imberbis*).
3. Herbs and low shrubs in higher well drained locations containing *Halosarcia indica*, the halophyte *Frankenia ambita* and *Hemichroa diandra*.
4. The most landward community which can tolerate the high salinity but not waterlogging, including *Neobassia astrocarpa, Trianthema turgidifolia* and some *Triodia* sp.

Mangroves

Mangroves grow between mean sea level and neap high tide and require some degree of protection from wave attack. In Western Australia no mangroves grow along the Bight and South Coast, through they do grow extensively on the South Australia side of the Bight, and as far south as 38° in Victoria. This description is largely based on the excellent mangrove handbook by Semeniuk, et al. (1978). The first mangroves on the west coast are located in Leschenault Inlet and adjacent Bunbury Harbour (33°S) with the single species *Avicennia maritima*. Moving up the coast, warmer water and, further north, more favourable habitats lead to a continual increase in mangroves area (Fig. 1.18), number of species and size or biomass.

Figure 1.18 Extensive mangroves just south of Carnarvon.

At Mangrove Bay (22°S) on Exmouth Peninsula three species occur: *Avicennia maritima, Rhizophora stylosa* and *Bruguiera exaristata*. By the DeGrey River mouth (Breaker Inlet) 20°S seven species are present: *Aegialitis annulata, Aegiceras corniculatum, Avicennia marina, Bruguiera exaristata, Ceriops tagal, Rhizophora stylosa,* and

Table 1.9 Distribution of mangroves species in Western Australia (from Semeniuk, et al., 1978)

Mangrove species	Central West	Pilbara	West Kimberley	North Kimberley
Avicennia maritima	x	x	x	x
Aegialitis annulata		x	x	x
Aegiceras corniculatum		x	x	x
Rhizophora stylosa		x	x	x
Ceriops tagal		x	x	x
Osbornia octodonta		x	x	x
Bruguiera exaristata			x	x
Camptostemon schultzii			x	x
Excoecaria agallocha			x	x
Sonneratia alba			x	x
Xylocarpus australasicus				x
Pemphis acidula				x
Lumnitzera racemosa				x
Avicennia eucalyptifolia				x
Bruguiera parviflora				x
Scyphiphora hydrophylacea				x

Excoecaria agallocha. In Willies Creek at Broome (17.5°S) eight species are recorded with some reaching 11 m in height: *Avicennia* (3-4 m), *Bruguiera, Rhizophora* (5-8 m), *Aegiceras* (to 5 m), *Camptostemon* (9-11 m), *Ceriops* (to 3.5 m), *Osbornia* and *Excoecaria*. In the northern Kimberley up to 16 species occur along the coast (Table 1.9).

In addition to the increase in species number and height, the width of the mangrove fringe widens as tide range increases and intertidal sand and mud flats become more prevalent. As a consequence the area of mangroves increases exponentially from south to north. Australia-wide 96% of mangroves are located in the tropics north of the Tropic of Capricorn. Western Australia has 2,430 km^2 of mangroves along the mainland, with another 90 km^2 on islands, which in total comprise 22% of Australia's mangroves.

Seagrass Meadows

Seagrass grows along lower energy sections of the open coast and in most bays (Fig. 1.19a). It is restricted to the lower intertidal and particularly the shallow subtidal zone to several metres depth, the growth limit primarily determined by sunlight penetration. Seagrass meadows occur extensively right along the entire Western Australian coast, with temperate species occurring in the south, tropical species in the north and a mixture in the centre at Shark Bay. The Shark Bay seagrasses cover 4000 km^2 and are the largest meadows in the world. In the Bight-South East they occur along sections of the open coast protected by low nearshore gradients. Along the South and Central West they grow to the lee of the shore parallel calcarenite reefs, which produce lower wave energy along much of this coast. They are less prevalent from Carnarvon north along the Pilbara and Canning coasts, while they again flourish in King Sound and around the Kimberley coast.

Three species dominate the temperate south. The grassy *Zostra* grows between mean sea level and low tide, while the taller *Posidonia australis* (fibre ball reed) and *Amphibolis antarctica* (sea nymph or wireweed) grow generally below low tide to as deep as 10 m. They occur

extensively across the Bight, and up the west coast, peaking in Shark Bay, the most extensive area of seagrasses in the world. Shark Bay straddles the tropical and temperate seagrass domains resulting in 12 species, with many occurring side by side. The most abundant are the temperate species *Amphibolis antarctica* and *Posidonia australis*, in addition to the tropical *Halodule uninervis*, which is a preferred food for the large dugong population in the bay.

In the Broome-King Sound region tropical species are more dominant, with *Halodule uninervis* the most prominent, particularly where the large tides expose the substrate. *Halodule pinifolia* and *Halophila ovata* prefer the intertidal and shallow subtidal. *Thalassia* sp. is associated with coarser sediments and *Thalassodendron ciliatum* grows directly on coral and calcarenite reefs. Muddy intertidal areas are favoured by *Halophila ovalis*. As in Shark Bay the tropical meadows are also grazed by dugongs and green turtles. Besides helping to stabilise the nearshore sands, the meadows support a rich epibiota and contribute a high proportion of red algae, foraminifera and bivalve fragments to the beach sediments, as well as seagrass roots and detritus (Fig. 1.19b). This material is reworked by the waves and tides to build the extensive tidal shoals, sub- to intertidal embankments, particularly in Shark Bay, and along all sections of coast is washed onshore to contribute sediments to the beaches and backing dunes. As a consequence they are a major, and in some areas an exclusive, source of beach sediments particularly along much of the South East and Central West coasts.

a.

b.

Figure 1.19 a) Seagrass meadows parallel the shore along Banky Beach, South Coast; b) Seagrass debris (right) piled 1 m high along Bilbunya Beach, South East Coast.

Coral Reefs

Western Australia has the world's most poleward coral reef systems forming the Houtman Abrolhos islands and reefs (29ºS). On the mainland fringing reefs become prominent north from Gnaraloo (24ºS) forming the Ningaloo fringing-barrier reef complex between Amherst Point and North West Cape. Between Exmouth and Cape Leveque the coast is dominated by generally low gradient beaches, tidal flats and mangroves, and as a result reefs

tend to occur off the coast as atolls and fringing islands. In the Kimberley however they dominate much of the coast, occurring as numerous fringing reefs, as well as fringing the numerous islands and forming atolls.

There are three major coastal coral reef systems in Western Australia. The most southern (21-24ºS) are the *Ningaloo* reefs, now incorporated in the Ningaloo Marine Park, which has been nominated for World Heritage listing. The 260 km long system ranges from 200 m to 7 km in width, averaging 2.5 km offshore, and forms barrier reefs, as well as the most extensive fringing reefs in Australia (Fig. 1.20).

The *Dampier Archipelago* between Cape Preston and Port Samson (20ºS) contains a mixture of large and small islands surrounded by barrier reefs, and some small patch reefs both on the islands and along rocky sections of the mainland.

The most extensive system surrounds much of the *Kimberley* coast and adjoining islands (15-17ºS). While most of the reefs lie off the coast surrounding the numerous island groups, there are extensive fringing reefs long the predominantly rocky coast, in addition to 163 usually small beaches fronted by fringing reefs.

Figure 1.20 Fringing coral reef along Pilgonaman Bay, Ningaloo Reef.

Beach Ecology

All tropical and temperate beaches house organisms living in the water above the beach and on and in between the sand grains. The basis of beach ecosystems is the microscopic diatoms that live in the water column (called phytoplankton) and microscopic meiofauna that live on and between sand grains: including bacteria, fungi, algae, protozoa and metazoa. Feeding on these are larger organisms that live in the sand, including meiofauna such as small worms and shrimps; and filter feeding benthos such as molluscs and worms. In the water column they are also preyed upon by zooplankton such as amphipods, isopods, mysids, prawns and crabs. At the top of the food chain are the fish and sea birds, and the occasional mammals such as dolphins, dugongs, turtles and whales. High energy south coast beaches can house a high diversity and density of organisms and export nutrients to the coastal ocean.

A number of hard-bodied organisms also contribute their skeletal material to the beach in the form of sediment. When all organisms die, their internal and/or external skeletons can be broken up, abraded and washed onto beaches by waves.

Shelf Biota

The biota of the southern Australian continental shelf makes a major contribution to the adjacent coast. The shelf contains a range of distinctive habitat assemblages grading from the inner, mid to outer shelf and including the upper continental slope (Fig. 1.21).

The shallow (30-70 m) *inner shelf* is dominated by wave action and contains a range of hard-bodied organisms, particularly molluscs, red algae, encrusting bryozoans,

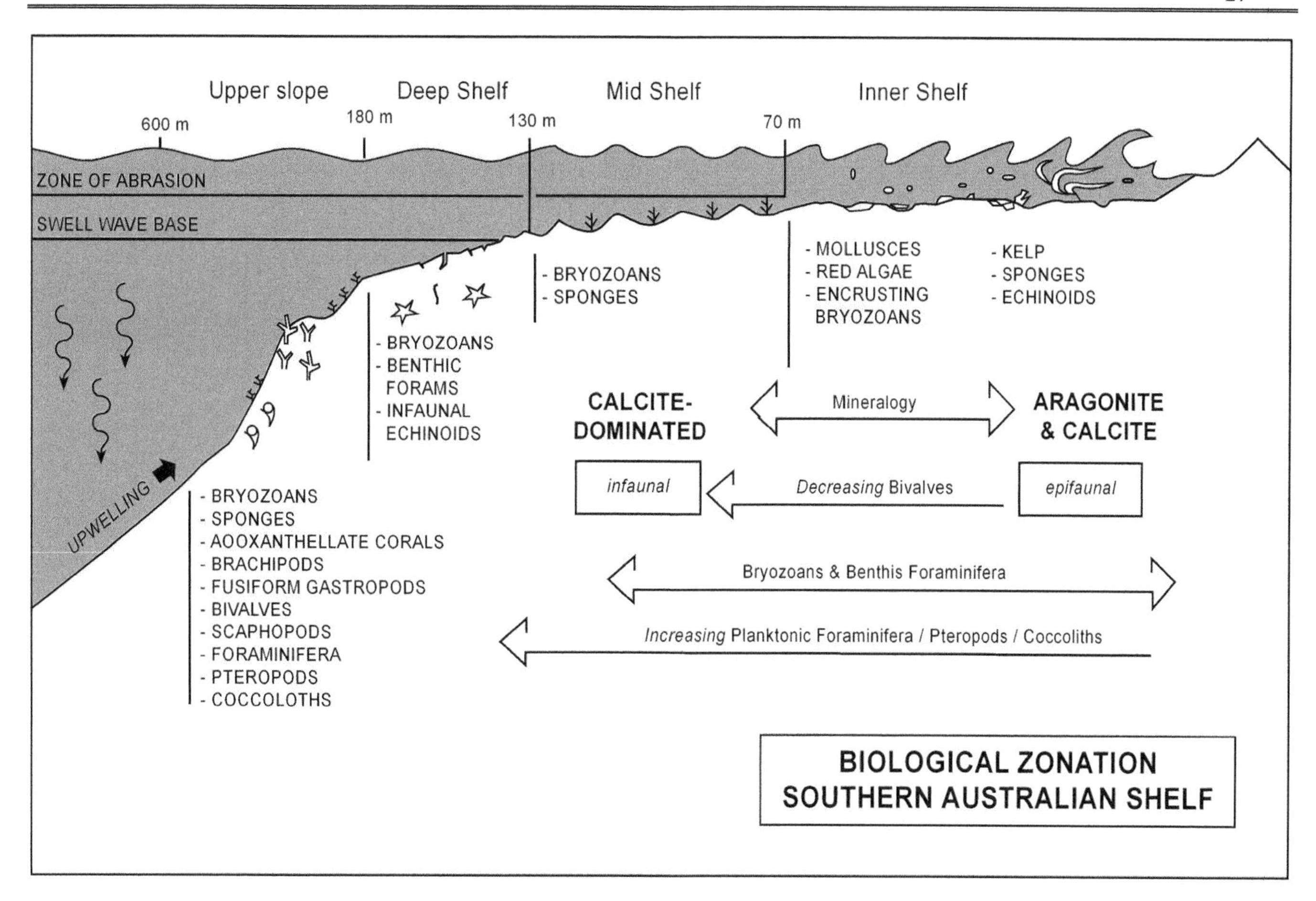

Figure 1.21 Typical shelf biota and zonation across a section of the South Australian continental shelf. Similar zonation exits across the high energy southern Australian continental shelf (Boreen, et al., 1993).

echinoids and soft kelps and sponges. Wave abrasion occurs across the inner shelf permitting erosion and transport of the carbonate detritus. The mid to outer shelf and slopes are dominated by a range of *Bryozoan* species. The *mid shelf* (70-130 m depth) is reworked by shoaling waves and is a zone of carbonate production and accumulation. The *outer shelf* (130-180 m depth) is only reworked during storm wave conditions and contains finer bioclastic sands. The top of the *continental slope* (180-350 m depth) has extensive bryozoan/sponge/coral communities which lead to the accumulation of muddy skeletal sands.

This massive area of carbonate production (shelf carbonate factory) is all the more important because during successive sea level regressions (falling sea level) and transgressions (rising sea level) massive volumes of carbonate detritus is reworked shoreward by waves during the rising sea level and deposited as the carbonate rich beaches and dunes that dominate most of the south coast. On average, approximately 20,000 to 70,000 m^3 of carbonate sand have been moved onshore for every metre of beach along the South East and South coasts of Western Australia. On exposed high energy beaches backed by massive dune systems, this figure can reach up to 100,000 m^3/m. In addition, at present sea level sediment can continue to be moved shoreward from the shallow inner shelf region continuing the supply of shelf carbonate sand to the beaches.

The Western Australian coast is exposed to three major marine ecosystems which contribute sediment to the shoreline. These include the massive shelf carbonate factory in the south, the temperate and tropical seagrass meadows which ring much of the coast, and the fringing coral reef systems in the northwest and north. As a consequence beach and dune sands around the coast are the only quartz rich in the southern region where there is sufficient rainfall to erode and supply terrigenous quartz sands to the shore, particularly during periods of lower sea level when it is dominated by carbonate material. Table 1.10 shows that carbonate material makes up the majority of beach sediments in the South East and up the entire West Coast including the Kimberley region. Only along the more humid South Coast does quartz sand dominate, as this is deposited on the innershelf. This material is subsequently reworked landward by the waves and rising sea level. Elsewhere in the state the general aridity minimises the supply of such material to the shore, contributing to the dominance of carbonate material.

In the more humid Kimberley region terrigenous mud and sands are supplied to the coast, however most of this material is deposited in the landlocked delta systems or deposited as inter and sub-tidal shoals, with insufficient wave energy to rework substantial proportions onshore.

Table 1.10 Regional proportion of carbonate beach sands

Region	Mean Carbonate %	Ω
South East	67	32
South	22	24
Central West-Canning	65	29
Kimberley	60	30

2 Beach Systems

Beaches through the world consist of wave deposited sediment and lie between the depth of wave activity and the upper limit of wave run-up or swash. The entire *beach system* therefore include the dry (subaerial) beach, the swash or intertidal zone, the surf zone and beyond the breakers, the nearshore zone (Fig. 2.1). Usually only the dry beach and swash zone are clearly visible, while bars and channels are often present in the surf zone but are obscured below waves and surf, and the nearshore is always submerged. The shape of any surface is called its morphology, hence *beach morphology* refers to the shape of the beach, surf and nearshore zone (Fig. 2.2).

Beach Morphology

As all beaches are composed of sediment deposited by waves, beach morphology reflects the interaction between waves of a certain height, length and direction and the available sediment; whether it is sand, cobbles or boulders, together with any other structures such as headlands, reefs and inlets.

Western Australian beaches can be very generally divided into three groups: wave-dominated, tide-modified and tide-dominated. The *wave-dominated* beaches occur along the open ocean south and southwest coasts. They are exposed to persistent ocean swell and waves and low tides (<2 m). The *tide-modified* and *tide-dominated* beaches occur in Shark Bay and in the northwest in areas of higher tide range and usually lower waves. The tide-modified beaches are usually exposed to the prevailing seas in the areas of higher tide range, while the tide-dominated are increasingly protected from waves and become increasingly dominated by the tides resulting in a mix of beach and tidal flats. The remainder of this chapter is devoted to a description of first the higher energy wave-dominated beaches, followed by the tide-modified and then the tide-dominated beaches.

Wave-dominated beaches

In two dimensions beaches consist of three zones: the subaerial beach, the surf zone and the nearshore zone, however the nature and extent of these zones vary considerably between the high wave energy beaches of the south coast and the sheltered low wave beaches of the northwest (Fig. 2.2).

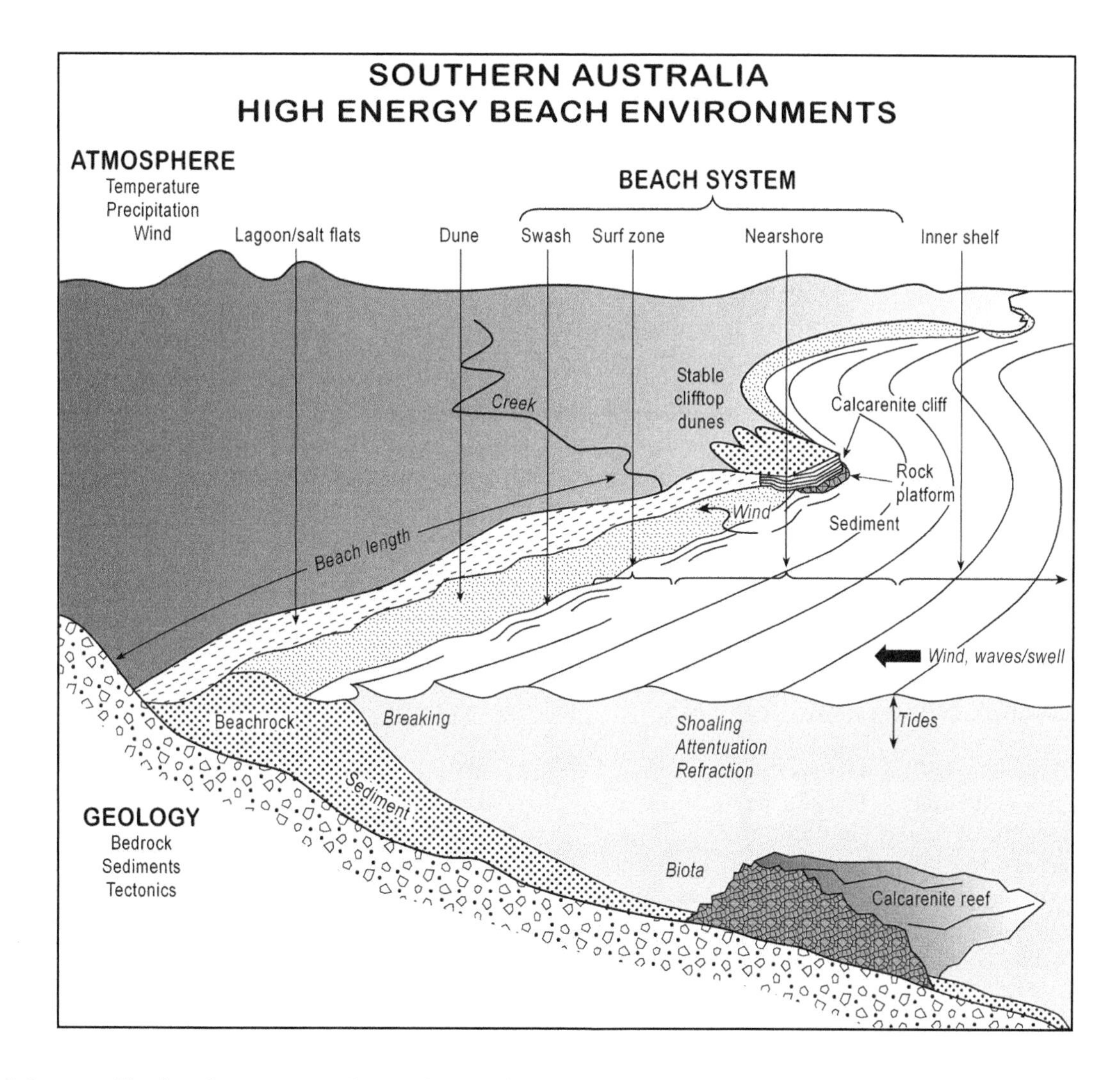

Figure 2.1 The beach system resides at the interface of the ocean, land and atmosphere, and is affected by all three together with input from the rich coastal biota.

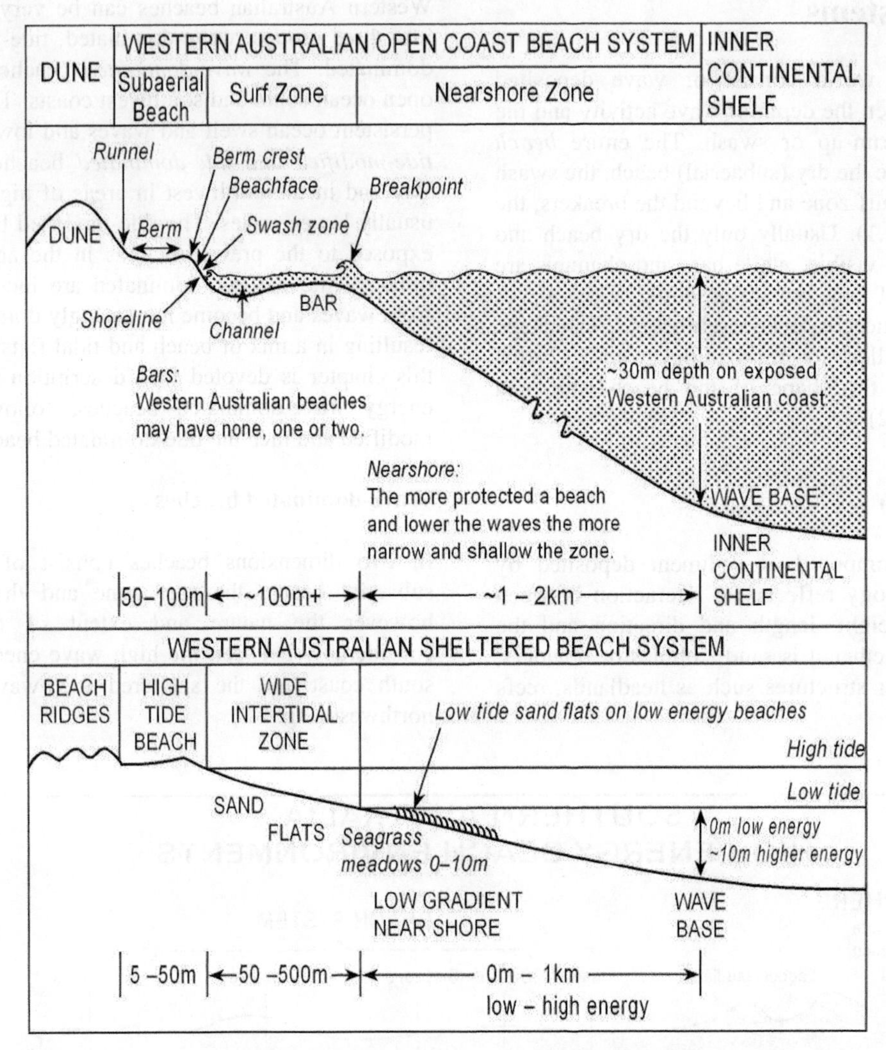

Figure 2.2 Examples of two typical Western Australian beach systems. The higher energy open coast beaches (upper) consist of the dry subaerial beach above the shoreline, the surf zone containing bars, troughs and breaking waves, and the nearshore zone which extends seaward of the breaker zone out to modal wave base. Wave base is the depth to which ocean waves can move beach sands. Seaward lies the inner continental shelf. The approximate width and depth of each zone are indicated. In the more protected northwest and some larger bays (lower) the lower waves and higher tide result in a lower gradient and shallower beach, intertidal and nearshore zone, with wave base as shallow as low tide.

Subaerial beach

The *subaerial beach* is that part of the beach above sea level, which is shaped by wave run-up or swash. It starts at the shoreline and extends up the relatively steep swash zone or beach face. This may be backed by a flatter berm or cusps, which in turn may be backed by a runnel, where the swash reaches at high tide. Behind the upper limit of spring tide and/or storm swash usually lies the edge of dune vegetation. The dry beach varies in width from tens of metres on high energy wave-dominated beaches, to a few metres on very low energy tide-dominated beaches. The subaerial, or dry, beach is that part which most people go to and consider 'the beach'. However, the real beach is far more extensive, in places extending several kilometres seaward, with the subaerial beach forming the figurative 'tip of the iceberg'.

Swash or intertidal zone

On wave-dominated beaches the swash zone connects the dry beach with the surf. The swash zone is the steeper part of the shoreline across which the broken waves run up and down. As wave height decreases and tide range increases, this zone tends to become flatter and considerably wider, and is termed the intertidal zone on tide-modified to tide-dominated beaches. In areas of higher tide range it may reach several hundred metres in width.

Surf zone

The *surf zone* extends seaward of the shoreline and out to the area of wave breaking. This is one of the most dynamic places on earth. It is the zone where waves are continuously expending their energy and reshaping the seabed. It can be divided into the area of wave breaking, often underlain by a bar, and, immediately shoreward, the area of wave translation where the wave bore (white water) moves toward the shoreline, transforming along the way into surf zone currents and, at the shoreline, into swash. Surf zones are up to 500 m wide on exposed southern Western Australian wave-dominated beaches, where waves average over 2 m and break across a double bar system. They decrease in width and bar number as wave height decreases. On tide-modified beaches the usually narrow surf zone is transient with the tide, while it is usually nonexistent on tide-dominated beaches.

Nearshore zone

On wave-dominated beaches the *nearshore zone* is the most extensive part of the beach. It extends seaward from the outer breakers to the maximum depth at which average waves can mobilise beach sediment and move it shoreward. This point is called the *wave base*, referring to the base of wave activity. On the high energy southern Western Australian coast, where waves average over 2 m and commonly reach several metres, it usually lies at a depth between 30 and 50 m and may extend 1, 2 or even 3 km out to sea. It decreases in depth and width as wave height decreases. On tide-modified beaches it is usually at most a few metres deep, while on tide-dominated beaches it may terminate at low tide. The fringe of the seagrass meadows, usually a metre or less below low tide, is a good indicator of its outer limit on sheltered lower energy shores.

Three dimensional beach morphology

In three dimensions, wave-dominated beaches become more complex. This is because most beaches are not uniform alongshore, but vary in a predictable manner.

Beaches vary longshore on two scales. First, they are usually swash-aligned, meaning they are aligned parallel to the crest of the dominant wave. If the wave crest is refracted or bent as it approaches the shoreline, owing to the presence of obstacles such as reefs, rocks and headlands, then the beach will also be shaped to fit the wave crest. The overall effect of the refracting wave crests is to cause a spiral or curvature in the shape of the beach, so that the curvature increases toward the more protected end. This bending of the wave crests is called *wave refraction* (Fig. 2.3a), while the loss of wave energy and height is called *wave attenuation* (Fig. 2.3b). The decrease in wave height along the beach results in lower breakers, a narrower and shallower surf zone, a narrower and often steeper swash zone and a lower and narrower subaerial beach.

In Western Australia beaches are usually bordered by headlands and reefs. They average 1.8 km in length south of Roebuck Bay and only 530 m in length around the Kimberley, with a State average of 1.3 km, close to the Australian average of 1.37 km. As a result there is considerable opportunity for these structures to influence the size and direction of waves arriving at many beaches.

a.

b.

Figure 2.3 a) Wave refraction around the main reefs at Margaret River focuses the waves on the reefs with lower waves to either side; b) Wave attenuation across shallow reefs off Gnarabup Beach, result in a low energy reflective shoreline.

The second scale of longshore beach variation relates to any and all undulations on the beach and in the surf zone, usually spaced from a few metres to as much as 500 m or more on very high energy beaches. Variable longshore forms produced at this scale include regular beach cusps located in the high tide swash zone and spaced between 20 and 40 m, and all variation in rips, bars, troughs and any undulations along the beach, usually with spacing of between 200 and 500 m. These features are associated with rip circulation and are known as rip channels, crescentic and transverse bars, and megacusp horns and embayments. Each of these and its associated beach type and characteristics are described in the section on Beach Types.

Beach Dynamics

Beach dynamics refers to the dynamic interaction between the breaking waves and currents and the sediments that compose the beach. In the long term (tens to thousands of years) this interaction builds all beaches and contributes sand to form backing dune systems. It can also ultimately erode these beaches. In the shorter term (days to months), changing wave conditions produce continual changes in beach response and shape, as sand is moved onshore (beach

accretion) and offshore (beach erosion), together with the associated movement of the shoreline, bars and channels.

Five factors determine the character of a beach. These are the size of the sediment, the height and length of the waves, the characteristics of any long waves present in the surf zone and the tide range. The impact of each is briefly discussed below.

Beach sediment

The size of beach sediment determines its contribution to beach dynamics. Unlike in air where all objects fall at the same speed, sediment falls through water at a speed proportional to its size. Very fine sediment, like clay, will simply not sink but stay in suspension for days or weeks, causing turbid, muddy water. Silt, sediment coarser than clay but finer than sand, takes up to two hours to settle in a laboratory cylinder (Table 2.1). As a consequence fine sediments usually stay in suspension through the energetic surf zone and tidal inlets and are carried out to sea to settle in deep water on the continental shelf.

Table 2.1 Sediment size and settling rates

Material	Size – diameter	Time to settle 1 m
clay	0.001-0.008 mm	hours to days
silt	0.008-0.063 mm	5 min to 2 hours
sand	0.063-2 mm	5 sec to 5 min
cobble	2 mm-6.4 cm	1 to 5 sec
boulder	> 6.4 cm	< 1 sec

Sand takes from a few seconds for coarse sand, to five minutes for very fine sand, to settle through 1 m of water. For this reason it is fine enough to be put into suspension by waves, yet coarse enough to settle quickly to the seabed as soon as waves stop breaking. In the energetic breaking wave environment, anything as fine or finer than very fine sand stays in continual suspension and is flushed out of the beach system into deeper, quieter water. This is why ocean beaches never consist of silts or mud. Most beaches consist of sand because it can be transported in large quantities to the coast and settle fast enough to remain in the energetic surf zone.

Twenty Western Australian beaches described in this book consist of cobbles and boulders. Cobbles and boulders require substantial wave or current energy to be lifted or moved and then settle immediately. They are therefore only rarely moved, such as during extreme storms, and then only very slowly and over short distances. Consequently, Western Australian beaches containing such coarse sediment always have a nearby source, usually an eroding cliff or bluff, while elsewhere they may lie at the mouth of gravel streams.

Depending on the nature of its sediment, each beach will inherit a number of characteristics. Firstly, the sediment will determine the mineralogy or composition of the beach. Sediment derived from the land via rivers and creeks is usually quartz sand or silica. In Western Australia, however, rivers and streams are few and most sand has been derived from the biotic detritus of the seagrass

meadows and inner continental shelf and is carbonate (algal, shell detritus, etc.) as discussed in Chapter 1. Secondly, the size of the sediment will, along with waves, determine beach shape and dynamics. Fine sand produces a low gradient (1-3°) swash zone, wide surf zone and potentially highly mobile sand. Medium to coarse sand beaches have a steeper gradient (4 to 10°), a narrower surf zone and less mobile sand. Cobble and boulder beaches are not only very steep (> 8°), but they have no surf zone and are usually immobile. Therefore, identical waves arriving at adjacent fine, medium and coarse sand beaches will interact to produce three distinctly different beaches.

Likewise, three beaches having identical sand size, but exposed to low, medium and high waves, will have three very different beach systems. Therefore, it is not just the sand or the waves, but the interaction of both, together with long waves and tides that determine the nature of our beaches.

Wave energy - long term

Waves are the major source of energy to build and change beaches. Seaward of the breaker zone, waves interact with the sandy seabed to stir sand into suspension and, under normal conditions, slowly move it shoreward. The wave-by-wave stirring of sand across the nearshore zone and its shoreward transport have been responsible for the delivery of all the sand that presently composes the beaches and coastal sand dunes of Western Australia.

Waves are therefore responsible for supplying the sand to build beaches. The higher the waves, the greater the depth from which they can transport sand and the faster they can transport it. Consequently, high wave energy beaches can deliver the largest volumes of sand and potentially supply sand to build the biggest dunes. Along parts of the South East and South coasts, massive amounts of up to 100 000 m^3 of sand have been transported onshore for every metre of beach. However, these same large waves can just as rapidly erode the very dynamic beaches they initially built. For this reason many high energy Western Australian beaches have left remnants of dunes on top of cliffs and bluffs, while the beaches in many cases have been completely eroded and removed.

Lower waves can only transport sand from shallow depths and at slower rates. Consequently, they build smaller barrier systems, usually delivering less than 10 000 m^3 for every metre of beach, as is typical of more sheltered parts of the open coast. However, these beaches are less dynamic, more stable and are less likely to be eroded.

Wave energy - short term

Waves are not only responsible for the long-term evolution of beaches, but also the continual changes and adjustments that take place as wave conditions vary from day to day. As noted above, a wave's first impact on a beach is felt as soon as the water is shallow enough for wave shoaling to commence, usually in less than 30 m water depth. As waves shoal and approach the break point, they undergo a rapid transformation, which results in the waves becoming slower and shorter, but higher, and ultimately breaking, as the wave crest overtakes the trough.

As waves break, they release kinetic energy, energy that may have been derived from the wind some hundreds or even thousands of kilometres away. This energy is released as turbulence, sound (the roar of the surf) and even heat. The turbulence stirs sand into suspension and carries it shoreward with the wave bore. The wave bore decreases in height shoreward, eventually collapsing into swash as it reaches the shoreline.

Breaking waves, wave bores and swash, together with unbroken and reformed waves, all contribute to a shoreward momentum in the surf zone. As these waves and currents move shoreward, much of their energy is transferred into other forms of surf zone currents, namely longshore, rip feeder and rip currents, and long waves and associated currents. The rip currents are responsible for returning the water seaward, while the long wave currents play a major role in shaping the surf zone.

It is therefore the variation in waves and sediment that produces the seemingly wide range of beaches present along the coast, ranging from the steep, narrow, protected beaches to the broad, low gradient beaches with wide surf zones, large rips and massive breakers. Yet every beach follows a predictable pattern of response, largely governed by its sediment size and prevailing wave height and length. The types of beaches that can be produced by waves and sand are discussed later.

Beach types

Beach type refers to the prevailing nature of a beach, including the waves and currents, the extent of the nearshore zone, the width and shape of the surf zone, including its bars and troughs, and the dry or subaerial beach. *Beach change* refers to the changing nature of a beach or beaches along a coast as wave, tide and sediment conditions change.

The first comprehensive classification of wave-dominated beaches was developed by the Coastal Studies Unit (CSU) at the University of Sydney in the late 1970s, followed by the first investigation of tide-dominated beaches at Cable Beach, Broome in 1980. The *wave-dominated* classification is now used internationally, wherever tide range is less than 2 m. In Australia it applies to most of the southern coast, from Fraser Island in the east around to Exmouth Peninsula in the west. In some large southern bays and gulfs, including Shark Bay, and across northern Australia, waves are lower to nonexistent and tides are generally higher, producing a different range of beach types. In the early 1990s the CSU undertook research along the central Queensland coast, which together with the earlier Broome investigations resulted in the identification of a range of *tide-modified* and *tide-dominated* beach types. Based on this work the full range of Western Australian wave-dominated, tide-modified and tide-dominated beach types is summarised in Tables 2.2 and 2.3, together with two additional types where rocks and/or reefs dominate the intertidal zone. In the following sections each of the beach types is described, together with examples and photographs of the fifteen beach types that occur along the Western Australian coast. Note this table and data relate only to the coast between Eucla and Roebuck Bay. The Kimberley coast and beaches will be covered in a later book on the beaches of Northern Australia.

Wave-dominated beach types

Wave-dominated beaches consist of three types: reflective, intermediate or dissipative, with the intermediate having four states as indicated in Figures 2.4 and 2.5. In Western Australia all wave-dominated beach types occur on the open coast (Table 2.2).

Dissipative beaches (D)

Dissipative beaches occur where a combination of high waves and fine sand ensures that they have wide surf zones and usually two to occasionally three shore-parallel bars, separated by subdued troughs. The beach face is composed of fine sand and is always wide, low and firm, firm enough to support a 2WD drive car. In Western Australia there is only one fully dissipative beach at South Beach (WA 1062, Dongara) and one which grades into a multi-bar dissipative system near Point Malcolm (WA 23). Their rarity on the coast is a product of the lack of fine sand, as there are numerous beaches with waves high enough for forming this beach type, but only if the sand is also fine enough.

On dissipative beaches wave breaking begins as spilling breakers on the outer bar, which reform to break again and perhaps again, on the inner bar or bars. In this way they dissipate their energy across the surf zone, which may be up to 300-500 m wide (Fig. 2.6). This is the origin of the name 'dissipative'.

In the process of continual breaking and re-breaking across the wide surf zone, the incident or regular waves decrease in height and may be indiscernible at the shoreline. The water and energy contained in the wave at the break point is gradually transferred in crossing the surf zone to a lower frequency movement of water, called a standing wave. This is known as red shifting, where energy shifts to the lower frequency, or red end, of the energy spectrum.

*Table 2.2 Western Australian (Eucla-Roebuck Bay) beach types by number and length (modal type in **bold**).*

	Beach Type	Number	Number %	Mean (km)	Ω	Total (km)	km (%)
	Wave-dominated						
1	Reflective	**830**	40.4	1.27	2.67	1052.5	28.6
2	Low tide terrace	239	11.7	2.47	6.7	595.3	16.1
3	Transverse bar & rip	220	10.7	3	8.1	660	18
4	Rhythmic bar & beach	85	4.1	1.65	2.8	140	3.8
5	Longshore bar & trough	0	0			0	0
6	Dissipative	1	0.2	2	0	8.2	0.2
	Tide-modified						0
7	R+low tide terrace	38	1.9	0.74	1.9	29.6	0.8
8	R+low tide rips	0	0	0	0	0	0
9	Ultra dissipative	**48**	2.3	6	11.8	288	7.8
	Tide-dominated						0
10	R+ridged sand flat	16	0.8	2.6	0	41.6	1.1
11	R+sand flat	**329**	16	1.74	2.4	572.5	15.5
12	R+tidal flat	139	6.8	1.44	1.83	200	5.4
13	R+mud flat	4	0.2	2.2	0	10.8	0.3
	Beaches+rock/reef flats						0
14	R+rock flat	60	2.9	0.65	1.2	39	1.1
15	R+coral reef	37	1.8	1.38	1.15	49.7	1.3
		2051	100.0			3687.3	100

Table 2.3 Western Australian beach types by coastal regions

Beach type[1]	South East	South	Leeuwin	Central West	***Southern WA***	Carnarvon	Pilbara	Canning	***Northern WA***	**Total WA**
1	6	165	75	340	***586***	171	46	27	***244***	**830**
2	20	117	14	38	***189***	51			***51***	**240**
3	14	173	23	13	***223***					**219**
4	6	79			***85***					**85**
5										**0**
6				1	***1***					**1**
7							38	1	***39***	**39**
8										**0**
9								48	***48***	**48**
10						13	3		***16***	**16**
11	2	3		16	***21***	227	51	30	***299***	**329**
12		1			***1***	70	38	30	***134***	**139**
13								4	***4***	**4**
14	4	22	2		***27***	5	12	15	***23***	**60**
15						37			***37***	**37**
Total	52	560	114	408	***1134***	574	188	155	***917***	**2047**

[1] See Table 2.2 for name of beach types.

At the shoreline, the standing wave is manifest as a periodic (every 60-120 seconds) rise in the water level (set-up), followed by a more rapid fall in the water level (set-down). As a rule of thumb, the height of the set-up is 0.3-0.5 times the height of the breaking waves (i.e. 1 - 1.5 m for a 3 m wave). Because the wave is standing, the water moves with the wave in a seaward direction during set-down, with velocities between 1 and 2 m/sec closer to the seabed. As the water continues to set down, the next wave is building up in the inner surf zone, often to a substantial wave bore, 1 m+ high. The bore then flows across the low beach face and continues to rise, as more water moves shoreward and sets up. This process continuously repeats itself every one to two minutes. Because of the fine sand and the large, low frequency standing wave, the beach is planed down to a wide, low gradient, with the high tide swash reaching to the back of the beach, often leaving no dry sand to sit on at high tide.

Dissipative beach hazards

The wide surf zones and high waves associated with dissipative beaches keep most bathers to the inner swash and surf zone. They are relatively safe close inshore, though not without some surprises, while the mid to outer surf zone is only for the fittest and most experienced surfers.

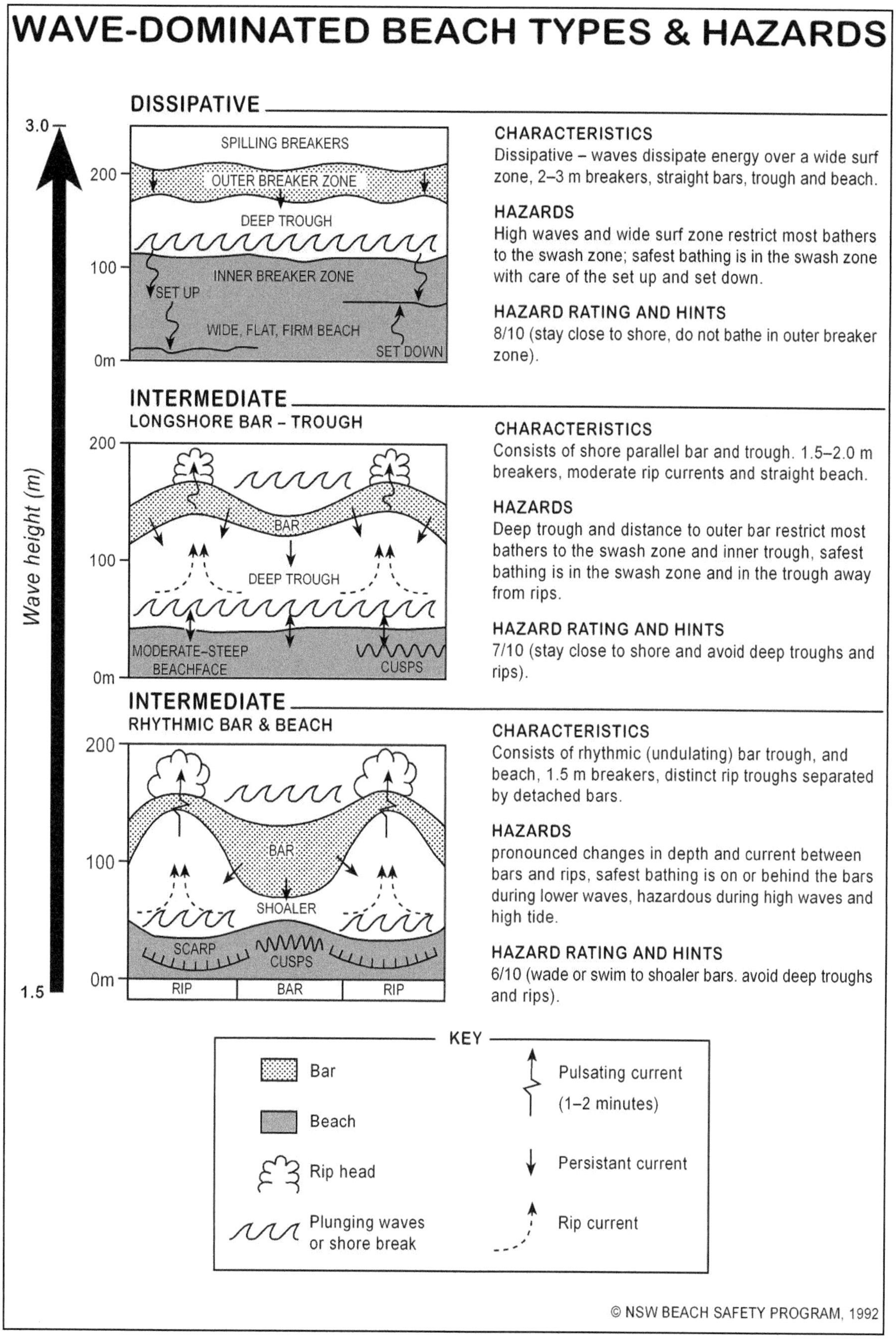

WAVE-DOMINATED BEACH TYPES & HAZARDS

DISSIPATIVE

3.0
200
100
0m

SPILLING BREAKERS
OUTER BREAKER ZONE
DEEP TROUGH
INNER BREAKER ZONE
SET UP
WIDE, FLAT, FIRM BEACH
SET DOWN

CHARACTERISTICS
Dissipative – waves dissipate energy over a wide surf zone, 2–3 m breakers, straight bars, trough and beach.

HAZARDS
High waves and wide surf zone restrict most bathers to the swash zone; safest bathing is in the swash zone with care of the set up and set down.

HAZARD RATING AND HINTS
8/10 (stay close to shore, do not bathe in outer breaker zone).

INTERMEDIATE
LONGSHORE BAR - TROUGH

200
100
0m

BAR
DEEP TROUGH
MODERATE-STEEP BEACHFACE
CUSPS

CHARACTERISTICS
Consists of shore parallel bar and trough. 1.5–2.0 m breakers, moderate rip currents and straight beach.

HAZARDS
Deep trough and distance to outer bar restrict most bathers to the swash zone and inner trough, safest bathing is in the swash zone and in the trough away from rips.

HAZARD RATING AND HINTS
7/10 (stay close to shore and avoid deep troughs and rips).

INTERMEDIATE
RHYTHMIC BAR & BEACH

200
100
0m
1.5

Wave height (m)

BAR
SHOALER
SCARP
CUSPS
RIP BAR RIP

CHARACTERISTICS
Consists of rhythmic (undulating) bar trough, and beach, 1.5 m breakers, distinct rip troughs separated by detached bars.

HAZARDS
pronounced changes in depth and current between bars and rips, safest bathing is on or behind the bars during lower waves, hazardous during high waves and high tide.

HAZARD RATING AND HINTS
6/10 (wade or swim to shoaler bars. avoid deep troughs and rips).

KEY

Bar
Beach
Rip head
Plunging waves or shore break
Pulsating current (1–2 minutes)
Persistant current
Rip current

Figure 2.4 A plan view of the dissipative, longshore bar & trough and rhythmic bar & beach beach types. As wave height increases between the rhythmic and dissipative beaches, the surf zone increases in width, rips initially increase and then are replaced by other currents, and the shoreline becomes straighter. The physical characteristics and beach and surf hazards associated with each type are indicated, as well as its beach hazard rating

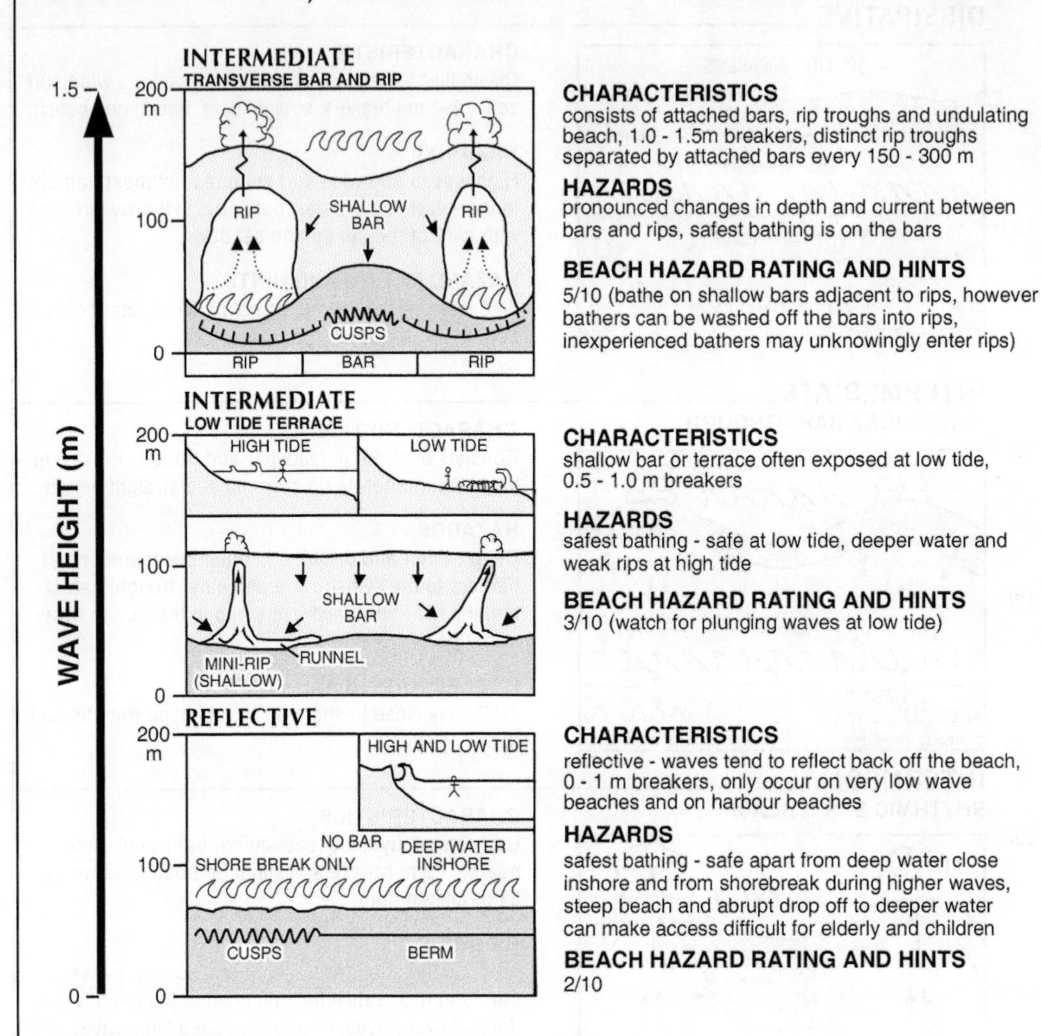

PLEASE NOTE:

This model represents average wave conditions on these beach types in micro tidal (< 2 m tide range) regions of southern Australia (south Queensland, NSW, Victoria, Tasmania, South Australia and southern Western Australia).

BEACH SAFETY IS INFLUENCED BY:

HEADLANDS - rips usually occur and intensify adjacent to headlands, reefs and rocky outcrops.

OBLIQUE WAVES - stronger longshore currents, skewed and migratory waves.

HIGH TIDE - deeper water and in some cases stronger rips.

RISING SEAS - eroding bars, stronger currents, strong shifting rips, greater set up and set down.

HIGH TIDE AND RISING SEAS - more difficult to distinguish bars and troughs.

STRONG ONSHORE AND ALONGSHORE WINDS - reinforced downwind currents.

MEGARIPPLES - large migratory sand ripples common in rip troughs can produce unstable footing.

LOW TIDE - rips more visible but normally more intensified due to restricted channel.

CHANGING WAVE CONDITIONS - (rising, falling, change in direction or length) - produce a predictable change in beach topography and type; the reason why beaches are always changing.

Figure 2.5 A plan view of the transverse bar and rip, low tide terrace and reflective beach types. As wave height increases between the reflective and transverse beaches, the surf zone and bar increase in width, rips form and increase in size, and the shoreline becomes crenulate. The physical characteristics and beach and surf hazards associated with each type are indicated, as well as its beach hazard rating.

Dissipative beach hazards

Most people do not venture far into dissipative surf zones as they are put off by their extremely wide surf and high outer breakers. However, if you do, this is what to watch out for:

- Outer surf zone - spilling breakers. Bigger sets break well seaward and catch surfers inside. Troughs - usually on/offshore currents, but chance of longshore and even rip currents, particularly under lower (< 1.5 m) wave conditions.
- Inner surf zone - watch for standing wave bores that can knock you over, fortunately shoreward. Set-down produces an often strong seaward flow, particularly closer to the seabed, which may also drag children off their feet.
- Swash zone/beach face - this is where most bathers stay and where most get into trouble, owing to the set-up and set-down. Be aware that water level will vary considerably between set-up and set-down, and currents will reverse from onshore to offshore. At best you will be knocked over by the incoming bore, at worst you might be dragged seaward by the set-down. Children in particular are most at risk.

Some young children, even babies in prams and parked cars, have been left on a seemingly safe part of the beach face, only to have a higher than usual set-up engulf them in water.

- **Summary:** Dissipative beaches are dangerous, however in Western Australia they only occur in a few locations, but are more frequent when the seas are very big, so most people don't consider swimming, or at least not beyond the swash zone. Definitely for experienced surfers only.

a.

b.

Figure 2.6 High energy dissipative beaches at a) Yokingup Bay, South East coast; and b) Shoal Point, Central West coast.

Intermediate beaches

Intermediate beaches refer to those beach types that are intermediate between the lower energy reflective beaches and the highest energy dissipative beaches. They tend to require waves greater than 0.5 m and can accommodate the highest waves along the South Coast, which can average over 2 m, combined with fine to medium sand. The most obvious characteristic of intermediate beaches is the presence of a surf zone with bars and rips. On the Western Australian coast 542 (26.4%) of the beaches are intermediate. They are the most common beach type along the more exposed South East and South coasts, secondary to the reflective beaches on the Leeuwin and Central West coasts, with only 51 occurring north of Kalbarri along the Carnarvon coast, and none along the Pilbara and Canning coasts where waves are low and tides increasingly high (Table 2.3).

As intermediate beaches are produced by waves between 0.5 m and 2 m plus, they exist in a wide range of waves and associated beach conditions. For this reason, intermediate beaches are classified into four beach states. The lowest energy state is called *low tide terrace*, then as waves increase, the *transverse bar and rip*, then the *rhythmic bar and beach*, and finally the *longshore bar and trough*. Each of these beaches is briefly described below.

Longshore bar and trough (LBT)

The *longshore bar and trough* beach type does not occur as a modal inner bar type in Western Australia and will not be further discussed. For more information on this beach type see descriptions in Short (1993, 1996, 2000 and 2001).

Rhythmic bar and beach (RBB)

The *rhythmic bar and beach* type is the highest energy beach type that commonly occurs on the open coast. In all, 80 beaches, averaging 3.9 km in length, are of this type. They occupy 134 km (3.6%) of the sandy coast. These energetic beaches require two primary ingredients for their formation, relatively fine to medium sand and exposure to the high deepwater waves. They occur where waves average at least 1.5 m and sand is fine to medium (Fig. 2.7).

Figure 2.7 High energy rhythmic bar and beach, Quallup beach, South Coast. Note the large rips (arrows).

Rhythmic beaches consist of a rhythmic longshore bar that deepens where the rips cross the breakers, and in between broadens, shoals and trends shoreward. It does not, however, reach the shore, with a continuous rip feeder channel feeding the rips to either side of the. The shoreline is usually rhythmic with protruding megacusp horns in lee of the detached bars and commonly scarped megacusp embayments behind the rips. The surf zone may be up to 100 to 250 m wide and the bars and rips are spaced every 250 to 1000 m alongshore.

The shallower sections of the rhythmic bar cause waves to break more heavily, with the white water flowing shoreward as a wave bore. The wave bore flows across the bar and into the backing rip feeder channel. The water from both the wave bore and the swash piles up in the rip feeder channel and starts moving sideways toward the adjacent rip embayment, which may be up to 200 m alongshore. The feeder currents are weakest where they diverge behind the centre of the bar, but pick up in speed and intensity toward the rip, particularly close to shore. In addition, the rip feeder channels deepen toward the rip.

In the adjacent rip channels, waves break less or often not at all. They may move unbroken across the rip to finally break or surge up the steeper rip embayment swash zone. The strong swash often causes slight erosion of the beach face and cuts an erosion scarp.

In the rip embayment, the backwash returning down the beach face combines with flow from the adjacent rip feeder channels. This water builds up close to shore (called wave set-up), then pulses seaward as a strong, narrow rip current. The currents pulse every 30 to 90 seconds, depending on wave conditions. The rip current accelerates with each pulse and persists with lower velocities between pulses. Rip velocities are usually less than 1 m per second (3.5 km/h), but will increase up to 2 m per second in confined channels and under higher waves.

To identify this beach type, look for the pronounced longshore beach rhythms, i.e. the shoreline is very sinuous. The shallowest, widest bars and heaviest surf lie off the protruding parts of the shore (the megacusps). Water flows off the bars, into the feeder channel, along the beach to the deeper rip embayment, then seaward in the rip current.

Rhythmic bar and beach hazards

This is the most hazardous beach type commonly occurring along the South and South East coasts. Most people are put off entering the surf by the deep longshore trough containing rips and their feeder currents. If you are swimming or surfing on a rhythmic beach, the following highlights some common hazards.

Rhythmic bar and beach hazards

- Bar - just to reach the bar requires crossing the rip feeder channel. This may be an easy wade at low tide or a difficult swim at high tide. Be very careful once the water exceeds waist depth, particularly if a current is flowing. Also, as you reach the bar, water pouring off the bar may wash you back into the channel.
- The centre of the bar is relatively shallow at low tide, but at high tide you run the risk of being washed into the rip feeder or rip channel.
- Rip feeder channel - depth varies with position and tide, both depth and velocity increase toward the rip.
- Rip - the rip channel is usually 2 to 3 m deep, with a continuous, but pulsating, rip current.
- High tide - deeper bar and channels, but weaker currents and rips.
- Low tide - waves break more heavily and may plunge dangerously, shallower bar and channels, but stronger currents and rips.
- Oblique waves - skew bar and rips alongshore.
- Higher waves - intensify wave breaking and strength of all currents.

- *Summary:* Caution is required by the young and inexperienced on rhythmic beaches, as the bar is separated from the beach by often deep channels and strong currents.

Transverse bar and rip (TBR)

The *transverse bar and rip* is the most extensive of the intermediate beach types. There are 222 TBR beaches, with a total length of 660 km (18% of coast) and an average length of 3 km. They occur primarily along the South Coast (173) on beaches composed of fine to medium sand and exposed to waves averaging between 1 and 2 m. TBR beaches receive their name from the fact that as you walk along the beach, you will see bars transverse or perpendicular to and attached to the beach, separated by deeper rip channels and currents (Fig. 2.8). The bars and rips are usually regularly spaced with a mean spacing in Western Australia of 350 m (Ω = 120 m). Their surf zones range from 50 to 150 m in width.

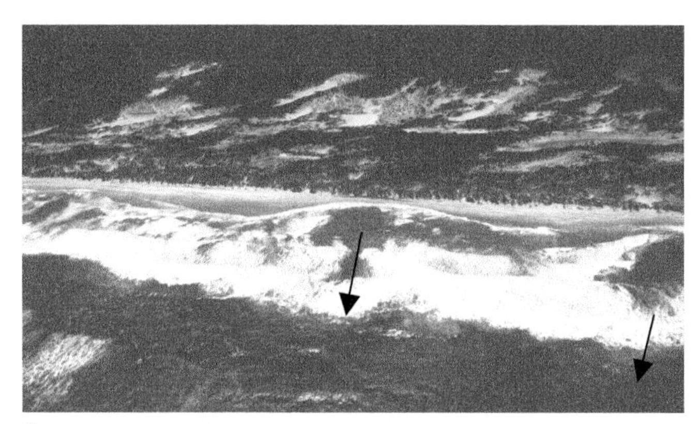

a.

b.

Figure 2.8 Transverse bar and rip beaches: a) Cliffy Head; and b) Sandy Bight, note the well developed rip heads, both South Coast.

Rip currents

Beach rips: Rip currents are a relatively narrow, seaward moving stream of water flowing over or through the sand bar (Fig. 2.9). They represent a mechanism for returning water back out to sea that has been brought onshore by breaking waves. They originate close to shore as broken waves (wave bores) that flow into longshore rip feeder troughs. This water moves along the base of the beach as rip feeder currents. On normal beaches, two currents arriving from opposite directions usually converge in the rip embayment, turn and flow seaward across the bar and through the surf zone. The currents usually maintain a deeper rip feeder trough close to shore and a deeper rip channel cut into the bar and through the surf zone.

The converging currents turn, accelerate and flow seaward through the surf zone, either directly or at an angle at speeds up to 1.5 m/sec. As the confined rip current exits the surf zone and flows seaward of the outer breakers, it expands and may meander as a larger rip head. Its speed decreases and it will usually dissipate within a distance of two to three times the width of the surf zone.

Rip currents will exist in some form on ALL beaches where there is a surf zone, particularly when waves exceed 1 m. In Western Australia they are most prevalent on the higher energy South Coast beaches. State-wide 1470 beach rips are operating on an average day along 520 km of the sandy coast (14%). They range in spacing from 200 to 700 m with a mean of 350 m (Ω = 120 m). This is the largest average beach rip spacing in Australia and among the largest in the world.

a.

b.

c.

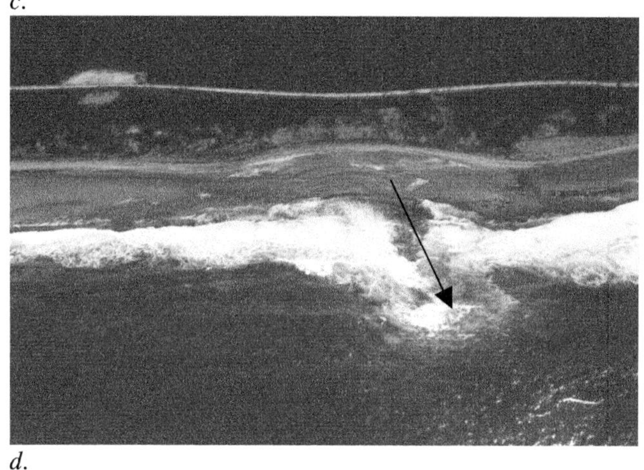

d.

Figure 2.9 a) Three beach rips on Surfers Beach, Esperance; b) strong topographic rips at Ledge Point; c) at Marlbemup; and d) at 12 Mile Beach, east of Hopetoun.

Topographic rips: Headlands and reefs in the surf can induce a strong seaward flow of water, called 'topographically controlled rip' or 'topo-rip'. During big seas these rips can expand to occupy an entire beach embayment and form large 'megarips', that is, large scale topographically controlled rips. Approximately 900 permanent topographic rips are located along the Western Australian coast, occurring primarily along the South Coast on all beaches with surf and headlands and/or surf zone reef (Fig. 2.9).

In total, Western Australia has about 2370 beach and topographic rip systems operating on a typical day, primarily on more exposed beaches in the southern half of the state.

Rip current spacing

- spacing approximately = surf zone width x 4
- in Western Australia rip spacing averages 350 m, ranges from 200-700 m
- also a function of beach slope, the lower the slope (hence wider the surf zone), the wider the rip spacing

TBR beaches are discontinuous alongshore as the alternation of shallow bars and deeper rip channels causes a longshore variation in the way waves break across the surf zone. On the shallower bars waves break heavily, losing much of their energy. In the deeper rip channels they will break less and possibly not at all, leaving more energy to be expended as a shorebreak at the beach face. Consequently, across the inner surf zone and at the beach face, there is an alternation of lower energy swash in lee of the bars and higher energy swash/shorebreak in lee of the rips.

This longshore variation in wave breaking and swash causes the beach to be reworked, such that slight erosion usually occurs in the embayments in lee of the rips and slight deposition in lee of the bars. This results in a rhythmic shoreline, building a few metres seaward behind the attached bars as deposition occurs and being scoured out and often scarped in lee of the rips. The rhythmic undulations are called megacusp horns (behind the bars) and embayments (behind the rips). Whenever you see such rhythmic features, which have a spacing identical to the bar and rips (200-700 m), you know rips are present.

The TBR surf zone has a cellular circulation pattern. Waves tend to break more on the bars and move shoreward as wave bores. This water flows both directly into the adjacent rip channel and, closer to the beach, into the rip feeder channels located at the base of the beach. The water in the rip feeder and rip channel then returns seaward in two stages. Firstly, water collects in the rip feeder channels and the inner part of the rip channel, building up a hydraulic head against the lower beach face. Once high enough, it pulses seaward as a relatively narrow accelerated flow, the rip. The water usually moves through the rip channel, out through the breakers and seaward for a distance up to twice the width of the surf zone.

The velocity of the rip currents varies tremendously. On a typical beach with waves less than 1.5 m, they peak at about 1 m per second, or 3.5 km per hour, about walking pace. However, under high waves they may double that speed. What this means is that under average conditions, a rip may carry someone out from the shore to beyond the breakers in 20 to 30 seconds. Even an Olympic swimmer going at 2 m per second would only be able to maintain their position, at best, when swimming against a strong rip.

Two other problems associated with rips and rip channels are their depth and their rippled seabed. The channel is usually 0.5 to 1 m deeper than the adjacent bar, reaching maximum depths of 3 m. Furthermore the faster seaward flowing water forms megaripples on the floor of the rip channel. These are sand ripples 1 to 2 m in length and 0.1 to 0.3 m high that slowly migrate seaward. The effect of the depth and ripples on bathers is to provide both variable water depth in the rip channels and a soft sand bottom, compared to the more compact bar. As a result, it is more difficult to maintain your footing in the rip channel for three reasons: the water is deeper, the current is stronger and the channel floor is less compact. Also, someone standing on a megaripple crest that is suddenly washed or walks into the deeper trough, may think the bottom has 'collapsed'. This may be one source of the 'collapsing sand bar' myth, an event that cannot and does not occur.

Transverse bar and rip beach hazards

Transverse beaches are one of the main reasons many Western Australian beaches have good surf. However, the good surf is also a hazard to the unwary swimmer and most drowning and rescues occur with this beach type. The shallow bars tempt people into the surf, while lying to either side are the deeper, more treacherous rip channels and currents.

Transverse bar and rip beach hazards

- Bars - the centres of the attached bars are the best place to swim. They are shallow, furthest from the rip channels and the wave bores move toward the shore.
- Rips - the rips are the cause of most surf rescues, so they are best avoided unless you are a very experienced surfer.
- Rip feeder channels - usually run along behind and to the sides of the bar, adjacent to the base of the beach. They carry water alongshore and deliver (feed) it to the seaward flowing rip current.
- In the rip embayment, the feeder currents converge and head out to sea. If you are not experienced, stay away from any channels, particularly if the water is moving and greater than waist depth.
- Children on floats must be very wary of feeder channels as they can drift from a seemingly calm, shallow, inner feeder channel, located right next to the beach, rapidly out into a strong rip current.
- Breakers - waves will break more heavily on the bar at low tide, often as dangerous plunging waves or dumpers. In the rip embayment, the shorebreak will be stronger at high tide.

- Higher waves - when waves exceed 1 to 1.5 m, both wave breaking and rip currents will intensify.
- Oblique waves - skew both the bars and rips alongshore and may make the rips more difficult to spot.
- Low tide - rip currents are more confined to the rip channel and as a result intensify at low tide.
- High tide - rip currents are weaker and may be partially replaced by a longshore current, even across the bar.
- *Summary:* It is relatively safe on the bars during low to moderate waves, but beware, as many hazards, particularly rips, lurk for the young and inexperienced. Stay on the bar/s and well away from the rips and their side feeder currents.

Low tide terrace (LTT)

Low tide terrace beaches are the lowest energy intermediate beach type and the most common intermediate type in Western Australia occurring on 239 beaches (12%) and occupying 595 km (16%) of the sandy coast. They occur on the open coast where sand is fine to medium and wave height averages between 0.5 and 1 m. They are in all coastal regions south of North West Cape, with most occurring along partly sheltered beaches of the South Coast (117) particularly where nearshore reefs and headlands lower waves to less than 1 m at the shore (Fig. 2.10).

Low tide terrace beaches are characterised by a moderately steep beach face, which is joined at the low tide level to an attached bar or terrace, hence the name - low tide terrace. The bar usually extends between 20 and 50 m seaward and continues alongshore, attached to the beach. It may be flat and featureless, have a slight central crest, called a ridge, and may be cut every several tens of metres by small shallow rip channels, called *mini rips*.

At high tide when waves are less than 1 m, they may pass right over the bar and not break until the beach face, behaving much like a reflective beach. However, at spring low tide, the entire bar is usually exposed as a ridge or terrace running parallel to the beach. At this time, waves break by plunging heavily on the outer edge of the bar. At mid tide, waves usually break right across the shallow bar.

Under typical mid tide conditions, waves break across the bar and a low surf zone is produced. Waves are less than 1 m and most water appears to head toward the shore. In fact it is also returned seaward, both by reflection off the beach face and via the mini rips, even if no rip channels are present. The rips, however, are usually weak, ephemeral and shallow.

a.

b.

Figure 2.10 A continuous low tide terrace beach at a) Barrens Beach; and b) Alexander Bay.

Low tide terrace hazards

Low tide terrace beaches are the least hazardous of the intermediate beaches, because of their characteristically low waves and shallow terrace. However, changing wave and tide conditions do produce a number of hazards to swimmers and surfers.

Low tide terrace beach hazards

- High tide - deep water close to shore; behaves like a reflective beach.
- Low tide - waves may plunge heavily on the outer edge of the bar, with deep water beyond. Take extreme care if body surfing or body boarding in plunging waves, as spinal injuries can result.
- Mid tide - more gently breaking waves and waist deep water, however weak mini rips return some water seaward.
- Diving - be very careful diving into the surf as the water is usually shallow and can result in head and spinal injuries.
- Higher waves - mini rips increase in strength and frequency and may be variable in location.
- Oblique waves - rips and currents are skewed and may shift along the beach, causing a longshore and seaward drag.
- Most hazardous at mid to high tide when waves exceed 1 m and are oblique to shore, such as during a strong summer sea breeze.

- **Summary:** One of the safer beach types when waves are below 1 m high, at mid to high tide. Higher waves, however, generate dumping waves, strong currents and ephemeral rips, called *side drag, side sweep* and *flash rips* by lifesavers. Use care when surfing or diving under the waves.

Reflective beaches (R)

Reflective sandy beaches lie at the lower energy end of the wave-dominated beach spectrum. They are characterised by relatively steep, narrow beaches usually composed of coarser sand. On the Western Australian open coast, sandy beaches require waves to be less than 0.5 m to be reflective. For this reason they are also found inside the entrance to bays, at the lower energy end of some ocean beaches and in lee of many of the calcarenite reefs and rock platforms that front many Bight and south west coast beaches.

In Western Australia there are 830 reflective beaches, making them the most common beach type (40%). They are however also the shortest of the beaches with a mean length of 1.27 km as they tend to form in protected pockets in lee of reefs and headlands, even along high energy sections of coast. As a result they have a total shoreline length of 1052 km, which represents only 28.5% of the sandy beach coast.

Reflective beaches are a product of both coarser sand and lower waves. Consequently, all 20 Western Australian beaches composed of gravel, cobble and boulders are always reflective, no matter what the wave height.

Reflective beaches always have a steep, narrow beach and swash zone. Beach cusps are commonly present in the upper high tide swash zone. They have no bar or surf zone as waves move unbroken to the shore, where they collapse or surge up the beach face (Fig. 2.11).

Reflective beach morphology is a product of four factors. First, low waves will not break until they reach relatively shallow water (< 1 m); second, the coarser sand results in a steeper gradient beach (5-10°) and relatively deep nearshore zone (> 1 m); third, because of the low waves and deep water, the waves do not break until they reach the base of the beach face; and finally, because the waves break at the beach face, they must expend all their remaining energy over a very short distance. Much of the energy goes into the wave swash and backwash, the rest is reflected back out to sea as a reflected wave, hence the name reflective.

The strong swash, in conjunction with the usually coarse sediment, builds a steep, high beach face. The *cusps,* which often reside on the upper part of the beach face, are a product of sub-harmonic edge waves, meaning the waves have a period twice that of the incoming wave. The edge wave period and the beach slope determine the edge wavelength, which in turn determines the cusp spacing. On the Western Australian coast cusp spacing can range from 20 to 40 m.

a.

b.

Figure 2.11 Steep, cusped reflective beaches at a) Hammer Head; and b) Little Tagon Beach.

Another interesting phenomenon of most reflective beaches is that all those containing a range of sand sizes have what is called a *beach step*. The step is always located at the base of the beach face, around the low water mark. It consists of a continuous band containing the coarsest material available, including rocks, cobbles, even boulders and often numerous shells. Because it is so coarse, its slope is very steep, hence the step-like shape. They are usually a few decimetres in height, reaching a maximum of perhaps a metre. Immediately seaward of the step, the sediments usually fine markedly and assume a lower slope.

The reason for the step is twofold. The unbroken waves sweep the coarsest sediment continuously toward the beach and the step. The same waves break by surging over the step and up the beach face. However, the swash deposits the coarsest, heaviest material first, only carrying finer sand up onto the beach, then the backwash rolls any coarse material back down the beach. The coarsest material is therefore trapped at the base of the beach face by both the incoming wave and the swash and backwash.

Reflective beach hazards

The low waves and protected locations that characterise reflective beaches usually lead to relatively safe swimming locations. However, as with any water body, particularly one with waves and currents, there are hazards present that can produce problems for swimmers and surfers.

Reflective beach hazards

- Steep, soft beach face - may be a problem for toddlers, the elderly and disabled people.
- Relatively strong swash and backwash - can knock children and unwary people off their feet.
- Step - causes a sudden drop off from shallow into deeper water.
- Deep water - absence of bar means deeper water close into shore, which can be a problem for non-swimmers and children.
- Surging waves and shorebreak - when waves exceed 0.5 m, they break increasingly heavily over the step and lower beach face. They can knock unsuspecting swimmers over. If swimming seaward of the break, swimmers may experience problems returning to shore through a high shorebreak.
- Most hazardous when waves exceed 1 m and shorebreak becomes increasingly powerful.
- Where fronted by a rock platform or reef, additional hazards are associated with the presence of the rock/reef.
- **Summary:** Low hazards under low wave conditions, so long as you can swim. Watch children as deep water is close to shore. Hazardous shorebreak and strong surging swash under high waves (> 1 m).

Determining wave-dominated beach type

The type of wave-dominated beach that occurs on the coast is a function of the modal wave height, which has a maximum of 2 m+, the wave period, which averages 12 seconds, and finally the sand size. While the wave period is essentially constant for the southern half of the state, the wave height is at a maximum along the South Coast, decreasing northward into the Bight and up the West coast (Fig. 1.15), as well as locally decreasing into more protected environments. Finally, sand size averages medium while it can range from fine to coarse sand.

Figure 2.12 provides a method for determining the predicted beach type based on these three parameters. Only the highest energy beaches, with waves averaging over 2 m, produce dissipative beaches in Western Australia. The same waves on beaches composed of slightly coarser (fine to medium) sand produce high-energy intermediate beach systems, in particular longshore bar and trough, rhythmic bar and beach and transverse bar and rip. Where waves are reduced slightly to average between 1 and 1.5 m, they form moderate energy transverse bar and rip beaches, while low tide terrace beaches occur where waves have been reduced to around 1 m and reflective beaches where waves are lowered still further to average about 0.5 m.

Tide-modified beaches

While the southern half of the state is dominated by high waves and low tides resulting in wave-dominated beaches, the opposite occurs in the northern half of the state. Waves are low in Shark Bay and north from Exmouth, while tide range increases progressively from Exmouth peaking at 9 m by Broome. As a consequence in Shark Bay and along the Pilbara and Canning coasts the beaches become increasingly tide-modified and tide-dominated. On the Western Australian coast 89 beaches (4.2%) are tide-modified and 472 (23%) tide-dominated, all but 21 occurring in the northern half of the state. They occupy a total of 1,093 km or 30% of the sandy coast.

By definition, tide-modified beaches occur when the tide range is between 3 and 15 times the wave height. Where tides remain low (i.e. 1 m or less) such as in Shark Bay, the waves must be less than 0.3 m to be considered tide-modified, while along the Pilbara and Canning coasts increasing tide range, as well as lower wind wave, produces these conditions on the open coast. Tide-modified beaches consist of three beach types - ultradissipative, reflective plus bar and rips and reflective plus low tide terrace (Fig. 2.13).

Ultradissipative (UD)

Ultradissipative beaches occur in higher energy tide-modified locations, where the beaches are also composed of fine to very fine sand. They are characterised by a very wide (200-400 m) intertidal zone, with a low to moderate gradient high tide beach and a very low gradient to almost horizontal low tide beach. Because of the low gradient right across the beach, waves break across a relatively wide, shallow surf zone as a series of spilling breakers (Fig. 2.14). This wide, spilling surf zone dissipates the waves to the extent that they are known as 'ultradissipative' beaches. During periods of higher waves (>1 m), the surf zone can be well over 100 m wide, though still relatively shallow.

There are 48 ultradissipative beaches in Western Australia all located along the Canning coast, north of Cape Keraudren, with the best example being the long Eighty Mile Beach. On these beaches the combination of tide exceeding 7 m, combined with the low westerly wind waves (~1 m) and fine sand beaches, produces the wide, low gradient ultradissipative systems. These are also the longest average beach type in the state, averaging 6 km in length and occupying 288 km (7%) of the sandy coast.

Basically the fine sand induces the low gradient, while the tide range moves the higher waves backwards and forwards across the wide intertidal zone every six hours. The two act to plane down the beach, while the lack of stationarity or stability of the surf zone and shoreline precludes the formation of bars and rips.

Hazards: The major hazards associated with ultradissipative beaches are their usually higher waves, the relatively deep water off the high tide beach, the long distance from the shore to the low tide surf and the often considerable distance from the shoreline out to beyond the breakers. Currents run along the beach when waves arrive at

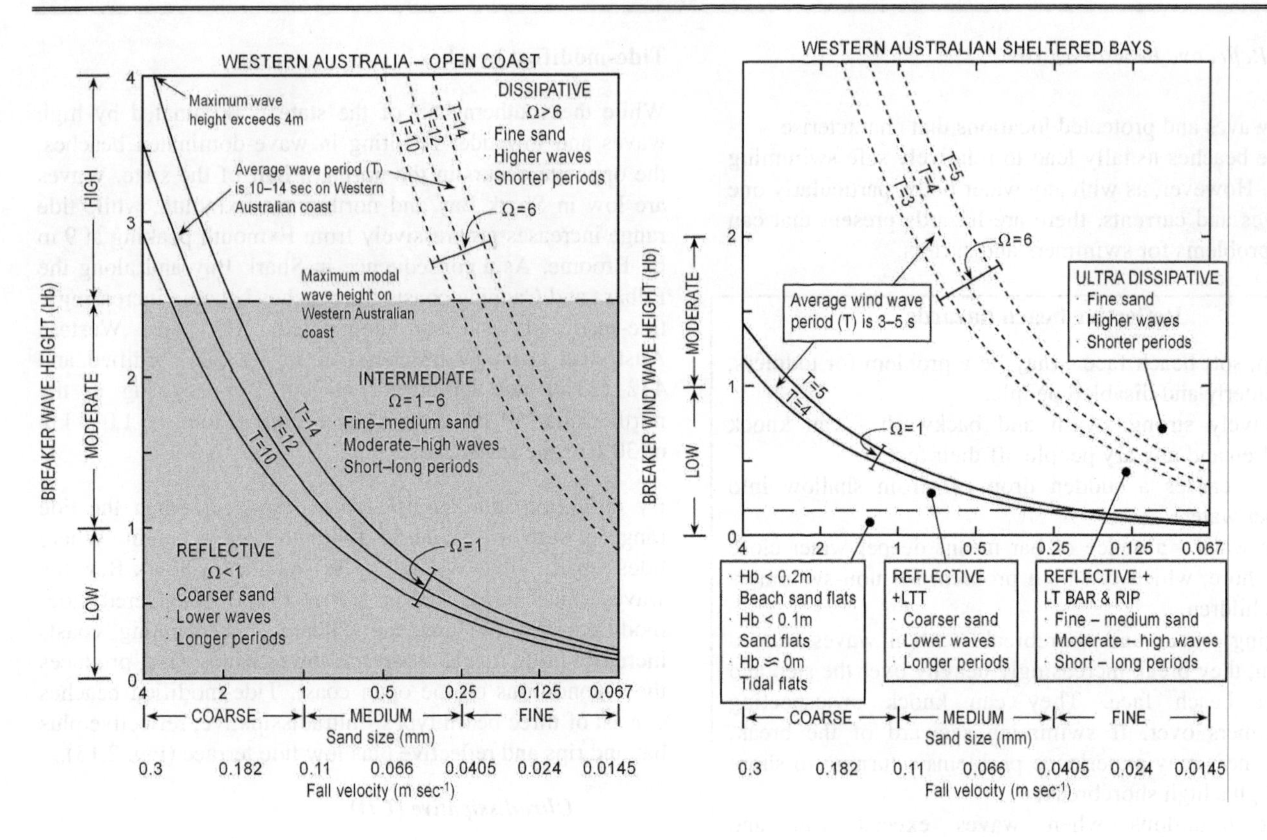

Figure 2.12 A plot of breaker wave height versus sediment size, together with wave period, that can be used to determine approximate Ω and beach type for high energy open coast (a) and sheltered beaches (b). To use the chart, determine the breaker wave height, period and grain size/fall velocity (mm or m/sec). Read off the wave height and grain size, then use the period to determine where the boundary of reflective/intermediate, or intermediate/dissipative beaches lies. Ω = 1 along solid T lines and 6 along dashed T lines. Below the solid lines Ω < 1 and the beach is reflective, above the dashed lines Ω > 6 and the beach is dissipative, between the solid and dashed lines Ω is between 1 and 6 and the beach is intermediate.

angles, however strong rip currents are generally absent. Seaward of the breakers, however, shore parallel tide currents also increase in strength.

Reflective plus bar & rips (R+LTR)

Reflective beaches fronted by low tide bars and rips occur in similar environments to the ultradissipative, only on beaches with fine to medium sand. There are however no R+LTR beaches in Western Australia, a result of the lack of waves high enough to maintain an energetic low tide surf zone on the high tide range beaches in the north of the state.

Reflective plus low tide terrace (R+LTT)

The lowest energy tide-modified beach is the reflective high tide beach fronted by a low tide terrace. A total of 40 such beaches occur, all but two on the Pilbara coast.

At high tide, waves surge up the steep beach face. This continues as the tide falls, until the shallower water of the terrace induces wave breaking increasingly across the terrace. At low tide, waves spill over the outer edge of the terrace, with the inner portion exposed and dry during spring low tide (Fig. 2.15). If rips are present, they will cut a channel across the terrace and are only active at low tide.

Tide-dominated beaches

When the tide range begins to exceed the wave height by between 10 and 15 times, then the tide becomes increasingly important in the beach dynamics, at the expense of the waves. In Western Australia these conditions in Shark Bay are due to very low waves and along the Pilbara and Canning coasts to both lower waves and the increasing tide range. Basically there are four beach types - three with a high tide beach and sand flats, while the fourth is essentially a transition from sand flat to a tidal mud flat (Fig. 2.16). These 'beaches' receive sufficient wave energy to build the sandy, and often shelly, high tide beach, while wide low gradient sand flats extend seaward of the base of the high tide beach. The intertidal sand flats grade from the higher energy ridged flats, to flat featureless sand flats, to sand flats with tidal flow and drainage features, to finally mud flats, still with the high tide sand beach. Beyond these pure intertidal flats dominate with no beach.

TIDE-MODIFIED BEACH TYPES & HAZARDS

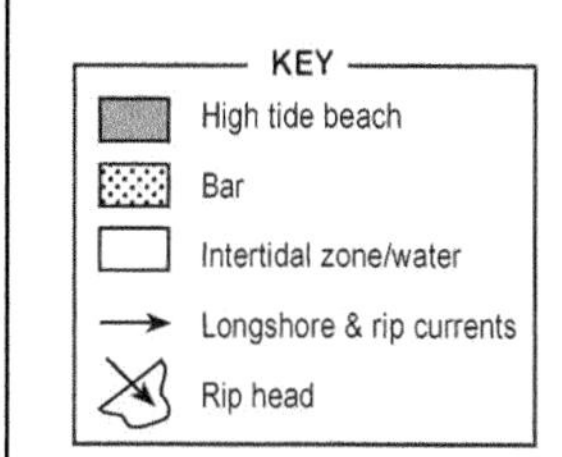

PLEASE NOTE:

High tide range beaches, with tides greater than 2m, extend across Northern Australia from Exmouth Cape in the west to Hervey Bay in the east and dominate north-west Western Australia, the Northern Territory and north Queensland. Additional high tide range beaches also occur in northern Tasmania, central Victoria and the South Australian gulfs.

Northern Australian high tide range beaches have three characteristics which differentiate them from southern Australian beaches, in terms of their nature and hazards to bathers:

1. They all lie in areas of higher tide ranges, beginning at 2m and reaching 8m in Queensland Broad Sound and 10m in parts of the north-west of Western Australia.
2. They receive lower waves, owing to lower energy wave sources, coupled with the protective impact of the Great Barrier Reef, offshore islands and low gradient nearshores.
3. They mostly lie in areas inhabited by a number of biological hazards, including box jellyfish and crocodiles.

ADDITIONAL HAZARD FACTORS

PHYSICAL:

WAVE HEIGHT – usually increases with onshore wind velocity.

STRONG WINDS – can rapidly build waves to over 1 m. Strong longshore winds will induce longshore currents.

WAVES – over 1 m may induce rip currents on all the beaches at low tide.

AT LOW TIDE – bathers may be located 100s of metres from the dry beach, therefore be very watchful if supervising young children; walk down to the water with them.

TIDAL CREEKS:

IF TIDAL CREEKS are located near the beach, stay well clear, especially on a falling tide.

STRONG TIDAL CURRENTS will flow in on the rising tide and seaward on the falling tide.

TIDAL CREEKS are also a preferred crocodile habitat and the source of many marine stingers.

REFLECTIVE + LOW TIDE TERRACE (+RIPS) (R+LTT)

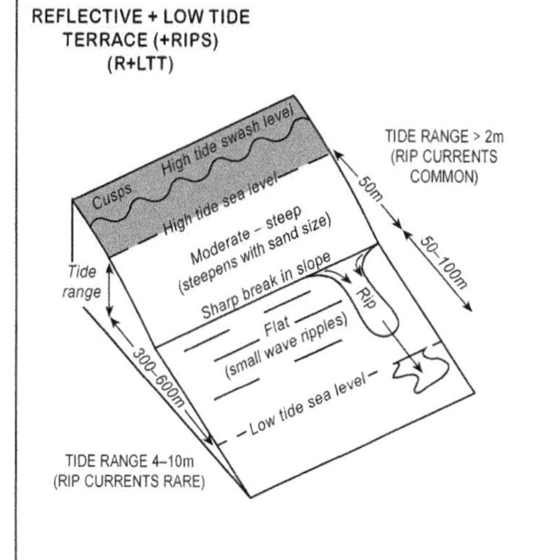

CHARACTERISTICS

Steep beach composed of medium to coarse sand; fronted by flat bar exposed at low tide, composed of finer sand; waves 0.5 to 1 m high.

High Tide	Low Tide
Steep beach (3-10°), waves surge at base of beach, no surf or surf zone, deep water (2–4m) immediately off beach.	Flat sand bar exposed at spring low tide, waves > 0.5m plunge on outer end of bar, waves > 1 m may generate rip currents and cut rip channels across the terrace.

HAZARDS

High Tide	Low Tide
Deep water close inshore, steep beach face may cause a problem for the elderly and young children.	Stay on bar inside breakers, if rip channels are present they will contain seaward moving rip currents, plunging waves can dump heavily on bar.

BEACH HAZARD RATING

High Tide	Low Tide
2/10	3/10 (no rips) 4/10 (with rips)

REFLECTIVE + BARS & RIPS (R+LTR)

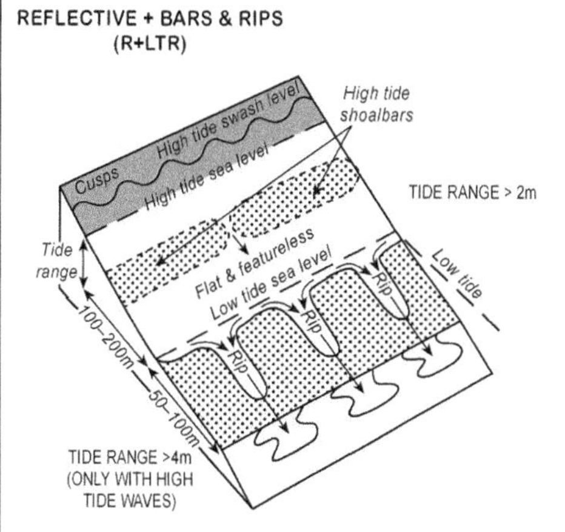

CHARACTERISTICS

Narrow, steep high tide beach face, wide, moderate gradient intertidal zone, low gradient low tide bar and rips, breaker height 0.5 to 1.5m (height increases with onshore wind velocity).

High Tide	Low Tide
Moderate beach face slope (2-5°), waves break across narrow continuous surf zone.	Wider surf zone, alternating bars and rips, waves spill to plunge on bar, rip currents in deeper rip channels.

HAZARDS

High Tide	Low Tide
Stay inshore of breaker zone, deep water beyond breakers.	Waves may plunge on shallow bar; deeper rip channels, seaward flowing rip currents, strongest at low tide.

BEACH HAZARD RATING

High Tide	Low Tide
2/10	4/10

ULTRA- DISSIPATIVE (UD)

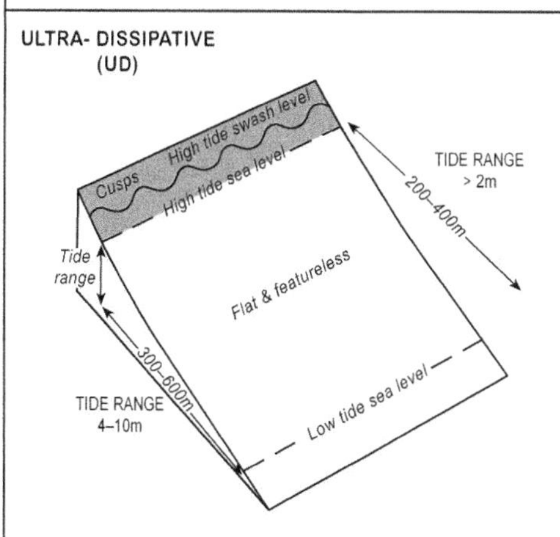

CHARACTERISTICS

Wide, featureless, low gradient beach—intertidal zone and surf zone; fine sand, spilling breakers, breaker height 0.5 to 1.5m (height increases with onshore wind velocity)

High Tide	Low Tide
Low to moderate beach slope (1-3°), wide zone of spilling breakers.	Wide, low gradient intertidal and breaker zone, very wide zone of spilling breakers.

HAZARDS

High Tide	Low Tide
Stay inshore of breaker zone, deep water beyond breakers.	Stay inshore of breaker zone, water depth gradually increases beyond breakers.

BEACH HAZARD RATING

High Tide	Low Tide
2/10	3/10

Figure 2.13 A three dimensional sketch of the three tide-modified Western Australian beaches: reflective beach and low tide terrace, reflective beach with low tide bar and rips, and ultradissipative.

Figure 2.14 *Two sections of Eighty Mile beach showing the wide, low gradient ultradissipative beach system.*

a.

b.

Figure 2.15 *Two steep high tide reflective beaches fronted by narrow low tide terraces, both in the Pilbara: a) Cape Lambert; and b) Cape Cossigny.*

Hazards: This beach type undergoes a marked change in morphology between high and low tide. At high tide hazards are associated with the waves surging against the steep high tide beach, together with the deeper water off the beach, while at low tide, waves spill across the broad, shallow terrace, with hazards associated with the deeper water off the terrace and the rips, if present.

Reflective plus ridged sand flat (R+RSF)

High tide beaches fronted by multiple sand ridges occur on only 16 beaches primarily on the Pilbara, with three on the Canning coast. They occupy a total of 42 km of the sandy coast.

These systems usually have a moderate to steep high tide beach, with shore parallel, sinuous, low amplitude, evenly spaced sand ridges extending out across the inter- to sub-tidal sand flats (Fig. 2.17). The ridges are not to be confused with the higher relief (sand) bars of the higher energy beaches. The beach is only active at high tide when either very low waves or calms prevail. The intertidal zone and ridges are usually inactive. The exact mode of formation is unknown, though it is suspected that the ridges are active and formed during infrequent periods of higher waves acting across the intertidal ridged zone.

In Western Australia the number of ridges averages 11 and ranges from 4 to 22, but can reach 40 in other Australian locations, while the sand flats range from 300 to 1500 m in width, averaging 700 m. The ridges are very low amplitude, no more than a few centimetres to a decimetre from trough to crest and average about 50 m in spacing. They tend to parallel the coast, but can at times lie obliquely to the shore, while at other places they merge into more complex patterns.

Hazards: The major hazard with these low energy beaches is the relatively deep water off the high tide beach and, in places, associated tidal currents. Low tide is dominated by the wide, shallow to exposed ridges and sand flats.

Reflective plus sand flats (R+SF)

The most common tide-dominated beach is the high tide beach fronted by very low gradient, flat, featureless sand flats (Fig. 2.18). There are 321 such beaches primarily in Shark Bay on the Carnarvon coast, as well as spread along the Central West, Pilbara and Canning coasts. In Western Australia the sand flats range from 10 to 5,000 m in width, with an average width of 1000 m (Ω=970 m). The beaches

TIDE-DOMINATED BEACH TYPES & HAZARDS

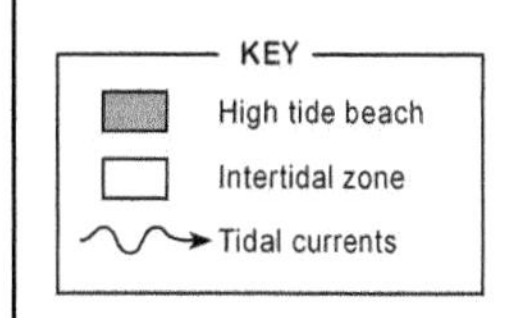

PLEASE NOTE:

High tide range beaches, with tides greater than 2m, extend across Northern Australia from Exmouth Cape in the west to Hervey Bay in the east and dominate north-west Western Australia, the Northern Territory and north Queensland. Additional high tide range beaches also occur in northern Tasmania, central Victoria and the South Australian gulfs.

Northern Australian high tide range beaches have three characteristics which differentiate them from southern Australian beaches, in terms of their nature and hazards to bathers:

1. They all lie in areas of higher tide ranges, beginning at 2m and reaching 8m in Queensland Broad Sound and 10m in parts of the north-west of Western Australia.
2. They receive lower waves, owing to lower energy wave sources, coupled with the protective impact of the Great Barrier Reef, offshore islands and low gradient nearshores.
3. They mostly lie in areas inhabited by a number of biological hazards, including box jellyfish and crocodiles.

ADDITIONAL HAZARD FACTORS

PHYSICAL:

WAVE HEIGHT – usually increases with onshore wind velocity.

STRONG WINDS – can rapidly build waves to over 1 m. Strong longshore winds will induce longshore currents.

WAVES – over 1 m may induce rip currents on all the beaches at low tide.

AT LOW TIDE – bathers may be located 100s of metres from the dry beach, therefore be very watchful if supervising young children; walk down to the water with them.

TIDAL CREEKS:

IF TIDAL CREEKS are located near the beach, stay well clear, especially on a falling tide.

STRONG TIDAL CURRENTS will flow in on the rising tide and seaward on the falling tide.

TIDAL CREEKS are also a preferred crocodile habitat and the source of many marine stingers.

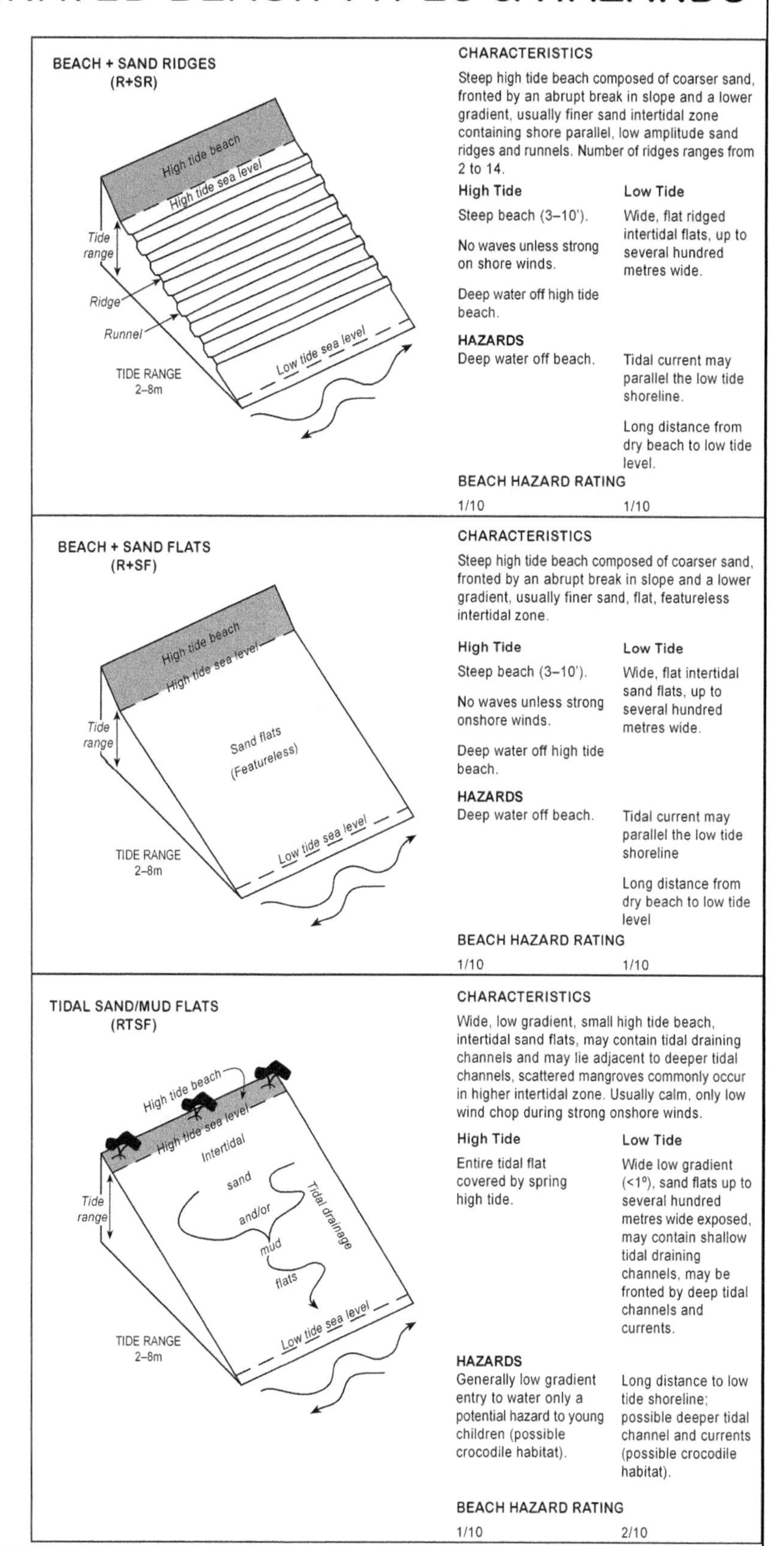

Figure 2.16 A three dimensional sketch of three tide-dominated Western Australian beaches: reflective+ridged sand flats, reflective+sand flats and tidal sand/mud flats.

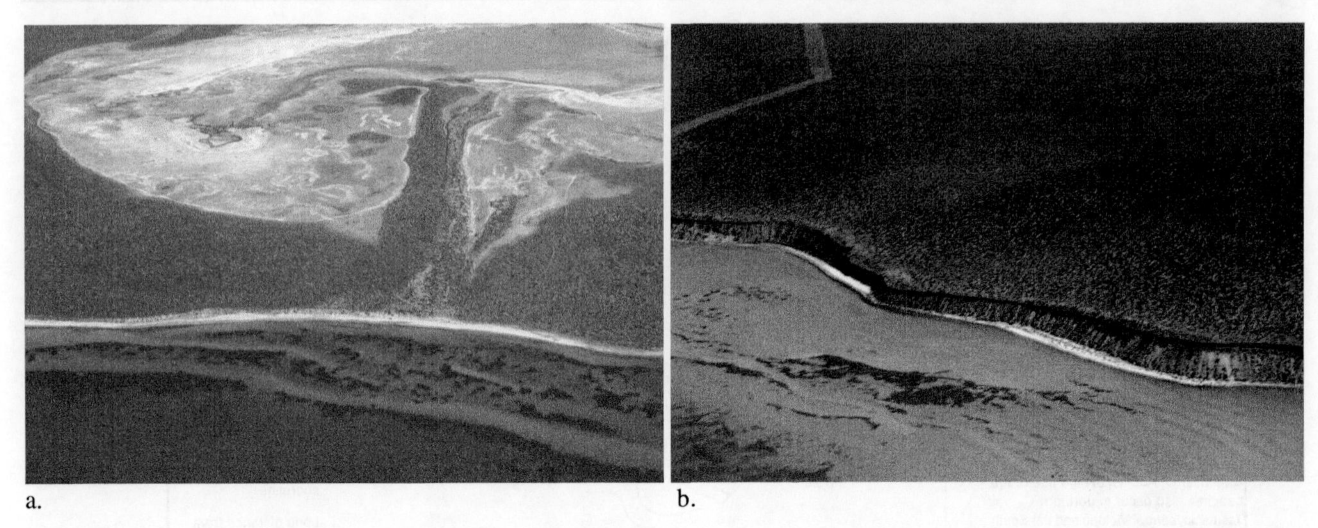

a. b.

Figure 2.17 Low energy reflective high tide beaches fronted by intertidal ridged sand flats at: a) Cattle Well; and b) Herald Bluff, both in Shark Bay.

a.

b)

Figure 2.18 Low energy high tide reflective beach, fronted by wide intertidal sand flats, both at Eagle Bluff, Shark Bay.

a.

b.

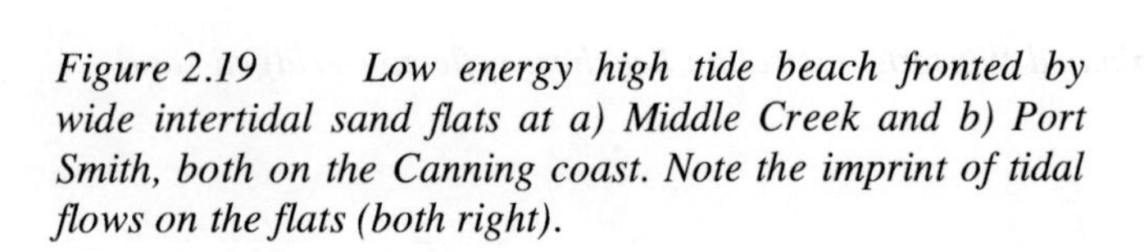

Figure 2.19 Low energy high tide beach fronted by wide intertidal sand flats at a) Middle Creek and b) Port Smith, both on the Canning coast. Note the imprint of tidal flows on the flats (both right).

average 1.7 km in length and occupy 558 km (15%) of the sand coast.

Hazards: The only hazards associated with these beaches are the deeper water off the high tide beach and the slight chance of tidal currents off the beach. Low tide reveals a wide, shallow to exposed tidal flat.

Reflective plus tidal sand flats (R+TSF)

Tidal flats are not by definition 'beaches', however they are included here for two reasons. First, these tidal flats represent a gradation from the true high tide beaches of the above two beach types, to the mangrove vegetated tidal flats that dominate the tide-dominated sections of the coast. Secondly, a number of sand flats are labelled 'beaches' and known locally as such. The main difference between the reflective+sand flats and reflective+tidal sand flats is that the tidal flats may be vegetated with seagrass and even scattered mangroves. The tidal sand flats differ from the above sand flats in that tide-generated drainage channels and other subdued features imprint themselves on the flats (Fig. 2.19). There are 139 beaches of this type, all occurring in the very low energy sections of Shark Bay on the Carnarvon coast and along similar sections of the Pilbara and Canning coasts. The sand flats range from 50 to 6000 m in width, averaging 890 m (Ω=1020 m). The beaches average 1.4 km in length and occupy 195 km (5%) of the sand coast.

Hazards: The only physical hazards associated with tidal sand flats are the deeper water over the flats at high tide, and the increased likelihood of tidal currents moving across, especially in drainage channels, or parallel to the flats and their wide distance at low tide, which increases the chance of being caught in the intertidal zone by the rising tide.

Reflective plus mud flats (R+MF)

There are only four beaches with an intertidal mud flat extending seaward from the base of the sandy high tide beach, all located on the Canning coast, adjacent to river mouths (Fig. 2.20). They are similar to the tidal sand flats, with mud, rather than sand, occupying the intertidal zone.

Beach plus rock flats (R+rock flats)

There are 59 beaches fronted by intertidal rock flats, about half located in the south, primarily on the South Coast, and half in the north spread along the Carnarvon, Pilbara and Canning coasts. The rock flats range from 20 to 400 m in width averaging 80 m. In each situation waves break at the edge and over the rocks, lowering waves at the shoreline, while the usually steep beach is only active at high tide (Fig. 2.21a). These are a more hazardous beach type, owing to the presence of the rocks in the sub- to intertidal zone, especially when exposed to moderate to high waves. Extreme care should be used in attempting to wade or swim off such a beach.

Figure 2.20 A steep sandy high tide beach, grading abruptly into very wide low gradient intertidal mud flats, at Hearson Cove, Dampier Peninsula.

Beach plus coral reef (R+coral reef)

There are 37 beaches fronted by fringing coral reefs. They are all located along the Ningaloo Reef system on the Carnarvon coast. The reefs extend seaward from the base of the high tide beach (Fig. 2.21b), which is usually steep and composed of coral and shell debris. The reef flats average 1400 m in width (Ω=1,200 m) and substantially lower waves at the shore, even at high tide, with no waves reaching the beach at low tide. While waves are usually low at the shore, care should be taken in wading or swimming off these beaches owing to the irregular and often sharp coral surface.

Calcarenite

Calcarenite is a lithified (cemented) sand composed predominantly of calcareous material. The 'calc' stands for the calcareous and 'arenite' means sand. The Western Australian coast is dominated by two types of calcarenite - aeolian or dune calcarenite, when sand dunes have been cemented, commonly called *dunerock* (Fig. 2.22a), and beach calcarenite, where the intertidal beach has been cemented, commonly called *beachrock* (Fig. 2.22b).

When calcareous rich dune sands are exposed to long periods (thousands of years) of pedogenesis (soil forming processes), the sand can undergo transformation into massive calcrete, normally within 1 m of the surface. When this occurs, the original sand grains and structures are no longer present. Below this surface the sand undergoes partial cementation, which leaves the individual grains and structures visible, though cemented. In addition, towards the surface the vegetation interacts with the sand to produce a range of lithocasts (limestone cast around a trunk or root) and lithoskels (limestone cast inside a trunk or root). What all this means is that if the lithified beach and dune are subsequently exposed to wave attack, much will remain as a resilient rock or reef structure, which exerts considerable influence on the present coast.

a.

b.

Figure 2.21 a) High tide sandy beach and rock flats at Cape Bossut; and b) fringing coral reef at Maggies Reef, Ningaloo reef.

Calcarenite occurs around the entire Western Australian coast and in places dominates the shoreline and inner shelf. It is prolific along the South East coast, particularly in the Bight where it helps lower waves along the shore between Eucla and Twilight Cove. It occurs sporadically along the South Coast, while it dominates the Leeuwin coast, where it occurs offshore as reefs and onshore as massive dune calcarenite, within which the cave systems have evolved.

Along the Central West coast it dominates the inner shelf occurring as submerged shore parallel Pleistocene barriers which form reefs, and occasionally as islands, including Rottnest Island, as well as forming many low headlands. The reefs and islands are responsible for the lower wave conditions along much of this coast. Dune calcarenite dominates the Carnarvon Coast forming the massive

a.

b.

Figure 2.22 a) One hundred metre high cliffs composed of multiple layers of dune calcarenite, Zuytdorp Cliffs; b) shore parallel beachrock reef causing heavy wave breaking and a backing more protected lagoon at Munglinup Beach.

Zuytdorp Cliffs together with Dirk Hartog, Doore and Bernier islands, and outcrops along much of the shore between Lake MacLeod and North West Cape. On the Pilbara and Canning coasts it occurs as the core of many of the Pleistocene barrier islands and systems, that form and/or back sections of the modern coast.

Large Scale Beach Systems

The beach systems described above are all part of larger scale beach and barrier systems, the barriers including backing beach and foredune ridges and, in places, sand dunes, as well as adjoining tidal creeks and inlets and headlands and reefs. Figure 2.23 provides a schematic overview of the typical arrangement of some of these beach and associated barrier systems. In general the lowest energy systems receive low waves and are backed by low beach ridge plains, as in parts of Shark Bay. Moderate energy systems are backed by prograding foredune ridge plains (Fig. 2.24a). Such plains occur on many sandy forelands along the Central West coast. Moderate to higher energy beaches tend to have

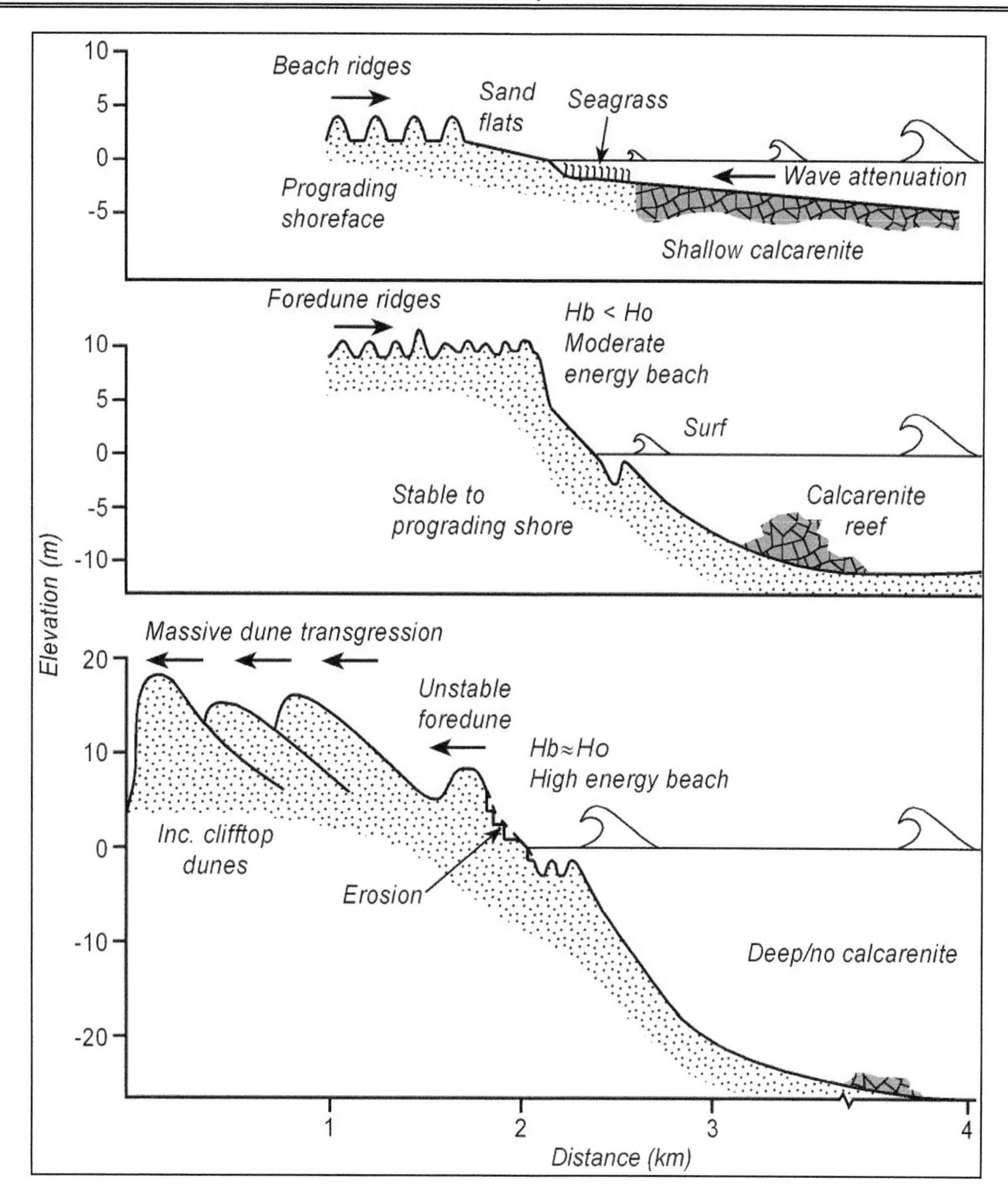

Figure 2..23 Schematic cross-section of typical low, moderate and high energy Western Australia beach-dune-barrier systems. As wave height increases the width and depth of the surf and nearshore zone increase, as do the size and instability of the backing dune systems.

backing blowouts and parabolic dunes, which often transgress over former foredune ridges (Fig. 2.24b), or climb backing bluffs and cliffs (Fig. 4.24c). In areas of extensive bare sand transverse dune are formed perpendicular to the prevailing wind (Fig. 2.24d). The highest energy systems are backed by massive dune transgression, as along the South Coast's Warren Beach (Fig. 2.24d) and at Eucla (Fig. 2.24e).

There is therefore a relationship between the level of wave energy at the shore, which is in part determined by the nearshore topography as indicated in Figure 2.23, and the type, size and extent of the backing dune system. Other contributing factors are the orientation of the coast, the direction and strength of the dominant onshore winds, generally strong south to southwest in southern Western Australia, and the supply of sand to the beach and dunes.

Lower energy systems tend to have prograded well vegetated foredune ridges. Moderate energy systems often have foredune ridges overrun by later parabolics dunes. The highest energy beach systems are usually backed by the most extensive dune systems consisting of parabolics dune, clifftop dunes and in places massive sand sheets. Australia's longest active parabolic dune extends for 38 km northward, to the lee of Steep Point, while Australia's most extensive dune system extends 105 km inland in lee of Kaniaal Beach in the Bight.

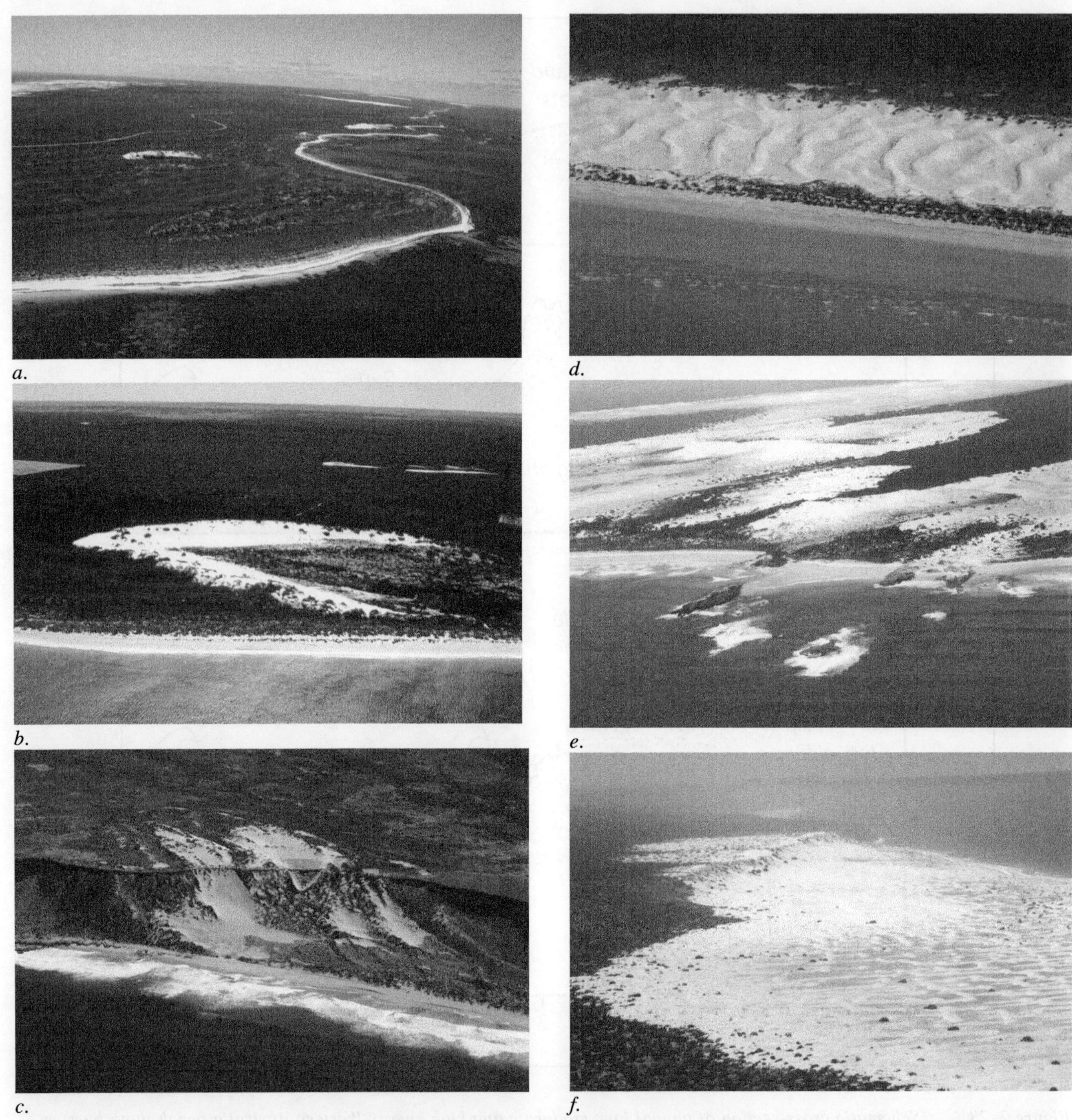

Figure 2.24 Dune types a) Foredune ridges overrun by some vegetated parabolics, Kangaroo Pt; b) Well developed active parabolic dune transgressive over foredune ridges, Ronsard Bay; c) Active sand ramp and clifftop dune, Red Bluff; d) Transverse dunes, Cape Keraudren; e) Massive dune fields, Point Jedacorrudup; and f) Massive dune sheet at Eucla.

3 Beach Hazards and Safety

All beaches contain variable water depth, deep water, breaking waves and currents, all of which can present hazards to beach users. This chapter reviews the types of beach hazards and how they can be mitigated and reduced through beach safety.

Beach usage

The Australian coast remained untouched until the arrival of the first Aborigines some tens of thousands of years ago. The coast and beaches that they found, however, no longer exist. Probably crossing to Australia during one of the glacial low sea level periods, they not only reached a far larger, cooler, drier and windier continent, but one where the shoreline was some tens of metres below present sea level. Hence the coast they walked and fished now lies out below sea level on the inner continental shelf.

The present Australian coast only formed some 6,500 years ago, when sea level rose to approximately its present position. As it was rising, at about 1 m every 100 years, the Aborigines no doubt followed its progress by slowly moving inland and to higher ground. Therefore, we can assume that usage of the present Western Australian beaches began as soon as the beaches began forming some 6,000 years ago and continued in the traditional way until the 1800s.

Following the initial European settlements in Albany (1826) and Perth (1829) the beaches and coast were largely ignored by the new settlers, considered a barren wasteland for farming purposes, and a place to avoid for the ships sailing the reef-strewn coast. It was not until the late 19th century that people began looking to the coast as a source of recreation, and in the process their attitude to beaches and coastal real estate changed dramatically, a process that continues today as more and more Australians choose to work and live near the coast.

The Surf Lifesaving Movement

The history of surf life saving in Western Australia is well documented by Jaggard (1979) and this summary is based on his excellent book. Ed Jaggard and a team of authors are presently writing the history of Surf Life Saving Australia (SLSA) for its centenary in 2007.

To understand the formation of the surf lifesaving clubs (SLSCs) and the broader Surf Life Saving Australia is to realise two things. Firstly, in their rush to the surf, most beachgoers could not swim and had little or no knowledge of the surf and its dangers. Secondly, the open coast of Western Australia is as dangerous as it is inviting to the unprepared.

Only in the early 1900s did beach and surf swimming become a popular recreational pastime, in both the east and west of the nation. In Perth it followed the opening of the Perth-Fremantle railway in 1880 and the development of Cottesloe, Perth's first beach suburb, from 1886. By the early 1900s up to a few thousand people were using Cottesloe Beach on hot Sundays. In 1905 one of the first recorded rescues took place on Cottesloe, prompting the placement of four lifebelts along the beach. As crowds continued to grow, a caretaker to patrol the beach was appointed in 1906 and a group of volunteers began patrolling in 1908. Finally in 1909 the Cottesloe Life Saving and Athletic Club was formed. This was followed by North Cottesloe SLSC in 1912. Between the wars the movement expanded in Perth with City of Perth SLSC formed in 1924, Scarboro SLSC in 1928, Swanbourne Nedlands SLSC in 1932 and Fremantle SLSC in 1934. To the south the third club in the state was City of Bunbury SLSC in 1915, with Geraldton SLSC the second country club forming in 1930.

Following World War 2 the surf life saving movement continued to expand in Perth, with Floreat SLSC established in 1948, followed by Trigg Island SLSC in 1954, which was able to assist neighbour Scarboro' patrol Perth's most dangerous section of beach, then Sorrento SLSC in 1958 and Mullaloo SLSC in 1960. In the country Albany SLSC started in 1956 and Denmark SLSC, patrolling the most dangerous beach in the state, in 1958.

There was then a lull in the formation of new clubs until the expansion of coastal communities and population north and south of Perth led to the foundation of Secret Harbour SLSC in 1981 and Quinns Mindarie SLSC in 1982, while up north Broome SLSC was founded in 1988 following the 'discovery' of Cable Beach as one of Australia's premier tourist destinations, and down south Esperance Goldfields SLSC started in 1990 to provide a safer patrolled beach along the hazardous Esperance coast.

The continuing population growth in the state, particularly its concentration and spread along the coast, have been matched by the formation of new clubs, all in areas of rapidly growing coastal population, with Yanchep SLSC formed in 1991 and the probationary Mandurah SLSC in 1996. Since 2002 there has been the biggest single increase in club numbers, with four new clubs (Binningup SLSC, 2002; Broadwater Bay SLSC, 2003; Coogee Beach SLSC, 2003; Dongara-Denison SLSC, 2004) and two probationary clubs (Margaret River SLSC, 2003; Port Bouvard SLSC, 2004), all but Coogee Beach in country areas. Table 3.1 lists all 26 Western Australian surf lifesaving clubs by their year of formation.

All beaches with surf lifesaving clubs are patrolled during the summer months by volunteer surf lifesavers (Table 3.2). In addition several beaches have professional lifeguards during the peak summer months, and two also have beach inspectors. For full and up-to-date details on all Western Australian surf lifesaving clubs and their patrol periods, as well as other information, visit Surf Life Saving Western Australia's official web site at www.mybeach.com.au.

Table 3.1 Western Australian Surf Lifesaving Clubs

Surf Life Saving Club	*Year*
Cottesloe	1909
North Cottesloe	1912
City of Bunbury	1915
City of Perth	1924
Scarboro'	1928
Geraldton	1930
Swanbourne Nedlands	1932
Fremantle	1934
Floreat	1948
Trigg Island	1954
Albany	1956
Sorrento	1958
Denmark	1958
Mullaloo	1960
Secret Harbour	1981
Quinns Mindarie	1982
Broome	1988
Esperance Goldfields	1990
Yanchep	1991
Binningup	2002
Broadwater Bay	2003
Coogee Beach	2003
Dongara-Denison	2004
Probationary Affiliated Clubs	
Mandurah	1996
Margaret River	2003
Port Bouvard	2004
Dalyellup	2005

The first surf lifesaving clubs and their national association (SLSA), formed in 1907, had embarked upon the establishment of an organisation that has become an integral part of Australian beach usage and culture, and through which it is so readily identified internationally.

Now, more than a century after the initial rush to the beaches and the foundation of the early surf lifesaving clubs, both beach usage and the 303 surf lifesaving clubs around the coast are accepted as part of Australian beaches and beach life. However, at the beginning of the 21st century, beach usage is undergoing yet another surge as the Australian population and visiting tourists increasingly concentrate on the coast. This is resulting in more beaches being used, more of the time, by more people, many of whom are unfamiliar with beaches and their dangers. In addition, while this has led to a rapid expansion of surf lifesaving clubs in Western Australia, still the vast majority of beaches have no surf lifesaving club or patrols.

There is now a greater need than ever to maintain public safety on these beaches, a service that is provided on patrolled beaches by volunteer lifesavers and professional lifeguards. This book is the result of a joint

Table 3.2 Western Australian volunteer surf lifesaving patrol periods (south to north). Note: based on 2005 periods, which may vary in future. ***Bold*** *indicates also professional lifeguards most of which patrol weekdays as well. Check with local surf life saving club or council for exact dates, days and patrol times. Lifeguard indicated lifeguard only.*

SLSC	Oct	Nov	Dec	Jan	Feb	March	April
Esperance Goldfields		Sun	Sun	Sat/Sun	Sun	Sun	
Albany			Sun	**Sat/Sun/Mon**	Sat/Sun/Mon	Sat/Sun/Mon	
Denmark			Sun	**Sat/Sun**	Sun	Sun	
Smiths Beach				**Lifeguard**			
Yallingup				**Lifeguard**			
Binningup			Sun	Sun	Sun	Sun	
Dalyellup			Sun	Sun	Sun		
Bunbury	Sat/Sun	Sat/Sun	Sat/Sun	Sat/Sun	Sat/Sun	Sat/Sun	
Port Bouvard	Sun	Sun	Sun	Sun	Sun	Sun	
Mandurah		Sun	Sun	Sun	Sun		
Secret Harbour	Sun	Sun	**Sat/Sun**	**Sat/Sun**	Sat/Sun	Sat/Sun	
Coogee Beach	Sun	Sat/Sun	Sat/Sun	Sat/Sun	Sat/Sun	Sat/Sun	
Fremantle	**Sun**	Sat/Sun	Sat/Sun	Sat/Sun	Sat/Sun	Sat/Sun	
Cottesloe*	**Sat/Sun**	**Sat/Sun**	**Sat/Sun**	**Sat/Sun**	**Sat/Sun**	**Sat/Sun**	**Lifeguard**
North Cottesloe	Sun	Sat/Sun	Sat/Sun	Sat/Sun	Sat/Sun	Sat/Sun	
Swanbourne Nedlands	Sun	Sat/Sun	Sat/Sun	Sat/Sun	Sat/Sun	Sat/Sun	
City of Perth*	Sat/Sun	Sat/Sun	**Sat/Sun**	**Sat/Sun**	**Sat/Sun**	Sat/Sun	
Floreat	**Sun**	**Sat/Sun**	**Sat/Sun**	**Sat/Sun**	**Sat/Sun**	**Sat/Sun**	
Scarboro	Sat/Sun	Sat/Sun	Sat/Sun	Sat/Sun	Sat/Sun	Sat/Sun	
Trigg Island	Sat/Sun	Sat/Sun	Sat/Sun	Sat/Sun	Sat/Sun	Sat/Sun	
Sorrento	Sun	Sat/Sun	**Sat/Sun**	**Sat/Sun**	**Sat/Sun**	Sat/Sun	
Hillarys Boat Harbout			**Lifeguard**	**Lifeguard**	**Lifeguard**		
Mullaloo	Sat/Sun	Sat/Sun	**Sat/Sun**	**Sat/Sun**	**Sat/Sun**	Sat/Sun	
Quinns Mindarie	Sun	Sun	**Sun**	**Sun**	**Sun**	Sun	
Yanchep	x	Sun	**Sun**	**Sun**	**Sun**	Sun	
Dongara-Denison		Sun	Sun	Sun	Sun	Sun	
Geraldton		Sun	**Sat/Sun**	**Sat/Sun**	Sat/Sun	Sun	
Broadwater Bay		Sun	Sun	Sun	Sun	Sun	

Surf Life Saving Australia, Surf Life Saving Western Australia and the Coastal Studies Unit project that is addressing this problem. The book is designed to provide information on each and every beach in Western Australia, between Eucla and Roebuck Bay, including Rottnest Island, in all, 2051 mainland beaches and 63 on Rottnest Island. It contains information on all 26 patrolled beaches including a map of each and their general characteristics and hazards. It also provides a description of every beach including its general characteristics, access and suitability for swimming and surfing. In this way, swimmers may be forewarned of potential hazards before they get to the beach and consequently swim more safely.

If you are interested in joining a surf lifesaving club or learning more about surf lifesaving, contact Surf Life Saving Australia, your state centre (listed below) or your nearest surf lifesaving club, or check www.mybeach.com.au.

Physical beach hazards

Beach hazards are elements of the beach-surf environment that expose the public to danger or harm (Fig. 3.1). *Beach safety* is the mitigation of such hazards and requires a combination of common sense, swimming ability and beach-surf knowledge and experience. The following section highlights the major physical hazards encountered in the surf, with hints on how to spot, avoid or escape from such hazards. This is followed by the biological hazards.

There are seven major physical hazards on Western Australian beaches:

1. water depth (deep and shallow)
2. breaking waves
3. surf zone currents (particularly rip currents)
4. tidal currents
5. strong winds
6. rocks, reefs and headlands
7. water temperature

In the surf zone, three or four hazards, particularly water depth, breaking waves and currents, usually occur together. In order to swim safely, it is simply a matter of avoiding or being able to handle the above when they constitute a hazard to you, your friends or children.

Surf Life Saving Australia Locked Bag 2 Bondi Beach NSW 2026 Phone: (02) 9130 7370 Fax: (02) 9130 8312 Email: info@slsa.asn.au Web: www.slsa.asn.au	**Surf Life Saving Western Australia** PO Box 1048 Osborne Park Business Centre WA 6916 Phone: (08) 9244 1222 Fax: (08) 9244 1225 Email: mail@mybeach.com.au Web: www.mybeach.com.au
Surf Life Saving New South Wales PO Box 430 Narrabeen NSW 2101 Phone: (02) 9984 7188 Fax: (02) 9984 7199 Email: experts@surflifesaving.com.au Web: www.surflifesaving.com.au	**Life Saving Victoria** PO Box 2288 Oakleigh VIC 3166 Phone: (03) 9567 0000 Fax: (03) 9568 5988 Email: mail@rlssav.org.au Web: www.lifesavingvictoria.com.au
Surf Life Saving Queensland PO Box 3747 South Brisbane QLD 4101 Phone: (07) 3846 8000 Fax: (07) 3846 8008 Email: slsq@lifesaving.com.au Web: www.lifesaving.com.au	**Surf Life Saving Tasmania** GPO Box 1745 Hobart TAS 7001 Phone: (03) 6231 5380 Fax: (03) 6231 5451 Email: slst@slst.asn.au Web: www.slst.asn.au
Surf Life Saving Northern Territory PO Box 43096 Casuarina NT 0811 Phone: (08) 8941 3501 Fax: (08) 8981 3890 Email: slsnt@austarnet.com.au Web: www.slsnt.org	**Surf Life Saving South Australia** PO Box 82 Henley Beach SA 5022 Phone: (08) 8356 5544 Fax: (08) 8235 0910 Email: surflifesaving@surfrescue.com.au Web: www.surfrescue.com.au

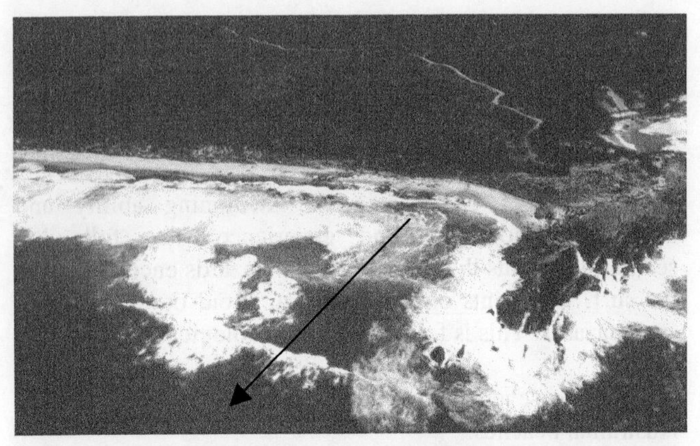

a,

b.

Figure 3.1 *Western Australian physical beach hazards include high waves, rocks, rips and tidal inlets, as illustrated in this series of photographs. (a) Surf, rocks and topographic rip (arrow) at Warrenup; and (b) beachrock reefs, surf and a strong central rip at Ten Mile Lagoon, Esperance.*

1. Water depth

Any depth of water is potentially a hazard.

- *Shallow water* is a hazard when people are diving in the surf or catching waves. Both can result in spinal injury if people hit the sand head first.
- *Knee depth* water can be a problem for a toddler or young child.
- *Chest depth* is hazardous to non-swimmers, as well as to panicking swimmers. In the presence of a current, it is only possible to wade against the current when waist deep or less. Therefore be careful when water begins to exceed waist depth,
particularly if younger or smaller people are present and if in the vicinity of longshore, rip or tidal currents.

Water depth

- Safest: knee deep - can walk against a strong rip current
- Moderately safe: waist deep - can maintain footing in rip current
- Unsafe: chest deep - unable to maintain footing in rip current

Remember: what is shallow and safe for an adult can be deep and distressing for a child.

Shallow water hazards

Spinal injuries are usually caused by people surfing dumping waves in shallow water, or by people running and diving into shallow water.

To avoid these:
- Always check the water depth prior to entering the surf.
- If unsure, WALK, do not run and dive into the surf.
- Only dive under waves when water is at least waist deep.
- Always dive with your hands outstretched.

Also
- Do not surf dumping waves.
- Do not surf in shallow water.

2. Breaking waves

As waves break, they generate turbulence and currents which can knock people over, drag and hold them under water, and dump them on the sand bar or shore. If you do not know how to handle breaking waves (most people don't), stay away from them. stay close to shore and on the inner part of the bar.

If you are knocked over by a wave, remember two points - the wave usually holds you under for only two to three seconds (though it may seem like much longer), therefore do not fight the wave, you will only waste energy. Rather, let the wave turbulence subside, then return to the surface. The best place to be when a big wave is breaking on you is as close to the seafloor as possible. Experienced surfers will actually 'hold on' by digging their hands into the sand as the wave passes overhead, then kick off the seabed to speed their return to the surface.

If a wave does happen to gather you up in its wave bore (white water) it will only take you towards the shore and will quickly weaken, allowing you to reach the surface after two to three seconds and usually leave you in a safer more shoreward location than where you started.

Breaking waves and wave energy

Surging waves - low hazard when low
- break by surging up beach face - usually less than 0.5 m high
- can be a problem for children and elderly, who are more easily knocked over
- become increasingly strong and dangerous when over 0.5 m high

Spilling waves - moderately hazardous
- break slowly over a wide surf zone
- are good body surfing waves

Plunging (dumping) waves - the most hazardous waves
- break by plunging heavily onto sand bar
- strong wave bore (white water) can knock swimmers over
- very dangerous at low tide or where water is shallow

- waves can dump surfers onto sand bar, causing injury
- most spinal injuries are caused by people body surfing or body boarding on dumping waves
- to avoid spinal injury, do not surf dumping waves or in shallow water. If caught by a wave do not let it dump you head first, turn sideways and cover your head with your arms
- only very experienced surfers should attempt to catch plunging waves

Wave energy ≈ square of the wave height and is proportional to wave period.
Wave energy represents the amount of power in a wave of a particular height.

0.3 m wave = 1 unit wave energy/power
1.0 m wave = 11 units
1.5 m wave = 25 units
2.0 m wave = 44 units
2.5 m wave = 70 units
3.0 m wave = 100 units

Therefore, a 3 m wave is 10 times more powerful than a 1 m wave and 100 times more powerful than a 0.3 m wave. Likewise a 10 sec wave will have twice the energy of a 5 sec wave of the same height.

3. Surf zone currents and rip currents

Surf zone currents and particularly *rip currents* are the biggest hazards to most swimmers. They are the hardest for the inexperienced to spot and can generate panic when swimmers are caught by them. See the later section on how to identify and escape from rip currents.

The problem with currents, particularly rip currents, is that they can move you unwillingly around the surf zone and ultimately seaward (Fig. 3.2). In moving seaward, they will also take you into deeper water and possibly toward and beyond the breakers. As mentioned earlier, currents are manageable when the water is below waist level, but as water depth reaches chest height they will sweep you off your feet.

Rip and surf zone current velocity

Breaking waves travel at 3-4 m/sec (10-15 km/hr)

Wave bores (white water) travel at
1-2 m/sec (3-7 km/hr)
Rip feeder and longshore currents travel at
0.5-1.5 m/sec (2-5 km/hr)
Rip currents under average wave conditions
(< 1.5 m high) attain maximum velocities of
1.5 m/sec = 5.4 km/hr

(Note: Olympic swimmers swim at about 7 km/hr) An average rip in a surf zone 50 m wide can carry you outside the breakers in as little as 30 seconds.

Figure 3.2 *There are approximately 1470 beach rips and 900 permanent (topographic) rips operating at any one time along the more exposed sections of the Western Australian coast. This view of Whalebone Beach shows a central beach rip, with topographic rips (arrows) to either end.*

4. Tidal currents

Tidal currents are generally weak around the southern half of the state and only impact beach processes near tidal inlets. However, amongst the higher tides of the Pilbara, Canning and Kimberley coasts the combination of shallower, constricted water and in places increasing tide range can generate strong tidal currents. Where such constrictions and stronger currents occur, tides must flow in and out twice a day and in so doing generate strong reversing currents, which also maintain deeper tidal channels (Fig. 3.3a). These strong currents and deep channels are a very real hazard on all such beaches. They are particularly hazardous on a falling tide as the currents flow seaward. Ocean Beach at Denmark is located next to the mouth of Wilson Inlet and is a good example of a beach with a low tide range (0.6 m), being impacted by a constricted and strong tidal flow (Fig. 3.3b).

When swimming or even boating in or near tidal creeks and inlets, always check the state and direction of the tide and be prepared for strong currents. Swimmers should not venture beyond waist deep water.

5. Strong winds

The Western Australian coast is exposed to predominantly westerly winds in the south associated with both the subtropical high and the passage of cyclonic cold fronts, southerly winds and sea breezes along the Central West, and a mix of seasonal winter easterlies and summer westerlies in the north. The strongest winds are associated with the southerly cold fronts, which tend to hit from the west and gradually swing more to the south, and with strong summer sea breeze. As a consequence wind strength and direction must be factored into the hazard level whenever beaches are exposed to moderate to strong winds (Table 3.3).

a.

b.

Figure 3.3 a) *Deep tidal inlet flowing along beach at Ocean Beach, Denmark; and (b) a deep tidal inlet with strong tidal currents at the mouth of the Maitland River.*

Table 3.3 Wind hazard rating for Western Australian beaches, based on wind direction and strength. Winds blowing on and alongshore will intensify wave breaking and surf zone currents, with strong longshore winds capable of producing a strong longshore drag. Their impact on surf zone hazards and beach safety is indicated by the relevant hazard rating, which should be added to the prevailing beach hazard rating.

	Light	*Mod*	*Strong*	*Gale*
Longshore	*0*	*1*	*3*	*4*
Onshore	*0*	*1*	*2*	*3*
Offshore	*0*	*1*	*1*	*2*

Along much of the Central West coast, where reefs filter the ocean waves, including most of the Perth beaches, the strong southerly sea breezes are responsible for a significant proportion of the wave energy at the shoreline (Table 3.4). They produce waves arriving obliquely from the south, which combined with the following wind produce a strong northerly current along the shore. This current can feed into mobile or flash rips flowing seaward of the surf zone, which are a major hazard during such conditions.

Whenever strong winds occur, their direction determines the impact they will have on beach and surf conditions, as follows:

- *Longshore winds,* particularly strong westerly winds, will cause wind waves to run along the beach, with accompanying longshore and rip currents also running along the beach. The waves and currents can very quickly sweep a person alongshore and into mobile rips on lower energy beaches or into deeper rip channels and stronger currents on higher energy beaches.
- *Onshore winds* will help pile more water onto the beach and increase the water level at the shore. They also produce more irregular surf, which makes it more difficult to spot rips and currents.
- *Offshore winds* tend to clean up the surf. They are generally restricted to hot summer northerly conditions and the winter Trades winds in the north. However, if you are floating on a surfboard, bodyboard, ski or wind surfer, it also means it will blow the board offshore. In very strong offshore winds, it may be difficult or impossible for some people to paddle against this wind.

Table 3.4 In waters sheltered from ocean swell sea wave height depends on strength of on or alongshore winds. Maximum sea height can reach about 3 m along the West coast. These seas are short and steep, compared to the longer ocean swell of the same height.

On-alongshore wind velocity	Max. sea wave height (m)*
Light	0.3
Moderate	0.5
Strong	1.0
Gale	3.0-4.0

*wave period 2-5 sec

Wind generated waves and currents

Winds in the Southern Ocean are responsible for the high swell that arrives at the coast, while the same winds also drive the great circumpolar ocean current. Closer to shore and at the coast winds generate seas, rather than swell, and can generate local wind driven currents.

Along the Central West coast the southerly sea breezes are responsible for the generation of southerly wind wave (seas) particularly during summer, with stronger northwest winds in winter. In addition these same winds along the open coast will generate a sea on top of any existing swell, hence the weather forecast for sea and swell conditions. The locally generated seas are by definition short, steep and more irregular (some say confused) compared to swell.

If winds are of sufficient strength and duration they can generate locally wind-driven currents at the coast. In Western Australian open waters, most wind-driven currents set to the east in the south following the larger ocean currents to the south. Changes to this pattern occur during strong northerly winds, which push water offshore, causing upwelling, while strong onshore winds push water onshore, causing downwelling. Along the Central West coast the southerly sea breeze causes coastal waters to move offshore generating cooler upwelling at the shore, while the winter northerlies will drive water shoreward resulting in warmer downwelling. In the North West the offshore easterlies will tend to drive surface waters to the west, while the onshore summer westerlies will drive the water to the east.

6. Rocks, reefs and headlands

Most open coast Western Australian beaches have some rocks, calcarenite reefs and headlands. These pose problems on higher energy beaches in the south of the state because they cause additional wave breaking, generate more strong topographic rips, and have hard and often dangerous surfaces. When they occur in shallow water and/or close to shore, they are also a danger to people walking, swimming or diving because of the hard seabed and the fact that they may not be visible from the surface.

Rocks, reefs and headlands

- If there is surf against rocks or a headland, there will usually be a rip channel and current (topographic rip) next to the rocks. There are 900 topographic rips around the Western Australian coast.
- Rocks and reefs can be hidden by waves and tides, so be wary.
- Most calcarenite reefs will be submerged at high tide.
- Do not dive or surf near rocks, as they generate greater wave turbulence and stronger currents.
- Rocks often have sharp, shelled organisms growing on their surface, which can inflict additional injury.
- If walking or fishing from rocks, be wary of being washed off by sets of larger waves

7. Water Temperature

Water temperature becomes an increasing hazard as the temperature falls, particularly below 20^{0}'s and into the teens. The cooler temperature reduces body heat and when exposed to low water temperature or even moderate temperature for a prolonged period it will cause hypothermia. For this reason most surfers where wetsuits in winter and cooler summer days.

Safe swimming

The safest place to swim is on a patrolled beach between the red and yellow flags, as these indicate the safest area of the beach and the area under the surveillance of the lifesavers. If there are no flags then stay in the shallow inshore or toward the centre of attached bars, or close to shore if water is deep. However, remember that rip feeder currents are strongest close to shore and rip currents depart from the shore. The most hazardous parts of a beach are in or near rips and/or rocks, outside the flags or on unpatrolled beaches and when you swim alone.

Remember these points:

- DO swim on patrolled beaches.
- DO swim between the red and yellow flags.
- DO swim in the net enclosure (where present).
- DO observe and obey the instructions of the lifesavers or lifeguards.
- DO swim close to shore, on the shallow inshore and/or on sand bars.
- ALWAYS have at least one experienced surf swimmer in your group.
- NEVER swim alone.
- DO swim under supervision if uncertain of conditions.
- DO NOT enter the surf if you cannot swim or are a poor swimmer.
- DO NOT swim or surf in rips, troughs, channels or near rocks.
- DO NOT enter the surf if you are at all unsure where to swim or where the rips are.
- BE AWARE of hypothermia caused by exposure to cold air and water, particularly on bare skin and with small children. Wind will add to the chill factor.
- DO check the tides in the north, as there may be no water at low tide
- DO be aware of strong tidal currents near inlets and tidal creeks

Patrolled beaches

- swim between the red and yellow flags
- obey the signs and instructions of the lifesavers or lifeguards
- still keep a check on all the above, as over 500 people are rescued from patrolled beaches in Western Australia each year

Unpatrolled beaches

- always look first and check out the surf, bars and rips
- select the safest place to swim, do not just go to the point in front of your car or the beach access track
- try to identify any rips that may be present
- select a spot behind a bar and away from rips and rocks
- on entering the surf, check for any side currents (these are rip feeder currents) or seaward moving currents (rip currents)
- if these currents are present, look for a safer spot
- it's generally safer to swim at low tide, if you avoid the rips

Children

- NEVER let them out of your sight
- ADVISE them on where to swim and why
- ALWAYS accompany young children or inexperienced children and teenagers into the surf
- REMEMBER they are small, inexperienced and usually poor swimmers and can get into difficulty at a much faster rate than an adult

Beach Hazard Rating

The *beach hazard rating* refers to the scaling of a beach according to the physical hazards associated with its beach type under normal wave conditions, together with any local physical hazards. It ranges from the low, least hazardous rating of 1 to a high, most hazardous rating of 10. It does not include biological hazards, such as sharks and crocodiles; these are discussed later in this chapter. The beach characteristics and hazard rating for wave-dominated beaches are shown in Figures 2.4 and 2.5, for tide-modified beaches (Fig. 2.14) and tide-dominated beaches (Fig. 2.16).

The modal beach hazard rating indicates the level of hazard under typical or modal wave conditions for each beach type. Figure 3.4 lists the six wave-dominated three tide-modified and three tide-dominated beaches, together with their beach hazard rating under waves between less than 0.5 m and greater than 3 m. The modal wave height and modal hazard rating are indicated in **BOLD**. The rating ranges from a low of 1 on most tide-dominated and some tide-modified beaches, to a high of 10 on high energy dissipative beaches. The figure also indicates how the hazard rating will change under changing wave and beach type conditions, together with the more generalised hazards associated with each.

The *prevailing beach hazard rating* refers to the hazard rating prevailing at a given time as a result of the prevailing wave, tide wind and beach type conditions. Figure 3.3 can be used to determine the prevailing hazard based on beach type and wave height, with Table 3.3 providing the additional wind hazard. If the beach also has local hazards such as rocks, reefs, headlands and inlets an additional 1 is added to the rating. The prevailing beach hazard rating is therefore a function of:

wave height+beach type+wind hazard+other local hazards

What this implies is that beach hazards are a function of some permanent features such as rocks and reefs, and some low energy beach types, as well as more variable factors such as waves, tides and wind, as well as changing beach types, particularly on more energetic beaches. It also means that the hazard rating will change both between beaches as well as over time on any particular beach or part of a beach. These changes can occur quickly as wave, tide and wind conditions change.

Beach Hazard Ratings

1 - least hazardous beach
10 - most hazardous beach

Beach hazard rating is the scaling of a beach according to the physical hazards associated with its beach type and local beach and surf environment.

Modal beach hazard rating is based on the beach type prevailing under average or modal wave conditions, for a particular beach type or beach.

Prevailing beach hazard rating refers to the level of beach hazard associated with the prevailing wave, tide, wind and beach conditions on a particular day or time.

Table 3.4 summarises the rating for all Western Australian beaches.

Table 3.4 provides an overview of beach hazard ratings around the entire Western Australian coast. Table 3.4a covers the beaches contained in this book. It highlights three aspects of the coast. First, the full range of hazards (1-10), a function of the wide range of wave-beach conditions around the coast. Second, the predominance of lower hazard ratings, a function of the sheltered Central West coast and lower wave energy Carnarvon, Pilbara and Canning coasts; and third, a still substantial number of beaches (1192) rating 5 and above, a product of the high wave conditions along much of the south and parts of the west coast, including 41 rating an extremely hazardous 10, the greatest number for any state. In comparison the Kimberley coast (Table 3.4b) has no beaches rating greater than 5. This is a result of the generally low to very low waves and absence of strong surf and rips.

Compared to other Australian states (Table 3.5) Western Australia has the greatest number of beaches (3411), it has the greatest number of beaches with a hazard rating 8, 9 and 10, and the greatest number rating 5 and higher. In other words it has the greatest concentration of hazardous beaches on the Australian coast, with the more hazardous beaches spread along the South East, South, and parts of the Leeuwin and Central West coasts. At the other extreme it also has the greatest number of beaches with a low hazard rating of 1 and 2, most of these located in Shark Bay and along the Carnarvon, Pilbara, Canning and Kimberley coasts.

Three factors contribute to the high number of hazardous beaches in the state. *First* is the prevailing high southerly swell, which averages 2 m and more year round (Fig. 1.15). The high waves themselves are a hazard as well as the dangerous surf and rips they generate. *Second* is the prevalence of rips on all exposed high energy beaches, with a total of 1470 beach rips usually present around the coast, particularly along the South East and South coasts, but also present along the Leeuwin, Central West and Carnarvon coasts. These rips have an average longshore spacing of 350 m, the largest in Australia. *Third* is the presence of rocks, reefs and headlands along or at the boundaries of many beaches. These features are a hazard in themselves, as well as generating topographic (or headland) rips where surf breaks against the rocks and the associated currents are deflected seaward. In all there are 900 such rips around the coast, again primarily along the South East, South and Leeuwin coasts. When the beach and topographic rips are combined, on an average day 2370 rips are operating around the coast. This book identifies every beach where rips occur and provides information on their location and general spacing.

BEACH HAZARD RATING GUIDE

Impact of changing breaker wave height on hazard rating for each beach type

Wave Dominated Beaches

BEACH TYPE \ WAVE HEIGHT	< 0.5 (m)	0.5 (m)	1.0 (m)	1.5 (m)	2.0 (m)	2.5 (m)	3.0 (m)	> 3.0 (m)
Dissipative	4	5	6	7	8	**9**	**10**	10
Long Shore Bar Trough	4	5	6	**7**	**7**	**8**	9	10
Rhythmic Bar Beach	4	5	6	**6**	**7**	8	9	10
Transverse Bar Rip	4	4	**5**	**6**	7	8	9	10
Low Tide Terrace	3	**3**	**4**	5	6	7	8	10
Reflective	**2**	**3**	4	5	6	7	8	10
Tide Modified Beaches (at high tide - at low tide add 1)								
Ultradissipative	1	2	4	6	8	10	10	
Reflective + Bar & Rips	1	2	3	5	7	9	10	
Reflective + LTT	1	1	2	4	6	8	10	
Tide Dominated Beaches (at high tide - at low tide add 1)								
Beach + Sand Ridges	1	1	2	Waves unlikely to exceed 0.5 - 1m				
Beach + Sand Flats	1	1	*Note: if adjacent to tidal channel, beware of deep water and strong tidal currents.*					
Tidal Sand Flats	1							

BEACH HAZARD RATING

Least hazardous: 1 - 3
Moderately hazardous: 4 - 6
Highly hazardous: 7 - 8
Extremely hazardous: 9 - 10

KEY TO HAZARDS

- (blank) Water depth and/or tidal currents
- (hatched) Shorebreak
- (light shading) Rips and surfzone currents
- (dark shading) Rips, currents and large breakers

NOTE: All hazard level ratings are based on a swimmer being in the surf zone and will increase with increasing wave height or with the presence of features such as inlet, headland or reef induced rips and currents. Rips also become stronger with falling tide.

BOLD gradings indicate the average wave height usually required to produce the beach type and its average hazard rating.

Figure 3.4 *Matrix for calculating the beach hazard rating for wave-dominated, tide-modified and tide-dominated beaches, based on beach type and prevailing wave height and, on tide-modified beaches, state of tide. The* ***modal*** *hazard rating is based on average wave height and beach type for the beach. The* ***prevailing hazard*** *rating is based on wave height and beach type prevailing at time of observation.*

*Table 3.4 Western Australian physical beach hazard rating, by number of beaches and their length (excluding wind hazard) for (a) the mainland coast between Eucla and Roebuck Bay, (b) the Kimberley coast and (c) the entire Western Australian coast. Modal conditions in **Bold**.*

a. WA: Eucla-Roebuck Bay

Beach hazard rating	Number	%	Mean length (km)	σ	Total length (km)	Total %
1	**550**	26.8	1.93	4.1	**1060**	28.6
2	247	12.0	1.49	2.05	368	10.0
3	**331**	16.1	2.06	5.4	**682**	18.4
4	216	10.5	1.78	5.0	386	10.5
5	163	7.9	1.65	8.3	268	7.3
6	119	5.8	2.54	4.3	302	8.2
7	137	6.7	1.57	2.6	215	6.0
8	171	8.3	1.97	4.1	337	9.1
9	76	3.7	0.56	0.6	42	1.1
10	41	2.0	0.68	1.6	28	0.8
	2051	99.8			3688	100

b. Kimberley coast

Beach hazard rating	Number	%	Mean length (km)	σ	Total length (km)	Total %
1	621	45.7	0.40	0.9	246	34.4
2	**633**	46.5	0.55	1.1	**348**	48.7
3	84	6.2	0.52	0.8	44	6.2
4	10	0.7	5.30	3.9	54	7.6
5	12	0.9	1.85	3.0	22	3.1
	1360	100			714	100

c. Western Australia (entire coast)

Beach hazard rating	WA number	%	WA km	%
1	**1171**	34.4	**1306**	28.6
2	880	25.8	716	10.0
3	415	12.2	726	18.4
4	226	6.6	440	10.5
5	175	5.1	290	7.3
6	119	3.5	302	8.2
7	137	4.0	215	6.0
8	**171**	5.0	**337**	9.1
9	76	2.2	42	1.1
10	41	1.2	28	0.8
	3411	100	4402	100

*3.1.1 Rip identification - how to spot rips

To the experienced surfer rips are not only easy to spot, but they are the surfer's friend, providing a quick way (and at times the only way) to get through the surf and 'out the back' beyond the breakers, as well as carving channels to produce better waves. To the inexperienced however, rips are not only unknown or 'invisible' to them, but if caught in one it can be a terrifying and even fatal experience. Most recreational swimmers and visitors do not have the time or desire to become experienced swimmers and surfers. In order to assist them, a check list of features that indicate a rip or rips are present on the beach is noted below:

Rip Current Spotting - Check List

Note: any one of these features indicates a rip, but not all will necessarily be present.
* indicates always present
\+ indicates may be present

* A seaward movement of water (Fig. 3.2) either at right angles to or diagonally across the surf. To check on currents, watch the movement of the water or throw a piece of driftwood or seaweed into the surf and follow its movement.

Table 3.5 *Beach hazard rating of all Australian beaches, by number of beaches.*
***Bold** indicates modal hazard rating/s. (Source: Short, 1993, 1996, 2000, 2001)*

Beach Hazard Rating	Qld	NSW	Vic	Tas	SA	WA	NT	Aust. Number	Aust. %
	325	0	61	**118**	**320**	**1171**	544	2539	23.8
2	**748**	45	36	101	271	880	**738**	**2819**	**26.4**
3	473	103	90	197	206	415	132	1616	15.1
4	58	134	**92**	242	154	226	54	960	9.0
5	13	85	66	**263**	125	175	11	738	6.9
6	**23**	**232**	109	140	93	119	9	725	6.8
7	9	112	**148**	103	**137**	137		646	6.0
8	1	7	77	78	117	**171**		451	4.2
9			11	23	20	76		130	1.2
10		3	2	4	11	41		61	0.6
Total	1650	721	692	1269	1454	3411	1488	10 685	100.0

* Rips only occur when there are breaking waves seaward of the beach. If water is moving shoreward, it must return seaward somewhere.
* Disturbed water surface (ripples or chop) above the rip, caused by the rip current as it pushes against incoming waves and water. May be difficult to spot.
+ Longshore rip feeder channels and/or currents running alongshore, hard against the base of the sloping beach face. Rips are usually supplied by two rip feeder channels converging from either side of the rip.
+ Rhythmic or undulating beach topography, with the rips located in the indented rip embayments.
+ Areas where waves are not breaking, or are breaking less across a surf zone, owing to the deeper rip channel.
+ A deep channel or trough, usually located between two bars or against rocks. The channel may contain inviting, clear, calmer water compared to the adjacent turbulent surf on the bars. However, do not be fooled. In the surf, calm water usually means it's deeper and contains currents.
+ A low point in the bar where waves are not breaking, or break less. This is where the rip channel exits the surf zone.
+ Turbid, sandy water moving seaward, either across the surf zone and/or out past the breakers.
+ In the rip feeder and rip channel, the stronger currents produce a rippled seabed. These ripples are called megaripples and are sandy undulations up to 30 cm high and 1.5 m long. If you see or can feel large ripples on the seabed, then strong currents are present, so stay clear.
+ If you see one rip on a long beach, there will be more if wave height remains the same along the beach. Rip spacing can vary from 200 to 700 m, depending on wave-beach conditions.
+ If there is surf and rocks, reef or a headland, a rip will always flow out close to or against the rocks. These are called **topographic rips**, in that they are controlled by the topography of the headland or reef. These rips are often permanent and have local names like *the express, the accelerator, the garbage bowl* and *the alley.*

Rips - how to escape if caught in one

- If the water is less than chest deep, adults should be able to walk out of a rip. Conversely, avoid going into deeper water. So if you are in any surf current, become very careful once the water exceeds waist depth. Get out while the water is shallow.
- Most people rescued in rips are children. Never let them out of your sight and if they get into difficulties, go to them immediately while the water is still relatively shallow.
- As long as you can swim or float, the rip will not drown you. There is no such thing as an undertow associated with rips, or for that matter, with surf zone currents. Only breaking waves can drive you under water. Most swimmers who drown when caught in rips do so because they panic. So stay calm, tread water, float and conserve your energy.
- If there are people/lifesavers on the beach, raise one arm to signal for assistance.
- Do not try to swim or wade in deep water directly against the rip, as you are fighting the strongest current. There are easier ways out.
- Where possible, wade rather than swim, as your feet act as an anchor and help you fight the current.
- If it is a relatively weak and/or shallow rip, swim or wade sidewards to the nearest bar. Once on the bar, walk or let the waves or wave bores return you to shore.
- If it is a strong and/or deep rip, go with it through the breakers. Do not panic. When beyond the breakers, slowly and calmly swim alongshore in the direction of the nearest bar, indicated by heavier wave breaking. If you are not a surfer, simply wait for a lull in the waves, swim into the break and allow the waves to wash you to shore. Do not dive under the wave,s stay near the surface so the broken wave can wash you shoreward.
- To summarise: stay calm, swim sideways toward breakers or the bar and let the broken waves return you to shore. Raise an arm to signal for help if people are on the beach.
- If rescued, thank the rescuer.

Surf safety

Surfing, as opposed to swimming, requires the surfer to go out to and beyond the breakers, so he or she can catch and ride the waves, in other words, go surfing. This can be done using just your body (body surfing) or a range of surfboards, bodyboards and skis.

Surfing safely requires a substantially greater knowledge of the surf, compared to swimming on the bar or close to shore. The following points should be observed before you begin to surf:

1. You must be a strong swimmer.
2. You must also be experienced at swimming in the surf.
3. You must be able to tell if and when a wave will break.
4. You must know the basics of how breaking waves and the surf zone operate. You should be able to spot rip currents.
5. Equally, you should know what hazards are associated with the surf, including breaking waves, rips, reefs, rocks and so on.
6. You must only use equipment that is suitable for you, i.e. the right size and level.
7. You must know how to use your equipment, whether it be flippers, bodyboard, surfboard or wave ski.
8. You should use safety equipment as appropriate, including a legrope or handrope, wetsuit, flippers and in some cases a helmet.
9. You should ensure your equipment is in good condition, with no broken fibreglass, frayed legrope, etc.

Some tips on safe surfing

- Remember surfing is fun, but it is also hazardous.
- Never surf alone. If you get into trouble, who will help you?
- Before you enter the water, always look at the surf for at least five minutes. This will enable you to first, gauge the true size of the sets, which may come only every few minutes; and secondly, besides picking out the best spot to surf, you can also check out the breaker pattern, channels, currents and rips; in other words, the circulation pattern in the surf. This is important as you can use this to your advantage.
- On patrolled beaches, observe the flags, surfboard signs and directions of the lifesavers. Do not surf between the red and yellow flags or near a group of swimmers.
- If you are surfing out the back, the safest, quickest and usually the easiest way to get out is via a rip. This is because the water is moving seaward, making it easier to paddle; the rip flows in a deeper channel, resulting in lower waves; and the rip will keep you away from the bar or rocks where waves break more heavily.
- Once out past the breakers, particularly if paddling out in a rip, move sidewards and position yourself behind the break.
- Buy and read the Surf Survival Guide, published by Surf Life Saving Australia and available at all newsagents.
- Obtain an SLSA Surf Survival Certificate from your school.

Some general tips

Surfers conduct many rescues around the Australian coast, so be prepared to assist if required. Remember, if you are on a surfboard, bodyboard or wave ski and have a legrope and wetsuit, you are already kitted out to perform rescues. The board is a good flotation device that can be used to support someone in difficulty. The wetsuit will keep you warm and buoyant and thereby give you more energy and flotation to assist someone in distress; and the legrope or board can be used to tow someone in difficulty, while you paddle them toward safety.

The simplest way to get someone back to shore is to lay them on your board while you swim at the side or rear of the board and let the waves wash the board, patient and you back to shore.

Some surfing hazards to watch out for when paddling out:

* Heavily breaking/plunging waves, particularly the lip of breaking waves.
* Rocks and reefs.
* Strong currents, particularly in big surf.
* Other surfers and their equipment.

... when you are surfing:

* Other surfers - the surfer on the waves has more control and is responsible for avoiding surfers paddling out, or in the way.
* Heavily breaking waves.
* Your own and other surfboards. They can and do hit you and can knock you out.
* Shallow sand bars.
* Rocks and reefs.
* Close-out sets and big surf.

... when returning to shore:

* Heavy shorebreaks.
* Rocks and reefs.
* Strong longshore/rip feeder currents.

Remember: The greatest danger to surfers is to be knocked out and drown. Most surfers are knocked out by their own boards or by hitting shallow sand or rocks. This can be avoided by always covering your head with your arms when wiped out, by wearing a wetsuit for flotation, by wearing a helmet and by surfing with other surfers who can render assistance if required.

Biological Hazards - Sharks and crocodiles

This book is not designed to deal with biological hazards. However some mention must be made of sharks and crocodiles, as there have been a number of fatal shark attacks since 2000 in Western Australia at Cottesloe and Cowaramup and in South Australia at Cactus and Elliston. There have also been several fatal crocodile attacks in Western Australia, all in the Kimberley region.

There is no way of avoiding sharks once you enter their territory. If you are concerned about sharks then it is best to stay out of their domain. However all surfers and divers and many swimmers are prepared to spend some time in the ocean with the knowledge that the chances of being attacked are extremely small. On average only one person is fatally attacked in all Australia waters each year, and as unfortunate as they are the above four attacks maintain this average. If you are at all concerned then swim only at patrolled beaches during patrol periods, where lookouts and at times aerial surveillance is used to spot and warn swimmers and surfers of the presence of sharks.

Likewise with crocodiles, the best way to avoid them is to stay well clear of their territory, which not only includes all creeks and estuaries in northern Australia but also many freshwater streams and river banks. While crocodile attacks on open beaches are rare, crocodiles commonly visit and come ashore on ocean beaches. I have seen crocodiles and particularly fresh crocodile tracks on many beaches around the Kimberley coast. I strongly recommend anyone working, fishing or recreating in the Kimberley coast region to visit the crocodile park in Broome or elsewhere to see crocodiles first hand and learn about their behaviour. Do this before venturing into their domain.

The best reference for **biological hazards** is:

Venomous and Poisonous Marine Animals - a Medical and Biological Handbook, 504 pp.
edited by J A Williamson, P J Fenner, J W Burnett and J F Rifkin
published by University of New South Wales Press, Sydney, 1996
ISBN 0 86840 279 6
Available through UNSW Press or the Medical Journal of Australia.

This is an excellent and authoritative text, which provides the most extensive and up-to-date description and illustrations of all hazardous marine animals and the treatment for their envenomation.

Rock fishing safety

Rock fishing is a very popular pastime around most of the Western Australian coast (Fig. 3.5). However, the rocky sections are the most hazardous part of the coast, with most fatalities due to fishermen drowning after being washed off the rocks.

Figure 3.5 Rock fisherman with heavy clothing unsuitable for swimming.

Rock fishing is hazardous because:

- Deep water lies immediately off the rocks, often containing submerged reefs and rocks and heavily breaking waves.
- Occasional higher sets of waves can wash unwary fishermen off the rocks.
- Most fishermen are not prepared or dressed for swimming, as they are often wearing heavy waterproof clothing, shoes and tackle, rather than buoyant and protective wetsuits.
- Many fishermen are not experienced surf swimmers and many cannot even swim.

To minimise your chances of joining this distressing statistic, two points must be heeded. Firstly, avoid being washed off; and secondly, if you are washed off, make sure you are prepared and know how to handle yourself in the waves until you can return to the rocks or await rescue.

The biggest problems usually occur when inexperienced fishermen are washed off rock platforms. To compound the problem, they either cannot swim or are not prepared for a swim. You only need to watch experienced board, bodyboard and body surfers surfing rocky point and reef breaks, to realise rocks are not a serious hazard to experienced and properly equipped surfers.

So the rules are:

1. Before you leave home:
- Check the weather forecast. Avoid rock fishing in strong winds and rain.
- Phone the boat or surf forecast and check the wave height. Avoid waves greater than 1 m.
- Check the tide state and time. Avoid high and spring tides.
- Are you suitably attired for rock fishing, particularly footwear?
- Are you suitably attired in case of being washed off the rocks?
- A loose fitting wetsuit is both comfortable and warm, and it will keep you afloat and protected if washed off the rock platform.

2. Before you start fishing:

- Check the waves for ten minutes, particularly watching for bigger sets.
- Choose a spot where you consider you will be safe.
- When choosing a spot to fish: if the rocks are wet, then waves are reaching that spot; if the rocks are dry, waves are not reaching them, but may if the tide is rising or wave height is increasing.
- Ensure you have somewhere to easily and quickly retreat to, if threatened by larger waves.
- Place your tackle box and equipment high and dry.

3. When you are fishing:

- Never turn your back on the sea, unless it is a safe location.
- Watch every wave.
- Be aware of the tide: if it is rising, the rocks will become increasingly awash.
- Watch the waves, to check for:
  - increasing wave height, leading to more hazardous conditions;
  - the general pattern of wave sets: it is the sets of higher waves that usually wash people off rocks.
- Remember, 'freak waves' exist only in media reports. No waves are freak, all that happens is that a set of larger waves arrives, as any experienced fisherman or surfer can tell you. These larger sets are likely to arrive every several minutes.
- Do not fish alone, two can watch and assist better than one.
- If you see a larger set of waves approaching - retreat. If you cannot retreat, lie flat and attach all your limbs to the rock. Forget your gear, you are more valuable. As soon as the wave has passed, get up and retreat.
- Wear sensible clothing. A wetsuit provides warmth, protection and safety, particularly if you are washed off or knocked over. Sandals with cleats to prevent slipping are also popular.

4. If you are washed off, here are some hints:

- If you have sensible clothing, that is, clothing that will keep you buoyant, such as a wetsuit or life jacket, then you should do the following:
  - Head out to sea away from the rocks, as they are your greatest danger.
  - Abandon your gear, it will not keep you afloat.
  - Take off any shoes or boots and you will be able to swim better.
  - Tread water and await rescue, assuming there is someone who can raise the alarm.
  - If you are alone, or can only be saved by returning to the rocks, try the following:
  - Move seaward of the rocks and watch the waves breaking over the rocks in the general area, then:

Choose a spot where there is either:

a channel - this may offer a safer, more protected route;

a gradually sloping rock - if waves are surging up the slope, you can ride one up the slope, feet and bottom first, then grab hold of the rocks as the swash returns;

or a steep vertical face with a flat top reached by the waves - swim in close to the rocks, wait for a high wave that will surge up to the top of the rocks, float up with the wave, then grab the top of the rocks and crawl onto the rock as the wave peaks. As the wave drops, you can stand or crawl to a safer location.

4 Western Australia Beaches - Eucla to Roebuck Bay

This chapter provides a description of every Western Australian beach between Eucla and Roebuck bay, in all 2051 beaches. The coast is divided into seven regions, which are listed in Table 4.1 and illustrated in Figures 1.1 and 4.1. The beach descriptions begin at Eucla and continue clockwise around the coast to Roebuck Bay, with beaches numbered sequentially from WA 1 to WA 2051. Each beach is located by number on the regional maps. Maps are also provided of every patrolled beach and several other popular beaches, while 356 beach photographs illustrate many of the beaches. Beaches can be located by region, number and name (see Beach Index at rear).

Table 4.1 Western Australian coastal regions and beaches

	Region	Boundaries	Beaches	No. beaches	km	km
1	South East[1]	Eucla-Cape Palsey	1-52	52	0-637	638
2	South[1]	Sandy Bight-Cape Leeuwin	53-617	565	638-1818	1181
3	Leeuwin[1]	Cape Leeuwin-Cape Naturaliste	618-730	113	1818-1930	112
4	Central West[1]	Cape Naturaliste-Kalbarri	732-1134	403	1931-2734	804
5	Carnarvon[1]	Kalbarri-Cape Preston	1135-1708	574	2734-4770	2036
6	Pilbara[1]	Cape Preston-Port Hedland	1709-1896	188	4770-5236	466
7	Canning	Port Hedland-Roebuck Bay	1897-2051	155	5236-5834	598
	sub-total[1]			**2050**		**5835**
8	Kimberley[2]	Broome-WA/NT border	1-1360	1360	0-4340	4340
	Total		3411	**3411**		**10 193**

[1] this book

[2] in Short (in prep) "Beaches of the Northern Australian Coast"

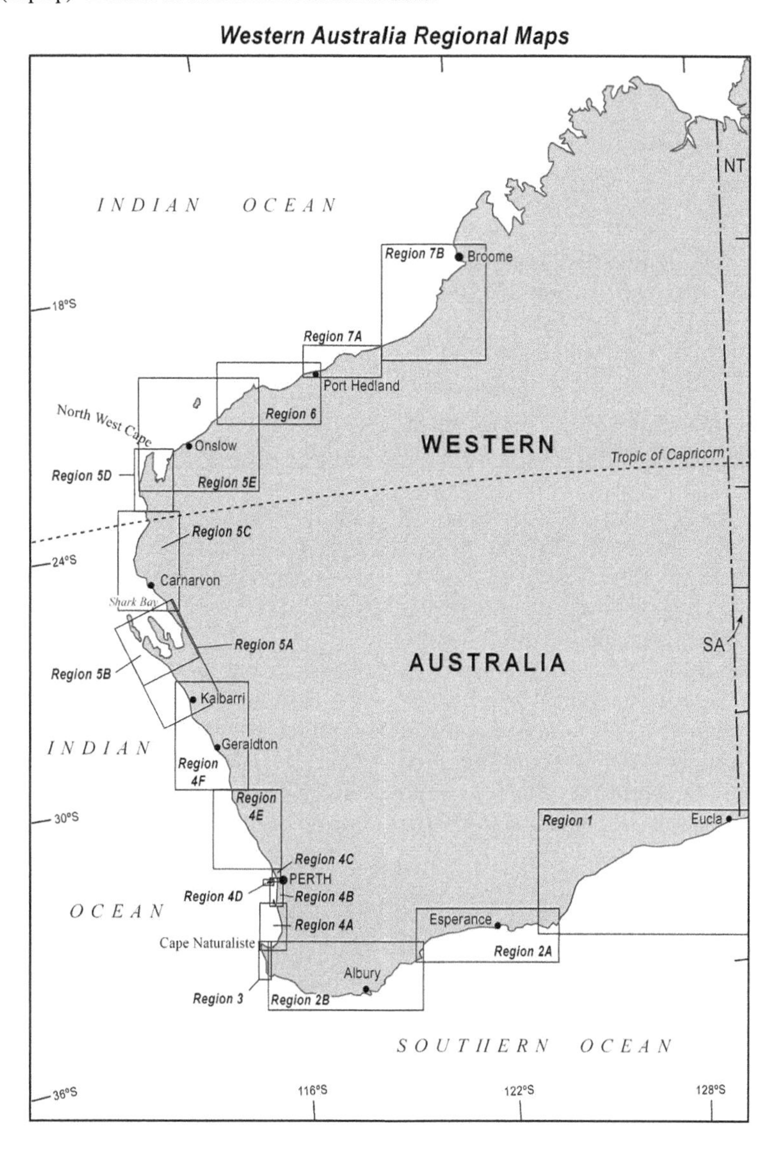

Figure 4.1 Location of the seven coastal regions (1-7) and 18 maps that cover the entire coast between Eucla and Roebuck Bay and locate by number each of the 2051 beaches between Eucla and Roebuck Bay.

Beach name, number and length

Most beaches in Western Australian do not have a name and many have poor or no vehicle access. In order to compile a database and description of each beach a number of methods are used in order to accurately located and identify each beach. These include the CSIRO kilometre distance around the Australian coast to locate the beginning and end of each beach; the central latitude and longitude of each beach obtained from Mapinfo, and the name if known. In addition the Australian Beach Safety and Management Program database contains a digital image of all or part/s of each beach.

Once identified each beach is numbered beginning with WA 1 at Eucla continuing clockwise for 5810 km to WA 2051 at Sandy Point on the southern shores of Roebuck Bay. The Kimberley region contains another 1360 beaches and 4340 km of coast, bringing the total number of Western Australian beaches to 3411, spread along 10 193 km of coast.

Each beach is named, when the name in known, or more usually given a name based on the nearest named feature, often a headland, reef, island or creek. The location and length of most beaches are well defined by headlands or prominent boundaries. However there are parts of the coast, including the Eucla to Twilight Cove section where a near continuous undulating stretch of sand extends for 308 km. Rather than divide the shore into a very few, very long beaches or many very small units located between every single undulation, it has to be divided into sections that usually incorporate a number of the smaller undulations, with division based on the occasional larger shoreline protrusion, usually in lee of an inshore calcarenite reef. The number or location of the beaches along this, and some other sections of coast, is therefore somewhat arbitrary and endeavors to achieve a balance between too few long sections, and too many small sections along and essentially uninhabited and for the most part unnamed coast.

Region 1 South East Coast - western Great Australian Bight (Eucla to Cape Palsey)

South East Coast

Length of Coast: 637 km (0-637 km)
Number of beaches: 52 (WA 1-52)

National Parks:	km	length	beaches	number
Eucla National Park	0-10	10 km	WA 1	1
Nuytsland Nature Reserve:	157-623	466	WA 8-43	36
Cape Arid National Park	623-637	14	WA 44-51	8
Total		490 km(77%)		45 (88%)

Coastal settlements: Eucla, Eyre Bird Observatory, Israelite Bay
Other locations (km inland): Mundrabilla (19 km), Mandura (32 km), Cocklebiddy (26 km), Caiguna (30 km)

The southeast coast of Western Australia extends for 638 km from the 90 m high limestone cliffs at Wilson Bluff on the WA-SA border, to Cape Palsey (Fig. 4.2). The cape and backing 164 m high conical Mount Palsey is the first major outcrop of non-limestone bedrock on the coast west since Cape Adieu (830 km to the east). In between is the world's largest limestone deposit, the 250 000 km^2 Nullabor Plain. The southern boundary of the plain is a 790 km long, 60-100 m high limestone escarpment, which includes the Nullabor (Bunda) Cliffs, the Hampton escarpment, the Baxter Cliffs and Wylie Scarp.

This entire coast is essentially uninhabited apart from the highway settlements/road houses of Border Village, Eucla, Mundrabilla, Mandura, Cocklebiddy and Caiguna all located a 20-30 km inland. The only permanent habitation is at the Eyre Bird Observatory (2 people) and a few shack owners at Red Rock, plus the seasonal cray fishermen at a few coastal locations south of the roadhouses, and at Israelite Bay. In addition 76% the coast is given over to the reserves, beginning with the Eucla National Park, occupying the eastern 10 km, then Nuytsland Nature Reserve (466 km of coast), which in the west abuts the Cape Arid National Park, which has 14 km of coast between the reserve boundary and Cape Pasley.

Eucla is one of Australia's most isolated communities, yet also one of the most frequented given its central location on the 1,674 km long Eyre Highway. Besides being a refuelling stop on the long drive across the Nullabor Plain, the small community is located just 12 km inside the Western Australia border and the first stop for those entering Western Australia from the east. Its location atop the 100 m high Wilson Bluff also provides commanding views of the Roe Plain and Southern Ocean. Many people also stop to visit the ruins of the Eucla Telegraph Station (1877-1927), partly buried in the dunes 4 km south of the settlement. A few people also walk or drive the additional 1 km across the active bare dunes to the old Eucla Jetty and Western Australian's first beach (WA 1).

Region 1: South East Eucla – Cape Pasley

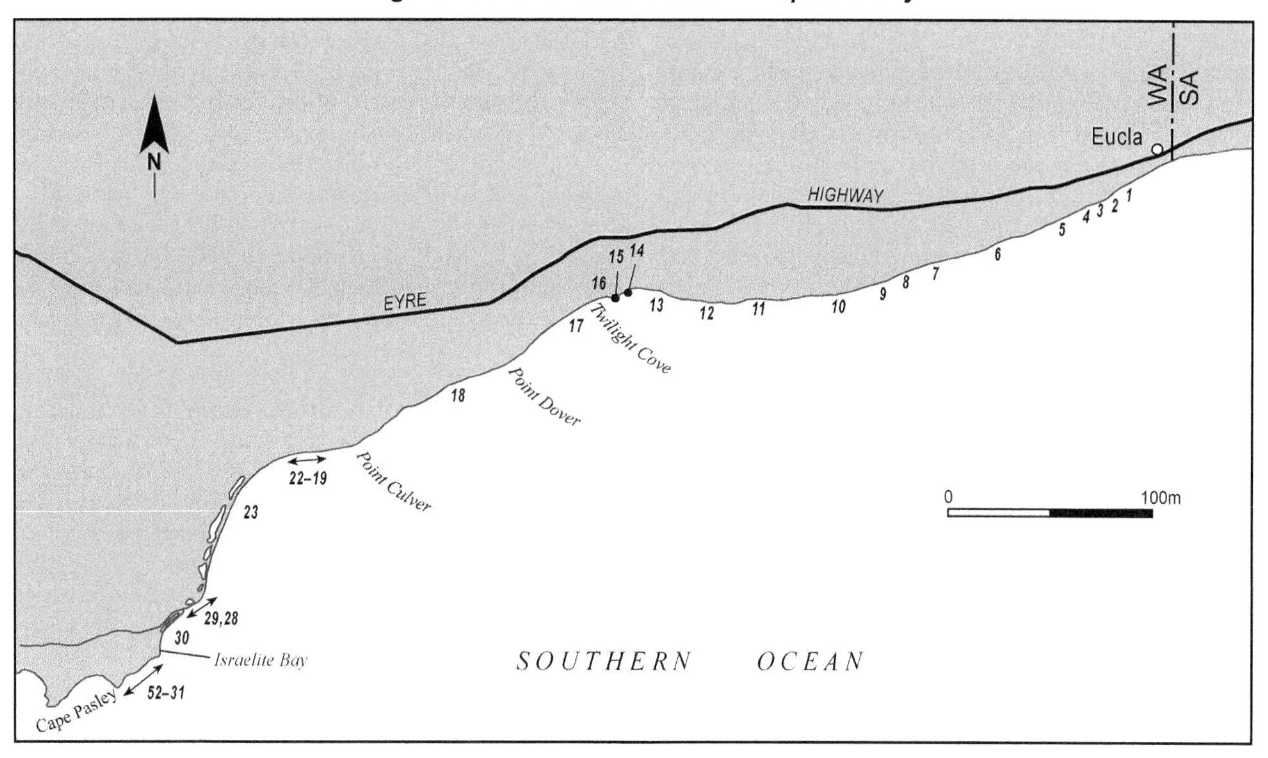

Figure 4.2 South East regional map: Eucla to Cape Pasley, beaches WA 1-52.

The beach (WA 1) begins at the state border below the 100 m high limestone cliffs of Wilson Bluff (Fig. 4.3). The bluffs also marks the western boundary of the 210 km long **Nullabor (or Bunda) Cliffs**. At the border the exposed sea cliffs turn inland and continue west as a sloping 80-100 m high escarpment (**Hampton Bluffs**), for 190 km to Mandura. The bluffs are fronted by the low Roe Plain, a former shallow calcareous seafloor, which reaches 45 km in width at Mandura. Past Mandura the escarpment continues on for another 110 km to Twilight Cove, which marks the western end of the Roe Plain and reemergence of the limestone seacliff, here called the Baxter Cliffs (Table 4.2).

Figure 4.3 Wilson Bluff (foreground) marks the Western Australia-South Australia border, with the Roe Plain and Eucla beach extending to the west.

Table 4.2 Southern boundaries of the Nullabor Plain

	Name	Location (coastal plain)	Elevation (m)	Length (km)
1	Nullabor Cliffs	Head of Bight-Wilson Bluff	60-90	210
2	Hampton Bluffs	Wilson Bluff-Twlight Cove (Roe Plain)	80-100	300
3	Baxter Cliffs	Twilight Cove-Pt Culver	100-120	160
4	Wylie Escarpment	Point Culver-lee of Pt Malcolm (Bilbunya barrier)	70-120	120
			Total	**790**

The **Baxter Cliffs** continue southwest for 160 km to Point Culver, where they again trends inland for another 120 km as the steep 100-120 m high **Wylie Scarp**. The scarp runs in lee of the Bilbunya dunes and salt lakes for 120 km to 14 km west of Israelite Bay, finally degrading into a lower gradient 70-90 m high slope which terminates 17 km west of Point Malcolm. In all the Nullabor – Wilson – Hampton – Baxter - Wylie escarpment extends continuously for 790 km and marks the southern boundary of the Nullabor Plain.

The Nullabor Cliffs are best viewed at the several lookouts along the South Australia side of the plain, while the Eucla-Mandura section of the highway parallels the base of the scarp and provides good views of the sloping Hampton Bluffs. The scarp continues west to Twilight Cove, with the gravel Burnabbie track following the base of the escarpment for 70 km between Mandura and the turn off to Eyre.

The **Roe Plain** lies between the escarpment and the coast. It is a low scrubby plain 290 km long and 40 km wide at its widest, with an area of approximately 6400 km^2. The plain is essentially an uplifted part of the seafloor composed of calcareous marine sediments. It extends from the scarp to the modern shoreline and continues seaward as a shallow calcareous seafloor. The emerged section is densely vegetated with low scrubby vegetation and mallee, while the submerged section forms a low gradient nearshore zone covered in seagrass and with numerous shallow limestone reefs, which substantially lower waves at the shore. As a consequence between Eucla and the Cove the shoreline is an near continuous stretch of low energy reflective carbonate sandy beaches, with numerous reef-induced undulations, a very few limestone outcrops, and numerous vegetated and active dune systems, the largest extending 105 km inland across the flat plain. This dune system commences in lee of Kaniaal Beach near Eyre, and extends 100 m inland, the longest coastal dunes in Australia, and perhaps the world. The **Eyre Highway** crosses the Nullabor Plain between Nullabor and Eucla, then descends Wilson Bluff at Eucla and traverses the Roe Plain to Mandura, 100 km to the west. It then climbs back up Mandura Pass onto the top of the Hampton Tableland (Nullabor Plain) again. In this description of the plain and its 52 beaches, all distance refer to shoreline distance from the WA/SA border at Wilson Bluff.

Eucla National Park

Area: 3560 ha
Length of coast: 10 km (0-10 km)
Beaches: part of beach WA 1

Eucla National Park is a relatively small 3, 560 ha wedge of park bounded by the WA-SA border, the highway and Southern Ocean. It was gazetted to include Wilson Bluff and the Delisser Sand Dunes which originate along the shore and in part drape the bluffs. It also provides a small western continuation of the larger South Australian Nullabor National Park.

For information: South Coast Regional Office
120 Albany Highway
Albany WA 6330
(08) 9742 4500

WA 1 **EUCLA**

No.	Beach	Rating	Type	Length
WA 1	Eucla	5→3	TBR→LTT	41.5 km

Spring & neap tidal range = 1.0 & 0.4 m

Eucla beach (**WA 1**) is a generally low energy sandy shoreline that commences below 100 m high Wilson Bluff and trends to the west-southwest for 41.5 km to a more prominent reef-induced sandy foreland. There is access to coast at the Eucla jetty immediately south of the Old Telegraph Station (Fig. 4.4) and at 15, 23 and 38 km to the west. The jetty was used to supply the station until 1929. The partial survival of the 150 m long timber jetty attested to the usually low wave energy along this section of the Bight (Fig. 4.5). There are no facilities or development along the continuous white sand, apart from a few backing fence lines, an old shed at the end of the 23 km track, and small boats, which are sometimes moored off the 15 km track. The highway is closest to the beach at Eucla. The gap then widens and is 10 km inland by the western end of the beach.

The first few kilometers of the beach have a deeper nearshore and receive moderate waves averaging over 1 m which produce a low gradient 100 m wide surf zone and strong rips every few hundred metres. Towards Eucla jetty the nearshore shoals and waves decrease to about 0.5 m, with low wave conditions dominating the remainder of the beach. The lower energy section consists of a low gradient, fine white sand beach with a continuous 50 m wide low tide bar, fringed by seagrass meadows. Swash deposited piles of seagrass debris is a prominent feature of the upper swash zone the entire beach, and most of the beaches between Eucla and Twilight Cove.

The beach is backed initially by the active **Delisser dunes** which have spread 2-3 km inland and in the east have climbed up the slopes of 90 m high Wilson Buff and then blown several hundred metres inland (Fig. 2.24f). These dunes, which extended further inland during the Pleistocene, were apparently reactivated by rabbits and sheep grazing during the latter part of the 19th century. West of 20 km the dunes become more stable and form a low foredune system, backed by discontinuous salt flats up to a few hundred metes wide, then older, low vegetated early Holocene dunes, which extend a further 3-4 km inland.

Swimming: The Eucla jetty is a moderately safe beach to swim at so long as waves are less than 1 m. Higher waves will strengthen the rips and gouge out rip channels. Remember this is a remote and usually empty beach, so make sure you have company and use caution if you swim here.

Surfing: Usually a low beach break.

Fishing: The beach is one of the few accessible spots on the Nullabor, and the jetty is the only one between Fowlers Bay and Esperance, a distance of 800 km. Unfortunately the jetty is in a bad state and not safe for fishing. The track provides 4WD access to the beach either side of the jetty.

Summary:This is a remote generally low energy ocean beach, made more interesting because it is one of the few beaches reasonably accessible from the Eyre Highway. It also has the historic jetty and backing ruins.

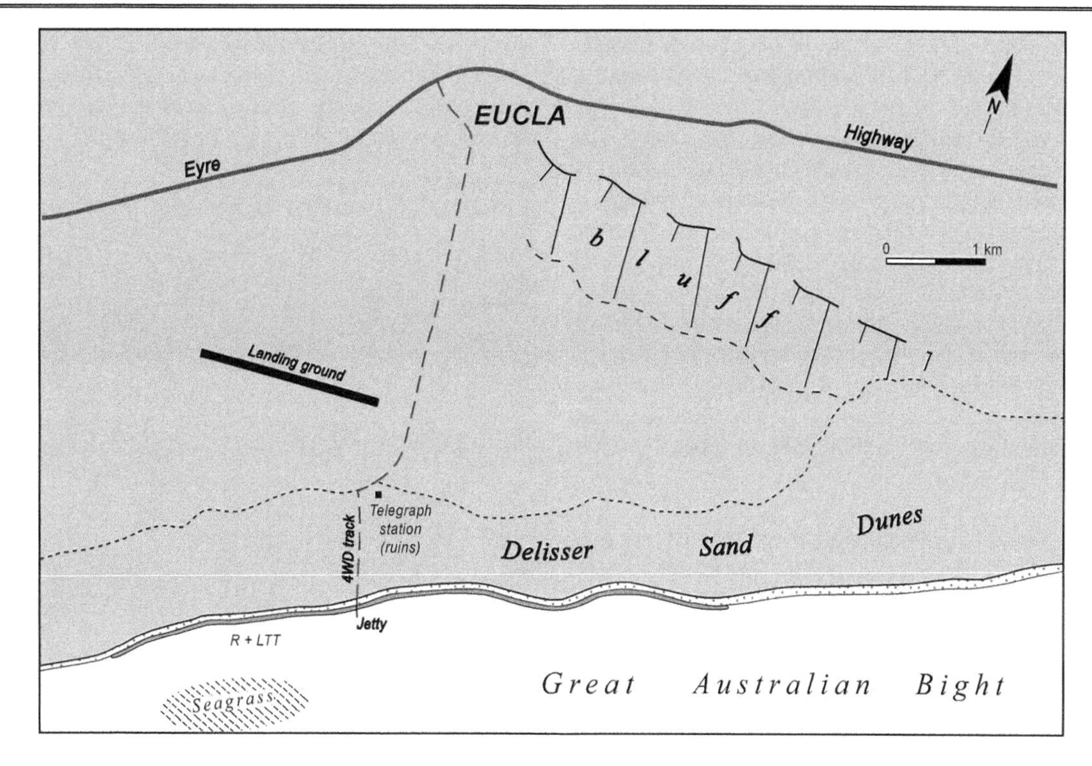

Figure 4.4 The eastern section of Eucla beach showing the access track from Eucla to the old jetty and beach.

Figure 4.5 Eucla jetty and typical low wave conditions.

WA 2-4 EUCLA-LOW PT

No.	Beach	Rating	Type	Length
WA-2	Eucla (W)	3	LTT	18.5 km
WA-3	Hearder Hill	3	LTT	15 km
WA-4	Mundrabilla	3	LTT	8.5 km
WA-5	Low Pt	3	LTT	3 km
			Total	45 km
Spring & neap tidal range = 1.0 & 0.4 m				

West of Eucla beach the coast trends to the west-southwest as a continuous crenulate low energy sandy beach for 45 km to Low Point. The beach can be divided into four beach sections (WA 2-5) each consisting of a low energy, low gradient beach and continuous attached low tide bar, fronted by shallow seagrass covered nearshore, and backed by a mixture of active and stable dune fields. Vehicle access to the shore via is available at 55 km, 68 km (via Hearder Hill Repeater Station) and 83 km from the Mundrabilla track. All tracks are only suitable for 4WD vehicles, and some cross salt flats, which flood following rains.

Beach **WA 2** is an 18.5 km long section of crenulate shoreline containing three longer (≈ 5 km) embayments, together with a series of smaller (≈ 1 km) crenulations, all the crenulations tied to low points in lee of calcarenite reefs. The beach is backed by a continuous series of low foredunes and in places backing salt flats, averaging about 1 km in width, together with four areas of active transverse sand dunes extending up to 1 km inland, with heights generally less than 20 m. There is access to beach at 56 km. The track crosses a series of vegetated foredunes and a deflation basin, before reaching the beach with seagrass piled high along the crest of the beach. The beach has a low gradient beach face, continuous shallow bar, with seagrass growing 100-200 m offshore.

A more prominent cuspate foreland separates beaches WA-2 and 3. Beach **WA-3** continues west for 15 km and contains three eastern 'embayments' containing smaller crenulations and occupying the first 11 km, and a straighter section making up the western 4 km. A vehicle track via the repeater station reaches the back of a 500 m wide series of active transverse dunes and deflation basin, where there is an informal campsite. The dunes are 5-8 m high with a wavelength of 80 m and have buried casuarina woodland. The beach has a low gradient, with seagrass debris forming a high tide berm, and shallow sand bar-shoals extending off the beach to the seagrass meadows. There are two areas of dune actively along the eastern half of the beach, with the entire beach backed by low vegetated transgressive dunes extending 2-3 km inland.

Beach **WA-4** lies at the end of the 16 km track that winds in from **Mundrabilla**. This is a more popular spot with a fishing camp located in behind the dunes, as well as remnants of an old well and stockyards close to the coast. During the crayfish season fishermen operate from this camp and leave their dinghies on the beach with the larger boats moored offshore. At the camp area is a low gradient beach and narrow bar with seagrass meadows offshore. It is backed by a deflation basin then the low vegetated dunes. The beach extends west for 8.5 km with two longer embayments either side of a central sandy foreland, together with shorter crenulations right along the shore. Low vegetated dunes extend 1-2 km inland, with stabilising areas of dune activity along about half the shoreline.

Beach **WA-5** is a 3 km long embayment bordered by two cuspate forelands, the western called Low Point, a more prominent sandy foreland at the eastern end of an 80 km long relatively straight section of shore. The beach is backed by generally stable 3-5 m high foredunes, and discontinuous salt lakes up to 500 m wide, then older vegetated dunes extending up to 2-3 km inland. There is no direct vehicle access to the beach.

WA 6-7 NOONAERA-RED ROCK

No.	Beach	Rating	Type	Length
WA-6	Noonaera	3	LTT	55 km
WA-7	Red Rock	3	LTT	16.5 km
WA-8	Red Rock Pt	3	R/LTT	1.5 km
Spring & neap tidal range = 1.0 & 0.4 m				

To the west of Low Point the coast trends to the west-southwest for 74 km to Red Rock Point, a low red sandstone point, and the only non-calcareous bedrock between Cape Adieu and Israelite Bay. Like the coast to the east the terrain remains generally low, with low energy beaches covered in seagrass debris and fronted by a shallow nearshore and seagrass meadows. Two beaches (WA-6 & 7) occupy most of the shore between the two points, with beach WA-9 tucked in lee of the point. There is access to the coast at 100 km, the old Noonaera homestead (119 km) and at Red Rock (150-160 km) where there are a few shacks, the only 'permanent' habitation on this section of coast.

Beach **WA 6** commences to the west of Low Point and trends west as a crenulate, low energy, low gradient, beach and bar fronted by seagrass meadows usually 100-200 m offshore, while it is backed by low foredunes. Much of the first 33 km to **Noonaera** is backed by low (<5 m) relatively stable foredune ridge plains varying in width from 100-400 m, then a near continuous 100-300 m wide linear salt flats, occupying the back barrier depression, with some older dunes and ridges extending up to 1 km further inland (Fig. 4.6). In total an undulating 1-2 km wide Holocene barrier forms the coastline. West of Noonaera the same type of coast originally extended to the boundary with beach WA-7, narrowing to the west. In places however dune activity has reactivated the foredune which has blown inland across the salt flats and backing dunes, with the active dunes ranging from 1-2 km in width and occupying about 50% of the western 22 km of the beach.

Figure 4.6 Noonaera beach and backing foredune ridges, salt flats and older dunes.

Beach **WA 7** commences on the eastern side of a small embayment, where a straight north-south track reaches the shore. The beach continues west for 16.5 km to the prominent low foreland 2 km east of **Red Rock**, which also marks the approximate boundary of Nuytsland Nature Reserve. The beach remains crenulate and low energy and is backed by generally stable low foredunes, which form a narrow (<500 m wide) band in front of older higher transgressive dunes. At the eastern end of the beach the transgressive dunes extend up to 16 km inland. The Red Rock track winds through a 2 km wide section of 30 m high vegetated dunes before reaching the coast. The Red Rock track is well used and in generally good condition. It services a few shacks located between the beach access area and the point, as well as occasional fishers. Tracks run east of the access point down to the well dug by John Eyre in 1841, and west for 10 km to the shacks and the point area. The beach remains a low gradient low energy system, with higher waves breaking on the outer reef. Towards the point the beach becomes more moderate gradient but is increasingly dominated by seagrass debris, as well as beachrock, which outcrops towards the western end.

Beach **WA 8** is located between the cuspate foreland formed in lee of a more prominent offshore reef, and Red Rock Point. It is a curving 1.5 km long very low energy beach, which is usually completely covered with dense piles of seagrass, which reach 2 m in height and hinder access to the shore. The seagrass berm continues right around the point. The point itself is also covered with clifftop dunes originating from neighbouring Middini Beach. The vehicle track from Red Rock runs around the point to the Middini Beach dunes, providing access to the dunes and beach.

NUYTSLAND NATURE RESERVE

Area:	625 332 ha	
Coast length:	466 km	(157-623 km)
Baxter Cliffs	145 km	(307-465 km)
Beaches:	31	(WA 8-43)

The Nuytsland Nature Reserve is the longest coastal reserve in Australia covering 466 km of continuous shoreline. The reserve includes 311 km of beach systems, as well as the 145 km long Baxter Cliffs. Much of the reserve is a 10 km wide coastal strip, which includes the beaches and dunes, and the cliffs. It does widen in the west to include the Cocklebiddy and Rankin Plain caves, and in the far west to abut the 65 km long, eastern boundary of Cape Arid National Park.

WA 9-11 MIDDINI BEACH-SCORPION BIGHT

No.	Beach	Rating	Type	Length
WA 9	Middini Beach	3	LTT	20 km
WA 10	Mandura	3	LTT	49.5 km
WA 11	Scorpion Bight	3	LTT	8.5 km
Spring & neap tidal range = 1.0 & 0.4 m				

At **Red Rock** the trend of the shoreline turns slightly more to the west and gradually curves round for 75 km to Scorpion Bight by which time the coast is trending due west. There are three long beach along this section (WA 9-11), with vehicle access only in the east via Red Rock (152 km) and the Middini track (171 km), and in the centre down from Mandura (202 km). To the west of Mandura there are no tracks until the Eyre Bird Observatory at 279 km. This entire section is within the Nuytsland Nature Reserve and apart from the tracks and the Observatory, there is no development or habitation along this section.

Middini Beach (WA-9) commences at Red Rock and trends west-southwest for 20 km to a southerly inflection in the shore. The first 10 km is dominated by an active dune system, which is 1 km wide in the west widening to 2 km at Red Rock, where it partly covers the rocks (Fig. 4.7). The active system has a narrow vegetated foredune, 100-200 m wide deflation basin, then the active transverse dunes. This beach receives slightly higher waves, which may contribute to the more active dunes. Some dune calcarenite outcrops at 172 km forming a small point, while red sandstone reef is located off the beach at 171 km. The rocks have been eroded and deposited as storm cobble-boulder ridges, which are now exposed in the deflation basin. The beach maintains a low gradient, with a continuous low tide bar, usually free of rips. The western half of the beach is more protected by shore parallel reefs lying 200-300 m offshore and tends to have stable vegetated foredune and older backing dunes extending up to 4 km inland.

Beach **WA 10** is centered on the track from **Mandura**, 50 km to the north. The beach extends for 20 km to the

Figure 4.7 The Middini Beach transgressive dunes, extending west of Red Rock.

east and 28 km to the west. Where the track reaches the beach it crosses 5 km of elevated dunes, which rise to 32 m. However the present 50 km of shoreline is dominated by low wave conditions and generally stable foredune, with only a few areas of dune activity, the largest at 192 km, extending about 500 m inland. However earlier (mid-Holocene ?) transgressive dunes have extended on average 2-3 km, and in places up to 4-5 km inland. These are now vegetated and stable. The beach maintains a low gradient, with a continuous low tide bar, and seagrass meadows further out. Most wave energy is attenuated across the continuous shallow nearshore zone (Fig. 4.8).

Figure 4.8 Wave attenuating across the shallow seabed in front of Mandura beach.

Beach **WA 11** occupies **Scorpion Bight**, named after the cutter *Scorpion* wrecked here in 1876. The bight is a 2 km deep indentation in the run of the shoreline, with the 8.5 km long beach occupying the relatively straight eastern arm, and a protruding western section. Wave energy is slightly higher along the bight and dunes are more active. On the eastern arm dunes have extended 500 m inland while out on the protrusion a 4 km long dune system is up to 1 km wide. All the dunes tend to have a stable foredune, deflation basin, then active transverse dunes spilling onto the backing scrubland. In addition early Holocene dunes, part of the Burremul Sand Patch extend up to 7 km in from the bight, with still more extensive dunes originating from Kaniaal Beach, 50 km to the west lie between 12 and 25 km inland. These are discussed in the next beach description. The beach continues as an undulating, low gradient beach face, with a high tide seagrass debris berm, and a continuous low tide bar, then seagrass meadows.

WA 12-16 **BURREMUL-EYRE-TWILIGHT COVE**

No.	Beach	Rating	Type	Length
WA 12	Burremul	3	LTT	26 km
WA 13	Kaniaal Beach (Eyre)	3	LTT	29 km
WA 14	Kaniaal (W)	5	R+reefs	6.5 km
WA 15	Twilight Cove (E)	3	LTT	4 km
WA 16	Twilight Cove	3	LTT	3 km
Spring & neap tidal range = 1.0 & 0.4 m				

To the west of Scorpion Bight the shoreline trends due west along the Burremul Sand Patch for 26 km, before trending west-northwest for 29 km along Kaniaal Beach, which includes the Eyre Bird Observatory. It then protrudes southeast for 14 km around the bulge in the shoreline that terminates at Twilight Cove against the 100 m high northern edge of the Baxter Cliffs. This 70 km long section of coast is one of the more geologically interesting on the Australian coast. It is backed by the largest, single mainland, coastal dune system in Australia, which extends up to 105 km inland. It also marks the western boundary of the Roe Plain and the beginning of the spectacular Baxter Cliffs, with most of the cliffs draped in both lithified Pleistocene and Holocene dunes, including massive clifftop dunes, that in places have blown up, along and back over the 100 m high vertical cliffs. The coast is accessible only at Eyre (280 km) and Twilight Cove (305 km).

Burremul beach (**WA 12**) is named after the **Burremul Sand Patch** which has originated from the 26 km long beach. The dune system extends inland between 3 km in the west to 8 km in the east, and is in turn backed by the larger Wurrengoodyea dunes. The Burremul dunes have blown eastward at an acute angle to the coast and have travelled in places 30 km from the beach source. There is evident of three periods of dune activity. First is the inner, most extensive dune field, then a second now stable system with a more northerly orientation extending 2-3 km inland, and the modern active dunes, which in the east extends 200-500 m inland, decreasing is size and occurrence to the west. The beach is protected by the shallow nearshore zone with waves shoaling and breaking across the reefs to average about 0.5 m at the shoreline, together with numerous low amplitude reef-induced undulations alongshore.

Kaniaal Beach (WA 13) commences at the slight inflection in the shoreline and continues west-northwest for 12 km to Eyre, then another 5 km to Wonglabilla Well before curving round for 7 km to terminate against the beginning of the calcarenite shore adjacent to Rigbys shack, with a total length of 29 km. The old telegraph station and the Eyre Bird Observatory are located 1 km inland at 279 km. The beach has acted as the source of sand for the largest single dune system in Australia. The dunes trend east-northeast, sub-parallel to the coast, for a maximum distance of 105 km. They are 15 km wide at the coast gradually narrowing to the east. They have a total area of approximately of 1000 km^2, and with an average height of approximately 10 m, the dunes have a volume of 10 km^3 of quartz and calcareous sands. It is second only to Fraser Island in the size ranking of Australian dune systems. The beach system continues to be protected by the shallow nearshore zone, with wave attenuating across the calcarenite reefs and averaging about 0.5 m at the shore where they maintain a 50 m wide low tide bar, with seagrass offshore, and seagrass debris dominating the upper beach.

Beach **WA 14** marks change in the shoreline from one backed by sand dunes, to a calcarenite section of coast backed by the southern end of the 80-90 m high Hampton escarpment. The Pleistocene dunes reached the base of the escarpment and in places climbed to the crest. These conditions continue west for the 14 km to the Baxter Cliffs. The first 6.5 km long section is backed by sloping vegetated dune calcarenite rising within 1.5 km to the crest of the escarpment. The calcarenite also dominates the shoreline with outcrops along the beach and off the beach as inter to subtidal reefs. Wave averaging over 1 m break over the reef, forming numerous reef-induced crenulations alongshore. As a result the beach is generally impassable for vehicles. Beach **WA 15** is a 4 km long continuation of this system, and is backed by more active dune systems extending up to 500 m inland, that have deflated down to the Pleistocene surfaces exposing the red soil horizons. The beach has a low gradient, waves about 0.5 m high and a continuous low tide bar and lower beach face, with seagrass debris on the upper beach.

Twilight Cove (WA 16) is named after the cutter *Twilight* wrecked on the beach in 1877, one of three known 19th century wrecks at the beach. The cove has a continuous moderately steep quartz-rich sand beach that trends west for 3 km to the base of the 100 m high cliff line (Fig. 4.9). The beach receives waves averaging 0.5-1 m, which maintain a continuous low tide bar, usually free of rips. The system both undulates in response to offshore calcarenite reefs, as well as increasingly oscillating towards the western end, where the shoreline may move backwards and forwards more than 100 m over a period of years. The beach is backed by a low foredune area, then a deflation basin that includes Pleistocene calcarenite, which has been scarped by waves in the past, then an active dune system, which is transgressing over the backing casuarina woodland, and finally the older Pleistocene dunes, which abut the escarpment. In the far west the Holocene dunes have reached the base of the cliffs. There is vehicle access to Twilight Cove both along the beach from Eyre, and via tracks from Cocklebiddy, 26 km to the north. The dunes behind the beach are used from informal camping. The remains of Carlisle's hut are located on the Pleistocene slopes towards the western end of the dunes. Carlisle and his family of eleven children lived in this remote location during the mid parts of the 20th century.

Figure 4.9 Twilight Cove beach terminates against the beginning of the 160 km long Baxter Cliffs.

WA 17-18 **BAXTER CLIFFS**

No.	Beach	Rating	Type	Length
WA 17	Baxter Cliffs (329 km)	10	TBR+rocks	150 m
WA 18	Toolinna	8	TBR+rocks	100 m
Spring & neap tidal range = 1.1 & 0.5 m				

The **Baxter Cliffs** commence at Twilight Cove and trend to the southwest for 158 km to the eastern end of Bilbunya beach and the beginning of the Wylie Scarp. The cliffs are named after John Baxter who was murdered at 355 km in 1841, while the Wylie scarp, was named after John Eyre's final companion in his historic walk across the Great Australian Bight. The sheer cliffs are composed of Tertiary marine limestone and average 60-80 m high. They run uninterrupted for the entire length, with just two small pockets of sand located in small coves at the base of the cliffs (WA 17 & 18).

Beach **WA 17** is located at 329 km, 25 km southwest of Twilight Cove. The narrow 150 m long strip of sand has accumulated at the base of a 500 m long indentation in the run of the cliffs. The beach is backed by vertical 80 m high cliffs, with no safe land access. Boulder debris from the cliffs litters the back of the beach, and the moderate to high waves wash over the entire beach at high tide, leaving no dune. A permanent rip drains the beach against the eastern rocks.

Toolinna Cove beach (**WA 18**) lies another 90 km along the cliffline at 419 km. The cove occurs at an inflection in the run of the cliffs as they turn and trend more to the south. The beach is approximately 100 m long, but is in part covered by a massive slopes of boulder debris, leaving only a 50 m long patch of sand towards its southern end. It has a usually detached bar cut by a central and two permanent boundary rips, with some rocks also in the surf (Fig. 4.10). A vehicle track leads to the top of the beach, and a flying fox has been set up in a steep gully toward the southern end to provide access to and transfer of fishing gear and fish, to and from the beach. Access to the beach however remains extremely hazardous. Extending to the northeast of the beach and adjacent

Figure 4.10 Toolinna Cove is a remote, difficult to access and hazardous beach.

cliffline is a 10 km long area of clifftop dunes, covering an area of 20 km^2.

Point Culver clifftop dunes: Clifftop dunes extend 7 km inland and for 45 km to either side of Point Culver covering an area of 250 km^2. The dunes have originated from sand blown up from the eastern end of Bilbunya beach, up and over the 100 m high Wylie Scarp, and then spread to the east for up to 45 km. The dunes average 8 m in height and become increasingly longitudinal to the east.

WA 19-23 **BILBUNYA**

No.	Beach	Rating	Type	Length
WA 19	Bilbunya (E 4)	9	RBB	150 m
WA 20	Bilbunya (E 3)	9	RBB	400 m
WA 21	Bilbunya (E 2)	9	RBB	600 m
WA 22	Bilbunya (E 1)	9	RBB	600 m
WA 23	Bilbunya	6→3	RBB→LTT	103.2 km
Spring & neap tidal range = 1.1 & 0.5 m				

Bilbunya beach (WA 23) is one of the longest and most spectacular beaches on the Australian coast and also one of the least known and least visited. It commences as a series of discontinuous beach remnants (WA 19-22) located along the base of 100 m high cliffs, 10 km west of Point Culver. It then trends to the west, curving round to eventually trend south for 103 km, terminating 5 km east of Point Lorenzen. The entire system is part of the Nuytsland Nature Reserve and apart from a few access tracks is undeveloped. The 100-140 m high Wylie Scarp lies about 10 km inland and parallels the entire beach. Between the scarp and the beach are Holocene and Pleistocene barrier and dune deposits, including the 120 m high active Bilbunya sand dunes, the highest active dunes in Australia.

Beach **WA 19** is a 150 m long strip of sand located at the base of 100 m high cliffs. It occupies a slightly indented cavern in the cliffs, with the upper part of the cliff overhanging the beach, as well as considerable rock debris at either end of the beach. It has a small patch of dune in the cavern, but is otherwise awash at high tide. It is fronted by an energetic 150 m wide surf zone that continues west to the main Bilbunya beach.

Beach **WA 20** lies 1 km to the west and is a narrow 400 m long strip of sand at the base of the cliffs, and separated from beach WA 21 by a 100 m long debris slope. Beach **WA 21** continues on in a similar fashion for 600 m, with steep debris slopes and the cliff backing the central 400 m of the beach. Beach **WA 22** commences 200 m to the west and is a narrow discontinuous 600 m strip of sand, that is cut in places by rock debris and slight protrusions in the cliff line. It terminates as the cliff abruptly turns inland and becomes the landlocked Wylie Scarp. This point also marks the beginning of the main Bilbunya beach (WA 23). All four beach share a surf zone dominated by a usually detached bar cut by large rips spaced every 500 m.

Bilbunya beach (**WA 23**) initially faces south and trends west, gradually curving round to the southwest at 20 km and south by 55 km. The scarp initially parallels the back of the beach 1 km inland, then as the beach swings the gap widens to 5 km, before the scarp turns and trends southwest essentially paralleling the beach its entire length, with the gap gradually widening to 20 km in the south. The 103 km long beach not only changes orientation but also experiences a gradual decrease in wave energy from north to south, which manifests itself in the nature of the surf zone, beach and backing barrier systems and indicated in Table 4.3.

The first **30 km** of the beach faces southeast and receives high waves averaging over 1.5 m which maintain a rip dominated single bar system and a generally a moderately steep beach face, fronted by a deep trough and the bar. Rips are spaced about every 500-700 m. This section is backed by a wide deflated surface with some inner active transverse dunes. Between **30 and 40 km** the shoreline turns more to the southwest and waves decrease slightly with a low tide terrace forming along the base of the beach and an outer rip dominated bar system, with rips maintaining a similar spacing. In lee of this section are the large Bilbunya dunes, massive coalescing transverse dunes that have formed a star dune that reaches about 120 m height. Between **40-55 km** the shoreline continues to curve slightly more to the south and waves decrease to 1 m resulting in a continuous low tide bar, with a straight outer bar and usually no rips. Along this section the northern bare dunes gradually stabilise with vegetation and decrease in height. By **55 km** the beach is trending south-southwest and wave height is decreasing to less than 1 m. As grain size decreases (Table 1) the beach and surf zone gradient also decreases resulting in three parallel bars. This system transforms to a very low gradient system with four parallel bars by **70 km**, which continue for 30 km to the curving convex end of the beach. A the same time as wave height and depth decreases, seagrass meadows increases along the beach and the amount of seagrass on the beach increases, particularly south of 55 m (Fig.1.19b). There is also a small outcrop of eroded rock slabs (possibly beachrock) at 57 km.

The barrier system south of 55 km is characterised by low vegetated dunes, apart from the active dunes at Wattle Camp (60 km). On the beach the seagrass dominates the upper and intertidal beach all the way to the end, with the seagrass berm up to 50 m wide and 1-2 m high. For this reason the beach access track runs along behind the beach south of about **90 km**, as it is literally impossible to drive on the large 'soft' hummocky seagrass berm. The beach sediments continue to fine and become rich in carbonate at 100 km (Table 1) forming a very low gradient, flat wide white sandy beach, fronted by multiple bars (Fig. 4.11) and backed by the massive piles of seagrass. The beach finally terminates at a low rock attached foreland, by which time it has recurved round to face southwest again, but now is fronted by a shallow 2 km wide barless sand flats, then the seagrass meadows.

Figure 4.11 The southern end of Bilbunya beach is composed of very fine sand which contributes to the wide, very low gradient 3-4 bar dissipative surf zone.

The beach is backed by a continuous Holocene barrier system, then a near continuous depression occupied by salt flats. The barrier is widest in the north where the active dunes and deflation basin extend up to 3 km inland, backed by a 500 m wide salt flats. South of the dunes the lower vegetated barrier averages 1-2 km in width and is backed by more irregular 2-3 km wide salt flats. At Wattle Camp the barrier widen to 3 km and salt flats are 3 km wide. To the south the barrier narrows to 1-2 km, and the salt flats vary from 2-5 km in width.

Table 4.3 Bilbunya beach characteristics

Beach length (km)	WA coast (km)	Orient-ation (deg)	Breaker wave height (m)	Sand size (mm)	% Ca CO_3	Beach slope (deg)	Bar number (1, 2,3,4)- beach type	Surf zone width (m)	Barrier type comments
0-30	465-495	160	1.5-2	0.33	3	5	1- RBB	150	deflation surface & transgressive dunes
30-40	495-505	140	1.5	0.25	11	4	1-LTT 2-RBB	200	deflation surface & transgressive dunes
40-55	505-520	130	1	0.20	2	4	1-R/LTT 2-Diss	250	vegetated low dunes some instability
55-70	520-535	110	0.7	0.23	5	3	1-LTT 2-LBT	400	vegetated low dunes seagrass berm increasing
				0.22	12		3-Diss		along beach (rocks at 522 km)
70-97	535-562	110	0.5	0.14 0.16 0.19	7 4	1.5	1-LTT 2-Diss 3-Diss 4-Diss	500	vegetated low dunes large seagrass berm over beach
98-101	562-565	95	<0.5			1	1-LTT 2-Diss 3-Diss 4-Diss 5-Diss	500	vegetated low dunes- washover flats large seagrass berm over beach
101-104	565-569	150	<<0.5	0.10	79	1	sand flat	2000	vegetated washover flats large seagrass berm over beach

WA 24-30 **ISRAELITE BAY-PT MALCOLM**

No.	Beach	Rating	Type	Length
WA 24	Pt Lorenzen (E3)	1	R+sand flats	1.7 km
WA 25	Pt Lorenzen (E2)	2	R	800 m
WA 26	Pt Lorenzen (E1)	4	LTT	700 m
WA 27	Pt Lorenzen	3	R	1.3 km
WA 28	Israelite Bay	3→1	R→R+sandflats	5.2 km
WA 29	Israelite Bay spit	1	R+sand flats	1.8 km
WA 30	Pt Dempster-Malcolm	6→1	TBR-R+sandflats	23 km

Spring & neap tidal range = 1.1 & 0.5 m

At the southern end of the long Bilbunya beach the shoreline curves sound to trends west and terminates at a small section of low metasedimentary rocks. These are the first non-calcareous rocks on the coast since of Red Rock, 410 km to the east. Between here and Point Malcolm 30 km to the southwest are seven low to moderate energy beaches. All are accessible by vehicle.

Beach **WA 24** commences on the western side of the rocks and curves round for 1.7 km trending south against the next more substantial 200 m long low rocky section. Despite the southerly orientation of most of the beach, it is well protected by a continuation of the 1-2 km wide low gradient sand flats that extends around the southern end of Bilbunya beach extending the length of beach WA 24, narrowing to a few hundred metres against the western rocks. Waves attenuate across the sand flats, and seagrass debris is piled wide and high along much of the low gradient beach. The beach is backed by a low vegetated 100-200 m wide foredune, then a shallow lagoon. There is vehicle access to the western end, with the main track running along behind the lagoon.

Beach **WA 25** commences on the southern side of the low massive metasedimentary rocks and trends southwest for 800 m, curving round to face east against the next section of rocks. The beach is reflective, with a low gradient while the nearshore zone is a little deeper and dominated by seagrass meadows with seagrass debris dominating the upper beach and foredune. The beach is initially backed by a low foredune, which increases in height to the south as 10-20 m high vegetated transgressive dunes from beach WA 26 spill over onto the southern half of the beach.

Beach **WA 26** is a relatively straight 700 m long, white sand beach that commences amongst the sloping, 5 m high, massive gneiss rocks at its eastern end, trending essentially west to a straight 500 m long rocky section that separates it from beach WA 27. The beach receives waves averaging about 1 m, which maintain a continuous 50 m wide low tide bar, with permanent rips between the bar and the boundary rocks at either end. It is backed by generally stable transgressive dunes that extend up to 300 m inland and rise to 20 m. Beach **WA 27** continues to the west on the other side of the rocks before curving round to face east in lee of Point Lorenzen. The point is a low linear 300 m long section of genesis, with beaches WA 27 and 28 forming a narrow sandy 200 m long tombolo to its lee (Fig. 4.12). The beach is composed of medium quartz and receives waves less than 1 m, which decrease towards the point and maintain a reflective beach, with is backed by vegetated parabolic dunes extending up to 1 km inland, with an access track through the dunes to the rocks between beaches WA 26 and 27.

Israelite Bay beach (**WA 28**) commences on the western side of the Point Lorenzen tombolo and initially trends west before curving round to face east at the old jetty site. The beach is 5 km long and receives low waves in the north, decreasing in height to the south, with sand flats lying off the southern end of the beach. The waves are sufficiently low to permit the operation of the jetty, which was used to supply the telegraph station (1887-1927), located 1 km southwest of the jetty. Today the sandy Fisheries Road track terminates at the southern end of the beach near the ruins of the old jetty. There is a fishing camp behind the beach as well as an informal camping area, with fishing boats sometimes moored off the beach.

Figure 4.12 Point Lorensen is a low rock reef tied by a tombolo formed by beaches WA 26 and 27.

Beach **WA 29** is a slowly migrating recurved spit that has migrated across the southern few hundred metres of Israelite Bay beach enclosing a small salty lagoon, which drains across the boundary of the two beaches (Fig. 4.12). The spit originated in lee of Point Dempster as a curving 1 km long very low energy beach, which continues on for another 800 m as the spit, which curves back into the main Israelite Bay beach. The beach and spit are fronted by 300 m wide sand flats and then shallow seagrass meadows, all protected by the point and rock reefs extending 2 km east of the point. Seagrass debris dominates both beaches.

Beach **WA 30** commences on the southern side of the dune-capped, 10 m high, 400 m wide Point Dempster. It gently curves to the southwest and finally south for 23 km, terminating in lee of Point Malcolm. The northern section of the beach receives waves averaging over 1 m which maintain a 50 m wide bar initially cut by rips. After a 1-2 km, as wave height begins to decrease, the bar remains continuous to 5 km south of the point, beyond which the beach is reflective and barless to about 16 km south, where decreasing wave energy permits 100-200 m wide sand flats and seagrass meadows to form along the southern several kilometers of beach. At this point the beach also become crenulate alongshore, induced by shore-attached sand waves. The northern 6 km of beach is backed by some partially active dune transgression, which widens to 2 km in lee of the point. To the south 200-300 m wide vegetated foredunes, then beach ridges back the beach, which are in turn backed by 1-2 km wide, salty Lake Daringdella. The only direct access to the beach is 4 km north of Point Malcolm, near the junction of the Telegraph Line track, and near some ruins of an early homestead. The vehicle track follows the back of the beach to 1 km north of Point Malcolm where there is a small parking and picnic area.

Figure 4.13 Israelite Bay beach showing the recurved spit and lagoon.

WA 31-37 **POINT MALCOLM (W)**

No.	Beach	Rating	Type	Length
WA 31	Pt Malcolm	7	TBR/LTT	1.3 km
WA 32	Pt Malcolm (W1)	5	LTT→R	800 m
WA 33	Pt Malcolm (W2)	3	R	600 m
WA 34	Pt Malcolm (W3)	8	TBR+rocks	500 m
WA 35	Pt Malcolm (W4)	7	TBR	1.5 km
WA 36	Pt Malcolm (W5)	9	R+rock platform	200 m
WA 37	Pt Malcolm (W6)	9	R+rock platform	500 m
Spring & neap tidal range = 1.1 & 0.5 m				

Point Malcolm is a 5-10 m high, dune-capped, gneiss headland, with large tabular slabs of gneiss thrown back on the exposed southern face. While seagrass and sand flats dominate the north side of the point, the coast to the south and west is exposed to the high southerly swell and energetic beaches and rocky coast dominate (Fig. 4.14). To the west of the point is a cuspate foreland at 2 km and the next major inflection at 8 km. In between are seven moderately exposed beaches separated by generally low bedrock points and backed by moderately active dunes. The Telegraph Track runs 2-3 km inland, with no defined access to the beaches.

Beach **WA 31** commences on the western side of the 500 m long point and continues west for 1.3 km to a 100 m band of dune-capped bedrock. While the beach is well exposed to southerly waves they are lowered slightly by inshore bedrock reefs and Franklin Rock, a small islet 2 km to the south. The beach receives waves averaging about 1.5 m high which maintain permanent rips at either end and 2-3 beach rips crossing the 100 m wide bar. During calmer periods the beach rips can infill and a continuous bar front the beach. There is foot access across the dunes from the picnic area at the southern end of beach WA 30. The beach is backed by a vegetated frontal dune area, then a 10 ha area of active dunes that have transgressed up to 500 m inland, with older vegetated dunes extending up to 1 km inland.

Figure 4.14 Point Malcolm with a low energy seagrass fringed beach to the north, and an exposed high energy beach to the west.

Beaches WA 32 and 33 lie ether side of a sandy foreland formed in lee of a low, linear 1 km long reef lying up to 1 km offshore. Beach **WA 32** commences on the western side of the rocks that separate it from beach WA 31. It is 800 m long, faces southeast and terminates at the tip of the foreland. Wave height decreases along the beach to the foreland owing to the protection afforded by the islet and other rock reefs. The beach initially has a narrow bar grading to a reflective beach on the foreland. Beach **WA 33** commences at the foreland and trends to the west for 600 m are a reef protected reflective beach. The beaches are backed by a 10 m high vegetated sandy foreland.

A 200 m long series of discontinuous rock outcrops form a sand foreland and separate beaches WA 33 and 34. Beach **WA 34** is a curving 500 m long beach with boulders and rock reefs defining either end, as well as large rocks outcropping along the beach and in the surf. It is also more exposed to the waves, which average about 1.5 m generating three permanent rock-controlled rips. It is backed by generally vegetated transgressive dunes extending up to 500 m inland.

Beach **WA 35** extends from the western rocks for 1.5 km to the next low rocky point. It is well exposed with moderate waves, decreasing slightly to the west. They maintain a 100 m wide surf zone cut by permanent rips to either end and usually one large central beach rip. It is backed by a foredune, deflation hollow, and a few parabolic dunes in amongst vegetated transgressive dunes, which extend up to 500 m inland.

Beaches WA 36 and 37 are two rock-dominated strip of sand. Beach **WA 36** is a 200 m long strip of high tide sand, bordered and fronted by low bedrock platforms, with and energetic rip-dominated rock and sandy surf zone off the platform. Beach **WA 37** lies 1 km to the west and is a similar, but longer, 1 km long series of sand pockets and strips bordered, separated and fronted by the supratidal bedrock, with rock reefs off the beach. Rough vehicle tracks reach the rear of both beaches.

WA 38-46 **BELLINGER ISLAND**

No.	Beach	Rating	Type	Length
WA 38	Bellinger Is (E4)	7	RBB→LTT	4.5 km
WA 39	Bellinger Is (E3)	9	R+rock platform	600 m
WA 40	Bellinger Is (E2)	5	LTT+rocks	1.2 km
WA 41	Bellinger Is (E1)	4	LTT→R	1.5 km
WA 42	Bellinger Is	7	TBR	500 m
WA 43	Bellinger Is (W)	8	TBR+rocks	150 m
WA 44	Marlbemup (W2)	8	RBB	1 km
WA 45	Marlbemup (W1)	4	LTT	600 m
WA 46	Marlbemup	8	RBB→LTT	3.5 km
Spring & neap tidal range = 1.1 & 0.5 m				

Bellinger Island is a linear 2 km long granite island located midway between Point Malcolm and Cape Pasley. The shoreline protruded in lee of the island, which is located approximately 1 km off the coast. Beaches WA 38-41 are located on the mainland coast to the east of the protrusion, with beach WA 42 to the lee of the island and beaches WA 43-46 west of the island. The Telegraph Track is located 3 km north of the coast with no formed access to the shore. The boundary of the Nuytsland Nature Reserve and Carpe Arid National Park bisects beach WA 44.

Beach **WA 38** is separated from beach WA 37 by a low rocky point that extends offshore for 200 m as a rock reef. The beach trends to the southwest for 4.5 km, with a slight foreland at 3 km to the lee of a rock reef located 500 m offshore. Waves average over 1.5 m in the east, decreasing slowly to the west, particularly past the foreland. The higher waves maintain a 100 m wide surf zone cut by strong beach rips along the first 2 km, which give way as waves decrease to a continuous attached bar. The beach is backed by moderately active dunes climbing to 20 m and extending 100-200 m inland. The Telegraph Track passes within 500 m of the rear of the beach.

Beaches WA 39-41 are located along the eastern side of the protrusion in lee of Bellinger Island and associated rock reefs. Beach **WA 39** is a 600 m long strip of high tide sand located behind a near continuous 20 m wide band of intertidal rocks (possibly beachrock?), with a 50 m wide surf zone breaking over additional rocks. Beach **WA 40** extends for 1.2 km between low rocky points, with rocks also outcropping in the centre and forming a slight foreland. Waves decrease towards the island with a continuous bar occupying the sandy sections. Beach **WA 41** extends from a clump of rocks to a low rocky point at the tip of the protrusion. It is 1.5 km long and sufficiently protected by its orientation and the island to maintain a barless reflective beach. All three beaches are backed by bare deflated surfaces and active dunes that have migrated from beaches WA 43-45 up to 3 km across the protrusion.

Beach **WA 42** is located directly behind **Bellinger Island.** It is a curving 500 m long beach that is tied to a discontinuous line of rocks in the east and a 200 m wide low rock headland in the west. Waves average over 1 m and maintain a central bar with permanent rips to either side flowing out against the rocks. It is backed by a vegetated foredune, then a generally deflated dune surface (Fig. 4.15).

Figure 4.15 Active transgressive dunes link beaches WA 42 (foreground) and WA 41 to the lee.

Beaches WA 43 and 44 occupy the western side of the protrusion. Beach **WA 43** is a 150 m long pocket of sand wedged in between the headland and a low linear rocky point. The beach faces south with waves averaging over 1.5 m, which maintain a strong permanent rip running out against the western rocks. A continuous boulder beach backs the sand beach, then a foredune and the deflated dune surface. Beach **WA 44** commences on the other side of the rocks and trends west for 1 km to the next low bedrock headland. This is a high energy exposed beach dominated by a 200 m wide surf zone with two large permanent rips flowing out at either end of the beach (Fig. 2.9c). It is backed by a moderately active foredune, deflated surface and then active dunes extending 2-3 km across to beaches WA 39-41.

Beach **WA 45** lies between two low rock headland located 300 m apart. The beach curves round between the two headlands for 600 m and receives waves lowered by the headland and adjacent rock reefs to less than 1 m. These maintain a narrow continuous attached bar, with a stable vegetated foredune behind the beach, then vegetated transgressive dunes from beach WA 46.

Beach **WA 46** commences on the western side of the 500 m wide headland and trends west, then southwest for 3.5 km to the next low headland. Waves are high in the east where they maintain a rip-dominated surf zone, gradually decreasing to the west, to form a continuous attached bar along the western half of the beach. Vegetated transgressive dunes, including 18 m high Marlbemup Hill, back the beach and extend east for up to 4-5 km to link with the dunes from beaches WA 43 and 44.

CAPE ARID NATIONAL PARK

Area:	279 832 ha	
Coast length:	101 km	(622-723 km)
Beaches:	47	(WA 43-89)

Cape Arid National Park is named after the cape located in the centre of the southern coastal boundary of the park. The park contains a the spectacular 100 km long coastline dominated by Cape Pasley, Cape Arid and Tagon Point with Sandy Bight and Yokinup Bay between the capes. In addition the coast contains 46 beach systems and backing massive dune systems. The park also extends up to 80 km inland and includes Mt Ragged. There is vehicle access to the coast and camping areas at Poison Creek in Sandy Bight and Thomas River in Yokinup Bay, with access across the park to the Nuytsland Nature Reserve and Israelite Bay via the Fisheries Road/track and the Telegraph Track.
For information:

Cape Arid National Park
PO Box 185, Esperance WA 6450
Phone: (08) 9075 0055

WA 47-52 **CAPE PASLEY**

No.	Beach	Rating	Type	Length
WA 47	Cape Pasley (W3)	5	R+rocks	150 m
WA 48	Cape Pasley (W2)	6	TBR	500 m
WA 49	Cape Pasley (W1)	6	TBR	500 m
WA 50	Cape Pasley (1)	10	R+rocks	100 m
WA 51	Cape Pasley (2)	10	R+rocks	200 m
WA 52	Cape Pasley (3)	10	R+rocks	100 m

Spring & neap tidal range = 1.1 & 0.5 m

Cape Pasley is a prominent 5 km wide bedrock headland covered in dense low scrubs and dominated by 164 m high Mount Pasley. It is exposed to the full force of the westerly gales and high waves and has a very hazardous shoreline, with the waves washing over the sloping rocks and numerous rocks and reefs along the shore. While most of the cape has a rocky shore, there are three partly protected beaches on the western side of the cape (WA 47-49), and three exposed pockets of sand on the southern tip of the cape (WA 50-52).

Beach **WA 47** is a 150 m long pocket of east-facing sand, wedged in a small rocky cove. The beach itself is relatively protected with waves averaging about 0.5 m. However rock outcrops on the beach and rocks and reefs narrow the cove entrance to about 50 m.

Beaches **WA 48** and **49** share a 1 km long curving embayment. The are both 500 m long and are separated by a group of low rocks in the centre of the bay. They face southeast and receive waves averaging less than 1.5 m. These each maintain a 50 m wide bar cut by a central beach rip, with a strong permanent rip at the northern end of beach WA 48. They are backed by the dense vegetated slopes of Mount Pasley.

Beaches WA 50-52 are located in three adjoining gaps in the rocky coast along the southern tip of the cape. Beach **WA 50** is a 100 m long sandy beach backed by a low vegetated foredune. It is bounded in the east by a 300 m long low rocky point and the main cape shore to the west. A single strong permanent rip drains the cove, with rock and reef also outcropping in the 200 m wide surf zone. Beach **WA 51** lies 1 km to the west and is a similar 200 m long beach cut in two by a large rock outcrop. It is backed by a grassy foredune, then older transgressive dunes that have blown across the backing rocks for 500 m. It also faces into a 300 m deep rocky cove with a strong central rip draining the 200 m wide surf zone (Fig. 4.16).

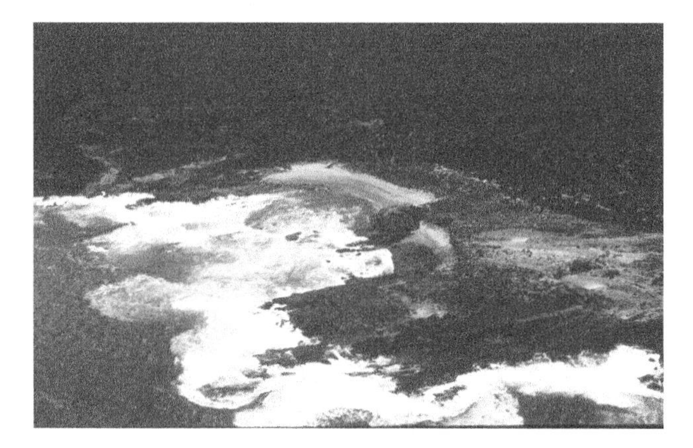

Figure 4.16. Beach WA 51 is an exposed hazardous beach located 1 km west of Cape Pasley.

Beach **WA 52** lies another 1 km to the west, near the tip of the cape. It is a 100 m long high tide sand beach and backing foredune and some dune transgression. The sand beach grades into rocks flats and rocks, then a rock-dominated 200 wide surf zone, bordered by low rocky points and also drained by a single rip against the eastern rocks.

Region 2 South Coast –Cape Pasley to Cape Leeuwin

South Coast

Length of Coast: 1181 km (637-1818 km)
Number of beaches: 565 (WA 53-617)
See Figs 4.17 & 4.79

National Parks:	km	length	beaches	number
Cape Arid	637-723	86 km	53-89	37
Cape Le Grande	775-853	78	115-141	27
Butty Habour (Nat. Res.)	890-923	33	156-168	13
Stokes	956-984	28	183-200	18
Fitzgerald River	1083-1159	76	257-297	41
Fitzgerald River	1226-1229	3	316	1
Waychinicup	1375-1390	15	364-365	2
Two Peoples Bay	1407-1431	24	374-382	9
Torndirrup	1460-1498	38	410-421	12
West Cape Howe	1510-1530	20	429 433	5
William Bay	1562-1573	11	450-464	15
Walpole-Nornalup	1598-1651	53	491-533	43
D'Entrecasteaux	1651-1756	105	534-587	54
D'Entrecasteaux	1762-1772	10	593-600	8
Total		580 km (49%)		285 (50%)

Towns: Esperence, Hopetoun, Bremer Bay, Albany, Denmark, Walpole, Augusta
Coastal settlements: Duke of Orleans Bay, Peaceful Bay, Windy Harbour

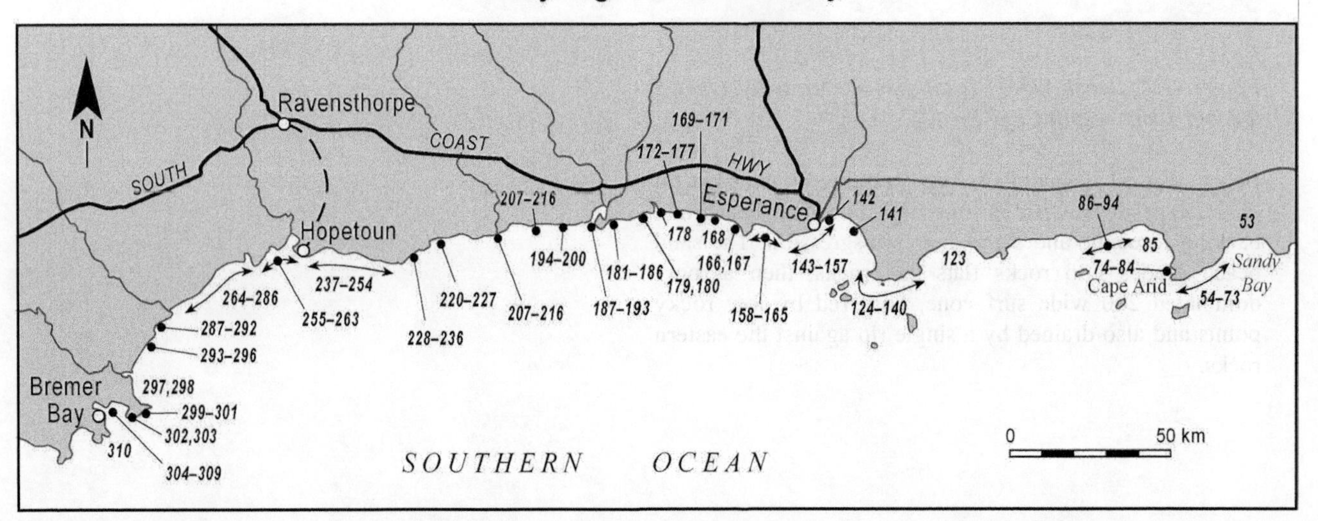

Figure 4.17 South Coast regional map 2A, between Sandy Bight and Bremer Bay, beaches WA 53-310.

The South Coast of Western Australia extends for 1181 km in a protruding arc from Cape Pasley in the west to Cape Leeuwin at the 'heel' of the state. This is a relatively remote, partly developed section of coast containing ten national parks, which occupy 580 km (49%) of the coast. The structure of the coast is dominated by ancient intrusive Proterozoic rocks, with most of the 565 beaches located between gneiss or granite headlands. Many of the beaches are also well exposed to the prevailing southwesterly waves and winds, resulting in energetic, rip-dominated beaches, many backed by extensive dune systems. Figure 4.16 illustrates the eastern section of the South Coast.

WA 53-58 **SANDY BIGHT**

No.	Beach	Rating	Type Inner Outer bar	Length
WA 53	Sandy Bight	8	TBR RBB	14 km
WA 54	Sandy Bight (W)	6	TBR	500 m
WA 55	Fern-Poison Ck	4	LTT/R	1.2 km
WA 56	Seal Ck (N)	4	LTT	1 km
WA 57	Seal Ck	4	LTT	600 m
WA 58	Seal Ck (S)	4	LTT	50 m
WA 59	Jorndee Ck	3	R	400 m
Spring & neap tidal range = 0.8 & 0.4 m				

Sandy Bight is a 15 km wide, southwest-facing embayment bordered by Cape Pasley to the east and a 20 km long rocky section of coast terminating at Cape Arid to the west. One long beach (WA 53) occupies the base of the bay, with a series of smaller embayed beaches along the western side of the bight. The entire bight lies within the Cape Arid National Park, with vehicle access to most of the western beaches, and camping areas at Seal and Jorndee creeks.

Sandy Bight beach (**WA 53**) is an exposed 14 km long, gently curving, southwest-facing beach backed by sand dunes that have blown 10-15 km behind Cape Pasley to the eastern shoreline. The beach receives waves averaging 2 m, which together with the fine to medium sand maintain a double bar 300 m wide surf zone. The inner bar has rips spaced every 200-300 m, while the outer bar has more widely spaced rips (Fig. 4.18). In addition there are permanent rips against the boundary rocks and the slight foreland in lee of a rocky islet located just off the centre of the beach. The beach is backed by a series of active blowouts-parabolic extending from a few hundred metres to 1 km inland (Fig. 2.8b), and then a far more extensive older, vegetated series of longwalled parabolics which reaches the shoreline at beaches WA 46 and 47, up to 15 km to the east. In the west an isolated rocky point separates it from beach **WA 54**, a 500 m long extension of the main beach. The beach does however receives slightly lower waves and has a single bar usually cut by strong permanent rips to either end, and an occasional central rip. It is backed by a vegetated foredune, then a narrow elongate wetland that extends for 1 km behind the western end of the main beach.

Fern-Poison Creek beach (**WA 55**) commences 300 m to the west and occupies a 1 km wide southeast-facing embayment. The beach curves for 1.2 km between the two low vegetated headlands, with Fern Creek deflected the length of the beach to link with Poison Creek and breakout across the southern end of the beach. The beach is moderately exposed in the north, with wave energy decreasing into the southern corner, where the beach curves round to face north. A few rips occupy the northern half of the beach, with a continuous and narrowing bar along the southern half. There is vehicle access along Poison Creek to the creek mouth, where the beach is used to launch small fishing boats.

Figure 4.18 The central section of Sandy Bight (WA 53) showing the wide high energy surf zone and spiraling rip head in the centre (arrow)

The **Seal Creek** beaches occupy the next embayment, 1.5 km to the southwest. The two beaches lie either side of a cuspate foreland formed in lee of two small rocky islets (Fig. 4.19). Seal Creek is blocked by the northern beach and breaks out only following winter rains across the very northern end of the beach. The northern beach (**WA 56**) faces southeast and curves round for 1 km, while the main beach (**WA 57**) is 600 m long and faces south. Both beaches receives waves averaging about 1 m which maintain a continuous attached low gradient bar, with rips only forming during higher wave conditions. There is an active foredune behind both beaches and dune transgression extending up to 200 m in behind the northern beach. The main access road runs along behind the dunes to the southern end of the main beach, where there is also an outcrop of granite at the end of the beach, which separates it from a 50 m long pocket beach **WA 58**. The beach has a continuous bar and is usually free of rips.

Figure 4.19 The Seal Creek beaches (WA 56 and 57) lie either side of a sandy cuspate foreland formed to the lee of the low islets.

Jorndee Creek beach (**WA 59**) lies in the next small embayment 2 km to the south. The beach is 400 m long and curves round the base of the embayment to face out into the bight. It is relatively protected by its orientation and a southern low rocky point and rock reefs with waves averaging less than 1 m, which maintain a relatively steep reflective beach. The creek drains across the beach and rocks at the southern end of the bay. This is the most protected and least hazardous of the bight beaches. There is a vehicle access to the beach, with the car park and small camping area behind the centre of the beach.

WA 60-68 **SANDY BIGHT (W)**

No.	Beach	Rating	Type	Length
WA 60	Sandy Bight (W1)	3	R/LTT	750 m
WA 61	Sandy Bight (W2)	3	R/LTT	100 m
WA 62	Sandy Bight (W3)	3	R/LTT	100 m
WA 63	Sandy Bight (W4)	4	LTT	80 m
WA 64	Sandy Bight (W5)	3	R/LTT	300 m
WA 65	Sandy Bight (W6)	3	R/LTT	250 m
WA 66	Sandy Bight (W7)	3	R	1.2 km
WA 67	Sandy Bight (W8)	3	R	500 m
WA 68	Sandy Bight (W9)	3	R	600 m
Spring & neap tidal range = 0.8 & 0.4 m				

To the south of Jorndee Creek massive sloping gneiss and granites dominate the shoreline for the next 20 km to Cape Arid. In amongst the rocky shoreline are several generally small beaches afforded varying degrees of protection by the rocks, reefs and in some cases a more easterly orientation. Beaches WA 60-68 occupy a 6 km long section of the rocky shore commencing 3 km south of Jorndee Creek. There is no vehicle access to these beaches.

Beaches **WA 60** is a curving east-facing, 750 m long beach bounded by a 1 km long sloping northern headland, and a smaller southern headland, with a rock reef extending 200 m into the embayment, as well as some scattered reefs across the bay mouth. As a consequence the beach receives waves lowered to less than 1 m which maintain a usually steep reflective beach, backed by densely vegetated foredune and some older minor dune activity extending up to 200 m inland, all covered and backed by dense coastal heath.

One kilometre to the south is beach WA 61, which is the first of a series of five pocket beaches (WA 61-65) that occupy gaps in the next 2 km of sloping rocky shoreline. Beach **WA 61** is a 100 m long pocket of sand wedged at the rear of a 200 m deep gap in the massive gneiss that rises to 20 m. Rock reefs also extending part way across the mouth of the gap resulting in low waves and a reflective beach backed by a small stable foredune, then coastal heath. Beach **WA 62** lies 300 m to the south and is a similar 100 m long pocket of sand located at the base of a narrow 250 m deep rocky gap that narrows to less than 100 m at the mouth, which is also strewn with rock reefs. Usually low wave to calm conditions prevail at the beach, which is backed by a stable foredune, then low coastal heath that links with an ephemeral lagoon 300 m to the rear. Beach **WA 63** lies 200 m to the south and is an 80 m long wedge of sand at the base of a 150 m deep, 50 m wide rocky gully, with rock reefs across the mouth. Bare rock lines the sides of the gully, with a small foredune and coastal heath to the rear.

Beach **WA 64** occupies the rear of a 200 m wide break in the rocky shore. The bay is 400 m deep and widens to accommodate the curving 300 m long beach. Waves are usually low at the shore and surge up the steep reflective beach. It is backed by a low foredune, then dense low coastal heath which links with a 10 ha ephemeral lagoon 300 m to the west.

Beach **WA 65** lies 500 m further south, at the base of a 200 m wide gap between the rocky shore and a 10 ha low rocky islet, that is tied to the shore by a tombolo formed by beaches WA 65 and 66 (Fig. 4.20). Beach WA 65 curves between the mainland rocks and islet for 250 m. It faces east and receives low waves, which surge up the steep reflective beach face. The beach is narrow, low and overwashed along the tombolo section, then backed by a 10-20 m high unstable foredune, originating from the eastern end of beach WA 66. Beach WA **66** is a curving 1.2 km long south-facing white sand beach, that is partly protected by Cape Arid and its rocky islands, as well as inshore rock reefs. Waves average just over 1 m along the centre of the beach and maintain a continuous attached bar, with rips only forming during periods of higher waves. It is backed by a moderately stable foredune, then densely vegetated older blowouts and transgressive dunes trending northeast to the rear of beach WA 65.

Figure 4.20 Beaches WA 65 and 66 lie either side of a small tombolo-tied rock islet.

A 200 m long sloping rocky point separates beaches WA 66 and 67. Beach **WA 67** is a relatively straight 500 m long beach, bounded by sloping gneiss, with rock reefs off each end. It receives waves averaging less than 1 m, which maintain a steep reflective beach. It is backed by a vegetated hummocky foredune and remains of some earlier minor dune transgression. Beach **WA 68** lies on the southern side of a protruding 300 m long gneiss point. It is a slightly curving 600 m long beach bordered in the south by a low, sloping gneiss point, which is awash in storms. Rocky reefs and exposed rocks extend from the western point to off the centre of the beach, lowering waves and resulting in a steep reflective beach face. It is backed by a partly stable foredune, then a deflated surface associated with dune transgression originating from beach WA 69.

WA 69-73 **THOMAS FISHERY-CAPE ARID**

No.	Beach	Rating	Type	Length
WA 69	Thomas Fishery (E3)	6	TBR	900 m
WA 70	Thomas Fishery (E2)	3	R+rocks	250 m
WA 71	Thomas Fishery (E1)	4	R/LTT	1.3 km
WA 72	Thomas Fishery	3	R+rocks	300 m
WA 73	Cape Arid (W)	2	R	20 m
Spring & neap tidal range = 0.8 & 0.4 m				

Gulch Island is a 2 km long, 20 m high granite island that lies 3 km due south of beaches WA 69-72. The island is part of a series of granite islands, islets, rocks and reefs at extend up to 17 km south of Cape Arid, the largest 90 ha Middle Island. They are all part of the Recherche Archipelago and afford varying degrees of protection to the backing shoreline. There is vehicle access to the beaches via the Thomas Fishery track.

Beach WA 69-72 lie between 7-9 km east of the cape. Beach **WA 69** is a curving 900 m long southwest-facing beach bordered by a generally bare 20 m high domed gneiss headland to the east and a low bedrock shore to the west. The beach is well exposed to the southwest winds, while partly protected from the waves by Cape Arid and some of the island. The waves average about 1.5 m in the centre and usually maintain 3-4 central beach rips and a permanent rip against the eastern rocks, with waves decreasing towards the western end. The beach is backed by a relatively stable foredune, then a deflated area in the east that extends up to 2 km northeast to beach WA 68 (Fig. 4.21). The western half is backed by a generally vegetated longwalled parabolic with vehicle access to the western end of the beach.

Figure 4.21 Beach WA 69 is bordered by a low gneiss headland and backed by active dunes that extend east to beach WA 68.

Beach **WA 70** commences at the southern end of a 300 m long section of bedrock that separates it from beach WA 69. It is a narrow curving, east-facing reflective beach wedged in lee of the islet. Bedrock dominates the northern half of the beach, with only a small southern sandy access to the ocean. It is backed by a 10 m high vegetated foredune that has originated on beach WA 71. There is vehicle accessed to the northern end of the beach.

Beach **WA 71** commences on the western side of the vegetated tombolo and curves round for 1.3 km to the next bedrock section of shore. It is moderately protected from waves by inshore islands and nearshore rock reefs, including some rocks close to shore. Waves average less than 1 m and maintain a steep surging reflective beach. It is backed by some minor dune activity in the 20 m high foredune, then densely vegetated dunes extending up to 400 m inland in the east. The beach can be access from a track that terminates at the western rocks.

Thomas Fishery beach (**WA 72**) commences at the southern end of a 1 km long section of bedrock that separates it from beach WA 71. The beach is 250 m long, faces east and curves between the bedrock and a small, partly vegetated, rocky point and reefs. The beach is protected by its orientation, the point and inshore rock reefs, and consist of a narrow reflective beach dominated by the bordering low sloping bedrock and extensive bedrock reefs which fill the small bay. The vehicle track leads to the eastern end of the beach where a few campsites are located. Mount Arid rises to 357 m, 2 km northwest of the beach.

Beach **WA 73** lies 4 km to the west and 3 km to the east of Cape Arid. The beach is a speck of sand just 20 m long in a 100 m wide, 200 m deep bedrock bay surrounded on three sides by bare sloping granite rising to 40 m. The beach lies in the northwestern corner of the bay, and provides the only sandy access to the rocky bay. It is backed by rocks and dense scrub and is awash at high tide and during high waves.

Cape Arid was sighted and named by Admiral D'Entrecasteaux "Cap Arride" in 1792. Matthew Flinders subsequently anglicised it to "Cape Arid" in 1802.

WA 74-77 **ARID BAY**

No.	Beach	Rating	Type	Length
WA 74	Arid Bay (1)	3	R	400 m
WA 75	Arid Bay (2)	6	TBR	2 km
WA 76	Arid Bay (3)	6	TBR	1 km
WA 77	Barrier Anchorage	3	R	700 m
Spring & neap tidal range = 0.8 & 0.4 m				

Arid Bay is a 3.5 km wide, southwest-facing, bedrock bay bordered by the protruding granite of Cape Arid to the south, and the gneiss of Barrier Island and its backing headland to the north. In addition three other islands and several rock reefs lie across the entrance to the 2 km deep bay. Mount Arid lies 3 km to the east. Four beaches (WA 74-77) dominate the bay shore, with extensive dunes backing the western three. There is no formal vehicle access to the bay region.

Beach **WA 74** is located on the northern side of the 2 km long spur that terminates at Cape Arid. The beach faces west and receives protection from the cape, its northern rocky point and the islands, with waves averaging less than 1 m. The beach is reflective and composed of yellow sand, unlike the white sand that dominates its neighbours.

It is bounded by bare granite, which rises to 40 m in the east and is backed by low coastal heath.

Beaches WA 75-77 are part of a continuous 3.7 km long strip of sand, which is broken into the three beaches by two low wave-washed rocky points. Beach **WA 75** is a curving 2 km long, west-facing moderately exposed beach. Waves are lower in the east in lee of a low rocky island and increase towards the west, where they average about 1.5 m. These maintain an initially continuous bar that is cut by a few central and western beach rips, with the western end of the beach curving out to link with the low gneiss point. Beach **WA 76** continues on the western side of the point for 1 km as a similarly exposed, rip-dominated, beach and surf, with a permanent rip against the eastern rocks, and rocks outcropping in the surf towards the western end of the beach. It terminates at a small 10 m high rocky point, with beach **WA 77** continuing for 700 m to a 60 m high domed headland. This beach faces southwest and is protected by the headland and Barrier Island, with wave averaging less than 1 m and reflective conditions usually prevailing (Fig. 4.22). The protected western end of the beach and bay is the anchorage after which the 'Barrier Anchorage' is named. All three beaches are linked alongshore and backed by a common dune sheet that extends up to 2 km inland climbing to 60 m on the lower slopes of Mount Arid. The dune area is largely deflated, with scattered knobs of vegetation. Older vehicle tracks leading to the landing are now closed to the public.

Figure 4.22 Barrier Anchorage is a partly sheltered bay containing the reflective beach WA 77, which terminates at the 60 m high granite headland.

WA 78-79 **JENAMULLUP CK**

No.	Beach	Rating	Type	Length
WA 78	Jenamullup Ck (1)	4	LTT	250 m
WA 79	Jenamullup Ck (2)	6	TBR	1.5 km
Spring & neap tidal range = 0.8 & 0.4 m				

Beaches WA 78 and 79 are contained within a 1.5 km long, west-facing bay, bordered by the 1 km long Barrier Anchorage headland to the south and a 40 m high granite headland to the north. Jenamullup Creek drains the northern slopes of Mount Arid and dissipates into the dunes behind beach WA 79, only breaching out across the western end of the beach during heavy rain. A second creek also breaks out across the eastern end of the beach. There is vehicle access to the northern headland, though access is restricted.

Beach **WA 78** is located at the northern base of the southern headland and trends north for 250 m to a small sloping granite outcrop, which separates the upper beach from beach WA 79. The beach is moderately protected by the headland and receives waves averaging about 1 m, which maintain a continuous 50 m wide bar, usually free of rips. It is backed by deflated clifftop dunes. Beach **WA 79** continues on from the rocky protrusion for 1.5 km to a steep, 1 km long, northern headland. Wave energy increases along the beach and a 50-100 m wide, rip dominated surf zone prevails, with up to five beach rips and a strong permanent rip against the northern rocks. The beach is backed by a vegetated southern section, and a partly vegetated northern section extending 500 m inland. These in turn are backed by 3-4 km of densely vegetated longwalled parabolic dunes, which link with the more extensive Yokinup Bay dunes.

WA 80-84 **PT JEDACORRUDUP**

No.	Beach	Rating	Type	Length
WA 80	Pt Jedacorrudup (S5)	7	RBB	1 km
WA 81	Pt Jedacorrudup (S4)	4	LTT	100 m
WA 82	Pt Jedacorrudup (S3)	4	LTT	300 m
WA 83	Pt Jedacorrudup (S2)	6	TBR	800 m
WA 84	Pt Jedacorrudup (S1)	7	RBB	1.4 km
Spring & neap tidal range = 0.8 & 0.4 m				

Point Jedacorrudup is a low vegetated outcrop of gneiss that separates the bedrock-controlled beaches WA 80-84, from the longer Yokinup Bay beach (WA 85). Beaches WA 80-84 extend 4 km to the southeast of the point to the Jenamullup Creek headland. The five intervening beaches are separated by three central outcrops of gneiss (Fig. 2.24e), which are in turn fronted by rock reefs and islets. There is no public vehicle access to the beaches.

Beach **WA 80** commences on the northern side of the headland and trends northwest for 1 km to the first of the low bedrock outcrops. The beach faces into the southwest swell, which is reduced by a central rock reef, to the lee of which is a small sandy foreland. Waves average about 1.5 m and maintain an 80 m wide surf zone across the fine low gradient white sand. Permanent rips are located to either end of the beach and on the central foreland. To the lee of the beach active dunes extend 2 km inland.

Beaches WA 81, 82 and 83 are clustered in amongst the central rock outcrops. Beach **WA 81** is a 100 m long pocket of sand bounded by two low, 100 m long, gneiss points. The beach is partly protected by offshore reefs and islets with waves averaging about 1 m, which break across a 40 m wide bar, with a single rip against the southern rocks. Beach **WA 82** lies immediately to the north and is a 300 m long low gradient beach tied to small gneiss outcrops. It lies to the lee of a 200 m long islet, which is tied to a central sandy foreland at low tide. Waves average about 1 m and maintain a continuous bar south of the islet, with a permanent rip on the northern side of the foreland. Beach **WA 83** extends for another 800 m from the outcrop to a sand foreland formed in lee of a low 5 ha vegetated islet. The beach is well exposed to the southerly wave, which average over 1.5 m. The waves maintain a permanent rip against the southern rocks and a central beach rip, with waves decreasing slightly to the lee of the islet. All three beaches are backed by active dunes extending 2 km inland.

Beach **WA 84** commences on the northern side of the islet-foreland and trends northwest for 1.4 km to the foreland formed in lee of Point Jedacorrudup. Waves average over 1.5 m and maintain a 100 m wide surf zone, dominated by a central beach rip and a permanent rip against the point. The beach is backed by a degraded foredune, a 2 km wide deflation surface, then a series of three pulses of active dunes extending up to 4 km inland and reaching 60 m in height (Fig. 4.23). All the dunes to the lee of beaches WA 81-84 link with the larger Yokingup Bay dunes, which are discussed with beach WA 85.

Figure 4.23 Beach WA 84 is a high-energy beach backed by massive active dune transgression.

WA 85 **YOKINGUP BAY (THOMAS RIVER)**

No.	Beach	Rating	Type Inner Outer bar	Length
WA 85	Yokingup Bay (Thomas River)	7	TBR RBB	14 km
Spring & neap tidal range = 0.8 & 0.4 m				

Yokingup Bay occupies the northwestern half of the 20 km wide bay bordered by Cape Arid and Tagon Point. Yokingup Bay beach (WA 85) extends from Point Jedacorrudup to the Thomas River mouth. The beach is 14 km long and initially faces southwest, curving round to face south at the river mouth. The entire beach is backed by both active and vegetated dunes, which increase in size and width from north to south, with active dunes extending up to 10 km inland. More extensive vegetated dunes, most likely Pleistocene in origin extend up to 60 km inland, blanketing much of the terrain north of the cape and covering an area of about 2 000 km^2. These include some area of reworking by glacial era westerly winds.

The beach is well exposed to the southwest swell, which combines with the fine white beach sand to maintain a 200 m wide, double-barred surf zone (Fig. 2.6a). The inner bar is dominated by beach rips spaced every 500 m, with more widely spaced rips crossing the outer bar. Waves average over 2 m in the southern-central section of the beach, decreasing to 1.5 m at **Thomas River** (Fig. 4.24). A strong permanent rip and deep channel flows out between the western end of the beach and a small rock islet, which is exacerbated when the river mouth is open. There is vehicle access to Thomas River mouth and access along the low gradient beach when the mouth is closed, as is usually the case. Parking and two camping areas are located next to the river mouth.

Figure 4.24 Thomas River mouth, showing the typical high surf, closed mouth and headland rip (arrow).

Surf: *Thomas River* is the first readily accessible surfing break west of Eucla. It provides persistent moderate, to occasionally high south swell, breaking across a wide surf zone. The rip against the rocks provides access to the outer breaks. Further down the beach the surf zone widens as waves and rips increase.

WA 86-88 **TAGON**

No.	Beach	Rating	Type	Length
WA 86	Dolphin Cove	4	R	500 m
WA 87	Little Tagon	3	R	500 m
WA 88	Tagon	6	TBR	2.3 km
Spring & neap tidal range = 0.8 & 0.4 m				

To the west of Thomas River the coast for the next 100 km to Esperance, is dominated by outcrops of Precambrian granite both at the shore where they form prominent headland, and as hundreds of rock reefs, islets and islands that lie up to 25 km offshore, all part of the **Recherche Archipelago**. Along the shore are scores of generally white sandy quartz-rich beaches with varying exposure to the prevailing southerly waves and winds. Beaches WA 86-88 occupy parts of the first 6 km of shore, with each beach bound by granite headlands. These are the westernmost beaches in Cape Arid National Park, with vehicle access from Thomas River to all three. The ranger station is located on the beach access track.

Dolphin Cove (**WA 86**) lies 1 km southwest of the Thomas River mouth and is a 500 m long southeast-facing moderately protected beach, bounded by bare sloping granite headlands and backed by dense coastal heath. The beach receives waves averaging less than 1 m and which usually maintain a moderately steep reflective white sand beach. There is a car park on the southern headland, with a 3-minute walk to reach the usually secluded beach. The car park provides good views of the beaches and bay and is used to view whales during the breeding season.

Little Tagon (**WA 87**) lies in the next cove 500 m to the south and is a 300 m long east-facing beach, afforded slightly more protection, with waves averaging about 0.5 m which surge up a moderately steep, reflective beach face (Fig. 2.12a). This is the least hazardous beach in this section of the park. However be careful, as while waves are usually low, the water is deep right off the beach. It is backed by dense coastal heath and bordered by partly vegetated granite slopes. It shares the car park with Dolphin Cove, with a 2-minute walk down the slopes to the beach.

Tagon beach (**WA 88**) is located 1 km to the west, with a vehicle track to the rear of the foredune. There is 4WD access in the centre of the beach, however beware as the beach is sloping with relatively soft sand. So be prepared if driving on the beach, as bogging is easy. The 2.3 km long beach faces south and is bounded by a low wave-washed granite point to the east and 70 m high Tagon Point in the west that protrudes 1.5 km to the south. The exposed area in between is known both as Tagon Bay and Tagon Harbour. It remains however an exposed bay with waves average over 1.5 m which usually maintain an 80 m wide surf zone with five beach rips and permanent rips to either end. The beach is popular for beach and rock fishing, but relatively hazardous for swimming. It is backed by two blowouts in the foredune and vegetated older dunes extending up to 1 km inland in the east, while active dunes from beach WA 89 extend across the western end of the beach and up to 2 km inland

WA 89-94 **TAGON-ALEXANDER PTS**

No.	Beach	Rating	Type	Length
WA 89	Tagon Pt (W1)	7	TBR	5.1 km
WA 90	Tagon Pt (W2)	6	TBR	1.7 km
WA 91	Kennedys beach	5	TBR/LTT	1.2 km
WA 92	Kennedys beach (W)	5	LTT+rocks	100 m
WA 93	Taylor Boat Hbr (E)	4	LTT	2.3 km
WA 94	Taylor Boat Hbr	3	R/LTT	1.1 km

Spring & neap tidal range = 0.8 & 0.4 m

Beaches WA 89-94 occupy the next 14 km of south-facing coast between the prominent Tagon Point and the lower Alexander Point, the latter backed by 104 m high granite dome of Alexander Hill. The six beaches are each separated by granite outcrops, while Taylor and Inshore islands lie close inshore, and several islands, islets and many rock reefs lie up to 10 km offshore, including the 7 km long Twin Peak islands. The western boundary of the national park lies 2 km west of Tagon Point. There is however, no formal vehicle access to this section of coast.

Beach **WA 89** extends due west of Tagon Point for 5.1 km to a low 200 m wide, bare domed granite headland. The beach is exposed to waves averaging over 1.5 m and to the full force of the southwesterly winds. The waves maintain a 100 m wide rip-dominated surf zone, with permanent rips against the point and rocks, located 500 m west of Tagon Point, together with up to 15 beach rips. It is backed by active dune transgression, which widens from 200 m in the west to 3 km in the east, where earlier now vegetated dunes have dammed the linear Boolenup Lake.

Beach **WA 90** is a relatively straight 1.7 km long beach bounded by the eastern granite headland and by a 600 m long low irregular granite section of shore to the west. A 15 km long 4WD track that originates at the Merivale Road reaches the western end of the beach. Waves average about 1.5 m and maintain permanent rips against the boundary rocks with usually five beach rips in-between. In addition Blackboy Creek drains across the beach just to the west of Tagon Point and is at times deflected a few hundred meters along the back of the berm. Active dunes extend 200-300 m inland, with older vegetated dunes up extending up to 1 km inland and climbing the slopes of the backing Tertiary siltstone.

Kennedys Beach (**WA 91**) extends for 1.2 m to the west-southwest and is bounded by low sloping 1 km long section of rocky shore. Wave height drops to less than 1.5 m owing to sheltering of offshore islands. They maintain a 50 m wide surf zone usually cut by two beach rips, together with permanent boundary rips. It is backed by an active foredune and blowouts extending up to 100 m inland, then dense coastal heath, an a low granite dome behind the western end.

Beach **WA 92** is a transitory pocket of intertidal sand, which accumulated in a gap in the rocks that extend for 1 km west of beach WA 92. The beach is centred on the narrow 20 m wide gap in the rocks, which is backed by a small but active foredune. The beach expands along the front of rocks for 100 m during periods of below average waves. Periods of high waves however wash much of the beach away. While waves average about 1 m, which is a hazardous place to swim owing to the rocks.

Beach **WA 93** is a curving 2.3 km long, south-facing beach, bounded by the rocks of beach WA 92 and 30 m high granite domed headland to the west. This headland is the southern extension of Alexander Hill, which lies 2 km to the north. The beach receives increasing protection from the offshore islands, with waves dropping to about 1 m. These maintain a narrow continuous bar, grading to reflective conditions following periods of low waves, with rips only present during and following periods of higher waves. It is backed by a 1 km wide area of dune instability in the east, bounded by a small blocked creek, then dense coastal heath in the centre, and steeply sloping, partly vegetated granite slopes to the west.

Beach **WA 94** commences on the western side of the domed headland and trends southwest for 1.1 km to the low domed Alexander Point. The area between the beach and the islands is known as Taylor Boat Harbour. The beach is composed of fine white quartz sand, which combines with waves averaging less than 1 m to maintain a narrow low gradient bar, usually free of rips. The beach is backed by a stable foredune, and then transgressive dunes that have originated from the western beach WA 95 in Alexander Bay (Fig. 4.25).

Figure 4.25 The sheltered beach WA 94 extends east of the domed Alexander Point.

WA 95-97 **ALEXANDER BAY**

No.	Beach	Rating	Type	Length
WA 95	Alexander Bay (1)	7	TBR	1.7 km
WA 96	Alexander Bay (2)	6	TBR	6.6 km
WA 97	Alexander Bay (3)	4	LTT	1.1 km
Spring & neap tidal range = 0.8 & 0.4 m				

Alexander Bay is an 8 km wide, south-facing embayment bounded by Alexander Point in the east and the headland in lee of Ben Island in the west. A curving near continuous strip of sand occupies the northern bay shore and contains beaches WA 95-97. There is vehicle access to the recreation and camping reserve, located at the western end of the bay, which offers basic facilities.

Beach **WA 95** commences on the western side of the low sloping granite of Alexander Point and trends northwest for 1.7 km to a low 100 m long granite outcrop. The beach is well exposed to the southwest waves and winds, with a 100 m wide rip dominated surf zone (Fig. 2.11b), and backing massive active and vegetated transgressive dunes, the latter extending up to 6 km inland. Several beach rips and permanent rips against the boundary rocks, as well as a rock reef near the centre of the beach result in hazardous swimming conditions.

Beach **WA 96** commences on the western side of the outcrop and continues to the northwest then west for 6.6 km to the lee of a small linear rock outcrop. Most of the beach receives wave averaging over 1.5 m, decreasing close to the western end. The higher waves result in a rip-dominated surf zone, with up to 25 rips present along the beach. Active dunes maintain a series of blowouts behind the beach, with older vegetated transgressive dunes extending several kilometers to the northeast, blocking several small upland creeks in the process. The small Alexander River does reach the shore towards the western end of the beach and breaks out following heavy rain. Just to the east of the river entrance rock outcrops at the shore and in the surf.

Beach **WA 97** extends from the western side of the islet for 1.1 km, curving round in lee of the western granite headland to face east. Waves average about 1 m and break across a low gradient bar and beach face, decreasing into the western corner in lee of the point and Ben Island. The vehicle track reaches this lower energy end of the beach with access to the beach. Several informal campsites are located in the coastal heath behind the beach, which is used to launch small boats. Seagrass meadows lie just off the beach and seagrass debris also increases in presence towards the western end of the beach where it can from a low berm. It is backed by dense coastal scrub that covers the siltstone slopes, which rise to 50 m.

WA 98-104 **BEN ISLAND-TABLE ISLAND**

No.	Beach	Rating	Type	Length
WA 98	Ben Island (W1)	6	TBR	1.5 km
WA 99	Ben Island (W2)	4	LTT	1.8 km
WA 100	Ben Island (W2)	7	TBR+rock	300 m
WA 101	Menbinup beach	4	R/LTT	4.2 km
WA 102	Menbinup (E 1)	4	R/LTT	150 m
WA 103	Menbinup Ck (E2)	6→5	TBR-LTT	300 m
WA 104	Dailey River	6→2	RBR→R	5.7 km
Spring & neap tidal range = 0.8 & 0.4 m				

To the west of Ben Island the coast trends due west for 12 km before curving round to the lee of Table Island. In between is 16 km of predominately sandy shoreline occupied by seven beaches (WA 98-104), each separated by relatively small granite headlands and rocks. The granite also outcrops offshore as numerous small islands, the largest being the Mart Islands located 12 km south of Table Island. These islands, as well as the large protrusion of Hammer Head, act to increasingly lower waves from east to west along the beaches. There is rough vehicle access to beach WA 98 and formed access to beaches WA 101 to 104.

Beach **WA 98** is a curving, south-facing, 1.5 km long beach, bounded by a low granite headland and rock reefs in the east, and a domed granite headland that protrudes 500 m south, to the west. The beach receive moderate waves averaging about 1.5 m in the centre where they maintain a few beach rips, with lowers waves and a continuous bar to either end. It is backed by a well developed scarped foredune, then a series of well vegetated parabolic dunes extending up to 1 km to the northeast. A vehicle track from the Alexander Bay access road reaches the western end of the beach.

Beach **WA 99** is a slightly curving, south-facing beach that extends from the western side of the granite headland for 1.8 km to the next smaller granite outcrop. It receives protection from offshore islands, with waves averaging about 1 m, which maintain a continuous narrow bar along the beach, with rips only forming during periods of higher waves. It is backed by a 1 km wide area of revegetating dune transgression in the east, and a high scarped foredune along the centre and western section of beach, backed by vegetated transgressive dunes extending up to 500 m inland.

Beach **WA 100** is a 300 m long straight strip of sand bounded by low outcrops of granite, with some rocks also outcropping in the surf. Waves average over 1 m and maintain three permanent rips against the boundary and central rocks. It is backed by a steep, high vegetated foredune. **Menbinup Beach** (**WA 101**) commences immediately to the west and is a relatively straight, 4.2 km long, south-facing beach, which curves around at its western end in lee of two granite outcrops that extend 1 km to the south. It receives waves averaging about 1 m, which maintain a continuous narrow bar usually free of rips. The beach is backed by four stabilising blowouts in the east and centre, with vegetated older transgressive dunes in between, extending up to 1 km inland. Mungliginup Creek reaches the western end of the beach, where it is usually deflected to the east and blocked during the summer months. There is formed vehicle access to the eastern end of the beach, with tracks leading off to a basic camping area and the river mouth.

Beaches WA 102 and 103 occupy parts of the granite headland. Beach **WA 102** is a 150 m long pocket beach located between the two eastern fingers of the point (Fig. 4.26). The beach faces south, but receives waves lowered to less than 1 m, which maintain a narrow bar. Rocks occupy part of the intertidal zone, and at high tide there is a rock pool formed on the eastern half of the beach. It is partly surrounded by the granite rocks and backed by 20 m high partly scarped foredune. Beach **WA 103** lies immediately to the west between the main point and a smaller outcrops located 300 m to the west, with rocks also outcropping in the surf towards ether end, leaving an open 150 m long central section. This section is drained by a permanent rip toward the eastern end. It is backed by a partly vegetated 20 m high foredune, with the road access and car park to the rear of the foredune, as well as a vehicle access track at the eastern end of the beach.

Beach **WA 104** commences on the western side of the 150 m long outcrop of granite and curves round for 5.7 km to the west and finally south in lee of Table Island where it forms a narrow overwashed tombolo, which attaches it to the coast. Wave height decreases from less than 1.5 m in the east to less than 0.5 m at the island. As the waves drop the beach grades from a continuous bar cut by a few rips, to no rips and finally a low gradient reflective beach at the island, with seagrass meadows close to shore, and seagrass debris piles on the upper beach. There is vehicle access and a car park in lee of the tombolo, as well as a rough track along the rear of the beach. Two stabilising blowouts are located towards the eastern end, with generally stable foredune to the centre and west, with the Dailey River deflected for 2 km along the eastern end of the beach to exit across the beach at the tombolo.

Figure 4.26 Two low granite headlands separate beaches WA 101,102 (centre) and 103.

WA 105-107 **DUKE OF ORLEANS BAY**

No.	Beach	Rating	Type	Length
WA 105	Duke of Orleans Bay	2-1	R-R+sand flats	2.8 km
WA 106	Duke of Orleans Bay (S)	2	R+rocks	50 m
WA 107	Nares Island	2	R+seagrass	1.4 km
Spring & neap tidal range = 0.8 & 0.4 m				

Duke of Orleans Bay offers the first commercial facilities on the coast east of Fowlers Bay, 950 km to the east in South Australia. A spacious and shady caravan park and store are located at the southern end of the main beach, with the official name of the settlement called Wharton. The bay area is dominated by 168 m high granite dome of Mount Belches and its adjoining granite points and islands. The bay faces east and is sheltered by its orientation, headlands and islands, with low wave to calm conditions usually prevailing at the three beaches (WA 105-107).

Duke of Orleans Bay beach (**WA 105**) is a semicircular, south to east-facing 2.8 km long beach that spirals round from the tombolo, in the lee of Table Island, to a 500 m long sloping granite headland that rises to 60 m. In the north low waves lap against the low gradient reflective beach, while towards the more protected southern end, seagrass covered sand flats reach the beach, with seagrass debris becoming increasingly prominent along the upper beach. It is backed by a steep, vegetated 10-20 m high foredune, with a solitary blowout at the northern apex of the beach. The main road runs around the back of the dune to the caravan park. The southern section of the beach is used to launch small boats, with dinghies often stored on the beach. Rock outcrops and shallow sand flats dominate the very southern end.

Beach **WA 106** is a 50 m long pocket of sand located on the granite headland, with access via a walking track from the caravan park. The high tide sand beach is bordered by granite slopes, and fronted by inter to sub-tidal granite rocks.

Beach **WA 107** faces northeast and curves for 1.4 km between the headland and the granite point in lee of Nares Island. It is a very low energy reflective beach, with seagrass growing to the shore and usually covered in seagrass debris. There is vehicle access to a small car park and camping area at the eastern end of the beach. It is also used by professional fishermen for launching and storing small boats.

WA 108-114 **HAMMER HEAD**

No.	Beach	Rating	Type	Length
WA 108	Hammer Head (E1)	2	R	300 m
WA 109	Hammer Head (E2)	2	R	550 m
WA 110	Hammer Head (W1)	2	R	200m
WA 111	Hammer Head (W2)	2	R	100m
WA 112	Hammer Head (W3)	2	R	250 m
WA 113	Hammer Head (W4)	2	R	80 m
WA 114	Hammer Head (W5)	2	R	300 m
Spring & neap tidal range = 0.8 & 0.4 m				

Hammer Head is a prominent 75 m high granite dome located 2 km south of Mount Belches. Between the head and mount are three other granite outcrops, which between them impound seven small, relatively protected, pure quartz beaches. There is vehicle access to the northern two beaches (WA 108 and 114), with the remainder only accessible on foot. Beaches WA 108 and 109 occupy the eastern side of the head, while beaches WA 110 to 114 are located along the western side extending to the slopes of Mount Belches..

Beach **WA 108** faces east is 300 m long and is located between sloping granite headlands. The beach is composed of golden quartz sand. It receives waves averaging about 0.5 m which usually maintain a series of regularly spaced beach cusps on the moderately steep beach face, with deep water off the beach (Fig. 2.12b). It is backed by the vegetated transgressive dunes, which originated from beach WA 112. There is vehicle access and limited parking at the northern end of the beach.

Beach **WA 109** is bordered by a 40 m high granite headland to the north and Hammer Head to the south, the latter extending 500 m to the east. The beach is 550 m long and composed of white quartz sand, with low waves and cusps dominating the beach face and deep water close inshore. It is backed by active dune transgression originating from its western neighbours, beaches WA 110 and 111. The three beaches form a tombolo, linking three granite outcrops (Fig. 4.27).

Figure 4.27 Beaches WA 109, 110 and 111 are both sheltered by and link the granite rocks of Hammer Head. Note the seagrass growing close to shore.

Beaches WA 110 and 111 are located either side of a 20 m high granite outcrop, which together with beach WA 109 link Hammer Head and mainland. Beach **WA 110** is a curving 200 m long pocket of sand with sloping headlands extending 500 m westward. The headlands and scattered rock reefs lower waves to less than 0.5 m at the shore, with seagrass also growing in the bay between the heads and reaching to within 50 m of the shore. The beach is usually cusped and moderately steep, and backed by an unstable foredune, with sand blowing across the 100 m wide isthmus to beach WA 109. Beach **WA 111** occupies then next small bay. It is 100 m long, bordered by sloping granite, with rock reefs off the beach. It faces west but receives only low waves, with no cusps and seagrass growing almost to the shore. It is however backed by an unstable foredune, which also supplies sand to beach WA 109.

Beach **WA 112** occupies the next small bay 200 m to the north, with the northern slopes of Mount Belches forming its northern boundary. The beach is a curving 250 m long southwest-facing strip of sand, protected by the headlands, reefs and Lorraine Island, with waves usually less than 1 m. The beach spirals into a protected tidal pool at the northern end. It is backed by a scarped but stable foredune, then densely vegetated transgressive dunes, which extend for 500 m across the backing valley to beach WA 108.

Beach **WA 113** is an 80 long pocket of sand located 300 m to the west in a small valley at the base of Mount Belches. The beach faces due west and while protected from waves by Lorraine Island it is exposed to the westerly winds, which have reactivated an older dune to maintain a 300 m long blowout behind the beach. Beach **WA 114** lies 100 m to the west and is a 300 m long strip of sand located along the southern slopes of Mount Belches. The beach lies in lee of Lorraine Island and protrudes as a slight foreland with seagrass growing just off the beach. It is accessible by vehicle down the backing granite slopes and used to launch boats across the beach while the beach and western rocks are used as a car park.

CAPE LE GRANDE NATIONAL PARK

Area: 31 578 ha
Coast length: 78 km (775-853 km)
Beaches: 27 (WA 115-141)

Cape Le Grande National Park contains some of the most spectacular coastal scenery in Australia. The shoreline is dominated by towering granite domes connected by curving pure white sandy beaches and crystal clear waters, all backed by diverse coastal heathland. There is good vehicle access in the east at Wharton Beach, at Dunn Rocks in the centre and between Rossiter Bay and Le Grande beach in the west. Camping areas are located Lucky Bay and Le Grande Beach and there are several coastal walks between Rossiter Bay and Le Grande Beach. In 1772 Le Grande was an officer on the French ship *L'Esperance* which explored this coast under the command of Admiral D'Entrecasteaux.

For information:

Cape Le Grande National Park
PO Box 706, Esperance WA 6450
Phone: (08) 9075 9022

WA 115-118 WHARTON-VICTORIA HABROUR

No.	Beach	Rating	Type	Length
WA 115	Wharton Beach	6	TBR	4.5 km
WA 116	Cheyne Point	2	R	250 m
WA 117	Victoria Hbr (1)	6	LTT→TBR	1.5 km
WA 118	Victoria Hbr (2)	7	TBR	400 m

Spring & neap tidal range = 0.8 & 0.4 m

To the west of Mount Belches the coast trends west and continues to be dominated by granite outcrops forming both headlands and islands. Beaches WA 115-118 occupy the first 6 km of coast with the four beaches bordered by the prominent granite headlands of Mount Belches, Cheyne Point and the 120 m high spur at the western end of Victoria Harbour. Wharton Beach also marks the eastern boundary of the spectacular Cape Le Grande National Park.

Wharton Beach (WA 115) is located 2 km west of Duke of Orleans Bay and is the most accessible and most popular surfing beach in the area. The beach faces south and curves round in a semi-circle for 4.5 km between the lower slopes of Mount Belches and the 50 m high spur of Cheyne Point. The car park above the eastern end of the beach provides an excellent view up the beach (Fig. 4.28) as well as 4WD access to the beach. Waves average about 1.5 km in the west and centre, decreasing to the lee of Cheyne Point, adjacent Station Island and some scattered rock reefs. A rip-dominated 100 m wide surf zone dominates the eastern half of the beach. Up to 12 rips usually form along this section of the beach, including a permanent rip below the car park with the rips usually clearly visible from the car park. To the west waves decrease and a continuous narrowing bar fronts the beach, usually free of rips. It is backed by vegetated dunes in the east grading into active dunes to the west. Two creeks are impounded by the dunes at times forming shallow lagoons in the deflation basins. They only flow to the shore following heavy rain.

Figure 4.28 View along the eastern end of Wharton Beach.

Beach **WA 116** is located in a 100 m wide gap in the rocks of Cheyne Point. It is a curving, almost circular 250 m long beach protected by the headlands, island and reefs with waves averaging less than 0.5 m. Seagrass grows almost to the shore and seagrass debris is usually piled on the upper beach.

Beach **WA 117** commences on the northern side of Cheyne Point and is initially protected by an island just off the point with waves averaging about 1 m. However a strong current can flow though the 50 m wide rocky channel between the point and island. Waves increase up the 1.5 km long beach with the central-northern section averaging 1.5 m high waves and an 80 m wide surf zone usually containing three beach rips and a permanent rip against the northern low granite point. Beach **WA 118** commences on the north side of the 100 m long point and is a 400 m long exposed beach, bounded to the north by a 120 m high granite spur. The beach receives waves averaging over 1.5 m, which maintain a single permanent rip against the northern rocks. A small creek drains out across the southern end of the beach. The two beaches are backed by both active and vegetated transgressive dunes. Active dunes and deflated areas extend for 2.5 km east of beach WA 117 reaching the back of Wharton Beach (Fig. 4.29) and climbing Cheyne Point to form a 50 m high clifftop dune which descends down onto the small point beach (WA 115). There is 4WD access across the deflation surfaces to all four beaches.

Surf: *Wharton Beach* is the most popular beach in the region with usually moderate to occasional high swell breaking over the bars, with access out the back via the rips.

WA 119-123 VICTORIA HBR-DUNN ROCKS

No.	Beach	Rating	Type	Length
WA 119	Victoria Hbr (W1)	7	TBR	1.1 km
WA 120	Victoria Hbr (W2)	7	TBR	250 m
WA 121	Victoria Hbr (W3)	7	TBR/RBB	1.7 km
WA 122	Victoria Hbr (W4)	7	TBR	250 m
WA 123	Dunn Rocks	7	TBR/RBB	9.7 km
Spring & neap tidal range = 0.7 & 0.1 m				

Figure 4.29 Cheyne Point (right) is bordered by Victoria Harbour (foreground) and Wharton Beach, with deflated dune surfaces extending between the two.

To the west of the 120 m high Victoria Harbour headland the coast trends west in a gentle curve for 13 km to the prominent Dunn Rocks, a 50 m high granite headland that protrudes 1 km seaward. In between are five south-facing beaches all exposed to moderate to high waves averaging 1.5-2 m, which maintain a 50-100 m wide rip-dominated surf zone. Active and vegetated transgressive dunes back the five beaches. The only vehicle access to the coast is at Dunn Rocks.

Beach **WA 119** is a straight 1.1 km long beach located between the large Victoria Harbour head and a western 30 m high, 100 m wide spur of bare granite that protrudes 200 m seaward. The waves maintain an 80 m wide surf zone dominated by boundary rips and up to three central beach rips. Beach **WA 120** continues on the western side of the spur for 250 m to a low wave-washed finger of granite. The high tide beach is only 100 m long, with the intertidal beach and surf also extending east in front of the sloping rocks. Two permanent rips drain each end of the small beach.

Beach **WA 121** extends west of the spur for 1.7 km to a 100 m long outcrop of granite. It has an 80 m wide surf zone dominated by boundary rips and up to six beach rips. Beach **WA 122** lies between the granite outcrop and a 20 m high dome of granite that is tied to the shore by a tombolo formed by the ends of beaches WA 122 and 123. The beach is only 250 m long and curves round to the lee of the outcrop where it faces east. Waves decreases slightly towards the tombolo, however the eastern rocks and a smaller central outcrop in the surf maintain a strong permanent rip draining out the eastern end of the beach, with a small rip against the western outcrop.

Beaches WA 119-122 are all backed by some minor foredune instability, including a few blowouts extending up to 300 m inland. Vegetated transgressive dunes then extend up to 9 km inland and to heights of 60 m.

Beach **WA 123** commences on the western side of the outcrop and curves round slightly to the west for 9.7 km to Dunn Rocks, where it faces southwest in lee of the rocks. Waves remain moderate to high the length of the beach, which results in a 100 m wide, low gradient surf zone, dominated by up to 30 beach rips, as well as a strong permanent rip against the western outcrop and adjoining rocks, and a weaker rip against Dunn Rocks, where waves reduce to about 1 m. The entire beach is backed by pure white active dune transgression, extending up to 1 km inland, then vegetated older transgressive dunes that have spread up to 19 km inland. Several small creeks drain into the dune field and work their wave across the deflation surface to reach the beach during heavy rain (Fig. 4.30).

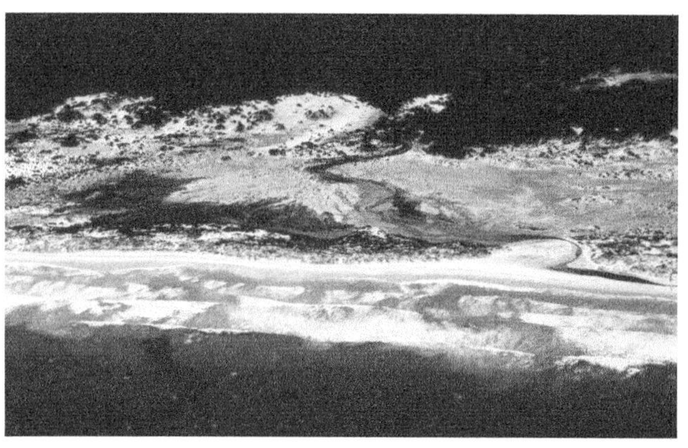

Figure 4.30 Dunns Rocks (WA 123) beach has high waves, a rip dominated surf zone and creeks draining across the deflated dune surface.

WA 124-126 ROSSITER BAY

No.	Beach	Rating	Type	Length
WA 124	Rossiter Bay	6→3	TBR→R	9 km
WA 125	Rossiter Bay	3	LTT+rocks	100 m
WA 126	Rossiter Bay	3	R+rocks 1	50 m
Spring & neap tidal range = 0.7 & 0.1 m				

Rossiter Bay is named after Captain Rossiter who commanded the whaler "Mississippi' which in 1841 provided assistance to Eyre and Wylie on their epic trek to Albany. The name of the ship is also attached to a point and a hill. The bay is an open 9 km wide, southeast-facing embayment, bounded by Dunn Rocks to the north and Mississippi Point to the south. In addition several small islands lie off the western half of the bay, as well as the six Troy Islands, 3 km southeast of Mississippi Point. The result is a moderate energy shoreline in the east, with decreasing waves to the west.

Beach **WA 124** occupies most of the bay shoreline. It commences against Dunn Rocks and trends to the southwest for 9 km, with a slight central foreland in lee of two islands, finally curving round to face east against the southern rocks. Moderate waves and rips dominate the first few kilometres of shore (Fig. 4.31), then as the waves drop along the foreland the bar narrows and remains continuous. It finally grades into a low energy reflective beach along the southern kilometre, with seagrass growing close to shore, and seagrass debris piled up on the beach. The higher energy western section is backed by some moderate dune transgression extending up to 500 m inland, with stable vegetated transgressive dunes backing the remainder of the beach and all trending to the northeast, along the rear of the beach. Further inland are west-trending vegetated parabolics which were active during the glacial maxima, and whose sands originated from La Grande Beach, 20 km to the west. The Lucky Bay road provides access to the southern end of the beach, with 4WD access to the beach where boats are launched in the southern corner.

Figure 4.31 *View along Rossiter Bay beach from Dunn Rocks.*

Beaches WA 125 and 126 are two pockets of sand located just south of the beginning of the southern boundary rocks. Beach **WA 125** lies on the southern side of a 50 m wide, 150 m long spur of granite that separates it from beach WA 124. It is a 50 m long pocket of high tide sand, widening in front of the rocks to 100 m. It is bordered by bare granite with some rocks and seagrass lying off the beach. Beach **WA 126** is located 300 m to the south and is a 150 m long reflective beach wedged between a bare granite points and rocks. It is backed by dense wind swept vegetation, and fronted by a seagrass and rock littered foreshore, including some extensive rock reefs. Both beaches can be accessed on foot from the southern end of the bay access road.

WA 127-128 **MISSISSIPPI PT**

No.	Beach	Rating	Type	Length
WA 127	Mississippi Pt	4	R+rocks	300 m
WA 128	Mississippi Hill	4	LTT	500 m
Spring & neap tidal range = 0.7 & 0.1 m				

Mississippi Point is a prominent 60 m high granite headland, located 2 km east of 178 m high Mississippi Hill. This substantial granite outcrop forms the eastern boundary of a 17 km long section of high sloping granite shore, backed by a series of granite domes reaching 250 m at Frenchman Peak and 345 m at Mount Le Grande, and terminating in the west at Cape Le Grande. In between are 12 rock bound beaches. The first two (WA 127& 128) lie either side of Mississippi Point, and are accessible only on foot.

Beach **WA 127** lies 1 km northeast of the point and is a 300 m long, southeast-facing, white sand reflective beach, protected by the point and a 30 m high islet off the eastern head. Waves are usually less than 1 m and surge up the steep beach face and amongst a cluster of granite boulders along the eastern end of the beach. It is backed by a steep, scarped 15 m high foredune then dense coastal heath filling the valley between the two headlands. Beach **WA 128** lies 1 km to the west and is bounded by a steeply sloping, 50 m high, eastern head and a steep vegetated western head. The beach faces south and extends for 500 m between the rocks. It receives waves averaging over 1 m, which maintain a continuous narrow bar, with a small rip usually located against the eastern rocks. It is backed by a densely vegetated foredune, then dense coastal heath, with the backing slopes rising to a 100 m high granite ridge line, which extend to Mississippi Hill in the west.

WA 129-132 **LUCKY BAY**

No.	Beach	Rating	Type	Length
WA 129	Lucky Bay (1)	4	LTT	150 m
WA 130	Lucky Bay (2)	7	TBR	200 m
WA 131	Lucky Bay (3)	7	TBR	80 m
WA 132	Lucky Bay	6→2	TBR→LTT	3.4 km
Spring & neap tidal range = 0.7 & 0.1 m				

Lucky Bay is where the lucky Eyre and Wylie meet the whaler 'Mississippi', thus ensuring their continued survival on their long trek across the southern coast. Today the bay area is one of the most popular destinations and camping areas in the park. The entire bay is a 2 km wide, southwest-facing embayment, with granite headland extending 2-3 km off either end (Fig. 4.32). Its orientation exposes the central-eastern half of the bay to moderate south swell, while the western half becomes increasingly protected as it spirals round to the lee of the western rocks. Four near continuous beaches (WA 129-132) occupy the northern shore of the bay.

Beach **WA 129** is a 150 m long beach, wedged in between two rugged granite headlands rising steeply to 40 m, together with partly dune-covered steep vegetated slopes rising towards Mississippi Hill. The beach is slightly protected and receives waves averaging about 1 m, which maintain a narrow bar. Beach **WA 130** lies 200 m to the west. It is a 200 m long white sand beach with steeply rising bare granite bordering the eastern side and a wash-washed finger of granite at the western boundary. Waves average over 1.5 m and maintain permanent rips to either end of the beach. Beach **WA 131** is an 80 m long sliver of sand located between two fingers of granite, which extend 150 m offshore. In between is a 100 m wide surf zone, with a strong solitary rip draining the entire pocket system.

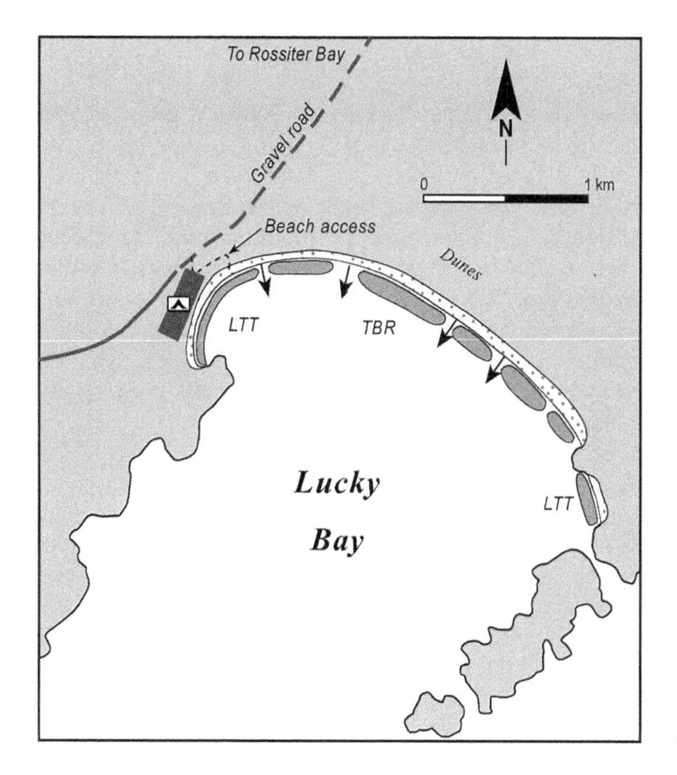

Figure 4.32 Lucky Bay curves round between two granite headlands, which provide shelter to the western side of the bay where the camping area is located.

The main **Lucky Bay** beach (WA 132) commences on the western side of the granite finger and trends for 3.4 km to the northwest, before spiraling round to finally face east in lee of the 80 m high western headland The beach is initially exposed to waves averaging over 1.5 m which maintain a 100 m wide surf zone, cut by several strong beach rips. Along the western kilometre of beach waves decrease with a continuous bar narrowing (Fig. 4.33). and becoming reflective in the western corner. The beach is backed by a relative stable well vegetated 10 m high foredune with one area of instability at the northern apex of the bay. These are backed by well vegetated transgressive dunes that extend up to 3 km inland, climbing the lower slopes of Mississippi Hill in the east. The main car park and camping area is located on the slopes above the western corner, with the beach also used to launch small coast and provide 4WD access along the shore to the remainder of the beach. Water seeps out of the western rocks and drains across the beach.

Swimming: The western corner is the safest spot, with usually low waves, no rips and a shallow bar. Be careful toward the centre of the bay where higher waves and strong rips dominate.

Figure 4.33 The western side of Lucky Bay usually has moderate waves breaking across a continuous bar.

Surfing: The centre of the bay usually has beach breaks across the bars and rips.

Fishing: This is a popular destination for boat fishers who launch off the beach, together with beach fishing in the central rips, and rock fishing off the granite boulders than form the headlands. Beware on the rocks however as waves have washed fishers off.

Summary: An attractive crescentic bay with white beaches and usually clear sparkling water and surf. A popular camping and fishing spot.

WA 133-134 **THISTLE COVE**

No.	Beach	Rating	Type	Length
WA 133	Thistle Cove	5	LTT	1 km
WA 134	Thistle Cove (W)	4	R/LTT	200 m
Spring & neap tidal range = 0.7 & 0.1 m				

Thistle Cove has a 600 m wide entrance between two 100 m high granite headlands, the eastern headland shared with Lucky Bay. The cove opens to a 1.2 km wide bay containing two moderately protected beaches (WA 133 & 134). Access is via a road terminating at a small car park at the eastern end of the main beach.

The Thistle Cove beach (**WA 133**) is a curving 1 km long beach that faces southeast out of the bay mouth. It receives moderate waves averaging about 1 m, which maintain a continuous low tide bar, usually free of rips, apart from a small permanent rips against the eastern rocks, just below the car park. The beach is backed by a 20-30 m high, unstable foredune, which has impounded a 4 ha lake and backing wetlands. It is bordered by sloping granite rocks, with a 100 m long western outcrops separating it form beach **WA 134** (Fig. 4.34). This pocket beach is 200 m long, faces east across the bay and receives waves averaging less than 0.5 m, which break across a narrow attached bar. A few boulders also litter the beach and seagrass grows to within 50 m of the shore. It can only be reached on foot along the main beach.

Figure 4.34 Thistle Cove contains two moderately sheltered beaches.

WA 135-136 **HELLFIRE BAY**

No.	Beach	Rating	Type	Length
WA 135	Hellfire Bay (E)	3	R/LTT	200 m
WA 136	Hellfire Bay	4→6	LTT→TBR	600 m
Spring & neap tidal range =0.7 & 0.1 m				

Hellfire Bay lies 2.5 km to the west of Thistle Cove, with 178 m high Boulder Hill separating the two. Hellfire Bay is a 1 km deep, south-facing bay with a 1 km wide mouth and shoreline dominated by steeply rising granite slopes. Two beaches are located at the mouths of valleys on the northern side of the bay (WA 135 & 136). A gravel road leads to the eastern end of the main beach (WA 136) with a 500 m walk around the back of the rocks required to reach beach WA 135.

Beach **WA 135** is a 200 m long pocket of sand, which faces west across the bay. It receives low waves and is a moderately steep reflective beach, with a string of rocks extending off the southern end. It is backed by a densely vegetated foredune wedged in between the bordering granite slopes.

The main Hellfire Bay beach (**WA 136**) extends for 600 m along the northern shore of the bay and faces south out the bay entrance. It receives waves averaging 1 m in the east increasing to about 1.5 m at the western end. The beach commences with a permanent rip against the eastern rocks, then a continuous 50 m wide bar widening into a rip-dominated system in the west, together with a permanent rip against the western rocks. A well vegetated 15 m high foredune backs the beach. It impounds a linear wetland and creek, which breaks out across the western end of the beach.

WA 137-140 **CAPE LE GRANDE**

No.	Beach	Rating	Type	Length
WA 137	Cape Le Grande (E2)	4	R/LTT	900 m
WA 138	Cape Le Grande (E1)	3	R+rocks	200 m
WA 139	Cape Le Grande (W)	3	R/LTT	600 m
WA 140	Cape Le Grande (N)	3	R/LTT	400 m
Spring & neap tidal range = 0.7 & 0.1 m				

Cape Le Grand forms the western end of the spectacular 17 km long section of coast bordered in the east by Mississippi Point. Like the point it is also backed by a prominent granite peak, in this case 345 m high Mount Le Grande, the highest outcrop in the park. Around the base of the point is 12 km of irregular granite shoreline containing four beaches, all of which are only accessible by foot around the base of the mount, with the easiest access via boat when waves are low.

Beach **WA 137** lies 1.5 km due south of the peak and 2 km west of Hellfire Bay. It is a 900 m long white sand beach. It faces southwest but is moderately protected by offshore islands resulting in waves averaging 1 m at the eastern end, decreasing to 0.5 m in the northern corner. These maintain a cusped beach face, with a narrow bar in the east grading into a fully reflective beach to the west. It is bordered by vegetated granite slopes and backed by a densely vegetated foredune and parabolic dunes extending up to 300 m inland. These impound a small creek, which drains out at the western end of the beach.

Beach **WA 138** occupies the mouth of a rocky gully 700 m to the southwest. It is an irregular, discontinuous 200 m long beach composed of yellow sand and bordered, backed and fronted by bare granite slopes, rocks and reefs. As a result of the reefs waves are low at the shoreline, with a small lagoon formed between the shore and reef. A creek drains across the eastern part of the beach obliterating the foredune, with coastal heath then climbing the backing granite slopes.

Beach **WA 139** is located 2 km north of Cape Le Grande and is a slightly curving, 600 m long south-facing beach. It receives however only low waves owing to the protection of inshore islands. Waves average between 0.5-1 m and break across a continuous narrow bar, with seagrass growing to within 50 m of the shore and some debris on the beach. It is backed by an active 150 m long blowout in the west, and dune-draped granite to the east, with steep granite slopes to either side (Fig. 4.35).

Beach **WA 140** lies 1 km to the north, in a steep valley on the northern side of granite point. The 400 m long beach faces north along the large Le Grande Beach. It is relatively sheltered and receives waves averaging about 0.5 m, which surge up a moderately steep beach face. It is bordered by steeply rising granite slopes, and backed by a vegetated foredune rising into the coastal heath of the valley.

WA 141 **LE GRANDE-WYLIE BAY**

No.	Beach	Rating	Type	Length
WA 141	Le Grande Beach	6→4	TBR→LTT	20.7 km
WA 142	Wylie Bay	4→3	LTT→R	1.3 km
Spring & neap tidal range = 0.7 & 0.1 m				

Le Grande Beach (WA 141) is one of the longer beaches on the south coast. It commences below the northern slopes of Mount Le Grande and spiral to the north-

northwest and finally west for 20 km to terminate at a small granite outcrop 1 km west of Wylie Head. The western boundary of the national park is located 8 km up the beach. There is good access at either end of the beach. In the national park, at the southern end of the beach there is a car park and popular camping area, and at Wylie Head at the northern extremity there is a car park and the sheltered corner of the beach which is used to launch small boats. In between is 22 km of natural beach backed by some massive dune transgressions (Fig. 4.36). The entire beach can however be accessed via 4WD from either end.

Figure 4.35 Beach WA 139 is a moderately sheltered beach bordered by steep granite headlands, with a dune blowout at its western end.

Along the central-southern section of the beach waves average over 1.5 m and maintain a 150 m wide, low gradient rip dominated surf zone, with a low gradient firm beach face. Up to 70 rips can operate along the beach spaced about every 300 m. Waves decrease slightly between 13-16 km up the beach in lee of the small Lion Island and its associated reefs, which have formed a slight foreland formed in their lee. Waves pick up again north of the foreland, only decreasing significantly close to its northern boundary.

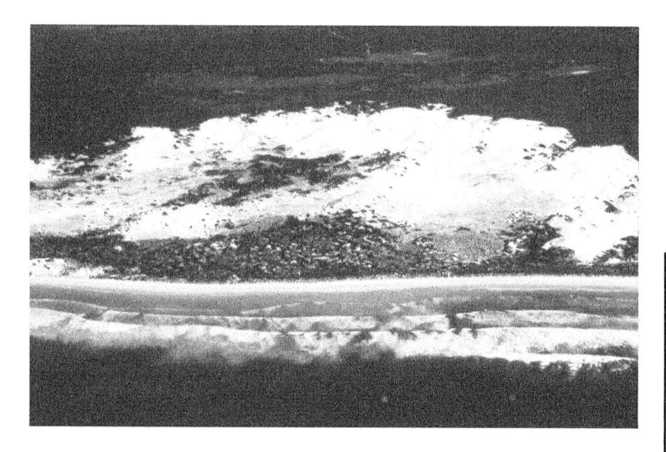

Figure 4.36 The central section of Le Grande Beach is typified by a wide rip dominated surf zone and backing active and vegetated transgressive dunes.

Wylie Bay (WA 142) continues to curve round from the rock outcrop for 1.3 km to the lee of Wylie Head where it faces east. Waves decrease from about 1 m to less than 0.5 m and the beach transforms from a continuous bar with a 50 m wide surf zone to a sheltered reflective beach used to launch the boats (Fig. 4.37). In addition in the southern corner there is a large boulder just north of the southern tip of the beach, and seagrass debris often accumulated on the high tide beach.

Figure 4.37 Wylie Head and Bay form the northern lower energy end of the 21 km long Le Grande Beach system.

The two beaches are backed by a series of active blowouts extending up to 2 km inland, and then by vegetated Holocene parabolics dunes that reach up to 5 km inland, and then by massive vegetated Pleistocene transgressive dunes averaging 60-80 m in elevation that in places extend 35 km to the west, reaching the shores of Rossiter Bay and beyond. Many of the inner dunes were reactivated during the glacial maxima (approximately 18 000-20 000 years ago) and have a more westerly orientation together with numerous water-filled deflation hollows. The entire system covers and area of about 350 km^2, and surround some of the inner granite peaks.

Surf: *Le Grande Beach* is the first surfing beach east of Esperance and offers spilling breaker across a variety of beach breaks You can drive up the beach and have a choice of dozens of breaks, including a reef break located between Lion Island and the shore, 8 km from Wylie Head.

WA 143-151 WYLIE HEAD-ESPERANCE

No.	Beach	Rating	Type	Length
WA 143	Wylie Head	7	LTT+rocks	150 m
WA 144	Picnic Beach	6→4	TBT→LT	4.4 km
WA 145	Brandy Ck (E)	2	R	150 m
WA 146	Brandy Ck (W)	2	R	150 m
WA 147	Castletown Beach	4→3	LTT	3.5 km
WA 148	Esplanade Beach	3	R	1 km
WA 149	Yacht Club Beach	3	R	400 m
WA 150	Harbour Beach	2	R	400 m
WA 151	Breakwater Beach	5	LTT	100 m

Spring & neap tidal range = 0. & 0.1 m

ESPERANCE is a growing coastal town with a population of about 10 000. It is a major service centre, port and tourist destination. It is located at the western end of Esperance Bay where a relatively safe anchorage lead to the development of the port in the 1890's to service the booming Kalgoolie goldfields, 400 km to the north. The port has been upgraded and enhanced with the construction of a major breakwater, a series of groynes and a 700 m long jetty. These facilities, together with the more recent Brandy Creek boat harbour, have however dissected the once continuous 10 km beach, that ran from Wylie Head to Dempster Head, into eight beaches (WA 144-151). Today the town provides a wide range of accommodation and services and is an excellent base of visiting the surrounding coast line, as well as the many islands in the Recherche Archipelago. Cape Arid and Le Grande national parks lie to the east, Stokes National park to the west, while in town is the spectacular scenic coastal drive out to Observatory Point.

Wylie Head is a 40 m high granite knob which anchors the northern end of the long Le Grande beach, and the beginning of the once 10 km long beach that sweeps west to Dempster Head and Esperance. Beach **WA 143** is located in between the southern side of the 200 m wide head and a smaller wave-washed granite outcrop. The beach is 150 m long and with beach WA 144 forms the eastern side of a tombolo that links the rocks to the head. The beach faces southeast and receives moderate protection from the bordering rocks and rock reefs, with waves breaking over the reefs, and reducing to less than 1 m at the shore. While a continuous bar and lower wave dominate the beach, the small embayment is drained by a permanent rip that runs out through the reefs. There is a car park at the base of the head adjacent to the northern end of the beach.

Picnic Beach (WA 144) commences at the small tombolo adjacent to Wylie Head and trends west for 4.4 km to the eastern groyne and breakwater of Brandy Creek boat harbour. There is a slight bulge in the centre produced by wave refraction around Low Rock, located 2 km south of the beach. The beach receives waves averaging 1.5 m in the east decreasing slowly to the west. The waves maintain a 70 m wide surf zone with rips spaced every few hundred metres all the way to Brandy Creek, with permanent rips against the tombolo rocks and the breakwater just east of the boat harbour. The beach is backed by vegetated dunes, with an area of active dunes 1-2 km inland. The Wylie Head road runs along the back of the beach with access at either end.

Brandy Creek boat harbour was constructed during the 1990's and straddles the original Brandy Creek mouth. The small boat harbour has been dredged with converging breakwaters constructed either side of the mouth as well as groynes inside the harbour to stabilise the shore and miminise sand transport into the harbour (Fig. 4.38). As a result of the construction there are two sandy beaches (WA 145 & 146) just inside the harbour entrance. Beach **WA 145** extends from the eastern breakwater for 150 m into the harbour and has a central and terminal groyne. On the opposite side of the harbour is beach **WA 146** a similar 150 m long beach bounded by the western breakwater, a central and terminal groyne. Both beaches are usually calm and face into the harbour channel.

Figure 4.38 Brandy Creek boat harbour.

Castletown Beach (WA 147) commences on the western side of the harbour and curves slowly to the west then southwest for 3.5 km (Fig. 4.39), finally terminating at the now disused Esperance jetty. Wave height decreases along the beach, averaging about 0.5 m at the jetty. The beach commences with a continuous low tide bar and a few rips, grading into a continuous narrow low tide bar at the jetty. Castletown Road then residential development backs the western half of the beach, the first beachfront houses in Western Australia. In addition between the junction with the highway and the jetty are is a seawall and four rock groynes that have each slightly reoriented sections of the beach.

Figure 4.39 View east of Esperance along Castletown Beach.

Beaches **WA 148** commences on the western side of the jetty and trends southwest for 1 km to a 150 m long groyne, with a rock seawall backing the entire beach. The beach faces southeast and is partly protected by the port breakwater that extends for 1 km north of Dempster Point. Waves average about 0.5 m and surge up a moderately steep narrow beach, backed by a foreshore reserve then the Esplanade road. Beach **WA 149** continues on the southern side of the groyne for 400 m to the Yacht Club breakwater and jetty. The attached breakwater extends 200 m seaward curving at its tip to provide additional shelter to a floating jetty to its lee. In addition there is wharf and a beach boat ramp on the north side. A second boat ramp is located toward the centre of the beach. Waves are usually low to calm along the beach. A seawall and foreshore reserve backs both beaches, then the Esplanade and town centre.

Beach **WA 150** is located between the Yacht Club breakwater and the beginning of the continuous harbour seawall, with a rock groyne also crossing the beach. The beach is 250 m long with a small central rock groyne and faces and east along the main harbour wharves and towards the breakwater. Small boats are moored at the jetty in lee of the curving breakwater at the northern end of the beach, while a range of boating facilities are located in the reserve behind the beach.

Beach **WA 151** is a beach that has formed out on the harbour breakwater following the construction of a 100 m long groyne to trap sand and prevent it from entering the harbour. The groyne has been effective enough to impound a beach between the groyne and breakwater, which is up to 100 m long at low tide, and awash at high tide. It receives waves averaging about 1 m which maintain a 50 m wide bar and rip running out off the groyne.

WA 152-155 LOVERS-WEST/SECOND-BLUE HAVEN-SALMON

No.	Beach	Rating	Type	Length
WA 152	Lovers Beach	7	LTT	150 m
WA 153	West-Second Beach	8	TBR+reef	1.2 km
WA 154	Blue Haven	5	LTT+rock	1.4 km
WA 155	Salmon Beach	7	LTT+rock	400 m
Spring & neap tidal range = 0.7 & 0.1 m				

Immediately west of Esperance is one of the more scenic coastal drives in Australia. Commencing at 90 m high Wireless Hill, the drive runs along for 25 km to Eleven Mile Beach, then turns inland to return to Esperance via the Pink Lake. It provides excellent views of the coast and access to all the beaches (Fig. 4.40). Most of the beaches are exposed to moderate to high waves and numerous rocks and reefs induce additional rips and hazards. The only suitable location for swimming is the more protected western end of Twilight Beach where the Esperance-Goldfields Surf Club is located. The other beaches are however popular for beach and rock fishing, and some have excellent surf. Beaches WA 152 to 155 occupy the first 5 km of the drive. Some of the beaches have names, others are numbered consecutively by the mileage, with Second, Fourth and Fifth beaches, as well as 9 Mile, 11 Mile and 13 Mile beaches.

Rock fishing – the rocks and headlands along this section of coast provide numerous excellent spots for rock fishing. However care must be taken on the sloping granite rocks as the persistent moderate to high southerly swell can wash high up on the rocks. Such waves have swept many fishers into the sea and caused fatalities along these beaches, so be watchful and beware.

Lovers Beach (**WA 152**) is located at the base of Wireless Hill, with Dempster Head forming its steep eastern boundary. It is bordered by steep granite slopes and backed by a steep vegetated valley, partly filled with a climbing vegetated dune, with a walking track through the bushes to the beach. The 150 m long beach is exposed to moderate waves, which maintain a 50 m wide bar, drained by a permanent rip against the eastern rocks.

West-Second Beach (**WA 153**) extends from the steep, bare western side of Wireless Hill for 1.2 km to a 30 m high granite Chapmans Point. The beach is exposed to waves averaging over 1.5 m, which break across a discontinuous shore-parallel beachrock reef, lying up to 80m offshore, and fronting a central sandy foreland, that separates West from Second beach. The reefs provides some excellent surf with the water returning out via four strong permanent reef-controlled rips (Fig. 4.41) that drain the both ends of the beach. There are three blufftop car parks above the beaches and two sets of steps leading down to the beach. While a popular beach for fishing and surfing it is very hazardous for swimming owing to the rocks and rips. Up until the 1980's active clifftop dunes and deflection surfaces covered the 30 m high calcarenite bluffs behind the beach. These have since been covered by residential development.

Blue Haven beach (**WA 154**) is as the name suggests is a little more protected than its neighbours. It commences on the western side of Chapmans Point and curves round to the southwest, then south for 1.4 km to a 50 m high granite headland. Its curvature and orientation results in waves averaging 1 m or less along the beach, and a continuous bar dominating the narrow surf zone, with beachrock paralleling the rear of the beach. The beach is backed by steep 20-30 m high vegetated bluffs, with the only access at the western end. The western corner is clear of beachrock and provides the best location for swimming.

Salmon Beach (**WA 155**) lies immediately to the west and is a curving, south-facing 400 m long beach bordered by two protruding wave-washed granite points. A continuous band of beachrock rims the beach, which while providing an excellent platform for beach fishers, it makes for hazardous swimming conditions. Waves average over 1 m coming into the beach and break heavily on the edge of the beachrock, with a solitary permanent rip draining the beach through the centre of the bay. A small car park is located behind the centre of the beach and provides good access to the shore.

Swimming: These are four exposed and potentially hazards beach owing to the higher waves, rocks, beachrock, reefs and strong permanent rips. The least hazardous location is at the western end of Blue Haven.

Surfing: *West* and *Second beaches* provide both left and right breaks over the edge of the reefs, as well as a right-hand break against *Chapmans Point*. There is also a heavy short right off the rocks at *Salmon Beach*. All these breaks are associated with shallow reefs and adjacent strong rips, so be careful if new to the area.

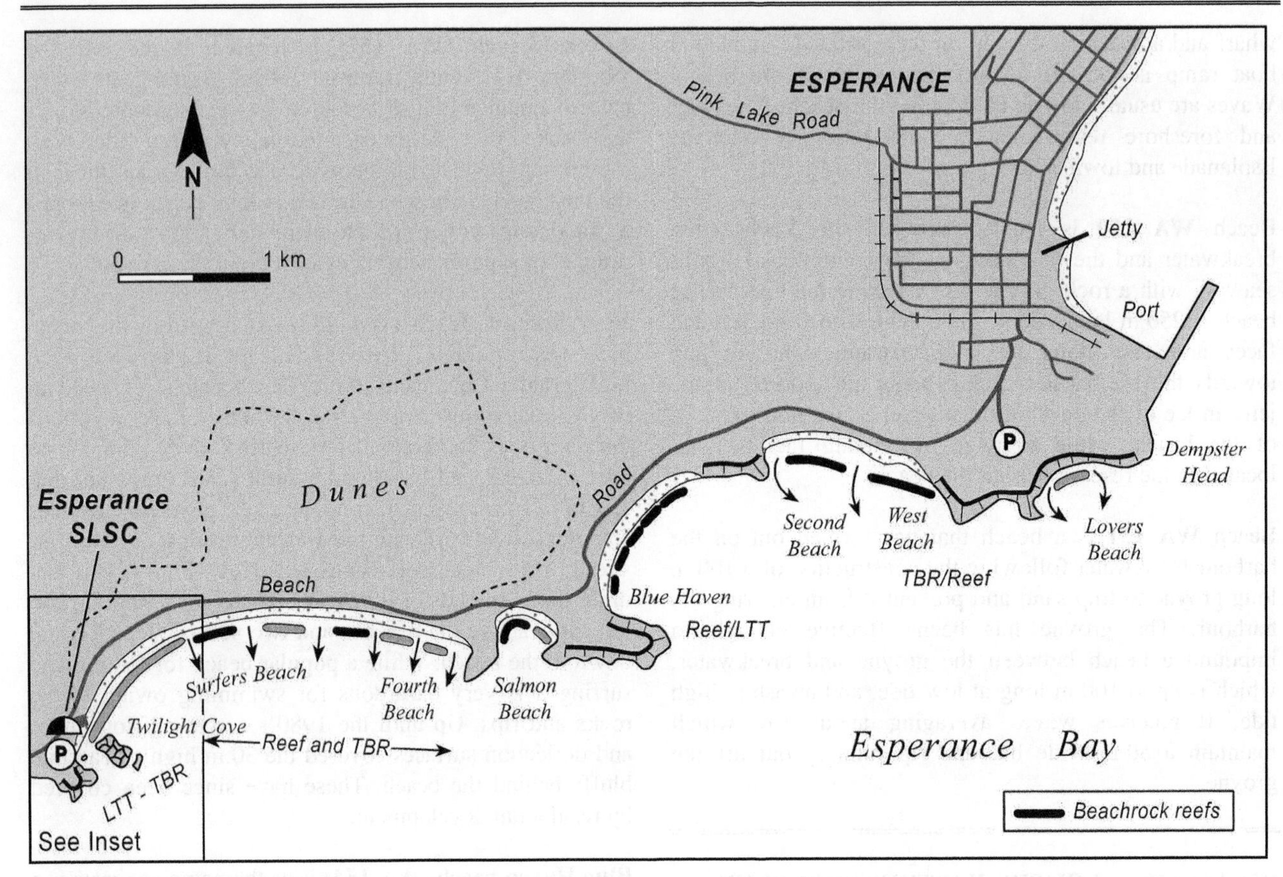

Figure 4.40 Esperance has a series of exposed surfing beaches extending west to Twilight Cove (see Fig. 4.42).

Figure 4.41 Second Beach is fronted by shore parallel beachrock reefs, which induce heavy wave breaking and strong rips (arrow) flowing between gaps in the reefs.

WA 156 FOURTH-SURFERS-TWILIGHT BEACH ESPERANCE-GOLDFIELDS SLSC

No.	Beach	Rating	Type	Length
WA 156	Fourth	6	TBR	1.2 km
	Surfers	6	TBR	1 km
	Twilight	6→4	TBR→LTT	1 km
		Total length		3.2 km

ESPERANCE-GOLDFIELDS SLSC
Patrol period: Sundays November-March
Saturday January
Spring & neap tidal range = 0.7 & 0.1 m
See Figure 4.42 for Beach Map

Twilight Beach (**WA 156**) lies 7 km and seven beaches west of Esperance. It commences immediately west of Blue Haven headland and trends west then southwest for 3.2 km. The beach grades from an exposed high energy rip-dominated system in the east where it is called initially Fourth, then Surfers (Fig 2.9a). As it curves to the southwest it becomes increasingly protected by its orientation and the rocks and islets off Twilight Cove. The Cove region is the site of the Esperance/Goldfields Surf Life Saving Club and offers the least hazardous swimming and surfing beach along this section of shore (Fig. 4.42). The club was founded in 1990 and patrols the beach on Sundays between December and March.

Fourth and **Surfers** beaches are exposed to high waves, which together with scattered beachrock reefs, induce strong permanent rips (Fig. 2.9a). This section is more popular with surfers and fishers. The road to the Cove runs along the bluffs behind the beach with several car parks and access points down to the beach.

The more protected **Twilight Beach** lies at the western end of the beach with two large car parks either side of the Surf Life Saving Club (Fig. 4.43). The beach faces southeast in the Cove and has rounded granite rocks forming the western headland, wave-washed granite islets just off the beach, as well as slabs on granite on the beach. The beach is composed of fine white sand, which combine with lower waves averaging 1 m, to produce a wide, flat beach and continuous shallow bar. Rips are usually absent in the western corner, but increase east of the Surf Life Saving Club as wave height picks up.

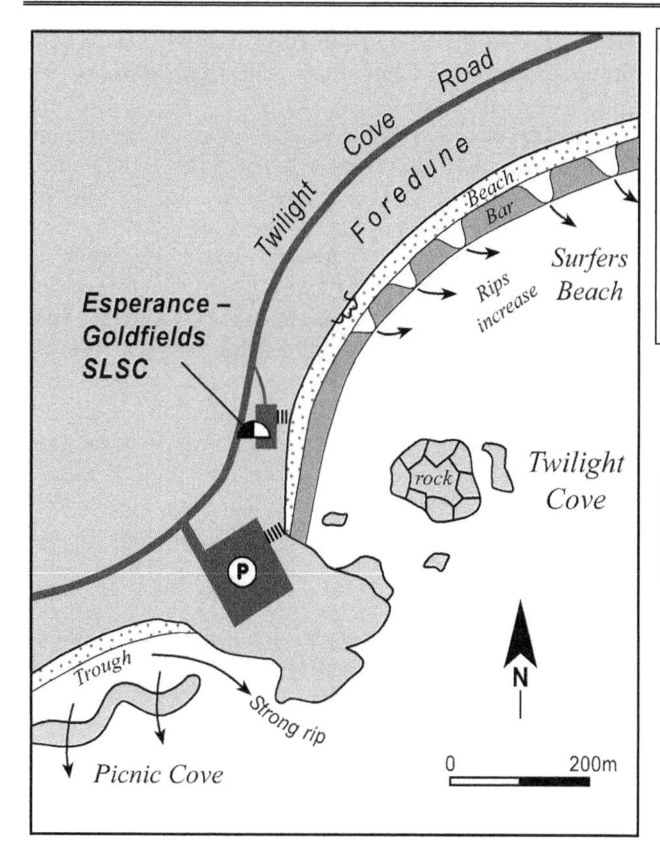

Figure 4.42 Twilight Cove is a moderately sheltered beach with exposed rip dominated beaches to either side. It is the site of the Esperance-Goldfields SLSC

Figure 4.43 Twilight Cove with the Esperance-Goldfields SLSC patrolling this more sheltered end of the beach.

Swimming: Twilight Beach is the safest swimming and surfing beach in the Esperance area, as well as being patrolled in summer. Still stay in the western patrolled area, close inshore and on the attached bar. Avoid swimming up the beach where waves and rips are more prevalent.

Surfing: Most surfers head for the high waves along *Fourth* and the aptly named *Surfers* beach where beach breaks and strong rips prevail, with Twilight more suitable for beginners.

BUTTY HARBOUR NATURE RESERVE

Coast length:	33 km	(890-923 km)
Beaches	12	(WA 156-168)

Butty Harbour Nature Reserve extends from half way along Twilight Beach west to Roses Beach. The reserve includes the spectacular coastline with granite points, calcarenite bluffs and intervening beaches, as well as the backing active and vegetated dune systems.

WA 157 PICNIC COVE (FIFTH BEACH)

No.	Beach	Rating	Type	Length
WA 157	Picnic Cove	8	TBR+rocks	2.8 km
Spring & neap tidal range = 0.7 & 0.1 m				

Picnic Cove beach (WA 157 also known as **Fifth Beach**) shares a 20 m high granite headland and large car park with Twilight Cove. The 2.8 km long beach faces southwest and is exposed to waves averaging over 1.5 m at the eastern end, decreasing closer to the western boundary, 70 m high Observatory Point. The eastern end of the beach is therefore dominated by high waves, a 100 m wide surf zone cut by strong rips, with a strong permanent rip right in front of the car park access. While the beach is excellent for beach fishing, it is unsuitable for swimming. In addition beachrock outcrops along the beach resulting in several strong permanent rips. Only in the far western corner do waves drop below 1 m and a continuous bar usually free of rips prevails.

The road runs along the top of the 50-60 m high calcarenite bluffs, which back the beach, with a car park and viewing area at Observatory Point, and the steep access down to the more sheltered western end of the beach.

WA 158-163 NINE MILE-PLUMPUDDING

No.	Beach	Rating	Type	Length
WA 158	Nine Mile Beach	8	TBR/RBB+rock	1.3 km
WA 159	Ten Mile Lagoon	8	TBR+rocks	4 km
WA 160	Eleven Mile Beach	8	TBR	6 km
WA 161	Butty Harbour (1)	7	TBR	250 m
WA 162	Butty Harbour (2)	7	TBR+rocks	200 m
WA 163	Plumpudding Beach	4	LTT→R	600 m
Spring & neap tidal range = 0.7 & 0.1 m				

To the west of Observatory Point is a 13 km long exposed beach system that terminates in the lee of Butty Head. Beachrock and bedrock outcrops divide the beach into six systems, three longer exposed beaches (WA 158-160) and three shorter increasingly protected beaches approaching Butty Head (WA 161-163). The Twilight Beach scenic drive continues on west of Observatory Point for 6 km to the boundary between beaches WA 159 and 160, with the remainder of the beaches not accessible by vehicle.

Nine Mile Beach (WA 158) commences on the western side of Observatory Point and trends northwest for 1.3 km to a protruding calcarenite point. The beach is bounded by a low linear granite point and reefs in the east and has a shore-parallel beachrock reef dominating the surf along the western half (Fig. 4.44). The high waves, rocks and reefs result in a high energy 150 m wide surf zone, with four permanent rips located adjacent to the points and gaps in the beachrock reef. While the western half has relatively calm conditions inside the reef, particularly at low tide, it is drained by a strong rip towards its eastern end. The entire system is backed by scarped Pleistocene dune calcarenite capped by Holocene clifftop dunes, reaching 90 m in height behind the beach. Generally well vegetated dunes then extend up to 12 km to the northeast, reaching as far as Esperance.

Figure 4.44 The western end of Nine Mile Beach is dominated by a shore parallel beachrock reef with rip-drained lagoon to the rear.

Ten Mile Lagoon (WA 159) extends from the calcarenite point for 4 km to the northwest, with the entire beach dominated by a near continuous shore parallel beachrock reef, which impounds the lagoon. While the inner lagoon can be relatively calm, three strong permanent rips exit through three gaps in the central-western section of the reef, and beachrock outcrops along the shore (Fig. 3.1b). Use caution if swimming here and stay close to shore and clear of the beachrock. The bluffs and dunes continue the length of the beach, with the northern boundary consisting of a small protruding section of calcarenite. There is a large car park on the bluffs behind the centre of the beach, and a vehicle track off the main road leading to the top of the boundary point, which provides views of Ten Mile and Eleven Mile beaches. The crest of the bluffs behind the beach is studded with large windmills, part of a wind power farm. This is the western most point of vehicle access. In addition it is also a 'free' or nude beach.

Eleven Mile Beach (WA 160) continues to the northwest for another 6 km with the western half called **Thirteen Mile beach**. The beach terminates as it begins to curve round to the southwest against the granite rocks of Butty Harbour. Most of the beach is well-exposed and dominated by a 200 m wide surf zone and strong beach rips, While some beachrock outcrops along the shore, the surf zone is relatively fee of rock. Calcarenite bluffs leading to 140 m high clifftop dunes back the beach, with the dunes extending 6 km inland to the Pink Lake.

Beaches WA 161-163 are located along an east-facing 1 km long section of shore in lee of Butty Head, called **Butty Harbour** (Fig. 4.45). Beach **WA 161** is a 200 m long strip of southeast-facing sand bounded by two sections of dune-capped granite rocks. The beach is exposed to waves averaging about 1.5 m, which maintain a 60 m wide surf zone with permanent rips to either end. Beach **WA 162** lies 200m to the south. It is 300 m long, faces east and is bordered by the northern granite, with granite outcropping in the surf the length of the beach and in two outcrops at the southern end. Waves average over 1 m and break across the rocks, which are drained by two permanent rips. Active dunes originating from beach WA 168 cascades down 30 m slopes onto the rear of the narrow beach.

Figure 4.45 Butty Harbour and beaches WA 161, 162 and 163.

Plumpudding Beach (WA 163) lies between the granite rocks and protruding granite point. The 600 m long beach curves between the two outcrops finally facing north against the southern rocks. It receives increasing protection to the south with waves decreasing to less than 0.5 m. The beach initially has a continuous bar grading south to reflective conditions. Active dunes cascade down onto the northern half, with 30 m high vegetated dunes on the point. There is a vehicle track leading to the point where a *righthand break* runs along the outside of the rocks.

WA 164-167 BUTTY HEAD

No.	Beach	Rating	Type	Length
WA 164	Butty Head (E 1)	10	R+rocks	200 m
WA 165	Butty Head (E 2)	10	R+rocks	250 m
WA 166	Butty Head (W 1)	10	R+rocks	200 m
WA 167	Butty Head (W 2)	8	TBR+rocks	300 m
Spring & neap tidal range = 0.7 & 0.1 m				

Butty Head extends to the southwest for 5 km from Plumpudding Rocks to the 40 m high granite head. Most of the bedrock is capped by active dunes from Roses Beach (WA 168) with vegetated clifftop dunes south of Plumpudding Beach. Between Plumpudding Beach and the head is 4 km of rocky southeast-facing shore containing two small pockets of sand (beaches WA 164 & 165), with beaches WA 166 and 167 located out on the headland. There is no vehicle access to the three outer head beaches.

Beach **WA 164** is located 500 m southwest of Plumpudding Beach and is a 200 m long strip of high tide sand fronted by continuous intertidal rocks and rock flats. Waves average 1.5 m and break across the rocks to feed a permanent rip at the western end. It is bordered by a 100 m wide rock platform and backed by steep vegetated clifftop dunes, with a vehicle track winding through the dunes. Continuous rock platforms continue to beach **WA 165**, which lies 1 km north of the head and is a 250 m long strip of east-facing sand also fronted by continuous 50 m wide rock flats and rocks, with waves averaging about 1-1.5 m. It receives sand form beach WA 167, which blows across the 200 m wide head.

Beaches WA 166 and 167 occupy parts of the 1 km long western side of the head. Beach **WA 166** is a 200 m long strip of high tide sand, fronted by 50-100 m wide rock platforms, with a *left-hand reef break* located off the beach. Beach **WA 167** lies immediately to the north and is a 300 m long sand beach, wedged between the rocks of the head and a 300 m long section of 50 m high calcarenite cliff that separates it from the long Roses beach (WA 168). A patch of calcarenite also outcrops on the centre of the beach. The beach however shares a 200 m wide surf zone with beach WA 168, and has strong permanent rips running out against the eastern rocks and in front of the western cliffs. Active dunes blow from the beach across the 20 m high rear of the head to beach WA 165.

WA 168 ROSES-QUALLUP BEACH

No.	Beach	Rating	Type	Length
			inner outer bar	
WA 168	Roses-Quallup	8	TBR RBB	13.2 km
Spring & neap tidal range = 0.7 & .1 m				

Roses-Quallup Beach (**WA 168**) is a 13 km long exposed, high energy southwest-facing beach located between Butty Head and Munroe Point, both prominent dune capped granite points. The entire beach and most of the backing massive dune system is contained within the Butty Head Nature Reserve. There is rough 15 km long flood-prone 4WD access via Murrays Road to Roses Beach and the western end of Quallup Beach, where a clifftop car park provides a view along the beach.

The beach is 13.2 km long and commences at the base of the calcarenite cliffs, which separate it form beach WA 167. It then trends northwest as Roses and finally Quallup beach, where it curves round for the last 1 km in lee of the base of Munroe Point, which extends another 3 km to the southwest. The entire beach is exposed to high southwest waves, averaging over 2 m. These combined with the fine white beach sand to maintain a 500 m wide, double bar system (Fig. 2.7). It has a rip dominated inner bar, with rips spaced every 300-400 m and more widely spaced rips cross the outer bar. The beach is backed by dunes climbing to heights of 140 m (Fig. 4.46). The dunes overlap Pleistocene dune calcarenite and extend up to 8 km inland in the west, where they are backed by Lake Mortijinup, while to the east they continue for up to 13 km inland. The massive dune field includes three episodes of dune activity. The first along much of the beach and extending up to 2 km inland, a second up to 5 km inland, and the third up to 10-12 km inland, with vegetated dunes and deflated surfaces in between.

Figure 4.46 Roses-Quallup Beach is dominated by high waves, rips and backing massive transgressive dunes.

WA 169-171 MUNROE POINT

No.	Beach	Rating	Type	Length
WA 169	Munroe Pt (1)	9	TBR+rocks	250 m
WA 170	Munroe Pt (2)	9	R+rocks	200 m
WA 171	Munroe Pt (3)	10	R+platform	200 m
Spring & neap tidal range = 0.7 & 0.1 m				

Munroe Point is a 90 m high dune-capped, granite point, which extends from Quallup Beach for 2.5 km to the southwest. Beaches WA 169-171 are located along its rocky eastern shore. All three beaches are backed by steep vegetated Pleistocene calcarenite and Holocene dunes. There is 4WD access down to beach WA 169, with a track running along the top of the ridge behind beaches WA 170 and 171.

Beach **WA 169** is located halfway along the eastern shore of the point. It is a 250 m long sandy beach wedged in a small embayment, with a protruding rock platform forming its eastern boundary and the rocks at the base of the high bluffs to the west. Rocks also outcrop in the centre of the beach. It is backed by a crenulate, partly unstable foredune, then the vegetated bluffs. It has a 200 m wide surf zone, with one strong central rip draining the entire system.

Beach **WA 170** lies 500 m east of the tip of the point and is a curving 200 m long strip of high tide sand, fronted by a rock and reef strewn high energy surf zone. Rock reefs dominate the eastern side, with rock platforms and bluffs the western side. A strong rip drains out the centre of the small embayment (Fig. 4.47).

Figure 4.47 Monroe Point is surrounded by the hazardous beaches WA 170 (foreground), 171 and 172 (background).

Beach **WA 171** lies at the tip of the point, immediately west of beach WA 170. It is 200 m long, convex in shape, curving round the point, with an irregular rock platform then rocks and rock reef extending 500 m seaward of the beach where they are exposed to high waves. There is an area of sandy surf access on the western side of the beach, however it leads to a strong permanent rip. It is backed by an unstable foredune, from which sand blows across the point to beach WA 170.

WA 172-178 SHELLY BEACH-BARKER INLET

No.	Beach	Rating	Type	Length
WA 172	Shelly Beach (1)	8	TBR+rocks	3.7 km
WA 173	Shelly Beach (2)	5	R+rocks	100 m
WA 174	Warrenup Beach (1)	8	TBR+rocks	2 km
WA 175	Warrenup Beach (2)	8	TBR+rocks	250 m
WA 176	Warrenup Beach (3)	5	R+rocks	300 m
WA 177	Warrenup Beach (4)	5	R+rocks	600m
WA 178	Barker Inlet	8→5	TBR+rocks-LTT	7.8 km
Spring & neap tidal range = 0.7 & 0.1 km				

Between Munroe Point and the headland at Barker Inlet is 14 km of generally south-facing coast containing the two Shelly beaches (WA 172 & 173), the four Warrenup beaches (WA 174-177) and a longer beach that terminates at Barker Inlet (WA 178). All seven beaches also located within a nature reserve, which also incorporates much of their backing dune systems. The is vehicle access to the coast at Shelly and Warrenup beaches via the winding 15 km long Crisps track, and to Barker Inlet via a track from Quagi Beach. There are however no facilities at any of the beaches.

Beach **WA 172** commences immediately west of Munroe Point and trends west-northwest for 3.7 km to a dune-capped, 400 m long rocky section of shore that separates it from beach WA 173. The beach is well exposed to southerly waves, which average about 2 m and break across a 200 m wide surf zone. The western half is dominated by rocks in the surf and has four permanent reef-controlled rips, while the western half has fewer rocks and usually has six beach rips spaced about every 400 m. The entire system is backed by densely vegetated 70 m high, dune-capped, calcarenite bluffs with vegetated dunes extending another 3-4 km inland. The Crisps track reaches the western end of the beach and then runs along the top of the bluffs.

Beach **WA 173** is tucked in the western corner of the beach, between the rocky shore and a 300 m long gneiss headland. The beach is 100 m long, faces east and is protected by the headland and rock reefs, with waves averaging about 1 m at the shore where there surge up a reflective beach. The small bay is, however, drained by a permanent central rip.

Beaches WA 174-177 occupy the next 3 km of shore between the two more prominent gneiss headland. Beach **WA 174** trends west for 2 km to a 300 m long rock platform that separates it from beach WA 175. The beach receives waves averaging less than 2 m, which break across a sandy surf zone with several rock outcrops. The rocks to control the location of four permanent rips, the largest against the eastern headland (Fig. 3.1a). Beach **WA 175** is a 250 m long continuation of the beach, with a reef across the inner surf zone, and a strong permanent rip draining the beach. Both beaches are backed by densely vegetated dune-capped, calcarenite bluffs, rising to 70 m with the dunes then extending 1-2 km inland. Vehicle tracks reach either end of the two beaches.

Beach **WA 176** and **177** are two adjoining curving beaches 300 m and 600 m long respectively. They faces southeast and are protected by bordering rocky points and rock reefs lining the points, with waves breaking heavily on the reefs. At the shore the waves average less than 1 m and surge up moderately steep reflective beaches. The vehicle track runs along the top of the 40 m high vegetated bluffs that backs the beaches. Both embayments are drained by a single permanent rip, against the boundary between the beaches (Fig. 4.48).

Figure 4.48 View east along beach WA 177, with the deep channel in foreground leading to a permanent rip against the western rocks.

Barker Inlet beach (**WA 178**) commences on the western side of the 60 m high gneiss headland that separates it from beach WA 177. The beach curves slightly to the west for 7 km to the usually closed 100 ha inlet, finally curving round to face west in the corner south of the inlet (Fig. 4.49). The beach receives high waves in the east, which gradually decrease to the west owing to offshore islands and reefs, with wave averaging about 1 m at the inlet. Strong rips dominate the eastern beach, with the central section consisting of a high tide beach and near continuous rock platform, while at the inlet a sandy low tide terrace prevails. The entire beach is backed by 40-90 m high densely vegetated, calcarenite bluffs, with vegetated dunes then extending 3-6 km inland. The access track reaches the western corner, where there is a shady informal camping area, and 4WD access to the beach. There is also a *right-hand break* off the western point.

WA 179-183 QUAGI BEACH

No.	Beach	Rating	Type	Length
WA 179	Quagi Beach (1)	8	TBR+rocks	3.5 km
WA 180	Quagi Beach (2)	7	TBR+rocks	2.2 km
WA 181	Quagi (W 1)	7	LTT+rocks	800 m
WA 182	Quagi (W 2)	8	TBR+rocks	250 m
WA 183	Quagi (W 3)	8	TBR+rocks	650 m
Spring & neap tidal range = 0.7 & 0.1 m				

Quagi Beach is a readily accessible beach, 9 km due south of the highway. It has a small camping area in the western corner, a picnic area on the point and is a popular spot for beach fishing, swimming and the occasional surf. There are five beaches (WA 179-183) accessible by 4WD from the camping reserve.

Figure 4.49 The usually blocked Barker Inlet.

The eastern half of the main beach (**WA 179**) commences immediately west of the 50 m high dune capped Barker Inlet headland and trends west for 3.5 km before a continuous band of beachrock separates it from the western half. The beach receives moderate to high waves, which break across a 150 m wide surf zone, dominated by discontinuous beachrock reefs in the inner surf, resulting in hazardous conditions and several small but strong permanent rips amongst the reefs. Boulder beaches lie behind the beachrock, abutting 30-40 m high vegetated calcarenite bluffs, including two clifftop blowouts, with the track running along the crest.

The main **Quagi Beach** (**WA 180**) extends from the beachrock for another 2.2 km to the west finally curving round to face east in lee of the western 20 m high headland. Wave height averages 1.5 m in the east decreasing to less than 1 m in the corner. Only the far corner is suitable for swimming, as higher waves, beachrock reefs and rocks, and small permanent rips dominate the rest of the beach. The best *surf* is usually along this section of beach. There are three car parks behind the western end of the beach and a picnic area on the rocks at the end of the main track.

Beaches WA 181-183 are accessible by a 4WD track that follows the predominately rocky shoreline to each of the beaches. They are three exposed, hazardous rocky beaches. Beach **WA 181** is an 800 m long south-facing beach with beachrock reef and rocks dominating the lower beach, with a sandy bar beyond. Waves average about 1.5 m and maintain three permanent rips. It is bordered by gneiss headland and backed by 30 m high vegetated calcarenite bluffs.

Beach **WA 182** lies 1 km to the west and is an exposed 250 m long headland-bound beach. It has a 150 m wide surf zone and a permanent rip draining out against the eastern rocks, with a rock reef also off the eastern end of the beach. Beach **WA 183** lies immediately to the west and is a similar 650 m long beach, with a patch of rock in the centre. It has a strong eastern rip, and second rip running out off the central rocks, with waves decreasing slightly towards the western end in lee of rock reefs. The beginning of the beach marks the eastern boundary of Stokes National Park.

STOKES NATIONAL PARK

Area:	10 667 ha	
Coast length:	28 km	(956-984 km)
Beaches:	17	(WA 183-200)

Stokes National Park is centered on Stokes Inlet, a 700 ha coastal lagoon which enters the sea 4 km west of Shoal Cape, and is feed by the Lort and Young rivers. The park has 28 km of coastline containing 17 beaches, separated by both bedrock gneiss and beachrock reefs, the latter forming extensive shore parallel reefs between the cape and inlet. Well vegetated dune systems rising to 100 m back the beaches. There is vehicle access to the coast and camping areas at Fanny Cove, Stokes Inlet and Skippy Rocks. The ruins of the historic Muir Homestead are located 2 km in from Fanny Cove.

WA 184-189 FANNY COVE-STOKES INLET

No.	Beach	Rating	Type	Length
WA 184	Fanny Cove	7	TBR+rocks	6.1 km
WA 185	Shoal Cape (E2)	6	R+rocks	150 m
WA 186	Shoal Cape (E1)	4	R+rocks	1 km
WA 187	Shoal Cape (W1)	4	R+reefs	1.2 km
WA 188	Shoal Cape (W2)	7	R+reefs	1.5 km
WA 189	Stokes Inlet (W)	6	R+reefs	900 m
Spring & neap tidal range = 0.7 & 0.1 m				

The eastern half of **Stokes National Park**, east of Stokes Inlet contains seven beaches (WA 183-189), all bordered by low gneiss headlands and rocks, with dune-draped Shoal Cape occupying a central position. Shore-parallel beachrock reefs dominate between the cape and inlet. There is 4WD access off the Quagi Road, via the sandy 10 km long Farrells track to Fanny Cove and Shoal Cape, with a basic camping area in lee of the sheltered western corner of Fanny Cove.

Six kilometre long beach **WA 184** commences against a 100 m long section of low gneiss rocks, and trends west for 5 km, before curving round in **Fanny Cove** to finally faces north. The exposed eastern section of beach receives waves averaging about 2 m which maintain a 200 m wide surf zone, with both beach rips as well as permanent rips associated with beachrock outcrops in the surf and along the beach. In total about 15 strong rips are usually present. The western 1.5 km receives increasingly lower waves, with waves in the cove averaging less than 0.5 m. Seagrass grows just offshore in the cove and some seagrass debris piles upon the beach. The beach is backed by vegetated calcarenite bluffs rising 70-90 m, with the dunes extending up to 10 km inland. There is a sheltered camping area in lee of the cove beach, with a picnic area on the point.

The Muir brothers established a sheep farm in lee of Fanny Cove in 1888 and used the cove to ship their wool and sandlewood. The cove was also used by miners enroute to the Dundas and Norseman goldfields in the 1890's.

Beach **WA 185** lies 500 m south of the cove and is steep, curving 150 m long east-facing shelly beach, bordered by gneiss rock platforms, together with rock reefs off the beach. Waves averaging over 1 m break heavily across the rocks and reefs, but are relatively low at the shore. However a permanent rip drains the entire beach and flows out across the reefs. The beach is backed by vegetated slopes rising to 40 m, with a small car park at the southern end of the beach. There is a *right-hand surf break* over the rocks at the southern end of the beach.

Beach **WA 186** lies immediately to the east of the 20 m high Shoal Cape. It is a curving 1 km long southeast-facing reflective beach (Fig. 4.50), which receives refracted waves, lowered to less than 1 m. These maintain a moderately steep, cusped reflective beach, backed by partly vegetated sand dunes originating from beach WA 187. **Shoal Cape** forms the southern boundary of the beach. The cape is a low gneiss point, capped by partly vegetated dunes rising to 10-20 m.

Figure 4.50 Shoal Point (left) with the sheltered beach WA 186 and exposed beach WA 187 to either side.

To the west of the cape is a series of three shore-parallel, beachrock reefs, perhaps the best developed multiple series of beachrock reefs in southern Australia. The outer reef lies up to 500 m offshore, while the inner reef is attached to the shore. The reefs represent prior positions of the shoreline, where the sandy beach has been cemented as beachrock. Waves break heavily on the outer reef and are progressively lowered by the second and third reefs. The vehicle track leads to the western end of beach WA 187, where there are two campsites and wooden steps leading down to the beach. Beach **WA 187** is the most protected of the three western beaches. It is 1 km long and forms three sandy crescents between inner beachrock outcrops along the shore (Fig. 4.51). Waves averaging about 0.5 m at the shore and a reef formed 'lagoon' lies between the beach and second reef. However beware of currents flowing through gaps in the beachrock and out of the lagoon. An 800 m long section of beachrock separates it from beach **WA 188**, a curving 1.2 km beach, with wave breaking heavily on the inner reef and strong currents exiting from both ends of the beach. Beach **WA 189** curves to the west for another 900 m, terminating at Stokes Inlet when open, and linking with beach WA 190 when the inlet is closed (Fig. 4.52). It is fronted by more discontinuous reefs, which lower waves at the shore. However rocks and two permanent rips maintain hazardous conditions. All three beaches are backed by scarped vegetated dunes rising to 30 m, together with four blowouts, the largest between beaches WA 187-188 extending 1.5 km inland.

WA 190-191 DUNSTER CASTLE BAY

No.	Beach	Rating	Type	Length
WA 190	Stokes Inlet (W)	6	R+reefs	600 m
WA 191	Dunster Castle (1)	8	TBR+reefs	400 m
WA 192	Dunster Castle (2)	8	TBR+reefs	3.7 km
WA 193	Dunster Castle (3)	5	R+reefs	150 m
Spring & neap tidal range =0.7 & 0.1 m				

Figure 4.51 Shoal Point beach (WA 187) is fronted by three successive beachrock reefs (arrows), which substantially lower waves at the shoreline.

Figure 4.52 Stokes Inlet with beaches WA 189 & 190 to either side.

Dunster Castle Bay occupies an open 4.5 km long, south-facing embayment between Stokes Inlet and a western 30 m high headland. In between are four beaches (WA 190-193) bounded by beachrock reef and calcarenite in the east and a rocky gneiss headland to the west. There is 4WD vehicle access via Stokes Inlet to the inlet mouth, and in the west to the Dunster Castle headland. All four beaches are within the Stokes National Park, with the nearest camp site back on the western shores of the inlet.

Beach **WA 190** commences on the western side of the inlet mouth and is joined to beach WA 189 when the inlet is closed. The beach trends to the southwest for 600 m to a sandy foreland formed in lee of a shore-parallel beachrock reef. This reef and a second outer reef parallels the beach, reducing waves to less than 1 m at the shore, which maintain a moderately steep reflective beach. Beware when the inlet is open, as strong tidal current flow past the beach, together with water flowing offshore through gaps in the reefs.

Beach **WA 191** extends from the cuspate foreland for 400 m to the west where it is bordered by scarped 60 m high calcarenite bluffs fronted by a rock platform. Waves average over 1.5 m and reach the central-western part of the beach, where they maintain a strong permanent rip that drains eastward along the beach to the exit against the reef off the foreland. The beach is backed by calcarenite bluffs decreasing in height to the east and a 700 m long blufftop blowout. Vegetated blufftop dunes extend anther 1-3 km to the western shores of Stokes Inlet.

Beach **WA 192** is the main bay beach and extends for 3.7 km from the western side of the 300 m long bluffs, initially as a very narrow beach along the base of the bluffs, which then widens to the west as the bluffs become increasingly vegetated apart from two blowouts. Waves average about 2 m and maintain a 200 m wide surf zone with up to six well developed beach rips, as well as a permanent rip amongst the rocks off the western end of the beach. This rip leads out to a *right-hand surf break* over the boundary rocks.

Beach **WA 193** is located 300 m to the south at the end of the vehicle track to the bay. It is a pocket 150 m long strip of high tide sand, bounded by gneiss points, with rock reefs linking the two points. Waves break over the reef, with a calmer rock-filled 'lagoon' off the beach. The beach is backed by steep, vegetated calcarenite bluffs rising to 30 m.

WA 195-200 SKIPPY ROCKS-MARGARET COVE

No.	Beach	Rating	Type	Length
WA 194	Skippy Rocks (W3)	7	TBR+rocks	2.5 km
WA 195	Skippy Rocks (W2)	6	R+rocks	100 m
WA 196	Skippy Rocks (W1)	5	R+rocks	350 m
WA 197	Skippy Rocks	8	R+rocks	100 m
WA 198	Torradup R (W)	7	TBR+rocks	2 km
WA 199	Torradup R (E)	4	LTT→R	800 m
WA 200	Margaret Cove	4	R+rocks	300 m
Spring & neap tidal range = 0.7 & 0.1 m				

The western section of Stokes National Park commences at Dunster Castle headland and trends west to Skippy Rocks and Margaret Cove, a total distance of 8 km. In between are eight beaches (WA 194-200) and the narrow mouth of the Torradup River. There are 4WD access tracks in the east to Dunster Castle headland, to the Skippy Rocks and the river mouth, and to Margaret Cove. There are camping sites at Skippy Rocks, but otherwise no facilities.

Beach **WA 194** commences on the western tip of the Dunster Castle headland, and trends west for 2.5 km to the first of the low gneiss boulders of Skippy Rocks. Rocks also outcrop in the inner surf and on the beach. Waves average about 1.5 m and maintain a 100 m wide surf zone, usually cut by six permanent beach rips (Fig. 4.53). A vegetated foredune, then a densely vegetated slopes rise to a 50 m high vegetated calcarenite ridge line, with older well vegetated dunes extending up to 4 km inland.

Figure 4.53 View along the moderate energy beach WA 194, west of Skippy Rocks.

Beach **WA 195** is a 100 m long pocket of sand wedged between the first two outcrops of the Skippy Rocks. It is fronted by continuous rock reefs and boulders, which lower waves to less than 1 m at the shore, with a small 'lagoon' against the steep, narrow sandy beach. The lagoon is however drained by a permanent rip through the reefs. Beach **WA 196** commences immediately to the west and is a curving 350 m long sandy beach, bounded by low gneiss points, with rock reefs also off the beach, forming a quieter 'lagoon', with waves averaging less than 1 m at the shore (Fig. 4.54). There are *left and right-hand surf breaks* off the rocks at either end. The main car park and camping area are located at the eastern end of Skippy Rocks headland.

Figure 4.54 The sheltered 'lagoon' beach WA 196 adjacent to Skippy Rocks.

Beach **WA 197** is located at the southern tip of the small headland, and consists of a rock bound 100 m long beach. It is backed by 20 m high vegetated bluffs and fronted by continuous inter to sub-tidal rocks, with waves averaging 1.5 m on the outer rocks, but less at the shore.

Beach **WA 198** commences on the western side of the head and trends west for 2 km to the narrow, deflected mouth of the Torradup River. The first kilometre of beach has shore-parallel beachrock reef, with a more continuous bar toward the river mouth. Waves average less than 1.5 m and maintain a 50 m wide surf zone, with three permanent rip to the east, and occasional beach rips to the west. A 500 m long blowout backs the central section of the beach.

Beach **WA 199** commences on the western side of the narrow river mouth, and is joined to beach WA 198 when the river is closed. The beach curves to the southwest for 800 m with waves decreasing from about 1 m to 0.5 m in the southern corner, which lies in lee of a rock reef. There is a *right-hand surf break* off the end of the reef. The reef is attached to a cuspate foreland, with beach **WA 200** extending from the southern side of the foreland for 300 m. The beach curves round between the reef and a boulder strewn southern point, with rock reefs also off the beach. Waves break across the reefs with a calmer inner lagoon at the shore (Fig. 4.55). Low vegetated terrain backs both beaches.

Figure 4.55 Margaret Cove beach (left) and Lake Shaster behind.

LAKE SHASTER NATURE RESERVE

Area: 10 505 ha
Coast Length: 36 km (986-1006 km, 1008-1024 km)
Beaches: 22 (WA 201-215, 217-222)

Lake Shaster is a 1000 ha lake in lee of Munglinup beach. It is the largest in a series of lakes and wetlands formed in the backbarrier depression behind Munglinup and adjoining beaches. The reserve includes 36 km of coast and 22 near continuous beach systems, together with the backing 1 to 3 km wide coastal dune systems, most consisting of vegetated Holocene dunes over Pleistocene dune calcarenite. There is vehicle access to the park at Munglinup Beach.

WA 201-206 **PINCER POINT**

No.	Beach	Rating	Type	Length
WA 201	Pincer Pt (E5)	9	TBR+reef	1.8 km
WA 202	Pincer Pt (E4)	10	R+platform	600 m
WA 203	Pincer Pt (E3)	7	TBR+rocks	1.2 km
WA 204	Pincer Pt (E2)	7	TBR+rocks	1.6 km
WA 205	Pincer Pt (E1)	6→5	TBR+reef→LTT	3.5 km
WA 206	Pincer Pt	6	R+rocks/reef	400 m

Spring & neap tidal range = 0.7 & 0.2 m

Pincer Point is a 20 m high dune-capped, metasedimentary headland, that extends south about 1 km. It lies 7 km west of the Margaret Cove headland, and between the two are five near continuous, south-facing beaches (WA 201-205), with a small beach on the point (WA 206). There is 4WD access to the point, but no facilities. All the beaches and backing dune areas are located within the Lake Shaster Nature Reserve.

Beach **WA 201** commences just west of the Margaret Cove headland and trends west for 1.8 km to a band of bedrock. The beach receives waves averaging over 1.5 m, which break across discontinuous beachrock reefs present in the surf the entire length of the beach. While the beachrock forms calmer 'lagoon' sections close to shore, it also maintains four permanent rips draining the surf and lagoons, as well as rocks present through the surf. It is backed by a steep narrow beach, a high partly unstable foredune, and then vegetated Holocene dunes over Pleistocene calcarenite, all extending 1-2 km inland to Margaret Cove and the Torradup River.

Beach **WA 202** is a 600 m long, narrow strip of high tide sand, fronted by a continuous, though irregular rock platform averaging about 50 m in width, with some rock reef further out. Waves break heavily on the outer rocks and reefs and flow shoreward across the platform. It is backed by steep, vegetated bluffs rising to 50 m. Beach **WA 203** commences immediately to the west. It curves slightly for 1.2 km between two rock outcrops, with small forelands attached to each outcrop. Permanent rips are located at either end, with a solitary large central beach rip. It is backed by a partly unstable foredune, then vegetated Holocene dunes, draped over 40-50 m high Pleistocene calcarenite extending about 2 km inland.

Beach **WA 204** is a 1.6 km long beach located between the next two foreland bound rock outcrops. Waves break over reefs off each foreland and a small creek is usually dammed behind the eastern foreland. The central section of beach usually has 2-3 beach rips, with permanent rips to either end. It is backed by a partly active foredune, then the creek, which winds though a 1 m long section of vegetated Holocene dunes overlying the Pleistocene calcarenite.

Beach **WA 205** commences at the western foreland and curves to the west for 3.5 m terminating against the eastern rocks of Pincer Point. The eastern half of the beach receives waves averaging about 1.5 m and together with boundary rocks and some rock outcrops, has a 100 m wide surf zone drained by both strong permanent and beach rips. The western half becomes increasing protected and the bar more continuous, as it narrows into the western corner (Fig. 4.56). The beach is backed by steep, partly active 20 m high foredune, and then 2 km wide vegetated dunes. The access track reaches this beach 1 km east of the point, and then continues on to the point.

Beach **WA 206** is located out on Pincer Point. It is a rock and reef bound 400 m long strip of high tide sand, tied at each end to 300 m wide rock platforms and reefs, with a subtidal reef in between. Waves break heavily on the outer reefs with relatively calm conditions at the shore.

Figure 4.56 View along the western half of beach WA 205 towards Pincer Point.

WA 207-214 **PINCER ROCKS-MUNGLINUP**

No.	Beach	Rating	Type	Length
WA 207	Pincer Rocks (W1)	7	TBR	5.5 km
WA 208	Pincer Rocks (W2)	8	R+rocks	50 m
WA 209	Pincer Rocks (W3)	5	LTT	2.7 km
WA 210	Munglinup (E5)	9	R+platform	700 m
WA 211	Munglinup (E4)	5	R+reef	800 m
WA 212	Munglinup (E3)	4	R+rocks	300 m
WA 213	Munglinup (E2)	7	TBR+reefs	800 m
WA 214	Munglinup (E1)	4	R+rocks	700 m
Spring & neap tidal range = 0.7 & 0.2 m				

To the west of Pincer Point is an open 10 km wide south-facing embayment, with an eastern sandy shore containing three beaches (WA 207-209) and a more rocky western shore with beaches WA 210-214. It terminates at the Munglinup beach headland. All the beaches are within the Lake Shaster Nature Reserve and there is vehicle access via the Munglinup beach road to beaches WA 209 to 214.

Beach **WA 207** is a slightly curving, south-facing 5.5 km long beach that extends from the western side of Pincer Point to a 500 m long outcrop of metasedimentary rocks. The beach is well exposed to southerly waves, which average about 2 m and maintain a 250 m wide rip dominated surf zone, with up to 15 beach rips spread along the beach, together with two permanent boundary rips (Fig. 4.57). It is backed by three blowouts in the east, the largest extending 1.5 km inland, and a moderately active foredune to the west. The dunes are transgressing over vegetated dunes 500 m wide in the west, widening to 2 km in the east.

Beach **WA 208** is a 50 m long pocket of sand located out on the metasedimentary headland. It is wedged in at the base of two bedrock arms, with a strong permanent rip draining the small embayment.

Figure 4.57 Well developed boundary rips (arrows) on beach WA 207 west of Pincer Point.

Beach **WA 209** commences on the western side of the rocks and trends to the southeast for 2.7 km curving round in the south to face east. The beach is moderately protected with waves averaging less than 1.5 m and decreasing to the south. Rips occasionally form along the eastern section of beach, with a more continuous narrow bar to reflective conditions along the western section. It is backed by a low narrow vegetated foredune and bounded in the south by a 700 m long low rocky platform. Beach **WA 210** lies along the rear of the platform. It is backed by dune-covered rocks, and fronted by the irregular metasedimentary platform and some rock reefs.

Beaches WA 211 and 212 occupy a 1 km wide southeast-facing embayment, immediately to the west. Beach **WA 211** is a curving 800 m long steep, reflective beach, separated from its neighbour by a low rock outcrop, with rock reefs paralleling the northern half of the beach. Waves break over the reefs, with a calmer 'lagoon' against the shore, drained by a permanent rip flowing out though the reef. Beach **WA 212** lies in the southern corner, and is a curving 300 m long steep reflective beach, bordered by rocky outcrops. Both beaches are backed by 20 m high scarped dunes, the dunes originating from beach WA 215.

Beaches WA 213 and 214 occupy the next 1.3 km wide embayment that terminates at the Munglinup headland. Beach **WA 213** is a curving 800 m long, southeast-facing, moderately exposed beach that receives waves averaging about 1.5 m, which break over rock reefs. Three permanent rips drain out either end and through a central gap in the reefs. A small, rock outcrop separates it from beach **WA 214**, a curving, more protected, east-facing, 700 m long beach. Waves are lowered by refraction around the point and some inshore reefs, and surge up a steep reflective beach face. Both beaches are backed by 20 m high scarped dunes.

WA 215-217 **MUNGLINUP**

No.	Beach	Rating	Type	Length
WA 215	Munglinup	3	R+reef	600 m
WA 216	Munglinup (W1)	3	R+reef	1.2 km
WA 217	Munglinup (W2)	7	RBB	2.8 km
Spring & neap tidal range = 0.7 & 0.2 m				

Munglinup is a relatively popular beach area as it is one of the few accessible beaches along this section of coast. In addition to 2WD access, it has a car park, a small camping reserve with water and amenities, beach vehicle access and boat launching from the beach. While the beach faces due south it is protected from the swell by a near continuous beachrock reef lying 50-100 m off the beach, with a usually calm lagoon between the reef and steep beach.

Munglinup Beach (WA 215) is a double crenulate, south-facing 600 m long beach lying in lee of a beachrock reef (Fig. 4.58) which extends from the adjoining eastern headland for 2.5 km to the usually blocked mouth of the Oldfield River. While waves break heavily on the reef, they average only about 0.5 m at the beach, increasing slightly at high tide. The beach is suitable for swimming close to shore, however be aware of increasing currents closer to the reefs. It is backed by 10-20 m high partly active dunes, with the car park and beach access behind the centre of the beach. Beach and rock fishing are also popular, however be careful on the headlands as the rocks slope seaward and large waves run over the point seaward of the reef.

Figure 4.58 Munglimup Beach is located to the lee of a near continuous beachrock reef, forming a more sheltered inner lagoon.

Beach **WA 216** continues on west of Munglinup for another 1.2 km to the Oldfield River mouth bar with the reef generally 100-200 m offshore, though in places it reaches the shore. The beach faces south and is slightly crenulate behind the reef, with waves remaining low at the shore (Fig. 2.22b). Ten to twenty metre high dunes back the beach. There is vehicle access to the eastern side of the river mouth and 4WD access to the beach.

Beach **WA 217** commences on the western side of the river mouth and trends to the west for 2.8 km, terminating at a low bedrock reef. Once past the river mouth the reef ceases and high waves averaging over 1.5 m reach the shore. These maintain a 200 m wide surf zone usually cut by strong rips every 400-500 m, including a permanent rip against the western reef. The beach is backed by partly stable foredunes and three blowouts extending up to 500 m inland, then by vegetated dunes up to 1 km inland. A rough track winds through the dune to reach the western end of the beach.

WA 218-222 LAKE SHASTER–SWAN HILL

No.	Beach	Rating	Type	Length
WA 218	Lake Shaster (1)	6	TBR-LTT	4 km
WA 219	Lake Shaster (2)	7	TBR+reef	4.7 km
WA 220	Parriup (1)	6	TBR/LTT+reef	2.2 km
WA 221	Parriup (2)	5	LTT+reef	1.7 km
WA 222	Swan Hill	5	LTT+rocks&reef	6.6 km
Spring & neap tidal range =0.7 & 0.2 m				

Lake Shaster is a shallow 800 ha lake located in lee of beaches WA 218 and 219, with a 1 km wide dune system separating the lake from the beaches. The surrounding reserve, which extends along the coast is named after the lake. The lake is the largest in a series of shallow lakes and wetlands that occupy the backbarrier depression between Munginlup Beach and Emu Hill, 20 km to the west. All are part of the reserve.

Beach **WA 218** extends west of the gneiss reef that separates it from WA 219. The beach is 4 km long, faces south and is slightly crenulate in lee of scattered reefs lying up to 4 km offshore. The reefs lower waves to about 1.5 m, which maintain a usually attached 100 m wide bar, cut by rips on the more exposed sections and continuous in the more protected sections. A partly active foredune and several larger blowouts back the beach, with stable dunes up to 30 m high extending 1 km inland to the lake shore. The beach terminates at the most prominent cuspate foreland with beach **WA 219** continuing to west for another 4.7 km to terminate on the next foreland in lee of an inshore bedrock reef. A solitary 4WD track crosses the dunes and reaches the western end of the beach. The beach is partly protected by the scattered offshore reefs, with waves averaging about 1.5 m. The waves break across a 50-100 m wide usually attached bar, with some section of beachrock in the surf. The result in a combination of both beach rips and rips against the rocks, resulting in more hazardous conditions. The beach is backed by a more continuous 20 m high foredune containing numerous small blowouts, then 1 km of vegetated dunes.

Beach **WA 220** commences on the western side of the reef-tied foreland, and curves to the southwest for 2.2 km to terminate at a prominent reef foreland, with larger reefs extending up to 2 km offshore. The slight eastern orientation of the beach and the reef lower waves towards the western end where they average about 1 m. The beach is cut by rips in the east, grading to a continuous low tide terrace in the west. The backing 10-20 m high foredune, while steep and scarped in places, is generally moderately stable and backed by 500 m wide densely vegetated dunes. A series of four circular lakes (Parriup) occupy the 1.5 km wide back-barrier depression behind the beaches WA 220 and 221. Beach **WA 221** extends for 1.7 km due west of the foreland (Fig. 4.59). The entire beach is fronted by scattered shallow reefs, with deeper reef offshore. These lower waves to about 1 m, producing a continuous low tide terrace, inter-fingering with the reef, and some small rips against the reef. The entire beach is backed by a steep vegetated 10 m high foredune, then 500 m wide scrub covered dunes.

Figure 4.59 Beach WA 221 a lower energy open beach sheltered by shallow inshore reefs.

Beach **WA 222** commences at the reef-bound foreland that separates it from beach WA 221. The beach initially extends west for 2 km before curving round to the southwest for a total length of 6.6 km. It is backed by a 10-20 m foredune, then a high 500 m wide band of lower, scrub-covered dune, then a 1-2 km wide area of shallow lakes and wetlands. Swan Hill is a 13 m high dune towards the western end of the beach. There is vehicle access to the beach via a rough track in the centre, and from the track that runs north from Starvation Boat Harbour in the west. The beach becomes increasingly protected by Powell Point to the west with wave height decreasing from about 1 m in the east to less than 0.5 km in the west. The waves maintain a continuous low tide terrace apart from a few small rock outcrops, terminating against the low sloping gneiss below 11 m high Emu Hill.

WA 223-227 STARVATION BOAT HARBOUR

No.	Beach		Rating	Type	Length
WA 223	Emu Hill	1	3	LTT+rocks	100 m
WA 224	Emu Hill	2	3	LTT+rocks	70 m
WA 225	Emu Hill	3	3	LTT/R	250 m
WA 226	Starvation Boat Hbr	2	3	R	600 m
WA 227	Starvation Boat Hbr	2	3	R	50 m
Spring & neap tidal range = 0.7 & .2 m					

Powell Point is a low gneiss headland, capped by 29 m high calcarenite Lime Hill. The point forms one of the more significant inflections in the shoreline between Esperance and Hopetoun. Immediately east of the point the shoreline trends north for 3 km before resuming its easterly trend. The east-facing Starvation Boat Harbour lies to the lee of the point, followed by a 2 km long section of low gneiss shoreline, which at its northern end forms the western boundary of beach WA 222. Between this boundary and the harbour are six protected, low energy east-facing beaches (WA 223-227). A gravel road reaches the harbour beach, with a 4WD track leading back along the crest of the 10 m high dune-covered rocky shore to beach WA 222 and provides access to the other beaches.

Beach **WA 223** is an east-facing 100 m long strip of high tide sand located in amongst the rocks that from the boundary with beach WA 222. It is backed by 10 m high vegetated dunes, with vehicle access to either end. The highest point of the dune is known as Emu Hill. Immediately north is a 70 m long pocket of sand (**WA 224**) wedged in between the sloping gneiss and fronted by shallow rock reefs. The rocks between the two beaches were used to locate the southern end of an earlier dingo fence, with remnants of the fence and an old jetty visible on the rocks and in the water. Waves average about 0.5 m at both beaches.

Beach **WA 225** commences 100 m to the south and is a 250 m long sand beach, initially fronted by rocks, then a continuous reflective beach, bordered by low sloping gneiss rocks and shallow reefs. It is backed by a well vegetated 10 m high foredune then the access track.

Starvation Boat Harbour beach (**WA 226**) is a relatively straight 600 m long east-facing reflective beach, backed by a low vegetated foredune. These is vehicle access to the centre of the beach, with a car park and small camping area at the northern end, as well as vehicle access for beach boat launching (Fig. 460). A fishing shack is located behind the centre of the beach and a lifebuoy is maintained at the beach. Beach **WA 227** is a 50 m long pocket of sand located 200 m east of the boat launching area. It is separated from the main beach by Tide Gauge Rock, a low gneiss rock and reef. Dune capped Powell Point backs the beach, with the road running out to a car park and lookout on the point.

Figure 4.60 Starvation Boat Harbour beaches.

WA 228-230 **POWELL PT**

No.	Beach	Rating	Type	Length
WA 228	Powell Pt	6	TBR	6.7 km
WA 229	Powell Pt (W)	2	R	900 m
WA 230	Mason Bay (E 3)	2	R+rocks	2.2 km
Spring & neap tidal range =0.7 & 0.2 m				

At Powell Point the shoreline turns and sweeps to the southwest for 13 km to Mason Bay. Between the two are nine beaches ranging from exposed to very protected. The gravel Southern Ocean Road runs along behind the backing dunes and together with the Mason Bay Road provides reasonable access to all the beaches. The first three beaches (WA 228-230) occupy the first 10 km of shore.

Beach **WA 228** commences hard against the sloping gneiss of Powell Point and curves slightly to the southwest for 6.7 km to a reef-tied tombolo that separates it from beach WA 229. The beach is initially exposed to waves averaging 1.5 m, which maintain a rip dominated 80 m wide surf zone (Fig. 4.61). This continues for 5 km before inshore reefs lower waves resulting in a low tide terrace, then reflective conditions dominating the western 1.5 km. The beach is backed by a foredune which matches the beach types, being scarped and dominated by small blowouts rising to 25 m height in the east, while becoming increasing stable, lower and well vegetated to the west. There is a car park and viewing area on Powell Point overlooking the eastern corner of the beach, and a track from the road across the foredunes to the beach 5 km west of the point. The beach is hazardous for swimming, with a permanent rip right below the point car park. There are some rock pools on Powell Point and close inshore against the point providing calmer places to swim. A lifebuoy is located at the car park.

Figure 4.61 The moderately exposed rip-dominated beach WA 228 extends west of Powell Point.

Beach **WA 229** commences on the western side of the tombolo and curves round to the southwest for 900 m to a more rounded cuspate foreland that separates it from beach WA 230. Gneiss reefs and seagrass meadows occupy the nearshore between the points, with usually calm conditions at the shore, and a steep reflective beach and low stable foredune. There is no vehicle access to the beach, however it can be reached from the car park in lee of beach WA 230, 500 m to the west.

Beach **WA 230** initially trends west of the foreland before curving to the south with a total length of 2.2 km. The entire embayment off the beach is filled with shallow gneiss reefs and rocks, with rocks also outcropping along the western half of the beach. The beach is usually calm with a steep, narrow reflective beach backed by a low densely vegetated foredune, which diminishes to the west. There is a small basic camping area with amenities and water behind the eastern end of the beach, with vehicle access to the beach for boat launching. However beware of the steep, soft beach and rocks and reef right off the launching site. A second rough access track also reaches the western end of the beach.

WA 231-238 MASON BAY

No.	Beach	Rating	Type	Length
WA 231	Mason Bay (E2)	5	R+rocks	50 m
WA 232	Mason Bay (E1)	5	R+rocks	250 m
WA 233	Mason Bay	7	TBR/LTT+rocks	2 km
WA 234	Mason Bay (W1)	2	R	150 m
WA 235	Mason Bay (W2)	3	R+rocks	250 m
WA 236	Mason Bay (W3)	3	R+rocks	80 m
Spring & neap tidal range = 0.7 & 0.2 m				

Mason Bay is an open south-facing, 3 km wide embayment, bordered by 30 m high dune-capped gneiss headlands. The bay shore is dominated by one longer exposed beach (WA 233), with small protected, rock-bound pocket beaches dominating both headlands. The gravel Mason Bay Road reaches the shore in lee of beach WA 235, providing foot access to the remaining beaches.

Beaches WA 231 and 232 occupy the eastern headland. Beach **WA 231** is a 50 m long pocket of southeast-facing sand wedged in a gap in the steep dipping beds of gneiss. The rocks extend up to 200 m seaward of the beach resulting in usually calm conditions at the shore. Immediate to its west is beach **WA 232** a curving 200 m long beach bordered by low gneiss rocks and reefs, together with reefs lying immediately off the beach. The reefs lower the waves to about 0.5 m at the shore. Thirty meter high vegetated dunes blanket the headland and a rough vehicle track terminates on the crest of the dunes and provides foot access to both beaches.

Beach **WA 233** occupies more than half the bay shore. The 2 km long beach faces south and receives waves averaging up to 1.5 m, which break across a 50 m wide rip dominated surf zone, with usually 4-5 active beach and reef bound rips. A 20 m high partly stable foredune backs the beach, which is bordered by low gneiss points, as well as some rock outcropping along the beach.

Beach **WA 234** is a 150 m long pocket of sand located between the rocks that from the western boundary of beach WA 223 and the more continuous rocks along the western side of the bay. The steep reflective beach faces south with wave reduced by the boundary rocks and reefs. Beaches **WA 235** and **236** are located 250 m to the south. They are adjoining 250 m and 80 m long curving beaches tied by a central tombolo. Both are bordered by prominent rounded gneiss rocks and boulders, with rock reefs dominating the nearshore, while steep vegetated dune sand rocks back the beaches. There is a *lefthand surf break* over the reef off beach WA 236. The Mason bay Road terminates at a car park behind beach WA 235 proving good access to all three beaches.

WA 237-239 FIFTEEN MILE BEACH

No.	Beach	Rating	Type	Length
WA 237	Fifteen Mile	6	LTT+rocks/reefs	4.2 km
WA 238	Fourteen Mile	6	LTT+rocks/reefs	2.7 km
WA 239	Thirteen Mile	6	LTT+rocks/reefs	1.6 km
WA 240	Twelve Mile (E4)	6	LTT+rocks/reef	300 m
Spring & neap tidal range = 0.7 & 0.2 m				

To the east of Hopetoun is a 30 km long gently curving south-facing sweep of coast that terminates at the southern Mason Bay headland. The once continuous sand beaches has been eroded and now consists of 16 beach systems (WA 237-253) each separated by bedrock and/or calcarenite rocks and reefs, and all backed by a continuous generally steep 20-30 m high foredune, for the most part overlying Pleistocene calcarenite. Gneiss rocks and reefs in the east give way to extensive beachrock reefs in the centre, while towards the west the reefs, rocks and forelands define the boundary of the beaches. The gravel Southern Ocean East Road parallels the back of the dunes and provides access to a several of the beaches. Some of the beaches are named according to their distance from Hopetoun (Two, Five and Twelve Mile), the remainder have been unofficially named in this publication in a similar fashion. While all the beaches faces south they are protected to varying degrees by bedrock and calcarenite reefs, resulting in waves usually less than 1.5 m and in places less than 1 m at the shore. The first four beaches (WA 237-240) are the longest and occupy the first 9 km.

Beach **WA 237** commences at the tip of the eastern headland and is initially fronted by irregular rock reefs. These give way to a continuous, though crenulate 4.2 km long beach, with irregularities due to outcrops of beachrock on and along the shore. Waves average just over 1 m, which maintain a continuous 30 m wide low tide bar, usually only cut by rips adjacent to rocks and reefs. It is backed by a steep vegetated foredune reaching up to 40 m in height, with one large blowout-parabolic towards the eastern end extending 1 km inland (Fig. 4.62). Elsewhere the dunes average 500 m in width, with the road 1-2 km inland. The only vehicle access track reaches the rocks at the southern end of the beach.

A 100 m long strip of beachrock separates beaches WA 237 and 238, with beach **WA 238** continuing west for 2.7 km as a relatively straight beach interrupted by rocks and reefs along the shore. It has a similar 30 m wide continuous bar with rips against the rocks, and a backing 20-30 m high dune system that narrows from 600 m in the east to 200 m in the west. Access tracks reach the centre and eastern end of the beach.

Figure 4.62 Fifteen Mille Beach has waves lowered by inshore reefs, as well as beachrock outcropping along the shore.

Beach **WA 239** is separated from beach WA 238 by a 1 km long section of scarped calcarenite, with the dune continuing behind. The beach is 1.6 km long with waves averaging about 1 m. They break across a near continuous beachrock reef that lies at and off the shoreline, forming elongate lagoon when off the beach, with some strong rips exiting through gaps in the reefs. A scarped 20 m high foredune backs the beach. At the western end of the beach is an 80 m long section of rocks that separates it from 300 m long beach **WA 240**. This beach is bordered by rocks, with some outcropping on the beach. Waves remain about 1 m and break across a 30 m wide bar. A vehicle access track reaches the back of the dunes between the two beaches and provide foot access to both.

WA 241-246 TWELVE-EIGHT MILE BEACH

No.	Beach	Rating	Type	Length
WA 241	Twelve Mile (E3)	5	R+reef	200 m
WA 242	Twelve Mile (E2)	4	R+parallel reef	400 m
WA 243	Twelve Mile (E1)	4	R+reef	150 m
WA 244	Twelve Mile	4	R+parallel reef	2.2 km
WA 245	Eight Mile	4	R+parallel reef	1.6 km
Spring & neap tidal range =0.7 & 0.2 km				

The 4.5 km long section of shore between beaches WA 241 and 245 consists of a relatively straight south-facing shoreline, which has however been segregated into the six beaches by the dominance of near continuous shore parallel beachrock reefs, the reefs forming lagoons in places. Waves average just over 1 m on the reefs, but are reduced at the shoreline, resulting in relatively calm conditions in the lagoon segments. All the beaches are backed by steep, in places scarped 20-30 m high foredunes, which are backed by vegetated dunes extending from 200 m to 1 km inland. The road tends to run along the rear of the dunes with the main beach access from the car park above beach WA 245.

Beach **WA 241** is a 200 m long pocket beach wedged in between a low calcarenite point and reefs, with a 100 m long central bar section and a steep scarped dune behind. Beach **WA 242** commences immediately to the west and while the boundary 50 m wide reef continues straight, the backing beach curves round forming a 400 m long lagoon, between the usually calm beach. The lagoon is up to 50 m wide, with the beach usually calm and backed by steep, vegetated 20-30 high dunes (Fig. 4.63).

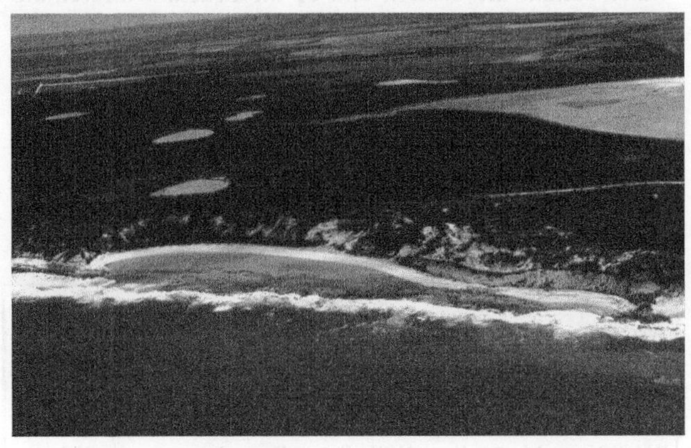

Figure 4.63 A beachrock enclosed eastern section of Twelve Mile Beach.

Beach **WA 243** commences where the reef rejoins the shore and is a 150 m long strip of high tide sand fronted by continuous, but irregular intertidal calcarenite reef, with steep vegetated dunes behind. Immediately to the west, beach **WA 244** separates from the reef again forming an elongate lagoon between the relatively straight near continuous reef and the crenulate shore. The beach and lagoon are 2.2 km in length, with the lagoon reaching up to 100 m in width. Blowouts in the east and scarped dunes in the west back the beach, with the road running just behind the western half of the beach.

Beach **WA 245** is locally knows as the Twelve Mile Beach (12 miles or 18 km from Hopetoun) and has two car parks on the dunes above the beach. The beach is 1.6 km long and like its neighbours consists of a triple crenulate low energy beach in lee of a near continuous beachrock reef. This impounds a lagoon up to 100 m wide, then the calm shoreline backed by steep 20-30 m high dunes. While the lagoon is relatively calm, beware of a strong permanent rip that drains water out of the lagoon toward the eastern end of the beach, directly below the first car park (Figs. 2.9d, 4.64).

Surf: There are a few reef breaks along the outer side of the beachrock reefs, with the best and more accessible adjacent to the gaps in the reef.

Figure 4.64 A strong permanent rip (arrow) drains the beachrock-enclosed lagoon in front of Twelve Mile Beach.

WA 246-249 **SEVEN-FOUR-MILE BEACH**

No.	Beach	Rating	Type	Length
WA 246	Seven Mile	5	LTT+rocks	3.4 km
WA 247	Six Mile	4	R+reefs	300 m
WA 248	Five Mile	4	R+reefs	1.3 km
WA 249	Four Mile	5	LTT	2.3 km
Spring & neap tidal range = 0.7 & 0.2 m				

Beaches WA 246 to 249 occupy a 7 km long section of near continuous sandy shoreline, one however still dominated by calcarenite reefs and outcrops. The beaches are backed by a 10-20 m high, 200-400 m wide vegetated dune system, with the road running along the crest of the dunes and providing access to the central Five Mile Beach (WA 248).

Beach **WA 246** is a relatively straight 3.4 km long, south-facing beach that receives waves averaging about 1 m. Along the eastern half they break across a 30 m wide continuous bar, while to the west beachrock increasingly dominates the shoreline, and a second reef parallels the shore resulting in a rock dominated surf zone. Beach **WA 247** is located immediately west, and consists of a 300 m long high tide beach fronted by a raised beachrock ridge, then a parallel reef in the surf. The road clips the western end of the beach with an access track down to the beach.

Beach **WA 248** is a 1.3 km long crenulate beach in lee of a shore parallel beachrock reef that lies up to 60 m off the beach, forming an elongate lagoon (Fig. 4.65). Raised beachrock also outcrops on the beach. Waves break heavily on the reef, with calmer conditions in the lagoon, while the water drains out through a gap in the reef forming a strong permanent rip, located just below the dune top car park. Be wary of the currents and the rip if swimming in the lagoon.

Beach **WA 249** commences at the western end of the lagoon and continues west for 2.3 km to the broad cuspate foreland that separates it from Two Mile Beach (WA 250). The beach has beachrock fringing the shore along its eastern half, while it is free of rock with a continuous 30 m wide low tide bar along the western half, with waves averaging about 1 m. It is backed by a 10-20 m high 200 m wide dune then the road.

Figure 4.65 Five Mile Beach is sheltered in lee of a near continuous beachrock reef.

WA 250-253 **TWO MILE-HOPETOUN**

No.	Beach	Rating	Type	Length
WA 250	Two Mile (E)	5	LTT	3.6 km
WA 251	Two Mile	4	R	200 m
WA 252	Two Mile (W)	5	R+rocks	200 m
WA 253	Hopetoun	4	R+reef	2.3 km
Spring & neap tidal range = 0.7 & 0.2 m				

Immediately east of Hopetoun are the first four in the string of 16 beaches that extend east for 30 km to Mason Bay. The four beaches (WA 250-253) consist of two longer beaches separated by two central pockets of sand. All four faces south-southeast and receive waves that gradually decrease from 1 m in the east to 0.5 m at Hopetoun.

Beach **WA 250** is a 3.6 km long gently curving beach that lies in lee of deeper reefs. Waves average about 1 m and break across a narrow continuous low tide bar, which decreases in width to the west. It is backed by a generally stable 10 m high foredune. Beach **WA 251** occupies the western end of the beach, which is defined by a low shore perpendicular gneiss reef that protrudes 200 m off the beach, with a dune covered gneiss headland forming the western boundary. Waves average less than 1 m in height and maintain a usually reflective beach. Beach **WA 252** runs along the front of the headland for 200 m. It is a narrow beach backed by a 1 m high vegetated dune covered headland and fronted by ridges of gneiss extending up to 200 m offshore, in addition to rock reefs off the beach. There is a gravel road and car park above the western end of the beach..

Beach **WA 253** commences on the western side of the headland and curves gently to the west-south-west for 2.3 km to terminate against the Hopetoun jetty. The eastern half of the beach is backed by densely vegetated 10 m high dunes. The dunes continue to the jetty, with a foreshore reserve then the small town of Hopetoun and backing the dune along the western half of the beach. Wave height decreases slowly to the west with the beach usually narrow, steep and reflective, with the shallow nearshore dominated by reefs and seagrass meadows.

Hopetoun is a small town of about 300, located right on the coast at Mary Ann Haven. The town owes it origin to the shelter afforded by the reefs off the haven, which lead to its establishment in 1901. This was followed by the development the original jetty in 1906 and the railway in 1908 to service the copper and gold mine at Ravensthorpe. Following World War 1 the mine closed and the town stagnated. In the past decade has it begun to reestablish itself as a tourist and retirement destination. It offers a service station, store, hotel and post office, as well as a range of accommodation. It provides ready access to the Fitzgerald River National Park to the west, as well as the numerous beaches to the east. The old timber jetty was replaced by a large rock jetty in 1984, which also serves as a boat ramp.

WA 254-258 **HOPETOUN (W)**

No.	Beach	Rating	Type	Length
WA 254	Mary Ann Haven	2	R	700 m
WA 255	Whale Bay	2	R	5.9 km
WA 256	Four Mile (E)	7	TBR+reef	1.6 km
WA 257	Four Mile (W1)	7	TBR	1.5 km
WA 258	Four Mile (W2)	3	R+reef	400 m
Spring & neap tidal range = 0.7 & 0.2 m				

The Hopetoun jetty forms an artificial boundary between the long sweep of beaches to the east and the five beaches (WA 254-258) that extend for 10 m to the west across the mouth of Culham Inlet, to terminate against the steep slopes of East Mount Barren (Figs. 4.66 & 4.67). The beaches are initially sheltered by the extensive reefs that lie off Mary Ann Haven, becoming more exposed along Four Mile Beach. Hopetoun and Hopetoun caravan park then the Whale Bay subdivision back the first 3 km of beaches, with relatively natural conditions extending to the west.

Mary Ann Haven beach (**WA 254**) faces due south and while exposed to strong winds is sheltered from ocean waves by a shallow 2 km wide reef system that lowers waves to less than 0.5 m at the shore. The beach commences on the western side of the Hopetoun jetty and extends west for 700 m to a low but prominent sand foreland, formed in lee of the reefs. Shallow reefs and a mixture of sandy substrate and seagrass meadows extend seaward of the beach. Hopetoun backs the eastern half and the caravan park the western half of the beach. While there is no surf at the shore there is a heavy lefthand break, called *Crazies* out on the reef edge.

Whale Bay beach **WA 255** commences at the foreland and curves round to the northwest then west for 5.9 km to a sandy foreland. For the most part the beach is well protected by shallow reefs and sandy sea floor, resulting in a steep reflective beach, together with some rocks outcropping toward the western end. It is backed by a well vegetated 10 m high foredune, backed by a few foredune ridges. The eastern section of the beach was subdivided in the late 1990's.

Four Mile Beach (**WA 256**) extends from the foreland past a section of intertidal rocks for 1.6 km to the usually blocked mouth of Culham Inlet. Wave height increases towards the inlet resulting in a 50 m wide surf zone usually cut by a few rips. The inlet marks the eastern boundary of the Fitzgerald River National Park. Beach **WA 257** continues west on the other side of the inlet for another 1.5 km. It is free of rocks and usually has 3-4 beach rips, before wave height decreases into the western corner, in lee of a linear reef that projects eastward and separates it from the next beach. Both beaches are backed by a 10-20 m high foredunes. The inlet mouth is deflected along the western kilometer of beach WA 256, and breaks out across a 100 m wide section of bare sand that separates the two beaches.

Beach **WA 258** is a 400 m long strip of southeast-facing sand that is bordered in the west by the linear reef, and in the west fronted by a 200 m long section of small rock reef, leaving a 100 m long open water section in between. It is backed by a series of well vegetated foredune ridges. There is a car park behind the reef that separates the two beaches while the national park ranger station and Four Mile Beach camp site are located behind the beach.

FITZGERALD RIVER NATIONAL PARK

Area: 329 589 ha
Coast length: 79 km (1083-1159 km;1226-1229 km)
Beaches: 42 (WA 257-297; part WA 316)

Fitzgerald River National Park is covers a 79 km long section of coast and the backing rugged terrain extending up to 50 km inland, with a total area of 3,300 km^2. The park marks the beginning of a 40 km long section of steep rugged coast dominated by resilient Proterozoic metasedimentary rocks. The park is renown for its rugged terrain, wild flowers (in spring), whale watching (in August-November), bush walks and spectacular coastal scenery, especially in the east. The park has remained largely undeveloped and vehicle access is limited to the Hammersley Drive in the east and Point Ann access roads in the west, with no through access and a number of tracks only accessible by 4WD. There are campsites in the east at Four Mile Beach, Hammersley Inlet, Whalebone Beach and Quoin Head, the latter two accessible by 4WD only. In the west there is camping at Fitzgerald Inlet, St Mary Inlet and Point Ann, only the latter accessible by car. A 1000 ha section of the park is located behind the eastern end of Dillon Beach 3 km south of Bremer Bay township.
For information:
Ranger, Fitzgerald River National Park
PO Box 44, Jerramungup WA 6337, phone:(098) 355 043
Rangers are also located at East Mount Barren
(098) 383 060 and Murray Road (098) 371 022.

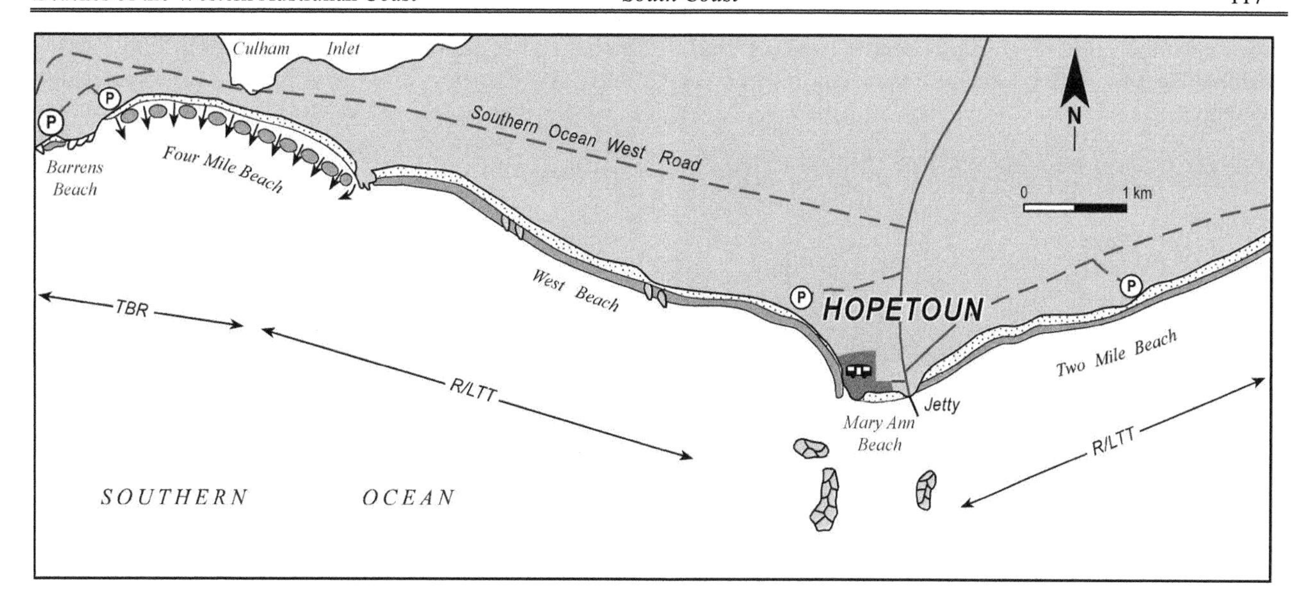

Figure 4.66 Hopetoun and regional beaches.

Figure 4.67 Hopetoun with the jetty. Two Mile Beach is to the right and Mary Ann Haven to the left.

WA 259-260 **BARRENS**

No.	Beach	Rating	Type	Length
WA 259	Barrens Beach	4	LTT	200 m
WA 260	Barrens (W)	3	R+reefs	50 m
Spring & neap tidal range = 0.7 & 0.2 m				

Barrens Beach (WA 259) is a 200 m long south-facing beach wedged in between the base of East Mount Barren in the west and the rocks and reefs that separate it from beach WA 258 in the east. It is partly protected by its orientation and the reefs and receives waves averaging about 1 m, which maintains a 60 m wide continuous shallow bar (Fig. 2.11a). It is usually free of beach rips, through higher waves induce rips against the boundary rocks. It is backed by a 10 m high foredune and a partly stable blowout extending about 200 m inland. There is a small car park at the western end of the beach.

Beach **WA 260** is located 300 m to the south and occupies the mouth of a steep valley descending from the mountain. It is a 50 m long, east-facing pocket of sand backed by a steep vegetated valley and bordered by massive quartzite rocks, with rocks and reefs almost encasing the small beach. Waves are low and the beach usually reflective. It can only be access around the rocks from Barrens Beach.

WA 261-263 **MYLIES-WEST BEACH**

No.	Beach	Rating	Type	Length
WA 261	Mylies Beach	6	TBR	1.8 km
WA 262	Caves Pt	4	R+rocks	20 m
WA 263	West Beach	7	TBR+rocks/reef	300 m
Spring & neap tidal range = 0.7 & 0.2 m				

To the west of Barrens Beach steep metasedimentary terrain rising as high as 300 m dominates the coast for the next 40 km to Dempster Inlet. These rocks are the southern edge of the Yilgarn block, the ancient core of continental crust that underlies much of southern Western Australia. Between East Mount Barren and Edwards Point, the next major headland 7 km to the southwest, are three beaches (WA 261-263).

Mylies Beach (WA 261) is a slightly curving, 1.8 km long, south-facing beach bordered by the southern 80 m high slopes of East Barren in the east and 40 m high Cave Point in the west. The beach receives waves averaging just over 1.5 m which maintain a 100 m wide surf zone usually cut by 2-3 beach rips along the eastern-central section, with wave height decreasing towards the western end (Fig. 4.68). The numerous beach breaks are known as *Mylies*. Two creeks break out across the beach in the centre and towards the west, with a largely stable 10-20 m high 100 m wide foredune backing the beach. There is vehicle access to either end of the beach.

Caves Point extends 2 km south of the western end of Mylies Beach. Midway along the steeply sloping 50-60 m high point is a 20 m long wedge of sand (**WA 262**) at the base of a narrow winding gully. The beach faces east and is fronted by 100 m wide rocks and reefs. A vehicle track runs along the crest of the point with a walking track down to the beach. It is suitable for fishing but not for swimming.

Figure 4.68 Mylies Beach offers the best surf in the Hopetoun region.

West Beach (WA 263) is bounded by Cave Point in the east and the beginning of Edwards Point to the west. The beach is 300 m long, faces south and receives waves averaging about 1.5 m. These break across an 80 m wide surf zone and maintain permanent rips against the rocks at either end of the beach, as well as breaking across rocks in the centre of the beach. The beach is backed by a foredune extending 100 m inland rising up over the backing bedrock. There is a small car park on the rise above the centre of the beach containing a bar-b-que and shelter. The beach is not recommended for swimming owing to the strong rips and rocks, but is used by surfers.

Surf: *Mylies* offers the best beach breaks in the region. It picks up most swell and usually has something to offer. *West Beach* does have surf, but beware of the numerous rocks.

WA 264-265 **HAMMERSELY INLET**

No.	Beach	Rating	Type Inner outer bar	Length
WA 264	Hammersely Inlet	8	TBR RBB	2.2 km
WA 265	Hammersley Inlet (W)	8	TBR+reef	600 m
Spring & neap tidal range = 0.7 & 0.2 m				

Hammersely Inlet is an elongate 2 km long inlet occupying a drowned valley feed by the Hammersely River. The inlet is located 3 km west of Edwards Point and is usually blocked by Hammersely Inlet beach (WA 264). The inlet and beach are accessible by car with a camping area located besides the inlet.

Beach **WA 264** is a curving, south-facing, 2.2 km long, exposed beach located between Edwards Point and rocks on the western side of the inlet (Fig. 4.69). The beach receives waves averaging 2 m which break across a 200 m wide surf zone containing an inner bar cut by 4-5 rips and an outer bar with 3-4 rips, as well as a permanent rip flowing out against the eastern headland. Partly active transgressive dunes have climbed up over the backing bedrock extending up to 3 km inland and to elevations of 70 m. Today there are more than 300 ha of bare active sand dunes.

Figure 4.69 The exposed high energy beach WA 264 curves round to the usually blocked entrance to Hammersley Inlet.

Beach **WA 265** occupies the lee of the rocks and reefs to the west of the inlet. The 600 m long south-facing beach is exposed to high waves, which break across the reefs and inshore rocks. The water returns seaward as a strong permanent rip flowing through a gap in the reefs. It is backed by a dune-draped 40 m high bedrock scarp, with a 4WD track running along the crest.

These are two accessible but very hazardous beaches for swimming. Use extreme care if entering the water and stay well clear of the strong rips.

WA 266-273 **WHALEBONE BEACH**

No.	Beach	Rating	Type	Length
WA 266	Whalebone (E)	8	RBB+reef	1 km
WA 267	Whalebone	8	RBB+reef	1.7 km
WA 268	Whalebone (W1)	8	RBB+rocks	150 m
WA 269	Whalebone (W2)	8	RBB+rocks	60 m
WA 270	Whalebone (W3)	8	RBB+reef	50 m
WA 271	Whalebone (W4)	8	RBB+rocks	150m
WA 272	Whalebone (W5)	8	RBB+rocks	250 m
WA 273	Whalebone (W6)	8	RBB+rocks	200 m
Spring & neap tidal range =0.7 & 0.2 m				

Whalebone Beach is part of a series of high energy beaches that extend for 5 km west of Hammersley Inlet. All face south into the high southerly waves and are dominated by a 200 m wide energetic, rip-dominated surf zone, together with numerous outcrops of rocks and reef, resulting in a very hazardous section of coast. There is 4WD access to the top of the 30-50 m high bluffs that runs along behind the beaches, with steep tracks down to the beaches.

Beach **WA 266** is located immediately east of the main beach. It is a 1 km long high energy beach dominated by four permanent rips that flow out between the rocks and reefs scattered along the beach and in the surf (Fig. 4.7-). It is backed by a 40 m high dune-draped bedrock escarpment, then low heath land. It is separated from the main Whalebone Beach (**WA 267**) by 500 m of jagged quartzite. This beach is 1.7 km long, faces south. The eastern half is dominated by two permanent boundary rips and a large central beach rip (Fig. 3.2), while rocks and reef increasingly dominate the western half and maintain three permanent rips. In addition rocks are scatted along either end of the beach. A 20-40 m high vegetated bluff backs the beach and increasingly protrudes onto the western section of the beach.

Figure 4.70 Whalebone Beach is an exposed, high energy beach with a hazardous rock-strewn surf zone.

Beaches **WA 268** and **269** are sections of the main beach truncated by the protruding steep 30 m high bluff. They are 150 and 60 m long respectively, backed by the steep, partly vegetated bluff, with rock debris at their base. They are separated by the steep point, which can be rounded at low tide. Both receive high waves, which break across the continuous 200 m wide surf zone, interrupted by the intervening rocks.

Beaches **WA 270** and **271** lie immediately to the west and consist of two 50 and 150 m long pockets of sand at the base of steep bluffs with a rocky point separating the two. Both are backed by a small foredune, then the bluffs. The continuous energetic surf zone extends 200 m seaward of the shore and a permanent rip drains beach WA 271.

Beach **WA 272** lies 200 m to the west at the mouth of an elongate drowned valley. The 250 m long beach is skewed along the steep narrow embayment, with a small creek draining out at its eastern end. Waves break across a 300 m wide surf zone with a strong rip running along the base of the western rock boundary. A 4WD track reaches the end of the headland between the two beaches.

Beach **WA 273** lies 300 m to the southwest in the next small embayment. It is a narrow curving 200 m long cobble beach, backed by steep partly vegetated slopes, and bordered by steeply rising quartzite headlands, with rocks and reef doming the surf zone. A single strong rip drains out the northern side of the bay.

All eight Whalebone beaches are exposed to high waves and dominated by strong usually permanent rips, with rock and reefs strewn along the beaches and in the surf. They are extremely hazardous and not recommend for swimming.

WA 274-275 QUOIN HEAD

No.	Beach	Rating	Type	Length
WA 274	Quoin Head	6	TBR	300 m
WA 275	Quoin Hd (W1)	8	RBB+reef	200 m
WA 276	Quoin Hd (W2)	5	LTT+rocks	100 m
Spring & neap tidal range =0.7 & 0.2 m				

To the west of beach WA 273 is a 17 km long section of rugged rocky coast composed of steeply dipping metasedimentary rocks which rise steeply to over 100 m. Rocky shoreline dominates with only a few small pockets of sand, cobbles and boulders located at the mouth of a few of the steep creeks. The first three beaches (WA 274-276) are located either side of Quoin Head.

Quoin Head is a steep 50 m high metasedimentary headland located 4 km southwest of Whalebone Beach. It is the westernmost accessible beach in this part of the park, with a 4WD track winding down the steep gradient to the rear of the beach where a camping site is located next to the small dammed lagoon.

Quoin Head beach (**WA 274**) is a 300 m long pocket beach and foredune that partly dams an elongate V-shaped valley, forming a narrow 200 m long lagoon (Fig. 4.71), which occasionally breaks out across the northern end of the beach. The beach is 300 m long and bordered by jagged rock platforms and backing steep 50 m high headlands. Its southeast orientation and the headland afford moderate protection resulting in waves averaging less than 1.5 m. These break across a 50 m wide surf zone, which is drained by a permanent rip against the northern rocks.

Figure 4.71 The small headland-bound Quoin Head beach.

Beach **WA 275** lies on the southern side of Quoin Head 200 m to the south. It is a more exposed south-facing 200 m long beach. It is wedged in between steep jagged rock slopes, with linear rock outcrops crossing the beach. Waves average up to 2 m and break across a 150 m wide surf zone with a strong permanent rip running along the western rocks. It is backed by a clump of dunes that have climbed up onto the saddle that separates it from the main beach. It is only accessible on foot via the saddle.

Beach **WA 276** lies 600 m to the south and is a 100 m long sand beach located at the base of steep vegetated slopes, with a small creek crossing the southern end of the beach. It is bordered by rugged rocky slopes, which narrow the embayment to 50 m. Waves are reduced to about 1 m and break over the rocks and across a low tide terrace. A rock-controlled rip drains the small embayment.

WA 277-279 MARSHES BEACH

No.	Beach	Rating	Type	Length
WA 277	Marshes Beach	7	Boulder	80 m
WA 278	WA 278	8	Boulder+rock flats	150 m
WA 279	WA 279	9	R boulder+TBR	200 m
Spring & neap tidal range = 0/7 & 0.2 m				

Marshes Beach (**WA 277**) is located 4 km southwest of Quoin Head and is a curving 80 m long boulder beach located at the based of a steep V-shaped valley, with the valley walls extending seaward as 50 m high cliffs, composed of steeply dipping metasedimentary rocks.

Beach **WA 278** is located 3 km to the west and consists of a 150 m long high tide boulder beach, fronted by a series of irregular shore parallel rock reefs, which extend up to 100 m off the beach. The reefs form a quieter lagoon at the beach, with difficult access to the ocean through a small gap at the eastern end of the reefs. A 4WD track winds down the backing slopes to reach the rear of the beach.

Beach **WA 279** lies another 1.5 km to the west and is a south-facing 200 m long beach composed of a boulder high tide beach, fronted by a low gradient sandy beach and 150 m wide energetic surf zone. It is wedged in between a steep eastern headland and a lower rocky point, with rips running out along each boundary. It is backed by a vegetated foredune that has partly filled the entrance to the small valley, with a creek bed running along the western side of the valley to the beach.

WA 280-285 TWIN BAYS

No.	Beach	Rating	Type	Length
WA 280	Twin Bays (E1)	8	RBB+rocks	200 m
WA 281	Twin Bays (E2)	8	RBB+rocks	500 m
WA 282	Twin Bays (E3)	8	RBB+rocks	150 m
WA 283	Twin Bays (E4)	8	RBB+rocks	150 m
WA 284	Twin Bays (W)	8	RBB+rocks	800 m
WA 285	Twin Bays (W1)	8	R-boulders	150 m
Spring & neap tidal range = 0.7 & 0.2 m				

Twin Bays is on open southeast-facing bay located 2 km east of 440 m high Mid Mount Barren. It occupies the mouth of a steep valley bordered by the mount to the west and 500 m high Thumb Peak, 4 km to the east. A vehicle track runs down the heath-covered valley to reach the rear of the eastern bay beaches.

The eastern side of Twin Bays consists of four near continuous beaches (WA 280-283) each separated and partly dominated by fingers of protruding metasedimentary rocks. The beaches are linked by a continuous 150 m wide, high energy, rip-dominated surf zone. Beach **WA 280** is a 200 m long beach bordered by continuous steep rocks to the east and a 100 m wide finger point to the west. A permanent rip runs out against the eastern rocks, while a steep, vegetated 30 m high bluff backs the beach. Beach **WA 281** lies immediately to the west and is a 500 m long continuation, with some rocks also outcropping along the beach, and a large low outcrop forming the western boundary. It has permanent rips to either end and is backed by both vegetated and deflated climbing dunes, the deflated dunes exposing sloping bedrock. The vehicle track reaches the rear of this beach.

Beach WA 282 and 283 lie amongst the rocks that dominate the western side of this half of the twin bay. Beach **WA 282** is a 150 m long pocket of sand bordered by low irregular rocky points, with numerous rocks also outcropping along the beach. Beach **WA 283** is a similar 150 m long beach bordered in the west by a sloping 30 m high bedrock point that divides the twin bay in two. Both beaches are fronted by the continuous surf zone and backed by largely vegetated dunes that have climbed up onto the backing bedrock.

Beach **WA 284** occupies the western half of Twin Bays. It is an 800 m long southeast-facing high energy beach, dominated by a 150 m wide surf zone containing two permanent rips against the boundary headlands and 2-3 beach rips. The beach is backed by a 200-300 m wide bedrock deflation surface, fringed by boundary dunes, then the heath land of the backing valley, while the western 60 m high headland rises to Mid Mount Barren 1.5 km to the north (Fig. 4.72). A vehicle track winds down the backing slopes to reach the southern end of the beach.

Figure 4.72 The western Twin Bays beach at the base of Mid Mount Barren, receives usually high waves and is backed by an active dune field.

Beach **WA 285** is located in the central section of the 3 km of steep rocky coast that separates Twin Bay from Dempster Inlet (Beach WA 286). The beach is a 150 m long high tide boulder beach located at the mouth of two small valleys. It is fronted by an irregular 50 m wide rock platform, then by a deeper 100 m wide sandy surf zone.

WA 286 DEMPSTER-FITZGERALD INLETS

No.	Beach	Rating	Type	Length
WA 286	Dempster-Fitzgerald Inlets	7→4	RBB→TBR	6 km
Spring & neap tidal range = 0.7 & 0.2 m				

Dempster and Fitzgerald inlets are two elongate shallow coastal lagoon that drain Copper Mine Creek and the Fitzgerald River respectively and reach the sea that either end of beach WA 286. Both inlets are usually blocked only breaking out following heavy winter rains.

Beach **WA 286** is a 6 km long southeast-facing beach, that curves to the east in lee of the westerly Point Charles and Dempster Inlet (Fig. 4.73). The beach is well exposed along its central-eastern section receiving waves averaging up to 2 m which break across a 200 m wide rip-dominated surf zone, with rips spaced every 300-400 m. Wave height gradually decreases along the curving western section of the beach as the bar narrows and ultimately becomes a continuous low tide terrace to the west of the Fitzgerald Inlet mouth. Besides the two inlets the beach is backed by active massive dune transgression in the centre that has climbed up over the backing 50-60 m high sloping bedrock and extend up to 1 km inland. Much of the dune area has been deflated down to the bedrock. Blowouts dominate the western section together with a 1 km long dune sheet emanating from the unstable mouth of the Fitzgerald Inlet. There is a 4WD track to the eastern side of Fitzgerald Inlet where a basic camping area is located.

Figure 4.73 Beach WA 286 extends west of Point Charles (right) and the blocked Dempster Inlet, as a high energy rip-dominated beach.

WA 287-292 POINT CHARLES BAY

No.	Beach	Rating	Type	Length
WA 287	Pt Charles Bay	7	RBB→TBR	4.7 km
WA 288	St Mary River (W)	6	TBR	1.8 km
WA 289	St Mary River (E)	6	TBR	400 m
WA 290	Pt Ann	4	LTT	250 m
WA 291	Pt Ann (1)	5	LTT+rocks	80 m
WA 292	Pt Ann (2)	5	LTT+rocks	100 m
Spring & neap tidal range = 0.7 & 0.2 m				

Point Charles Bay is an 8 km long, open southeast-facing bay located between Point Charles and Point Ann. A dammed salt lake backs the centre of the beach, while the small Saint Mary River breaks out towards the southern end during winter rains. There is a gravel road out to Point Ann, the first 2WD vehicle access to the coast west of Hammersley Inlet, 40 km to the east.

Beach **WA 287** is the main bay beach. It commences 1.5 km west of 60 m high Point Charles and curves gently to the southwest for 4.7 km, to a small outcrop of inter and subtidal rocks. Waves average over 1.5 m and maintain a 150 m wide surf zone cut by rips every 400-500 m. It is backed by active climbing parabolic dunes rising over the backing bedrock to 60 m in the centre-east. They extend up to 500 m inland with older vegetated dunes located up to 1 km inland. The dunes increase in stability and decrease in size to the west as the shoreline swings more to the south. There is vehicle access to the slopes above the boundary rocks.

Beach **WA 288** commences on the southern side of the rocks and continues south for 1.8 km to a low rocky outcrop on the southern side of the inlet mouth. The beach faces east-southeast and waves decrease from 1.5 m in the north to about 1 m at the rocks. These maintain a low gradient 70 m wide rip dominated surf zone, backed by a foredune abutting sloping vegetated bedrock, then the usually blocked inlet mouth. Vehicle tracks from Point Ann reach the southern side of the inlet and overlook the beach.

Beach **WA 289** extends from the rock outcrop for 400 m to the first rocks of Point Ann. It faces east and receives waves average about 1 m which during higher wave conditions produce rips across the 50 m wide surf zone. Beach **WA 290** is the main Point Ann beach (Fig. 4.74). It is a 250 m long pocket of sand located between the small 15 m high northern point and the more continuous rocks that extend for 500 m east to Point Ann. The beach receives waves averaging less than 1 m, which usually break across a narrow low tide terrace. The Point Ann road terminates here and runs down the slopes to the beach, which is used to launch small boats. In addition there is a camping area with an amenities block and a bar-b-que.

Figure 4.74 The sheltered Point Ann beach in foreground, with the blocked mouth of St Mary River (top).

Figure 4.75 Beach WA 293 is located along the base of the Cheadanup Cliffs.

Point Ann is composed of folded red schists, conglomerates and metasedimentary rocks that rise to 30 m in height at the point. Between beach WA 290 and the point are two narrow strips of sand located at the base of the sloping rocks. Beach **WA 291** is located immediately south of the main beach and is an 80 m long strip of sand backed by steep 20 m high vegetated slopes and fronted by inter to subtidal rocks. Beach **WA 292** lies 50 m to the south and is a similar 100 m low energy long strip of sand and rocks. The beaches are only accessible on foot from the main beach.

WA 293-296 CHEADANUP CLIFFS

No.	Beach	Rating	Type	Length
WA 293	Cheadanup Cliffs (1)	7	TBR+rocks	300 m
WA 294	Cheadanup Cliffs (2)	6	R+rocks/reef	150 m
WA 295	Cheadanup Cliffs (3)	6	TBR	300 m
WA 296	Cheadanup Cliffs (4)	7	TBR+rocks	300 m
Spring & neap tidal range = 0.7 & 0.2 m				

The **Cheadanup Cliffs** are steep rocky section of shore extending for 4 km southeast of Point Ann. The cliffs rise to 50-60 m and become increasing dissected to the south. The Point Ann road runs a few hundred metres inland with a few vehicle tracks leading to the top of the cliffs. Beaches WA 293-296 are located along the central section of the cliffs with vehicle tracks leading to the top of beaches WA 293 and 295.

Beach **WA 293** is a 300 m long pocket of sand wedged in at the base of steep 50 m high vegetated slopes. It receives waves averaging about 1.5 m which breaks across a mixture of sand and rocks, with a single rip draining the surf during higher waves (Fig. 4.75). Beach **WA 294** lies 200 m to the south and is a protruding rock-bound beach, with steep slopes behind and an irregular rock platform fronting much of the beach and only a 20 m wide open water section at the northern end.

Beach **WA 295** commences at the southern end of the 300 m long rock platform and is a 300 m long beach located at the mouth of a V-shaped valley, impounding a small elongate lagoon. The beach is bounded by the rock platform to the north and a low rock outcrop to the south, with a 50 m wide surf zone. Beach **WA 296** lies on the southern side of the rock outcrop and extends south for 300 m terminating against a steep 30 m high vegetated headland, with a rock platform at its base, while rock reefs lie off the southern half of the beach. Climbing foredunes have developed against the base of the bluffs behind beaches WA 294, 295 and 296.

WA 297-298 TRIGELOW-TOOREGULLUP

No.	Beach	Rating	Type	Length
WA 297	Trigelow Beach	6	TBR	11.4 km
WA 298	Tooregullup Beach	6→5	TBR→LTT	8 km
Spring & neap tidal range = 0.7 & 0.2				

To the south of Point Ann is an east-facing Doubtful Bay. The open bay extends for 23 km to Point Hood, which together with the Doubtful Islands, projects 11 km to the east, thereby protecting the southern section of the bay from southerly waves. Much of the bay shore is taken up with two long continuous beaches (WA 297 & 298) with five small beaches located around Point Hood.

Trigelow Beach (**WA 297**) commences 4 km south of Point Ann, at the end of the Cheadanup Cliffs, and curves gently to the south-southwest for 11.4 km to the usually blocked mouth of Gordon Inlet. The inlet also marks the southern boundary of the national park. The beach receives waves that are slightly reduced by refraction around Hood Point, to average about 1.5 m. These break across a low gradient white sand beach and 100 m wide surf zone, usually cut by rips every 400 m. The beach is backed by a 200-300 m wide field of generally vegetated foredune-blowouts. The small Boondadup River breaks out 2 km down the beach, while the larger Gordon Inlet forms the southern boundary. There is vehicle access to the north end of the beach from the Point Ann road.

Tooregullup Beach (**WA 298**) commences at Gordon Inlet and continue almost due south for 8 km to the northern rocks of Whalebone Point, with wave height gradually decreasing down the beach. Rips commonly occur along the northern section grading into a more continuous low tide terrace along the southern section. The beach is backed by a few blowouts in the north extending up to 200 m inland, grading to a more stable narrower foredune to the south. Kellys Creek flows out across the beach 2 km south of the inlet, with vehicle access to the creek mouth and to the southern end of the beach. Several beachfront shacks are located on the slopes behind the southern 1 km of the beach.

WA 299-303 POINT HOOD

No.	Beach	Rating	Type	Length
WA 299	House Beach	2	R	150 m
WA 300	Corner Cove	2	R	250 m
WA 301	Whalebone Pt	4	R/LTT	2.4 km
WA 302	Pt Hood (1)	3	R	80 m
WA 303	Pt Hood (2)	3	R	80 m
Spring & neap tidal range =0.7 & 0.2 m				

Whalebone Point and Point Hood are part of a 12 km long band of folded Proterozoic gneiss that extends east to the Doubtful Islands. The bedrock forms a ridge that rises to 88 m behind Whalebone Point and up to 100 m along the southern ridge.

House Beach and Corner Cove are located in a 700 m long northeast-facing bay immediately west of Whalebone Point. **House Beach** (**WA 299**) is a 150 m long pocket of sand wedged in at the base of steeply descending bedrock slopes and bordered by steeply sloping headlands (Fig. 4.76). The only safe access is via boat. The beach receives low waves and has a steep reflective shore. Dense seagrass meadows lie 200 m offshore.

Figure 4.76 House Beach is a classic pocket beach surrounded by steep granite slopes and fringed by seagrass meadows.

Corner Cove (**WA 300**) is located 500 m to the southeast at the other end of the bay. It is a very protected 250 m long, northwest-facing beach, bordered by sloping rocky point and backed by vegetated bedrock slopes. Waves are often calm and seagrass grows almost to the shore. There is a vehicle track to the beach, with a house behind the centre, and a clearing for boats at the western end. The beach is used to launch small fishing boats.

Beach **WA 301** curves towards the southeast between Whalebone Point and Point Hood. The beach is 2.4 km long with low vegetated rocky slopes to either end. It faces northeast and receives low refracted waves around Point Hood. These maintain a continuous low tide bar, with seagrass lying 300 m offshore (Fig. 1.19a). The beach is backed by a low 200-300 m wide barrier, with a circular lake behind the centre of the beach. Vehicle tracks run along the barrier to a few fishing shacks located behind the centre and southern end of the beach.

Beaches WA 302 and 303 are two small south-facing pockets of sand located on the southern side of Point Hood, directly south of Whalebone Point beach. Beach **WA 302** is an 80 m long pocket of sand bordered by steeply rising vegetated slopes, with a small grassy foredune at the rear. Beach **WA 303** lies in the next small bay 300 m to the west and is a similar 80 m long beach which is used for launching small fishing boats. A vehicle track leads down the steep valley to the beach and several small shacks are located on the slopes behind the beach. Both beaches are fronted by a pocket of clear sand then seagrass growing 100 m offshore.

WA 304-310 BREMER BAY

No.	Beach	Rating	Type	Length
WA 304	Peppermint Beach	4→6	LTT→TBR	2.5 km
WA 305	Bremer Beach	4→6→4	LTT→RBB→LTT	6.3 km
WA 306	John Cove	3	LTT	500 m
WA 307	Back Beach	4	R	3 km
WA 308	Fishery Beach	2	R	250 m
WA 309	Short Beach	4	R+reef	500 m
Spring & neap tidal range = 0.7 & 0.2 m				

Bremer Bay is an 8 km wide southeast-facing bay that curves up to 6 km to the north, with a total shoreline of 25 km (Fig. 4.77). It is bordered by 80 m high Point Hood to the east and 60 m high Black Point to the west. The curving shoreline is interrupted by several prominent headlands that separate the bay into six beaches (WA 304-309), which total 13 km in length. The small town of Bremer Bay with a population of about 200 is located on the shores of Wellstead Inlet, which is feed by the Bremer River. The estuary breaks out across John Cove in the northwestern corner of the bay. The only development on the bay shore is the golf course behind Back Beach and a small marina at Fishery Beach.

Peppermint Beach (WA 304) is located on the eastern side of the bay and faces west-southwest across the bay. The curving 2.5 km long beach is protected along its southern section by the protruding rear of Point Hood. Waves average about 1 m in the southern corner, where they break across a continuous low tide terrace, with wave height increasing up the beach to average about 1.5 m along the northern half, where a 100 m wide rip-dominated surf zone prevails, with a permanent rip against the northern rocks. The beach is backed by generally vegetated parabolic dunes that have extended up to 1.5 km inland, climbing up over the backing bedrock to heights of 80 m. A large partially active parabolic dune is located behind the northern end of the beach, with a vehicle access immediately in front, as well as access at the southern end of the beach.

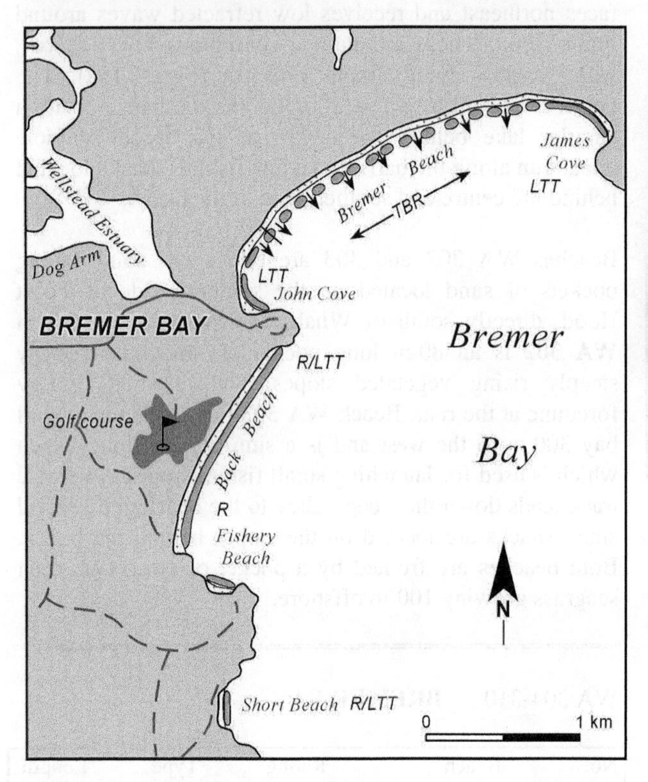

Figure 4.77 Bremer Bay contains a series of partly sheltered beaches.

Bremer Beach (WA 305) is the main bay beach. It is a curving 6.3 km long south-southeast-facing beach, bordered by a protruding 65 m headland in the east and the mouth of Wellstead Inlet and a 20 m high headland in the east (Fig. 4.78). The headlands protect both ends of the beach where lower waves and a low tide terrace prevail, and where access tracks reach the beach. These protected areas are known as James and John cove respectively. The central few kilometers receive higher waves averaging 1.5 m which maintain a 100 m wide rip dominated surf zone, with several rips located along this section. The beach is backed by a 1000 ha transgressive dune field, with an area of inner active dunes and a vegetated deflation surfaces. The dunes have extended up to 3.5 km inland, transgressing over low bedrock and reaching heights of 40 m. The Hunter River is partly blocked by the dunes and forms an elongate lagoon. It then winds across the deflated surface to break out in the centre of the beach.

John Cove beach (**WA 306**) is a 500 m long section of sand that fronts the Wellstead Estuary, terminating against the rocks of the western headland, where low wave to calm conditions usually prevail. The beach is partially washed away following break out of the lagoon and overwashed by high waves. There is road access to the southern corner of the beach, which is also used to launch fishing boats. This is the main boat launching beach and in season fish are netted from the beach. A Surf Life Saving Club was established in Bremer Bay in 1962, patrolling this beach. Unfortunately it only survived a few years.

Figure 4.78 The mouth of Wellstead Inlet and Johnny Cove at Bremer Bay.

Back Beach (WA 307) extends south of the John Cove headland for 3 km to the northern base of 162 m high Tooreburrup Hill. The beach faces east-southeast, looking out the bay mouth. It is protected by its orientation and Black Point, with refracted waves averaging about 1 m. These combine with the medium to coarse sand to maintain a steep reflective beach, with a surging break and high tide cusps. A 1 km long section of shore parallel beachrock occupies the centre of the beach. A vegetated foredune parallels the beach, with the Bremer Bay golf course located behind the centre of the beach. There is vehicle access to either end.

Fishery Beach (WA 308) is a small beach located on the headland at the southern end of Back Beach. The 250 m long beach is bordered by sloping rocky points and faces north resulting in usually low wave to calm conditions that maintain a low energy reflective beach with seagrass growing close to shore. An attached 150 m long curving breakwater was constructed in the mid-1990's to further protect the beach (Fig. 4.80). The breakwater includes a boat ramp and a jetty together with several boat moorings between the breakwater and beach.

Short Beach (WA 309) lies 1 km to the south and is a 500 m long east-facing beach bordered and backed by steeply rising vegetated slopes and headlands. A rock reef lies about 50 m off shore and parallels most of the beach, resulting in low waves at shore and a steep reflective beach. A gravel rocks descends the southern slopes to a beachfront car park at the end of the beach.

Surfing: The best surfing is along *Bremer Beach* where there is several kilometres of beach breaks over the wider bar.

Region 2B SOUTH COAST: Point Henry – Cape Leeuwin

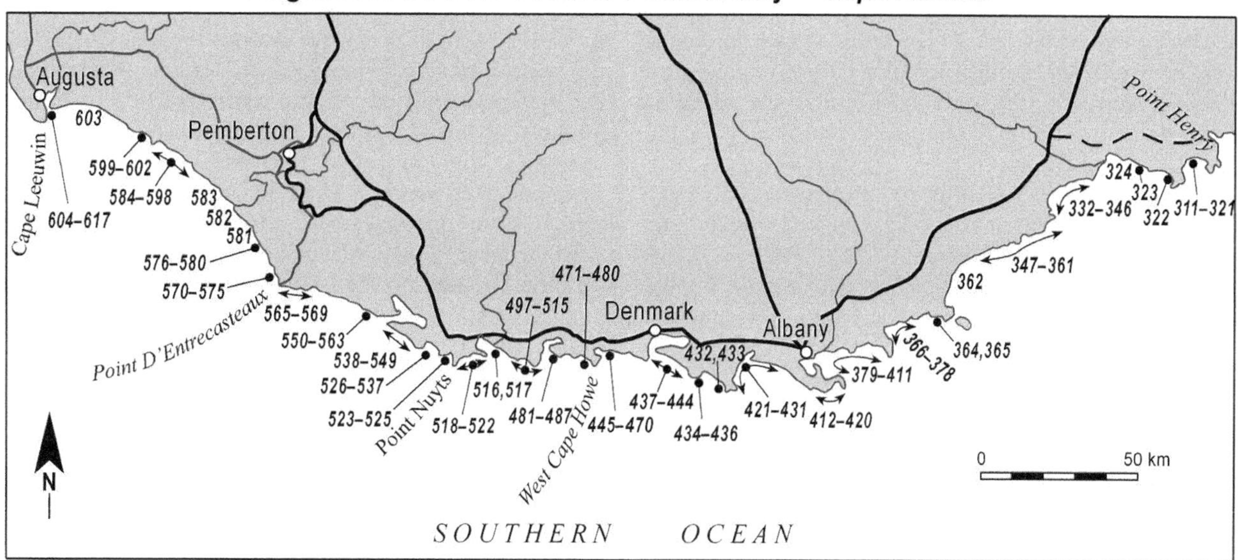

Figure 4.79 South Coast regional map 2B, between Point Henry and Cape Leeuwin, beaches WA 311-617.

Figure 4.80 Fishery Bay with its breakwater and boating facilities.

WA 310-311 **BLACK PT-PT HENRY**

No.	Beach	Rating	Type	Length
WA 310	Banky Beach	8	TBR+rocks	350 m
WA 311	Pt Henry	4	boulder	70 m
Spring & neap tidal range = 0.7 & 0.2 m				

Between Bremer and Dillon bays is 10 km of steep, irregular bedrock coast dominated by Black Point in the east and Point Henry to the south. Most of the shoreline is steep rugged gneiss rising to 60-80 m. Only one pocket of sand and one boulder beach are found along the shore.

Banky Beach (WA 310) lies just outside Bremer Bay and 2 km west of Black Point. The 350 m long beach is wedged in between two 40 m high headlands, with steep 20-30 m high active dune-capped scarps behind. There is vehicle access to the top of the bluff, with a steep climb down to the beach. This is an exposed, southeast-facing beach, which receives waves averaging over 1.5 m which break across a 100 m wide surf zone and a rock reef off the western end. Two permanent boundary rips and a third central rip against the reef dominate the surf and result in hazardous conditions.

Beach **WA 311** is located on the southern side of Point Henry and consists of a 70 m long, pocket boulder beach, located at the base of a steep narrow valley. An equally steep 4WD track descend the side of the valley to a small turning area just above the beach.

WA 312-321 **DILLON BAY**

No.	Beach	Rating	Type (Inner Outer bar)	Length
WA 312	Little Boat Harbour	3	R	100 m
WA 313	Little Boat Harbour (N)	3	LTT	50 m
WA 314	Blossoms Beach	6	TBR	750 m
WA 315	Native Dog Beach	7	RBB	300 m
WA 316	Dillon Beach	7	TBR RBB	6.7 km
WA 317	Dillon Beach (W)	6	TBR→LTT	1.1 km
WA 318	Stream Beach (N)	4	Boulder+LTT	100m
WA 319	Stream Beach	4	R/LTT	450 m
WA 320	Stream Beach (S 1)	4	LTT+rocks	30 m
WA 321	Stream Beach (S 2)	4	boulder+LTT	80 m
Spring & neap tidal range = 0.7 & 0.2 m				

Dillon Bay is a U-shaped south-facing 5 km deep bay bordered by Point Henry to the east and Cape Knob to the west. The bay entrance is 6 km wide and contains 17 km of shoreline, including 10 beaches, which occupy 10 km of the shore, the remainder composed of steeply sloping granite and gneiss. Three kilometres of the shoreline and backing dunes are part of the Fitzgerald River National Park. Apart from some 4WD access tracks and a couple of shacks there is no development in the bay.

Little Boat Harbour (WA 312) is located on the north side of Point Henry. The 100 m long beach faces northwest into the bay and combined with the protection afforded by the point receives low waves, which maintain a steep reflective cusped beach, with seagrass meadows

30 m off the shore (Fig. 4.81). The beach is bordered and backed by steep vegetated slopes, with vehicle access to the eastern end of the beach. The beach is used to launch small fishing boats and there are remains of an old winch system at the rear of the beach, which relates to the name of the beach.

Beach **WA 313** lies 100 m to the north around a jagged rocky point. It is a 50 m long pocket of sand backed and bordered by a small sloping vegetated valley. Rocks border each end as well as a few outcropping on the beach. Waves are slightly higher and usually maintain a narrow low tide terrace, though seagrass meadows are close inshore.

Figure 4.81 Little Boat Harbour beach is a protected beach in lee of Point Henry that is used for launching small boat.

Blossoms Beach (**WA 314**) lies 500 m to the north. This is a more exposed 750 m long, west-facing beach that receives waves that increase up the beach from about 1 m to 1.5 m. The waves maintain a continuous low tide terrace in the south grading to a wider bar with rips in the north, and a permanent rips against the northern rocks. The beach is backed by a mixture of vegetated dunes, areas of dune deflation and a central active parabolic dune, that have all climbed up over the backing bedrock and have blown up to 3 km inland almost to the leeward shore of Bremer Bay. A gravel road runs down to the northern boundary of the beach with a 4 WD track then leading to the centre of the beach.

Native Dog Beach (**WA 315**) is located 200 m to the north. The 300 m long southwest-facing beach is bordered by a 20 m high headland in the south and a rugged rock platform in the north. It shares the same access road as Blossoms Beach, the road terminating on the southern headland with foot access down to the beach. Waves average over 1.5 m and break across a 150 m wide surf zone, which is drained by permanent rips against each headland. Steep vegetated transgressive dunes extend up to 2 km inland and link with the Blossoms Beach dunes.

Dillon Beach (**WA 316**) is the central and main bay beach. It occupies the northern shore of the bay, faces south-southeast and curves round for 6.7 km between an eastern 50 m high headland and a protruding 30 m high bluff that separates it from beach WA 317. The beach receives relatively high waves averaging over 1.5 m for most of its length, the wave decreasing slightly towards the west. The waves break across a 300 m wide double bar systems along the eastern-central section, with rips dominating both bars. Towards the west the surf zone narrows to an 80 m wide single bar, also cut by rips every few hundred metres. Massive active and vegetated transgressive dunes extend northeast from the beach reaching up to 3.5 km inland and to height of 60 m. These decrease in width to the west, where Bitter Water and a second creek drain through the dunes and across the beach. A 4WD access track follows Bitter Water Creek to the shore.

Beach **WA 317** commences on the western side of the 50 m long bluffs and curves to the southeast for 1.1 km, terminating against a 50 m high headland. Wave height decreases slowly down the beach, with a continuous low tide terrace only cut by rips towards the northern end. The entire beach is backed by steep 20-30 m high bluffs, with a vehicle track along the top of the bluffs. It descends the southern end where a solitary shack in located behind the beach.

Beaches WA 318-322 are four pocket beaches located along a 3 km section of steep rocky coast that extends south of beach WA 317. Waves average about 1 m or less along the shore. Beach **WA 318** is a 100 m long high tide boulder beach wedged into a 200 m wide cove. It has some sand immediately off the boulders which at times forming a narrow low tide terrace. **Stream Beach** (**WA 319**) is a 450 m long east-facing sand beach bordered and backed by 50-100 m high steep vegetated slopes, together with some rocks outcropping on the steep reflective beach.

Beach **WA 320** lies 300 m to the south and is a 30 m long sliver of sand wedged in at the base of a steep narrow valley. It forms a low tide terrace with some rocks scattered over the beach. Beach **WA 321** is located another 300 m to the southeast and consists of an 80 m long high tide boulder beach fronted by a narrow sand low tide terrace. A vehicle track leads down to the beach where a single shack is located.

WA 322-324 FOSTER & REEF BEACHES

No.	Beach	Rating	Type	Length
WA 322	Foster Beach	8	TBR	6.5 km
WA 323	Reef Beach	8	TBR	8.4 km
Spring & neap tidal range = 0.7 & 0.2				

To the west of Cape Knob is a 28 km long relatively open south-facing embayment bounded by Groper Bluff in the west. In between are two very exposed eastern beach (WA 322 & 323), then a longer spiraling near continuous Wray Bay beach (WA 324-325) which gradually decreases in energy to the west, together with beaches WA 326-331) located amongst the east-facing rocky shoreline immediately north of the Bluff. The entire shoreline is undeveloped with 4WD access at Reef Beach and in the west at Pallinup and Groper Bluff.

Foster Beach (WA 322) is one of the highest energy beaches on the south coast. The 6.5 km long beach faces directly into the southwest waves, which average over 2 m. These combine with the fine-medium sand to maintain a 400-500 m wide surf zone dominated by 12 large beach rips spaced every 500 m together with boundary headland rips. It also faces directly into the prevailing southwest winds which have blown sand up through valleys and over the backing bedrock slopes, which rise rapidly to 100 m, then blown for up to 7 km inland to heights of 180 m, in places reaching the leeward slopes of Dillon Bay and sliding down onto the east-facing rocks. The backing transgressive dune sheet is only partly vegetated, with two active sheets including extensive deflated areas on the slopes, occupying about 700 ha, and the entire sheet covering 3500 ha. The 4WD access track, called the Minarup 'Road', reaches the western end of the beach where there is a solitary fishing shack.

A 3 km long, 140 m high dune-capped granite headland separates Foster from Reef Beach. **Reef Beach** (WA 323) is a second 8.4 km long exposed high energy beach, which also faces to the southwest. However in addition to the high waves the beach as a discontinuous beachrock reef paralleling much of the shore. The reef forms five major shoreline crenulations, which occupy the central 5 km of the beach (Fig. 4.82). Waves break on the reefs, reform in lagoons between the reef and the shore and break heavily on the steep beach, with strong permanent rips draining the five embayments. More normal beach and surf conditions occur to either side, with four beach rips to the east and three to the west, though some reefs still outcrop in the surf. The entire beach is backed by some of the most active dunes on the southern coast, with no foredune the length of the beach. Rather, sand blows directly from the beach into a dune sheet of transverse dunes that climb the backing 50-100 m high slopes and extend for 5.5 km inland. Earlier vegetated dunes have blown up to 9 km inland and to heights of 220 m, the entire sheet covering 7500 ha. There is 4WD vehicle access to the beach via the eastern Warramurrup 'Road' and in the west via the rough Reef Beach 'Road', which terminates at a small fishing camp and informal camping area. The beach is the base for the Offshore Angling Club.

Figure 4.82 Large beachrock-induced shoreline crenulations along Reef Beach, and backing active dunes.

WA 324-331 WRAY BAY

No.	Beach	Rating	Type	Length
WA 324	Wray Bay (1)	6	TBR	8.7 km
WA 325	Wray Bay (2)	10	R+platform	4.8 km
WA 326	Pallinup Beach	6→5	TBR→LTT	2.2 km
WA 327	Groper Bluff (N5)	5	R+reef	200 m
WA 328	Groper Bluff (N4)	5	boulder+LTT	80 m
WA 329	Groper Bluff (N3)	7	TBR+platform	400m
WA 330	Groper Bluff (N2)	5	boulder	100 m
WA 331	Groper Bluff (N1)	5	boulder	100 m
Spring & neap tidal range =0.7 & 0.2 m				

The western half of **Wray Bay** commences at the western headland of Reef Beach and trends west for 14 km before finally curving round to face east at Beaufort Inlet. Three kilometres of rocky shoreline extends south of the inlet to Groper Bluff the western boundary of the bay. The entire bay is essentially undeveloped apart from a few 4WD access tracks and a few shacks at the southern end of Pallinup Beach.

Beach **WA 324** commences 1 km west of Reef Beach and curves gently to the east for 8.7 km to the beginning of an exposed bedrock section. The beach is free of bedrock for the most part, with waves averaging over 1.5 m in the east by decreasing slowly to the west. The waves usually maintain a continuous 100 m wide bar cut by rips, particularly to the east. The beach is backed by a vegetated dune sheet that has extended up to 8 km to the northeast, with one active blowout located 4 km along the beach and extending 500 m inland. The dunes blanket both Pleistocene calcarenite as well as bedrock. The Reef Beach 'Road' crosses the backing dunes, with a track leading also to the eastern end of the beach.

Beach **WA 325** commences as a continuous 50 m wide strip of beachrock that forms the shoreline for 4.8 km. The beachrock has an active intertidal platform, as well as a 1 m higher inactive platform, then sand forming a high tide beach, as well as lying seaward of the rock. The beach faces south-southeast, with waves averaging less than 1.5 m, breaking heavily on the rocks and only reaching the beach at high tide. The beach is backed by a well vegetated foredune, then vegetated dune transgression extending 1-2 km inland. There is an access track over the 30 m high dunes to the centre of the beach, which is only suitable for rock fishing.

Pallinup Beach (**WA 326**) is the western extension of beaches WA 324 and 325, commencing as the beachrock creases. The sandy 2.2 km long beach curves increasingly to the south and blocks the mouth of **Beaufort Inlet**, a 400 ha lagoon that is feed by the Pallinup River. The beach terminates at the base of scarped vegetated 20 m high bluffs, adjacent to the northern rocks of Groper Bluff. Waves average 1 m in the north decreasing to the south, as a 50 m wide bar cut by occasional rips gives way to a continuous low tide terrace. The beach is backed by the low overwashed inlet mouth, then the bluffs, with vehicle access to either side of the mouth, and several shacks located towards the southern end of the beach.

Groper Bluff is a 50 m high 1 km long south-facing gneiss headland, with 4 km of generally east-facing steep rocky shore lying between the bluff and the southern end of Pallinup Beach. Five small rock controlled beaches (WA 327-331) are located along this section, with vehicle access to two of the beaches.

Beach **WA 327** is a 200 m long reflective sand beach wedged in between east-west tending fingers of gneiss, with some rocks outcropping on and off the beach. Beach **WA 328** lies 50 m to the south and is an 80 m long pocket beach consisting of a high tide boulder beach and a mid to low tide sand terrace, with rock bordering either side and some on the beach. Both beaches are backed by steep vegetated bedrock, with an access track reaching the rear of beach WA 328.

Beaches WA 329-331 occupy a 1.2 km wide east-facing bay located immediately north of Groper Bluff. Beach **WA 329** is a 400 m long east-facing sand beach, with an open northern half dominated by a 50 m wide bar and permanent rip, while beachrock forms the shoreline along the southern half. A vehicle track winds down the ridge of the steep backing slopes to reach the centre of the beach. Beach **WA 330** lies 400 m to the south and is a curving 100 m long narrow, pocket boulder beach backed by steep partly vegetated slopes. Beach **WA 331** lies 150m to the south and is a similar 100 m long boulder beach grading into an irregular rock platform at its southern end, with some seagrass growing in the protected lee of Groper Bluff, which forms the southern headland.

WA 332-339 **GROPER BLUFF-BOAT HARBOUR**

No.	Beach	Rating	Type	Length
WA 332	Groper Bluff (W1)	6	TBR	2.1 km
WA 333	Groper Bluff (W2)	9	R+platform	3.2 km
WA 334	Groper Bluff (W3)	6	TBR	400 m
WA 335	Groper Bluff (W4)	9	R+platform	700 m
WA 336	Boat Harbour (E3)	6	TBR	900 m
WA 337	Boat Harbour (E2)	7	TBR+reef	150 m
WA 338	Boat Harbour (E1)	7	TBR+reef	200 m
WA 339	Boat Harbour	4	R/LTT	1 km

Spring & neap tidal range = 0.6 & 0.2 m

At Groper Bluff the shoreline turns and trends due west for 7 km before curving round to the south for 3 km, along a rocky section of coast that terminates at Boat Harbour. This is a moderately exposed shoreline containing eight beaches (WA 332-339) of varying length, each bordered by beachrock and bedrock reefs. There is 4WD access to the coast at beach WA 334 and Boat Harbour (WA 339).

Beach **WA 332** is a straight, 2.1 km long south-facing sandy beach that commences against the western end of Groper Bluffs and extends west for 2.1 km, terminating at the start of the beachrock reef. The beach receives waves averaging about 1.5 m which maintain a 50 m wide bar usually cut by 3-4 beach rips, as well as a permanent rip against the bluffs. It is backed by steeply rising, 50 m high vegetated calcarenite bluffs, then a few hundred metes of vegetated transgressive dunes, which overlie bedrock. Beach **WA 333** commences at the start of the beachrock and continues west for 3.2 km as a sandy high tide beach fronted by the near continuous beachrock shoreline, with a 50 m wide sandy surf zone beyond (Fig. 4.83). There are a few gaps in the reef through which permanent rips run seaward. Vegetated calcarenite bluffs continue along the rear of the beach and are scarped in lee of the gaps in the reef.

Figure 4.83 Beach WA 333 west of Groper Bluff is dominated by the beachrock reef, with a strong permanent rip flowing out through the gap in the reef.

Beach **WA 334** occupies a 400 m long break in the beachrock at the mouth of the largest creek to reach the beach. A 10 ha lagoon lies 500 m inland with the creek only breaking out following heavy rain. Beachrock and vegetated bluffs fringe both ends of the beach, which is backed by a low foredune with the creek crossing at the eastern end of the beach. Waves average about 1.5 m and break across a 50 m wide bar. A vehicle track reaches the western end of the beach. Beach **WA 335** extends for another 700 m to the west in lee of the remaining beachrock. It terminates in lee of a low rounded granite outcrop, with a 50 m long section of open beach between the granite and beachrock. Sixty metre high vegetated calcarenite bluffs back the beach.

Beach **WA 336** commences on the western side of the 500 m long granite point and continues west for 900 m, to a low rocky point. Rocks and reefs are also located off either end of the beach. It receives waves averaging about 1.5 m, which maintain a 50 m wide rip-dominated surf zone containing two central beach rips and permanent rips to either end. It is backed by steep 70 m high, vegetated calcarenite bluffs.

Beaches WA 337 and 338 are located between two low gneiss points, with a central attached reef separating the two beaches. Beach **WA 337** is 150 m long with a 100 m long central sand section dominated by a 50 m wide surf zone and permanent rip flowing out against the western rocks. Beach **WA 338** is a similar 200 m long beach, with a rip against its eastern rocks. Both are backed by steep 70 m high, vegetated bluffs.

Boat Harbour beach (**WA 339**) is a curving 1 km long southeast to east-facing beach, protected by its orientation and protruding gneiss headlands, including the southern Black Point. Waves are lowered to about 1 m decreasing toward the southern end of the beach. The beach in turn grades from a low tide terrace to reflective in the southern corner (Fig. 4.84). A creek breaks out across the northern end of the beach, with an informal camping area on the shore of the small backing lagoon. There is 4WD access to the beach with the track running down to the southern end of the beach where a few fishing shacks are located between the bluffs and beach. This more protected end of the beach is used to launch small fishing boats.

Figure 4.84 View south along Boat Harbour beach.

WA 340-346 **LONG BEACH-CHEYNE INLET**

No.	Beach	Rating	Type	Length
WA 340	Long Beach (1)	6	TBR	3.3 km
WA 341	Long Beach (2)	4	TBR+rocks	600 m
WA 342	Long Beach (3)	4→3	LTT→R	5 km
WA 343	Long Beach (4)	4	TBR	80 m
WA 344	Schooner Beach	4	LTT/R	1.7 km
WA 345	Schooner (S)	3	R+rocks	200 m
WA 346	Cheyne Inlet	4	LTT/R	500 m
Spring & neap tidal range = 0.6 & 0.2 m				

Cheyne Bay lies to the west of Black Point with the bay shoreline trending to the southwest then gradually curving round to the south and finally southeast in lee of Cape Riche. The 12 km of generally east-facing bay shoreline contains seven beaches (WA 340-346), which receive increasing protection to the south. There is a gravel road to the southern beaches and 4WD access to the remainder.

Long Beach (**WA 340**) commences at 70 m high Black Point and trends to the southwest for 3.3 km to a slightly protruding 20 m high section of calcarenite bluffs. The beach is exposed to waves averaging about 1.5 m at Black Point, which slowly decrease to the west. The waves maintain a 50 m wide surf zone, with a combination of beach rips and occasional sections of intertidal beachrock inducing permanent rips. The beach is backed by a generally stable foredune that onlaps older scarped Pleistocene dunes that extend up to 200 m inland and in turn overlie bedrock. There is 4WD access to Black Point and along the crest of the bluffs backing the eastern half of the beach.

Beach **WA 341** occupies a narrow 600 m long strip of sand that runs along the base of the bluffs extending west of Long Beach to the mouth of a dammed creek. Steep 20-30 m high, dune-capped bluffs back the beach, together with some protruding sections, as well as rock debris on the beach and in the 40 m wide surf zone. A continuous bar fronts the beach and is cut by rips during higher waves.

Beach **WA 342** commences at the creek mouth and curves for 5 km to the south-southwest, then south to the mouth of Black Sand Creek. Waves height continues to decrease down the beach averaging less than 1 m in the south. The waves maintain a continuous narrow bar in the north grading into a cusped reflective beach in the south. The beach is backed by both scarped and vegetated 20-30 m high calcarenite bluffs, with considerable bluff debris at the base of the narrow beach. It terminates 300 m south of the creek against a protruding 20 m high bluff. There is vehicle access to along the rear of the beach to both creek mouths.

Beach **WA 343** is an 80 m long pocket of sand located in an amphitheater in the boundary bluffs. It is backed by curving 20 m high vegetated bluffs, with a rocky point on its northern side grading into reefs, and a continuation of the bluffs to the south. It receives low waves and is usually reflective.

Schooner Beach (**WA 334**) commences 200 m to the south on the southern side of the bluffs. It is a low energy 1.7 km long east-facing beach with either a narrow continuous bar or reflective conditions, and patchy seagrass meadows growing 50-100 m offshore. A reef lies 50 m off the centre of the each causing a slight foreland. It is backed by a steep vegetated 20 m high scarp with some rock debris on the beach. The main gravel road runs along the top of the bluffs. Beach **WA 335** is located immediately south of the beach, where the bluffs are replaced by a low rocky shore. The 200 m long beach is bounded by the rocks, as well as a central outcrop, with seagrass reaching to within 50 m of the shore. There are a few fishing shacks on the northern boundary bluffs and a council camping and caravan area behind the beach, with the southern end of the beach used to launch small boats. Thirty metre high Cheyne Island lies 1 km off the beach.

Cheyne Inlet beach (**WA 336**) is the southern-most beach in the bay and is located at the mouth of the usually blocked Eyre River (Fig. 4.85). The beach is 500 m long, faces east and is bordered by a low vegetated rocky point to the north and slopes rising to 30 m high Mount Catherine in the south. The beach is backed by a low foredune grading landward into an overwash plain, with the river flowing along the southern side to cross the beach in the southern corner. Waves are usually less than 1 m high and maintain a narrow bar to reflective conditions. There is vehicle access to the northern end, and a 4WD crossing of the river mouth to private land on the southern side.

Figure 4.85 The sheltered Cheyne Inlet beach is located across the mouth of the small Eyre River.

WA 347-350 **CAPE RICHE-WILLYUN BEACH**

No.	Beach	Rating	Type	Length
WA 347	Cape Riche (S)	6	TBR	300 m
WA 348	Ledge Point (W)	7	TBR/RBB	600 m
WA 349	Willyun Beach	6→4	TBR→LTT	3 km
WA 350	Willyun (S)	4	LTT	200 m
Spring & neap tidal range = 0.6 & 0.2				

Cape Riche is a prominent, steep sloping vegetated headland dominated by 139 m high Mount George and 123 m high Mount Belches. The cape protrudes 3 km to the east and forms the southern boundary of Cheyne Bay, and the eastern boundary of a 45 km long southwest-trending section of coast that terminates at Lookout Point next to Cheyne Beach. The first 15 km of coast is dominated by steep vegetated cliffs and bluffs rising over 100 m and in places to 160 m. The slopes are dissected by several steep streams. Four beaches (WA 247-350) only occupy 4 km of the shoreline. All are backed by private farmland with restricted private access.

Beach **WA 347** lies at the mouth of a small steep valley 3 km southwest of Cape Riche. It consists of a 300 m long pocket of sand that narrows to the rear, with rocks, boulders and step grassy sloping behind rising to 100 m. It faces southeast and receives waves averaging about 1.5 m, which maintain a 60 m wide surf zone with permanent rips to either end. The beach is backed by farmland with a steep walking track down the southern slopes.

Beach **WA 348** is located another 6 km to the southwest, 1.5 km west of Ledge Point and its reef system, called the Giants Causeway. Rocky points backed by steeply rising slopes back either end with valley sides rising to 90 m to the rear and a small creek that breaks out across the centre of the beach. The 600 m long beach faces south and is exposed to high waves averaging about 2 m. These break across a 150 m wide surf zone drained by permanent rips to either end, with a very prominent and strong western rip (Fig. 2.9b). It is backed by an unstable foredune, then an active dune that is climbing up the 120 m high eastern valley side. Older Holocene dunes have previously climbed the slopes and transgressed 1 km across the rear of Ledge Point. A vehicle track runs along the top of the backing slopes and winds down to the western side to the beach.

Willyun Beach (WA 349) is a slightly curving, southeast-facing beach, located 3 km to the southwest. The 3 km long beach spreads across the mouth of Willyun Creek in the east and continues west along the base of 50 m high vegetated bluffs, with a 100 m high headland bordering the eastern end and low rocks to the west. Waves average up to 1.5 m in the east decreasing to the west in lee of the western headland and 70 m high Haul Off Rock, located 2 km south of the beach. A 70 m wide rip dominated bar dominates the eastern half of the beach, with the bar becoming continuous, narrower and free of rips to the west. The central-eastern half of the beach is backed by an active sand sheet, which has blown 500 m inland and partly up the northern 70 m high valley sides. In addition the creek winds its way through the western side of the dunes to reach the centre of the beach. Cleared farmland occupies the backing valley and vehicle tracks reach the bluffs about the western section of the beach, with a narrow track also winding around the western headland and along the base of the bluffs to reach two fishing shacks at the southern tip of the beach.

Beach **WA 350** is a 200 m long pocket of sand located 100 m south of Willyun Beach on the southern side of an irregular low bedrock point. Rocks lie to either side and outcrop on the beach dividing it into three high tide pockets of sand connected by a bar at low tide. Eighty metre high bluffs back the beach, with the track to the shacks running down the western headland and the length of the beach. Waves average about 1 m and break across the 40 m wide, rock-interrupted bar.

WA 351-355 **SALMON BAY**

No.	Beach	Rating	Type	Length
WA 351	Salmon Bay	10	TBR+reef	300 m
WA 352	Salmon Bay (W1)	10	LTT+reef	400 m
WA 353	Salmon Bay (W2)	7	TBR+reef	200 m
WA 354	Salmon Bay (W3)	8	RBB+reef	250 m
WA 355	WA Beach 355	7→6	RBB→TBR	1.2 km
Spring & neap tidal range =0.6 & 0.2 m				

Salmon Bay lies on the western side of the southern Willyun Beach headland. The 100 m high headland continues west as 100 m high calcarenite cliffs, rising steeply as 130 m high slopes for 3.5 km to the west. In between are four small rock-dominated, south-facing exposed beaches (WA 351-354). The cliffs are draped in Pleistocene calcarenite and early Holocene dunes both extending a few hundred metres inland to the edge of cleared farmland. There is an access track along the top of the bluffs and a steep track descending between beaches WA 351 and 352.

Beach **WA 351** is a narrow 300 m long strip of sand located at the base 90 m high dune-draped cliffs. It is bordered by the headland to the east and beachrock fronted cliffs to the west. The entire bay is exposed to high waves, which break across a 150 m wide surf zone drained by permanent rips to either end of the beach. The only access is via the track to the shack then along the beachrock at the base of the cliffs, to the beach. Beach **WA 352** lies 600 m to the west and is a similar 400 m long narrow beach located at the base of sheer, eroding calcarenite cliff rising to 80 m. The beachrock reef lies 60 m off the eastern half of the beach forming a small lagoon, with the water flowing out a permanent rip at the open western half of the beach and through the 200 m wide reef strewn surf zone. The fishing shack is located midway between the two beaches at the base of the steep vehicle track.

Beaches WA 353 and 354 occupy the western end of Salmon Bay. Beach **WA 353** is a 200 m long, curving southeast-facing beach, located between the base of the steep 120 m high vegetated slopes and a lower vegetated point that protrudes 200 m to the south. Waves average about 1.5 m and break across a 70 m wide bar, with a permanent rip against the western rocks and some rocks in the surf. Beach **WA 354** lies 100 m away on the western side of the point. It is a more exposed, south-facing 250 m long beach bordered by the point and south-trending 120 m high steep vegetated slopes. Waves averaging up to 2 m break across a 150 m wide surf zone, with a major permanent rip against the eastern rocks, and a small rip against the western bluffs. An active dune is transgressing 200 m across the point to the rear of beach WA 353. Earlier vegetated Holocene dunes have climbed the backing 130 m high eastern slopes and up onto the top of the cliffs and transgressed up to 500 m inland, to the edge of the farmland.

Beach **WA 355** is located 1 km to the west in an open south-facing, cliff-rimmed 1.5 km long embayment. The 1.2 km long beach receives high waves at the eastern end, which decrease to less than 1.5 m at the western end, with the surf zone decreasing from 150 m to 80 m in width. There are usually three beach rips, together with permanent rips to either end, the eastern being the more substantial as well as flowing along an irregular rock platform. The beach is backed by 100 m high vegetated slopes, which are blanketed by earlier dunes along the eastern half of the beach. A vehicle track reaches the base of the eastern headland and provides foot access along the rocks to the beach.

WA 356-360 CORDINUP-WARRIUP

No.	Beach	Rating	Type	Length
WA 356	Cordinup (E)	8	TBR+reef	350 m
WA 357	Cordinup	8	RBB+reef	1.6 km
WA 358	Cordinup (W)	8	TBR+rocks	100 m
WA 359	South Warriup (E 1)	10	R+platform	100 m
WA 360	South Warriup (E 2)	9	RBB+rocks	50 m
WA 361	South Warriup	8	TBR/RBB	1.9 km
Spring & neap tidal range -0.6 & 0.2 m				

The Cordinup River reaches the coast 8 km west of Salmon Bay and 2.5 km west of Beach WA 355. Between the river and South Warriup Creek 4 km to the west are six relatively exposed south-facing beaches, each bordered and backed by steep slopes rising from 50-100 m.

Beach **WA 356** lies 1 km east of the river mouth. It is a 350 m long beach bordered by a rocky point and 70 m high slopes to the east and a low rock outcrop to the west (Fig. 4.86). It is exposed to waves averaging about 2 m and partly fronted by intertidal beachrock and a 150 m wide surf zone drained by permanent rips to either end. **Cordinup Beach (WA 357)** is 1.6 km long and commences on the western side of the 50 m wide rocks, the sand linking the beaches behind the rocks, with beachrock continuing for 300 m west along the beach. The centre of the beach is backed by the Corindup River and active dunes, with bluffs rising to 80 m along the western half. The bluffs terminate at a gneiss headland that protrudes 400 m seaward affording some protection to the western end of the beach. Waves average 2 m in the east decreasing to 1.5 m in the west, with a continuous rip-dominated surf zone that narrows from 200 m in the east to 50 m in the west. Active dune transgression from the river mouth area and the eastern half of beaches WA 356 and 357, has blown up over the backing 50 m high bluffs and up to 1.5 km to the east. These overlie earlier dunes that extended up to 2 km east almost reaching beach WA 355. Farmland commences 1 km in from the river mouth with a vehicle track running along the top of the western bluffs.

Figure 4.86 The eastern end of the rip-dominated Cordinup beach is bordered by rocky points and backed by steep climbing dunes.

Beach **WA 358** is located 500 m west of the boundary western headland, and consists of a curving 100 m long pocket of sand wedged in at the base of 50 m high vegetated sand-draped slopes. It is bordered by steep slopes to the west and rear and a low rocky point to the east, with a vehicle track running down the steep slopes to reach the rocks. Waves averaging about 2 m break off the point and the surf runs into the small embayment containing the beach, with a single rip running out along the base of the western bluffs.

Beaches WA 359 and 360 are two rock-bound beaches located 100 and 300 m east of South Warriup Beach. Beach **WA 359** consists of a 100 m long high tide beach, backed by steep dune-draped 50 m high slopes, and fronted by a continuous 30 m wide bedrock platform, with sand and a 100 m wide surf zone beyond. Beach **WA 360** is a 50 m pocket of sand open to the 200 m wide surf, but bounded by rock platforms and backed by steep rocky slopes.

South Warriup Creek beach (**WA 361**) blocks the small creek mouth. The beach is 1.5 km long and faces south-southeast, with a sloping rocky point separating it from beach WA 359, and a 100 m high headland protruding 400 m to the south at the western end. The beach receives waves averaging between 1.5-2 m, which break across a 100 m wide rip-dominated surf zone, with usually 4-5 beach rips. The creek mouth is blocked by a plug of active dunes, which in the past have climbed up the eastern 100 m high slopes and transgressed up and over the eastern headland and down onto beach WA 359, 500 m to the east. A vehicle track runs down the slide of the southern bluffs to reach the beach and a solitary fishing shack located half way up the bluffs.

WA 362 HASSELL-CHEYNE BEACH

No.	Beach	Rating	Type Inner outer bar	Length
WA 362	Hassell Beach	6→4	TBR→LTT LBT	22 km
Spring & neap tidal range = 0.6 & 0.2 km				

Hassell Beach (**WA 362**) is one of the longer beaches on the South Coast. It commences against the 100 m high, 1 km long, migmatite headland that separates it from South Warriup beach (WA 361) and curves for 22 km to the southwest, then south, finally spiraling to face northeast in the southern corner at Cheyne Beach (Fig. 4.87), in lee of Lookout Point and Bald Island. The beach is exposed to moderately high waves, which gradually decrease to the south. They average over 1.5 m in the north where they maintain a double bar system, the outer shore parallel longshore bar that extends for 20 km to the south, usually lying about 150-200 m off the shore. The inner bar is a 100 m wide transverse bar and rip system backed by a relatively steep beach in the north. The inner bar narrows, but continues inside the outer bar, as rips diminish in size down the beach. Only along the southern 2 km, as wave height drops substantially, does a single continuous bar, then reflective conditions prevail, with a patch of seagrass off the very southern end of the beach. The southern corner has finer sand, a low slope and is firm for driving on.

The beach is exposed to strong winds along the central-northern section, which have generated both active and earlier now vegetated dune transgressions. There are active dunes extending up to 1.5 km inland backing the northern 12 km of the beach, with some older dunes extending up to 2 km inland. In the south the dunes stabilise and average 1 km in width, with dunes only diminishing along the final north facing 2 km of shore. The small Bluff River and two smaller creeks occasionally break out across the beach just south of the main active dune area. Farm land backs most of the dunes areas with a few vehicle tracks to the beach.

Figure 4.87 The sheltered Cheyne Beach and small settlement lies at the southern end of the longer Hassell Beach.

There is 4WD access in the north near Swan Lake, the track crossing a 1.5 km wide section of dunes and deflated surfaces to reach the beach. In the south the sealed **Cheyne Beach** road reaches the small settlement of the same name that backs the southern end of the beach. At the very southern end, against the gneiss rocks of 70 m high Lookout Point is a large car park and beach boat launching area, adjacent to a camping and picnic area. There is a caravan park with a small store located on the slopes behind the southern end of the beach.

Surfing: East of *Cheyne Beach* wave height increases and a double bar system provides a variety of beach breaks along *Hassell Beach.*

WAYCHINCUP NATIONAL PARK/NATURE RESERVE

Area:	approx 4000 ha	
Coast length:	18 km	(1373-1390 km)
Beaches:	3	(WA 363-365)

Waychincup National Park occupies the rugged coastline west and south of Cheyne Beach, extending along the coast for 18 km from the southern end of Cheyne Beach to 2 km west of Waychinicup Inlet. It also extends inland for up to 7 km and included areas of lower coastal heathland. Four wheel drive tracks lead to the bluffs above the Mermaid Point beaches and a small camping area on the western side of the inlet.

WA 363-365 **LOOKOUT-MERMAID PTS**

No.	Beach	Rating	Type	Length
WA 363	Lookout Pt	7	TBR	250m
WA 364	Mermaid Pt (W1)	9	TBR+reef	200 m
WA 365	Mermaid Pt (W2)	10	TBR+reef	400 m
Spring & neap tidal range = 0.6 & 0.2 m				

Lookout Point marks the southern boundary of Hassell-Cheyne Beach and the beginning of a steep rocky section of shore that trends west for 17 km to Norman Beach. It is composed of Proterozoic granite that forms several bare-domed peaks, the highest the massive 500 m high Mount Many Peaks, whose steep crests are located 1.5 km from the rocky shore. The 26 km of rocky shoreline contains just three small beaches (WA 363-365) totalling 850 m in length.

Beach **WA 363** is the most accessible of the three beaches. It is located 1.5 km east of Cheyne Beach at the end of a sandy 4WD track. The beach faces east towards Bald Island. It is bordered by the steep rocky Lookout Point to the north and Channel Point to the south, with rock then vegetated slopes rising to 120 m and backed by a steep heath covered valley sides. The vehicle track terminates on the slopes above the centre of the beach. Wave average about 1.5 m and maintain a 50 m wide surf zone with a permanent rip running out against the northern rocks (Fig. 4.88).

Figure 4.88 Lookout Point beach is located between the granite rocks of the point, with a permanent rip located against the northern rocks.

Beaches WA 364 and 365 are adjoining rock-dominated, south-facing beaches located 3 km southwest of Cheyne Beach and immediately west of 150 m high Mermaid Point, with a rough vehicle track leading to the top of both beaches. Beach **WA 364** is a 200 m long exposed beach located at the base of steep, vegetated 100 m high slopes, with a steep foot track to the beach. The beach is wedged in between sloping granite, with some rocks also in the inner surf zone. Waves break across a 100 m wide surf zone, with a permanent rip exiting against the western rocks. Beach **WA 365** lies 300 m to the west and is a 400 m long high tide sand beach, fronted by a continuous intertidal band of beachrock, then a 100 m wide surf zone, with a permanent rip against the eastern point. It is backed by steep, vegetated 60 m high slopes, with the vehicle track terminating on the centre of the slopes, and a steep foot track down to the beach.

Four kilometres west of the beach is the downed mouth of the **Waychinicup River**. The rocky mouth is 100 m wide at its entrance. It winds inland for 600 m as a deep inlet and anchorage. A vehicle track terminates a camping area on the western side of the river.

WA 366-370 **NORMANS-BETTY BEACHES**

No.	Beach	Rating	Type	Length
WA 366	Normans Beach	6	TBR	2.5 km
WA 367	Bettys Beach	4	LTT	1 km
WA 358	Bettys (S1)	3	R	50 m
WA 369	Bettys (S2)	2	R	150 m
WA 370	Bettys (S3)	2	R	100 m
Spring & neap tidal range = 0.6 & 0.2 m				

Normans and Bettys beaches (WA 366 & 367) are two adjoining sand beaches divided by the usually closed entrance to Normans Lake (Fig. 4.89). Normans Beach commences as the rocky shoreline surrounding Mount Many Peaks then gives way to a series of embayed beaches and high granite headlands that continue to Albany, 30 km to the west.

Figure 4.89 Norman Lake divides Normans (right) from Bettys (left) beach.

Normans Beach (WA 366) is 2.5 km long, faces southeast and receives waves averaging 1.5 m in the north decreasing slowly to the south. **Bettys Beach (367)** extends south of the inlet for another 1 km, to the base of 170 m high Boulder Hill, with waves averaging 1 m. A 50 m wide bar cut by several beach rips occupies Normans Beach, with a more continuous low tide terrace free of rips along Bettys Beach. They are backed by 20-30 m high foredune blocking the narrow valley and 15 ha Normans Lake in the centre, with valley sides rising steeply to either side. The Normans Beach Road provides access to the beach, where camping is permitted through no facilities are provided. A house is located on the western headland overlooking the beach.

Four hundred meters to the south on the northern side of North Point are three protected, east-facing pocket beaches (WA 368-370) (Fig. 4.90). Beach **WA 368** is a 50 m long pocket of sand bordered by rounded granite rocks and boulders and backed by a storm boulder beach and steep vegetated slopes. Beach **WA 369** lies 50 m to the south and is a curving 150 m long reflective beach set is a 100 m wide semi-circular granite embayment, which it shares with 100 m long beach **WA 370**. A 20 m wide pile of large boulders separates the two. There is vehicle access via the Bettys Beach Road to the rear of beach WA 370 where there are several fishing shacks used by professional fishermen who launch their boats from the steep beach.

WA 371-375 TWO PEOPLES BAY

No.	Beach	Rating	Type	Length
WA 371	Two Peoples Bay (N 3)	3	R+rocks	100 m
WA 372	Two Peoples Bay (N 2)	4	LTT	200 m
WA 373	Two Peoples Bay (N 1)	4	LTT	150 m
WA 374	Two Peoples Bay	5	TBR/LTT	4.5 km
WA 375	Two Peoples Bay (S)	3	R+seagrass	200 m
Spring & neap tidal range = 0.6 & 0.2 m				

Two Peoples Bay is a semi-circular, 3 km wide east-facing embayment bordered by steeply rising granite slopes of Boulder Hill in the north and 380 m high Mount Gardner in the south. The bay shore is dominated by the main beach (WA 374) with smaller beaches located along the rocky shore to either end. There is access to the bay in the north off the Bettys Beach Road and in the south along the Two Peoples Bay Road.

Figure 4.90 The three small Bettys beaches are a popular boat launching and sheltered swimming area.

Beaches WA 371-373 occupy a 300 m wide, shallow south-facing bay at the western base of Boulder Hill. Low swell averaging 1 m reaches the bay entrance and refracts into the small bay. Beach **WA 371** is located on the eastern side of the bay and is a 100 m long narrow strip of sand set at the base of steep vegetated slopes with rocks outcropping to either end and off the shallow beach. It can only be accessed on foot around the rocky shore of the bay. Beach **WA 372** lies 100 m to the west and is a similar 200 m long strip of sand backed by steep slopes, with numerous rocks outcropping along and off the low energy beach. Waves break across a 20 m wide low tide terrace. Beach **WA 373** is the main bay beach. It is clear of rocks and receives waves averaging less than 1 m, which break across a 30 m wide low tide terrace. A short vehicle track leads off the road to the top of the backing 40 m high bluffs with a steep walking track down to the beach.

Two Peoples Bay beach (**WA 374**) occupies the eastern shore of the bay and curves for 4.5 km between the steep granite slopes to either end. It faces due east out of the bay mouth and receives waves averaging over 1 m in the centre, decreasing to either end. The waves combine with the fine white sand to maintain a continuous low tide terrace, only cut by rips in the centre during periods of higher waves. The beach is backed by a 10 m high foredune and low vegetated 500 m wide barrier, with Angrove and Gardner lakes occupying the 1 km wide back barrier depression. The Two Peoples Bay access road runs between the two lakes to the beach.

Lake Gardner is connects by a small deflected creek to the very southern end of the beach, where seagrass grows to the shore and piles upon the beach. Beach **WA 375** lies 100 m east of the creek mouth. It extends for 200 m along the base of vegetated slopes, with granite rocks bordering each end and 40 m high South Point lying 200 m to the east. This is a very low energy reflective beach with seagrass growing to the shore and rocks scattered off the beach .

TWO PEOPLES BAY NATURE RESERVE

Area:	approx 3,500 ha	
Coast length:	21 km	(1407-1428 km)
Beaches:	6	(WA 374-380)

Two Peoples Bay Nature Reserve compasses a 3,500 ha area of rugged granite coastal cliffs, backing Holocene and Pleistocene dune systems and Gardner and Montes lakes. The Two People Bay Road terminates at the southern end of Two Peoples Bay Beach where there is a visitor centre and picnic area and boat launching permitted from the beach. Four wheel drive tracks also lead through the reserve to the cliffs above Stinkers Reef and the centre of Nanarup Beach.

WA 376-378 WATERFALL BEACH

No.	Beach	Rating	Type	Length
WA 376	South Point	3	R+platform	100 m
WA 377	Waterfall Beach	2	R	200 m
WA 378	Little Beach	3	R+rocks	80 m
Spring & neap tidal range = 0.6 & 0.2 m				

Five hundred metres east of South Point are three pockets of protected sand containing beaches WA 376-378. A gravel road runs out from Two Peoples Bay to a small car park on the slopes above Waterfall Beach (Fig. 4.91). Beach **WA 376** lies 100 m north of the car park. It is a curving 100 m long high tide beach, fronted by an irregular 50-100 m wide rock platform with a tidal pool along the eastern half of the beach connected by a narrow western channel to the bay.

Figure 4.91 The sheltered South Point, Waterfall and Little beaches.

Waterfall Beach (WA 377) is located 200 m to the east and directly below the car park. It is a curving, north-facing 200 m long beach, bordered by rounded granite points and boulders, with a large boulder also in the centre. A steep narrow valley and creek, the source of the waterfall, crosses the centre of the beach. Waves average about 0.5 m and surge up a moderately steep reflective beach. This is a relatively popular swimming beach. **Little Beach (WA 378)** is located 50 m further east on the other side of a granite point. It is a straight 80 m long pocket of reflective sand with granite boulders to either end and off the center of the beach. Steep vegetated slopes back both beaches.

WA 379-380 MT GARDNER

No.	Beach	Rating	Type	Length
WA 379	Vancouver Bay	5	R+rocks	60 m
WA 380	Inner Island	5	boulder+rock	50 m
Spring & neap tidal range = 0.6 & 0.2 m				

The eastern and southern base of Mount Gardner are dominated by steep, massive granite slopes and generally deep water close to shore. The only beaches are located deep inside two narrow inlets on the southern side of the mountain. Beach **WA 379** is a 60 m long pocket of sand located inside a V-shaped 1 km deep bay, 1 km north of Cape Vancouver. The bay narrows to 100 m in the north with the 60 m long low energy reflective beach wedged in at the end, with rocks outcropping on and off the beach. It is surrounded by steep vegetated slopes rising to 250 m. Beach **WA 380** is located 2 km to the northwest, 1 km to the lee of Inner Island, in a 500 m deep V-shaped, west-facing embayment. The beach consists of a low energy 50 m long pocket of boulders, bordered and backed by steep valley sides. The only reasonable access to the beaches is by boat.

WA 381-384 NANARUP BEACH

No.	Beach	Rating	Type	Length
WA 381	Rocky Pt	9	LTT+reef	300 m
WA 382	Nanarup Beach	8→6	RBB+reef→TBR	4.2 km
WA 383	Nanarup (W)	6	TBR	400 m
WA 384	Islet Point	3	R	100 m
Spring & neap tidal range = 0.6 & 0.2 m				

Between False Island and Herald Point is a 13 km wide, relatively open, south-facing bay. The bay is afforded some protection by Michaelmas and Breaksea islands in the west, with higher waves dominating the central-eastern section. Rocky shore with beachrock reefs dominate the eastern half of the bay along a section known as Stinkers Reef. Beaches WA 381-384 extend west of Rocky Point to Islet Point, followed by a series of embayed beaches (WA 385-388) between the point and Herald Point. The only vehicle access is to the bluffs above Stinkers Reef and at Nanarup Beach.

Stinkers Reef is a 2.5 km long section of shore-attached beachrock reef, containing an intertidal active platform and higher relic platform cut into the beachrock. While there are some patches of high tide sand along the rear of the reef it does not classify as a beach. There is 4WD access to the eastern end of the 50 m high bluffs that overlook the reef, with a steep climb down to the reef. Waves averaging over 1.5 m break heavily on the reef, which is backed by calcarenite bluffs and vegetated clifftop dunes.
Beach **WA 381** commences at Rocky Point, a 30 m high calcarenite bluff. The 300 m long beach is wedged in between the base of the steep eroding bluffs and a continuation of the beachrock reef, lying 10-50 m offshore. Waves averaging about 2 m break heavily on the reef resulting in very hazardous conditions. They then reform in a small lagoon between the reef and beach, with relatively calm conditions at the shore.

Nanarup Beach (WA 382) commences 100 m to the west and curves slightly west-southwest for 4.2 km to the usually closed mouth of **Taylor Inlet**. The first 1 km of beach has scattered beachrock reef lying 50 to 100 m offshore, which combine with the high waves to produce a crenulate shoreline with three strong permanent rips. As the reefs cease, the waves maintain a 100 m wide rip-dominated surf zone, that extends to the inlet, with waves decreasing slightly to the east. The beach is backed by scarped 30-40 high calcarenite bluffs in the east and unstable foredune and blowouts in the west, with vegetated and active dunes transgressing up to 3 km inland. There is vehicle access via Nanarup and along the western shore of the inlet to the inlet mouth, with 4WD access to the firm low gradient beach.

The western end of Nanarup beach (**WA 383**) commences at the inlet mouth and curves to the southwest for 400 m to the 20 m high granite boulders of Islet Point (Fig. 4.92). The point is connected to the shore by a small tombolo between Nanarup and the point beaches. Waves continue to decrease along Nanarup Beach averaging about 1 m by the point. The beach grades from a lower energy transverse bar and rip to a low tide terrace against the point.

Figure 4.92 Nanarup Beach extends from the usually blocked mouth of Taylor Inlet to Islet Point.

Beach **WA 384** is a curving 100 m long pocket beach wedged in between the tombolo-point and the sloping granite of a western point. The two points converge to form a reef-strewn 50 m wide opening to the beach, with waves averaging about 0.5 m at the shore, resulting in a low energy reflective beach. The beach is suitable for swimming inshore, but beware of the reef and rocks where the points converge offshore.

Surfing: *Nanarup Beach* is worth checking out if you have a 4WD to head up along the beach to the numerous beach breaks including reefs breaks at the eastern end.

WA 385-390 **BEN DEARG-HERALD POINT**

No.	Beach	Rating	Type	Length
WA 385	Ben Dearg Beach (E)	4	R	500 m
WA 386	Ben Dearg Beach (W 1)	5	R+rocks	50 m
WA 387	Ben Dearg Beach (W 2)	4	R	250 m
WA 388	Herald Pt	4	R	150 m
WA 389	Herald Pt (W 1)	4	R/LTT	150 m
WA 390	Herald Pt (W 2)	5	LTT+rocks	100 m

Spring & neap tidal range = 0.6 & 0.2 m

Mount Taylor rises to 200 m, 1.5 km west of Islet Point. The steep slopes of the mount dominate the next 5 km of shoreline to Ledge Point, with 150 m high Mount Martin 3 km further west. The first 3 km contains six small beaches (WA 385-390) all bordered by granite points and backed by the densely vegetated slopes. There is no formal access to any of the beaches, though most can be accessed via rough 4 WD tracks.

Ben Dearg Beach occupies a 1 km long, southeast-facing, open embayment that is divided into of three sections by some central rocks The eastern part (**WA 385**) is 500 m long and bordered by the eastern granite point and the central rocks. It is composed of medium sand, which maintains a steep reflective beach with waves averaging about 1 m. The central rocks contains a 50 m long pocket of sand (**WA 386**) fronted by a narrow low tide terrace. Immediately to the west is the 250 m long western section (**WA 387**) which grades from a narrow low tide terrace near the rocks to reflective towards the lower energy western end. Steep vegetated slopes back all three sections of the beach. There is a large house above the eastern section with 4 WD tracks to the top of the western slopes.

Herald Point, a 60 m high vegetated headland, lies 2 km to the southwest, with beach **WA 388** located on the eastern side of the point. It a 350 m long southeast-facing beach bordered by sloping granite points and backed by a scarped 20 m high foredune, with several 4WD tracks reaching the rear of the dune. Waves are lowest in the west where the beach is reflective, increasing to about 1 m at the western end where they maintain a narrow bar.

Beach **WA 389** is located 500 m to the west in a small embayment on the western side of the point. The bay has converging sloping rocky shores and a central sandy section the 150 m long beach. It is protected by the offshore islands with waves averaging about 1 m, which surge up a reflective beach face. A 4 WD track reaches the back of the beach. Beach **WA 390** is located 200 m further west and consists of a 100 m long pocket of southwest-facing sand located at the base of 30 m high bluffs, with rocks outcropping to either end of the beach.

KING GEORGE SOUND is an 8 km wide, east-facing embayment bordered by Herald Point to the north and the Flinders Peninsula, terminating at Bald Head to the south. The sound has two major tributaries, Oyster Harbour in the north and Princess Royal Harbour in the west, with the city of Albany located along its northwestern shores. This is the largest city on the South Coast with a population of 25 000. The 30 km of shoreline around the Sound is relatively protected by its orientation, the Flinders Peninsula and the larger Michaelmas and Breaksea islands, as well as several smaller islands. The sound has a total of 17 beaches all either low energy reflective to low tide terrace, with the highest waves averaging about 1 m. Albany Surf Life Saving Club, the southern most in the state, is located at Middleton beach.

WA 391-396 **LEDGE-EMU PT**

No.	Beach	Rating	Type	Length
WA 391	Ledge Pt	7	LTT/TBR+reef	1.8 km
WA 392	Ledge Bay	4	R/LTT	2.2 km
WA 393	Cheyne Head (E)	5	R+rocks	100 m
WA 394	Cheyne Head (W)	5	R+rocks	50 m
WA 395	Mount Martin (1)	4	R	300 m
WA 396	Mount Martin (2)	4	R+rocks	300 m

Spring & neap tidal range = 0.6 & 0.2 m

To the west of Herald Point the steep granite slopes and intervening headlands continues for 9 km to the entrance to Oyster Harbour, below Mount Martin. Two longer beaches (WA 391 & 392) are followed by four smaller, rock-dominated systems (WA 393-396). Gravel roads run out to the two main beaches however there is no development along this section of coast.

Beach **WA 391** is a 1.8 km long, south-facing beach bordered by Herald and Ledge points. The beach receives waves averaging just over 1 m, which break across a discontinuous intertidal beachrock reef. The reef and wave combine to form a mixture of reef-controlled rips, a central reef-bound lagoon and a western section of sand beach. The beach is backed by a 20 m high scarped foredune, with generally vegetated transgressive dunes extending up to 1 km inland and climbing up to 100 m up the backing slopes. A large climbing blowout is located towards the eastern end of the beach. There is car access to the crest of 30 m high Ledge Point where the car park provides views of the beach (Fig. 4.93).

Ledge Bay beach (**WA 392**) commences on the western side of Ledge Point and curves round to the west for 2.2 km to the base of 120 m high Cheyne Head. The beach is initially protected by the point maintaining a cusped reflective beach, which grades in the west to a narrow low tide terrace as waves pick up to about 1 m. The beach is backed by a well vegetated 20 m high foredune, then the shallow 50 ha Gull Rock Lake and elongated wetland, which occasionally flows across the far western end of the beach. A gravel road reaches the western end of the beach terminating at a small car park with no facilities.

Figure 4.93 Ledge Point is a prominent granite headland separating Ledge Bay beach (right) from Cheyne Head beach (left).

Cheyne Head is a steep 120 m high, densely vegetated, rounded headland surrounded by lower energy rocky shore. In amongst the rocks are two small pockets of sand (WA 393 & 394). Beach **WA 393** is located on the eastern side of the head. It is a 100 m long sliver of intertidal sand, backed and bordered by high tide rocks. Waves break across a narrow bar and against the rocks at high tide. Beach **WA 394** lies 1 km to the east in a small embayment, on the eastern side of the head. It is a 50 m long pocket of high tide sand almost encased in inter to low tide boulders and rocks. It is also backed by steeply rising vegetated slopes.

Mount Martin is a 150 m high, densely vegetated, granite dome that forms the eastern boundary of Oyster Harbour. Two small beaches are located along its western base. Beach **WA 395** is a 300 m long, southwest-facing reflective beach bordered by granite rocks, with a few rocks outcropping on the centre of the beach. It is backed by dense coastal heathland. Beach **WA 396** commences immediately to the west facing west across the entrance channel to Oyster Harbour and lies directly opposite Emu Point. The beach is 300 m long, backed by steep vegetated slopes, with numerous large rounded granite boulders littering the beach. The high tide beach is reflective, but grades within a few tens of meters into the deep entrance channel, with its strong tidal currents. While these beaches have no vehicle access, their location opposite Emu Point makes them a popular destination for boat fishers and picnickers.

WA 392 MIDDLETON BEACH (ALBANY SLSC)

No.	Beach	Rating	Type	Length
WA 397	Emu Point (N)	1	R+tidal sand flat	400 m
WA 398	Emu Beach	4	R	400 m
WA 399	Middleton Beach	3→4	R→LTT	4.5 km

Albany Surf Life Saving Club

Patrol Period:	**Saturday**	**January**
	Sunday	**December-March**
Lifeguard	**Mon-Fri**	**January**

Spring & neap tidal range = 0.6 & 0.2 m

Emu Point forms the western boundary of the 150 m wide entrance to Oyster Harbour. The 1500 ha coastal lagoon is feed by the King and Kalgan rivers. A deep tidal channel occupies the entrance, however scour caused by the waves and tides has resulted in the construction of a 200 m long, rock training wall that parallels the channel at the tip of Emu Point. There are two beaches either side the point, one inside the harbour (WA 395) and one just outside (WA 396).

Beach **WA 397** is a sheltered beach located just inside the entrance to Oyster Harbour, on the northern side of Emu Point. The beach is 400 m long extending from the eastern training wall to the Emu Point marina. The beach receives no ocean waves, only low winds waves from the harbour. It faces north into the harbour and is fronted by 200 m wide shallow tidal sand flats grading into seagrass meadows. A fenced tidal pool is located towards the eastern end, while there is a dredge hole associated with the marina at the western end. The popular beach is backed by Emu Point park, picnic area, playground, car park and boat ramp.

Emu Beach (WA 398) commences at the southern end of the training wall, faces south and curves for 400 m to a sand foreland in the lee of a second breakwater that attaches it to the shore. In between is a usually calm curving reflective beach. However beware of strong tidal current flowing in the deep channel outside the breakwaters. The eastern end of the beach is backed by the Emu Point park, with beachfront houses behind the western end.

Middleton Beach (WA 399) (Fig. 4.94) is the main beach for the Albany, a city of 25 000 people. The city is located on the protected shores of King George Sound, with the main anchorage inside the almost landlocked tributary Princess Royal Harbour. The Sound is 5 km wide at the entrance, with Middleton Beach located 10 km inside the entrance in the northwest corner. The beach is 4 km long, faces the southeast and receives low refracted waves. Sandy Emu Point forms the northern boundary and granite Wooding Point, capped by 100 m high Mount Adelaide, the southern boundary. Waves are usually 1 m or less and maintain a low energy beach fronted by a continuous shallow bar, usually free of rips (Fig. 4.94). The southern end is called Middleton Beach, which is an eastern suburb of Albany and site of the Albany Surf Life Saving Club (Fig. 4.95 & 4.96). The Surf Club, which was established in 1956, is located in a grassy foreshore reserve, and patrols the beach on Sundays between January and March. The well laid out reserve also provides parking, a kiosk and other amenities. In the southern corner, called Ellen Cove, is a wooden walkway that runs out past the jetty to Wooding Point. The small jetty is popular with swimmers and fishers. Right behind the reserve is a resort hotel and shopping centre, with a caravan park just up the beach.

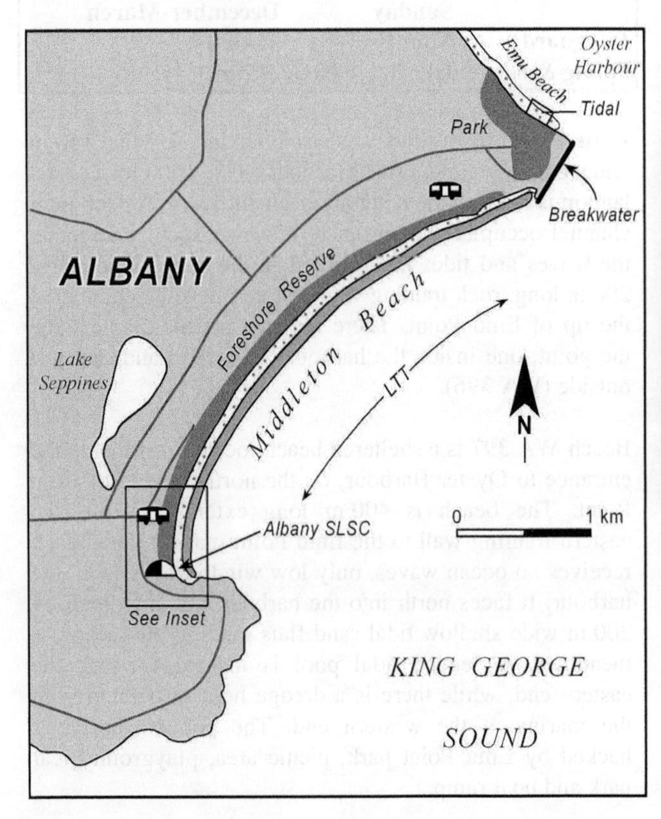

Figure 4.94 Middleton Beach, site of the Albany SLSC. See Fig. 4.95 for insert.

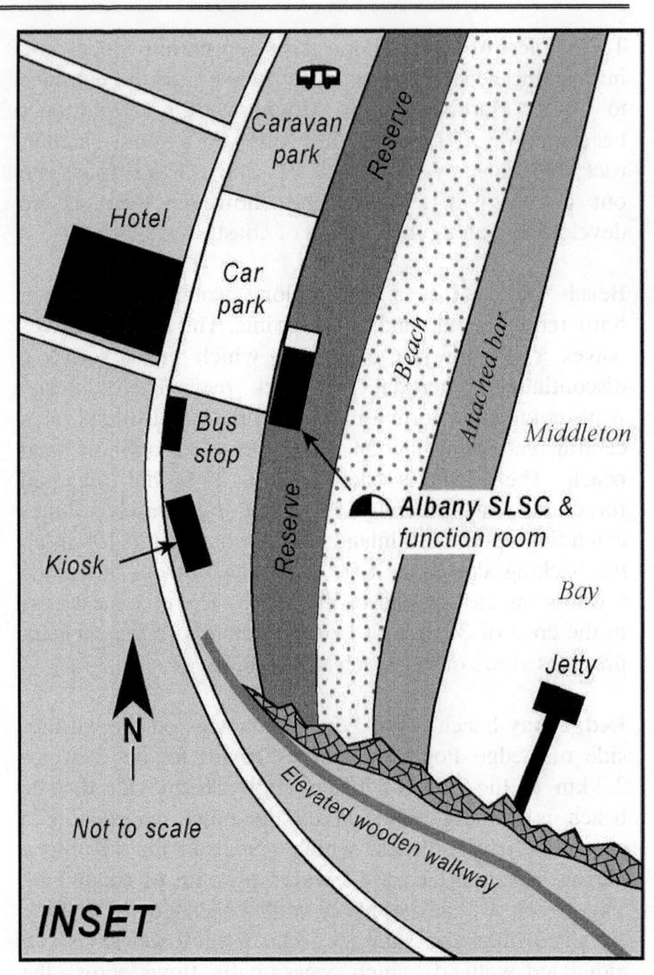

Figure 4.95 The Albany Surf Life Saving Club is located at the southern end of Middleton Beach.

Figure 4.96 The southern end of Middleton Beach and site of the Albany Surf Life Saving Club.

The beach is backed by a 500 m wide foredune ridge plain which has been largely developed. The central section is occupied by Albany golf course, with the northern half called Emu Beach, which has a large caravan park and the northern suburb of Emu Point. The Albany Surf Life Saving Club patrols the southern end of the beach (Figs. 4.95 & 4.96). The point forms the boundary between Oyster Harbour and the Sound, with a narrow inlet linking the two.

Swimming: This is a relatively safe beach with waves averaging 1 m and less, with surf breaking across a shallow bar usually free of rips. There is deep water and tidal currents in Emu Point inlet off beach WA 398.

Surfing: Usually just a low beach break.

ALBANY is the largest town on the south coast with a population of 25 000. It is located on the protected shores of Princess Royal Harbour, with Middleton beach representing its more exposed eastern shore. The town is dominated by several granite domes including 150 m high Mount Melville in the west, and 180 m Mount Clarence in the centre. The city offer all the facilities and services of a major regional centre.

WA 400-403 BARKER BAY-WHALING COVE

No.	Beach	Rating	Type	Length
WA 400	Barker Bay	4	R	700 m
WA 401	Whaling Cove (1)	4	R+rocks	200 m
WA 402	Whaling Cove (2)	4	R+rocks	200 m
WA 403	Whaling Cove (3)	4	R+rocks	100 m
Spring & neap tidal range = 0.6 & 0.2 m				

Albany is located on the northern shores of Princess Royal Harbour, which is connected to King George Sound by a 500 m wide entrance between the prominent King Point to the north and Possession Point to the south. Procession Point marks the beginning of the southern shoreline of the Sound which trends south then west along the Flinders Peninsula to Bald Head, a shoreline distance of 17 km. There are nine beaches located along this section, the first four (WA 400-403) in the first two bays south of Possession Point.

Barker Bay beach (WA 400) occupies the western shore of 1 km wide east-facing Barker Bay. The 700 m long beach links 40 m high Possession Point with the mainland. The beach receives refracted waves averaging about 1 m, which usually surge up a steep cusped reflective beach. This is backed by a steep 10-20 m high foredune, which is scarped in the centre. The foredune is part of a 200 m wide barrier that faces into Princesses Royal Sound on its low energy western shore. There is a gravel road along the crest of Vancouver Peninsula to Geak Point at the southern end of the western shore. It terminates at the entrance to Camp Quaranup, a sport and recreation camp.

Whaling Cove is located 500 m southwest of the bay and is a 700 m wide northeast-facing bay. It contains three low energy near continuous beaches, with wave height decreasing from less than 1 m at the western end of the bay to near calm at the more protected eastern end. Remnants of the past whaling days are located in the southern corner. Beach **WA 401** is the western-most and more exposed of the three beaches. It is a 200 m long reflective beach bordered by a low granite point in the west and a low rock outcrop to the east, with seagrass growing 50-100 m offshore. Beach **WA 402** commences on the eastern side of the rocks and is a similar 200 m long reflective beach, with a flat granite point in the east and several large boulders on and off the beach. The flat rocks separates it from beach **WA 403** a relatively steep, narrow 100 m long pocket of sand, with a few rounded granite boulders on the beach and seagrass growing almost to the shore. The Quaranup Road runs along behind the beaches, with a car park on the 20 m high bluffs between beaches WA 402 and 403.

WA 404-408 FRENCHMANS BAY

No.	Beach	Rating	Type	Length
WA 404	Goode Beach	4	R	3 km
WA 405	Whalers Beach	4	R	700 m
WA 406	Cheyne Beach	4	R	600 m
WA 407	Whaling Station	4	R	80 m
WA 408	Misery Beach	4	R	200 m
WA 409	Flat Rock	3	R	50 m
WA 410	Isthmus Hill (W	4	R	400 m
WA 411	Isthmus Hill (E)	3	R	500 m
Spring & neap tidal range = 0.6 & 0.2 m				

The southwestern corner of King George Sound is located between Mistaken Island in the west, the old Whaling Station in the south, with the 5 km long Flinders Peninsula forming the southern arm. Between Mistaken Island and the Whaling Station are five beaches (WA 404-408), with three more beaches (WA 409-411) occupying the next 1.5 km, then steep cliffs dominating the remaining 5 km of the northern side of the peninsula.

Goode Beach (WA 404) occupies the western shore of 3.5 km wide Frenchman Bay. The beach faces due east and receives 1 m high refracted waves, which surge up a steep cusped reflective white sand beach. Seagrass meadows lie 50-100 m offshore (Fig. 4.97) and seagrass debris litters parts of the beach. It is backed by a well vegetated 10-20 m high foredune, which is part of a 400-500 m wide barrier that links the Vancouver Peninsula, and in turn Possession Point, with the mainland. The low energy western side is fronted by 500 m wide tidal sand flats. There is vehicle access and a small car park at the northern end of the beach and a few beach front houses and a car park at the very southern end, with the dunes in between.

Whalers Beach (WA 405) is a curving 700 m long northeast-facing reflective beach located between the southern rocks of Goode Beach and 40 m high Waterbay Point. Wave energy decreases east along the beach, with seagrass 30 m offshore in the west reaching the shore to the east, and seagrass debris also increasing to the east. It is backed by houses on the northern point, and a caravan park and car park at the southern end, with vegetated 30-40 m high bluffs in between. The southern corner is used to launch small boats.

Figure 4.97 Goode Beach is a curving sheltered white sand beach paralleled by seagrass meadows.

Cheyne Beach (WA 406) is located on the eastern side of Waterbay Point and curves to the east for 600 m to the rocks of the old Whaling Station. It receives slightly higher waves averaging almost 1 m which maintain a steep reflective beach, with the seagrass 50 m offshore in the west and close to shore in the west. It is backed by scarped 10-20 m high dunes that have transgressed across the peninsula from Jimmy Newhills Harbour 2 km to the south. There is a sealed road to the southern end of the beach where the car park for the whaling station is located.

The **Whaling Station** operated from 1952 to 1978. Since it closed it has been redeveloped as a whaling museum, called *Whaleworld*, with most of the original station, including a whaling ship on display. There are two beaches at the station. Beach (**WA 407**) is a curving 80 m long pocket of sand located between the car park and the old whaling ship. It receives low waves and is backed by a low bluff and the Whaling museum (Fig. 4.98).

Misery Beach (**WA 408**) extends 200 m east of the whaling station to a low granite point. It is a low energy reflective beach with patchy seagrass offshore. It is backed by cleared land, with a gravel road and car park on the eastern point, which it shares with beach WA 409. The ramp for the whaling station is located at the western end of the beach, and when the station operated with beach used to be covered in the blood and debris from the butchered whales, perhaps the reason for the name. Flat Rock beach (**WA 409**) is separated from beach WA 408 by a 50 m wide elongate flat granite point. The 50 m long beach is wedged in between the point and higher granite slopes to the east, with rocks also off the beach and the two boundary points converging to form a narrow entrance. The backing road terminates at a small car park, with a concrete boat ramp located at the western end of the beach. Seagrass grows to the shore and litters the beach.

Figure 4.98 The Albany Whaling Station bordered by Cheyne, Whaling Station and Misery beaches.

Beach **WA 410** lies 200 m to the east. It is a north-facing 400 m long steep, cusped sandy beach bordered by a 40 m high granite rocks to the west and a sheer, steeply sloping granite rising 120 m to Isthmus Hill to the east. The beach receives waves averaging about 1 m, which surge up the reflective beach face. There is a gravel road out to a car park above the beach and a walking track down to the eastern end of the beach. Beach **WA 411** is located on the eastern side of Isthmus Hill and is a curving 500 m long lower energy beach with waves averaging about 0.5 m. These permit seagrass to grow close inshore, with seagrass debris littering much of the beach. It is bordered and backed by steeply rising vegetated slopes composed of calcarenite covered granite. A walking trail runs along the crest of the 100-150 m high backing ridge, with a track down to the rocks at the eastern end of the beach.

TORNDIRRUP NATIONAL PARK

Established:	1918	
Area:	approx. 4000 ha	
Coast length:	38 km	(1460-1498 km)
Beaches:	11	(WA 410 421)

Torndirrup National Park is one of the oldest in the state being first gazetted in 1918. It was not however named until 1969. It encompasses some of the most spectacular coastal cliffs and scenery in Australia. The park includes most of Flinders Peninsula which rises to 230 m, as well as the ancient rugged granites and more recent calcarenite cliffs between the end of the peninsula at Bald Head to the cliff backed Sand Patch Beach 30 km to the west. Set in amongst the cliffs and points are 12 beaches. These range from the quieter north-facing beaches below Isthmus Hill, to the more exposed and often rocks and reef-dominated, beaches along the southern shore. Be very careful if walking close to the rocky shore, as occasional higher waves washover the rocks and have washed people into the sea.

WA 412-420 TORNDIRRUP BEACHES

No.	Beach	Rating	Type	Length
WA 412	Isthmus Bay (E)	8	TBR	800 m
WA 413	Isthmus Bay (W)	10	LTT+reef	1.5 km
WA 414	Salmon Pools	5	R/LTT+reef	300 m
WA 415	Peak Head (N)	5	LTT+rocks	70 m
WA 416	Jimmy Newhills Hbr	5	boulder	50 m
WA 417	Black Head (W)	8	R+rocks	80 m
WA 418	Cable Beach (E)	8	TBR+reef	300 m
WA 419	Cable Beach (W)	8	R/LTT+reef	600 m
Spring & neap tidal range = 0.6 & 0.2 m				

The southern shores of **Torndirrup National Park** faces into the high waves and strong winds of the Southern Ocean. The waves are reduced slightly by the offshore Eclipse islands, as well as some inshore rocks and islets, and closer to shore by some of the headlands. The winds however remain strong and have blanketed the entire southern shoreline with multiple layers of Pleistocene dune calcarenite, anchored by the ancient massive granite that forms the base of all the headland and points. The calcarenite has blown up and over the granite and up to 6 km inland in lee of Sand Patch Beach. It averages over 100 m in height and reaches a maximum height of 230 m at Limestone Head on Flinders Peninsula. The first 16 km of shoreline between Bald Head and Cave Point contains eight beaches (WA 412-419).

Isthmus Bay occupies most of the southern side of Flinders Peninsula, with Bald Head forming the eastern boundary. The exposed south-facing, 2.5 km long bay contains two beaches (WA 412 & 413). The beaches can only beach reached on foot along the Isthmus Hill track, with a steep descent down to each beach. Beach **WA 412** is an 800 m long beach wedged in between calcarenite capped-granite points and backed by steep scarped calcarenite rising to 150 m. The beach receive slightly reduced waves averaging about 1.5 m, which maintain a 50 m wide surf zone breaking across both sand and a discontinuous beachrock reef. Four reef-controlled rips drain the surf. Beach **WA 413** is located 200 m to the west and occupies the western 1.5 km of the bay. It is a similar beach with a more continuous beachrock reef that replaces the beach along the central-western section. Waves break heavily on the reef, with two permanent rips at the eastern end of the beach, while the rocks dominate the rest. It is also backed by the steep slopes rising to 100 m.

Salmon Pools (**WA 414**) is located in the western corner of Isthmus Bay, where the cliffs turn and trend south for 3 km to Peak Head. The 300 m long beach faces southeast, and consists of a double-crenulate, steep sandy beach, fronted by a beachrock reef along the central-western section, with an open eastern end (Fig. 4.99). Permanent rips run out the eastern end and at the western tip of the beachrock reef. The reef encloses a relatively calm lagoon along the western half in lee of the reef, however the water flows out through the western rip, so be cautious. A car park is located on top of the 30 m high bluffs behind the beach, with wooden steeps leading down to the beach, which is a popular fishing spot and also offers some right and lefthand *surfing breaks* running off the reef.

Figure 4.99 View east along the beachrock fringed Salmon Pools beach.

Peak Head from the southernmost tip of this section of coast, with an irregular section of shoreline occupying the next 8 km of shoreline to Sharp Point. Beach **WA 415** is located in a small west-facing embayment on the western side of the head. The 70 m long beach is bordered and backed by 70-120 m high calcarenite slopes, with basal granite and boulder beaches bordering either side. The beach receives refracted 1 m high waves, which break across a 50 m wide bar, with a permanent rip against the northern rocks. There is no safe land access to this beach.

Jimmy Newhills Harbour (**WA 416**) is a 500 m deep cleft in the 100-150 m high calcarenite capped-granite, with boulders dominating the more protected inner shoreline, where waves average less than 1 m. While the boulders dominate the shore there is some low tide sand amongst the boulders at the apex of the bay. A clifftop car park is located above the western side of the harbour with a steep descent down to the shore. It is used by fishers and surfers who ride a peaky righthand *surf break* off the western harbour entrance.

Black Head is located 1 km west of the harbour and is a 100 m high calcarenite-capped, granite headland. The headland is the site of the Blowholes and a road runs out to the tip of the head, with a walking track down to the holes. Beach **WA 417** is located in a small southwest-facing gap on the western side of the head, near the blowholes. The 80 m long beach faces west and consists of a high tide strip of sand and small foredune, backed by steep vegetated slopes rising to 130 m. It is bordered and fronted by granite rocks and boulders, as well as a beachrock reef off to the west. A permanent rip exists the small bay.

Cable Beach occupies most of the 2 km wide embayment between Black Head and Cave Point. It consists of two beaches separated by a section of protruding vegetated calcarenite. The eastern side of the beach (**WA 418**) is a 300 m long strip of high tide sand and parallel beachrock reef. Waves averaging over 1.5 m break over a 50 m wide sand bar, then the reef. The reef forms the shoreline along the eastern half of the beach, and lies 10-30 m offshore along the western half, where a permanent rip drains the 70 m wide surf zone. The western half of the beach (**WA 419**) is a curving 600 m long high tide beach, fronted by a continuous 40 m wide beachrock reef, which forms a small lagoon in the east, while rocks are piled to the rear in the west. The Gap Road runs along the bluffs to western end of the beach, to a 70 m high clifftop car park with steep foot access down to the beaches. The Gap and Natural Bridge are located 500 m west of the beach.

WA 420-422 **SAND PATCH**

No.	Beach	Rating	Type	Length
WA 420	Sharp Pt (E)	9	R/LLT+reef	1 km
WA 421	Sand Patch	10	TBR/RBB+reef	9.5 km
WA 422	Hanging Rock	8	RBB+rocks	80 m
Spring & neap tidal range = 0.6 & 0.2 m				

At Cave Point the shoreline turns and trends west-northwest for 20 km to Hanging Rock headland. In between is one of the more exposed, high-energy sections of the South Coast, with high waves breaking on reef-dominated beaches at the base of 100 m high cliffs.

Beach WA 420 is located between **Sharp Point** in the west, with the Natural Bridge to the east. The beach is 1 km long, faces south and receive slight protection from the Green islands located 1 km offshore. The waves average about 1.5 m and break across a 50 m wide bar, then a 30 m wide shore-attached beachrock reef that dominates the shoreline. This is backed by a sandy high tide beach, then steep vegetated slopes rising to 100-180 m. The gravel Eclipse Island road terminates high above the western end of the beach, however then is no safe access down to the beach.

Sand Patch beach (**WA 421**) commences 5 km west of Sharp Point. The discontinuous 9.5 km long beach represents the remnants of a once more robust and continuous beach that extended 17 km from Sharp Point to Hanging Rock. Loss of sand to the backing high dunes and offshore have eroded the former beach leaving the discontinuous pockets of sand and high tide remnants. The beach faces southwest and receives the full forces of the high southwest swell (Fig. 1.13), which averages over 2 m. These maintain a rip-dominated 200-300 m wide surf zone, with rip spaced every 500-600 m, and flowing up to 1 km seaward. The inner surf zone and shoreline is dominated by discontinuous beachrock reef, with an intertidal active platform and higher inactive platform. While it is possible to walk along most of the 'beach' section, there are rock fall and much rock debris along the high tide beach. This is an extremely hazardous beach and unsuitable for swimming. This beach picks up any swell and there is always a wave on the numerous beach breaks, often too big. Even with a board this is a hazardous beach, so be cautious and talk to the locals. There is vehicle access to the top of the cliffs on the Sand Patch Road, with a steep climb down to the shore.

Beach **WA 422** is located in a curving 100 m wide embayment at the western end of the cliffs. It consists of a curving 80 m long beach located at the base of 70 m high unstable slopes, with Hanging Rock headland forming the western boundary. It is fronted by a 150 m wide surf zone, drained by a permanent rip against the western rocks. There is 4 WD access to the top of the cliffs and a steep descent down to the beach.

WA 423-428 **PORT HUGHES-HARDING**

No.	Beach	Rating	Type	Length
WA 423	Port Hughes	6	TBR	2 km
WA 424	Perkins Beach (N)	6	TBR	1.7 km
WA 425	Perkins Beach (S)	6→5	TBR→LTT	300 m
WA 426	Port Harding	4	LTT	1.6 km
WA 427	Cosy Corner	4	R	200 m
WA 428	Migo Island	3	R	300 m
Spring & neap tidal range = 0.6 & 0.2 m				

Port Hughes and Port Harding are part of a 5 km wide southeast-facing embayment bordered by Hanging Rock and Shelter Island to the north and Forsyth Bluff to the south. The northern half of the bay is relatively exposed, while to the south a series of small islands extend 3 km to the east and afford increasing protection to the shore, which was used as an anchorage in the 19th century and is still used by fishing boats today. Between the lee of Shelter Island and Migo Island to the south are six near continuous beaches. Most are accessible by car and cleared farmland lies to the west.

Port Hughes beach (**WA 423**) is a south-facing, curving 2 km long beach that commences in lee of Shelter Island (also Mutton Bird Island), the only potential anchorage in the 'port', and curves around to the mouth of Torbay Inlet in the west (Fig. 4.100). It receives waves averaging about 1.5 m, which maintain a 70 m wide rip-dominated surf zone, with usually 5-6 beach rips occurring. It is backed by bluffs, then vegetated transgressive dunes, that have extended up to 2 km inland up onto the rear of Hanging Rock, where they reach elevations of 60 m. There is access along Mutton Bird Road to the northern tip of the beach. The local surfers drive along the beach, which can offer a range of *beach breaks*.

Figure 4.100 The usually closed Torbay Inlet separates Port Hughes beach (right) from Perkins Beach.

Perkins Beach (**WA 424**) commences on the southern side of the often blocked inlet mouth and trends southwest for 1.7 km, with a central foreland formed due to wave refraction around Seagull Island. It terminates at a 100 m long, 20 high metasedimentary point that separates it from the southern part of the beach. The main beach receives waves averaging 1.5 m in the north decreasing to just over 1 m at the point. Beach rips dominate most of the beach, grading to a low tide terrace close to the point. The best *surf* is up the beach as the waves, rips and quality of the beach breaks improve. It is backed by a well vegetated 10-20 m high, 200-300 m wide foredune system, and then farmland. There is access to the beach along a 4 WD track that runs off the Perkins Beach road.

Beach **WA 425** commences on the southern side of the point and trends southwest for 300 m to a prominent cuspate foreland formed in lee of an inshore reef, with the beach just attached to the inner reef (Fig. 4.101). The beach receives waves averaging about 1 m, which break across a 30 m wide low tide terrace, which decreases in width towards the foreland. The beach is bordered by the point and reef and backed by a densely vegetated 10 m high, 100 m wide foredune then farmland. The Perkins Beach Road terminates on the southern side of the foreland at a small car park. The Kennedy Youth Camp is located in lee of the foreland.

Figure 4101 A small rocky islet and sandy foreland separates the southern end of Perkins Beach (right) from Port Harding beach.

Port Harding beach (**WA 426**) commences at the foreland and curves to the southwest, then south for 1.6 km to a small rock outcrop. Wave height averages about 1 m slowly decreasing down the beach. The waves break across a continuous 20 m wide low tide terrace. A small creek, feed by a drainage system, crosses the beach just south of the foreland adjacent to the car park, while a low vegetated foredune backs the beach. The Cosy Corner Road terminates that the southern end of the beach. Beach **WA 427** extends from the southern side of the rocks for another 200 m to a 200 m long section of low rocky coast. The beach faces east and is sheltered by inshore reefs and rocks with waves averaging less than 1 m and reflective conditions prevailing. Patchy seagrass meadows grow close to shore. A low foredune and solitary fishing shack back the beach.

Beach **WA 428** is located to the lee of Migo Island and extends for 300 m between the small northern point and the northern rocks of Forsyth Bluffs. The beach protrudes slightly in the centre in lee of a small reef. Waves average about 0.5 m and surge up a reflective beach face, with seagrass growing close to shore. Fishing boats moor in lee of the island and a steep vehicle track descends to the centre of the beach, which is used to launch small dinghies to access the boats.

WEST CAPE HOWE NATIONAL PARK

Area:	3517 ha	
Coast length:	20 km	(1510-1530 km)
Beaches:	4	(WA 429-433)

West Cape Howe is an 80 m high calcarenite-capped, granite headland, that together with Torbay Head forms a prominent headland that is the southernmost tip of the Western Australian mainland at 35°00'08"S. The national park encompasses much of the massive Pleistocene and Holocene dune systems that have blown up over the basement granite to form the 100-150 m high calcarenite cliffs, and then blown inland for 2-3 km reaching elevations of 250 m. The entire system is now vegetated in dense coastal heath. There is access to the park coast at Shelley Beach, where camping is permitted, and along the William Road to the 100 m high cliffs above the eastern end of Bornholm Beach, one of the highest energy beaches on the south coast.

WA 429-431 DINGO-HORTON-DUNSKY

No.	Beach	Rating	Type	Length
WA 429	Dingo Beach	4	R	900 m
WA 430	Horton Beach	4	R	800 m
WA 431	Dunsky Beach	3	R	100 m
Spring & neap tidal range = 0.6 & 0.2 m				

The eastern shores of **West Cape Howe National Park** extends from Forsyth Bluff to Torbay Head, with West Cape Howe located 2 km to the west. The eastern shore consists of 8 km of steep 100-150 m high calcarenite-capped, metasedimentary rocks grading to granite at the Head. Three protected beaches (WA 429-431) are located below the steep slopes.

Dingo Beach (**WA 429**) lies on the southern side of Forysth Bluff and is a 900 m long southeast-facing steep reflective beach, backed by densely vegetated slopes rising from 40-90 m. The slopes consist of transgressive dunes that have blown across the cape to this leeward shore. Waves average about 1 m and surge up the steep beach faces. The Torbay Beach Road terminates on the crest of Forysth Bluff, with a steep walking track descending down to the beach.

Shelley Beach (**WA 430**) is located 1 km to the south and is an 800 m long, east-facing beach, also known as Horton Beach. It is backed by densely vegetated dune-draped granite slopes, which rise from 100 to 150 m. A steep narrow valley backs the northern end of the beach from which a small creek emerges to cross the beach during heavy rain. The beach receives low reflected waves averaging about 1 m, which surge up the steep reflective beach (Fig. 4.102). The Horton-Shelly Beach Road runs down the northern side of the valley to the beach where there is a small camping area. Commercial fishermen also use the beach as a base to launch small boats, while a hang and kite gliders launch off a northern ridge top platform to glide along the face of the slopes.

Figure 4.102 Shelly Beach is a sheltered steep reflective beach.

Dunksy Beach (WA 431) lies on the northern side of Torbay Head and faces north. It is a 100 m long pocket of sand, backed by scarped vegetated transgressive dunes that have moved over the Head. Low waves surge up the steep narrow beach, with seagrass growing 50 m off the shore. A rough steep 4 WD track descends to the rear of the beach.

WA 433-436 BORNHOLM-LOWLANDS

No.	Beach	Rating	Type	Length
WA 432	Golden Gate Beach	9	TBR+rocks	200 m
WA 433	Bornholm Beach	9	TBR/RBB+rocks/reef	2.5 km
WA 434	Bornholm (W)	10	RBB+rocks/reef	100 m
WA 435	Lowlands (1)	7	TBR	100 m
WA 436	Lowlands (2)	7	TBR	200 m
Spring & neap tidal range =0.6 & 0.2 m				

Between West Cape Howe and Knapp Head is an 11 km wide, open southwest-facing embayment, for the most part exposed to high waves and strong onshore winds. Most of the original beach systems have been eroded leaving two remnants along the southwest-facing shore (WA 432 & 433) and the two Lowlands beaches (WA 435 & 436) tucked in a small embayment at the western end.

Golden Gate Beach (WA 432) is a 200 m long strip of sand located at the based of 50 m high densely vegetated slopes and bordered by rounded granite, capped by calcarenite. It receives waves averaging about 2 m, which break across a 100 m wide surf zone, with rocks littering the beach and inner surf. A strong permanent rips runs out against the sloping western granite rocks. A 4 WD track called the 'Golden Road' reaches the top of the beach.

Bornholm Beach (WA 433) is an exposed, high energy 2.5 km long southwest-facing beach. It is located at the base of 80-100 m high, steep vegetated slopes. It commences in the east against an irregular granite point that protrudes 200 m seaward, and continues west, wedged in between the base of the slopes and some patches of beachrock reef. It receives waves exceeding 2 m on average which break across a 200 m wide rip-dominated surf zone, with usually three large central beach rips, and permanent reef-controlled rips to either end. There is 4 WD access via the rough Lake William 'road' to the eastern end of the beach, and via the Shepherds Lagoon 'road' to the centre of the beach, with a rough 4 WD track leading down onto the beach. While this is a relatively popular beach for fishing, it is unsuitable for swimming, and should only be surfed by those experienced with such conditions.

Beach **WA 434** is a 100 m long pocket of sand located at the western end of the calcarenite cliffs, which continue for 3.5 km west of Bornholm beach. The beach is backed and bordered by 50 m high cliffs and littered in cliff debris. It receives waves averaging about 2 m, which break across a 150 m wide surf zone, including a beachrock reef, with one strong rip draining out the western side of the small embayment. A steep foot track leads down the western slopes to the beach.

The **Lowlands** beaches (WA 435 & 436) are located 500 m to the west and share a 700 m wide southeast-facing, semi-circular embayment (Fig. 4.103). The northern beach (**WA 435**) is a 100 m long pocket of sand wedged in a small south-facing embayment, with granite points extending 100 m seaward on either side, and a small creek draining across the western end of the beach. It receive waves averaging about 1.5 m which break across a 100 m wide surf zone and drain out via strong permanent rip that exits along the lower eastern granite point. The gravel Tennessee Road terminates at a small car park on the eastern slopes overlooking the beach. This is an accessible beach from which to check the surf and reasonably popular with locals.

Figure 4.103 The two adjoining Lowland beaches are popular surfing locations.

The southern Lowlands beach (**WA 436**) is located 300 m to the west in a second small embayment. It is a 200 m long southeast-facing beach, backed and bordered by smooth vegetated, dune-draped 20-30 m high slopes. The beach receives slightly higher waves, which break across a 150 wide surf zone, with permanent rips running out along both boundary headlands. Four wheel drive tracks reach the backing slopes, with foot access down the eastern headland.

WA 436-443 KNAPP HD-RATCLIFFE BAY

No.	Beach	Rating	Type	Length
WA 437	Knapp Hd (W1)	9	LTT+reef	250 m
WA 438	Knapp Hd (W2)	9	LTT+rocks/reef	250 m
WA 439	Knapp Hd (W3)	9	LTT+rocks/reef	100 m
WA 440	Anvil (E)	8	LTT+rocks/reef	300 m
WA 441	Anvil Beach	8	TBR+reef	1 km
WA 442	Ratcliffe Bay (1)	8	R+reef	150 m
WA 443	Ratcliffe Bay (2)	8	TBR/RBB	2 km
Spring & neap tidal range = 0.6 & 0.2 m				

Ratcliffe Bay is an open, southwest-facing bay, located between Knapp Head and Wilson Head, 15 km to the west. Most of the bay shore is exposed to the full force of the southerly waves and winds, which throughout the Pleistocene have built massive calcarenite dunes and cliffs the length of the bay shore. The dunes average 100 m in height and extend up to 3 km inland. The early Holocene beach/es that ran from Knapp Head to Wilson Inlet and supplied the dune system, has been largely eroded with only a few pockets of sand in the east (beaches WA 427-439), while a little more sand remains around the inlet mouth (beaches WA 440-444) as the beach becomes slightly sheltered by Wilson Head. Most of the shore between the head and inlet is dominated by 100-150 m high calcarenite cliffs.

Beach **WA 437** is located 1.5 km northwest of Knapp Head and is a 250 m long strip of sand at the base of 150 m high cliffs and steep slopes, with a near continuous beachrock reef lying 10-40 m offshore. Waves, averaging over 2 m break on the reef and reform in the lee 'lagoon' to reach the beach as lower waves. A strong permanent rip drains out the eastern end of the lagoon. There is no safe foot access to the beach.

Beach **WA 438** lies 1 km to the west and is a similar 250 m long narrow strip of high tide sand backed by 160-180 m high cliffs, with a major land slide immediately east of the beach and rock debris littering the beach and inshore. A beachrock reef lies 100 m off the beach, with waves breaking heavily on the reef and reforming in the backing lagoon, where they break again across the rock-littered inner bar. There is a steep walking track down to the rocks 200 m west of the beach.

Beach **WA 439** is a 100 m long strip of sand at the base of densely vegetated steep slopes rising to 140 m. It lies to the lee of a continuous beachrock reef located 20-30 m offshore, with an irregular lagoon between the reef and beach. The eastern end of the beach and backing slopes protrude seaward to attach to the reef, with an elongate lagoon extending to the west. Rock debris litters the beach and the high waves break on the reef.

To the west of beach WA 439, 100 m plus high calcarenite cliffs extend for 9 km west-northwest to beach WA 440. Beach **WA 440** is the first of the five near continuous beaches located at the western end of Ratcliffe Bay. It is a 300 m long strip of high tide sand cut into three 100 m long sections by rock debris in the east and a slight protrusion of the bluffs in the west. A continuous beachrock reef lies 50-100 m off the beach, with a rock and sand-filled lagoon in between. High waves reach the eastern end of the beach decreasing to the west as the lagoon widens. The beach is backed by steep vegetated 20-30 m high bluffs, with a 4 WD track reaching the top of the western bluffs.

Anvil Beach (WA 441) lies immediately to the west and is a curving 1 km long beach that lies in lee of beachrock reef, with a 200 m wide central gap (Fig. 4.104). Waves break heavily on the reef and pass through the gap to break again inside the 'lagoon', which is 300 m wide in the centre. A large permanent rip drain out from the centre of the beach and through the gap while a second beachrock reef also parallels the central shoreline of the beach. The beach is backed by well vegetated 20-30 m high bluffs in the east grading to a lower foredune in the west. Massive vegetated transgressive dunes then extend 3 km northeast to the shores of Wilson Inlet. The same 4 WD track terminates above the eastern bluffs.

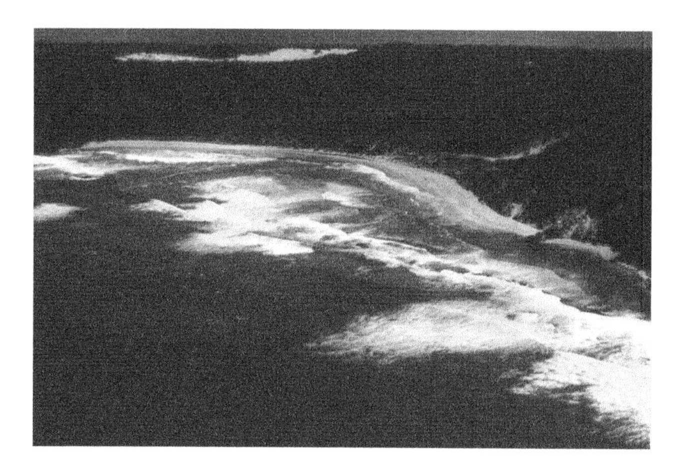

Figure 4.104 Anvil Beach receives high surf, which breaks across a beachrock reef and feeds into a strong central permanent rip.

A 200 m long calcarenite headland separates beach WA 441 from **WA 442**, which is a curving 150 m long strip of high tide sand located in a gap between two headlands. High waves break across a bar 200 m off the beach then a beachrock reef, with a strong permanent rip draining out from the centre of the small beach. It is backed by partly exposed calcarenite bluffs rising to 40 m.

Beach **WA 443** commences 100 m to the west and is a curving, double crenulate 2 km long beach. It is attached to a beachrock reef and calcarenite headland in the east, with a smaller reef inducing a central foreland, then sand extending to the variable mouth of Wilson Inlet in the west, where the beach curves round to face southeast as a bare sand spit. Usually 4-5 large beach rips dominate the 150-200 m wide surf zone. Two active blowouts and vegetated longwalled parabolics back the beach and extend up to 2.5 km northeast to Wilson Inlet.

WA 444 OCEAN BEACH (DENMARK SLSC)

No.	Beach	Rating	Type	Length
WA 444	Ocean Beach	7	TBR+inlet	500 m

DENMARK SLSC
Patrol period: Saturdays January
Sundays December-March
Lifeguard Mon-Fri mid Dec. -January
Spring & neap tidal range = 0.6 & 0.2 m

Ocean Beach (WA 444) is the surfing beach for the town of Denmark and site of the Denmark Surf Life Saving Club (Fig. 4.105). The town is located 5 km north of the beach on the banks of the Denmark Inlet, which flows into Wilson Inlet. The inlet in turn flows into the sea at Ocean Beach and its usually closed mouth forms the northern boundary of the beach. The southern boundary is the 100 m high Wilson Head. A sealed road from Denmark runs along the western shore of the inlet right to the beach, with two car parks on the bluffs at the southern end above the Surf Life Saving Club. A caravan park is located 500 m to the west just inside the inlet.

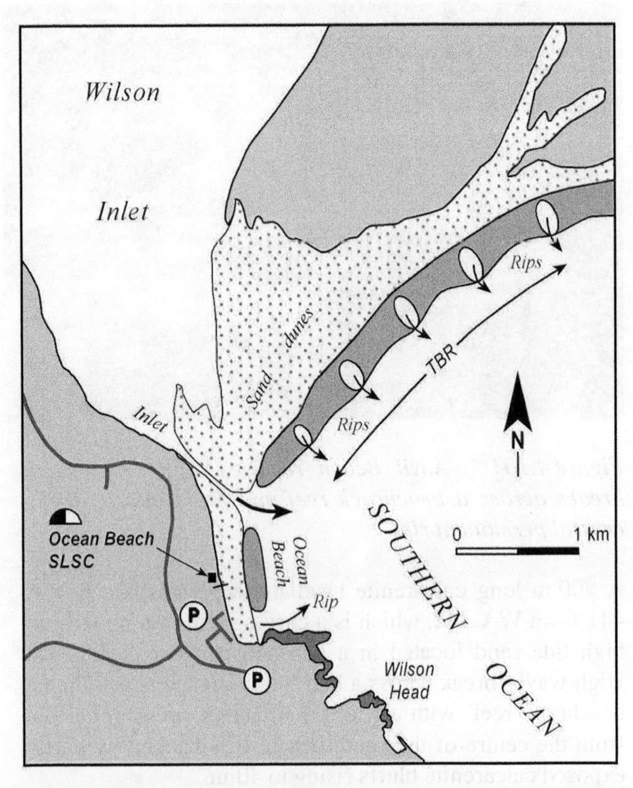

Figure 4.105 Ocean Beach is located at the mouth of Wilson Inlet and is one of the more hazardous beaches on the South Coast. Beware of strong rips and the tidal flow through the inlet.

The beach is about 500 m long, the length depending on the state of the inlet entrance. It faces east and receives moderate protection from the dominant southwesterly waves. It is backed by moderately steep, vegetated slopes rising to 20-30 m, with the Surf Life Saving Club perched on the slopes and the car park to the rear. Waves average over 1 m in front of the Surf Club, but rapidly increasing in height toward the inlet. The waves are sufficient to maintain a 100 m wide usually attached bar, with a permanent rip against the southern rocks. The bar detaches from the beach and the rips increase in size and intensive east of the inlet. When the inlet is open strong tidal currents usually flow just off and sometimes along the beach (Fig. 3.1c), so be careful as a double tragedy in the southern rip lead to the formation of the Surf Club in 1958.

Swimming: This is a potentially hazardous beach, with the least hazardous swimming on the inner portion of the bar in front of the Surf Club. Be very careful if waves exceed 1 m, as strong rips are generated, including in the southern corner against Wilson Head. Do not swim near the inlet, particularly if it is open. Strong rips prevail from the inlet east.

Surfing: Ocean Beach is a popular surfing spot with beach breaks right along the beach and east past the inlet.

WA 445-450 POINT WALTON

No.	Beach	Rating	Type	Length
WA 445	Pt Walton (W1)	9	TBR+reef	400 m
WA 446	Pt Walton (W2)	9	TBR+reef	400 m
WA 447	Back Beach	9	RBB+reef	1.8 km
WA 448	Pt Walton (W4)	9	TBR+reef	250 m
WA 449	Pt Walton (W5)	9	TBR+reef	200 m
WA 450	Pt Walton (W6)	9	TBR+reef	100 m
WA 451	Pt Walton (W7)	9	R/LTT+reef	200 m
WA 452	Pt Walton (W8)	9	LTT+reef	150 m

Spring & neap tidal range =0.6 & 0.2 m

Point Walton is a 120 m high headland formed of calcarenite-capped Proterozoic gneiss. It forms the boundary between Ratcliffe Bay to the east, and the next beach system to the west, which commences at the point and extends 5 km west to Light Beach. The once continuous beach has been eroded and now consists of eight, southwest-facing, rocks and reef bound remnants (WA 445-452). The beaches are backed by a continuous densely vegetated Holocene transgressive dune system that extends up to 3 km northeast to reach the shores of Wilson Inlet and Ocean Beach. The dunes overlie Pleistocene calcarenite, which outcrops along the beach as beachrock reef and calcarenite bluffs. The only vehicle access is in the west at beach WA 450.

Beach **WA 445** commences immediately west of Point Walton, with steep, vegetated 100 m high slopes backing the beach. Rock debris litters the eastern half, while beachrock reefs dominate the surf zone. Waves averaging up to 2 m break across a 200 m wide sand and reef surf zone, with a strong permanent rip existing towards the western end of the beach. There is no vehicle access to the beach, with foot access only along the shore from beach WA 447.

Beach **WA 446** lies 300 m to the west and is a 400 m long strip of high tide sand fronted by near continuous beachrock reef at the shoreline, as well as scattered reefs up to 100 m offshore. High waves break across a 200 m wide reef and sand surf zone, with a central permanent rip draining the beach. Steep vegetated slopes rising up to 70-100 m back the beach, with no vehicle access.

Beach (WA 447) is the main beach at 1.8 km in length. It is also the highest energy with waves averaging about 2 m and breaking across a 200-300 m wide surf zone, which includes large beach rips as well as a central reef extending up to 500 m offshore. A 4 WD track reaches the top of the 50 m high vegetated bluffs in lee of the central reef, with foot access down to the beach. Five beach and reef-controlled rips dominate the surf and a mixture of beachrock reef and rocks are scattered along the beach and in the inner surf zone. It gradually narrows to the west terminating at a 200 m long section of rocky bluffs.

Beach **WA 448** commences on the western side of the bluffs and continues west for 250 m. It is initially fronted by a continuous beachrock reef, which lies off the beach enclosing a 100 m long tidal pool. The western half of the beach is open and both sections are fronted by a 200-300 m wide surf zone, with rips to either end. It is backed by steep 50 m high vegetated bluffs, with access along the shore from the adjacent beaches.

Beach **WA 449** lies 100 m to the west and is a 200 m long beach bordered by a long beachrock point to the east and low rounded granite to the west. The eastern beachrock also lies in the inner surf and some rocks outcrop on the centre of the beach. It receives waves averaging over 1.5 m, which break across a 200 m wide surf zone, with rips to either end. It is backed by a 20 m high foredune, then steep, vegetated bluffs rising to 40 m. It can be accessed from the neighbouring beach WA 450.

Beach **WA 450** is a 100 m long pocket of sand wedged in between a low, massive granite point that extends 100 m into the 200 m wide surf zone. The surf drives a permanent rip that runs out along the western point. A gravel road terminates at a car park on top of the 20 m high western bluffs and provide a good view east along the beaches.

Beaches WA 451 and 452 extends west of the car park and are dominated by granite outcrops (Fig. 4.106). Beach **WA 451** is a curving 200 m long south-facing beach that is bounded by two 100 m wide granite reefs, both forming sandy forelands to their lee, with the beach linking the forelands. Wave enter a 100 m wide gap between the reefs and are lowered to about 1 m at the shore where they surge up a steep reflective to low tide terrace beach. This in backed by a steep 15 m high foredune, with the car park on the bluffs above the eastern end of the beach.

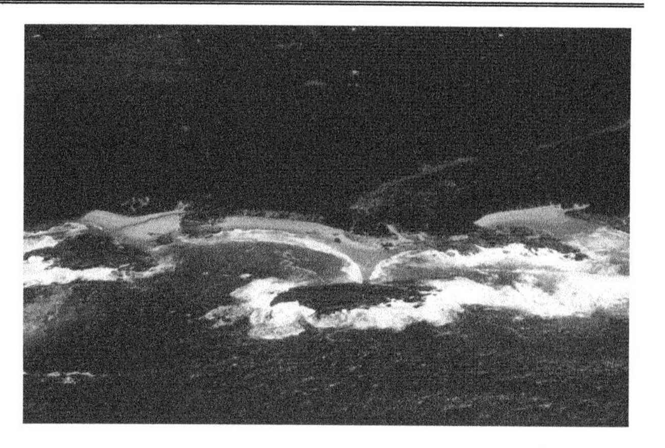

Figure 4.106 Beaches WA 451 and 452 are dominated by granite reefs and islets, which result in small curving beaches and salients.

Waterfall beach (**WA 452**) lies immediately to the west. It consists of a 150 m long beach formed in lee of the western granite reef to which it attaches at low tide. A creek flows down over Pleistocene coffeerock to form a small waterfall, which flows out across the centre of the beach. To the east of the reef the beach has lower waves and is reflective, while to the west it is more exposed with waves breaking into a 50 m wide, 100 m deep bedrock channel, which is drained by a strong permanent rip.

WILLIAM BAY NATIONAL PARK

Area:	approx. 1300 ha	
Coast length:	11 km	(1562-1573 km)
Beaches:	15	(WA 450-464)

William Bay National Park is a relatively small park that includes Edward Point and adjacent Mazzoletti Beach. The beach is backed by an area of densely vegetated Holocene and Pleistocene transgressive dunes, and includes one inland patch of active dunes. The shoreline around the low point is dominated by numerous granite boulders, reefs and islets, that together with the intervening sands produce a number of small attractive beaches and bays, including Madfish Bay, Smooth Pool, Elephant Rocks Beach and Green Pool. Mazzoletti Beach extends west of the point, terminating at the entrance to Parry Inlet, with the park incorporating the eastern side of the inlet.

WA 453-459 **LIGHTS BEACH-MADFISH BAY**

No.	Beach	Rating	Type	Length
WA 453	Waterfall (W)	7	LTT+reef	100 m
WA 454	Lights Beach (E 2)	5	R+reef	100 m
WA 455	Lights Beach (E 1)	5	R+reef	30 m
WA 456	Lights Beach	4	R+reef	150 m
WA 457	Madfish Bay (1)	3	R	80 m
WA 458	Madfish Bay (2)	3	R/LTT	50 m
WA 459	Madfish Bay (3)	3	R+sand flats	50 m

Spring & neap tidal range = 0.6 & 0.2 m

The eastern side of **Edwards Point** extends northeast for 1.5 km to Lights Beach. The generally low shoreline is dominated by numerous outcrops of rounded granite boulders both at the shore and off the beaches, where they form reefs and small islets. A gravel road off the Williams Bay Road runs out to Lights Beach and Madfish Bay both popular locations for picnicking, swimming and fishing. Beach **WA 453** lies 200 m west of Waterfall Beach and is a 100 m long south-facing beach wedged in between low sloping granite points, with a narrow line of granite down the centre dividing it in two. Waves averaging 1.5 m break over granite reefs, 100 m off the beach , while the rip from Waterfall Beach flows between the reefs and past the beach. The beach can only be accessed along the shoreline from Lights Beach or Madfish Bay.

Beaches WA 454-456 are three small adjoining beaches located in a 500 m wide southeast-facing, reef-strewn embayment. There is a gravel road to a car park behind the western Lights Beach (WA 456), which provides foot access along the shore to the adjoining beaches. Beach **WA 454** is a 100 m long narrow strip of sand bordered by low granite points, with a small granite outcrop off the centre of the beach. Waves average about 1 m and break across a narrow continuous bar. Beach **WA 455** is a 30 m long pocket of sand, wedged in a gap in the centre of a low granite point that separates the Light beaches. Waves break at the entrance to the gap, and across a narrow bar where a small permanent rip drains the beach. **Lights Beach (WA 456)** is a steep 150 m long reflective beach, with waves averaging about 1 m, which produce a heavy surging shorebreak. Deeper water lies off the beach, with two granite reefs 50-100 m offshore, with other reefs and islets further out.

Madfish Bay occupies the next embayment. The bay and its three beaches (WA 457-459) are all located to the lee of a 500 m long, shore-parallel granite islet. The islet and other rocks and reef shelter the beaches with waves averaging less than 1 m at the shore. Beach **WA 457** is an 80 m long pocket of sand with a steep reflective beach. Beach **WA 458** lies 100 m to the west and is a 50 m pocket of sand with sloping granite to either side. It is fronted by a partly submerged 200 m wide sand spit, which connects the shore with the islet. Beach **WA 459** lies 300 m to the west and 100 m inside a gap in the rocks, into which flows a small rocky stream. The 50 m wide beach consists of exposed high tide sand and intertidal sand and creek flats, with rocks dominating the shoreline. Waves are very low to usually calm. A central car park and walking tracks lead to all three beaches.

WA 460-462 **EDWARDS PT**

No.	Beach	Rating	Type	Length
WA 460	Smooth Pool	3	R+sand flats+rocks	50 m
WA 461	Elephant Rocks	4	R+rocks	60 m
WA 462	Green Pool	5	R/LTT+rocks	750 m
Spring & neap tidal range = 0.6 & 0.2 m				

On the western side of **Edward Point** is a 2 km long section of granite shoreline dominated by massive sloping granite rocks and boulders and backed by moderately sloping dune-draped slopes. Three beaches occupy this section, two small pockets of sand (WA 460 & 461) and the longer Green Pool (WA 462). Access to the three beaches is via the Green Pool car park, with a 400 walk to Smooth Pool and 100 m walk to Elephant Rocks Beach.

Smooth Pool (WA 460) is a 50 m wide pocket of sand wedged in a small V-shaped gap in the rocks. It is protected by extensive granite rocks and reefs extending up to 150 m seaward of the shore with waves break heavily over the outer rocks. The 'pool' occupies the shallow, usually calm, sandy channel between the rocks.

Elephant Rocks Beach (WA 461) is named after the two large elephant-like granite boulders that occupy the upper part of the beach. The steep 60 m long beach is located in a 60 m wide gap in the bordering granite. At the shore it is fronted by more boulders, which narrow the small bay to about 20 m in width. It widens past the boulders into a protected deep pool in lee of more granite reefs extending 100 m seaward, with high wave breaking on the outer rocks.

Green Pool (WA 462) is the main beach located 100 m west of the car park. It is a 750 m long south-facing beach that is afforded protection from a string of granite rocks and reefs extending up to 500 m offshore. The protection is greatest in the east where a relatively calm 'green' pool lies between the steep reflective beach and the rocks. Wave height increases slowly to the west, where a combination of low tide terrace and rocks form the shore.

WA 463-467 **MAZZOLETTI-PARRY BEACHES**

No.	Beach	Rating	Type	Length
WA 463	Mazzoletti (E)	7	TBR+rock	700 m
WA 464	Mazzoletti Beach	7	TBR	5.7 km
WA 465	Parry Inlet	5→4	TBT→LTT	700 m
WA 466	Parry Beach	4	R/LTT	200 m
WA 467	Parry Beach (S1)	4	R+rocks	20 m
Spring & neap tidal range = 0.6 & 0.2 m				

William Bay is a 6 km wide south to southeast-facing exposed embayment bordered by 20 m high Edward Point to the east and 160 m high Point Hillier to the west. The shoreline is dominated by rocks to either end, with a central more sandy section occupied by the Mazzoletti and Parry beaches (WA 463-457). The shoreline and beaches between Edward Point and Parry Inlet are located within the national park. The only vehicle access is in the west along the Parry Road to the mouth of the inlet and Parry Beach.

Mazzoletti Beach consists of two parts (WA 463 & 464). Beach **WA 463** occupies the eastern 700 m of the beach and is a curving, south-facing, moderately exposed beach. It is tied to low granite rocks that separate it from Green Pool in the east and a sandy cuspate foreland formed in lee of granite reefs in the west. Rocks and reefs also lie off and along parts of the beach and lower waves to about

1.5 m at the shore. The waves maintain a 70 m wide surf zone, dominated by three permanent reef-controlled rips. As the waves pick up so does the surf, with reef, then beach breaks extending west to Parry Inlet. There is no formal access to the beach, though these is at least one 4WD track across the backing dunes to the shore.

The main **Mazzoletti Beach (WA 464)** is a slightly curving south to southeast-facing sandy beach, bordered by the sandy foreland to the east and the narrow entrance to Parry Inlet in the west. It receives waves averaging up to 2 m, particularly in the central-east, which break across a 150 m wide surf zone usually dominated by ten beach rips, the waves and rips decreasing slightly to the east. Vegetated transgressive dunes extend up to 2 km inland in lee of both beaches. The national park occupies a narrow strip along the beach and foredune, with private land behind, including one private residence. A few informal 4 WD tracks reach the beach, which is otherwise inaccessible by car.

Parry Inlet is a 10 ha coastal lagoon, feed by the Kordabup River. It breaks out across the junction of Mazzoletti and Parry beaches during the winter rains. Parry Inlet beach (**WA 465**) extends from the inlet mouth for 700 m to the south, with wave height decreasing southward from about 1.5 m to 1 m. The beach initially has a 50 m wide surf with occasional beach rips, grading into lower energy low tide terrace, terminating at a small sandy foreland tried to a low partly submerged granite rock. The Parry Road runs 300 m in from the beach, with a vehicle track just behind the beach.

Parry Beach (WA 466) commences on the southern side of the foreland and curves to the south for 200 m to the start of a 600 m wide granite point (Fig. 4.107). This beach is protected by the point and reefs, with waves averaging about 1 m, which break across a narrow low tide terrace backed by a low gradient beach face. There is a large car park, camping and caravan area behind the southern end of the beach, which is also used to launch small boats.

Figure 4.107 Parry Beach (centre) curves between granite rocks and is sheltered by the southern point.

Beach **WA 467** is located 100 m to the southeast and consists of a 20 m wide pocket of sand wedged in a gap in the sloping granite rocks of the point. The beach is almost encased in boundary rocks, with a small sand blowout behind the beach, perhaps exacerbated by pedestrian traffic from the car park.

WA 468-473 **POINT HILLIER**

No.	Beach	Rating	Type	Length
WA 468	Parry (S2)	5	LTT+rocks	150 m
WA 469	Pt Hillier (N)	5	R+rocks	200 m
WA 470	Pt Hillier	7	TBR+rock	1 km
WA 471	Pt Hillier (S)	6	LTT+rocks	300 m
Spring & neap tidal range = 0.6 & 0.2 m				

South of the small Parry Beach settlement is a 20 m high point that forms the western boundary of the Parry beaches, with the larger Point Hillier located 1.5 km to the south. In amongst and between the points are four beaches (WA 468-471). There is vehicle access to the first point and to a sand quarry high above the main Point Hillier beach (WA 470).

Beach **WA 468** lies at the eastern apex of the first point, and consists of a curving 150 m long, east-facing, steep reflective beach. It receives waves averaging about 1 m, which refract into the small bay and past a few granite reefs that lie on and just off the beach. The beach is backed by dune-draped granite, with foot access from the Parry Beach car park 300 m to the north.

Beach **WA 469** is located on the southern side of the first point and is a south-facing 200 m long curving reflective beach, located in between boundary granite rocks, with reefs extending 200 m offshore and lowering waves to less than 1 m at the shore.

Beach **WA 470** is the main 1 km long beach and connects the northern point with the first rocks of Point Hillier. The beach faces east and receives waves averaging 1.5 m in the north, decreasing due to wave refraction, to about 1 m in the south. Three permanent rips and a shore parallel section of beachrock dominate the northern 100 m wide surf zone, with a 50 m wide low tide terrace to the south. The northern rip is located below the 4WD access point and is the strongest of the rips.

Beach **WA 471** is separated from the main beach by a protruding section of dune-draped rocks. It consists of a 300 m long sliver of sand located at the base of the steep, vegetated slopes that rise 100 m to the lee of Point Hillier, the dune originating from the southern side of the point. The beach receives refracted waves averaging less than 1 m, which break across a 50 m wide sand and rock strewn surf zone.

WA 472-480 **PT HILLIER-BOAT HARBOUR**

No.	Beach	Rating	Type	Length
WA 472	Pt Hillier (W1)	9	LTT+reef	1.2 km
WA 473	Pt Hillier (W2)	9	TBR+reef	250 m
WA 474	Pt Hillier (W3)	9	TBR+reef	200 m
WA 475	Pt Hillier (W4)	9	TBR+reef	700 m
WA 476	Pt Hillier (W5)	9	RBB+reef	450 m
WA 477	Pt Hillier (W6)	9	RBB+reef	80 m
WA 478	Pt Hillier (W7)	9	RBB+reef	150 m
WA 479	Pt Hillier (W8)	7→6	RBB→TBR	1 km
WA 480	Boat Harbour	4	LTT	200 m
Spring & neap tidal range =0.6 & 0.2 m				

Point Hillier and Stanley Island form the eastern boundary of an open 7 km wide south-southwest facing embayment, with the Boat Harbour headland forming the western boundary. In between is an exposed high energy section of coast consisting of eight beach remnants at the base of calcarenite cliffs and slopes rising to 150 m and all fronted by shore parallel beachrock reefs. The small sheltered Boat Harbour bay offers the only protected section of the coast. The only formal vehicle access is via a 4 WD track to Boat Harbour.

Beach **WA 472** commences 2 km west of steep rocky Point Hillier. It is a relatively straight 1.5 km long high tide strip of sand, fronted by a near continuous beachrock reef along the shore, with high waves breaking on bars up to 70 m seaward of the reef. There are a few linear tidal pools between the reef and beach, with a permanent rip at the eastern end flowing through a gap in the reef. Following high waves the western end of the beach may be eroded back to the cliffs and separated from the main beach. It is backed by steep, vegetated cliffs rising to 130 m in the east and decreasing to 80 m in the west. The only access to the beaches is on foot from the 4 WD access track to beach WA 473.

Beaches **WA 473** and **474** are neighbouring 200 m long strips of sand, separated by a central outcrop composed of an elevated beachrock platform, with an active platform extending seaward. Both are bordered and partly fronted by beachrock points and reefs, with waves breaking over the reef and bars extending up to 100 m seaward. A permanent rip exits from beach WA 474. A 4 WD track descends the densely vegetated sloping dunes to reach the rear of beach WA 473.

Beach **WA 475** lies 400 m to the west. It consists of a 700 m long beach with gneiss outcropping in the east, including one reef tied to the beach by a small tombolo, and beachrock to the west. In addition there are scattered smaller reefs in the 150 m wide surf zone with permanent rips to either end. A 4 WD track from beach WA 474 runs along the base of the slopes to reach the eastern end of the beach. Beach **WA 476** commences immediately to the west and is a curving 450 m long beach bordered by a small calcarenite sea stack to the east and a protruding section of calcarenite slopes to the west. In between is a shore parallel beachrock reef lying 20-30 m off the beach and impounding an elongate lagoon, drained by a permanent rip at the eastern end, which flows past the stack. It is backed by steep eroding 100 m high dune-capped calcarenite. The only access is on foot from beach WA 475.

Beach **WA 477** is located on the western side of the small protrusion and is an 80 m long pocket of sand located at the base of steep, partly vegetated sandy slopes rising to 70 m. It is bordered by a beachrock reef in the east and the base of the cliffs to the west. Waves break up to 200 m offshore, with the surf drained by a permanent central rip. It can only beach reached on foot around the bluffs from beach WA 476.

Beach **WA 478** is an isolated 150 m long strip of sand located at the base of 80 m high partly vegetated calcarenite cliffs. Rock debris litters the beach and high waves break up to 200 m offshore. A skewed permanent rip drains the surf. The beach can only be reached by a climbing down the sloping eastern bluffs.

Beach **WA 479** is located towards the western end of the embayment. The 1 km long beach commences as a high energy beach fronted by a 200 m wide rip-dominated surf zone. As it curves to the southwest it is afforded increasing protection by the Boat Harbour headland, and waves decrease to less than 1.5 m and the surf narrows to about 50 m with four beach rips usually occupying the total surf zone. It is backed older well vegetated dunes rising to 40 m in the east and extend 2 km inland to bisect a coastal wetland. The dunes are breeched by two active blowouts, one extending 1 km inland up over the earlier dunes, then older dunes. A track off the Boat Harbour track leads out to the centre of the western dune-capped granite headland and provides a view up the beach, as well a 4 WD access onto the beach.

Boat Harbour breach (**WA 480**) is located in a 150 m wide gap between two rounded 20 m high granite points (Fig. 4.108). The curving 200 m long beach faces southeast and receives refracted waves averaging about 1 m. These break across a low gradient 40 m wide low tide terrace. The Boat Harbour track runs along the back of the beach with vehicle access to both points. There is a basic camping area, two fishing shacks and boat launching from the beach. Freshwater seeps from the backing slopes and drains across the beach.

WA 481-487 **OWINGUP**

No.	Beach	Rating	Type	Length
WA 481	Owingup (1)	5	R/LTT+rocks/reef	150 m
WA 482	Owingup (2)	8	TBR+reef	200 m
WA 483	Owingup (3)	8	TBR+reef	50 m
WA 484	Owingup (4)	8	R+reef	300 m
WA 485	Owingup (5)	6	R+reef	100 m
WA 486	Owingup (6)	8	TBR+reef	500 m
WA 487	Owingup (7)	8	TBR/RBB	700 m
Spring & neap tidal range = 0.6 & .2 m				

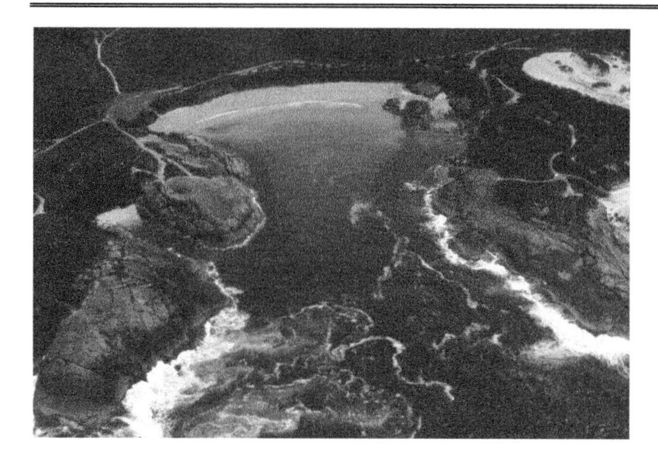

Figure 4.108 The sheltered Boat Harbour is well protected by the bordering granite points.

To the west of 60 m high Boat Harbour headland is a 4 km long south-facing open embayment bordered in the west by a 60 m high dune-capped gneiss headland. In between is a mixture of calcarenite cliffs rising to 60 m and seven predominately rock and beachrock reef-dominated beaches. The only access to the coast is via two 4 WD tracks from Boat Harbour that reach the top of the bluffs above beaches WA 481 and WA 484. The bluffs and beaches are backed by largely vegetated, longwalled parabolics that overlie the Pleistocene calcarenite that extends up to 2 km inland intruding part way into the Owingup Swamp.

Beaches WA 481 and 482 occupy a 600 m wide, 200 m deep, cliffed embayment 1 km west of Boat Harbour. Beach **WA 481** is located in the eastern corner. It is backed and bordered by vegetated calcarenite rising to 50 m, and fronted by massive granite rocks, including a large wave washed rock 100 m off the beach. The rocks lower waves to about 1 m at the beach, which usually has a narrow bar. While the water is relatively calm off the bar, a permanent rip does drain the small embayment between the islet and eastern rocks. There are 4 WD tracks to the crest of the eastern point and the backing bluffs.

Beach **WA 482** lies 200 m to the west in the western corner of the bay, and is a narrow 200 m long high tide beach, fronted by a continuous intertidal beachrock reef, then a sand and reef-strewn 200 m wide surf zone. A permanent rip drains out the eastern side to flow past the granite rock. It is backed by steep, scarped, 50 m high calcarenite cliffs, with a 1 ha patch of active clifftop dune above the eastern end of the beach.

Beach **WA 483** lies at the base of a 1.5 km long section of steep, vegetated 50-60 m high calcarenite cliffs. The beach is a 50 m patch of high tide sand littered with boulder debris, and fronted by a continuous strip of beachrock reef, then a predominately reef-strewn 100 m wide surf zone, drained by a strong central permanent rip. There is no safe access to the beach.

Beach **WA 484** is a 300 m long, narrow strip of high tide sand that runs along the base of densely vegetated dunes that gradually rise inland. The beach is littered with rocks and fronted by a continuous 50 m wide supratidal and intertidal beachrock reef, then a 100 m wide rip dominated surf zone. A 4 WD track winds through the dune to reach the centre of the beach.

Beach **WA 485** lies 100 m to the west and is a curving 100 m long pocket of sand located in lee of a 20 m wide gap in the beachrock reef. Waves break across a 100 m wide reef-strewn surf zone seaward of the gap, with a small calm lagoon between the gap and low energy sandy shoreline. The beach is backed by steep, partly vegetated 20 m high dunes and a 4 WD track that continues behind beach WA 486. It is bordered to the west by a small tombolo, with beach WA 486 linking the western side.

Beach **WA 486** extends from the tombolo for 500 m to the west, with more beachrock occupying the intertidal along the western 100 m of the beach. In addition scattered beachrock reefs extend up to 100 m offshore, with the high energy 200 m wide surf zone having multiple breaks over the reefs and patches of sandy seafloor. While waves are relatively low by the time they reach the shore, this is a very hazardous, reef-strewn surf zone drained by a strong central rip that flows out between the reefs.

Beaches WA 486 and 487 are separated by a rounded 200 m long, vegetated headland. Beach **WA 487** is a 700 m long southeast-facing sandy beach located in lee of the western gneiss headland. It receives waves averaging 1.5-2 m which break across a 100-150 m wide surf zone, with permanent rips to either end and a central beach rip (Fig. 1.2a). Wave height decreases slightly to the lee of the headland. It is backed by an unstable foredune, with active blowouts to either end, which periodically feed into a longwalled parabolic that has migrated 2.5 km inland to reach the shores of the Qwingup Swamp. There is a 4 WD track along the western headland.

WA 488-490 QUARRAM BEACH

No.	Beach	Rating	Type	Length
WA 488	Little Quarram	8	TBR/RBB+rock	500 m
WA 489	Little Quarram (W)	8	TBR/RBB+rock	50 m
WA 490	Quarram Beach	7	TBR/RBB	4.8 km
Spring & neap tidal range = 0.6 & 0.2 m				

Foul Bay is a curving 9 km wide, south to southeast-facing embayment, occupied by Quarram Beach to the east and the Peaceful Bay beaches to the west, with Irwin Inlet in the centre. The only access to Quarram is via 4 WD tracks in the east, while there is a sealed road to the small Peaceful Bay settlement.

Little Quarram Beach (WA 488) is a 500 m long, south-southwest-facing high energy beach, bordered by a dune-capped gneiss headland to the east and a small protruding gneiss bluff in the west, which separates it from beach WA 489 (Fig. 4.109). The beach is fronted by a 200 m wide surf zone with a strong permanent rip against the eastern headland and usually one central beach rip. It is backed by well vegetated dune-covered gneiss hills rising to 70 m. There is a 4 WD track to the eastern head and one running along the rear of the beach to reach beach WA 489.

Figure 4.109 Little Quarram Beach (foreground) is an exposed high energy rip-dominated system.

Beach **WA 489** is a 50 m long pocket of sand wedged in between two gneiss points. It shares the surf zone with its larger neighbours and has one permanent rip existing along the western boundary rocks. A 200 m wide vegetated sloping gneiss headland separates it from the main Quarram Beach (WA 490).

Quarram Beach (WA 490) is a 4.8 km long, south-southwest to south-facing, high energy beach. It receives waves averaging about 2 m, which maintain a 200 m wide rip dominated surf zone. It has a permanent rip against the eastern rocks and usually ten beach rips spaced about every 400 m. It terminates at the small, often closed, mouth of Irwin Inlet. The beach and its two eastern neighbours are backed by well vegetated, nested longwalled parabolic dunes, which have transgressed up to 4 km inland over Pleistocene dunes. There is one active blowout midway along the main beach, as well as dune instability within 1 km of the inlet mouth.

WALPOLE-NORNALUP NATIONAL PARK

Area:	21 500 ha	
Coast length:	53 km	(1598-1651 km)
Beaches:	43	(WA 491-533)

Walpole-Nornalup National Park covers 53 km of generally exposed high energy and rocky shoreline, containing 42 beaches, which occupy 30 km (60%) of the coast. Most of the beaches are backed by massive, vegetated transgressive dunes. There is road access to the park at Peaceful Bay, Conspicious and Mandalay beaches.

WA 491-496 **PEACEFUL BAY**

No.	Beach	Rating	Type	Length
WA 491	Peaceful Bay (N)	6→5	TBR→LTT	3.7 km
WA 492	Peaceful Bay	3	R	600 m
WA 493	Peaceful Bay (S1)	4	R+rocks	800 m
WA 494	Peaceful Bay (S2)	8	R+rock flats	150 m
WA 495	Peaceful Bay (S4)	7	R+rock flats	50 m
WA 496	Peaceful Bay (S5)	6	TBR+rocks	150 m
Spring & neap tidal range = 0.6 & 0.2 m				

Peaceful Bay is a small settlement containing several rows of houses and a large camping and caravan park right behind the main beach (Fig. 4.110). A sealed road runs out from the highway to the protected bay, which is a popular holiday and retirement destination. The settlement has a store and fuel, with a picnic area and toilets in the beachfront park that surrounds the main bay beach (WA 492).

Figure 4.110 The small Peaceful Bay settlement is fronted by the sheltered main beach.

This western side of Peaceful Bay faces southeast and is partly protected by its orientation and Point Irwin. It contains three main beaches (WA 491-493) and three smaller rock dominated beaches (WA 494-496) out towards the point. Irwin Inlet forms the eastern boundary of the beaches with a 1.5 km long channel connecting the main 1000 ha lagoon with the beach. The lagoon is feed by the relatively small Bow River.

Beach **WA 491** commences on the western side of Irwin Inlet and trends to the southwest, then south for 3.7 km to the northern foreland of Peaceful Bay. The foreland is formed in lee of series of gneiss reefs that extend a few hundred metres offshore. The beach correspondingly receives wave averaging about 1.5 m at the inlet, decreasing to less than 1 m at the foreland. The surf zone is 100 m wide with rips at the inlet, narrowing to a 20 m wide low tide terrace in the south. The beach is backed by several well vegetated foredune ridges, with some older transgressive dunes extending 1 km inland close to the inlet. A road runs along the rear of the beach to the inlet.

Peaceful Bay beach (**WA 492**) is a curving 600 m long east-facing beach, extending from the foreland in the north to a low dune-covered gneiss headland in the south. Several gneiss reefs also lie in the bay and partly block the entrance, resulting in low waves at the shore and a moderate gradient reflective beach. A few low well vegetated foredune ridges back the beach and converge at the foreland. The main road runs behind the ridges, with the caravan park then the settlement all located within 1 km of the shore. There is vehicle access to the beach in the south, where boats at launched, and a car park and picnic area on the foreland. Two navigation markers assist boats returning through the reefs to the beach, while there are surfable breaks over the reefs off the beach.

Beach **WA 493** commences on the southern side of the Peaceful Bay point and trends south for 800 m. It lies to the lee of one large central rock islet and several small gneiss rocks and reefs, resulting in waves averaging about 1 m at the shore. The waves surge up a relatively steep reflective beach, backed by moderately active dunes extending up to 500 m inland. The dunes were fenced in the 1990's and are in the process of being stabilized. There is 4 WD access to the beach across the point from Peaceful Bay.

Beaches WA 494 and 495 are two strips of high tide sand fronted by jagged gneiss rock platforms and located just south of beach WA 493. Beach **WA 494** is a 150 m long strip of sand fronted by a 50 m wide irregular supratidal platform, with a rock strewn 100 m wide surf zone. There is a 4 WD track to the northern end of the beach, which is used for rock fishing. Beach **WA 495** lies 100 m to the west and is a 50 m pocket of sand fronted by a small sandy tidal pool, then 50 m of gneiss platform, before the surf.

Beach **WA 496** lies another 100 m to the west in a small 100 m wide bay located 1 km north of Point Irwin. The southeast-facing beach is 150 m long, with rocks strewn along a 50 m wide low tide terrace backed by a sandy high tide beach. Waves average about 1 m and during higher wave generate a rip running out of the bay.

WA 497-500 **PT IRWIN-THE GAP**

No.	Beach	Rating	Type	Length
WA 497	Pt Irwin (W 1)	8	TBR+reef	150 m
WA 498	Pt Irwin (W 2)	8	TBR+reef	150 m
WA 499	The Gap(E)	8	RBB+rocks	100 m
WA 500	The Gap (W)	8	RBB+rocks	250 m
Spring & neap tidal range = 0.4 & 0.0 m				

At **Point Irwin** the shoreline turns and trends west for 6 km to Rame Head. In between is an exposed high energy, rock and reef-dominated shoreline containing 12 generally small beaches. There is access to the back of most of the beaches via a 4 WD track from Peaceful Bay. The first four beaches (WA 497-500) occupy the first two small rocky bays west of the point.

Beaches WA 497 and 498 are located 1 km west of 50 m high Point Irwin and consists of two small pockets of sand on an otherwise rocky shore. Beach **WA 497** is a 150 m long high tide sand beach, fronted by a rock-strewn inner surf zone. Rock reefs and a low gneiss point extend to either side of the 100 m wide outer surf zone with a permanent rip draining out the centre of the small embayment. It is backed by a blowout, then older vegetated dunes that have transgressed 1 km across the rear of the point to the Peaceful Bay shore. Beach **WA 498** lies 300 m to the west and is a similar 150 m long narrow high tide beach, bordered by irregular rock platforms, with a 100 m wide surf zone in between. It is backed by vegetated older dunes. There is 4 WD access to both beaches.

The Gap beaches occupy a 500 m wide south-facing embayment, located 2 km west of Point Irwin (Fig. 4.111). The eastern beach (**WA 499**) is a 150 m long, southwest-facing pocket of sand, wedged in between a gneiss point and a low rocky point that separates it from beach WA 500 with rocks also outcropping on the beach. Waves break on reefs 500 m offshore and along both sides of the narrow bay, with a strong permanent rip running out the centre. The western beach (**WA 500**) is a steep, slightly curving, 250 m long beach, with waves breaking on the same outer reefs and reforming to break heavily on the shore with a permanent rip running out the western side of the bay. A large blowout extending 500 m to the northeast backs the beach, together with vegetated transgressive dunes backing both beaches and extending 2 km towards Peaceful Bay. There is 4 WD access to both beaches.

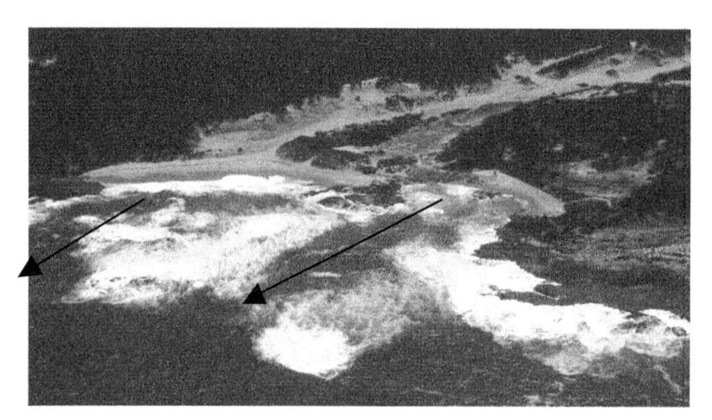

Figure 4.111 The Gap contains two exposed small beaches, both dominated by strong permanent rips (arrows).

WA 501-508 **HERRING ROCK-RAME HD**

No.	Beach	Rating	Type	Length
WA 501	Herring Rock (E3)	6	R/LTT+rocks	200 m
WA 502	Herring Rock (E2)	8	RRB+reef	1.8 km
WA 503	Herring Rock (E1)	6	LTT+platform	400 m
WA 504	Herring Rock	6	LTT/TBR+reef	450 m
WA 505	Herring Rock (W)	6	LTT/TBR+reef	150 m
WA 506	Rame Head (1)	7	TBR+reef	500 m
WA 507	Rame Head (2)	6	LTT+reef	150 m
WA 508	Rame Head (3)	7	TBR+reef	200 m
Spring & neap tidal range = 0.4 & 0.0 m				

Between The Gap and Rame Head are two curving embayments. The first terminates in lee of Herring Rock reefs and contains three beaches (WA 501-503), while the second curves round to Rame Head and contains five beaches (WA 504-508). All are exposed to high waves with considerable rock and reef presence in the surf zone and along the beaches. The entire shoreline is backed by generally vegetated transgressive dunes that have moved up to 2 km inland.

Beach **WA 501** is a curving 200 m long beach bordered by a gneiss platform to the east and tied to a small rock outcrop to the west, with reefs extending another 500 m offshore. High waves break on the reef, which lower waves reaching the shore to maintain a reflective to narrow low tide terrace beach. An active climbing blowout backs the beach extending 600 m inland and rising to elevations of 60 m.

Beach **WA 502** commences on the western side of the rock outcrop and curves round in a south-facing semi-circle for 1.8 km to the lee of **Herring Rock**. High waves break on a mixture of bars and reefs lying up to 300-400 m offshore, with usually two central beach rips and permanent rips to either end. It is backed by steep, vegetated calcarenite, which decreases from 60 m in the east to 20 m in the west, where a 4 WD track reaches the beach. The beach terminates at a vegetated 20 m high headland fronted by an irregular gneiss rock platform. Beach **WA 503** curves around the point for 400 m and is located between the base of the point and the platform. Waves break over the platform and reefs, resulting in generally low waves at the shore and a mixture of rocks and tidal pools at the base of the beach.

Beaches WA 504 to 508 occupy the shoreline of a curving 1.3 km wide southeast-facing bay between Herring Rock and Rame Head, with numerous rock reefs scattered across the bay. Beach **WA 504** is 450 m long with waves reduced by the reefs to about 1 m at the shore, which maintain a 30 m wide low tide terrace. It is backed by vegetated transgressive dunes, with a 4 WD track reaching the western end of the beach then running behind the point to reach beach WA 505. Beach **WA 505** lies 50 m to the west on the other side of a small gneiss point. It consists of a 150 m long pocket of sand wedged in between the low point and a 100 m wide western point. It also receives waves averaging about 1 m, which break across a low tide terrace. It is backed by steep vegetated dunes rising to 50 m.

Beaches WA 506-508 are located along the eastern side of 100 m high **Rame Head**. They can only be accessed on foot from beach WA 505. Beach **WA 506** lies on the western side of the point and is a curving 500 m long southeast-facing beach, located at the base of steep, vegetated 100 m high cliffs. It receiving waves averaging 1.5 m in the north, decreasing slightly to the south. These maintain a 70 m wide surf zone drained by two permanent rips. Beach **WA 507** is located immediately to the south and is a curving 150 m long beach bordered by two protruding low rocky points, with rocks also off the centre of the beach, which form a small calmer 'lagoon' off the narrow central section of the beach. Beach **WA 508** extends from the southern rocky point for 200 m to the eastern tip of Rame Head. It faces east and receives waves averaging about 1.5 m, which break across a 100 m wide surf zone, with rocks to either side, and a central permanent rip.

WA 509-517 **RAME HEAD-CONSPICIOUS-BELLANGER**

No.	Beach	Rating	Type inner	outer bar	Length
WA 509	Rame Hd (W1)	9	TBR+reefs		400 m
WA 510	Rame Hd (W2)	9	TBR+reefs		300 m
WA 511	Conspicious Beach(E)	7	TBR+rocks		600 m
WA 512	Conspicious Beach	8	RBB		900 m
WA 513	Conspicious (W1)	8	TBR+rocks		50 m
WA 514	Conspicious (W2)	8	TBR+rocks		80 m
WA 515	Conspicious (W3)	7	R+boulders		30 m
WA 516	Bellanger Beach	8	R/TBR	RBB	10 km

Spring & neap tidal range = 0.4 & 0.0 m

To the west of **Rame Head** are two exposed, 1.5 km wide south-facing embayments containing beaches WA 509-510 and 511-514 respectively, then the curving arc of 10 km long Bellanger Beach (WA 516) that terminates at the mouth of Nornalup Inlet in lee of Rocky Head, 11 km west of Rame Head. All the beaches are backed by generally vegetated transgressive dunes extending 1-3 km inland. The only vehicle access to this part of the national park is a gravel road to Conspicious Beach (WA 512) and 4 WD track to Blue Holes in the centre of Bellanger Beach.

Beaches WA 509 and 510 are located in a 900 m wide, southwest-facing open embayment, rimmed by 10 m high steep, vegetated calcarenite slopes with 100 m high Rame Head to the east and a 40 m high head to the west and a central rock reef, which is usually awash. Beach **WA 509** is a narrow 400 m long strip of sand at the base of the rock strewn slopes. It is fronted by a 200 m wide surf zone dominated by a large western and smaller eastern rips. Beach **WA 510** is a narrow 300 m long beach with a rock and reef strewn surf and which at times it is completely eroded back to the rocks. The waves and reefs maintain a stronger eastern rip, and weaker western rip against the point. A vehicle track leads to the bluffs above the beaches, with a steep climb down to the shore.

The **Conspicious** beaches (also known as Nornalup Beach) are located at the base of the steep vegetated 130 m high Conspicious Cliffs (Fig. 4.112). The eastern beach (**WA 511**) is a curving 300 m long west-facing beach that lies at the base of the cliffs. It terminates at a slight protrusion in the bluffs, with rocks at the base of the cliffs protruding 50 m into the surf. It receives waves averaging 1.5-2 m which break across a 150 m wide surf zone with usually a central bar drained by permanent rips at either end. The main beach (**WA 512**) extends west from the rocks for 900 m to a low granite point. It receives waves averaging about 2 m which maintain a 200 m wide surf zone usually dominated by two central beach rips, separated by detached bars, and a western permanent rip against the point. It is backed by initially scarped climbing, generally vegetated, longwalled parabolics, which extend up to 2 km inland with one central active parabolic extending 1 km to the northeast. The Conspicious Beach road reaches the western end of the beach where there is a small car park, picnic area, shelter and amenities, and a viewing platform with an excellent view of the entire beach. This is a relatively popular beach for surfing, however be aware of the strong rips.

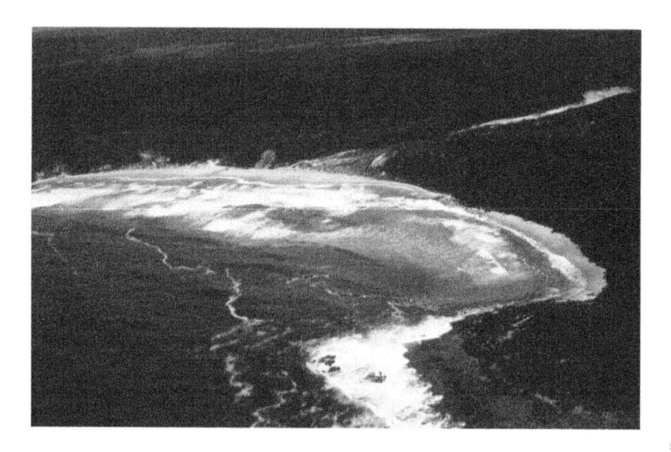

Figure 4.112 The Conspicious beaches (WA 511 & 512) are exposed to high waves and dominated by beach and topographic rips.

At the western end of Conspicious beach is a 1 km long southwest trending dune-capped granite headland, rising to 40 m. Three small beaches (WA 513-515) are located in pockets along its base. Beach **WA 513** lies 100 m past the main beach and is a 50 m pocket of high tide sand wedged in a gap in the rocks. It is fronted by intertidal rocks and a 100 m wide surf zone, which it shares with the main beach. Beach **WA 514** lies 100 m to the west and is an 80 m long pocket of sand located at the base of a 100 m deep gap in the rocks. Waves break from the entrance of the gap to the beach, with a permanent rip running out along the eastern side of the rocks. It is backed by three vegetated gullies, with beach sand washed into the base of each. Beach **WA 515** lies at the base of a 400 m deep, V-shaped gap on the southern tip of the headland. The gap narrows to 50 m with the 30 m long beach located at the very end. It consists of a patch of high tide sand, fronted by more than 50 m of intertidal rocks, then the deeper water of the rock-lined gap. Waves are lowered too less than 1 m at the beach, however the rocks make landing at the beach difficult.

Bellanger Beach (WA 516) commences on the western side of the point and curves in a 10 km long, south-facing arc to the mouth of Nornalup Inlet. Most of the beach receives waves averaging over 2 m which break across a 400 m wide double-barred surf zone, with strong beach rips spaced every 500 m along the inner bar, and more widely spaced rips on the outer bar. The inner bar is usually detached from the shore resulting in strong rip feeder currents. The entire beach is backed by multiple-longwalled parabolic dunes extending inland from 1 km in the west, to 3 km inland in the east where they reach up to 100 m in height and includes a few active blowouts (Fig. 4.113). The only access is via the central Blue Holes track, which leads to a section of beach with a few rocks, and the large channels of the beach rips forming the 'blue holes'. This beach is used for beach fishing and occasional surfing, however use care as it is a treacherous location owing to the high waves and strong rips. There is 4 WD access to the beach however beware as its steep and soft.

Figure 4.113 A section of the high energy Bellanger Beach with one of the large climbing blowouts behind.

WA 517-522 **ROCKY HD-PT NUYTS**

No.	Beach	Rating	Type inner	outer bar	Length
WA 517	East Pt	5	rocks+LTT		50 m
WA 518	Shelly Beach	7	TBR+rocks		1.1 km
WA 519	Shelly (W)	8	TBR+rocks		30 m
WA 520	Circus Beach	8	RBB	RBB	1.7 km
WA 521	Thompson Cove	4	R+reef		60 m
WA 522	Aldridge Cove	8	R+rocks		100 m
Spring & neap tidal range = 0.4 & 0.0 m					

The western side of Nornalup Inlet is bordered by a dune-covered rocky headland that extends 500 m southeast to 24 m high Rocky Head. The shoreline then trends west for 11 km to Point Nuyts, a 144 m high steep, vegetated point. Between the two rocky points is 15 km of predominately steep, rocky shoreline, containing the two larger Shelly and Circus beaches (WA 518, 520), and three small pockets of sand (WA 519, 521, 522). The only vehicle access is to the ridge high above Thompson Cove.

East Point is a vegetated dune-capped, granite point that forms the western boundary of Nornalup Inlet. On its eastern face is a 50 m pocket of intertidal sand and narrow bar at the base of the rocks (**WA 517**). The beach is relatively sheltered with waves usually less than 1 m. However the strong inlet tidal currents flow just off the point.

Shelly Beach (WA 518) is located on the western side of Rocky Head. It trends southwest for 1.1 km to a 100 m long granite point. The beach faces southeast into waves averaging 1.5-2 m, which break across the western Snake Ledge beachrock reef, as well as central granite outcrops.

The result is a 100-150 m wide surf zone with a permanent rip against Rocky Head and the central rocks and a third exiting the western end of the ledge. Dunes climbing to 50 m back the beach and extend 500 m to the inlet mouth.

Beach **WA 519** is located on the western side of the boundary point and consists of a 50 m long pocket of sand wedged in at the base of 60 m high steep slopes, and bordered and backed by rocks and rock debris. It is partly sheltered by Snake Ledge with waves averaging about 1.5 m, which break across a 70 m wide mixture of sand and rocks.

Circus Beach (**WA 520**) is located 1.5 km west of Nornalup Inlet and Rocky Point. It is a 1.7 km long, southwest facing beach exposed to high waves. These maintain a 400 m wide double barred surf zone, with usually two strong central rips, and two boundary rips, as well as some rocks in the eastern surf (Fig. 4.114). It is backed by climbing vegetated dunes that in the past have blown 1 km across the point to the shore of Nornalup Inlet. Two walking tracks lead from the inlet to either end of the beach, with no vehicle access.

Figure 4.114 Circus Bay is an exposed high energy rip-dominated hazardous beach.

Thompson Cove is located 3 km east of Point Nuyts. The narrow cove is cut in steep granite slopes that rise to 140 m with a 50 m wide pocket of sand (beach **WA 521**) at its base. Waves break over reefs and rocks outside the entrance to the cove resulting in low waves at the shore, where they maintain a steep reflective beach, crossed by a small creek. A 4 WD track reaches the crest of the western ridge above the beach. The cove could be used to land a small boat once the entrance reefs are negotiated.

Aldridge Cove is located 1 km to the west and is a similar small gap cut into the granite slopes that rise to 190 m in the west. A 100 m long strip of sand is located at the base of the cove, which narrows to about 50 m off the beach (**WA 522**) with waves breaking across the entrance. A small permanent rip drains out the centre. The sandy high tide beach is fronted by intertidal rocks and boulders, then low surf. Point Nuyts lies 2 km southwest of the cove.

WA 523-525 LOST & HUSH HUSH BEACHES

No.	Beach	Rating	Type	Length
WA 523	Lost Beach	7	TBR+rocks	250 m
WA 524	Hush Hush Beach	9	RBB+rocks/reef	1.4 km
WA 525	Hush Hush (W)	9	RBB+rocks	50 m

Spring & neap tidal range = 0.4 & 0.0 m

To the west of **Point Nuyts** is 14 km of exposed southwest-facing shore centred on Hush Hush and Mandalay beaches, with Long Point separating the two systems. The first system between Point Nuyts and Long Point consists of a 6 km wide open embayment containing Hush Hush and two smaller beaches, with steep granite and calcarenite slopes dominating most of the shore.

Lost Beach (**WA 523**) is located at the western base of Point Nuyts and consists of a 250 m long pocket of sand wedged in at the base of the 120 m high slopes of the point and equally high backing and western slopes with cliff debris also littering the beach. The beach receives waves averaging up to 2 m, which break across a 200 m wide surf zone, with one permanent rip draining the system.. There is a vehicle track to the ridge above the beach, with steep walk down to the shore.

Hush Hush Beach (**WA 524**) occupies the centre of the embayment. It is a relatively straight 1.4 km long southwest-facing high energy beach fronted by a 300 m wide rip-dominated surf zone, and backed by continuous cliffs rising to 120 m. The beach is narrow and in places eroded back to the base of the rock-littered slopes. Beach **WA 525** is a 50 m pocket of sand and rock debris located 300 m west of the end of the beach, at the base of 100 m high steep slopes. It shares the wide surf with the main beach. There is a vehicle track to the ridge above this beach with a steep descent down to the shore. *Hush Hush* is a remote, rippy and dangerous beach, which is occasionally surfed by locals.

WA 526-533 LONG PT-MANDALAY BEACH

No.	Beach	Rating	Type	Length
WA 526	Long Pt (W1)	7	LTT+reef	100m
WA 527	Long Pt (W2)	7	LTT+reef	100
WA 528	Long Pt (W3)	7	RBB+rocks/reef	200 m
WA 529	Long Pt (W4)	9	RBB+rocks	80 m
WA 530	Long Pt (W5)	9	RBB+rocks	300
WA 531	Long Pt (W6)	9	RBB+rocks/reef	1 km
WA 532	Long Pt (W7)	9	RBB+rocks	80 m
WA 533	Mandalay Beach	9	RBB+rocks	1.4 km

Spring & neap tidal range = 0.4 & 0.0 m

Long Point is a 200 m wide 40 m high, granite point that protrudes 700 m to the southwest and separates the Hush Hush beaches from a 4.5 km long section of eroded shoreline and remnant beaches (WA 526-532) backed by steep calcarenite slopes rising up to 120 m and terminating at Mandalay Beach (WA 533). All eight beaches are backed by a massive Pleistocene and Holocene dunes extending up to 7 km inland. There is a

4WD track to Long Point and a gravel road to Mandalay Beach.

Beaches WA 526-528 are three adjoining beaches located in an exposed 400 m wide embayment between Long Point and a western granite point that extends 200 m to the south (Fig. 4.115). They are separated by protruding calcarenite-capped granite bluffs, and share a common surf zone that breaks over a central calcarenite reef. Beach **WA 526** is a 100 m long pocket of sand located in lee of the reef, with most waves breaking 50-100 m offshore over the reef, and lower waves at the shore. Beach **WA 527** is located immediately to the west and is a similar 100 m long protected pocket of sand, with the reef reaching the shore at the western end of the beach. A permanent rip drains across the 'lagoon' fronting both beaches and out along the western side of Long Point. Beach **WA 528** occupies the western side of the small embayment and is a 200 m long beach with rocks strewn along the beach. It is bordered by the reef to the east and granite point to the west, with waves breaking up to 200 m offshore and a strong permanent rip draining between the reef and point. All three beaches are backed by 120 m high vegetated calcarenite slopes, with a few small parches of active climbing blowouts.

Figure 4.115 Beaches WA 526, 527 and 527 located immediately west of Long Point.

Beach **WA 529** lies 500 m to the west and is an 80 m pocket of sand wedged in a gap in the 100 m high calcarenite cliffs. The cliff narrows to the rear and the beach is littered in cliff debris. It is fronted by a 300 m wide surf zone, with a strong rip running out against the eastern rocks, The surf zone continues unbroken for 4 km to Mandalay Beach. Beach **WA 530** is a 300 m long strip of sand located at the base of the 100 m high cliffs. It protrudes slightly to the west in lee of a calcarenite reef, with some rocks also littering either end of the beach. The 300 m wide surf zone continues the length of the beach, with a permanent rip originating from the lee of the reef to drain out the centre of the surf.

Beach **WA 531** is a crenulate, discontinuous, eroding 1 km long beach that runs along the base of the cliffs. It is narrow, littered with cliff debris, with rocks and reefs also scattered along the 300 m long rip-dominated surf zone, the main reef forming permanent rips. Beach **WA 532** is located immediately to the west and is an 80 m long pocket of sand wedged in at the base of a gap in the 90 m high calcarenite bluffs. It is littered with rocks on the beach and in the surf, with a permanent rip draining out of the gap and across the 300 m wide surf zone.

Mandalay Beach (**WA 533**) commences at the western end of the high bluffs as they give way to the extensive dune field that has blown out of the beach. The beach is 1.4 km long faces southwest and is exposed to waves averaging over 2 m, which break across the rip-dominated 200-300 m wide surf zone. Scattered rocks and reefs lie off the eastern end, while the western end terminates at a wave-washed granite point. The surf is dominated by permanent rips to either end and usually two strong central beach rips. The backing dunes are actively extending up to 2 km inland, with older vegetated longwalled parabolics extending up to 7 km inland. A gravel road runs out to the western end of the beach, where there is a car park and a wooden platform that provides a commanding view along the beach, with a walkway down to the beach. The beach is popular for beach fishing and occasional *surfing*, however beware of the strong rips and western rocks. The Norwegian barque 'Mandalay' was wrecked on the beach in 1911.

D'ENTRECASTEAUX NATIONAL PARK

Area:	114 566 ha	
Coast length:	115 km	(1651-1756, 1762-1772 km)
Beaches:	62	(WA 534-587, 593-600)

D'Entrecasteaux National Park is the largest national park on the South Coast covering 114 566 ha and 115 km of exposed coastline, including 60 beaches. The beaches include the massive dunes centred on Warren Beach, which extend for 50 km between Black Head and Black Point. This system is one of the highest energy beach and dune systems in the state and one of largest dune systems in Australia. Conventional vehicles can only access the coast at Windy Harbour, with 4 WD tracks, often through bogging winter sections, reaching the coast at Fish Creek, Warren and Yeagarup Beach and Black Point.

WA 534-538 BANKSIA CAMP

No.	Beach	Rating	Type	Length
WA 534	Banksia Camp (E 2)	8	TBR/RBB	500 m
WA 535	Banksia Camp (E 1)	9	RBB+rocks	150 m
WA 536	Banksia Camp	8	RBB	400 m
WA 537	Banksia Camp (W)	7	TBR+rocks	100 m
WA 538	Cliffy Head	6	LTT/TBR+rocks	90 m
Spring & neap tidal range = 0.4 & 0.0 m				

To the west of Mandalay beach the shoreline trends to the southwest for 4.5 km to 115 m high Cliffy Head, with 10 ha, 182 m high Chatham Island located 1.5 km south of the head. There is 4 WD access to Banksia Camp beach, which marks the beginning of the 115 km long D'Entrecasteaux National Park.

Beach **WA 534** is a 500 m long south-facing beach, set in a semi-circular embayment, 400 m wide at the entrance. The beach receives waves up to 2 m, which break across a 300 m wide surf zone drained by strong permanent rips to either side of the bay. Only in the western corner in lee of a projecting, low wave-washed granite point, is there some protection. It is backed by a 20 m high scarped dunes, with older vegetated dunes extending 7 km inland as part of the Mandalay Beach dune system. A small creek drains across the eastern end of the beach with a 4 WD track to the western end. There is surf on the bar, but be careful of the strong western rip.

Banksia Camp occupies the next small embayment and consists of two adjoining beaches. In the east is a narrow 150 m long section of shore containing two patches of sand (**WA 535**) largely fronted by granite rocks and backed by steep vegetated bluffs rising to 40 m. Waves break on rock reefs 100 m off the shore and again on the rocks, with a permanent rip between the reef and shore. The main beach (**WA 536**) extends for 400 m to the west where it is bordered by steep 100 m high vegetated cliffs, with a central active 20 m high foredune, in front of higher vegetated bluffs. It receives high waves, which break across a 200 m wide surf zone, with permanent rips to either end (Fig. 4.116). The eastern rip provides easy access to the *righthand break* that runs into the rip.

Figure 4.116 Banksia Camp beach has a wide surf zone drained by a deep, permanent rip channel (arrows) that flows out against the western point.

Beach **WA 537** is located a few hundred metres to the southwest at the foot of the 150 m high cliffs that form the eastern side of the dune-draped Cliffy Head headland. In the past active dunes have blown down over the cliffs onto the beach. The beach consists of a 100 m long narrow strip of rock-littered sand at the base of the calcarenite cliffs, with a 300 m long surf zone that widens from 50 m against the southern granite point to 200 m in the north. There is no safe foot access to the beach.

Cliffy Head consists of dune and calcarenite-draped granite, and rises to 115 m in the west and up to 175 m to the east. Between the two ridges is a 200 m wide, 300 m deep gap within which is a 90 m long beach (**WA 538**). High waves at the entrance decrease to about 1 m at the beach where they maintain a 40 m wide attached bar, with a small permanent rip exiting the western side of the beach. A small foredune, then an amphitheatre of steep slopes backs and borders the beach. There is no formal land access to this remote beach.

WA 539-542 CLIFFY HEAD (W)

No.	Beach	Rating	Type	Length
WA 539	Cliffy Hd (W 1)	9	LTT+rocks/reef	800 m
WA 540	Cliffy Hd (W 2)	9	LTT+rocks/reef	200 m
WA 541	Cliffy Hd (W 3)	9	LTT+rocks/reef	400 m
WA 542	Cliffy Hd (W 4)	9	RBB+rocks/reef	1.7 km
Spring & neap tidal range = 0.4 & 0.0 m				

At Cliffy Head the shoreline turns and trends west-northwest for 48 km to Point D'Entrecasteaux and Windy Harbour. In between is an exposed section of high energy coast with numerous eroding beaches usually at the base of high calcarenite bluffs, and occasionally granite points. This is an isolated difficult to access stretch of coast, all backed by massive largely vegetated dune systems. There is limited and difficult 4 WD access to some of the beaches between Broke Inlet and Windy Harbour, the remainder, largely inaccessible except on foot or by boat across Broke Inlet.

The first section of the coast extends for 5.5 km west of Cliffy Head to a 160 m high calcarenite-capped granite point that protrudes 200 m seaward. In between is a continuous high energy surf zone breaking against the base of 100-160 m high calcarenite cliffs, with four sections of remnant, essentially inaccessible, sand beaches (WA 539-542).

Beach **WA 539** commences on the western side of Cliffy Head and trends northwest for 800 m. The high tide sand beach is fronted by a continuous 50 m wide intertidal beachrock reef, with a few narrow elongate tidal pools between the reef and the beach. Waves averaging over 1.5 m break off and over the reef. Steep vegetated 100 m high calcarenite cliffs back the beach and some cliff debris litters the western end of the beach.

Beach **WA 540** lies 500 m to the west and is a 200 m long narrow strip of high tide sand, littered with cliff debris and fronted by a continuous 50 m wide beachrock reef. Waves break seaward of the reef flowing over the reef to reach the beach. Beach **WA 541** is a similar 400 m long strip of high tide sand located 1 km further west. It is also littered in cliff debris and in the late 1990's was backed by fresh rock slides down the backing 150 m high calcarenite cliffs.

Beach **WA 542** is the most substantial of the four beaches. It extends for 1.7 km to terminate at the base of the western point. Beachrock reef fronts the eastern section of the beach then trends seaward, lying 50 m off the western end. The backing channel feeds a permanent rip that drains out through a gap between the end of the reef and the point. The beach is also backed by vegetated 100 m high cliffs as well as some fresh rock falls, which in places cut the beach.

WA 543-550 **BROKE INLET (E)**

No.	Beach	Rating	Type	Length
WA 543	Broke Inlet (E7)	9	TBR+rocks/beachrock	800 m
WA 544	Broke Inlet (E6)	9	LTT+beachrock	400 m
WA 545	Broke Inlet (E5)	9	TBR+rocks/beachrock	500 m
WA 546	Broke Inlet (E4)	9	TBR+rocks/beachrock	250 m
WA 547	Broke Inlet (E3)	9	TBR+rocks	700 m
WA 548	Broke Inlet (E2)	9	TBR+rocks	150 m
WA 549	Broke Inlet (E1)	9	TBR+beachrock	100 m
WA 550	Broke Inlet	9	LTT+tidal channel	700 m

Spring & neap tidal range = 0.4 & 0.0 m

Midway between Cliffy Head and Broke Inlet is a 160 m high calcarenite-capped granite headland. Between this headland and the inlet is 7 km of exposed southwest-facing shoreline, all backed by 100-190 m high calcarenite cliffs, with the base littered with cliff debris and fronted by near continuous beachrock reef. Scattered along the base of the cliffs are seven strips of sand (beaches WA 543-549), plus the sandy eastern side of Broke Inlet (beach WA 550). This is a hazardous, almost inaccessible section of cliff-backed, rocky shore. The only easy access is by boat to the inlet, while there are two 4 WD tracks that reach the top of the cliffs above beaches WA 543 and 547.

Beach **WA 543** is an 800 m long narrow strip of sand wedged in between the cliffs and shore parallel intertidal beachrock reef, with wave averaging over 1.5 m breaking 50 m seaward of the reef edge. There is a gap in the reef at the western end, where the beach curves round for 150 m and a strong rip runs out through a 150 m wide surf zone. Cliffs rising to 190 m and fresh rock slides back the beach.

Beach **WA 544** lies to the lee of deeper reefs extending 400 m offshore, which lower waves to about 1.5 m. The beach is 400 m long and protrudes slightly seaward in lee of a continuous beachrock reef, with backing 140 m high vegetated cliffs. Waves break seaward of the reef, washing over the reef to reach the beach.

Beach **WA 545** is a straight 500 m long strip of high tide sand, littered with cliff debris, and fronted by a discontinuous beachrock reef. Waves averaging over 1.5 m break over a 100-150 m wide sand and reef surf zone, with a strong permanent central rip.

Beach **WA 546** lies 100 m to the west and occupies a slight indentation in the backing 100 m vegetated high cliffs. A few patches of foredune have accumulated at the base of the cliffs. The beach is 250 m long, bordered at each end by the bluffs, with a near continuous broken beachrock reef in between. Waves averaging 1.5-2 m break 150 m seaward of the reef over a sandy seabed, with a strong rip originating from a small gap towards the western end of the beach and flowing out through the surf.

Beach **WA 547** consists of two arcs of sand totalling 700 m in length. The narrow rock strewn eastern arc links to the longer, wider western arc, which has a small foredune to the rear, then vegetated slopes rising to 120 m. While there are rocks littering the inner surf zone, a sandy sea bed dominates with two strong rips draining the 200 m wide surf zone.

Beach **WA 548** is a 150 m long pocket of sand bordered and backed by 100 m high vegetated slopes and cliffs. Calcarenite platforms border each side, with the 200 m wide surf zone between drained by a strong permanent rip against the western rocks. Beach **WA 549** lies 200 m to the west and consists of a 100 m long pocket of sand wedged at the base of a small embayment. It is backed by 90 m high steep, vegetated slopes. Wave break over 200 m off the beach and flow across a beachrock reef lying 50 m offshore. A strong rip flows west along the beach inside the reef and out against the 100 m high western high headland, which protrudes 200 m seaward.

Broke Inlet lies 1.5 km west of the headland. The inlet is one of the most exposed high energy lagoon opening on the Australian coast (Fig. 4.117). It closes during the drier summer season and during high waves, only opening during winter floods. The inlet drains the 4500 ha Broke Inlet lagoon, with the lagoon connected to the inlet mouth along a 200 m wide, 4 km long channel, bordered by 40-60 m high transgressive dunes. At the mouth calcarenite reefs lie off the entrance, which together with inlet tidal shoals cause waves to break across a treacherous 500 m wide surf zone. The beach itself (**WA 550**) extends from an eastern calcarenite point for 700 m to the northwest, with the reefs extending seaward. The length of the beach depends on the entrance conditions and when closed links with the beach on the western side of the inlet (WA 551). When open the entrance can be as narrow as a few tens of metres flowing straight into the high surf. While this beach offers great beach fishing, it is very hazardous for swimming.

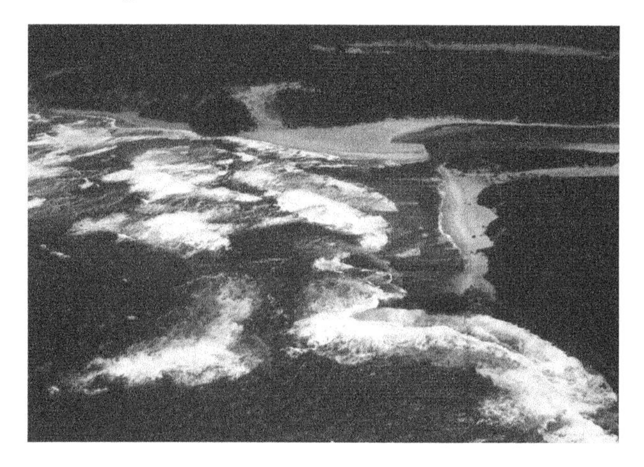

Figure *4.117 Broke Inlet is an exposed tidal inlet, which flows behind beachrock reefs to work its way through the high energy surf zone.*

WA 551-553 **BROKE INLET (W)**

No.	Beach	Rating	Type	Length
WA 551	Broke Inlet (W1)	9	TBR+tidal shoals	300 m
WA 552	Broke Inlet (W2)	9	RBB+tidal channel	750 m
WA 553	Broke Inlet (W 3)	9	RBB+rocks/reef	1.2 km
Spring & neap tidal range =0.4 & 0.0 m				

The western side of Broke Inlet consists of a 2 km long predominately sandy embayment containing three beach systems. The first two are fronted by the inlet reefs and shoals, and all three backed by massive dunes rising to 103 m at Sand Peak and extending up to 6.5 km to the northeast, well into Broke Inlet lagoon. There is a 4 WD track to the top of beach WA 553, with the only other access by boat across the lagoon to the inlet.

Beach **WA 552** forms the western side of the periodically open inlet, with its length depending on inlet conditions, while during floods the beach may be washed away. It is usually about 300 m long extending east from a 20 m high vegetated calcarenite knoll and terminating at a low sandy recurved spit. It faces south across the inlet channel, shoals and associated calcarenite reefs that dominate the inlet mouth. Waves average over 2 m and break up to 500m off the shore.

Beach **WA 553** extends west of the knoll for 750 m to a slightly protruding rock-littered section of calcarenite, surrounded by an intertidal reef. It is fronted by a 500 m wide surf zone, with the inlet channel running out the western side when open, and a strong permanent rip flowing out past the western reef (Fig. 4.118). The beach is backed by hummocky scarped calcarenite bluffs rising up to 60 m, with Sand Peak located 400 m inland. All the bluffs are backed by predominately well vegetated, longwalled parabolic dunes.

Figure 4.118 Beach WA 553 is an exposed rip-dominated beach backed by rocks and largely vegetated climbing dunes.

Beach **WA 554** commences on the western side of the rocks and continues for 1.2 km to a 40 m high calcarenite point that protrudes 400 m to the south. The beach, while relatively straight and predominately sandy, has several patches of beachrock reef and rocks littering its western half and intertidal zone. It has a high energy 400 m wide surf zone usually cut by two central rips and permanent boundary rips. It is backed by scarped calcarenite bluffs in the east grading into well vegetated bluffs to the west. Largely vegetated, longwalled parabolic dunes back the beach and extend to Broke Inlet lagoon. One active parabolic in the centre of the beach extends 3 km inland.

WA 554-563 **WEST CLIFF PT (W)**

No.	Beach	Rating	Type	Length
WA 554	West Cliff Pt (E9)	10	R+platform	150 m
WA 555	West Cliff Pt (E8)	10	R+platform	100 m
WA 556	West Cliff Pt (E7)	10	R+platform	100 m
WA 557	West Cliff Pt (E6)	9	TBR+reef	1.9 km
WA 558	West Cliff Pt (E5)	8	R+platform+reef	200 m
WA 559	West Cliff Pt (E4)	8	R+platform+reef	450 m
WA 560	West Cliff Pt (E3)	9	R+platform+reef	100 m
WA 561	West Cliff Pt (E2)	5	R+reef	250 m
WA 562	West Cliff Pt (E1)	6	R+reef	100 m
WA 563	West Cliff Pt (E)	5	R/LTT+reef	400 m
Spring & neap tidal range = 0.4 & 0.0m				

Two kilometres west of Broke Inlet is the beginning of a 6.5 km long section of southwest-facing shore dominated by 30-50 high calcarenite bluffs, fronted by intertidal beachrock platforms and offshore calcarenite reefs. Ten beaches of varying lengths, shape and exposure are located in amongst the bluffs, platforms and reefs. The bluffs are backed by generally longwalled parabolic dunes covered in coastal heath and extend up to 6 km to the northeast. There are a few 4 WD tracks that reach the top of some of the beaches, and a solitary shack above beach WA 561.

Beaches WA 554, 555 and 556 are three hazardous pockets of sand located in gaps along the calcarenite headland that commences 2 km west of the inlet. Beach **WA 554** is a 150 m long strip of high tide sand, backed by 20 m high bluffs and fronted by a 50 m wide intertidal beachrock platform, then irregular rocks and reefs, with waves breaking 80 m offshore. A 4 WD track reaches the bluffs above the beach, with a track down to the shore. Beach **WA 555** lies 300 m to the west and is a 100 m long pocket of sand bordered and backed by 20-30 m high calcarenite point and bluffs. It has an intertidal beachrock outcrop along its base then a 100 m wide rock and reef dominated surf zone. Beach **WA 556** lies 50 m to the west and is a similar 100 m long beach, with beachrock and reefs extending out into the surf. In addition to the surrounding bluffs, it has an outcrop cutting across the centre of the beach. The bluffs behind all three beaches are capped with parabolic dunes covered in coastal heath.

Beach **WA 557** is a crenulate 1.9 km long, narrow high tide beach, that runs along, between the base of 50-80 m high calcarenite cliffs, and an irregular, in places discontinuous beachrock platform, with scattered reefs offshore. Waves averaging 2 m break over the reefs and platform, with the water draining out through a number of gaps in both. A mixture of bare cliffs, cliff debris, steep vegetated slopes and one patch of foredune back the beach. A 4 WD track reaches the top of the bluffs towards the western end of the beach.

Beach **WA 558** is a 200 m long patch of high tide sand, separated by rocks falls from beach WA 557. It is fronted by a 100 m wide beachrock platform, which lowers waves at the shore, permitting the beach to widen to about 50 m. It is backed by 40 m high bluffs, which protrudes seaward at the western end as a narrow sloping point, which separates it from beach WA 559. Beach **WA 559** continues west for 450 m to a more prominent 200 m long calcarenite point. It is a narrow high tide beach backed by 40-50 m high bluffs, and fronted by beachrock platform-reef varying from 50-100 in width, together with scattered rocks and reefs. Waves break across an irregular 200 m wide surf zone, with low waves at the shore.

Beaches **WA 560** is a 100 m long strip of sand located on the western side of the protruding point. It is backed by steep, vegetated 50 m high bluffs, and fronted by a continuous 50 m width beachrock platforms, that breaks up as its extends up to 100 m seawards. Waves break across the reefs and platform, resulting in low waves at the shore. A 4 WD track reaches the top of the bluffs, with a steep decent to the western end of the beach.

Beaches WA 561 and 562 occupy a small 400 m wide embayment located 1.5 km east of West Cliff Point. Calcarenite reefs extending 600 m seaward of the shore, result in low waves within the embayment. Beach **WA 561** is a curving 250 m long, narrow strip of high tide sand, backed by bare eroding 30-40 m high calcarenite cliffs, with a shack located above the western end of the beach. Waves average less than 1 m at the shore. Beach **WA 562** lies 50 m to the west and is a 100 m long strip of high tide sand at the base of steep 50-60 m high, vegetated bluffs with rock debris on the beach and an irregular intertidal platform and rocks extending 50 m off the beach. Waves average over 1 m high and break across the rocks and reefs.

Beach **WA 563** occupies a curving 500 m wide embayment, bordered to the west by 50 m high West Cliff Point. The beach curves round the base of the embayment for 400 m and is bordered and backed by calcarenite bluffs, with a 200 m wide blowout behind the centre of the beach and extending 500 m inland. Scattered rocks and reefs lie across the bay and lower waves to about 1 m at the shore, where they both surge up the beach as well as break over a central section of rocks. There are several 4 WD tracks reaching the beach through the central blowout.

WA 564-560 CLIFF PT-WINDY HARBOUR

No.	Beach	Rating	Type inner	outer bar	Length
WA 564	Cliff Pt (W)	8	TBR	RBB	16 km
WA 565	Walbingup	7	LTT+inlet		250 m
WA 566	Ledge Islet	6	LTT/TBR		2.3 km
WA 567	Flat Rock	5	LTT/TBR		5.5 km
WA 568	Windy Harbour	4→3	LTT→R		1.5 km
WA 569	Windy Harbour (W)	5	LTT/TBR		800 m
Spring & neap tidal range = 0.4 & 0.0 m					

To the west of West Cliff Point the shoreline trends to the west for 22 km to Point D'Entrecasteaux, one of the major headlands and inflection points on the South Coast. The small settlement of Windy Harbour is located 2 km east of the point and is the only coastal settlement between Peaceful Bay, 85 km to the east, and Augusta, 95 km to the west. This section of shore begins with an exposed, high wave energy beach extending for 16 km west of West Cliff Point to the Gardner River. The remaining 6 km of shore west of the river mouth are protected to a moderate degree by shallow reefs extending up to 3 km offshore, as well as a few rocks, reefs, islets and islands, the largest 25 ha Sandy and 10 ha Flat islands. There is a sealed road to Windy harbour and 4 WD tracks to the remaining beaches.

Beach **WA 564** is one of the more exposed, higher energy and longer beaches on the South Coast. It commences at West Cliff Point where it is afforded slight protection by rock reefs extending up to 500 m offshore. However, for most of its 16 km it is fully exposed to waves averaging over 2 m, which break across a 500 wide double bar surf zone. The inner bar is usually attached to the shore and cut by strong beach rips every 500 m, while the outer bar has more widely spaced rips. In addition there is a permanent rip against West Cliff Point and a rock outcrop located midway along the beach. Between this rock outcrop and the western end of the beach wave height decreases slightly as the offshore shoals, however rips continue to the river mouth. The beach terminates at the narrow, occasionally blocked mouth of the Gardner River (Fig. 4.119). It is backed by generally well vegetated longwalled parabolic dunes extending up to 7 km to the northeast, then an irregular swamp that drains via Chesapeake Brook and the Shannon River in the east to Broke Inlet, and into Lake Maringup in the west. Four wheel drive tracks reach the beach in a number of locations, including at West Cliff Point.

Gardner Beach (WA 565) is located on the western side of the Gardner River mouth and when the mouth is closed it links to beach WA 564. The beach is usually about 250 m long and faces southeast across the river mouth shoals. It is tied at its western end to a low granite rock, with some ridges of granite also outcropping at the mouth, hence the name Granite Bar. The small shack settlement of Walbingup, is located on the western banks of the river, just inside the mouth. The beach is moderately protected by the rocks and shoals with waves averaging 1-1.5 m breaking across an attached bar. However be careful, as when the river is open, strong tidal currents flow along the beach. It is backed by a vegetated deflation area, then an active blowout, with a vehicle track through the blowout to the beach.

Figure 4.119 The small and often blocked mouth of the Gardner River reaches the sea between beaches WA 564 and 565.

Beach **WA 566** commences at the granite rocks and trends west as a double crenulate shoreline for 2.3 km to a cuspate foreland formed in lee of Ledge Islet, a small granite reef located 400 m offshore. The islet and generally shallow sea floor and reefs lower waves to less than 1.5 m. These maintain a continuous 50 m wide bar, which is only cut by rips during higher wave conditions. Reflecting the lower waves the beach has prograded seaward and is backed by a 500 m wide foredune ridge plain, which continue all the way to Windy Harbour.

Beach **WA 567** commences on the western side of the Ledge Islet cuspate foreland and trends essentially due west for 5.5 km to the next cuspate foreland that forms the eastern boundary of Windy Harbour. The beach is fronted by 3-4 km of shallow nearshore including Sandy Island located 3 km to the south. Waves break on some of the outer reefs and average just over 1 m at the shore. They maintain a continuous shallow low tide terrace only cut by rips during higher wave conditions. It is backed by the foredune ridges, and a solitary creek, which crosses the western end of the beach.

Windy Harbour beach (**WA 568**) is a semi-circular shaped, south-facing 1.5 km long beach bordered by the foreland to the west and a low series of granulite ridges to the west (Fig. 4.120). Waves average about 1 m in the west where they continue the low tide terrace, decreasing to less than a metre in the west, where the beach is steep and reflective. It is backed by a 500 m wide foredune ridge plain, with the small settlement built on the ridges. Beside the rows of shacks there is a grassy camping area and beach boat launching behind the rocks, with a few boats moored off the beach, but no other services. A newer subdivision is located in behind the original shacks.

Figure 4.120 Windy Harbour is located to the lee of Point D'Entrecasteaux, which affords moderate protection.

Immediately west of the rocky point is beach **WA 569** an 800 m long, southwest-facing beach, bordered in the west by the inner calcarenite slopes of Point D'Entrecasteaux. The beach is partly protected by Flat Island and receives waves averaging over 1 m, which break across a 50 m wide bar, which is occasionally cut by rips. It is backed by a 20 m high vegetated foredune crossed by two walking tracks, then a gravel road which runs out to a car park at the western end of the beach. This is the most sheltered section of the beach and is bordered by some large calcarenite outcrops called Cathedral Rock.

WA 570-575 SALMON-DOGGERUP

No.	Beach	Rating	Type inner	outer bar	Length
WA 570	Salmon Beach	9	RBB/TBR	RBB	1.4 km
WA 571	Salmon (W)	9	RBB/TBR	RBB	250 m
WA 572	Sandy Peak	10	TBR+rocks		100 m
WA 573	Wheatley (1)	10	RBB+rocks		80 m
WA 574	Wheatley (2)	10	RBB		70 m
WA 575	Doggerup	9	RBB/TBR	RBB	2.8 km

Spring & neap tidal range =0.4 & 0.0 m

At **Point D'Entrecasteaux** the shoreline turns and begins a 90 km sweep to the northwest, terminating at Augusta and Cape Leeuwin. In between is one of the highest energy sections of the Western Australia coast, and for that matter the Australian coast. The entire section faces directly into the dominant southwest swell and westerly winds. The waves have moved massive amount of sand to the coast to build high energy beaches, while the wind have blown the sand inland for up to 15 km as massive transgressive Holocene and Pleistocene dune sheets. Most of the coast is within the D'Entrecasteaux National Park and access is limited. The only car access to Salmon Beach in the far east, with the remainder restricted to a few 4 WD tracks. The first 9 km of coast between the point and Black Head contains the two longer Salmon and Doggerup beaches (WA 570 & 575), with four intervening pockets of sand (beaches WA 571-574) at the base of 100 m high calcarenite cliffs.

Salmon Beach (**WA 570**) is an exposed 1.4 km long, southwest-facing sandy beach, bordered by 100 m high cliffs and backed by a few new foredunes, then active dunes climbing up over the backing calcarenite and blowing several hundred metres inland. Earlier dunes have migrated 6 km inland and reached heights of 70 m. It receives waves averaging over 2 m, which maintain a 300 m wide, double bar surf zone. The inner bar is usually detached, with two large beach rips and two permanent rips to either end. A sealed road runs over the backing ridge to a lookout above the southern end of the beach and to a car park, picnic area and toilets behind the centre of the beach. Be very careful if swimming here as strong rips are always present. The beach picks up any swell and usually has several breaks out over the bars.

Beach **WA 571** is a 250 m long strip of sand separated from the main Salmon Beach by a 100 m section of protruding calcarenite. The steep 100 m high bluffs continue behind the beach with rock debris littering the back of the beach. The main surf zone continues past the bluffs and along the beach, with usually one large rip draining out from the beach. The only access to the beach is around the base of the rocks at low tide.

Beach **WA 572** is located 1 km to the northwest in a 100 m wide gap in the steep calcarenite cliffs, which reach 117 m high at Sandy Peak. The beach fills the gap, with rocks littering the beach and the inner surf zone. Waves break across rocks and sand up to 300 m off the beach, with a strong rip exiting the western side of the beach. They is no safe access to this hazardous beach.

Beaches WA 573 and 574 occupy adjoining gaps in the calcarenite cliffs at the eastern end of Doggerup beach. Both beaches share an eastern extension of the main 300 m wide surf zone and are equally hazardous. Beach **WA 573** is an 80 m long pocket of sand located at the base of the 100 m wide gap. High bluffs border the beach, which is backed by a sloping sand shute, which feeds a clifftop dune that has migrated up and over the 80 m high bluffs and 700 m inland. A permanent rip runs out along the eastern headland and through the wide surf zone. Beach **WA 574** occupies the next embayment 300 m to the west. It is a 70 m long pocket of sand surrounded by steep 60 m high bluffs, with rock debris littering the sides. Waves associated with the Doggerup surf zone break up to 300 m off the beach. The Wheatley Coast Road, a 4 WD track reaches the top of the bluffs behind the beaches, with the only access down the steep slopes.

Doggerup beach (**WA 575**) is also known as Pebbley Beach in its western corner. The beach is 2.8 km long and faces straight into the southwest waves and winds. High calcarenite bluffs back the eastern end, while the low dark granulite of Black Head border the western end. Waves averaging over 2 m breaks across a 300 m wide double bar surf zone, usually containing four large beach rips and a western boundary rip. It is backed by the **Doggerup Dunes**, which cover 400 ha and have moved up to 4 km inland and to heights of 60 m, while older vegetated parabolic dunes have reached 5.5 km inland. Doggerup Creek runs through the centre of the active dunes and drains across the eastern end of the beach.

WA 576-580 **BLACK HEAD**

No.	Beach	Rating	Type	Length
WA 576	Black Head (W1)	10	R+platform	200 m
WA 577	Black Head (W2)	10	R+platform	400 m
WA 578	Black Head (W3)	9	R+platform/TBR	300 m
WA 579	Black Head (W4)	10	TBR+rocks	50 m
WA 580	Black Head (W5)	9	RBB+rocks	100 m
Spring & neap tidal range = 0.4 & 0.0 m				

Black Head is located that the southern end of a low 4 km long west-facing outcrop of Proterozoic granulite, capped by Pleistocene calcarenite and more recent Holocene dunes. The resilient rocks form an irregular rocky shore containing five pockets of sand, remnants of the beaches that supplied the backing clifftop dunes. The Summertime 4 WD track runs along the edge of the coast providing access to all five beaches.

Beaches WA 576 and 577 occupy either end of a 700 m long west-facing open embayment located immediately north of Black Head. Beach **WA 576** commences 500 m north of Doggerup beach and is a curving 200 m long strip of high tide sand fronted by linear ridges of granulite forming a 50-70 m wide rock platform, and backed by moderately sloping, dune-covered rocks. In addition two small streams drain across the beach and rocks. Waves averaging 2 m break over the rocks to reach the beach. Beach **WA 577** is located 200 m to the north and is a double crenulate, 400 m long high tide beach fronted by a 50 m wide granulite rock platform, with wave breaking over sand and rocks 50 m seaward of the platform. Dune-covered rocks back the beach.

Beach **WA 578** is located in a small embayment on the northern side of a low point 200 m further north. It is a 300 m long beach consisting of a southern section located to the lee of a 50 m wide rock platform, while the northern half is open to the waves lowered to about 1.5 m by the southern point. The waves run up the steep beach face, with a permanent rip draining out of the small bay.

Beach **WA 579** lies 500 m to the north and is a 50 m pocket of sand that is awash at high tide. It is encased in 20 m high sloping granulite bluffs, as well as rocks on either side of the beach. Waves break 200 m offshore and run into the small gap, with a permanent rip running back out the centre.

Beach **WA 580** is located in a small south-facing embayment, 1 km south of the beginning of Meerup beach. The 100 m long pocket of sand faces straight into the high waves, which break on an outer bar over 100 m offshore, then reform to break heavily on the beach. A strong rip drains the narrow surf. Vegetated dunes rise to the east of the beach, with a few shacks located 300 m up the slopes, in lee of a vegetated dune. A walking track leads from the shacks to the beach.

WA 581-583 MEERUP-WARREN-YEAGARUP

No.	Beach	Rating	Type inner outer bar	Length
WA 581	Meerup beach	8	RBB RBB/LBT	10 km
WA 582	Warren beach	8	RBB RBB/LBT	7.2 km
WA 583	Yeagarup beach	8	RBB RBB/LBT	18.4 km

Spring & neap tidal range = 0.4 & 0.0 m

Black Head forms the southern boundary of the longest, straightest, highest energy section of beach on the South Coast, and one of Australia's highest energy beach and dune systems. Between Black Head and Black Point, 50 km to the northwest is a near continuous beach and surf system, backed by massive dune systems extending up to 15 km inland. In addition three rivers – the Meerup, Warren and Donnelley, penetrate the dunes to reach the shore. This is a wild, isolated and difficult to access section of coast, with 4 WD's required to negotiable the boggy swamps, high dunes and sandy beach tracks. Some of the tracks are flooded and closed during winter. The bulk of the coast is located within the D'Entrecasteaux National Park.

Meerup beach (**WA 581**) trends straight northwest from the northern rocks of Black Head for 10 km to the shifting and often deflected mouth of the Meerup River. The small river has to drain through 5 km of high dunes, rising to 190 m to reach the shore. The beach receives waves averaging 2-2.5 m, which break over an outer bar located 500 m offshore, and an inner bar cut by rips every 500 m and usually detached from the shore. These are backed by a continuous 1 km wide deflation basin, then the Meerup Dunes, which include active dunes (Fig. 4.121) extending up to 7 km to the west and to heights of 235 m as well as vegetated Pleistocene dunes up to 15 km inland. The only access track is via the Summertime Track to Black Head.

Warren Beach (**WA 582**) continues the trends of the shore and dunes from the mouth of the Meerup River to the larger Warren River, 7.2 km to the northwest. The river may be deflected as much as 1 km to the north or south, causing the actual beach length to accordingly. Sometimes the deflected sand spit forms a separate beach up to 2 km long known as Coolyarbup. The net result is a highly dynamic mouth responding to the high waves, strong winds and wind blown sand, episodic floods and tidal flows. The wide double bar system continues along the beach, with the rips spaced every 500 m, in places producing major rip embayments in the shore. The beach is backed by a continuous 1 km wide deflation basin, then the Callcup Dunes. These active dunes are a series of longwalled parabolics rising to 200 m and extending up to 3 km inland, while the older vegetated dunes reach 8 km inland. The highest dune is 236 m Callcup Hill located just off the Warren Beach track, on the crest of which is a lookout tower. A solitary fishing shack is located nearby. This beach is popular for beach fishing, however be very careful of the strong rips if entering the water.

Figure 4.121 The high energy Mererop beach and backing deflation basin and dunes.

Yeagarup beach (**WA 583**) commences at the Warren River mouth and continues straight to the northwest for 18.4 km to the beginning of a 1.5 km section of calcarenite bluffs, just south of the Donnelly River mouth. This is one of the longer, continuous high energy beaches in the state. Wave remains high and up to 30 large rips dominate the inner bar, some with large rip embayments (Fig. 2.8a). The deflation basin narrows to 500 m and continue north for 7 km before it is replaced by active dunes. The backing Yeagarup Dunes consists of a 0.5-2 km wide active zone, then largely vegetated dunes extending 10 m inland, with one large central active zone covering about 1800 ha. The dunes average 100 m in height rising to 163 m. Two access tracks wind through the dunes to the coast, with a few fishing shacks located in the dunes and a solitary shack on the coast 8 km west of the river mouth.

BLACK HEAD TO BLACK POINT

The Meerup-Warren-Yeagarup-Jaspar beach system beach extends essentially unbroken for 50 km between Black Point and Black Head. The only interruptions are the two river mouths and a few sections of calcarenite bluffs. The beaches share a continuous high energy surf zone which contains on average 86 large beach rips spaced every 500 m and 13 permanent rips in the western calcarenite sections, all draining across and a 400-500 m wide double bar surf zone. It is backed by a massive 38 000 ha dune system, including 8 700 ha of active dunes. The dunes average 100 m in height reaching well over 200 m in places, resulting in approximately 38 000 000 000 m^3 of sand in the dunes, delivered at an average rate of 750 000 m^3/m of beach, which if we assume all is Holocene sands, would require a rate of deliver of 105 m^3/m/yr during the past 7 000 years. The probably source of most of the predominately quartz sand (80-95% quartz) is the granites that dominates the river catchments. The sand would have been be delivered to the shelf at lower sea levels and reworked onshore during sea level transgressions and high stands, as at present.

BLACK HEAD-AUGUSTA

When the extra 40 km of shoreline between Black Point and Point Frederick, at Augusta, is included the near continuous system increases to 90 km in length, with the backing dune system totalling 46 000 ha, resulting in one of the largest beach-dunes systems in Australia.

WA 584-588 BROADWATER-DONNELLEY R

No.	Beach	Rating	Type inner	outer bar	Length
WA 584	Broadwater (E3)	9	RBB+rocks	RBB/LBT	100 m
WA 585	Broadwater (E2)	9	RBB+rocks	RBB/LBT	200 m
WA 586	Broadwater (E1)	9	RBB+rocks	RBB/LBT	150 m
WA 587	Donnelly R (E)	8	RBB	RBB/LBT	700 m
WA 588	Donnelly R (W1)	8	RBB	RBB/LBT	2.2 km
Spring & neap tidal range =		0.6 & 0.4 m			

The northern end of Yeagarup beach terminates at a straight 1.5 km long section of 20-30 m high calcarenite bluffs that continue to within 500 m of the Donnelly River mouth. While the bluffs break the continuity of the long beach system, the wide surf zone continues along the bluff to the river mouth and beyond. Tucked in along the base of the bluffs are three strips of sand (beaches WA 584-586), with beaches WA 587 and 588 straddling the river mouth.

Beach **WA 584** is a narrow 100 m long strip of rock-littered sand located at the base of the bluffs. It is awash at high tide and fronted by a deep inshore trough that feeds into a large rip system, with the waves breaking up to 500 m offshore. Beach **WA 585** lies 100 m to the west and is a similar 200 m long beach, a little wider and with patches of dry sand at high tide. It is also backed by dune-covered calcarenite bluffs, with an inner trough and wide surf zone. Beach **WA 586** is located another 80 m to the west and is a 150 m long pocket of sand backed by the bluffs and a small gully towards the western end. A trough feeds a permanent rip that runs out of the eastern rocks. It is bordered to the west by a 150 m wide calcarenite headland, with the Donnelly River beaches on the other side.

Beach **WA 587** extends from the northern side of the calcarenite bluffs for 700 m to the **Donnelly River** mouth. The first 400 m are backed by 20 m high partly vegetated bluffs, then a low partly vegetated dynamic spit that forms the eastern side of the river. The river mouth may be deflected up to a few hundred metres to the east or west, which will influence the nature and length of the beach. The 500 m wide double bar system continues along the beach, with 1-2 inner beach rips and then the outer bar. Beach **WA 588** extends from the river mouth, for 2.2 km to the west, to the next section of low calcarenite bluffs. The spit sections of both beaches are backed by partly active dunes extending up to 1 km inland, with older vegetated dunes paralleling the western side of the river for up to 7 km inland and rising to over 100 m. There is a shack settlement along the river banks extending from 1-3 km inland. The shacks are accessed by boats launched from a boat ramp located another 8 km upstream, at the end of the aptly named Boat Landing Road. The settlement area is also known as the Broadwater.

WA 589-598 JASPAR BEACH-BLACK PT

No.	Beach	Rating	Type inner	outer bar	Length
WA 589	Donnelly R (W2)	9	RBB+rocks	RBB	300 m
WA 590	Donnelly R (W3)	9	RBB+rocks	RBB	150 m
WA 591	Donnelly R (W4)	10	RBB+rocks	RBB	300 m
WA 592	Donnelly R (W5)	9	RBB+rocks	RBB	400 m
WA 593	Jasper Beach	8	RBB	RBB	4 km
WA 594	Jasper Beach (W)	8	RBB	RBB	1.3 km
WA 595	Black Pt (E4)	9	RBB+rocks		150 m
WA 586	Black Pt (E3)	9	RBB+rocks		250 m
WA 597	Black Pt (E2)	9	RBB+rocks		400 m
WA 598	Black Pt (E1)	9	RBB+rocks		200 m
Spring & neap tidal range =0.6 & 0.4 m					

To the west of the Donnelly River mouth the shoreline continues straight to the northwest for 13 km to the large Black Point. The point protrudes 1.5 km to the southwest and reaches heights of 60 m. In between is beach WA 588, then a series of ten beaches all backed by calcarenite bluffs and high vegetated dunes extending up to 6 km inland. Beaches WA 589-592 and WA 595-598 are dominated by the bluffs and rocks debris. The Jasper Beach track runs from Black Point through the backing high dunes to the highway and provides access to Jaspar Beach (WA 593) and beaches WA 590 and 591. A continuous double bar surf zone extends to beach 504, then as the bluffs protrude slightly seaward, one high energy bar dominates the remaining 2.5 km to Black Point. The western 7 km of shore is located within the D'Entrecasteaux National Park.

Beach **WA 589** is a 300 m long beach backed by low bluffs and scarped dunes rising to 20 m in the east, while rocks litters the western end of the beach. The 500 m wide double bar surf zone continues along the beach and past the low boundary bluffs, with a usually deep trough against the beach feeding a large beach rip. Beach **WA 590** lies 50 m to the west and is a 150 m pocket of sand located in a curve in the backing 20 m high vegetated bluffs. Calcarenite points border each end and a permanent rip drains out the western end of the beach and into the wide surf.

Beach **WA 591** lies 400 m further west along the bluffs and is a discontinuous 300 m beach consisting of four pockets of sand separated by rock debris and backed by low undercut bluffs, with numerous rocks littering the inner surf zone. There is usually a continuous trough between the beach and inner bar, which feeds a permanent rip towards the eastern end of the beach. A 4 WD track reaches bluffs above the eastern end of the beach. Beach **WA 592** commences immediately to the west and is a 400 m long slightly wider beach backed by dune capped bluffs rising to 20 m. Some rocks also litter the inner surf.

Jaspar Beach (**WA 593**) commences on the western side of a protruding 50 m wide calcarenite point and then continues straight for 4 km to the next calcarenite outcrop. It is backed initially by low bluffs, then a 3 km section of active dunes extending 200-300 m inland, backed in turn by the older vegetated dunes up to 5 km wide. The 500 m wide surf zone continues along the beach with several large beach rips crossing the inner bar, then the outer bar (Fig. 4.122). There is 4 WD access down the dunes to the western end of the beach where there is an informal camp site. Beach **WA 594** is separated by a few hundred metres of calcarenite bluff. It commences as a narrow beach littered with rock debris, then widens for most of its length where it is backed by a low bluff backed by well vegetated older dunes. The rip dominated inner bar and outer bar continues along the beach.

Figure 4.122 High energy double-barred Jasper Beach.

Beach **WA 595** is a 150 m long pocket of sand located at the base of 60 m high calcarenite cliffs 300 m to the west. The narrow beach is drained by a permanent rip which runs out against the western rocks, with the bar extending 300 m seaward. There is no access to the beach other than around the wave-washed rocks from beach WA 594.

Beaches WA 596 and 597 are adjoining pockets of sands at the base of 80 m high calcarenite cliffs. Beach **WA 596** is 250 m long and drained by a strong central rip with waves breaking 200-300 m offshore. The surf continues past the intervening 100 m wide protrusion in the cliffs to beach **WA 597**, a 400 m long strip of sand. It consists of a 200 m long wider beach, then a narrow rock littered western half. Rips exit from both ends of the beach, with a strong permanent rip against the western rocks. There is no safe access to either beach.

Beach **WA 598** is located in a 200 wide embayment on the eastern side of Black Point. Wave break 300-400 m off the beach with much of the surf draining out a strong permanent rip against Black Point, as well as a skewed rip exiting the eastern end of the beach. The beach is backed by low narrow foredune, then 80 m high basalt cliffs capped by calcarenite and dunes. A walking track from the Black Point car park leads to a lookout above the beach, with a steep track down to the rocks at the western end of the beach. There is a righthand point beach between Black Point and the main rip.

Black Point is a focus for visitors to this section of the national park and contains three camping sites, a day use area with basic facilities and a number of walking tracks. There are excellent views along the coast from the 60 m high point, and of the columnar basalt that composes the point, including the 'Stepping Stones' that lead down to the western beach (WA 599). There are also *reef breaks* to either side of the point.

WA 599-602 **BLACK-WHITE PTS**

No.	Beach	Rating	Type	Length
WA 599	Black Pt (W1)	8	RBB/TBR	1.3 km
WA 600	Black Pt (W2)	8	TBR+rocks	150 m
WA 601	Black Pt (W3)	7	TBR→LTT+rocks	4.5 km
WA 602	White Pt (E)	6	LTT/TBR+rocks	2 km

Spring & neap tidal range = 0.6 & 0.4 m

The near continuous beach system that begins on the western side of Black Point continues for another 40 km to Frederick Point at the mouth of the Blackwood River. The only minor inflection in this section is the low, small White Point located 9 km west of Black Point. Between Black and White points are four beaches (WA 599-602) with wave energy gradually decreasing to the west, as inshore reefs and a shoaler nearshore lower the waves.

Beach **WA 599** is bordered in the east by Black Point, a 1 km long 60 m high headland. The 1.3 km long beach faces straight into the southwest waves which average about 2 m and break across a 300 m wide surf zone usually dominated by two central beach rips, with a strong permanent rip running out against the point, and a lefthand surf breaking along the point. To the west the beach is bordered by rock reefs with waves breaking over reefs up to 400 m offshore. A small foreland has formed in lee of the reef beach, with the beach continuing on the western side of the foreland as a slightly lower energy beach for another 100 m to boundary 60 m high calcarenite cliffs. The beach is backed by three large blowouts that have climbed up to 80 m over the backing dunes and calcarenite and blown 1 km inland (Fig. 4.123). There is vehicle access to the point and a track down the Stepping Stones to the eastern end of the beach. The track terminates at a basalt boulder and cobble swash zone before reaching the sand beach. This is a hazardous high energy beach and dangerous for both for rock fishing and swimming.

Beach **WA 600** is located 300 m to the west in a 150 m gap in the cliffs. It consists of a pocket of sand bordered and backed by steep, vegetated 60-80 m high calcarenite slopes. It receives waves averaging less than 2 m, which maintain a 100 m wide surf zone, with rocks debris to either end and a central permanent rip. A 4 WD track terminates on the buffs above the western end of the beach, with a steep climb down to the shore.

Figure 4.123 Beach WA 599 to the west of Black Head has a rip-dominated surf zone, backed by three prominent climbing blowouts.

Beach **WA 601** commences 400 m to the west, as the calcarenite bluffs move slightly inland and the beach trends northwest, with the 40 m high bluffs backing the first 2 km of the 4.5 km long beach. The beach curves round in a slight arc to terminate at a beachrock reef. Waves average about 1.5 km in the east decreasing slightly towards the west as inshore reefs increase wave attenuation. The beach has a rip-dominated 80 m wide bar in the east grading to a more continuous low tide terrace to the west, with some beachrock also outcropping along the western shore. It s backed by 60-80 m high scarped vegetated dunes which extend up to 1.5 km inland peaking at 146 m at Dickson trig. There is one large active blowout towards the western end.

Beach **WA 602** extends 2 km from the beachrock reef to White Point, another low calcarenite reef, with scattered shallow reefs extending up to 500 m offshore the length of the beach. As a result of the reef waves height varies along the shore between 1 and 1.5 m, with the beach a mixture of some surf and rips, low tide terrace and a few reflective sections, together with scattered beachrock along the shore. It is backed by a partly unstable foredune, and three large blowouts extending up to 1 km inland, with older vegetated dunes up to 4 km inland. There is a good 4 WD track off the main track to White Point. A gate at the beginning of the track indicates this is private property, however the public access is permitted to the beach. The beach is a popular fishing location, as well as offering a few *surf breaks* over the inshore reefs.

WA 603 **WHITE-FREDERICK PTS**

No.	Beach	Rating	Type inner	outer bar	Length
WA 603	White-Frederick Pts				
	0-5 km	4	R		5 km
	5-12 km	7	TBR+reefs		7 km
	12-21 km	8	RBB	LBT	9 km
	21-30 km	8	TBR/RBB		9 km
				Total	30 km

Spring & neap tidal range = 0.6 & 0.4 m

Beaches **WA 603** is a curving, south-facing 29.6 km long, continuous stretch of sand between the low White Point in the east and low sandy Frederick Point at the mouth of the Hardy Inlet. The beach includes Ledge Point, a cuspate foreland formed in lee of offshore reefs, and the area between the foreland and the estuary called Flinders Bay. The beach represents the northwestern section of the near continuous 90 km long beach system that commences at Black Point. The dunes system continues the length of the beach, with Pleistocene dunes extending up to 4 km inland, while the Holocene dunes average about 1 km in width. There are several access tracks via the backing farmland to the shore. Some pass through private property and are not open to the public.

The beach is exposed to variable breaker wave conditions along its length and can the divided into four sections. The first **5 km** receive lower waves averaging about 1 m owing to offshore reefs. The waves maintain a steep reflective beach, with increasing intertidal beachrock outcrops to the west. The beach is backed by a series of active blowouts extending a few hundred metres inland. Some of the large deflation basins and blowouts have been fenced to assist dune stablisiation and revegetation. This is to prevent the dunes blowing onto the backing farmland, which extends to within 1 km of the shore.

Between **5-12 km** the waves average about 1.7 m and break over two parallel discontinuous beachrock reefs. The inner reef outcrops along the shore, while the outer lies 200-300 m out in the surf. The result is a more crenulate shoreline and a series of about 10 reef-controlled feeder currents and strong permanent rips flowing out the gaps in the reef. Active blowouts averaging up to 500 m in length continue along this section.

The next section from **12-21 km** receives the highest waves averaging about 2 m, which maintain a 300 m wide double bar system (Fig. 4.124). The inner bar is cut by beach rips, as well as occasional reef-controlled rips, while the outer bar remains more shore parallel. Active blowouts, including a few large blowouts back the first few kilometers, decreasing to the west.

Figure 4.124 Section of the long high energy beach WA 603.

The final section **(21-30 km)** commences 2 km east of Ledge Point and extends for 9 km to Frederick Point. Waves are lowered slightly by the Ledge Point reef, and to the west by Cape Leeuwin and the Saint Alouran Islands, which combined extend 12 km to the south. The result is a single bar 150 m wide surf zone dominated by strong beach rips spaced every few hundred metres. Generally stable 200-500 m wide dunes back the beach. The final 1.5 km is part of a narrow recurved spit that forms the eastern entrance to the Hardy Inlet into which flows the Blackwood River. The town of Augusta is located on the western shores of the inlet.

WA 604-608 **AUGUSTA**

No.	Beach	Rating	Type	Length
WA 604	Seine Bay	3	R+tidal channel	700 m
WA 605	Duke Head	6→5	TBR→LTT	1.3 km
WA 606	Storm Bay	5	LTT+rocks/reef	900 m
WA 607	Granny Pool	4	R+rocks	300 m
WA 608	Barrack Pt (S 1)	4	LTT	150 m
Spring & neap tidal range = 0.6 & 0.4 m				

Augusta is located at the southwestern tip of Western Australia, with a population of about 1000. It is located on the western shores of Hardy Inlet which enters the sea between two low mobile sandy spits, called Frederick and Duke heads. The town has protected estuary beaches inside the heads (WA 604), and more exposed ocean beaches on the shores of open Flinders Bay (WA 605 onwards), which runs for 7 km from the inlet to Cape Leeuwin, the south western tip of Australia. The Leeuwin Road runs along the coast to the cape, providing good access to the coast and beaches. The main swimming beaches are located between the inlet and Barrack Point (WA 604-608). These beaches are backed by a continuous foreshore reserve, with the access along Albany Terrace (Fig. 4.125).

Seine Bay (WA 604) is located just inside the Hardy Inlet and curves round to face northeast across the inlet. The beach is sheltered from ocean waves by the two heads and narrow inlet entrance, with low wind wave to calm conditions usually prevailing. The beach curves round for 700 m to the southeast to terminate on the inside of Duke Head. It is fronted by a mixture of migratory tidal sand shoals, and the deeper inlet channel, particularly towards the eastern end of the beach. While conditions are relatively safe at the shore, be very careful of the strong tidal currents, which flow off the beach. The beach is backed by a foreshore reserve, with all facilities, as well as the Turner Street jetty and boat ramp, and a caravan park behind the northern end.

Duke Head (WA 605) is a low sand spit that forms the western side of Hardy Inlet. The spit trends to the southwest, curving round for 1.3 km to terminate against the low granulite rocks that dominate the shore for 4 km to the south. Near the entrance the tidal currents and periodic high waves produce a dynamic inlet, with shifting channels and bars. Extreme care, local knowledge

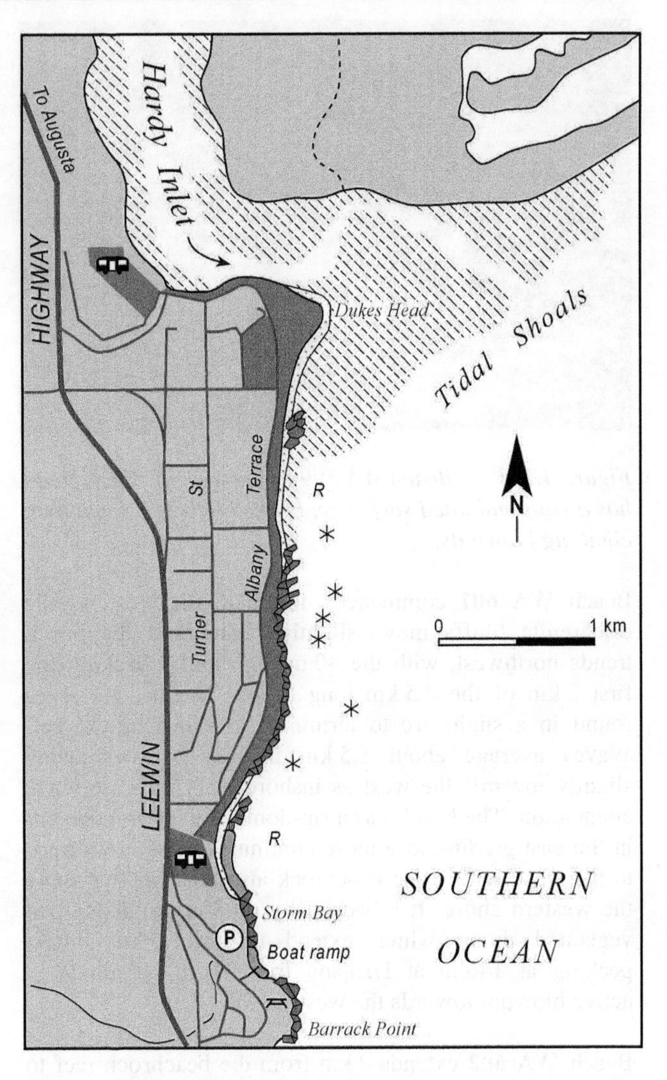

Figure 4.125 The main Augusta beaches are located between the mouth of Hardy Inlet at Duke Point and Barrack Point. All face east and are sheltered from the dominant southwest swell.

and low wave conditions are required to cross the entrance bar by boat. The beach near the entrance can vary considerably in width and only becomes more stable a few hundred metres south of the inlet, where rocks increasingly outcrop along and off the beach. Waves decrease from about 1.5 m at the inlet to less that 1 m at the southern end of the beach. The beach also changes from a 50 m wide bar with 2 to 3 beach rips in the north, to a continuous low tide terrace and occasional boulders south of Dere Reef. During higher swell and the right bar conditions the river mouth can provide some good *surf*. The beach is backed by a narrow foreshore reserve, Albany Terrace and a Lions Park, with good access to the entire beach. In lee of Dere Reef is The **Landing Place**, where the ship 'Emily Taylor' landed the first European settlers in 1830.

Storm Bay (WA 606) is a crenulate 900 m long, open east-facing bay, containing a moderately protected sandy beach, together with round granite rocks, boulders and reefs. The reefs lower waves to less than 1 m along most of the shore. The beach alternates between a low tide terrace and reflective conditions, with numerous rocks scattered along and off the shore. Seagrass meadows lie in the bay and seagrass debris often covers the beach. The bay is bordered by a caravan park and picnic area, with a large car park and boat ramp in the centre. Most boats going outside launch from here, rather than having to cross the bar. The area is also known as **The Whaling**, referring to its prior use as a whaling site.

Barrack Point is a low granulite point surrounded by rock platforms. It forms the southern boundary of Storm Bay and is occupied by a foreshore reserve and picnic area, with a large car park extending from the boat ramp onto the point (Fig. 4.126). Between the rocks and the point is a strip of high tide sand (beach **WA 607**), with the rocks enclosing a 100 m long tidal pool on the southern side of the point, know as **Granny Pool**. Low waves usually prevail breaking on the outer rocks with calm conditions in the pool.

Figure 4.126 Barrack Point is surrounded by three lower energy rock dominated beaches, including Storm Bay to the north and the small Granny Pool to the south.

Beach **WA 608** is located on the southern side of the point, between the pool area and the low granulite rocks that continue south. It is a 150 m long, south-facing, slightly more exposed sand beach. Waves average about 1 m and break across a 50 m wide low tide terrace, with a beach and small foredune behind, which are backed by a grassy reserve and the road. During higher waves a strong rip runs out along the western rocks.

Swimming: Seine Bay offers calm conditions, but beware of the tidal current flowing off the beach. Duke Head Beach has surf increasing in size towards the inlet, while the Granny Pool offers a quieter tidal pool, next to the usually low surf at Barrack Point.

WA 609-613 **BARRACK-MATTHEW PTS**

No.	Beach	Rating	Type	Length
WA 609	Barrack Pt (S2)	4	LTT+rocks	50 m
WA 610	Barrack Pt (S2)	4	LTT+rocks	100 m
WA 611	Pt Matthew (N3)	4	R+rocks/reef	150 m
WA 612	Pt Matthew (N2)	4	LTT+rocks	100 m
WA 613	Pt Matthew (N1)	3	R+reef	700 m
Spring & neap tidal range = 0.6 & 0.4 m				

Between Barrack Point and Point Matthew is a 3 km long section of low east to southeast-facing shoreline, dominated by granulite rocks to the north and the granite of Point Matthew to the south. In amongst the rocks are four small beaches (WA 609-612) and the longer Point Matthew beach (WA 613). The Leeuwin Road runs 100-200 m in from the beach, with small car parks and walking track providing access to the shore.

Beach **WA 609** is a 50 m long pocket on intertidal sand located 200 m south of Barrack Point and beach WA 608. It is backed and bordered by low rocks. Waves average about 1 m and break across a narrow bar. Beach **WA 610** is located 300 m to the south and is a similar 100 m long pocket of sand, backed by low rocks, with some rocks scattered across the 50 m wide sand bar. Waves average about 1 m and run across the bar to reach the rocks at high tide. A narrow vehicle track reaches the back of the beach.

Beach **WA 611** is 150 m long strip of high tide sand backed by a low foredune, then the Leeuwin Road. It is bordered by north-south trending ridges of rock, that run about 30 m off the shore partly enclosing a tidal pool. Waves average 1 m and break over the outer rocks and into the pool. Beach **WA 612** is located 100 m to the south and is a 100 m long high tide beach, bordered by low rocks, but open to the sea, and fronted by a 30 m wide low tide bar, with waves averaging less than 1 m. The road runs right behind the beach, with a near continuous car park backing of the beach.

Beach **WA 613** extends from the southern rocks of beach WA 612 for 700 m to the sloping granite of Point Matthew. The beach protrudes seaward in lee of shallow inshore reefs and seagrass meadows, with some seagrass debris usually littering the steep reflective beach. There is good access to the northern end and a vehicle track along the rear of the beach towards the point. The southern end of the beach in lee of the point is known as **Dead Finish Anchorage**. Small boats are launched from the beach and moored just offshore. The 15 m high point is named after Matthew Flinders, who anchored her in 1802.

WA 614-617 **GROPER, RINGBOLT & SARGE BAYS**

No.	Beach	Rating	Type	Length
WA 614	Groper Bay	5	R boulder	80 m
WA 615	Ringbolt Bay	4	R+rocks	150 m
WA 616	Sarge Bay	4	R+rocks/reef	500 m
WA 617	Sarge Bay (W)	4	R+rocks/reef	900 m
Spring & neap tidal range = 0.6 & 0.4 m				

The four southwestern-most beaches in Western Australia are located in a 2 km long rocks and reef strewn south-facing embayment between Point Matthew and Cape Leeuwin, the southwestern tip of the Australian mainland (Fig. 4.127). The four low energy beach face south across the shallow inshore granite reefs and a series of islands, islets and reefs that extend up to 6 km to the south. In addition to the protection afforded by the reefs, Cape Leeuwin extends 700 m to the south also shielding the beaches from westerly waves. The Leeuwin Road runs along the back of the beaches and out to Cape Leeuwin and the lighthouse.

Figure 4.127 View west across Groper, Ringbolt and Sarge Bay to cape Leeuwin, the southwest tip of the Australian continent.

Groper Bay is bordered in the east by Point Matthew and contains the small beach **WA 614**. This is a steep 80 m long low energy boulder beach wedged in between sloping, ridged granite rocks and backed by wind blasted vegetation rising to 20 m. Three hundred metres to the west is adjoining **Ringbolt Bay** and beach **WA 615.** It is a sandy 150 m long reflective beach, bordered by ridges of granite and backed by vegetated slopes. A vehicle track runs out from the road to the beach and a car park. If faces due south and receives waves averaging about 1 m.

Sarge Bay (**WA 616**) lies on the western side of a 100 m wide low granite point. The bay faces south, and curves round for 500 m between boundary points and scattered linear rock reefs, together with shallow reefs across the bay floor. Waves average less than 1 m and surge up a steep reflective beach face. A foredune, then vegetated slopes back the beach, with the road running 50 m behind the centre of the beach, where there is a car park on top of the foredune.

The western half of **Sarge Bay** commences 100 m to the west and is a double crenulate 900 m long steep reflective beach **(WA 617)**. Rock ridges and inshore reefs dominate the bay, with low waves at the shore. The road runs along the back of the beach, clipping the western end. The base of 20 m high Cape Leeuwin forms the western boundary, and extends 700 m to the south.

Region 3 Leeuwin Coast: Cape Leeuwin to Cape Nautraliste

Length of Coast: 112 km (1818-1930 km)
Number of beaches 113 (WA 618-730)

National Parks:	km	length	beaches	no.
Leeuwin-Naturaliste	1818-1993	103 km	WA 618-732	98

Towns: Augusta, Margaret River
Coastal settlements: Hamelin, Prevelly Park, Gracetown, Yallingup,

The South West tip of the state extends for 93 km from Cape Leeuwin for 93 km to Cape Naturaliste (Fig. 4.128)This is a west-facing shoreline exposed to the strong onshore southwest winds and high deepwater waves. It is also afforded varying degrees of protection by the near continuous string of calcarenite (limestone) islands, reefs and platforms, together with the ancient basement granulite and granite rocks. Consequently while there are some exposed high energy sections, for the most part the shoreline is sheltered from the full force of the waves, resulting in lower energy beaches at the shore. The crenulations induced by the rocks and reefs result in a shoreline length of 112 km containing 112 beaches. The combination of high bedrock and calcarenite coastline bordering numerous beaches and reefs, together with the persistent moderate to high southwest swell have combined to produce one of the more spectacular coastal sections of the world, deserving of its national park status.

Cape Leeuwin also represents the southwestern corner of the Australian continent. At the cape the shoreline turns and begins the long trend north to the Kimberley. It initially trends north for 1600 km to North West Cape and then northeast to Cape Londonderry, the northern most point in Western Australia, and 2700 km north of the Cape Leeuwin.

LEEUWIN-NATURALISTE NATIONAL PARK

Area: 20 000 ha
Coast length: 115 km (1818-1933 km)
Beaches: 114 (WA 618-732)

Leeuwin-Naturaliste

km	beaches	(No. beaches)
1818-1868	618-657	(39)
1875-1884	668-677	(9)
1886-1910	679-710	(31)
1913-1933	713-732	(19)
Total 103 km	98 beaches	

Leeuwin-Naturaliste National Park extends for 93 km from Cape Leeuwin north to Cape Naturaliste, two and prominent historic capes. In between is 115 km of shoreline containing 114 beaches, of which 103 km of shore and 98 beaches are located within the narrow park, which for the most part parallels the coast. The park was established in to preserve the spectacular coastline and coastal scenery between the capes. The shoreline and nearshore is dominated by Pleistocene calcarenite, overlying Precambrian granites and gneiss. Drowned Pleistocene barriers lie offshore as shore parallel series of calcarenite islets and reefs, and alongshore as numerous rocks and reefs, as well as the backing calcarenite cliffs and bluffs. The persistent and often high southwest swell breaks over the reefs producing some of the best surf in Australia, including the world renowned Margaret River break.

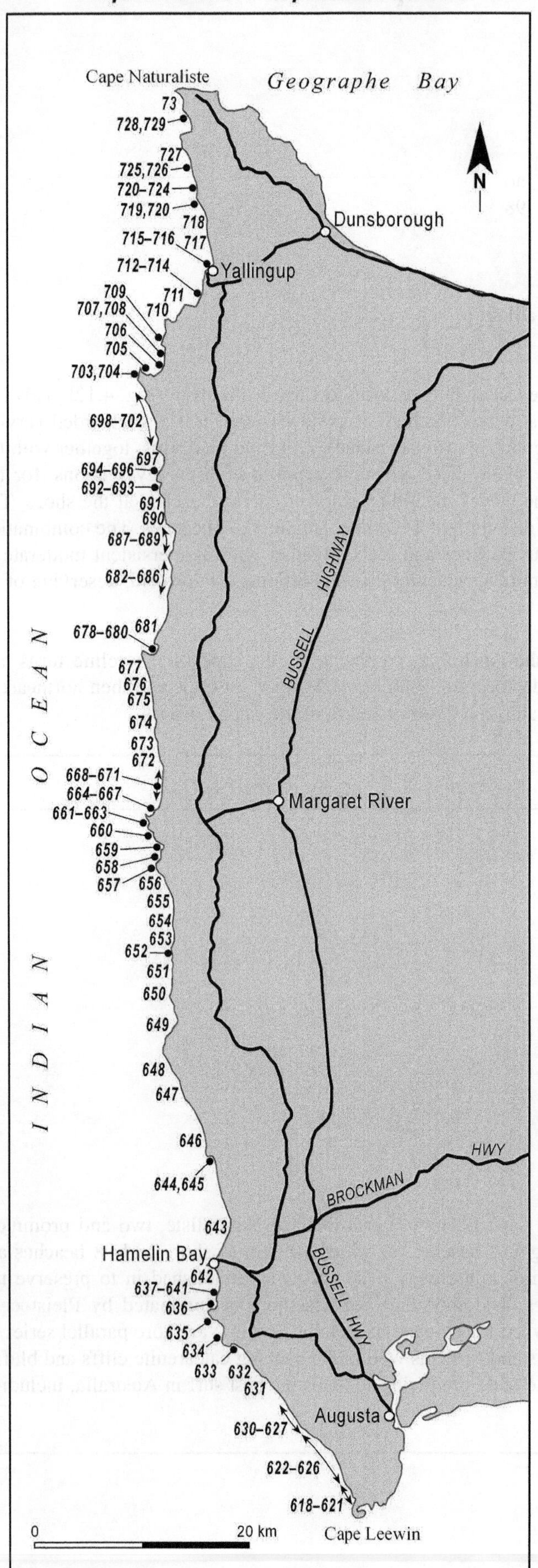

Figure 4.128 (left) The Leeuwin Coast, extends from Cape Leeuwin to Cape Naturaliste and contains 113 beaches.

WA 618-628 CAPE LEEUWIN (N)

No.	Beach	Rating	Type	Length
WA 618	Cape Leeuwin (W)	5	R+rocks/reef	70 m
WA 619	Quarry Bay	4	R+rocks/reef	250 m
WA 620	Skippy Rock (S2)	4	R+rocks/reef	50 m
WA 621	Skippy Rock (S1)	5	R+rocks/reef	150 m
WA 622	Skippy Rock (N1)	4→5	R→LTT	800 m
WA 623	Skippy Rock (N2)	5→6	LTT>TBR+rocks	700 m
WA 624	Skippy Rock (N3)	6	TBR+rocks	500 m
WA 625	Skippy Rock (N4)	7	TBR+rocks	650 m
WA 646	Skippy Rock (N5)	7	LTT/TBR+rocks	200 m
WA 627	Skippy Rock (N6)	7	LTT/TBR+rocks	40 m
WA 628	Skippy Rock (N7)	7	TBR+rocks	300 m
Spring & neap tidal range = 0.8 & 0.6 m				

Cape Leeuwin is a low 10 m high granite point, with much of the sloping rocks washed over by the regular southwest swell. Along the centre of the cape is a road to the old lighthouse keepers houses and the Cape Leeuwin lighthouse, sitting on the 22 m high crest of the cape. The shoreline trends northwest for 15 km to Cape Hamelin, with two broad arcs of sand beaches and calcarenite bluffs in between. The first arc extends for 7 km to a cuspate foreland formed in lee of Cumberland Rock and Jacks Ledge and contains beaches WA 618-628. These beaches are all located within the Leeuwin Naturaliste National Park. Only the southern four beaches are accessible via the Leeuwin and Skippy Rock roads. The Cape to Cape walking track commences/ends at the Skippy Rock car park and runs along the slopes behind these beaches.

The first beach (**WA 618**) on the west coast is patch of sand located 1 km north of the tip of the cape. The 70 m long pocket of sand lies immediately north of the car park for the Old Waterwheel, off the Leeuwin Road. The beach consists of a narrow strip of high tide sand bordered by large granite slabs and boulders, which almost encase the beach, with only a 30 m wide access between the rocks to the sea. Even in the entrance channel rocks form the seabed and extend seaward of the small embayment. Waves average about 0.5 m at the shore with the inner area forming a relatively calm rock pool during low wave conditions, while exposed conditions prevail outside the rocks.

Quarry Bay (WA 619) is located 200 m to the north and is a U-shaped, west-facing bay, from the shores of which rock was quarried for the lighthouse. The bay is 200 m across at the entrance with the 250 m long beach curving round the back of the bay. Waves break over granite rocks and reefs strewn across the mouth resulting in low wave to calm conditions at the shore. A gravel road, off the Skippy Rock road, leads to a car park at the back of the beach. This beach is usually suitable for swimming so long as you stay clear of the rocks.

Skippy Rock is a series of linear granite rocks that form a low islet that is tied to the shore by beaches WA 621 and 622. Beach **WA 620** is located in the small south-facing rocky bay located immediately east of the rock. The beach is a 50 m long south-facing patch of sand surrounded on three sides by 5-10 m high granite. It is well sheltered from most waves, with usually calm conditions at the shore. The Skippy Rock car park is located 100 m south of the beach, with a walking track past the beach.

Beach **WA 621** is a 150 m long south-facing beach that forms the southern side of the sandy foreland that ties Skippy Rock to the shore. The beach is sheltered by the rock and associated reefs, with waves usually low at the shore. Rocks also outcrop on and immediately off the beach.

Beach **WA 622** commences on the northern side of Skippy Rock and trends north for 800 m to a 20 m high, 100 m long section of slightly protruding calcarenite. Wave height increases from about 1 m at the rock to about to 1.5 m at the northern end, with the beach initially reflective grading into a low tide terrace in the north, with a permanent rip against the point. The rocky point and beach is initially backed by a deflated dune surface, then vegetated calcarenite bluffs rising to 40 m, with rising vegetated slopes behind. The beach can be accessed from the Skippy Rock car park.

Beach **WA 623** commences on the northern side of the calcarenite point and trends north-northwest for 700 m to the next narrow calcarenite point, which is capped by a clifftop blowout. Waves average about 1-1.5 m and maintain a low tide terrace with occasional rips forming, particularly amongst the rocks at the northern end of the beach. It is backed by vegetated 30-40 m high calcarenite bluffs.

Beach **WA 624** lies on the northern side of the point and is a 500 m long beach, which narrows to the north as it becomes increasingly dominated by the backing bluffs, with rocks littering the beach and calcarenite rocks and reefs in the intertidal. Waves averaging about 1.5 m break across a 50 m wide surf zone, with usually two rock-controlled rips. Beach **WA 625** lies 100 m to the north and is a similar 650 m long sand beach located at the base of 30-50 high vegetated bluffs, with rocks outcropping across the 50 m wide bar along most of the beach. A 50 m wide calcarenite point separates it from beach **WA 626**, a 200 m long pocket of sand bordered and backed by steep, vegetated calcarenite bluffs. Waves average less than 1.5 m and break across a 50 m wide bar and some rocks. The Cape to Cape walking track has a spur off to a lookout on the ridge, between beaches WA 625 and 626.

Beach **WA 627** lies 200 m to the north and is a wave-washed 40 m long pocket of sand, bordered and backed by 10 m high irregular calcarenite bluffs. Waves break off the boundary bluffs and across the small beach, with a permanent rip running out between rocks on the north side of the sand.

Beach **WA 628** commences 300 m to the north and continues northwest for 300 m to the rock-tied tip of a cuspate foreland (Fig. 4.129). The beach is initially narrow and backed by low bluffs. It widens towards the foreland, however calcarenite rocks and reef dominate the surf zone, with a permanent rip draining the northern half of the beach. It is backed by a deflation surface and active blowouts.

Figure 4.129 View north along beaches WA 628 and Deepende Beach.

WA 629-632 **CAPE HAMELIN (S)**

No.	Beach	Rating	Type	Length
WA 629	Deepende Beach (S2)	7	LTT/TBR+beachrock	400 m
WA 630	Deepende Beach (S1)	7	LTT/TBR+beachrock	1 km
WA 631	Deepende Beach	7	LTT→TBR	5.3 km
WA 632	Cape Hamelin (S)	5	R+rocks/reef	300 m
Spring & neap tidal range = 0.8 & 0.6 m				

The second arc north of Cape Leeuwin extends to the northwest from the Cumberland Rock cuspate foreland for 7.5 km along Deepende Beach to Cape Hamelin. The entire section lies within the national park, with the only vehicle access a 4 WD track off the Cosy Corner Road to Cape Hamelin. In the south Hillview Drive terminates at a 140 m high lookout, 2 km inland, that provides views of the coast. The Cape to Cape walking track run the length of Deepende Beach to the Cape Hamelin.

Beach **WA 629** commences on the northern side of the foreland, with a 150 m long section of beachrock separating it from beach WA 628. The beachrock has an inactive higher platform and an active intertidal platform. The beach trends north-northwest for 400 m to a 100 m long patch of calcarenite. Waves averaging over 1 m break over calcarenite reefs at either end with a central 50 m wide bar, drained by a central rip. A mixture of active and vegetated dunes rising to over 100 m back the beach. Beach **WA 630** continues north of the beachrock for another 1 km. The beach is narrow and for most of it length backed by low calcarenite bluffs, then well vegetated dunes, and fronted by intertidal calcarenite reefs. Waves break over a mixture of rocks and sand.

Deepende Beach (**WA 631**) commences on the northern side of the calcarenite and continues to the northwest and finally west to for 5.3 km to terminate in lee of Cape Hamelin. The southern half has extensive beachrock resulting in slightly lower waves, numerous tidal pools and some permanent rips along the shore. The northern half, while free of beachrock has more persistent beach rips, spaced about every 300-400 m (Fig. 4.130), including a small permanent rip against the northern boundary rocks, which provides the most accessible surf breaks on the beach. Wave energy increases up the beach, averaging just over 1.5 m along the northern few kilometres. Scattered reefs extending 2 km off Cape Hamelin, together with Geographe Reef and Minns Ledge contribute to the lower breaker wave heights.

Figure 4.130 Deepende Beach is typified by moderate waves and a rip-dominated surf zone.

The entire beach is backed by several blowouts, then generally well vegetated dunes, blanketing the calcarenite, and rising to a maximum of 173 m and extending up to 4 km inland. Turner Brook drains across the beach 1 km southeast of the cape. The Caves Road essentially runs along the eastern boundary of the dunes with a camping area is located in the dune area behind the northern end of the beach.

Beach **WA 632** is a 300 m long south-facing strip of sand located between the cape and the point at which the 4 WD and walking track reach the shore. The beach is entirely fronted by irregular lines of granite boulders and rocks extending up to 400 m offshore. Waves break over the rocks, resulting in some calmer tidal pools close to shore.

WA 633-635 **CAPE HAMELIN-COSY CORNER**

No.	Beach	Rating	Type	Length
WA 633	Cape Hamelin	5	R+rocks	50 m
WA 634	Cape Hamelin (N)	5	R+rocks	150 m
WA 635	Cosy Corner	5	R/LTT+beachrock	550 m
Spring & neap tidal range = 0.8 & 0.6 m				

Cape Hamelin consists of north-south trending ridges of granite and boulders. They rise to 20 m and extend inland where they are blanketed by dunes. Beside beach WA 632 on the southern side of the cape, beaches WA 633 and 634 are located amongst the rocks on the western side of the cape, with Cosy Corner located 1 km to the north.

Beach **WA 633** is a 50 m long pocket of south-facing sand located at the southwestern extremity of the cape. It is bordered by lines of granite boulders and fronted by submerged boulders, with waves averaging about 0.5 m at the shore. Beach **WA 634** lies 200 m to the north and is a 150 m long, west-facing strip of high tide sand, bounded by large granite boulders, which also outcrop along the shore and on the beach. The northern half usually receives low waves and has the only clear access to the water. A walking track reaches the northern end of the beach.

Cosy Corner beach (**WA 635**) is located on the northern side of the 1 km wide embayment formed between Cape Hamelin and Knobby Head. The beach is relatively straight, 550 m long and faces southwest (Fig. 4.131). It is however protected by the numerous rocks and reefs off the cape including the eroded calcarenite of the Honeycomb Rocks, which lie south of the beach resulting in waves averaging less than 1 m. The waves break across a mixture of straight intertidal beachrock and a central and western sandy low tide terrace. It is backed by rising vegetated dunes, with the Cosy Corner gravel road terminating at a car park on the 40 m high western headland.

Figure 4.131 Cosy Corner a southwest-facing beach sheltered by inshore reefs.

WA 637-642 **FOUL BAY**

No.	Beach	Rating	Type	Length
WA 636	Foul Bay (1)	7	TBT+rocks	750 m
WA 637	Foul Bay (2)	4	R+rocks/reef	200 m
WA 638	Foul Bay (3)	4	R+rocks/reef	150 m
WA 639	Foul Bay (4)	4	R+platform	200 m
WA 640	Foul Bay (5)	4	R+rocks/reef	300 m
WA 641	Foul Bay (6)	5	R+rocks/reef	400 m
WA 642	Foul Bay (7)	5	R+rocks	1 km
Spring & neap tidal range =0.8 & 0.6 m				

Foul Bay is a 3 km long, open west-facing bay bordered by Knobby Head and White Cliff Point. In between is 4 km of irregular rocky shoreline dominated by granite in the south and calcarenite in the north. It contains seven small beaches, all backed by generally well vegetated rising dunes, extending 1.5-2.5 km inland to the Caves Road. While the bay faces west the shore is moderately sheltered by a series of reefs extending 2 km south of Hamelin Island to South West Rock. There is access to either end via the Cosy Corner Road in the south and the Cape Hamelin Road to the north. The Cape to Cape walking track runs through the vegetated dunes behind the southern half of the bay and follows the shoreline along the northern beach (WA 642).

Beach **WA 636** is a relative straight 750 m long west-facing beach and the most exposed of the bay beaches, receiving waves averaging 1.5 m. The waves break across a mixture of rocks and a 70 m wide bar, with three rock-controlled rips located at either end and in the centre. It is backed by well vegetated dunes, which rise to 100 m, 500 m inland, with a lighthouse located on the highest dunes. There is a small car park off the Cosy Corner Road above the southern end of the beach.

Beaches WA 637 and 638 are located in the next small rocky embayment to the north. Beach **WA 637** is a 200 m long strip of high tide sand bounded by low granite points, linked by subtidal rocks off the centre of the beach. A spur off the Cape to Cape walking track reaches the southern point. Beach **WA 638** lies 150 m to the north and is a 150 m long strip of sand fronted by inter and supratidal lines of granite rocks. Both beaches receives low waves at the shore, and are backed by well vegetated rising dunes.

Beach **WA 639** is 200 m long and fronted by a continuous supratidal rocks, with subtidal rock reefs then extending a few hundred metres offshore. It is distinguished by an active blowout that backs the beach. Beach **WA 640** commences 200 m to the north and is a 300 m long strip of high tide sand fronted by linear granite ridges across the inter- and supratidal, with only one small area free of rocks at the southern end of the beach. It is backed by a series of small blowouts, then rising vegetated dunes. Beach **WA 641** lies 100 m to the north and is a straight 300 m long high tide sand beach fronted by a 100 m wide band of subtidal rock reefs, with waves breaking on the outer reefs, and relatively calm conditions at the shore. An unstable foredune then vegetated transgressive dunes back the beach.

Beach **WA 642** is the longest beach in the bay and extends for 1 km to the northwest terminating at the calcarenite basement rocks of White Cliff Point. The beach is tied to low granite rocks in the south, with a mixture of sand and calcarenite reef off the beach. The calcarenite forms an eroding 10 m high bluff at the northern end, the 'white cliffs' of the point. It receives waves averaging about 1 m, which surge up a moderately steep reflective beach. The beach can be accessed on foot from the Hamelin Bay car park, 400 m to the north.

WA 643-648 **HAMELIN BAY-CAPE FREYCINET**

No.	Beach	Rating	Type	Length
WA 643	Boranup Beach	4→6	LTT→TBR	7.8 km
WA 644	North Pt (S2)	5	R+rocks/reef	50 m
WA 645	North Pt (S1)	5	LTT+rocks	70 m
WA 646	North Pt (N)	5	R/LTT+rocks	3.2 km
WA 647	Cape Freycinet (S2)	7	TBR+rocks	2 km
WA 648	Cape Freycinet (S1)	7	TBR+rocks	900 m
Spring & neap tidal range = 0.8 & 0.6 m				

North of White Cliff Point is a 14 km long open west-facing section of coast, broken midway by North Point. In between are two long beaches (WA 643 and WA 646-647-648), together with two pockets of sand on the southern side of North Point (WA 644-645). This long section is only accessible in the south at Hamelin Bay and in the north at Cape Freycinet. In between the beaches are backed by predominately vegetated transgressive dunes extending 2-4 km inland, with an area of active dunes at Boranup and the Caves Road running along the rear. The Cape to Cape walking track runs along the beach from Hamelin Bay to North Point then turns inland for several kilometres, reaching the coast again at Contis Beach (WA 649).

Boranup Beach (WA 643) is one of the longer more continuous beaches in the national park. It commences at Hamelin Bay in lee of White Cliff Point and its numerous reefs, as a low energy reflective beach (Fig. 4.132), and trends north for 7.8 km to the southern rocks of North Point. Wave energy increases rapidly up the beach averaging 1-1.5 m and maintaining a rip-dominated bar that widens to 100 m in the north. The bar can produce some good *beach breaks* under the right conditions. The beach is backed by a series of mostly stable longwalled parabolics extending up to 4 km to the west and include the partly active Boranup Sand Patch, which reaches a height of 120 m near the Caves Road. At Hamelin Bay there is a large car park, caravan park, boat ramp and the remains of the old jetty. A 4 WD track from the Boranup camping area leads down to the northern section of the beach.

Figure 4.132 The dune-covered White Point and the sheltered southern end of Boranup Beach.

North Point is a 20-30 m high,, calcarenite-capped granite point that separates the two longer beaches. On the southern side of the point are two arcs in the calcarenite bluffs containing beaches WA 644 and 645. A 4 WD track runs along the top of the cliffs with no safe access down to the beaches. Beach **WA 644** is located 100 m north of the rocky end of Boranup beach. It is a 50 m long pocket of sand located at the base of 20 m high calcarenite cliffs. It is fronted by shallow reefs extending over 100 m offshore, with waves averaging less than 1 m at the beach. Beach **WA 645** lies 150 m to the west and is a similar 70 m pocket of sand at the base of steep vegetated bluffs. It is bordered by calcarenite to the east and calcarenite-capped granite to the west, with a 100 m wide bar extending off the beach and reefs to either side. Waves averaging over 1 m break over the bar and exit via two small rips to either side.

Beaches WA 646-647-648 are part of a near continuous 6.5 km long beach between North Point and Cape Freycinet, which is broken into the three beaches by calcarenite rocks (Fig. 4.133). Beach **WA 646** commences amongst the granite rocks and boulders on the northern side of North Point and trends north for 3.2 km to a section of broken beachrock, just south of the Hooley Road access point. The beach receives waves averaging 1-1.5 m, which maintain a usually narrow bar and occasional rips, including some rips against rock outcrops. It is backed in the south by several blowouts, then the vegetated dunes extending 3-4 km inland. Beach **WA 647** extends north of the beachrock for 2 km to a narrow finger of 10 m high calcarenite, which separates it from the northern beach. Waves average 1.5 m and maintain a 50-70 m wide bar cut by rips every few hundred metres, as well as a few rock outcrops. Beach **WA 648** continues on north of the rocks for 900 m to the southern granite rocks of Cape Freycinet. Waves increase slightly towards the cape, with a large permanent rip against the rocks. Beaches WA 647 & 648 are backed by generally well vegetated transgressive dunes, apart from an unstable foredune and large blowout at the southern end of beach WA 648.

Figure 4.133 Cape Freycinet and beaches WA 647 & 648.

WA 649-651 CONTIS BEACH-BOBS HOLLOW

No.	Beach	Rating	Type	Length
WA 649	Contis Beach	7	TBR+reefs	1.6 km
WA 650	The Fishing Place	4	R/LTT	700 m
WA 651	Bobs Hollow	4	R+rocks/reef	200 m
Spring & neap tidal range = 0.8 & 0.6 m				

To the north of Cape Freycinet the coastline continues roughly north for 15 km to the next major point at Cape Mentelle, with granite outcrops and calcarenite dominating the coast as reefs, points, headland and bluffs. The first 5 km north of the cape contains three beaches (WA 649-651).

Contis Beach (WA 649) is a relatively straight 1.6 km long west-southwest-facing beach exposed to waves averaging over 1.5 m, which maintain a 50-80 m wide bar. In addition beachrock outcrops along the centre of the beach and a few reefs lie off the southern and central sections, while it terminates at a sandy foreland tied to low slabs of granulite. The beach provides some good *beach breaks* and a heavy lefthand *reef break*. It is backed by steep, vegetated calcarenite bluffs rising to 100 m, with the vegetated calcarenite extending another 3-4 km inland. The Conto Road provides access to a car park at the top of the southern bluffs overlooking the beach as well as the granite shores of Cape Freycinet..

The **Fishing Place** is a 700 m long beach (**WA 650**) that extends north from the foreland to a 300 m long granulite point littered with large boulders. The beach is sheltered by inshore reefs with waves averaging about 1 m that surge up a steep reflective to low tide terrace beach. The shoreline has prograded up to 300 m out from the backing bluffs, and is covered with moderately unstable Holocene dunes including a few blowouts. A spur off the Bobs Hollow 4 WD track terminates on the bluffs above the northern end of the beach.

Bobs Hollow is located at the northern end of the rocky point and consists of a 200 m wide semicircular embayment dominated by a granulite shoreline capped by calcarenite. Two patches of sand (**WA 651**) totalling 200 m in length lie at the base of the small bay. The bay floor and mouth is dominated by rocks and shallow reefs, with low wave to calm conditions usually prevailing at the shore. The beach is littered with rock debris and fronted by the rocks and reefs. The Bobs Hollow track reaches the bluffs above the centre of the bay.

WA 652-656 ISAACS ROCKS-MARMADUKE PT

No.	Beach	Rating	Type	Length
WA 652	Redgate Beach (S)	8	TBR/RBB	350 m
WA 653	Redgate Beach (N)	4	R+reef	200 m
WA 654	Boodjidup (S)	4	R+reef	1.3 km
WA 655	Boodjidup Beach	8	TBR/RBB	1.9 km
WA 656	Grunters Beach	5	R/LTT+reef	300 m
Spring & neap tidal range = 0.7 & 0.5 m				

Isaacs Rocks and Marmaduke Point are two granulite points located 8 km apart. To either side and in between are five beaches with varying exposure to the high outside swell. There is a sealed road to a large car park on Issacs Rock, which overlooks Redgate Beach (WA 652), and anther road to the 30 m high calcarenite bluffs overlooking Grunters Beach (WA 656) in the north, together with a rough track to the rear of the northern end of Boodjidup Beach (WA 653). Elsewhere a mixture of both active and vegetated dunes back the beaches, then vegetated Pleistocene dunes extending up to 2 km inland.

Redgate Beach (WA 652) is located on the southern side of Isaacs Rock. It is a 350 m long, west-facing relatively exposed beach bordered by granulite rocks and boulders, and backed by three active blowouts which partially climb the backing 50 m high vegetated calcarenite bluffs (Fig 4.134). In addition Calgardup Brook drains out across the northern end of the beach. Waves average over 1.5 m and maintain an 80 m wide central bar, with two permanent rips draining out against the rocks to either end. There is a *righthand break* into the northern rip, as well as *beach breaks* across the bar.

Figure 4.134 Redgate Beach, with permanent rips to either end and the backing active climbing blowouts.

Beach **WA 653** is located on the northern side of Isaacs Rocks, just north of the car park. The beach is a curving 200 m long reflective beach, fronted by 500 m wide shallow reefs, and backed by 30 m high vegetated calcarenite bluffs. It is sheltered by the extensive rocks and reefs, which continue for 2.5 km to the north, lowering waves to less than 1 m at the shore..

Beach **WA 654** commences 100 m to the north and is a sheltered 1.3 km long reflective beach, also fronted by the 500 m wide band of shallow reefs, with some rocks outcropping along the shore. Waves are low at the shoreline, with a mixture of sand and rock along the intertidal. The beach is backed by several blowouts, which are climbing towards the backing 30 m high bluffs. The beach terminates at a small rock reef attached to the shore, beyond which wave increase in size along Boodjidup Beach.

Boodjidup Beach (WA 655) commences at the reef and continues north for 1.9 km to the rocks and boulders of Marmaduke Point. Waves average over 1.5 m and increase slowly up the beach. They maintain a well developed bar and rip system with usually 4-5 beach rips operating across the 50-100 m wide bar. The best *surf* is toward the northern end of the beach on the southern side of the point. The beach is backed by several blowouts set amongst predominately well vegetated dunes. Boodjidup Brook reaches the northern end of the beach and is then deflected south along the back of the berm for up to 1 km.

Grunters Beach (WA 656) is the back beach from Prevelly Park. The northern car park provides a good view of the beach and surf, with a second car park behind the southern end of the beach in lee of Marmaduke Point. The beach receives waves lowered by the offshore reefs and is usually steep and reflective with at times a heavy shorebreak (Fig. 4.135). However a strong permanent rip still drains out against Marmaduke Point. The best surf is off the northern reef, known as *Grunters*, and southern Marmaduke point, known as *Gas Bay* and *The Hollow*. Two vegetating blowouts back the centre of the beach, with the access road running behind.

Figure 4.135 Grunters Beach is a popular surfing beach, with reef breaks to either end.

WA 657-663 GNARABUP-MARGARET RIVER MARGARET RIVER SLSC

No.	Beach	Rating	Type	Length
WA 657	Gnarabup (S)	3	R+reefs	900 m
WA 658	Gnarabup Beach	3	R+reefs	1.6 km
WA 659	Margaret River car park	4	R+platform	300 m
WA 660	Margaret River mouth	5	R/LTT	500 m
Margaret River Surf Life Saving Club				
WA 661	Margaret River (N)	4	R+reef	250 m
WA 662	Cape Mentelle (S 2)	5	R+reef	150 m
WA 663	Cape Mentelle (S 1)	5	R+reef	100 m
Spring & neap tidal range = 0.7 & 0.5 m				

Margaret River is renowned for its surf. The small river enters the sea along a steep narrow valley set amongst the calcarenite bluffs and reefs just south of Cape Mentelle. The entrance is usually blocked by a 500 m long steep coarse sand beach. At the southern end of the beach is a 30 m high bluff in front of which reefs extending 300 m out to sea produce the famous surfing break. South of the bluff is the more protected Gnarabup Beach. Margaret River and Gnarabup beaches (Fig. 4.136) have been the traditional recreational beaches for the town of Margaret River, located 8 km inland. Now the surfing break between the two beaches, attracts surfers from around the world, and is the site of major surfing contests.

The road from town lead to six car parks, one at the river mouth behind Margaret River Beach (WA 660), a large one of the bluffs overlooking the surfing break (WA 659), two large car parks at the southern end of Gnarabup Beach (WA 658) and the two on either side of Grunters Beach (WA 656). The holiday settlement behind Gnarabup Beach is called Prevelly and has a large caravan park and store. There is also a small white Greek Church, dedicate to the people of the town of Prevelli in Greece, after whom the settlement is named

The 9 km long section of west-facing coast from north of Grunters Beach to Cape Mentelle is dominated by extensive offshore calcarenite reefs extending up to several hundred metres offshore, including the Margaret River surf break. Most waves break over the outer reefs resulting in low waves to often calm conditions at the shore. The six beaches in between are all dominated by the presence of the reefs and backed by the calcarenite.

Beach **WA 657** extends north from the northern side of Grunters Beach for 900 m to the southern boundary with Gnarabup Beach. It is backed by continuous 30-40 m high vegetated calcarenite bluffs, fronted by a 10-15 m high foredune. The beach extends in two arcs along the base of the bluffs, with shallow reefs extending a few hundred metres offshore. Waves break over the southern reefs (*Grunters*), while at the shore calm conditions usually prevail. There is access from the blufftop car parks at either end, while the national park terminates that the northern end of the beach.

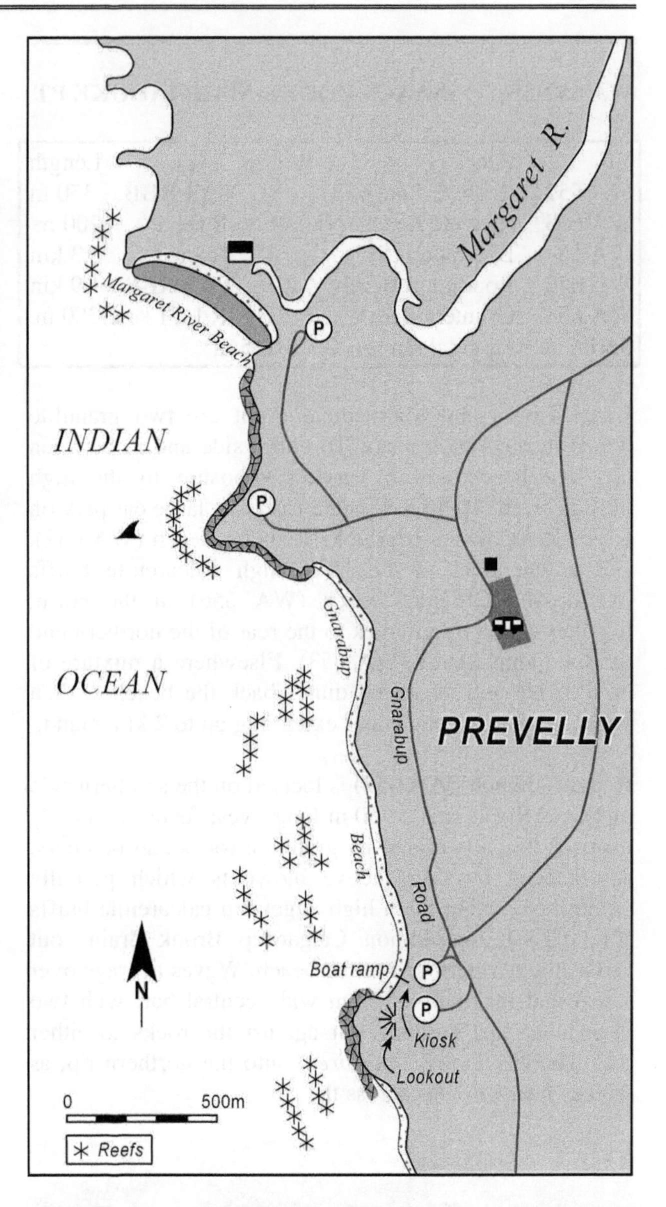

Figure 4.136 Gnarabup and Margaret River beaches are dominated by the numerous rocks and reefs that lie along the coast.

Gnarabup Beach (WA 658) is the main beach for Prevelly and is accessible at the southern end where there is a large car park and concrete boat ramp. It is 1.6 km long, faces due west, but is protected by near continuous reefs extending up to 500 m offshore which result in a double crenulate shoreline (Fig. 2.3b). Way out on the reefs is a break off the southern end known as *Boat Ramp*, and further north *The Bombie*, both of which hold big swell. Towards the northern end of the beach is another break called *Suicides*. After negotiating the reefs waves reaching the beach are usually less than 0.5 m, and lap up against the relatively steep beach. High foredunes back the beach with a road and houses behind.

Beach **WA 659** is located below the Margaret River blufftop car park. It is a narrow beach which surfers walk across to enter the channel to paddle out to the main break. The beach curves around the base of the 10-20 m high calcarenite bluffs for 300 m. It is fronted by intertidal calcarenite platform of varying width, with some open sandy patches. Waves are usually less than 1 m at the shore. However out on the reef wave refracting round and peaking on the reef produce the world famous main break (variously known as *Surfers Point/Main Break* and *Margaret River*, which hold lefts up to several metres, and a shorter righthanders known as *South Side*, which works up to 2 m (Fig. 2.3a).

Margaret River mouth beach (**WA 660**) lies across the mouth of the usually blocked river (Fig. 4.137). The beach faces southwest and curves for 500 m between boundary calcarenite bluffs of the old valley side to the south and a low reef to the north, where it forms a sandy foreland. It is backed by the dammed river which can break out and either end of the beach, resulting in a flat overwashed area behind the steep beach. The beach is partly protected by reefs, with waves increasing in height up the beach to about 1 m. Lower waves break right on the steep beach, with higher waves breaking as a heavy shorebreak or heavily on a narrow bar. The short heavy beach break, known as *Margaret River Mouth*, while a hazard to swimmers does attract surfers when waves exceed 1 m. The **Margaret River Surf Life Saving Club** was provisionally established at this beach in 2003 and patrols the beach on Sundays during the summer period.

Figure 4.137 Margaret River flowing across the river mouth beach.

Beach **WA 661** occupies the northern side of the sand foreland and curves for 250 m to the north, terminating at a 20 m high calcarenite bluff, which marks the start of Cape Mentelle. Waves average less than 1 m and usually lap against the steep beach face. On the reef that forms the foreland is a rarely surfed heavy righthand break known as *The Box.*

Beaches WA 662 and 663 are two narrow pockets of sand wedged in at the base of southern side of the cape. A walking track runs along the top of the bluffs above the beaches, with a steep climb required to reach the shore. Beach **WA 662** is a curving 150 m long, narrow rock littered, west-facing sand beach at the base of curving 40 m high calcarenite bluffs. It is fronted by shallow reefs, with low waves at the shore. Beach **WA 663** lies immediately to the west on the southern tip of the cape. It is a similar 100 m long curving strip of beachrock, with occasional patches of sand. Waves break on the boundary points and reefs with low waves at the shore.

Swimming: Margaret River beach is (WA 660) the more popular swimming beach and is relatively safe under normal conditions. However watch the rip against the southern rocks, and the shorebreak when waves exceed 1 m Gnarabup Beach (WA 658) usually has low waves lapping against the steep beach.

Surfing: *Margaret River* produces one of the world's great lefts on the inner and outer reef, the outer reef holding waves to several metres high. During average swell conditions there is also a shorter right across the inner reef. The river mouth beach can also produces a dumpy shore break.

WA 664-672 CAPE MENTELLE-GNOOCARDUP

No.	Beach	Rating	Type	Length
WA 664	Cape Mentelle (N1)	5	R+reefs	700 m
WA 665	Cape Mentelle (N2)	5	R/LTT+reefs	300 m
WA 666	Cape Mentelle (N3)	7	LTT/TBR+reefs	400 m
WA 667	Kilcarnup Beach (1)	7	LTT/TBR+reefs	150 m
WA 668	Kilcarnup Beach (2)	7	TBR+reefs/platform	500 m
WA 669	Kilcarnup Beach (N)	7	R+platform	100 m
WA 670	Dallip Springs	7	R+platform	300 m
WA 671	Gnoocardup Beach (1)	7	TBR+reef	200 m
WA 672	Gnoocardup Beach (2)	7	R+rocks/reef	750 m
Spring & neap tidal range = 0.7 & 0.5 m				

Between Cape Mentelle and Cowaramup Point is a relatively straight, 10 km long section of west-facing rock and reef-dominated shore containing 13 beaches, all influenced to varying degrees by the reefs, rocks and platforms, with the reefs extending up to 500 m offshore. Most of the shoreline is part of the Leeuwin-Naturaliste National Park, with formal access at Ellensbrook in the centre, together with 4 WD access to Kilcarnup in the south and *Big Rock* and *Cobblestones* surf breaks in the north. Partly active dunes cover the backing steep calcarenite slopes, the calcarenite rising to 100-200 m and extending 2-3 km inland to the Caves Road. The Cape to Cape walking track follows the beach from Cape Mentelle to Joey's Nose at the northern end of Kilcarnup Beach, then heads inland to reemerge on the coast at Ellensbrook.

On the north side of Cape Mentelle the shoreline trends to the north-northeast for 1.5 km and contains three moderately protected beaches (WA 664-666). Beach **WA 664** is a 700 m long relatively steep, narrow reflective beach fronted by shallow inshore reefs extending up to 500 m offshore, with waves averaging less than 1 m at the shore, and seagrass meadows just off the beach. The beach is bordered by a calcarenite platform surrounding the 20 m high cape in the south and a narrow shore perpendicular reef in the north. It is backed by a few private dwellings on cleared land, with no public access. Small boats are moored off the protected southern end of the beach. Beach **WA 665** commences on the northern side of the reef and continues to curve round towards the northeast for 300 m to a low point formed of granite boulders, with the sand continuing behind the rocks. The beach is fronted by the 500 m wide inshore reefs, with waves averaging about 1 m. Beach **WA 666** completes the string of beaches and is a curving 400 m long northwest-facing beach, tied to the boulders in the south and a low rocky northern point. It has 200 m of more open water in between permitting waves averaging 1-1.5 m to break across a central bar drained by rips to either end. The beach is backed by steep vegetated calcarenite bluffs rising to 50 m, with a rough winding 4 WD track reaching the beach from the north.

Kilcarnup Beach commences 200 m north of beach WA 666 and consists on two near continuous strips of sand (WA 667 & 668) and a northern pocket of sand (WA 669). The national park boundary crosses the centre of the beach, while a rough 4 WD track from Dallip Springs runs the length of the beaches. Beach **WA 667** is a 150 m long pocket of sand bordered by dipping metasedimentary rocks on the southern point and a low boulder northern point, together with shallow rock reefs across the front of the beach. Waves average between 1-1.5 m and break across the reefs and a wide shallow bar. The main Kilcarnup Beach (**WA 668**) commences on the northern side of the low rocks and trends north for 500 m to a low rocky point. It is initially fronted by a sand bar with wave breaking 100 m offshore, then by a shallow inshore reefs and rocks drained by two permanent rips during moderate to high waves. Beach **WA 669** is a 100 m long pocket of sand located immediately to the north. It is bordered by two protruding low rock points, with shallow reefs linking the points, resulting in low waves at the shore and a steep reflective beach. Continuous steep vegetated slopes back all three beaches, which are linked by the 4 WD track.

Dallip Springs lies 500 m to the north with the intervening shore dominated by straight bedrock, fronted by shallow inshore reefs, and backed by the steeply rising vegetated slopes. The Dallip Springs beach (**WA 670**) consists of two strips of sand totalling 300 m in length, which are wedged in between the slopes and the rocky platform. A spur off the 4 WD track reached the centre of the beach.

Gnoocardup Beach commences 400 m north of the springs and consists of two parts (WA 671 & 672) (Fig. 4.138). Beach **WA 671** occupies the first 200 m, extending north from the end of the longer rocky section to a small rock outcrop. Waves average about 1.5 m and break over a mixture of inshore reefs and sandy seabed drained by a permanent central rip. The main 4 WD access track descends part way down the 70 m high vegetated slopes behind the beach, with spurs off to the south and north to the main beach. Beach **WA 672** is the main Gnoocardup Beach and consists of a 750 m long sandy beach backed by several active blowouts extending 300 m up the backing slopes and climbing to 50 m. The beach is fronted by either continuous inshore or shoreline rocks and reefs, which lower waves at the shore to about 1 m.

Figure 4.138 Gnoocardup Beach is dominated by shallow inshore reefs, which lower waves at the shore.

WA 673-677 **ELLENSBROOK-COWARAMUP PT**

No.	Beach	Rating	Type	Length
WA 673	Ellensbrook (S)	7	R+platform/reef	200 m
WA 674	Ellensbrook	8	R+platform/reef	350 m
WA 675	Ellensbrook (N)	8	R+platform/reef	1.5 km
WA 676	Big Rock	8	R+platform/reef	800 m
WA 677	Cowaramup Pt (S)	8	R+platform/reef	500 m
Spring & neap tidal range = 0.7 & 0.5 m				

Ellensbrook is the site of the first European homestead on this section of coast. It was established in 1857 and is now restored and part of the national park. The coast to either side and up to Cowaramup Point remains dominated by the basement granite, with overlying calcarenite and patchy dunes. The Cape to Cape walking track runs along the shore from Ellensbrook to Cowaramup Bay and Gracetown. Apart from the valley carved by Ellen Brook, the entire 5 km of shoreline up to the point is backed by steeply rising, vegetated, calcarenite slopes that reach 190 m at Van Trips.

Beach **WA 673** is located 500 m south of Ellensbrook on the southwest-facing side of a dune-capped bedrock point. The beach is fronted by continuous intertidal bedrock rocks and reef, with waves break heavily on the outer reefs and some more protected tidal pools close to shore. Active dunes back the beach and climb to 30 m.

Ellensbrook Beach (**WA 674**) occupies the rocky shore to either side of the mouth of the small Ellen Brook. The beach is 350 m long and consists of a high tide beach fronted by near continuous intertidal calcarenite platform, together with reefs extending up to 100 m offshore. Waves averaging 1.5 m break over the outer reefs, with lower waves at the shore. It s backed by some active blowouts and vegetated foredune, the homestead, then older dunes extending 3 km inland to the Caves Road. Two surf breaks known as *Ellensbrook* and *The Womb* are located off the beach.

Beach **WA 675** commences 1 km north of Ellensbrook and continues north along the rocky shore for 1 km. The beach is fronted by a continuous intertidal calcarenite platform, with the high tide beach, and then sloping vegetated dunes behind. Waves break heavily on the outer edge of the platform, as well as over some scattered outer reefs. It terminates at a narrow indentation in the beach and backing bluffs, beyond which is beach WA 676. Beach **WA 676** continues north for another 800 m to the southern rocks of Cowaramup Point, with waves averaging over 1.5 m along this more exposed section. It is a similar high tide beach fronted by the intertidal rock platform, with some gaps towards the northern end. The beach can be accessed by gravel road to a large car park above the northern end of the beach, with two surf breaks called *Lefthander* and *Cobblestones* are located amongst the reefs south of the car park.

Beach **WA 667** is located along the southern side of Cowaramup Point. The 500 m long beach is fronted by continuous intertidal rock platforms, with reefs further out. Waves average over 1.5 m and break on some of the outer reefs, producing a heavy righthander called *Big Rock*, as well as breaking heavily on the platform. It is backed by two active blowouts that have climbed up the backing 30-40 m high slopes towards Gracetown (Fig. 4.139). A car park is located at the southern end of the beach and provides a good view of the surf.

Figure 4.139 Cowaramup Point and bay in the background.

WA 678-680 COWARAMUP BAY

No.	Beach	Rating	Type	Length
WA 678	Cowaramup Bay (S)	4	R+rocks/reef	100 m
WA 679	Cowaramup Bay	4	R+rocks/reef	800 m
WA 680	Cowaramup Bay (N)	4	R+reef	80 m
Spring & neap tidal range = 0.7 & 0.5 m				

Cowaramup Bay is a roughly semi-circular, 800 m wide, northwest-facing bay, bordered by the dune-capped granulite of Cowaramup Point to the south and a sloping granulite headland to the north. The bay provides one of the few anchorages on the coast, with usually a few fishing boats moored in the northern corner. The small settlement of Gracetown backs the southern half of the bay and rear of the point, with a foreshore reserve along the northern half of the bay. There are three beaches along the protected bay shore, two small ones to either end (WA 678 & 680) and the longer main beach (WA 679).

Beach **WA 678** is a narrow 100 m long strip of high tide sand located in lee of Cowaramup Point. It is backed by 20 m high crumbling calcarenite bluffs, with a blufftop car park overlooking the beach. Intertidal rocks and boulders front the beach and shallow reefs extend out into the bay, resulting in low wave to calm conditions at the shore. Surfers use the beach to paddle out to Cowaramup's *South Point*, a heavy lefthand point break.

The main bay beach (**WA 679**) curves along the centre of the bay shore for 800 m. Cowaramup Brook crosses the southern section of the beach, with a large car park and foreshore reserve along the northern half. Rocks outcrop along and off the beach, which combined with reefs in the bay lower waves to less than 1 m at the shore. The northern 80 m long beach (**WA 680**) is tucked in between rocks and the northern point and is located at a southern boundary of the national park A concrete boat ramp and small groyne are located at the eastern end of the sheltered beach. Out in the bay off the beach is the reef break called *North Point*, a potentially long heavy righthander.

WA 681-687 CULLEN ROAD

No.	Beach	Rating	Type	Length
WA 681	North Point	7	R+rocks/reef	50 m
WA 682	Cullen Road (S3)	7	R+platform/rocks/reef	70 m
WA 683	Cullen Road (S2)	7	R+platform	400 m
WA 684	Cullen Road (S)	7	R+reef	70 m
WA 685	Cullen Road	7	R+reef	1 km
WA 686	Cullen Road (N1)	7	R+platform	300 m
WA 687	Cullen Road (N2)	7	R+platform/rocks/reef	300 m
Spring & neap tidal range = 0.7 & 0.5 m				

The northern point of Cowaramup Bay forms the southern boundary of a relative straight west-facing section of coast that extends for 18 km to Cape Clairault. In between is a bedrock dominated section of coast, partly capped by calcarenite and containing 22 beaches (WA 681-704). All the beaches are bordered, and most dominated, by the granite, granulite and some calcarenite outcrops, together with offshore reefs. Access to the shore is restricted to a few gravel roads leading to 4 WD tracks. The first 6 km contains seven beaches (WA 681-687) most accessible off various tracks branching off the Cullen Road. The Cape to Cape walking track for the most part follows the top of the bluffs and beaches, providing an excellent view of this section of the coast.

Beach **WA 681** is located 500 m north of Cowaramup Bay's northern point, and in line with the main road into Cowaramup as it bends to turn into the bay with an informal car park on the slopes above the beach. The beach is a 50 m pocket of high tide sand bordered and fronted by granulite rocks and boulders, with rock reefs extending 200 m offshore.

Beach **WA 682** lies 2.5 km to the north at the end of a section of granulite shoreline and consists of a 70 m long pocket of southwest-facing sand located between the massive granulite. It lies at the base of 30 m high eroding calcarenite bluffs, with a 100 m wide calcarenite intertidal platform and rocks forming the northern boundary. Beach **WA 683** commences immediately to the north. It extends as a west-facing high tide beach for 400 m wedged in between the base of 20 m high vegetated calcarenite bluffs, and a flat but irregular intertidal calcarenite platform, up to 100 m wide. Wave break heavily over the platform with low waves at the shore (Fig. 4.140). A track running south from the Cullen Road leads to an informal car park above the centre of the beach, with a steep walking track down to the shore.

Beach **WA 684** is located immediately to the north, and consists of a 70 m long pocket of northwest-facing sand, wedged in at the base of steep 30 m high calcarenite bluffs, with narrow platforms to either side and a small central more open water section. It can only be reached on foot from the neighbouring beach. Beach **WA 685** commences 50 m to the north and is initially a narrow beach wedged between the bluffs and a 50 m wide calcarenite platform. The 1 km long beach then widens into a steep, sand-filled valley that parallels the end of the Cullen Road. The centre of the beach is fronted by shallow reefs, which include two surf breaks - *The Guillotine* to the south and *The Gallows* out front, just off the beach. As the names imply both are heavy reef breaks. The platform and bluffs resume along the northern half of the beach, which finally pinches out between the bluffs and platform.

Figure 4.140 Beach WA 683 is fronted by shallow reefs, which induce heavy wave breaking and result in low waves at the shore.

Beach **WA 686** continues on for 300 m further north, as a narrow sand beach between 50 m high calcarenite bluffs and the irregular 30-50 m wide platform. It terminates at a protruding section of bluffs, beyond which is beach **WA 687**, another 300 m long section of high tide beach wedged between the bluffs and platform. This section of bluffs and beach protrudes slightly seawards, and at its northern end the calcarenite retreats to expose 20 m high granulite cliffs

WA 688-691 WILYABRUP BEACH

No.	Beach	Rating	Type	Length
WA 688	Wilyabrup Beach	7	TBR	800 m
WA 689	Wilyabrup (N1)	7	TBR+platform	750 m
WA 690	Wilyabrup (N2)	8	LTT+rocks/reef	30 m
WA 691	Wilyabrup (N3)	8	TBR+reef	300 m

Spring & neap tidal range = 0.7 & 0.5 m

The central section of the coast between Cowaramup Bay and Cape Clairault is accessible by car only at Moses Rock, with a rough 4 WD track leading to the southern Wilyabrup Beach. Between Wilyabrup and the northern Moses Rock beaches is 4.5 km of bedrock shoreline, capped by steep dune-covered bluffs rising to 50 m, with a scattering of nine generally rock-dominated beaches (WA 688-696). The first four (WA 688-691) occupy the first 2.5 km of coast.

Wilyabrup Beach (WA 688) is located across the mouth of Wilyabrup Creek. The 800 m long beach fills the steep-sided valley mouth, with granulite points bordering each end (Fig. 4.141). It faces west and is one of the few beaches free of reefs. Wave average over 1.5 m and break across a 100 m wide bar with permanent rips draining out against the rocks at either end of the beach. The beach is backed by an unstable foredune, elongate lagoon and dunes climbing the southern valley side. The 4 WD access track terminates on the southern side of the dunes while the Cape to Cape track descends off the southern bluffs and runs along the beach.

Figure 4.141 Wilyabrup beach is located at the mouth of Wilyabrup Creek, with active dunes moving up the southern side of the valley.

Beach **WA 689** is located 300 m to the north and consists of a 750 m long generally narrow beach, fronted by bedrock and calcarenite platforms to either end, with a central 100 m long open water section. Waves averaging over 1.5 m break over a mixture of reefs to the side and a central bar, with a permanent rip draining out along the northern platform. It is backed by steep, partly vegetated 50 m high bluffs. A vehicle track reaches the top of the northern bluffs, with a steep foot track down to the beach.

Beaches WA 690 and 691 lie another 50 m to the north. Beach **WA 690** is a 30 m long pocket of sand that is wedged in lee of a narrow gap in the granulite platform. The beach widens to the rear where it is backed by steep 50 m high bluffs. Waves break seaward of the platform and surge into the narrow rocky gap to reach the beach. Beach **WA 691** lies immediately to the north and is an exposed 300 m long southwest-facing sand beach, bordered in the south by the rock platform, and tied to a small reef and low granulite point to the north. Waves average over 1.5 m and break across a 50 m wide bar and over the rocks and reef, with strong permanent rips forming against the rocks at either end of the beach. It is backed by an unstable foredune and climbing blowout extending 300 m inland and to heights of 60 m. The Cape to Cape walking track runs along the beach.

WA 692-696 **MOSES ROCK**

No.	Beach	Rating	Type	Length
WA 692	Moses Rock (S2)	9	R+rocks/reef	30 m
WA 693	Moses Rock (S1)	9	R+rocks/reef	200 m
WA 694	Moses Rock	9	R+rocks/reef	700 m
WA 695	Moses Rock (N1)	8	R+rocks/reef	60 m
WA 696	Moses Rock (N2)	8	R+rocks/reef	40 m
Spring & neap tidal range = 0.7 & 0.5 m				

Moses Rock is located 500 m inland on the crest of a 60 high ridge, with the name also applying to the adjacent rocky shoreline. Five beaches (WA 692-696) occupying a total of 1.5 km of shore are located either side of the end of the Moses Rock Road. The road terminates at three blufftop car parks overlooking some of the beaches.

Beaches WA 692 and 693 lie to the south of the road. Beach **WA 692** is a 30 m pocket of high tide sand fronted by a continuous granulite platform and rocks, with waves breaking up to 100 m offshore over rocky reefs. It is backed by steeply rising 50 m high vegetated bluffs. Beach **WA 693** commences 50 m to the north and is a 200 m long southwest-facing beach, fronted by a 150 m wide surf zone, which is a mixture of rocks and sand. It is bordered by irregular granulite points, and backed by an active dune system, which is ramped up against the backing 50 m high bluffs. A vehicle track terminates on top of the bluffs with a walking track leading to the southern end of the beach.

Beach **WA 694** commences on the northern side of the boundary rocky point and continues north for 700 m to a large rock outcrop. It is fronted by a continuous 50 m wide sloping, irregular rock platform, with some rock reefs further out, with the platform the site of the *Moses Rock* lefthand reef break. The beach is backed by sand-draped 30 m high bluffs, with a blufftop car park and track down to the beach and surf. While the beach offers good surf it is unsuitable for swimming.

Beaches WA 695 and 696 are two pockets of sand located immediately north of beaches WA 694. Beach **WA 695** is a 60 m long pocket wedged in a small gap in the backing bluffs. It is backed by sloping vegetated bluffs and fronted by intertidal rocks, with platforms to either side. Beach **WA 696** lies 80 m to the north and is a 40 m long pocket of high tide sand, also fronted by intertidal rocks, with boundary platforms.

WA 697 **QUININUP BEACH**

No.	Beach	Rating	Type	Length
WA 697	Quininup Beach	8	R+platform/rocks/reef	1.1 km
Spring & neap tidal range = 0.7 & 0.5 m				

Quininup Beach (**WA 697**) is a crenulate 1.1 km long beach that in part blocks the mouth of Quininup Brook. The beach extends to either side of the creek mouth, with continuous intertidal rock platforms fronting the beach, apart from an 80 m wide open section in front of the usually blocked mouth. Waves break over reefs off the platforms, with deeper water off the open section, where a permanent rip runs out the channel. The beach is backed by active dunes climbing the slopes to either side of the creek, in places rising to 70 m and extending 500 m inland. There is only foot access to the beach along the Cape to Cape track from Moses Rock. The track follows the beach, before returning to the bluffs that run north of the beach to Cape Clairault, 4 km to the north.

WA 698-704 **CAPE CLAIRAULT (S)**

No.	Beach	Rating	Type	Length
WA 698	Cape Clairault (S6)	8	R+platform/rocks/reef	300 m
WA 699	Cape Clairault (S5)	9	R+platform/rocks/reef	200 m
WA 700	Cape Clairault (S4)	9	R+platform/rocks/reef	600 m
WA 701	Cape Clairault (S3)	9	R+platform/rocks/reef	600 m
WA 702	Cape Clairault (S2)	9	R+rocks/reef	800 m
WA 703	Cape Clairault (S1)	9	R+platform/rocks/reef	250 m
WA 704	Cape Clairault	7	R+rocks/reef	50 m
Spring & neap tidal range =0.7 & 0.5 m				

North of Quininup beach is a relatively straight 4 km long section of 40-50 m high calcarenite bluffs, all fronted by intertidal rock platforms, rocks and reefs, with the bluffs terminating just south of Cape Clairault. In between are seven beaches wedged in between the bluffs and reefs exposed to waves averaging over 1.5 m and breaking heavily on the reefs and outer edge of the platforms resulting in lower waves and reflective conditions at the shore. The only access is via a 4 WD track that runs south from the cape along the top of the bluffs.

Beach **WA 698** is a 300 m long high tide sand beach, wedged in between the backing bluffs and irregular platform. The outer reefs lower waves at the shore. Beach **WA 699** lies immediately to the north and is a similar 200 m long beach, with deeper offshore reefs and higher waves breaking over the intertidal rocks that shelter the backing beach.

Beach **WA 700** consists of a 600 m long series of near continuous pockets of sand wedged in along a series of scallops in the 50 m high bluffs. The sand lies between the bluffs and irregular rock and reef strewn 50-100 m wide surf zone. Beach **WA 701** continues to the north for another 600 m as a more continuous high tide sand beach, backed by the 50 m high bluffs, with continuous intertidal rocks and reefs, and occasional rocks and rock outcrops on the beach. A track off the Quininup Road reaches the top of the bluffs behind the centre of the beach.

Beach **WA 702** is a curving 800 m long beach that runs initially behind a continuous intertidal platform, with a mixture of sand, rocks and reefs off the northern half of the beach. It terminates at the northern end of the bluffs, which give way to the large blowout behind beach WA 703. The *Windows* reef break is located on the reefs that form the southern boundary of the beach.

Beach **WA 703** is a 250 m long high tide beach fronted by a sloping intertidal platform containing some tidal pools, with waves breaking heavily on the outer platform. The *Wildcat* surf break is located on the reefs off this beach. An active blowout backs the beach and is transgressing right across the 600 m wide cape to Injidup Bay. Beach **WA 704** is a 50 m long pocket of sand located at the northern end of beach WA 703. It is bordered by a boulder beach to the south and the massive granite rocks of the cape to the north. Wave break over rocks and reef extending 100 m off the beach, resulting in low waves at the shore and form an exposed tidal pool.

WA 705-708 **INJIDUP**

No.	Beach	Rating	Type	Length
WA 705	Injidup Pt	6	R+rocks	350 m
WA 706	Injidup (1)	6	LTT→TBR	1.6 km
WA 707	Injidup (2)	6	TBR	250 m
WA 708	Injidup (3)	7	TBR	150 m
Spring & neap tidal range = 0.7 & 0.5 m				

The **Injidup** beaches are four near continuous sandy beaches (WA 705-708) located in on open northwest-facing bay to the lee of Cape Clairault, which is also called Injidup Point. The embayment is 2.2 km wide, with the curving shoreline containing the beaches anchored between the rocky granulite points. The sealed Clairault Road runs out to the main beach (WA 706) where there is a picnic and camping area besides Injidup Springs. The Cape to Cape walking track crosses the cape and runs along the main beach and up along the bluffs behind the northern beaches.

Beach **WA 705** is located on the rocky northern side of Cape Clairault and faces north, with only moderate refracted waves reaching the shore. The 350 m long beach has formed from dunes sands, which have blown the few hundred metres across the cape, and now drape an essentially low, irregular granulite shoreline. The waves breaking over the rock and reefs from a lefthand break known as *Injidup Point*. The beach terminates at a low rocky point.

The main **Injidup beach (WA 706)** commences on the western side of the small point and curves round to the northeast then north for 1.6 km to Mitchell's Rock (Fig. 4.142), a 200 m long north-south ridge of bedrock that forms the northern end of the beach. Waves average about 1 m at the southern end increasing to 1.5 m at the rock. The beach responds with a low tide terrace along the southern half, grading into a wider bar with a few rips towards the north. A few rocks are also located along the southern and centre of the beach. Waves breaking over the southern rock during higher swell produce the *Car Park* and *Pea Break* surf breaks. Vegetated slopes back the southern half as far as the car park, with a more unstable foredune against the backing slopes up to the rock.

Figure 4.142 Injidup Beach curves round to the lee of Injidup Point.

Beach **WA 707** commences on the northern side of Mitchell's Rock and continues north for 250 m past some attached reefs to a low rocky point. Waves averaging about 1.5 m break over a bar between the rocks and reefs, with permanent rips running out either side of the central reef. It is backed by an unstable climbing foredune in the south and vegetated bedrock slopes to the north. Beach **WA 708** lies immediately to the north and is a 150 m long strip of sand at the base of dipping 50 m high granulite bluffs. Waves average over 1.5 m and break across a 70 m wide bar drained by a permanent rip that runs out along the northern rocks.

WA 709-711 **WYADUP-CANAL ROCKS**

No.	Beach	Rating	Type	Length
WA 709	Wyadup	7	R/boulder+reef	100 m
WA 710	Canal Rocks	7	R/boulder+reef	200 m
WA 711	Canal Rocks (N)	7	R+rocks/reef	100 m
Spring & neap tidal range = 0.7 & 0.5 m				

Between Injidup and Smiths beaches is a 3 km long section of rocky coast consisting of massive granite and granulite rocks, including the scenic Canal Rocks, a 1 km long ridge of granite. In amongst the rocks are three small rock dominated beaches (WA 709-711). The Cape to Cape walking track follows the rocky shore to each of the beaches.

Beach **WA 709** is located at the mouth of Wyadup Brook and is a curving 100 m long north-facing boulder beach, contained in a small protected north-facing rocky embayment backed by the small valley carved by the brook. There is a patch of sand on the beach and a veneer of sand dune draping either side of the small valley, with cleared farm land behind. A spur off the Cape Clairault Road runs out to the western point overlooking the beach.

Canal Rocks beach (**WA 710**) lies in lee of the rocks. It consists of a protected 200 m long strip of north-facing boulders, with some low to subtidal sand and patchy rock reefs. It is backed by bluffs rising to the east and the Canal Rocks road, with the car park located at the western end of the beach.

One kilometre to the northwest is beach **WA 711**, a third north-facing beach set in lee of a small linear rocky point. It consists of a 100 m long high tide sand beach fronted by intertidal rocks and boulders, with rocks and reefs filling the small bay. Waves break over the reefs and average about 1 m at the shore. A vehicle track runs down the backing bedrock slopes to the point and rear of the beach.

WA 712-714 SMITHS BEACH

No.	Beach	Rating	Type	Length
WA 712	Smiths Beach	6	LTT→TBR	600 m
Lifeguard Patrols				
	Monday-Sunday		**January**	
WA 713	Smiths Beach (1)	6	TBR+reef	500 m
WA 714	Smiths Beach (2)	7	TBR+reef	200 m
Spring & neap tidal range = 0.7 & 0.5 m				

Smiths Beach is one of the more popular swimming and surfing beaches in the Leeuwin-Naturaliste National Park. The beach lies 2 km off the main road at the base of a steep drive down to the beach, which provides an excellent view of the beach, surf and coast to the north. At the base of the drive is a beach car park, with additional cars parks located toward the western end. There is also a large caravan park, the Canal Rocks Beach Resort and kiosk-cafe. However most of the beach and surrounding park is in a natural state with the Cape to Cape walking track running along the beach to the northern Torpedo Rocks and across the rocks to Yallingup (Fig. 4.143).

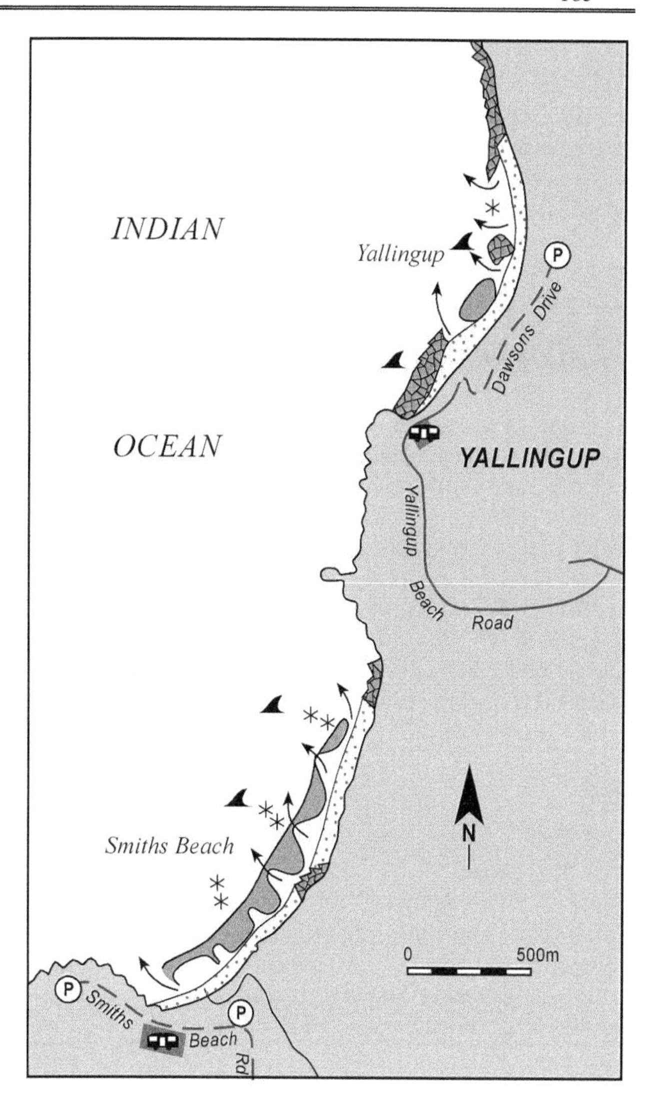

Figure 4.143 Smith and Yallingup beaches are two of the most popular surfing beaches on the Leeuwin coast. Both have good surf but are dominated by rips.

The beach faces west-northwest and has of three sections (WA 712-714) each separated by intertidal calcarenite reef. The main beach (**WA 712**) is 600 m long, faces northwest (Fig. 4.144). The 500 m long headland provides some protection at the southern end resulting in waves increasing from about 1 m in the southern corner to 1.5 m in the north, where they maintain 2-3 beach rips along the central-northern section of the beach. Gunyulgup Brook drains across the centre of the beach. It is backed by a moderately stable foredune south of the creek, then the resort development, while north of the creek are vegetated dunes climbing the backing 60 m high slopes. There is a *left break* off the point during

Figure 4.144 View north along Smith Beach.

higher waves and a variety of beach breaks over the reefs and bars.

A narrow calcarenite platform and backing bluffs separates the main beach from the central section. The beach (**WA 713**) is 500 m long and includes a 50 m wide central intertidal rock platform. Waves averaging over

1.5 m break on the reef producing the well known *Super Tubes* break. An exposed section of calcarenite separates it from the northern section (**WA 714**), a 200 m long beach bordered by the granulite slopes of Torpedo Rocks to the north, with a few reef patches off the beach. Waves break over a mixture of sand and reefs, the latter including a break known as *Torpedo Rocks*. Other breaks in the area are *Steeping Stones* and *Palisades*. Both beaches have strong rips between the reefs during higher waves. They are backed by dune-draped slopes climbing to a 90 m high ridge.

Swimming: South of the creek is the safer place to swim and is patrolled in summer, however strong waves and rips are still common, so use caution. North of the creek is increasingly hazardous owing to the reef, higher waves and stronger rips.

Surfing: This is one of the most popular surfing beaches in the area offering both consistent beach and several reef breaks. It also offers a little protection when the swell is too big elsewhere.

WA 715-716 YALLINGUP

No.	Beach	Rating	Type	Length
WA 715	Yallingup reef	5	R+platform/reef	200 m
WA 716	Yallingup	6→7	LTT→TBR+reef	1.3 km
Lifeguard Patrols				
	Monday-Sunday		**January**	
Spring & neap tidal range = 0.7 & 0.5 m				

Yallingup Beach has a long history of recreational use associated with the development of adjacent Caves House. Today Caves House remains a popular holiday location, while the beach area has developed into a holiday settlement, including a resort, caravan park and facilities for the numerous people who descend upon the town on weekends and in summer. There is a large car park over looking the protected southern end of the beach (WA 715), and a second park on the 30 m high slopes behind the more exposed centre of the main beach (WA 716).

Yallingup reef beach (**WA 715**) is located in lee of a 150 m wide intertidal calcarenite platform, which is bordered to the south by the sloping granulite rocks of Slippery Rock, while to the north it merges into the main beach. The reef beach has high waves breaking over the outer edge of the reef, where the famous long *Yallingup* left break occurs, with generally low wave to calm conditions at the shore. The inner tidal pools, located just below the main car park and facilities, are a popular spot for families, and are relatively safe so long as you stay close to the shore and well clear of the outer waves and currents. The reef is a marine reserve where rod fishing is permitted, but other methods of collected marine life prohibited.

The main **Yallingup Beach** (**WA 716**) (Figs. 4.143 & 4.145) extends north of the car park for 1.3 km to the beginning of a section of steep calcarenite bluffs and backing slopes, which rise to 173 m at Wardanup Hill. The curving beach faces west with wave height increasing up the beach. In the more protected southern corner waves average about 1 m and break across a low tide terrace. Wave height increases up the beach as the surf zone widens and is cut by strong permanent rips against the central reefs and usually a beach rip to either side The reef area provides good surf, but is hazardous for swimmers. The main settlement and rows of houses back the southern part of the beach, with the surfers car park in lee of the main central reef break. Some active dunes extend north of the car park before the bluffs and rock platforms begin to dominate the shore to the north.

Figure 4.145 Yallingup beach has a reef-fronted southern end, with good surf on the reef and the wave and rip-dominated main beach.

Swimming: The southern reef beach (WA 715) is the least hazardous location as it is usually free of waves and rips. Be careful if you swim north of the main car park (WA 716) as waves and rips increase in intensify up the beach.

Surfing: *Yallingup* is renown for its reef break off the southern end of the beach, locally know as *Car Park*. Depending on the size it hold lefts and right in a low swell, developing into a long left with bigger swell. There are also The beach and reef breaks accessible from the northern car park.

WA 717-727 YALLINGUP-KABBIJGUP-SUGARLOAF

No.	Beach	Rating	Type	Length
WA 717	Yallingup (N 1)	8	R+platform/reefs	300 m
WA 718	Yallingup (N 2)	8	R+platform/reefs	150 m
WA 719	Yallingup (N 3)	8	R+platform/reefs	70 m
WA 720	Yallingup (N 4)	8	R+platform/reefs	80 m
WA 721	Kabbijgup (S 3)	8	R+platform/reefs	30 m
WA 722	Kabbijgup (S 2)	8	R+platform/reefs	150 m
WA 723	Kabbijgup (S 1)	8	R+platform/reefs	70 m
WA 724	Kabbijgup Beach	8	R+platform/reefs	270 m
WA 725	Kabbijgup (N 1)	8	R+platform/reefs	60 m
WA 726	Kabbijgup (N 2)	8	R+platform/reefs	100 m
WA 727	Sugarloaf Rock (S)	9	R+rocks/reef	300 m
Spring & neap tidal range = 0.7 & 0.5 mm				

At the northern end of Yallingup beach is the beginning a straight 15 km long section of coast trending slightly west of due north to Sugarloaf Rock. The entire section consists of irregular calcarenite rock platforms, backed by steep 40-50 m high calcarenite bluffs, then vegetated slopes rising to a 190-200 m high calcarenite ridge. The lithified Pleistocene dune blanket the underlying Precambrian granite. The dunes extend on average 3 km inland to the cleared farmland of the eastern cape. Located along the base of the bluffs and usually wedged in between the bluffs and platforms are 11 strips of sand (beaches WA 717-727) exposed to waves averaging over 1.5 m. A vehicle track off the Hemsley Road runs along the top of the bluffs providing access to the bluffs above the beaches, while the Cape to Cape walking track runs along the top of the bluffs the entire way to Sugarloaf Rock.

Beach **WA 717** is located immediately north of Yallingup Beach and consists of a discontinuous 300 m long crenulate section of high tide sand wedged between the base of the 50 m high steep bluffs and an irregular 50 m wide platform, with reefs further out. Beach **WA 718** lies 50 m to the north and consist of two pockets of sand totalling 150 m in length and located either side of a protrusion in the bluffs, with a relict higher platform dividing the two sections.

Beaches WA 719 and 720 lie 2 km to the north and consist of two adjoining pockets of sand wedged in at the base of two scallops in the backing 50 m high bluffs. Beach **WA 719** is a narrow 70 m long strip of sand, with a steep gully cutting through the backing bluffs, while beach **WA 720** is a wider 80 m long high tide beach, with a patch of vegetated dune at the base of the bluffs. Both beaches are fronted by 50 m wide intertidal calcarenite platforms, with waves breaking over reefs further out.
Beaches WA 721 and 722 lies below the bluffs where the 4 WD track off the Hemsley Road first reaches the coast.

Beach **WA 721** is a 30 m long pocket of sand wedged in a narrow gap in the sloping 30 m high bluffs. A rock fall in the mid 1990's partly covered the rear of the beach. Immediately to the north is beach **WA 722**, with a vehicle track leading to the bluffs above the southern end of the beach. The beach is 150 m long and bordered and backed by steeply sloping 30 m high bluffs. Wave breaks up to 100 m offshore over the inner platforms and reefs off both beaches. The *Shivery Rocks* reef break is located off these beaches.

Kabbijgup Beach is located 1 km to the north. It has a southern neighbour (**WA 723**), which is a 70 m long pocket of sand contained is a semicircular gap in the backing 50 m high bluffs. A vehicle track terminates on the southern bluffs overlooking the beach. The main beach (**WA 724**) lies immediately to the north and is a 270 m long relatively straight steep sand beach backed by an unstable climbing foredune which has reached the top of the bluffs and blown 50 m inland (Fig. 4.146). The beach is fronted by a calcarenite intertidal platform then outer reefs, with the *Three Bears* surf break over the reefs. It is named after the three adjacent breaks, called *Mummy*, *Baby* and *Daddy Bear*.

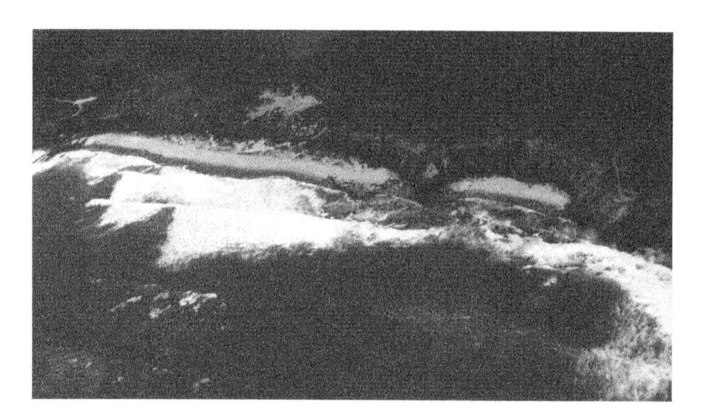

Figure 4.146 Kabbijgup Beach is backed by rocky bluffs and fronted by a reef dominated surf zone.

Five hundred metres to the north are two more beaches (WA 725 & 726). Beach **WA 725** consists of two adjoining pockets of sand totalling 60 m in length backed by two gaps in the 50 m high bluffs. Fifty metres to the north is its neighbour beach **WA 726** a 100 m pocket of sand in the next gap in the bluffs, with protruding bluffs bordering both beaches. The longer beach has some inner tidal pools, then the platform with wave breaking heavily over the outer platform and reefs. A vehicle track terminates on the northern bluffs.

Beach **WA 727** lies 1 km to the north and 1 km south of Sugarloaf Rock. It marks the end of the calcarenite platforms, as granulite rocks become exposed at the shore. The 300 long beach extends around a slightly protruding section of bedrock. It is irregular and wedged in between the backing vegetated bedrock and lines of granulite ridges that cross the beach and extend out into the surf. A spur off the vehicle track runs down to the rear of the centre of the beach.

WA 728-730 **GULL ROCK-CAPE NATURALISTE**

No.	Beach	Rating	Type	Length
WA 728	Gull Rock (S1)	10	R+rocks/reefs	300 m
WA 729	Gull Rock (S1)	10	R+rocks/reefs	500 m
WA 730	Cape Naturaliste	10	R+rocks/reefs	250 m
Spring & neap tidal range = 0.7 & 0.5 m				

Cape Naturaliste is one of the major capes on the Australian coast, forming the northern tip of the southwest corner of the state and continent. It also forms the northern boundary of the Leeuwin Naturaliste National Park, one of the more scenic coastal parks in the world, and one that offers a world class range of energetic surfing breaks over the many reefs. The northern 3 km of the coast between Sugarloaf Rock and the cape is dominated by massive granulite rocks. In amongst the rocks are three high tide strips of sand and backing dunes (beaches WA 728-730). The Cape Naturaliste Road runs up to the lighthouse on the crest of the 150 m high cape, with a gravel road down to Gull Rock.

Beach **WA 728** commences 1 km south of Gull Rock and curves round in a small rock-filled embayment for 300 m. A mixture of irregular rock platforms, rocks and reefs lie off the west-facing beach, with some tidal pools in against the shore and numerous rock outcrops along the beach. Waves average over 1.5 m and break across the rocks and reefs up to 100 m offshore. The *Windmills* surf break is located over these reefs. Beach **WA 729** continues on the northern side of the point for another 500 m. This is a similar more continuous high tide beach, with an inner surf zone dominated by a 50 m band of rocks and reefs. Both beaches are backed by generally vegetated dune-draped calcarenite, over the underlying granulite.

Beach **WA 730** is located on the northwestern side of Cape Naturaliste (Fig. 4.147). The 250 m long west-facing beach is bordered by low craggy bedrock and consists of a high tide sand beach, with numerous rock outcrops along the beach and extending into the inner surf zone. Surf breaks in this area include *The Other Side of the Moon* and *The Channels*. It is backed by a mixture of active sand and deflation surfaces, hence the reference to 'moonscape', which extend a few hundred metres across to the eastern side of the cape.

Figure 4.147 Beach WA 730 and Cape Naturaliste with its dune-deflated surface.

Region 4 Central West: Cape Naturaliste – Perth - Kalbarri

Length of coast: 802 km (1931-2734 km)
Number of beaches: 404 (WA 743-1134)

National Parks:

	shoreline km	no. beaches	page
Meelup Regional Park	6	6	191
Tuart National Park	-	-	-
Yalgorup	7	1	198
Shoalwater Is Marine Park	20	13	205
Marmion Marine Park	17	36	227
Nambung	21	15	246
Kalbarri	13	2	277

Towns/cities: Dunsborough, Busselton, Bunbury, Mandurah, Rockingham, Fremantle, Perth, Lancelin, Cervantes, Jurien, Green Head, Dongara, Geraldton, Kalbarri
Coastal settlements: Eagle Bay, Port Geographe, Binningup, Port Bouvard, Myalup, Avalon, Falcon, Madora, Singleton, Golden Bay, Secret Harbour, Quinns Rocks, Yanchep, Wreck Point, Guilderton, Ledge Point, Grey, Coolimba, Horrocks, Port Gregory.

The **Central West** section of the state from Cape Naturaliste up to Kalbarri is the most developed and populated in the state, with most of the population centred in and around Perth, and in the string of smaller cities, towns and settlements spread along the coast. The coast initially sweeps in a 60 km wide north-facing arc between Cape Naturaliste and Bunbury, then trends north for 140 km to Perth and then 510 km north-northwest up to Kalbarri. While most of the coast faces west, it is sheltered from the high deepwater waves by a near continuous series of submerged and some emerged Pleistocene calcarenite reefs, rocks, islets and islands, the latter including Rottnest Island off Perth. Further north the Houtman Abrolhos coral reef system lies 60 km off Geraldton and parallels the coast for 100 km. As a consequence the entire coast and beaches receive reduced waves and tend to have low to moderate energy conditions at the shore. Most of the beaches are backed by Holocene dunes produced by the strong southerly winds, often overlapping Pleistocene dune calcarenite.

WA 731-738 SHELLEY BEACH-EAGLE BAY

No.	Beach	Rating	Type	Length
WA 731	Cape Naturaliste (E)	5	R+rocks/reef	400 m
WA 732	Shelley Beach	4	R+rocks/reef	70 m
WA 733	Bunker Bay	4→5	LTT→TBR	2.3 km
WA 734	Rocky Pt (1)	4	R+rocks/reef	150 m
WA 735	Rocky Pt (2)	4	R+rocks/reef	50 m
WA 736	Eagle Bay (N)	3	R+rocks	750 m
WA 737	Eagle Bay	4	R→LTT	1.5 km
WA 738	Eagle Bay (N)	4	LTT	250 km
Spring & neap tidal range = 0.7 & 0.5 m				

At **Cape Naturaliste** there is a major change in the orientation and nature of the coast. The cape marks the southwestern boundary of the 45 km wide large Geographe Bay. This is an open north-facing bay, with 50 km of gently curving shoreline between the cape and Capel. The shoreline is protected from the direct attack of the southwest waves and winds, and generally low to very low energy conditions prevail, with refracted ocean waves only reaching many of the beaches during high outside wave conditions. The first 10 km of shoreline between the cape and Point Picquet contains a mixture of pockets of sand and two longer beaches (WA 731-738), all facing north to northeast and receiving only low refracted waves.

Beach **WA 731** is located 1.5 km due east of the cape and consists of a 400 m long strip of high tide sand fronted by an irregular granulite series of rocks and reefs, in places extending 100 m offshore, with deeper reef further out. *The Docks* reef break is located over the reefs and works during higher swell. The beach is backed by vegetated calcarenite bluffs rising to 40 m with a 4 WD track along the top of the bluffs.

Shelley Beach (WA 732) is a picturesque pocket beach located at the end of the Bunker Bay Road. There is a large car park at the end of the road and a short walk to the shady area around the beach. The beach is a curving 70 m long steep reflective beach, bordered and fronted by granite rocks and boulders, with calcarenite capping the western rocks. *The Quarries* surf break is located off the western headland.

Region 4A CENTRAL WEST: Cape Naturaliste – Mandurah

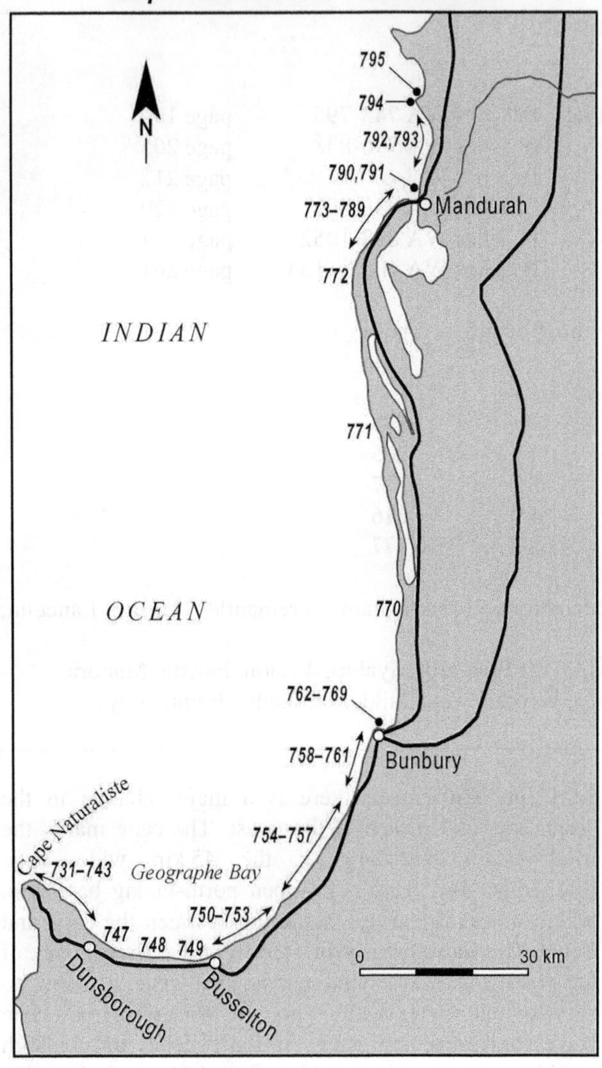

Figure 4.148 Region 4A extends from the southern boundary at Cape Naturaliste to Mandurah.

Bunker Bay is an open north-facing 2 km wide bay located between Shelley Beach and Rocky Point and contains a curving 2.3 km long sandy beach (**WA 733**). Waves energy increases eastward along the beach averaging just over 1 m at the eastern end. The beach usually has a low tide terrace extending along the western section to a central inshore reef, with usually several beach rips towards the east. This is the highest energy beach along this section of coast and popular with surfers looking for beach breaks known as *The Farm* and *Boneyards*. There is a car park at the western end with a walk along the beach to reach the better surf. It is backed by a vegetated foredune, two central dammed creeks, that bracket a farm, and rising vegetated slopes to either rend.

Rocky Point lies at the eastern end of a 600 m long section of granite shoreline. As the shoreline turns south at the point it permits two small pockets of sand to accumulate on the southern side of the point (beaches WA 734 & 735). Beach **WA 734** lies in the immediate lee of the point and is a curving 150 m long east-facing strip of high tide sand, fronted by inter- to subtidal boulders, with sandy sea floor further out. Beach **WA 735** lies 100 m to the south and is a 50 m long pocket of sand wedged in between low granite points which almost encase the small beach. Both beaches are backed by vegetated slopes rising to 30 m. There is a 4 WD track through private land to the point, which holds the reasonable *Rocky Point* lefthand break during high outside wave conditions.

The shoreline trends southwest from Rocky Point into **Eagle Bay**, an open 2.5 km wide bay, bordered in the east by the low granite Point Picquet. Three beaches occupy the bay shore (WA 736-738) (Fig. 4.149). Beach **WA 736** is located 500 m south of Rocky Point and is a sheltered east-facing low energy beach used to launch small boats. It is located in lee of a series of granite rocks and reefs and trends south for 750 m to the small mouth of Jingarmup Brook, beyond which is beach WA 737. It is backed by vegetated slopes containing a housing development off the Fern Road, including several beachfront houses.

Figure 4.149 Beaches WA 736 and 737 extend east of Point Picquet, with rocks dotted along the low energy shoreline.

Beach **WA 737** is the main bay beach and trends southeast for 1.5 km to a series of low granite rocks. Waves average less than 1 m and calm conditions are common. During higher swell, waves break across a narrow bar, which may be cut by beach rips towards the eastern end of the beach. The small Eagle Bay settlement is located behind the western end of the beach, with the Eagle Bay Road running along the back of the beach to Point Picquet, and a few car parks between the road and the beach. Beach **WA 738** is located between two series of two granite rock outcrops at the eastern end of the main beach. It is a 250 m beach, bordered by the low orange granite rocks. Waves are usually low, but a rip develops between the rocks during higher swell conditions. A car park off the Eagle Bay Road is located right behind the centre of the beach.

MEELUP REGIONAL PARK

Area: 500 ha
Coast length: 6 km (1939-1945 km)
Beaches: 6 (WA 737-743)

Meelup Regional Park occupies a 6 km long section of granite headland and rocks bordering six relatively protected northeast-facing beaches. The shore is backed by generally well-vegetated slopes rising to 90 m. The park contains the sheltered beaches, a coastal walk that runs the length of the park and the flora and fauna of the backing natural vegetation. It is popular for summer beach recreation, fishing from the shore and rocks, and walking.

WA 739-743 **MEELUP-CURTIS BAY**

No.	Beach	Rating	Type	Length
WA 739	Meelup Beach	4	R/LTT	400 m
WA 740	Castle Bay	4	R/LTT	400 m
WA 741	Castle Pt	4	R+rocks	100 m
WA 742	Castle Pt (S)	4	R+rock s	60 m
WA 743	Curtis Bay	4	R+rocks	60 m
Spring & neap tidal range = 0.7& 0.5 m				

Meelup Regional Park incorporates a series of low energy north-facing beaches, each bordered by sloping granite rocks and backed by densely vegetated slopes. The most popular of the beaches are Meelup and Castle Bay (WA 739 & 740), while a walking track is required to reach the more isolated Castle Point and Curtis Bay beaches (WA 741-743).

Meelup Beach (**WA 739**) is a northeast-facing 400 m long pocket of sand occupying the mouth of a small valley, bordered by the rising vegetated slopes in lee of Garnet Rock to the west and smaller Sail Rock to the east, both of which are fronted by boulder beaches. The main beach is often calm with a moderately steep beach fronted by a narrow bar then deeper water off the beach. There is road access to a large car park and shady picnic area immediately behind the beach, while Meelup Brook crosses the western corner of the beach.

Castle Bay (**WA 740**, also known as Little Meelup Beach) lies 500 m to the east in the next small valley and is a similar 400 m long northeast-facing low energy beach, with usually a moderately steep beach and narrow bar. Dolugup Brook flows across the western corner of the beach. It is bordered by the granite rocks of Sail Rock to the west and Castle Point to the east, which includes the prominent Castle Rock. Rocks also fringe either end of the beach, leaving an open 200 m long central section. During large outside waves there may be a reef break along the rocky eastern end of the beach. There is vehicle access to a small car park, but no facilities at the beach, which is backed by densely vegetated slopes rising to 90 m. A whaling station operated out of this little bay from 1845.

Castle Point lies immediately east of Castle Bay and can only be accessed on foot. In amongst the rocky shore are two small pockets of sand, both accessible off the walking track. Beach **WA 741**, is a crenulate 100 m long sand beach, bordered and partly fronted by granite rocks and boulders, which form tidal pools along the eastern half of the beach. There is only clear access to the sea at the western end. Waves are usually low to calm. Beach **WA 742** lies 200 further round the rocks and is a 60 m pocket of sand, bordered by granite rocks and fronted by a 30 m wide band of subtidal rocks with sand further out.

Curtis Bay is a slight indentation in the granite shoreline 1 km south of Castle Point. In the apex of the bay is a 60 m long pocket of sand (**WA 743**), bordered by intertidal granite rocks and boulders, and fronted by subtidal rocks, then the sand of the low energy bay floor.

WA 744-746 **DALLING PT-DUNSBOROUGH**

No.	Beach	Rating	Type	Length
WA 744	Pt Dalling	2	R+rocks/sand flats	150 m
WA 745	Pt Daking	2	R+rocks/sand flats	250 m
WA 746	Dunn Bay	1	R+sand flats	700 m
WA 747	Dunsborough	1	R+sand flats	7.5 km
WA 748	Toby Inlet	1	R+sand waves	5.3 km
Spring & neap tidal range = 0.7 & 0.5 m				

Dunsborough is the first coastal town in Geographe Bay and is located on the eastern end of the granite rocks that extend 10 km east of Cape Naturaliste. The growing town of about 1000 extends east behind the beginning of the long sandy shoreline that continues almost uninterrupted to Bunbury 60 km to the northeast. The Dunsborough beaches include two located in amongst the granite rocks (WA 744 & 745) and the two longer sandy sections (WA 746 & 747).

Beach **WA 744** is a curving 150 m long strip of high tide sand located in the small bay on the northern side of Point Dalling. Granite boulders lie along the shore, with only a 50 m clear section in the middle of the beach, from which boats are launched at high tide. Beach **WA 745** occupies the next small 250 m wide bay between Point Dalling and Point Daking. The 250 m long beach runs along the shore of the low energy bay, with granite boulders bordering each end and a 100 m long clear section towards the centre of the beach. Houses back the beach, which usually has a few small boats pulled up on the shore. Both beaches are fronted by 100-200 m wide intertidal sand flats and scattered rocks, then a 100 m wide band of subtidal seagrass meadows and shallow sandy sea floor beyond. As a result the boats can only be launched at high tide when the sand flats are covered.

Point Daking forms the end of the granite of the cape region and the beginning of the long sandy beaches of the Perth Basin, which extends 1000 km north to the Murchison River. **Dunn Bay** beach (**WA 746**) commences on the eastern side of the point and trends east for 700 m to a low sandy point. The beach faces northeast and receives only low wind waves, with calm conditions common along the shore. The narrow high tide beach is often covered or backed by seagrass debris and fronted by 200-300 m wide intertidal sand flats (Fig. 4.150), then subtidal seagrass meadows and deeper sand ridges. The beach is backed by a foreshore park in the western corner, then a caravan park and beach cottages.

Figure 4.150 Dunn Bay beach is a low energy high tide beach fronted by intertidal sand flats.

Dunsborough Beach (**WA 747**) commences on the eastern side of the sandy point and continues east for 7.5 km to Toby Inlet. The inlet drains a creek that runs behind the beach for 5 km to the east, as well as Annie Brook, a series of drainage canals that converge on the inlet, and drain the backing low farmland that occupy the interbarrier depression. The beach trends to the east-southeast as a series of subtle shoreline crenulations, each point tired to a shallow shore transverse sand wave that extend up to 1 km northwest into the bay. These sand waves are spaced about every 500 m and are slowly migrating to the east, and as they do so induce oscillations in the adjacent beach. Seagrass meadows occupy the swales in the between the sand wave crests. The beach itself is a low energy sandy beach, with the crests of the sand waves exposed at low tide. Waves only arrive during very high outside conditions, or strong northerly winds, the latter produce short wind waves. The Geographe Bay Road runs for 5 km along the beach, past sandy Point Templar, to a series of foreshore reserves and car parks, as well as houses on one or both sides of the road, with the elongate creek behind.

Beach **WA 748** commences on the eastern side of the often deflected mouth of Toby Inlet and continues much the same for 5.3 km to a rock groyne at the eastern end of Siesta Park (Fig. 4.151). The low energy beach is crossed by two drains, the small Molloy Ditch which drains Mary Brook, and the larger Carbunup River. Both drains close during the dry summer months and usually only have a low flow across the beach during the winter. The beach is backed by resort and residential development along much of its length and the Siesta Park cottages towards the eastern end.

Figure 4.151 Siesta Park beach and the beginning of the long Busselton beach are separated by a rock groyne, which has produced an offset in the shoreline.

WA 749 BUSSELTON

No.	Beach	Rating Type	Length
WA 749	Busselton Beach	1 R+sand waves	15.3 km
Spring & neap tidal range = 0.5 & 0.0 m			

Busselton is a sizable and growing coastal town of 17 000 people set on the southern shores of Geographe Bay. The town spreads for 6 km along the bay and beach WA 749 and is located on low beach and dune deposits that have built out into the bay over the past few thousand years. It is a major regional centre and summer holiday destination and has all the facilities and accommodation that go with a popular coastal town.

Busselton Beach (WA 749) is the longest continuous section of this part of the bay shore. It commences at the Siesta Park groyne and gently curves to the east for 15.3 km past the long Busselton jetty (Fig. 4.152) to the rock groynes of the Port Geographe development. The beach along this section is a relative continuous crenulate beach (Fig. 4.153), broken only by human structures, including the Buayanyup and New river drains and the 2 km long Busselton jetty, the longest in Australia. The length of the jetty is an indication of the extensive shallow sand flats, including sand waves and seagrass meadows that line the southern shores of the bay. The sand waves slowly migrate eastward along the bay, causing the adjacent shoreline to oscillate. Unfortunately structures and roads were built unwittingly in this zone of natural oscillation resulting in the need for seawalls and groynes to protect the road and property along parts of the beach. Most of the beach is backed by a wide foreshore reserve. It contains a bike path, and either side of the jetty numerous amenities, including an oceanarium and entertainment centre at the jetty, as well as boat ramps and sporting facilities.

The beach along the Busselton shoreline lies in the apex of north-facing Geographe Bay and initially faces northwest, then north and finally northeast against the eastern groyne. Owing to protection from Cape Naturaliste, it generally receives no to very low swell and only small local wind waves during northerly wind

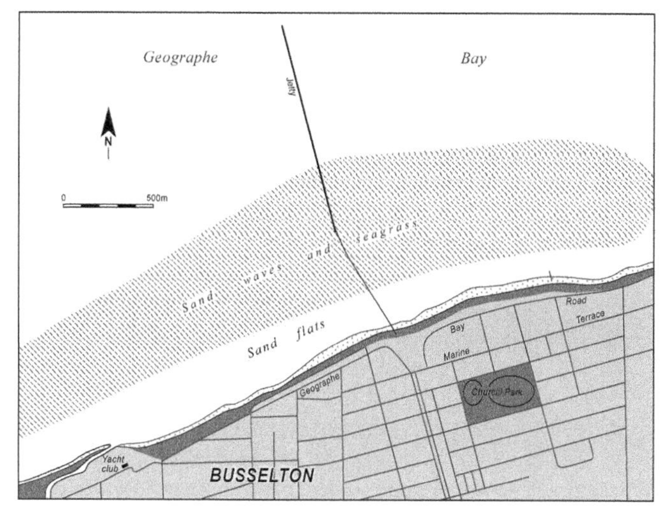

Figure 4.152 Busselton town centre, beach and jetty.

conditions. These result in a moderately steep narrow beach, fronted by the extensive sand flats and sand waves, with boats often moored off the beach. Seagrass is commonly washed up onto the beach, and at the far eastern end of the beach is causing major problems as it piles up against the Port Geographe groyne.

Figure 4.153 Section of Busselton beach north of the town centre.

Swimming: This is a relatively safe beach under normal conditions. A tidal swimming enclosure is located just west of the jetty, otherwise people swim all along the beach.

Surfing: Usually none.

WA 750-753 **PORT GEOGRAPHE**

No.	Beach	Rating	Type	Length
WA 750	Port Geographe (1)	2	R	80 m
WA 751	Port Geographe (2)	2	R	60 m
WA 752	Port Geographe (3)	1	R+sand flats	250 m
WA 753	Port Geographe (4)	1	R+sand flats	250 m
Spring & neap tidal range = 0.5 & 0.0 m				

Port Geographe is a marina and canal estate development, constructed between 1995 to 2003, on the Wonnerup wetland and beach area 5 km east of Busselton. In developing the canal estate the wetlands were dredged, and to connect the canals with Geographe Bay, a drainage-navigation canal and entrance training walls, and associated groynes were built. The structures cross a once continuous beach and include two training walls, two inner secondary walls and two downdrift groynes. Between the six shore perpendicular rock structures are four new beaches (WA 750-753) (Fig. 4.154), while the original beach is broken into beaches WA 749 and 754. The construction of such a structure has had a number of predictable impacts on a shoreline experiencing west to east longshore sand transport and with extensive seagrass meadows offshore. First and most obvious is the trapping of the seagrass debris against the updrift wall, along the eastern few hundred meters of beach WA 749, within the entrance canal on beaches WA 750 and 751, and towards the eastern ends of beaches WA 752 and 753. Second, has been the accumulation of sand on the same updrift side of the groynes and training walls causing the beach to build out tens of metres. Third, has been the starvation of sand from the downdrift Wonnerup Beach (WA 754) which is has lead to erosion of the beach back to the main Layman Road. As a result the council constructed first one, then a second rock wall to protect the road, and replace the former sandy beach and foredune .

Figure 4.154 Port Geographe under construction in 1997. The groynes and training walls have interrupted the northward movement of both seagrass debris and sand causing it to accumulate on the southern (right) side.

The four 'beaches' have all be artificially formed through the construction of the boundary groynes and training walls, then filled with sand. They may take some years to fully adjust to a more equilibrium states. The following description is based on site inspections undertaken in August 2001.

Beach **WA 750** is located on the western side of the navigation channel. It consists of a narrow low gradient 80 m of sand bordered by the long western groyne and the navigation channel training wall to the east. There is a large car park immediately east of the beach. The beach faces northwest along the 500 m long groyne to the 100 m wide entrance. Conditions are usually calm at the beach permitting seagrass to accumulate.

Beach **WA 751** is located on the eastern side of the navigation channel, between the eastern training wall and a secondary inner wall. The beach is 60 m long and acts as a natural seagrass trap. It is usually covered in thick seagrass debris and backed by a low grassy dune area.

Beach **WA 752** extends for 250 m east of the eastern training wall to a shore perpendicular rock groyne that extends 100 m into the bay. It is a narrow, moderate gradient sand beach, which in 2001 was eroding at its western end exposing rocks, while sand and seagrass was accumulating at the eastern side. Beach **WA 752** lies immediately to the west and is a near identical 250 m long beach, located between the central groyne and the final eastern groyne of the development. It too is eroding in the west with sand and much seagrass accumulating in the east. Car parks back both beaches.

Prior to the development **Wonnerup Beach (WA 754)** was the eastern end of beach WA 749. It now extends from the eastern groyne east for approximately 1.8 km to the migratory mouth of **Wonnerup Inlet** (Fig. 4.155) The western end of the beach has been severely impacted by the bay development and has been eroding since the groynes were constructed, even following sand nourishment of the beach. In 2001 it had eroded back approximately 100 m to Layman Road, resulting in the construction of rock seawalls to protect the road and backing houses. The central-eastern part of the beach is still in a relatively natural state, with a low energy beach fronted by the transverse migratory sand ridges. It terminates at the inlet, which is usually deflected to the east by the longshore current and sand transport.

Figure 4.155 Wonnerup Inlet is deflected northward by the transport of sand along the bay shore.

WA 755 FORREST-PEPPERMINT GROVE

No.	Beach	Rating	Type	Length
WA 755	Forrest-Peppermint Grove	2	R+sand ridges	14 km
Spring & neap tidal range = 0.5 & 0.0 m				

Forrest-Peppermint Grove Beach (WA 755) commences at Wonnerup Inlet and trends to the northeast as Forrest Beach, terminating at the channelised Capel River mouth as Peppermint Grove Beach. The beach represents a transition in the level of wave energy from less than 1 m in the south to about 1 km at the river mouth. It commences at Wonnerup Inlet with low waves and transverse sand ridges extending a few hundred metres offshore. Moving northwards along Forrest Beach the ridges decrease in width, and by the 10 km mark have moved in close to the beach, with the northern Peppermint Grove section having no ridges and a more typical reflective beach, usually dominated by high tide beach cusps and a deeper inshore. The backing barrier reflects this transition with the low foredunes that extends all the way to Dunsborough, increasing in height and width along Peppermint Grove beach, reaching 20 m by the river mouth. The dunes average a few hundred metres in width, with a 1 km wide interbarrier depression occupied by wetlands and lagoons, including the Wonnerup Estuary, separating the Holocene barrier from the inner Pleistocene barrier. The Turat Forest National Park occupies part of the Pleistocene barrier.

There is access to the beach at Forrest beach, with the Forrest Beach Road paralleling the back of the beach for 5 km, and in the north via the Peppermint Grove Road. Since the late 1990's Peppermint Grove has experienced considerable housing development in the backing foredunes and right up to the river. There are a number of beach access points, car parks and boat launching sites along the beach.

Waves are usually calm to low along the beach with a steep reflective beach face. During higher wave conditions a heavy shorebreak develops at the base of the beach and some times there are good waves over the *river mouth bars*. Both river mouths are usually narrow and shallow, and may close during the summer months. When open, channels and shoals form off the inlets.

WA 756 STIRLING BEACH

No.	Beach	Rating	Type	Length
WA 756	Stirling Beach	3	R+reef	10.9 km
Spring & neap tidal range = 0.5 & 0.1 m				

Stirling Beach (WA 756) commences at the small mouth of the Capel River and trends to the north-northwest for 11 km to the smaller, usually blocked mouth of Five Mile Brook Drain. The beach is moderately crenulate for most of its length owing to outcrops of beachrock on and off the shore, together with some shore parallel beachrock reefs, which induce shoreline crenulations, as well as lowering waves at the shore. The net result is a low energy reflective beach with variable inshore topography depending on the proximity to the rock outcrops and reefs. The dunes that had started to increase in size at Peppermint Grove again decrease to a lower, narrow moderately stable foredune for most of the length, through increasing substantially in size for the northern 2 km, which is free of the beachrock. The beach is backed by a generally drained wetland a few hundred metres wide, which is farmed, then the low Pleistocene barrier. There is public access to the beach at Dalyellup Road off the Minninup Road, 2 km south of the drain, and 4 WD access along the drain.

WA 757 **BUNURY BEACH**

No.	Beach	Rating	Type	Length
WA 757	Bunbury Beach	3	R	11.5 km
	Hungry Hollow Beach	3	R→LTT	500 m
	Ocean/Back Beach	3	R→LTT	500 m
		Total length		12.5 km

BUNBURY SLSC patrols Bunbury Beach
Patrol period: Saturday November-March
Sunday October–March

Spring & neap tidal range = 0.5 & 0.1 m

Bunbury is a thriving coastal city of 28 000 people and a major service and holiday centre located 160 km south of Perth. The city is situated on the southern shore of Koombana Bay, a natural harbour in lee of Casuarina Point (Fig. 4.156). The harbour has been heavily modified and improved for shipping. Modifications include a 1 km northward extension of Casuarina Point to the artificial McKenna Point and a 1.5 km long jetty inside the point.

The backing Leschenault Inlet has also been heavily modified. In the southwest corner of the bay a canal connects the now landlocked lower Leschenault Inlet to the bay. Two kilometers east is the new 150 ha Inner Harbour and finally 4 km to the northeast The Cut, provides a trained entrance into the main Leschenault Inlet.

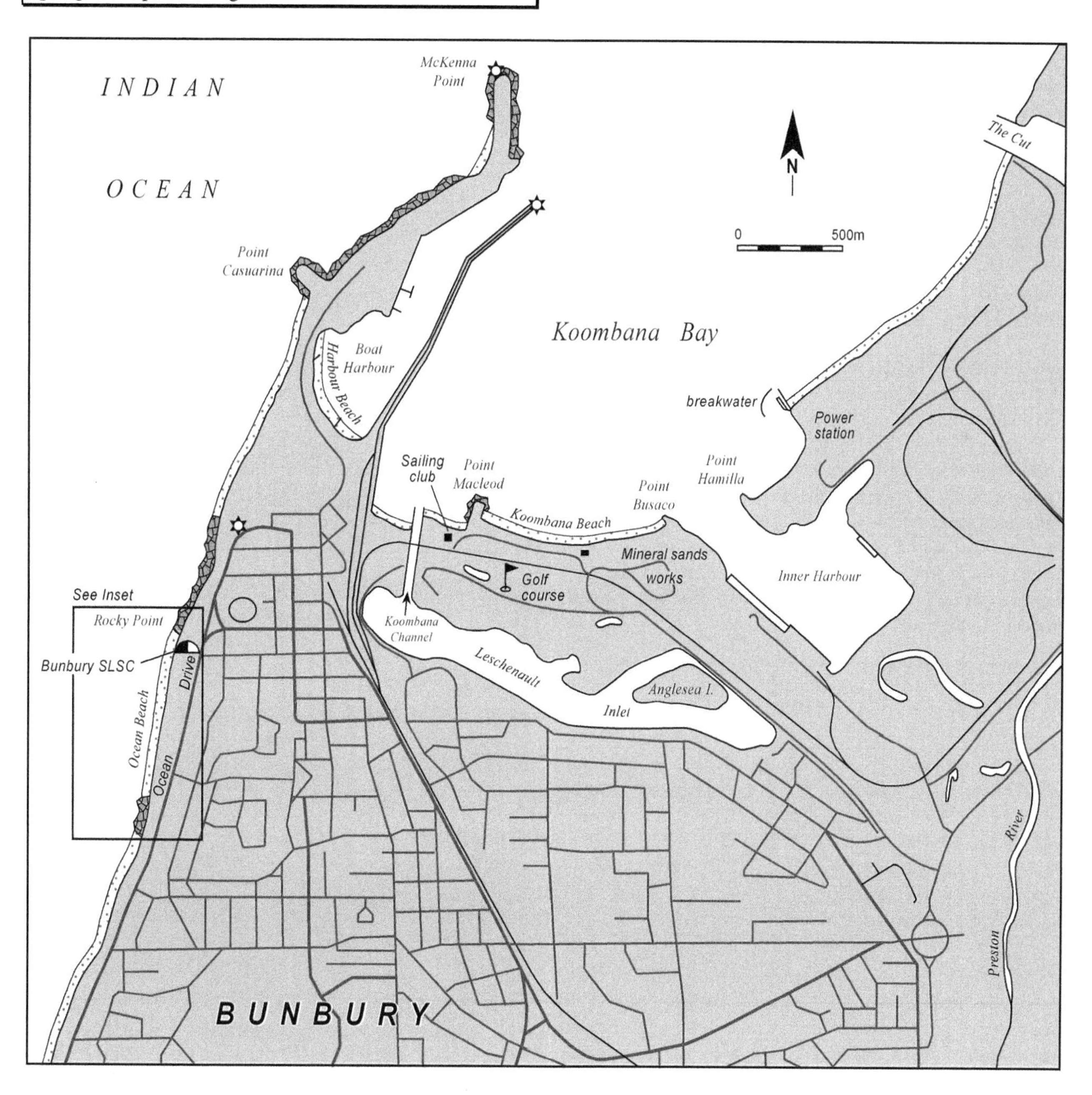

Figure 4.156 Bunbury is a large coastal town with exposed beaches on its western shore and sheltered beaches in Koombana Bay. See Fig. 4.155 for insert.

The eastern suburbs of Bunbury front the main surfing beaches of Bunbury and Casuarina Point (WA 757-760), which are backed by the Bunbury scenic ocean drive. Along the northern shore are the six harbour beaches of Koombana Bay (WA 761-766) which are all protected by Casuarina-McKenna points, and have all been heavily modified by harbour construction.

Bunbury Beach (WA 757) commences 12 km south of Bunbury city at the Five Mile Brook drains and trends to the north-northwest for 12.5 km to Rocky Point. The northern 4 km is known successively as Mindalong, Hasties St, Hungry Hollow and finally Ocean Beach. This is the main surfing beach for Bunbury and site of the **Bunbury Surf Life Saving Club**, which is located just south of Rocky Point (Fig. 4.157). The Surf Club is one of the oldest in Australia, and the third oldest in Western Australia and one of the oldest in Australia, having been founded in 1915. The beach is a near continuation of the long stretch of sand that extends north from Busselton. It receives waves averaging 0.5-1 m, which usually maintain a 50 m wide low tide terrace, cut by occasional rips (Fig.4.158). The beach commences at the Five Mile Brook drains and trends north for 12.5 km with little development, apart from one access road. It is backed by vegetated blowouts and parabolic dunes rising to 30 m and extending up to 1 km inland. Development of the coast commences 6 km south of the town centre, with Ocean Drive paralleling the shore for 4 km to the south of Rocky Point. This scenic drive provides good parking and access for 3 km south of the Surf Club. A number of rocky reefs are located along the beach, including either end of the Surf Club beach section. The Surf Club is surrounded by a long car park and fronted by a seawall.

The beach faces west and receives protection from both Cape Naturaliste 50 km to the southwest and beachrock reefs that lie off the beach. Consequently ocean waves average less than 1 m. These produce a relatively steep beach, often fronted by a low tide terrace. While the beach is usually free of rips permanent rips are located adjacent to the rocky sections.

WA 758-766 **PT CASUARINA**

No.	Beach	Rating	Type	Length
WA 758	Lighthouse Beach	5	R+platform	700 m
WA 759	Pt Casuarina	4	LTT	900 m
WA 760	Pt Casuarina (N)	3	R	50 m
WA 761	Pt McLeod (W	5	LTT/TBR	300 m
WA 762	Pt McLeod (E)	3	R	70 m
Spring & neap tidal range =0.5 & 0.1 m				

Rocky Point is a 700 m long section of exposed beachrock that continues north of Bunbury's Ocean Beach. The point and adjacent Casuarina Point form a major inflection in the shore, to the lee of which the Bunbury was established in 1841, with the first jetty constructed in 1864. It was gradually lengthen to 1.5 km by 1952. In the 1960's to provide more protection and improve the port Casuarina Point was extended 1.5 km northward a rock breakwater to form the outer harbour and in the process create a new shoreline that contains four additional beaches (WA 759-762). The inner harbour was opened in 1976.

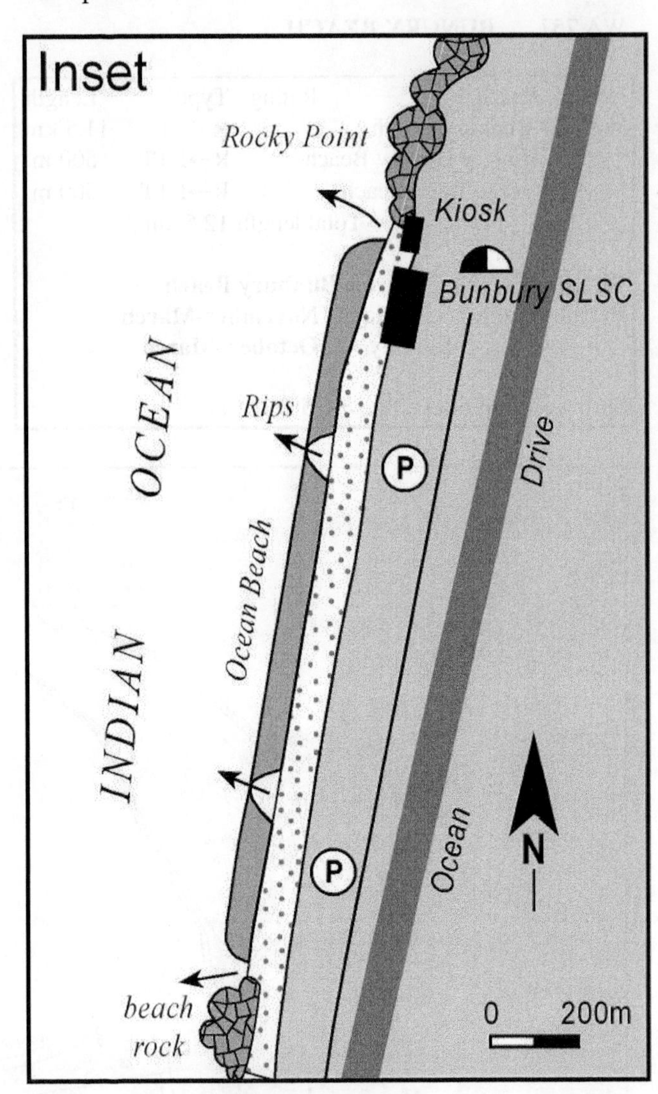

Figure 4.157 Bunbury SLSC is located on a straight section of beach immediately south of the low Rocky Point. The beach usually has a narrow bar cut by occasional rips.

Figure 4.158 View south from Bunbury SLSC with wave breaking across the low tide terrace.

Lighthouse Beach (also known as Symmonds St Beach) (**WA 758**) is a 700 m long crenulate strip of high tide sand fronted by near continuous inter- and supratidal beachrock rocks and reef. There are a few pockets of sandy shoreline, however the beach is most popular with rock fishers. Waves average about 1 m and break over the rocks at high tide. It is backed by Ocean Drive with parking along the rear of the beach.

Casuarina Point beach (**WA 759**) commences on the northern side of Rocky Point and continues north for 900 m to a 100 m long westward trending groyne. The groyne has been built to trap sand moving north along the breakwater, causing the beach to form. The beach is more exposed that the beaches to the south with waves averaging over 1 m and usually breaking across a low tide terrace. The beach is backed by port facilities and the port road. On the northern side of the groyne is a 50 m long pocket of north-facing sand (beach **WA 760**) wedged in between the groyne and the main breakwater. It is protected by the bordering rocks, however a sign indicate that swimming is prohibited at this beach, as it is on port land.

The breakwater continues northeast for another 1 km as **McKenna Point**. On the western side of the breakwater is an inflection within which sand has accumulated as 300 m long beach **WA 761**. The beach faces northwest and receives waves averaging over 1 m, which break across a 50 m wide bar with occasional rips. It is backed by the rocky breakwater, which also forms the boundaries as the beach pinches out to either end. To prevent sand moving round the breakwater and into the port, sand is from time to time removed from this beach. Swimming is also prohibited at this beach. Immediately west of the beach on the inner side of the breakwater is the final sediment trap, beach **WA 762**. This 70 m long north-facing beach is trapped between the main breakwater and an inner 100 m long north trending groyne. As sand accumulates in the trap the beach forms,. Like its neighbour this sand is periodically removed to prevent it moving further into the harbour.

WA 763-769 **BUNBURY HARBOUR**

No.	Beach	Rating	Type	Length
WA 763	Harbour Beach	2	R	600 m
WA 764	Harbour groyne	3	R	50 m
WA 765	Pt McLeod (W)	2	R	200 m
WA 766	Pt McLeod (E)	2	R	250 m
WA 767	Koombana Beach	2	R	800 m
WA 768	Pt Hamilla	2	R	300 m
WA 768	The Cut	2	R	1.5 km
Spring & neap tidal range = 0.5 & 0.1 m				

Bunbury Harbour has been heavily modified by the construction of breakwaters and groynes, which have been built to reduce wave height in the harbour, increase anchorage space and decrease sand moving into and shoaling the harbour. As a consequence all seven harbour beaches are artificial in that they have come into being due to harbour construction and modification. While some are located on the site of former beaches (WA 765-768), they have been modified by the overall reduction in wave height and construction of groynes and inlets, while two beaches (WA 763 & 764) have formed as a result of harbour construction.

Harbour Beach (**WA 763**) is the innermost of the harbour beaches and forms the southern shoreline of the recreational boat harbour. The beach faces northeast and curves round for 600 m. It is bordered by the main harbour jetty in the east and backed by the port road, port facilities and the main breakwater. The beach receives no swell, only low wind waves generate within the harbour. The main boat ramp is located at the northern end of the beach and numerous small boats are moored in the harbour off the beach. Beach **WA 764** is located out on the eastern side of the harbour jetty and consists of a 50 m pocket of north-facing sand wedged between the rock jetty and an angled groyne. It is backed by port facilities.

Between the jetty and the Point McLeod breakwater are two small harbour beaches (WA 764 & 765) , which are divided by the 70 m wide Koombana Channel and its training walls. The Channel connects Leschenault Inlet to the harbour. The inlet is home to the southernmost mangroves on the west Australian coast, and the first on the southern coast since Davenport Creek, South Australia, 1600 km to the east. Beach **WA 764** is 200 m long, faces north and is backed by a railway line. The beach is very sheltered and usually calm.

Beach **WA 765** is located on the eastern side of the Channel and extends east for 250 m to a 200 m long groyne. The Koombana Bay sailing club is located at the rear of the beach. Numerous yachts are moored off the beach and a small jetty crosses the sand. Calm conditions usually prevail at the shore.

Koombana Beach (**WA 766**) is the main harbour recreational beach. The 800 m long beach faces north out of the harbour, but is still sheltered by the main breakwater with usually calm conditions (Fig. 4.159). The beach is backed by a small foredune, three large car parks, a kiosk, picnic area, toilets and a Dolphin Discovery Centre at its eastern end. It also has a section for boat launching. This is the most suitable beach for swimming in the harbour region. The beach terminates at Point Busaco, a groyne that forms the western side of the newer Inner Harbour.

On the eastern side of the inner harbour entrance is Point Hamilla and two more harbour beaches (WA 767 & 768). Beach **WA 767** extends northeast of Point Hamilla for 300 m to the power station breakwater. The beach faces northwest toward McKenna Point and receives low refracted waves. However it is backed by port facilities and not suitable for swimming. Beach **WA 768** commences at the breakwater and continues to the northeast for 1.5 km to The Cut, the 200 m wide trained main entrance to Leschenault Estuary. The beach is backed by the power station at its southern end, while the

Fig. 4.159 Koombana Beach, the main harbour recreational beach.

northern end is reserved for the public with a park and boat ramp on the inner estuary side of Turkey Point.

Swimming: The main swimming beaches for Bunbury are Bunbury Beach, which has small swell and possible surf, and Koombana Beach in the more protected Koombana Bay. Swimming is not recommended on Point McLeod and in the Harbour.

WA 770 BINNINGUP BEACH

No.	Beach	Rating	Type	Length
WA 770	Binningup Beach	4	R	22 km

BINNINGUP SLSC
Patrol period: Sundays December-March

Spring & neap tidal range = 0.5 & 0.1 m

Binningup Beach is the name of a small beachfront settlement in the centre of a 22 km long beach (**WA 770**) that starts at The Cut at the mouth of the Leschenault Estuary and trends essentially due north to the small mouth of the Harvey River diversion drain, just below Myalup. The straight beach receives waves averaging about 1 m for most of its length, which maintain a steep reflective beach (Fig. 4.160). The entire beach is backed by moderately active 20-30 m high transgressive dunes, including blowouts and parabolics, extending up to 1 km inland, with vegetated dunes up to 1.5 km wide. The dunes are in turn backed by the 2 km wide Leschenault Estuary in the south and a swampy 1 km wide interbarrier depression for the remainder. The only access in the south is via the Buffalo Road around the top of the estuary to a 4 WD track across the dunes, and in the north at Binningup. The Binningup Surf Life Saving Club, founded in 2002 is located at the settlement and patrols the beach on Sundays between November and March..

The **Binningup** settlement extends for about 2 km through the dunes towards the northern end of the beach. The beach in this area has some outcrops of beachrock along and just offshore, resulting in a more crenulate

Fig. 4.608 Binningup community and beach access, and site of the Binningup SLSC.

shoreline. There is a large beachfront car park and boat launching area on a section of the beach partly protected by inshore reefs, while all the houses are located in behind the foredune.

WA 771 MYALUP-PRESTON-YALGORUP-CAPE BOUVARD

No.	Beach	Rating	Type	Length
WA 771A	Myalup Beach	4	R	22.5 km
WA 771B	Preston Beach	4	R	9 km
WA 772C	Yalgorup N.P.	4	R	7 km
WA 771D	Cape Bouvard	4	R	15 km
			Total	53.5 km

Spring & neap tidal range = 0.5 & 0.1 km

The 53.5 km of continuous sand between Myalup and Cape Bouvard is the longest beach (WA 771) in the southwest. The north-trending beach (WA 771A) commences at Myalup at the Harvey River diversion drain and trends almost due north for 32 km to the small Preston beach settlement (WA 771B), then another 7 km to the southern boundary of Yalgorup National Park, which occupies then next 7 km of shore (WA 771C), before the final 15 km which gently curves round Cape Bouvard (WA 771D). The beach terminates in the north at the first major beachrock outcrop (Fig. 4.161), located 1 km south of Tims Thicket.

The entire beach (**WA 771**) faces west, but is protected by shore parallel calcarenite reefs, including the Bouvard Reefs. The reefs are drowned Pleistocene beach and dune systems. They lie up to 5 km offshore and shoal to less than 10 m lowering the higher outside waves to about 1 m at the shore. These waves combined with the medium sand to maintain a relatively steep reflective beach the length of the long beach. The beach is backed by near continuous blowouts and parabolic dunes systems averaging about 1 km wide along Myalup and Preston beaches, but widening to 2-3 km along Yalgorup and Cape Bouvard sections. The dunes are moderately active with numerous blowouts located along the Myalup and Preston beaches. The Holocene dunes in turn are backed by an inner Pleistocene barrier which it overlaps to the north. There are a total of three inner shore parallel Pleistocene systems each separated by a series of lakes and wetland including Lake Preston and Lake Clifton.

Figure 4.161 The northern end of the 54 km long Myalup Beach at Cape Bouvard. The entire beach receives low waves and is steep and reflective.

The beach is accessible at **Myalup,** where there is a growing beachfront community and caravan park in the dune behind the beach and a large beachfront car park. The **Preston Beach** settlement, 20 km to the north, is located 500 m in from the beach and active dunes. It also has a large beachfront car park. The only other public road it the White Hill Road at **Cape Bouvard**, 22 km north of Preston Beach. This road leads to a 4 WD track to the beach, with other tracks leading to several houses located in the adjoining dune areas.

Waves average about 1 m along the entire beach and usually surge up the steep reflective beach face. During higher wave conditions a heavy shore break develops and may from an attached bar. Rips and surf are usually absent.

WA 772-776 TIMS THICKET-SURF BEACH

No.	Beach	Rating	Type	Length
WA 772	Tims Thicket (S)	5	R+reef	800 m
WA 773	Tims Thicket	4	R+reef	2.3 km
WA 774	Melros	5	R+reef	1.3 km
WA 775	Florida Bay	4	R	500 m
WA 776	Florida (N)	4	R/LTT	800 m
WA 777	Pyramids	4→5	LTT→TBR	1.4 km

Port Bouvard Surf Life Saving Club
Patrol period: Sunday October-March

Spring & neap tidal range = 0.7 & 0.5 m

The beachrock reefs that first outcrop at the southern end of Tims Thicket mark a change in the orientation and nature of the shoreline. The shoreline turns and trends north-northeast and then northeast for 18 km to Robert Point at Mandurah. In between is an increasingly developed section of coast including the new Dawesville Channel and adjacent developments and the growing city of Mandurah. The first 6 km between the Thicket and the Dawesville Channel contains six beaches (WA 772-777), all but the first undergoing considerable development during the 1990's and into the 2000's.

Tims Thicket is the name of a dense scrub area behind beaches WA 772 and 773. Beach **WA 772** commences at the beachrock outcrops that forms the northern boundary of the long beach WA 771. The beach trends north form 800 m, with a series of small outcrops along the shore inducing several crenulations. Waves break both over the reefs as well as a heavy shore break. The beach is backed a generally vegetated 20 m high foredune with a few blowouts, then older dunes extending up to 1 km inland. Numerous 4 WD tracks cross the dunes to the beach, and connect to Tims Thicket Road, the main access track.

A north-trending beachrock reef separates beaches WA 772 and 773. Tims Thicket beach (**WA 773**) begins in lee of the reef, initially curving round to the north as a 200 m long protected embayment. The beach continues north of the reef for a total of 2.3 km, with most of the beach exposed to 1 m high waves which surge up a steep reflective beach face, with a few outcrops of beachrock. A track off the end of the Tims Thicket Road reaches the centre of the beach, which is popular for offroad driving, beach fishing and occasional surf on the reefs. It is backed by moderately active dunes extending up to 300 m inland, with a sewerage treatment plant in the dunes.

Melros beach (**WA 774**) commences at the next major beachrock-induced inflection in the shore (Fig. 4.162). Wave break around the boundary reef generating a usually small lefthand *surfing break.* The beach extends for 1.3 km with a series of beachrock outcrops both on and off the shore, resulting in variable wave height and inshore topography. A foredune reserve, then the Melros settlement and caravan park backs the beach, with a large central beachfront car park and boat ramp.

Figure 4.162 Melros beach and backing community.

Florida beach consists of two parts (WA 775 & 776), which are both bordered by beachrock outcrops and adjoining reefs. The first beach (**WA 775**) is 500 m long and bordered by two more prominent beachrock reefs. Waves averaging 1 m surge up the steep beach face. It is backed by a foredune, then housing. Beach **WA 776** commences immediately to the north and continues for 800 m to a 100 m long intertidal reef at the northern end of the houses. A reef lying off the beach induces a central sandy foreland, with the area between known as Florida Bay. There is a large car park at the northern end, with a foredune to the north, and housing at either end.

Pyramids beach (**WA 777**) is a 1.4 km long beach that commences against the northern beachrock boundary of Florida beach and terminates at the 500 m long Dawesville Channel training wall (Fig. 4.163). The channel is part of the large Port Bouvard and Dawesville development, which cut the 1 km long, 200 km wide channel to connect the Harvey Estuary with the Indian Ocean. The beach is backed by the Pleistocene calcarenite dunes rising to 40 m, which are now covered by the southern Southport part of the development. The beach commences immediately north of the reef with a low tide terrace and waves averaging just over 1 m. The waves pick up in height toward the channel and the northern half has a centre beachrock section with a boundary rip, then a 50 m wide bar cut by a central beach rip. In addition there is a strong permanent rip against the training wall, and a small spur groyne off the wall. The spur causes the beach to spiral round to face south against the wall. There is surf on the small reef break at the southern end and *beach breaks* along the bar. The groyne has been built to trap sand at the northern end of the beach. This sand is then periodically pumped under the channel to the northern side, to assist in the longshore sand transport. This will need to be an ongoing process otherwise the northern beaches will be deprived of sand and continue to erode.

Figure 4.163 The Dawesville Channel, training walls and groynes with Pyramids beach, site of the Port Bouvard SLLC, to the south and the Port Bouvard beaches to the north.

The beach is patrolled by the **Port Bouvard Surf Life Saving Club**, founded in 2004. As there are usually a few rips along this beach, particularly towards the northern end, make sure to swim only when patrolled and in the patrol area.

WA 778-781 NORTHPORT-AVALON

No.	Beach	Rating	Type	Length
WA 778	Northport (1)	5	LTT+groynes	100 m
WA 779	Northport (2)	5	LTT+seawall	500 m
WA 780	Avalon	3	R+reefs	1.2 km
WA 781	Falcon	3	R+reefs	1.3 km
WA 782	Falcon Bay (1)	3	R+reefs	500 m
WA 783	Falcon Bay (2)	3	R+reefs	500 m
WA 784	Falcon (N 1)	4	R+reefs	1.3 km
Spring & neap tidal range = 0.4 & 0.0 m				

The northern side of the **Dawesville Channel** has a shorter training wall, just 50 m in length, with a 150 m long groyne located 100 m to the north and a seawall linking the two walls and continuing for 500 m to the north. The training wall has been built to control or train the flow through the channel, while the groyne is designed to presumably stop sand moving southward and into the channel. The seawall is a result of the severe beach erosion that has occurred since the channel and walls were constructed. While sand is pumped under the channel from Pyramids beach, erosion is continuing on the northern side, as it is starved of the sand that would normally move northward along the coast. This sand is part of a beach system that extends northeast for 11 km to Robert Point at Mandurah, where the sand moves around the point and into the next long sediment cell. In between are 13 near continuous beaches (WA 778-790) most separated by low beachrock reefs and calcarenite outcrops. The first seven beaches are located along the 5 km of northeast-trending shore between the channel and Falcon.

Between the northern wall and Falcon, 3 km to the north, are four beaches (WA 778-781). The first two have been heavily modified by the walls and groynes. Beach **WA 778** is a 100 m long pocket of sand wedged in between the large rocks of the training wall and the groyne. Waves averaging just over 1 m break across a 50 m wide bar, with a strong rip running back out the centre. Its neighbour (**WA 779**) varies in size depending on the amount of sand pumping. When small it consists of a wedge of sand in the southern corner between the groyne and backing seawall. During nourishment is may extend for 500 m along the seawall and link with the northern section, which is fronted by a beachrock reef. A road from Northport cuts through the backing dune and calcarenite to reach the rear of beach WA 789 and run out to the training wall.

Avalon beach (**WA 780**) commences on the northern side of the beachrock reef. The reef continues north as the shoreline curves to the east forming a protected 1.2 km long west-facing embayment (Fig. 4.164). Waves break on the reef, which lies up to 200 m offshore, and can produce a *lefthand break*. At the shore low wave to calm conditions usually prevail. The beach is backed by a foredune, then in the south the Northport development then a beachfront car park, while at the northern end is the older Avalon settlement, a car park and boat ramp. The beach width can vary considerable when sand is being pumped across the channel and onto beach WA 779.

Figure 4.164 Avalon beach located to the lee of a discontinous beachrock reef.

Falcon beach (**WA 781**) continues to the north for 1.3 km to the southern side of Falcon Bay. The beachrock reef continues off the beach partially enclosing a second protected embayment containing a low energy reflective beach. A foreshore reserve backs the southern corner, then a foredune backed by houses.

Falcon Bay consists of two adjoining beaches (WA 782 & 783). Beach **WA 782** is a curving 500 m long beach tied to two sand-capped intertidal calcarenite points. Reefs off the point and further out lower waves to less than 1 m at the shore. This is a popular beach with a large car park, reserve and boat ramp at the southern. A low foredune, road then houses back the rest of the beach. It terminates at a sandy foreland on the northern side of which is 300 m long beach **WA 783**. This is a similar curving beach, backed by a low foredune, road and houses in the south and a large active foredune to the north, backed by 200 m wide Rakoa Reserve.

Beach **WA 784** continues to the north for 1.2 km. The beachrock reef hugs the shoreline, with higher waves reaching the shore and breaking either over the reef or up the usually steep beach faces. Eldora Reserve backs the northern end of the beach with a small foredune and road, then houses behind the central 800 m of the beach. It terminates in the north at a 20 m high, 200 m long calcarenite bluff on the northern side of which is beach WA 765.

WA 785-790 **FALCON-ROBERT PT**

No.	Beach	Rating	Type	Length
WA 785	Falcon (N2)	4	R+beachrock	1 km
WA 786	Polleys Hole	4	R+beachrock	1.2 km
WA 787	Leighton Park	3	R+reef	2 km
WA 788	Blue Bay (S)	3	R+reef	800 m
WA 789	Blue Bay	3	R+reef	800 m
WA 790	Halls Head Beach	2	R+sand flats	800 m

Spring & neap tidal range = 0.4 & 0.0 m

The northern 6 km between the northern end of Falcon and Robert Point contains six near continuous beaches (Fig. 4.165), each separated by and most dominated by beachrock reef. Calcarenite bluffs back the beaches and are for the most part vegetated and/or covered by Holocene foredunes. A foreshore reserve runs the length of the shore, with housing increasing north of beach WA 786. These beaches form the western shore of Mandurah.

Figure 4.165 View south from Robert Point along the Blue Bay, Leighton Park and beyond.

Beach **WA 785** is a 1 km long beach bordered by 20 m high bluffs to either end, with dune-draped bluffs in between, then a 1 km wide vegetated reserve between the beach and the Old Coast Road. The beach is generally free of beachrock and has a steep reflective swash zone with waves averaging about 1 m. Beach **WA 786** continues on the northern side of the boundary bluff for 1.2 km to the next bluff-aligned inflection in the shore. Waves average about 1 m and break across beachrock located along the base of the steep beach, an area known as Polleys Hole. A vegetated 10-20 m high calcarenite bluff backs the beach, then the reserve that widens to 1.5 km.

Beach **WA 787** commences on the northern side of the reef-tied inflection and continues straight to the northeast for 2 km to a section of detached beachrock reef. Beachrock outcrops along much of the beach, which receives waves averaging about 1 m. It is backed by a continuous foreshore reserve that covers the 20-30 m high foredune, and then newer Leighton Park subdivision and a water treatment plant set in the dunes. There is a car park and good access at the southern end of the beach.

Beach **WA 788** starts in lee of a beachrock ridge, with a quiescent 'lagoon' for the first 100 m, then continues north for 800 m as steep reflective beach, to the next section of beachrock. The 100 m wide foredune-foreshore reserve continues along the beach, with Halls Head road backing the dune, then houses. A large car park is located at the southern end in lee of the reef.

Blue Bay (beach **WA 789**) commences in lee of the boundary reef. The beach receives low ocean waves averaging less than 1 m, which surge up the steep, narrow beach. Larger waves break over the narrow, continuous bar and patches of shallow calcarenite reef, with a *lefthand break* running off the southern reef. The beach trends northeast of the reef for 800 m, the first 400 m relatively free of rock, the northern 400 m fronted by continuous beachrock. Halls Head road runs along the top of the backing bluffs providing good access and views along the beaches. There is a car park, toilets and access at the southern ends, with the Blue Bay Caravan Park behind the centre of the beach. The road continues round Robert Point to beach WA 790 (Fig. 4.166). Halls Head subdivision backs all three beaches. The beach terminates at low sandy Robert Point, which sits atop an intertidal calcarenite reef.

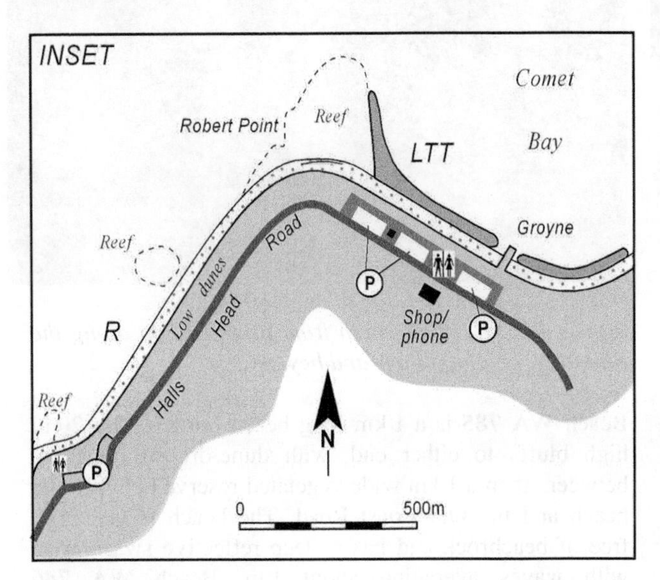

Figure 4.166 Detail of Blue Bay and Halls Head beaches. See Fig. 4.166 location.

Robert Point is a major inflection point in the coast and represents the northern end of the Bunbury-Mandruah 'embayment' and the beginning of the Mandurah-Becher Point 'embayment'. Sand moves northward to the point and accumulates on the 800 m long north-facing **Halls Head Beach (WA 790)** between Robert Point and the Peel Inlet entrance (Fig. 4.167). This beach is protected by the point and its northerly orientation and is usually calm. Because of the low waves, abundance of sand, plus the Peel Inlet ebb tide delta there are extensive ridged sand flats off the beach. The sand accumulates in the western side against the training wall where it is periodically removed to prevent it moving into the inlet entrance. The beach is backed by a foreshore reserve, with large car park, toilets, bar-b-que and picnic facilities, with a shop across the road, and the Mandurah Yacht Club at the eastern end. During higher swell a *lefthand break* runs off Robert Point into the beach.

Swimming: All the Mandurah beaches usually experience low waves and weak current with Robert Point being the most protected and popular. Rips only develop when waves exceed 1 m, particularly on Leighton and Blue Bay beaches. However water is relatively deep off the beach and care should be taken on the steep, surging beaches with young children.

Figure 4.166 Halls Head Beach located between Robert Point and the trained Peel Inlet entrance. This is a popular low energy north-facing beach.

Surfing: Usually only low beach breaks on the more exposed beaches. During larger swell there are good lefts at *Blue Bay* and *Halls Head* in higher swell, and a righthander just south of Blue Bay.

WA 791-794 **MANDURAH-SECRET HARBOUR**

No.	Beach	Rating	Type	Length
WA 791	Mandurah	2	R+sand flats	800 m
WA 792	Silver Sands	4	R	1.6 km
WA 793A	San Remo	4	R/LTT	4 km
Mandurah Surf Life Saving Club				
Patrol period:	**Sunday**		**November-February**	
WA 793B	Mandora	4	R/LTT	2 km
WA 793C	Singleton	5	LTT/TBR	2.5 km
WA 793D	Golden Bay	6	TBR	2.5 km
WA 793E	Secret Harbour	6→4	TBR→LTT	5.7 km
Secret Harbour Surf Life Saving Club				
Patrol period:	**Saturday**		**December-March**	
	Sundays		**October-March**	
Lifeguard:	**Mon-Fri**		**December-January**	
Spring & neap tidal range = 0.4 & 0.0 m				

Mandurah is a growing coastal city of over 40 000 people, located 75 km south of Perth. It straddles the 4 km long inlet that connects Peel Inlet with the ocean (Fig. 4.168). The entrance has been substantially modified by training walls to ensure a stable, permanent opening and a canal estate has been constructed in the former wetlands. The main part of the city lies to the east of the inlet, with beachfront suburbs to the west at Leighton Park and Halls Head, and extending north to Silver Sands. Halls Head and adjacent Robert Point also affords some protection to the inlet mouth, which is located 1 km east of the point. Mandurah is bordered by the two more exposed western beaches of Leighton Park,

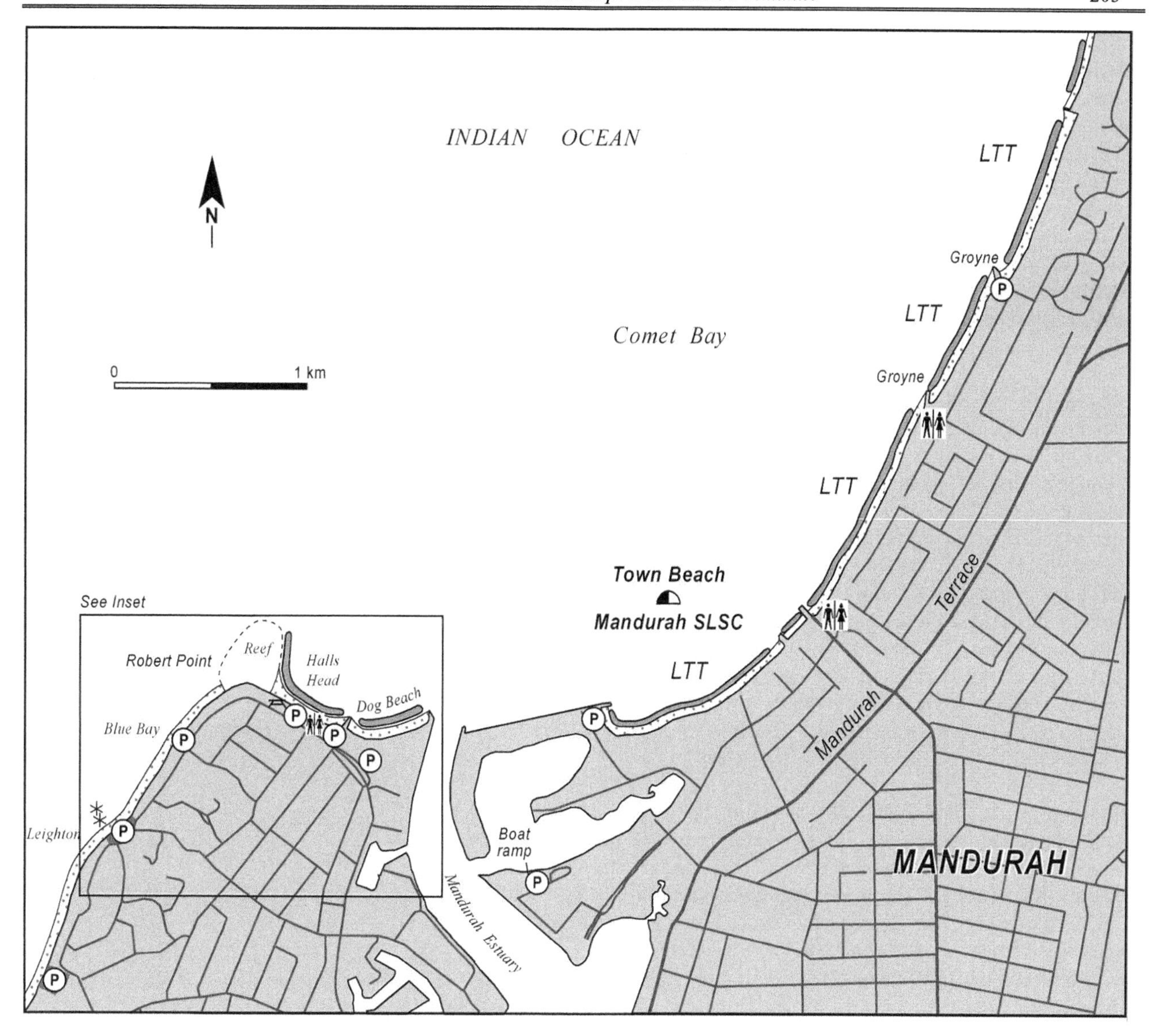

Figure 4.168   Mandurah has near continuous beaches fringing the shoreline. The Mandurah Surf Life Saving Club is located at Town Beach, San Remo.

Halls Head and Blue Bay (WA 788-789), south of Robert Point, the northern protected Robert Point beach (WA 790), and the long beach that begins at the inlet training walls at Silver Sands and continues north for 19 km to Becher Point (WA 791-793). In between are a series of separate beachfront communities of Silver Sands, San Remo, Mandora, Singleton, Golden Bay (formerly Peelhurst) and the newest, developed in the 1990's, called Secret Harbour.

Mandurah beach (**WA 791**) commences on the western side of the Peel Inlet training wall and breakwater that shelters the small boat harbour. The beach faces north-northwest and curves round for 800 m to the first of two rock groynes, located at the ends of Aileen and Henson streets. The beach is sheltered by its orientation, Robert Point, the training walls and the extensive tidal delta, which extends 200-300 m off the inlet. As a consequence the western half of the beach is usually calm with waves only begin to pick up along the eastern section of the beach. The western half is also backed by a wide foreshore reserve built on the reclaimed land, with the beachfront houses of Silver Sands backing the eastern half. The groynes and some seawalls have been built to protect these houses.

Silver Sands beach (**WA 792**) extends from the first groyne for 1.6 km, past one central groyne to the next major rock groyne located at the northern end of the Silver Sands subdivision at Wade Street. A road and beachfront houses back the entire beach. Scarping of the beach associated with natural sand movement and the inlet wall construction has lead to the construction of some seawalls and groynes, which cause the sand to build up on their southern sides, thereby slightly realigning the once straight beach. The beach receives waves averaging about 1 m, which surge up the usually steep reflective cusped beach face. They also generate small rips against the groynes.

A continuous 17 km long beach (WA 793A-E) extends north of Silver Sands to Becher Point (Fig. 4.169). The entire beach faces west however it is protected to varying degrees from ocean swell by a series offshore reefs, which lie up to 10 km offshore and lower the waves at the shore. Throughout, the beach is relatively steep, with a continuous attached bar, which in the more exposed central-northern section is cut by rips every 200 m when waves exceed 1 m. A 5 to 10 m high vegetated foredune backs the beach. The coast road lies 1 to 2 km inland and parallels the coast, with access to the beach at each of the settlements. All access points have a car park and usually toilets, with a Surf Life Saving Club at Secret Harbour.

San Remo Beach (**WA 793A**) is the site of the **Mandurah Surf Life Saving Club.** The beach extends from the northern Wade Street groyne, where it is called Watersun Beach, due north for 4 km, with the San Remo development occupying 3 km of the shoreline and now connected to the Silver Sands development. The beach receives waves averaging about 1 m, which maintain a reflective to at times narrow low tide terrace beach. A foredune reserve of varying width, then road and houses back the beach. There is good access at the central car park, which has a picnic area, playground and toilets.

Mandora Beach (or Mandora Bay Beach **WA 793B**) occupies the next 2 km of shoreline, with wave height picking up to over 1 m. These maintain a low tide terrace, with rips forming during higher wave conditions. The Mandora Foreshore Reserve parallels the beach and consists of 50-100 m wide fenced foredune, with car parks located in the south and centre and houses behind.

Singleton beach (**WA 793C**) extends north for 2.5 km. Waves continue to increase slightly averaging over 1 m and maintaining a continuous 50 m wide bar, usually cut by beach rips every 200 m. These conditions can produce a range of *beach breaks*. The beach is backed by a 150 m wide fenced foredune, with four car parks in the reserve, then a road and the subdivision.

Golden Bay (formerly Peelhurst) (**WA 793D**) occupies the next 2.5 km of the shoreline. This growing subdivision spreads along 1.5 km of the shore, with the 150 m wide Golden Bay Reserve between the houses and the beach. There is a large car park and viewing platform towards the northern end of the development. This section of the bay is more exposed with waves commonly reaching up to 1.5 m. These waves maintain an 80 m wide surf zone, with rips usually spaced every 200 m.

Secret Harbour (**WA 793E**) is the newest of the pocket developments and is located 2 km north of Golden Bay (Fig. 4.170). The beach occupies the remaining 5.7 km of the shore up to Becher Point. The beach also begins to curve round to face west-southwest and southwest at the point. The southern-central section of the beach is the highest energy in the bay with waves averaging up to 1.5 m. They maintain an 80 m wide surf zone with rips regularly spaced along the beach. The development is set well back from the shore, with a golf course, large foreshore reserve with several car parks, and protected foredunes backing the beach. The Golden Bay Surf Life Saving Club (founded in 1981) was moved here in 1995 and renamed the **Secret Harbour Surf Life Saving Club**. The club is located in the centre of the wide Secret

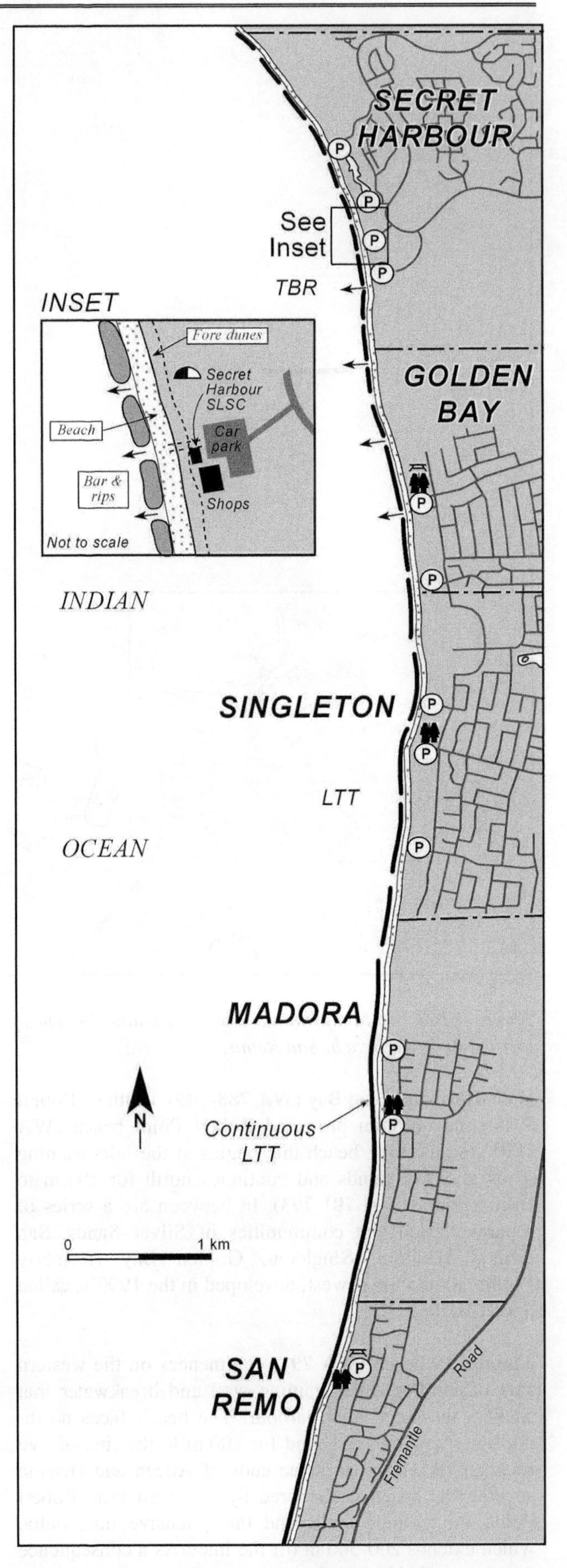

Figure 4.169 The 17 km long beach between San Remo, Madora, Singleton, Golden Bay and Secret Harbour. Wave energy gradually increase sup the beach, with surf and rips along the Secret harbour section.

Harbour Foreshore Reserve and patrols the beach on weekends between October and March. The beach continues for 2 km northwest of the development to terminate at Becher Point. The low sandy point is protected by the Murray Reefs located 2 km offshore and as a result waves decrease towards the point and the bar narrows and rips diminish. The backing low dune ridges between the harbour and the point is a scientific park, extending 2.5 km east to the Warnbro Sound Road.

Figure 4.170 Secret Harbour community with the Secret Harbour SLSC in the foreground and rip-dominated surf.

Swimming: This is a long moderately hazardous beach for swimming, with the major hazard being rips during higher waves. Under normal low wave conditions stay on the attached portion of the bar, close inshore and clear of rips and rip channels.

Surfing: When waves exceed 1 m there are beach breaks the length of the beach, with conditions depending on bar shape and wind direction.

SHOALWATER ISLANDS MARINE PARK

Area:	6545 ha	
Coast length:	20 km	(2139-2159 km)
Beaches:	13	(WA 794-807)

Shoalwater Islands Marine Park encompasses Warnbro Sound and the adjacent Rockingham shoreline around Point Peron to the Garden Island causeway (Fig. 4.171). It extends from 2-7 km offshore to include the chain of islands and reefs between Becher Point, Penguin Island and Point Peron, including Shag and Seal islands.

WA 794-797 **WARNBRO SOUND**

No.	Beach	Rating	Type	Length
WA 794	Becher Pt	1	R+sand flats	250 m
WA 795	Kennedy Bay	1	R+sand flats	1.3 km
WA 796	Warnbro Beach	3→4→3	R→LTT→R	8.3 km
WA 797	Safety Bay	1	R+sand flats	2.1 km

Spring & neap tidal range = 0.7& 0.5 m

Warnbro Sound is a 7 km wide west-facing semi-circular embayment, with a curving 11 km long shoreline. The bay has formed through the growth of two sandy cuspate forelands - Becher Point in the south in lee of Murray Reefs, and Mersey Point in the north in lee of Penguin Island. Between the two is a continuous chain of reefs, which represent the ridge of a drowned Pleistocene barrier that continues north to Cape Peron. As sand moved into the bay during the past 6000 years the shoreline has prograded seaward as much as 7.5 km

Region 4B CENTRAL WEST: Rockingham - Fremantle

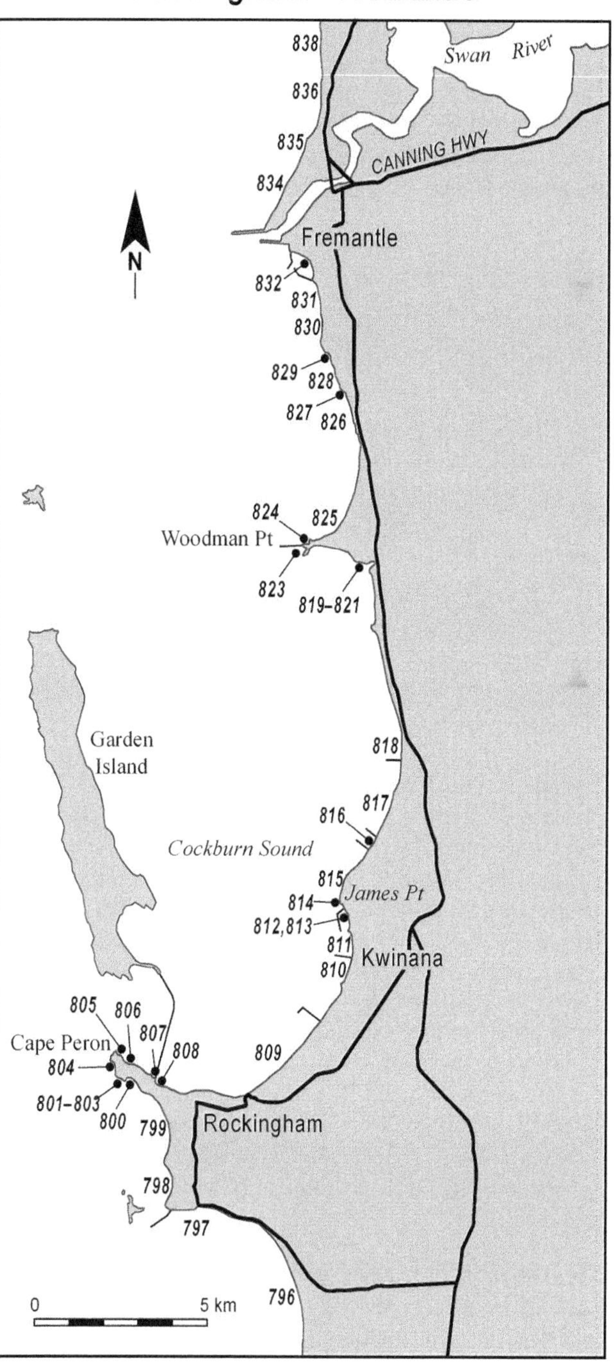

Figure 4.171 Region 4B extends from Rockingham to Fremantle.

through the development of up to 90 shore parallel foredune ridges. The ridges extend south of Becher Point as far as Mandurah and north of Mersey Point into Cockburn Sound. The entire system represents one of the largest areas of Holocene shoreline progradation on the Australian coast. Today however the ridges are rapidly being covered by housing estates, destroying this magnificent record of coastline evolution.

Becher Point (**WA 794**) is a 250 m long west-facing section of the foreland that links the southern beach WA 793, with the Sound beaches (Fig. 4.172). The point itself is a low energy beach backed by a low sandy foredune. The Murray Reefs lower waves at the shoreline to less than 1 m and calms are common. These conditions have permitted extensive sand flats to develop, which extend hundreds of metres off the beach out towards the reef. Seagrass meadows lie beyond the sand flats, and seagrass debris commonly accumulates on the low energy beach. It is backed by the 7 km wide low foredune ridge plain, and accessible by 4 WD tracks through the dunes and along the beach.

Figure 4.172 Point Becher marks the southern boundary of Kennedy Bay and Warnbro Sound.

Kennedy Bay beach **WA 795** extends for 1.3 km northeast from the point to Bridport Point. It is a curving north-northwest facing beach, and forms the southern arm of the sound. Sand flats continue along the beach and low wave to calm conditions usually prevail. The foredune ridges behind the eastern end of the beach were transformed in the late 1990's into a major development, with a gold course and housing estates, including beachfront houses, separated from the shore by a 50 m wide foreshore reserve. A large public car park is located at the western end of the development, towards the eastern end of the beach and a marina and boat harbour is proposed for the centre of the beach. It terminates at a gentle cuspate foreland formed in lee of the inner most sand shoal.

Warnbro Beach (**WA 796**) commences at the sandy foreland and curves round for 8.5 km to the northeast, then north and finally northwest, to terminate at the northern foreland, in lee of the inner northern sand shoal. Wave energy is low at both ends of the beach increasing towards the central section where it averages about 1 m maintaining a steep reflective beach. Towards the northern end the strong summer southwest sea breezes and associated wind waves can form a low tide terrace. The Port Kennedy golf course backs the southern 1 km, with housing development then extending all the way to Safety Bay. All the developments have a foreshore reserve covering the foredune and extending from 50-100 m inland, with the housing estates extending inland. The major developments are at Warnbro in the centre, where the once 500 m wide reserve has been further subdivided, and Waikiki to the north. There is a bike track along the reserve, together with numerous access points and car parks, with a boat ramp in the northern more protected corner.

Safety Bay beach (**WA 797**) extends west of the first cuspate foreland for 2.1 km to Mersey Point, with a second gentle foreland located in the centre (Fig. 4.173). Two shallow sand shoals are linked to both of the forelands, with seagrass meadows in between. The beach faces south with usually calm conditions, except during summer sea breezes conditions when low wind chop reaches the beach and seagrass debris is commonly washed up onto the shore. A foreshore reserve, road, then houses of Safety Bay back the beach all the way to the point. A boat ramp, jetty and yacht club are located in the Safety Bay Foreshore Reserve with boats moored off the jetty. Mersey Point is connected to Penguin Island by a 500 m long shallow sand shoal.

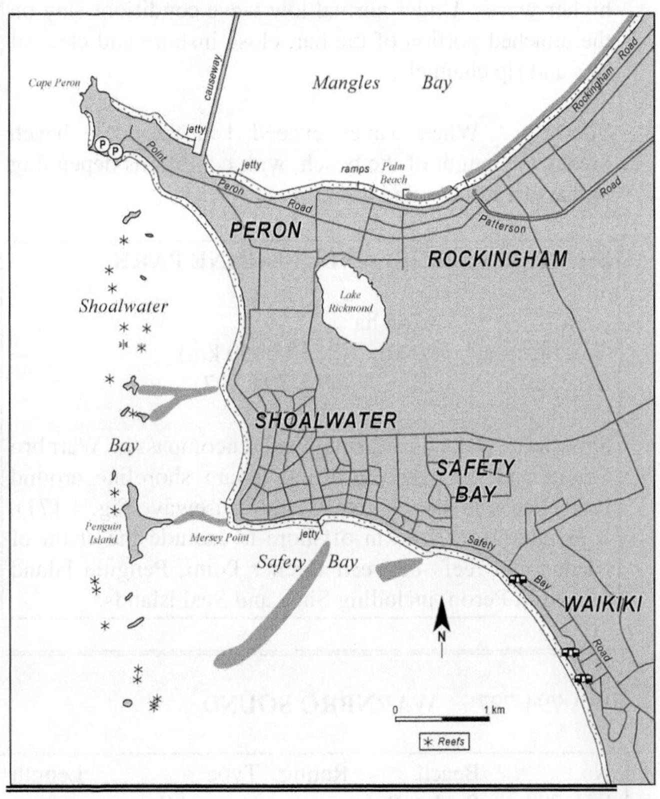

Figure 4.173 The Shoalwater-Rockingham area extends from Waikiki in the south around Cape Peron into Mangles Bay.

WA 798-804 **SHOALWATER BAY-CAPE PERON**

No.	Beach	Rating	Type	Length
WA 798	Shoalwater	3	R	1.5 km
WA 799	Peron	3	R	2.5 km
WA 800	White Rock	3	R	700 m
WA 801	Point Peron (S3)	4	R+reef	60 m
WA 802	Point Peron (S2)	4	R+reef	30 m
WA 803	Point Peron (S1)	4	R+reef	40 m
WA 804	Point Peron	4	R+reef	300 m
Spring & neap tidal range =0.4 & 0.0 m				

Shoalwater Bay is part of the 3 km wide cuspate foreland formed in lee of Penguin Island and a chain of reefs that lie 1-1.5 km offshore and extend 4 km north to Point Peron (Fig. 4.171). The bay as the name suggests is dominated by relatively shallow water, with extensive sand flats extending between Mersey Point and Penguin Island and out to Seal Island and Shag Rock. The near continuous line of reefs and small islands shield the bay from ocean waves, resulting in waves averaging only 0.5 m at the shore. There are three beaches along the bay shore (WA 798-800) (Fig. 4.174) and four located on the south to eastern side of Point Peron (WA 801-804).

Figure 4.174 Mersey Point divides Safety Bay beach (WA 797) from the Shoalwater Bay beaches (WA 798 and 799). All are located to the lee of extensive shallow calcarenite reefs and islets.

Shoalwater beach (**WA 798**) is a curving, west-facing, 1.5 km long low energy beach located between Mersey Point and a central sandy foreland. A shallow sands spit extends 1 km west from the tip of the foreland to Shag Rock and Seal Island. The low waves produce a relatively narrow steep beach, with seagrass meadows just off the shoreline, and the beach covered by seagrass debris following storms. The beach is backed by the 100-200 m wide Shoalwater Foreshore Reserve, with parking towards the southern end, then a road and houses of Shoalwater. Beach **WA 799** continues to curve north of the foreland for 2.5 km to the next gentle foreland located in lee of Bird Island, with a shallow sand ridge running 400 m out to the island. The beach is relatively steep and reflective, with seagrass meadows growing to the shore along much of the beach. It is backed by a 50 m wide foredune, which increases in height to the north, a central car park, road and then the houses of Peron in the south and a 1 km long series of holiday camps in the centre-north off Memorial Drive.

Beach **WA 800** occupies a curving 700 m long south-facing section of shore between the foreland and the first calcarenite of Point Peron. The beach is protected by the islands and shallow offshore reefs, with wave height decreasing to the west. The low waves result in a steep, narrow beach with seagrass debris usually littering the intertidal zone. It is backed by a generally vegetated foredune, with a central access area.

Beach **WA 801** lies immediately to the west and consists of a 60 m long pocket of sand wedged between the southern-most calcarenite outcrops of Point Peron (Fig. 4.175). The beach faces southwest, but is protected by a band on intertidal beachrock and reefs further out. At high tide low waves reach the shore, which is backed by a fenced 20 m high partly active foredune. A blufftop car park overlooks the beach, as well as providing access to the beaches **WA 802** and **803**, which are located 100 m to the west. These beaches consist of adjoining pockets of sand 30 m and 40 m long respectively, separated by a small calcarenite bluff. They are both bordered and backed by low calcarenite bluffs and fronted by a band of beachrock.

Figure 4.175 Point Peron is surrounded by several small sheltered beaches.

Beach **WA 804** extends from the tip of Point Peron north for 300 m to the 10 m high bluff in lee of Mushroom Rock. The beach is steep, narrow and curving, and is bordered by low calcarenite, with a central sandy section, fringed by a low tide beachrock reef. It is backed by a 20 m high vegetated foredune. Walking trials from the cape car parks run through the dunes and to the beach and out to the northern tip of the cape at John Point.

WA 805-809 **POINT PERON-ROCKINGHAM**

No.	Beach	Rating	Type	Length
WA 805	John Pt	4	R+reef	80 m
WA 806	John Pt (E)	2	R+sand flats	1 km
WA 807	Causeway	2	R	50 m
WA 808	Peron (N)	2	R+seagrass	1.7 km
WA 809	Rockingham	3	R/LTT	4.8 km
Spring & neap tidal range = 0.4 & 0.0 m				

Point Peron forms the southwestern boundary of Mangles Bay and the larger Cockburn Sound. The bay encompasses the southern shores of the Sound, which is backed by the city of Rockingham. The Sound extends north for 15 km to Woodman Point, with 10 km long Garden Island, 9 km to the west, forming the western boundary. Rockingham and the surrounding beaches have had a mixed history as port, holiday location and now a rapidly growing southern commuter suburb of Perth with a population of 60 000. Today the city also incorporates the adjoining seaside communities of Peron, Shoalwater, Safety Bay and Waikiki. The communities are located on a wide, low sandy coastal plain and are fronted by nearly 20 km of near continuous low energy beaches, with rocks only outcropping around Point Peron. The bay beaches (WA 805-809) begin at John Point on the northern tip of Point Peron and follow the generally curving north-facing shoreline for 8 km to the Kwinana Beach jetty.

John Point is a 10 m high narrow finger of calcarenite located at the northwestern tip of Point Peron. Immediately east of the point is a north-facing 50 m long strip of high tide sand (**WA 805**) bordered to the east by a vegetated calcarenite bluff, and fronted by 50-100 m wide calcarenite reefs, with sand further out. Waves are usually low to calm. Walking tracks from nearby car parks provide access to the entire cape area. Beach **WA 806** commences immediately to the east and extends for 1 km to the Garden Island causeway, as a double crenulate, low energy, north-facing beach. Extensive shallow reefs and sand flats lie off the western half of the beach, with seagrass extending to the shore along the eastern half. It terminates at a groyne located 200 m west of the causeway. Sand has moved east along the beach accumulating against the groyne causing up to 100 m of shoreline progradation. A central car park and vegetated low dune ridge backs the beach.

Beach **WA 807** is a south-facing 50 m long pocket of sand that has accumulated in lee of the groyne, and occupies part of the boat ramp facility constructed on the western side of the causeway. The channel for the boats runs between the eastern tip of the beach and the causeway. This is a quiet beach with most waves generated by boat wakes. It is backed by a large car park and the Sea Rescue headquarters.

Beach **WA 808** commences on the eastern side of the causeway and curves to the east for 1.7 km to two adjacent boat ramps. The entire beach faces north and has seagrass meadows growing to the shore, with waves usually low to calm. A foreshore reserve back the beach, which contains a yacht club with jetty and launching facilities with boats moored off the beach in the west. Some holiday camps are located behind the central section, and a road and houses back the eastern end of the beach.

Rockingham Beach (**WA 809**, also called Palm Beach) commences at the boat ramps and curves to the east, then northeast for 4.8 km (Fig. 4.176), terminating against the Kwinana jetty breakwater. The beach is backed by the continuous Rockingham Foreshore Reserve, which in the southern central Rockingham area has three jetties and the Cruising Yacht Club of Western Australia. The reserve widens to 200 m along the central-northern part of the beach, where it is crossed by the 700 m long Kwinana grain terminal jetty. Wave energy is low in the south and with boats moored off the yacht club. The waves pick up towards the north as it becomes more exposed to summer southwest sea breeze waves and occasional low swell. The beach consists of a steep reflective high tide beach fronted by a 50 m wide low tide terrace.

Figure 4.176 Rockingham Beach with the city in the background.

Swimming: Swimming is relatively safe along the generally low wave beaches of the three bays. Be careful if wading out on the sand flats at low tide, as people have been caught by the rising tide.

WA 810-818 KWINANA-CHALLENGER BEACH

No.	Beach	Rating	Type	Length
WA 810	Kwinana Beach (S)	2	R	1.3 km
WA 811	Kwinana boat hbr	2	R+sand flats	80 m
WA 812	Kwinana Beach (N)	2	R	600 m
WA 813	James Pt (S 2)	2	R	50 m
WA 184	James Pt (S 1)	2	R	70 m
WA 815	James Pt	2	R	1.2 km
WA 816	James Pt (N)	2	R	2.5 km
WA 817	Powerhouse	2	R	80 m
WA 818	Challenger Beach	2	R	2 km

Spring & neap tidal range = 0.7 & 0.5 m

Rockingham beach used to extend as a continuous 15 km long strip of sand from the north side of Point Peron around and up to Challenger Beach . The southern section is now divided into four beaches (WA 806-809) by the causeway and groynes, while the strip from Kwinana north is separated into eight beaches (WA 810-818) by groynes, jetties and other structures associated with the port and industrial facilities that back most of this section of Cockburn Sound. While most of the beaches here are sheltered by Garden Island and receive low waves, they are generally unsuitable for recreational activities owing to their proximity to the industrial and port facilities. The view north along the beaches is dominated by the high smoke stacks, grain loading facilities and jetties extending out into the Sound.

Kwinana Beach (WA 810) commences on the north side of the breakwater, jetty and boat ramp constructed at Wells Park, which separates it from Rockingham beach (WA 809) to the south. The beach extends north for 1.3 km to southern training wall of the oil refinery small boat harbour. The beach faces due west but is shielded by Garden Island from ocean swell, receiving only low wind waves and occasional low swell. There is a reserve at Wells Park providing a large car park and facilities as well as the fishing jetty and boat ramp. North of the park a fertilizer plant then the Kwinana oil refinery dominate the backshore, with no direct public access.

The **Kwinana boat harbour** consists of two shore parallel groynes spaced 100 m apart and extending 100 m into the sound, to a shore parallel breakwater, which all but enclose the harbour apart from a 30 m wide entrance. Sand has accumulated in the harbour forming a vegetated beach ridge, and curving 80 m long beach (**WA 811**) fronted by a sand flat, all of which half fills the harbour. The outer section is dredged to permit small boats to enter and moor. Vehicle access to the boat harbour is via the oil refinery.

The northern half of Kwinana Beach (**WA 812**) commences on the northern side of the boat harbour and continues north for 600 m to a small calcarenite outcrop on the southern side of James Point, adjacent to the centre of the oil refinery. Beach **WA 813** extends another 50 m from this outcrop to the solid base of the oil refinery jetty, which extends 400 m into the sound. Beach **WA 814** is wedged between the northern side of the jetty and a 200 m long groyne/water intake unit. The three beaches (WA 812-814) all face west-southwest and receive low waves. They are steep and narrow and backed by a strip of low vegetated beach, then the perimeter road and refinery.

Beach **WA 815** commences on the northern side of the long groyne and continues north for 1.2 km, extending to and around James Point. **James Point** is a low sandy foreland formed in lee of Garden Island with sand converging on the point. A series of seven small shore parallel breakwaters were constructed just off the beach and all have been attached to the shore by small salients, dividing the beach into a series of six small curving bays between each of the breakwaters (Fig. 4.177). The beach widens to the south as sand has been trapped by the longer groyne. It is backed by a low grassy dune, which widens to 100 m in the south, then the perimeter road and oil refinery. A water treatment plant is located behind the northern end of the beach, with a water discharge outlet flowing across the beach. The beach usually receives low wave to calm conditions, however during strong sea breeze conditions waves are sufficiently high to generate small rips within each of the embayments.

Beach **WA 816** continues to the north of the last breakwater for 2.5 km, past two 500 m long jetties of a steel mill, to the southern groyne/water inlet of the powerhouse. The beach receives slightly higher swell arriving around Garden Island and these usually maintain a narrow attached bar. It is backed by a 50 m wide grassy dune reserve, then a perimeter road with the steel mill in the south, and power house in the north. The small power station beach (**WA 817**) is an 80 m long strip of sand located between the two 100 m long groynes-water inlets of the power station. Water is taken in and discharged through these structures, which places the beach off-limits to the public. The large power station is located immediately behind the beach. There is access to the centre of the beach at the end of Barter Road.

Figure 4.177 James Point beach is dominated by the series of shore attached breakwaters.

Challenger Beach (WA 818) commences at the northern groyne and continues north past a 600 m long jetty for 2 km to where the beach and coastal plain merges with 10 m high calcarenite bluffs, the first bluffs north of Robert Point, 37 km to the south. This point also marks the northern boundary of the 8,000 ha coastal plain that commences at Mandurah and extends north through Warnbro Sound and Rockingham. The beach receives waves reaching up to 1 m, which maintain a narrow attached bar. It is backed by a 50 m wide grassy dune, then a large aluminum refinery. The Cockburn Road reaches the coast at the northern end of the beach where there is a car park, toilets and boat ramp off Sutton Road. A large linear shack settlement occupies the top of the bluffs to the north.

WA 819-825 WOODMAN POINT

No.	Beach	Rating	Type	Length
WA 819	Jervoise Bay (1)	1	R	400 m
WA 820	Jervoise Bay (2)	1	R+sand flats	100 m
WA 821	Jervoise Bay (3)	1	R+sand flats	80 m
WA 822	Woodman Pt (S)	2→3	R→R+reef	1.5 km
WA 823	Woodman Pt	3	R+reef	300 m
WA 824	Woodman Pt (N)	3	R+reef	400 m
WA 825	Coogee Beach	3	R+reef→R	3.7 km

Coogee Beach Surf Life Saving Club
Patrol period: Saturday November-March
Sunday October-March

Spring & neap tidal range = 0.7 & 0.5 m

Woodman Point is a 4 km long cuspate foreland tied to a low calcarenite reef 2 km west of the Pleistocene shoreline. The salient is composed of a series of low foredune ridges converging on the point. The point was used as an explosive area and quarantine station until the 1970's and has since been redeveloped as Woodman Point Regional Park. It is also the site of the large Woodman Point Caravan Resort and in the north the Coogee Beach Resort. Three beaches are located in the southern Jervoise Bay Boat Harbour (WA 819-821) and four surround the point (WA 822-825).

Beaches WA 819-821 are all located within the extensive breakwaters that were completed during the 1990s' to form the **Jervoise Bay boat harbour** (Fig. 4.178). The harbour is the site of a number of small ship building operations along its southern shore, with the three beaches occupying the northern corner. Beach **WA 819** commences at the northern end of the port works and curves to the north then northwest for 400 m. It is sheltered by the harbour walls and only receives low wind waves. It is backed by partly active foredune rising to 10 m in places, then the Cockburn Road. Beaches WA 820 and 821 are located either side of the two boat ramp areas, and are also bordered by groynes and a seawall. Beach **WA 820** is 100 m long, faces south and has a groyne to the east and the rocks of the boat ramp to the west. Beach **WA 821** is a similar 80 m long beach located on the western side of the boat ramps. They are sheltered by the harbour walls and only receive low wind waves. Both beaches are fronted by sand flats, which are occasionally dredged for navigation. Boats are common on and just off the beaches. A large car and boat trailer park, boat storage, the Cockburn Power Boat Association and Sea Rescue are all located on the flat area behind the beaches.

Figure 4.178 Jervoise Bay boat harbour and boat ramp, which includes beaches WA 819, 820 and 821.

Woodman Point beach (**WA 822**) occupies most of the southern side of the point. It is a curving, southwest to south-facing 1.5 km long low energy beach that extends from the harbour wall to low calcarenite outcrops on the point. In the east it is a sandy reflective beach, while to the west intertidal calcarenite reef extend 100-300 m of the shore. It is backed by a low grassy foredune, then the Woodman Point Road that runs out to the point.

Beach **WA 823** is located on the tip of the point and consists of a 300 m long sandy beach fronted by 100 wide calcarenite flats. It is bordered by a groyne at its southern end and the 600 m long Wapet groyne or Woodman Spit to the north. Seagrass debris is commonly piled up on the narrow low energy beach. Beach **WA 824** extends north of the groyne for 400 m to a rubble seawall that protects a shore facility, with a 100 m long jetty located at the northern end of the beach. The beach receives low waves and is fronted by patchy calcarenite reef and shallow sand flats.

Coogee Beach (**WA 825**) commences on the northern side of the seawall and gently curves to the north for 3.7 km. It forms the northern shoreline of the foreland and at its northern point reconnects with the inner calcarenite of the old Pleistocene shoreline called James Rocks. The southern part of the beach is within the regional park, with the Jervoise Bay Sailing Club located at the very southern end. It is then backed by a vegetated reserve and the old Quarantine station. The reserve continues to a small jetty backed by a recreational area with all facilities, a camping area and the old munitions depot, and finally the Coogee Beach Resort and a northern recreational jetty, with a terminal groyne just south of James Rocks. The beach is patrolled by the **Coogee Beach Surf Life Saving Club** founded in 2003.

WA 826-833 **SOUTH BEACH-FREMANTLE**

No.	Beach	Rating	Type	Length
WA 826	James Rocks	2	R	700 m
WA 827	Power station	2	R	50 m
WA 828	South Beach (1)	2	R	1.2 km
WA 829	South Beach (2)	2	R	300 m
WA 830	South Beach (3)	2	R	800 m
WA 831	South Beach (4)	2	R	400 m
WA 832	Boat Harbour	1	R	140 m
WA 833	Bathers Beach	3	R	250 m
Spring & neap tidal range = 0.7 & 0.5 m				

From James Rocks the shoreline trends north for 5.5 km to the southern training wall of Fremantle Harbour. In between is a once continuous sandy shore that has been heavily modified by structures and development when the area was backed by industry. Since the 1990's its backshore has undergone a transformation from a power station, railway yards and meat works, to in part, the Catherine Point Reserve, the South Beach Recreational Reserve and a large fishing and small boat harbour to the north. All the beaches (WA 826-832) face west and are partly sheltered by a string of reefs, including Carnac Island, that extend north of Garden Island. The reefs lies 5-6 km offshore and lower ocean waves to less that 1 m at the shore.

Beach **WA 826** commences on the northern side of the small groyne that separates it from Coogee Beach (WA 825). It initially trends north past the low calcarenite bluffs of James Rocks, then opens up with a backing 10 m high partly active foredune, for a total length of

700 m. It terminates at the seawall on the southern side of the old power station. The beach is narrow and relatively steep, with usually low waves. Beach **WA 827** is a 50 m pocket of sand wedged in between the power station seawall and a groyne built for the water intake. It faces southwest and receives low waves.

Beaches WA 828-830 are part of the **Catherine Point Reserve**. Beach **WA 828** extends north of the power station groyne for 1.2 km to the next groyne, which has induced a 100 m offset in the shoreline, as sand has accumulated on its northern side. The beach is backed by a low grassy dunes reserve, with access to the southern end off Robb Road. The reserve has been redeveloped as a passive recreation area, which includes picnic and bar-b-que areas and historic sites, including the old Robb Jetty Abattoir.

Beach **WA 829** commences on the northern side of the groyne offset and trends north for 300 m to the next groyne, which is located on a slight protrusion in the sandy shore called Catherine Point. Access to this beach is via Rollinson Road, which leads to a large car park. Beach **WA 830** continues north of the groyne for 800 m to the next groyne. This area is backed by a continuous foreshore reserve with extensive recreational facilities and a kiosk and large car park at the southern end. Beach **WA 831** forms the northern section of sand. It continues for another 400 m to the southern rock walls of Success Harbour. A low narrow grassy dune, then industrial development backs the beach, with car parking behind the southern groyne. All three beaches receive waves less than 1 m high and usually have a moderately steep reflective beach face.

Beach **WA 832** is located inside Success Harbour. It used to be a continuation of South Beach, but has been enclosed within the harbour. It consists of a 150 m long strip of sand, backed by a grassed strip, and backed and bordered by car parks. Jetties of the Fremantle Sailing Club and scores of boats lie at anchor off the protected beach.

Bathers Beach (WA 833) is a 300 m long pocket of sand located between the northern walls of the boat harbour and the southern training wall of Fremantle Harbour (Fig. 4.179). It faces west and receives waves averaging about 1 m. A rock structure in the centre almost cuts the beach in two, with the city of Fremantle immediately behind.

Figure 4.179 Bathers Beach is located between the southern training wall for Fremantle and the newer Success Harbour breakwater.

ROTTNEST ISLAND

Rottnest Island is an 11 km long, 1-5 km wide, east-west trending island, located 20 km due west of Fremantle. The island is the crest of a drowned Pleistocene calcarenite barrier system, and is located toward the outer edge of the calcarenite reefs and islands that extend north from Garden Island. While these reefs lower waves at the coast and on the lee eastern shores, higher waves are received on the more exposed southern and western shore of the island. The island's 36 km of shoreline is dominated by the calcarenite bluffs and reefs, which produce numerous bays and breaks in the shore, resulting in a total of 63 beaches (Fig. 4.180). The beaches range in length from 20 m to 1.3 km. There is one town on the island at The Settlement, the main landing site, as well as holiday accommodation in Geordie and Longreach bays, and an educational camp on Phillip Point. The remainder of the island is accessible by foot and bike, but offers no facilities. Private cars are prohibited on the island. The island is extremely popular holiday and day-tripper destination. It offers a wide range of attractive pocket beaches, numerous reefs and some good reef surf. It is probably the most popular and one of the most attractive islands on the Australian coast.

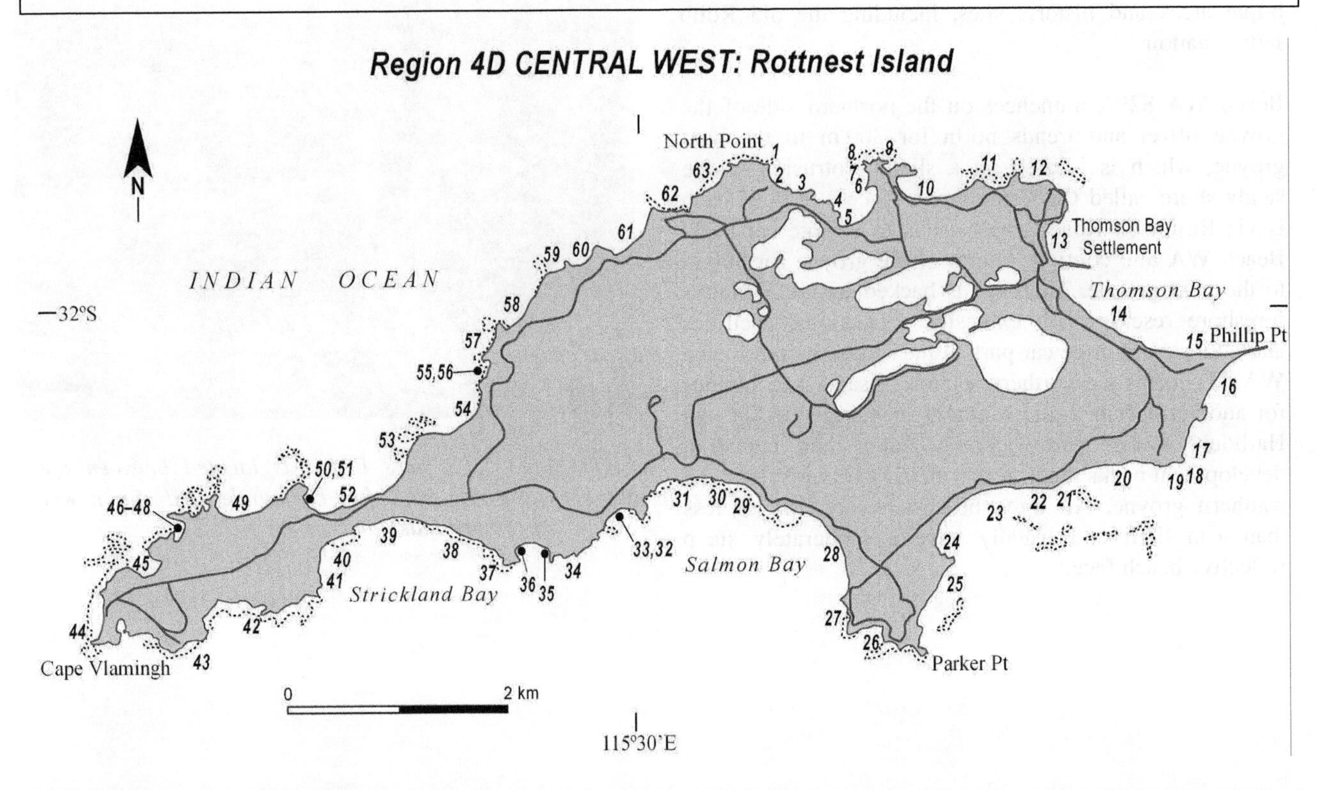

Figure 4.180 Rottnest Island is located 20 km offshore of Fremantle. The calcarenite, dune-capped island is surrounded by 63 generally small sheltered beaches.

ROTTNEST IS 1-6 **PARAKEET & GEORDIE BAYS**

No.	Beach	Rating	Type	Length
RI 1	Parakeet Bay	2	R+reef	300 m
RI 2	Little Parakeet Bay	2	R+reef	70 m
RI 3	Little Geordie Bay	2	R+reef	20 m
RI 4	Geordie Bay (W)	2	R+reef	60 m
RI 4	Geordie Bay	2	R+reef	450 m
RI 4	Geordie Bay (E1)	2	R+reef	50 m
RI 5	Geordie Bay (E2)	3	R+platform	60 m
RI 6	Geordie Bay (E3)	3	R+platform	30 m
Spring & neap tidal range = 0.4 & 0.0 m				

North Point is a 20 m high dune-capped calcarenite bluff that forms the northern tip of Rottnest Island and the western boundary of 600 m wide **Parakeet Bay**. Within the northeast-facing bay are two low energy beaches (RI 1 & 2). Parakeet Bay beach (**RI 1**) curves gently for 300 m between 20 m high bluffs. It is steep reflective beach fronted by patchy seagrass over a sand seafloor and backed by a scarped 10-20 m high foredune, with a small blowout at the eastern end. A car park is located behind the centre of the foredune. Little Parakeet Bay beach (**RI 2**) lies 50 m to the east and is a 70 m long pocket of sand bordered by platform fringed bluffs, with a 40 m open section of sand, then reefs further out. A car park overlooks the beach. Little Geordie Bay beach (**RI 3**) occupies a gap in the calcarenite bluffs, 100 m to the east, with a walking track from the car park to the top of the beach. It is a 20 m long pocket of sand bordered by the 15 m high bluffs, with reef across the front of the sandy beach.

Geordie Bay lies 300 m to the east and is a curving 400 m wide, northwest-facing bay, bordered by two 200-300 m long, 10-20 m high arms of calcarenite. The bay is a popular anchorage with numerous permanent moorings in the bay. Seagrass meadows extend to the usually calm shore. Beach **RI 4** is located towards the base of the western arm and is a 60 m long pocket of sand, bordered and backed by 15 m high calcarenite bluffs, with the road running along the top of the bluffs. Some rocks outcrop on the low energy beach with seagrass growing to the shore.

The main Geordie Bay beach (**RI 5**) is a curving 450 m long sandy beach that links the calcarenite bluffs. Seagrass grows to the shore and conditions are usually calm. The sandy beach is relatively steep and backed by a 5 m high foredune, then two rows of holiday apartments and an 80 m long jetty crossing the eastern end of the beach (Fig. 4.181). Beaches **RI 6, 7** and **8** are three pockets of sand 50 m, 60 m and 30 m long respectively, located on the eastern side of the bay, along the western shore of Point Clune. Each is bordered and backed by sloping vegetated bluffs rising to 15 m, and fronted by rock platforms and reefs, the reefs continuing most of the way across the 300 wide mouth of the bay.

Figure 4.181 Geordie Bay with Point Clune in the foreground.

RI 9-12 LONGREACH BAY

No.	Beach	Rating	Type	Length
WA 9	Fays Bay	2	R+reef	150 m
WA 10	Longreach Bay	2	R+reef	900 m
WA 11	The Basin	2	R+platform	200 m
WA 12	Pinky Beach	2	R+reef	250 m
Spring & neap tidal range = 0.4 & 0.0 m				

Longreach Bay is located on the northeastern tip of the island, 1 km north of The Settlement. The 1.5 km wide bay faces north, with a string of reefs almost linking across the bay mouth . Three beaches lie within the bay (RI 9-11) with Pinky Beach (RI 12) located on the eastern point.

Fays Bay beach (**RI 9**) is a 150 m long sandy beach bordered by 10 m high calcarenite bluffs on the western point of the Longreach Bay. Reefs and a sea stack extend across the mouth of the small embayment, as well as some rocks outcropping close to shore. The beach is backed by a 10 m high foredune, then walking tracks linking to adjoining Geordie and Longreach bays.

Longreach Bay beach (**RI 10**) us a curving 900 m long beach tied to a 20 m high calcarenite bluff in the west and a series of reefs in the east, with near continuous reefs lying 100-200 m off the beach. Waves are usually low to calm at the shore, which is backed by a 5-10 m high foredune, then holiday apartments on the western dunes. There is a narrow navigation channel through the reefs, which boats negotiate to moor of the western end of the beach. The reef reaches the shore at the eastern end, where a 100 m long detached calcarenite bluff separates it from The Basin.

The Basin (**RI 11**) is a popular picnic and swimming spot, backed by four Norfolk Island pine trees, shelters, a picnic area and toilet facilities. The narrow 200 m long beach is wedged in between a low calcarenite bluffs, including a protruding central section and an intertidal platform with several large tidal pools (Fig. 4.182). There is sand then reefs beyond the platform. The platform and outer reefs resulting usually calm conditions at the shore.

Figure 4.182 The Basin is a popular sheltered beach fronted by reefs and tidal pools.

Pinky Beach (**RI 12**) is located immediately to the east and is a 250 m long northeast-facing beach, with the Bathurst Lighthouse located above the southern end of the beach, and water treatment plant behind the northern end. It is bordered by calcarenite bluffs and backed by a grassy foredune. Scattered rocks, reefs and seagrass meadows lie off the low energy beach.

RI 13-15 THOMSON BAY

No.	Beach	Rating	Type	Length
RI 13	Settlement Beach	1	R+sand flats	500 m
RI 14	Hotel Beach	1	R+sand flats	1.3 km
RI 15	Kingston (N)	2	R	900 m
Spring & neap tidal range = 0.4 & 0.0 m				

Thomson Bay is the site of The Settlement, the sole town on the island and the centre of all ferry arrivals, activities, most accommodation and facilities. The 2 km long bay faces northeast and is bordered by Bathurst Point to the west and low sandy Phillip Point to the east. The once continuous bay beach has been divided into three (RI 13-15) by groynes and structures built across the beach.

The Settlement beach (**RI 13**) commences east of the low calcarenite rocks of Bathurst Point and occupies the first 500 m of the beach This is a sheltered section of the bay with seagrass growing almost to the shore and shallow sand flats between the beach and meadows, together with seagrass debris occasionally accumulates on the upper beach. Rows of holiday houses and cabins are located on the point and behind the beach, with the main Settlement commercial area behind the eastern end of the beach, which terminates at the main 300 m long ferry jetty. Conditions are usually calm at the beach, while a small jetty and swimming nets are located along the beach.

The Hotel beach (**RI 14**) extends from the eastern side of the jetty for 1.3 km terminating at a boat ramp. This is a similar low energy beach, with seagrass growing to the shore and numerous boats moored off the beach, as well as two smaller jetties crossing the western end. In the west it is backed by the large beachfront hotel, then holiday accommodation, with a grassy reserve along the eastern half of the beach.

Beach **RI 15** commences at the groyne protected boat ramp and continues east for 900 m to the narrowing sandy tip of Phillip Point (Fig. 4.183). Wave energy increases slightly along the beach and the seagrass lies 30-50 m off the shore. There are some calcarenite bluffs exposed half way along the beach, with a 10 m high foredune to either side. The land bordered by Phillip Point used to be the Kingston army base and is now the Kingston Environmental Education Centre.

Figure 4.183 View from Phillip Point along Hotel beach to The Settlement.

RI 16-24 PHILLIP-PARKER POINTS

No.	Beach	Rating	Type	Length
RI 16	Bickley Bay	2	R+seagrass	600 m
RI 17	Bickley Pt	3	R+reef	400m
RI 18	Jubliee Rocks (1)	2	R+seagrass	150 m
RI 19	Jubliee Rocks (2)	2	R+seagrass	150 m
RI 20	Patterson Beach	2	R+seagrass	400 m
RI 21	Henrietta Rocks (1)	3	R+reef	100 m
RI 22	Henrietta Rocks (2)	3	R+reef	200 m
RI 23	Porpoise Bay	2	R+seagrass	1 km
RI 24	Porpoise Bay (S)	2	R+seagrass	200 m
RI 25	Pocillopora Reef	2	R+seagrass	600 m
Spring & neap tidal range = 0.6 & 0.5 m				

The southeastern side of the island extends for 4 km from low sandy Phillip Point to calcarenite bluffs of Parker Point. In between are Bickley and Porpoise bays , sections of calcarenite bluff and nine low energy beaches (RI 16-25). All the beaches face southeast and are sheltered by Parker Point and extensive calcarenite reefs and islets extending up to 2 km offshore. Seagrass meadows grow close inshore off most of the beaches. The circum-island road runs along behind the shore provide good access to all the beaches.

Bickley Bay beach (**RI 16**) commences at the tip of Phillip Point, the eastern tip of the island and gently curves to the southwest for 1 km to the 20 m high calcarenite bluffs of Bickley Point, which has a World War 2 pillar-box on the tip of the point. The beach receives low waves in the north, which decrease down the beach enabling the seagrass to reach the shore along the southern half. As a result seagrass debris usually piles up on this section of the beach. The beach is backed 10-20 m high vegetated bluffs and then the Kingston Environmental Education Centre complex. A 100 m long section of calcarenite bluffs separates it from beach **RI 17**, which continues south for 400 m to Bickley Point. This is a narrow reflective beach located in lee of shallow reefs, which link with Twin Rocks Reef and Wallace Island. It is backed by dune-draped 10 m high bluffs.

Beaches RI 17,18 and 19 occupy a 1 km long south-facing section of shore known as Patterson Beach (Fig. 4.184). Beaches **RI 18** and **19** lie immediately west of Bickley Point along a section of rocky shore called Jubliee Rocks. They are two adjoining 150 m long pockets of sand bordered by 20 m high calcarenite bluffs, with grassy transgressive dunes rising to 30 m behind the beaches, with a pillar box on the highest dune. **Patterson Beach** (**RI 20**) continues to the west for another 400 m between bordering 20 m high calcarenite bluffs. It is backed by dunes rising to 40 m. All three beaches receive usually low wave to calm conditions, with seagrass growing to within 20 m of the shore. Dyer Island lies 500 m off the main beach.

Figure 4.184 Patterson beach consists of three pockets of sand fronted by seagrass meadows.

Henrietta Rocks lies off a 400 m long rocky section of shore that separates Patterson beach from Porpoise Bay. Two pockets of sand (RI 20 & 21) are located between the bluffs and the rocks. Beach **RI 21** is a 100 m long section containing four pockets of sand, each bordered by calcarenite bluffs and fronted by a discontinuous intertidal platform. Beach **RI 22** lies immediately to the west and is a 200 m long more continuous pocket of sand fronted by a near continuous, but irregular platform up to 50 m wide, with the remains of a wreck on the outer end of the platform. The main island road runs right behind both beaches, with a small parking area above the second beach providing a view of the wreck.

Porpoise Bay (**RI 23**) continues west of the beaches for 1 km, curving round towards the south. The beach is sheltered by offshore reefs with seagrass growing to the shore and its debris usually accumulating along the beach. The beach is backed by 10-20 m high vegetated bluffs, with the island road running along the top. Beach **RI 24** is located at the western end of the bay as the shoreline turns and trends southeast. It is a narrow 200 m long strip of sand wedged in between the backing 20 m high bluffs and shallow inshore reefs.

Pocillopora Reef is a prominent reef located 500 m off beach **RI 25**. The reef shelters the beach as well as providing a popular anchorage, with a few boats usually moored to the lee of the reef. The beach extends along the shore for 600 m between a 20 m high northern point and Parker Point, the southern most tip of the island. There is an active blowout behind the centre of the beach, which has climbed up onto the backing bluffs. A spur road off the main road runs out to a car park and toilets just south of the blowout, with steps leading down to the beach. During higher waves there is a left-hand surf break over *Pocillopora Reef*, also known as *Chicken R*eef.

RI 26-31 SALMON BAY

No.	Beach	Rating	Type	Length
RI 26	Little Salmon Bay	2	R+reef	60 m
RI 27	Salmon Pt	3	R+platform	80 m
RI 28	Salmon Bay	3	R+reef	1.1 km
RI 29	Fairbridge Bluff (1)	6	R+platform	400 m
RI 30	Fairbridge Bluff (2)	6	R+platform	100 m
RI 31	Salmon Bay (W)	7	TBR+reefs	400 m
RI 32	Nancy Cove	3	R+platform	300 m
RI 33	Green Island	3	R+platform	60 m
Spring & neap tidal range = 0.4 & 0.0 m				

Salmon Bay is a 3 km wide south-facing embayment that occupies the centre of the southern shore of the island. It is bordered to the east by Salmon Point with Kitson Point to the west. Between the points is 4.5 km of curving shoreline dominated by calcarenite bluffs, platforms and reefs, together with eight sandy beaches (RI 26-33). The island road follows the coastline and provides good access to all the beaches.

Little Salmon Bay (**RI 26**) is located 400 m east of Salmon Point in a 60 m wide gap in the bordering 20 m high calcarenite bluffs which continue east for 500 m to Parker Point. The steep reflective beach is fronted by 80 m wide shallow sandy seafloor, then patchy reefs and rocks, which lower waves to less than 1 m at the shore. The island road clips the rear of this popular swimming spot (Fig 4.185).

Figure 4.185 Little Salmon Cove is a sheltered popular swimming spot.

Beach **RI 27** is located 300 m north of Salmon Point and is an 80 m long crenulate strip of high tide sand backed by jagged calcarenite rocks and fronted by a mixture of sand and intertidal platforms. The platforms extend 50-100 m off the beach lowering waves at the shore. Out on the reef is a righthand surf break known as *Salmon Point.*

Salmon Bay beach (**RI 28**) is the longest beach on the south coast. It faces southwest and curves for 1.1 km between calcarenite bluffs. It consists of a moderately steep white sand reflective beach fringed with a 30 m strip of sandy seafloor, with seagrass meadows and patchy reef extending well into the bay, thereby reducing waves at the shore to less than 1 m. The western 100 m are fronted by a 50-100 m wide platform. Well vegetated foredune backs the beach and the island road runs along the rear providing good access to this popular swimming beach.

Fairbridge Bluff is an 800 m long section of dune-draped 20 m high calcarenite bluffs, which is fronted by irregular intertidal platforms extending up to 50 m seaward. Gaps in the platform permits waves averaging over 1 m to reach the shore in places. Between the platform and the bluff are two beaches (RI 29 & 30). Beach **RI 29** is a continuous 400 m long strip of high tide sand fronted by the platform with wave breaking over the reef and reaching the shore in places. The island road runs along the top of the bluffs providing views of the beach. Beach **RI 30** lies immediately to the west and is a 100 m long strip of sand bordered by two projecting sections of the bluffs. Platforms surround the base of both small headlands with an open section in between. Waves break along the platform to reach the shore and form a permanent rip as the water returns seaward. These are two hazardous beaches owing the higher waves, reefs and rips.

The western side of Salmon Bay consists of a 400 m long beach (**RI 31**), which extends from western side of Fairbridge Bluff to the beginning of a south-trending section of rocky shore. The beach has a fringing strip of beachrock along its base, with a sand bar beyond resulting in waves breaking over the bar before they reach the beachrock. This offers the only sandy surfing spot on the island and is a popular surfing location. However a permanent rip runs out against the rocks at both ends of the beach, which together with the beachrock requires caution if swimming at this beach. The beach is backed by vegetated slopes rising to 20 m, with access points at either end. The Rottnest Lighthouse is located 500 m north of the beach on Wadjemup Hill.

Nancy Cove is a 300 m wide, small southeast-facing embayment located 400 m southwest of Salmon Beach. The semicircular-shaped cove has a curving 300 m long sandy shoreline (**RI 32**), which is bordered, by a platform-fronted headland to the north and a sea stack to the south. The platforms and reefs lower waves to less than 1 m at the shore while vegetated 10-20 m high sand-draped bluffs backed the beach. A small jetty, connected to the shore by a rock groyne, crosses the southern end of the beach with a toilet block behind the jetty and dinghies sometimes pulled upon the beach. A navigation sign helps small boat navigate the channel between the platforms to reach the beach and jetty. **Green Island** is the name of the sea stack off the southern end of the beach, and to its lee is a 60 m long pocket of sand (**RI 32**). The beach is wedged between the base of vegetated 10 m high bluffs and a 50 m wide platform, with a small rock pool in the centre.

RI 34-42 KITSON PT-RADAR REEF

No.	Beach	Rating	Type	Length
RI 34	Kitson Pt	3	R+platform	30 m
RI 35	Mary Cove (E)	3	R+platform	50 m
RI 36	Mary Cove	3	R+reef	25 m
RI 37	Strickland Bay (1)	7	R+platform	250 m
RI 38	Strickland Bay (2)	7	R+platform	150 m
RI 39	Strickland Bay (3)	5	R+reef	400 m
RI 40	South Pt (1)	3	R+platform	200 m
RI 41	South Pt (2)	3	R+platform	80 m
RI 42	Wilson Bay	5	R+platform	80 m
RI 43	Radar Reef	7	R+platform	150 m

Spring & neap tidal range = 0.4 & 0.0 m

The southwestern shore of the island extends from Kitson Point west for 4.5 km to the western tip at Cape Vlamingh. In between is a predominately rocky section of shore dominated by calcarenite platforms and reefs and backed by bluffs rising between 20-30 m. In amongst the rocks are 10 generally small beaches, all protected to some degree by the rocks and reefs. The island road runs out to the cape with access to, or spurs off to, most of the beaches.

Kitson Point is a narrow spur of calcarenite, which forms the eastern boundary of Mary Cove. Tucked in on its eastern side below the bluffs is a low energy 30 m long pocket of sand (**RI 34**) fronted by shallow intertidal platform, with reefs further out. There is access to the bluffs above the small beach.

Mary Cove is a 300 m wide semicircular-shaped, south-facing embayment, bordered by platform-fringed Kitson Point and a western point, together with shallow reefs within the bay (Fig. 4.186). As a consequence low wave to calm conditions normally prevail at the shore. Beach **RI 35** is a 50 m long pocket of sand tucked in below the bluffs on the eastern side of the bay. It is fronted by a 100 m wide mixture of irregular platform and tidal pools. The main cove beach (**RI 36**) is a 250 m long sandy beach backed for 200 m by a vegetated foredune and bluffs, then narrowing as raised beachrock backs the beach. Patchy rocks, reef and sand front the beach and fill the centre of the bay. A walking track terminates above the western end of the beach.

Strickland Bay is a 1.6 km wide south-facing bay, located immediately west of Mary Cove. The bay receives some higher waves on the outer reefs and along the eastern shoreline. There are five beaches (RI 37-42) located along the bay shore. Beach **RI 37** is a 250 m long high tide beach wedged between a dune-draped 20 m high bluff and irregular 50-200 m wide platform. Waves up to 1.5 m high break over the platform and reach the beach at high tide. The Mary Cove track terminates above the western end of the each. Beach **RI 38** lies 200 m to the west and is a 150 m long narrow strip of high tide sand, with backing dune-draped bluffs, and a 70 m wide platform containing two large tidal pools along the beach. Waves break over the outer end of the platforms and

Figure 4.186 Mary Cove is a sheltered cove containing two small beaches.

reach the pools at high tide. A 300 m long walking track starts at the island road T-junction and runs out to the beach. Both beaches offer *reef breaks* off the shore.

The main Strickland Bay beach (**RI 39**) is a 400 m long sandy beach that forms the southern side of Narrowneck, an isthmus-like connection, which links Conical Hill and Cape Vlamingh with the main body of the island. The beach faces south but is protected by inshore and offshore reefs, with low waves at the shore. It has generally sandy seabed close to shore, which makes it a more suitable swimming location under low wave conditions. It is backed by 10-20 m high foredune, with the island road at the rear of the dune.

Beach **RI 40** is located 300 m to the southwest out towards South Point. It is a 200 m long, curving southeast-facing sandy beach, backed by densely vegetated slopes rising to 20 m, and fronted by a generally open sandy seafloor, with bordering rocky points and reefs. Beach **RI 41** lies 100 m to the south and consists of three pockets of narrow sand strips totalling 300 m in length. Each beach is backed and bordered by 20 m high vegetated bluffs and fronted by a continuous intertidal platform with reefs further out resulting in low waves at the shore.

Wilson Bay lies 800 m west of South Point. Beach **RI 42** is an 80 m long strip of high tide sand located in the centre of the bay, below the 200 m long walking track off the island road. The beach is backed by steep, vegetated 40 m high bluffs and fronted by a 100-150 m wide intertidal platform, with waves breaking over reefs off the platform. There are some tidal pools in the platform just east of the beach.

Radar Reef lies at the southern most point on the southwestern tip of the island. The exposed reef is a 100-200 wide intertidal platform, with relatively high waves breaking along the outer edge to form the lefthand *RadarReef* break. Beach **RI 43** consists of a narrow 150 m long strip of sand wedged in between the base of steep 40 m high vegetated bluffs and the platform. A spur off the island road terminates on the bluffs above the western end of the beach, with a walking track down to the shore.

RI 44-48 CAPE VLAMINGH-MABEL COVE

No.	Beach	Rating	Type	Length
RI 44	Cape Vlamingh	7	R+platform	120 m
RI 45	Eagle Bay	3	R+reef	250 m
RI 46	Mabel Cove (1)	3	R+platform	60 m
RI 47	Mabel Cove (2)	2	R+reef	250 m
RI 48	Mabel Cove (3)	3	R+platform	70 m
Spring & neap tidal range = 0.4 & 0.0 m				

Cape Vlamingh marks the southwestern tip of the island. From the Cape the shoreline trends roughly northwest for 2 km to Celia Rocks. In between is a rock and reef-dominated section of exposed shore containing five small beaches (RI 44-48). The island road terminates at the cape with spurs off the road providing good access to all the beaches.

Beach **RI 44** lies at the foot of the cape. It is a 120 m long high tide sand beach that commences beneath 20 m high bluffs and terminates at the more exposed rocks of the tip of the cape. It has a small dune along the base of the bluffs, while a continuous 70 m wide intertidal platform fronts the beach. Waves averaging about 1.5 m break on the edge of the platform. The right-hand *Bullet Reef* break lies off the beach.,

Eagle Bay is a 200 m wide, northwest-facing, semicircular-shaped bay located 600 m northeast of the cape. The bay is surrounded by 20 m high calcarenite bluffs, with platforms and reefs off each point. The reefs lower waves to about 0.5 m inside the bay, where a narrow 250 m long beach (**RI 45**) hugs the base of the bluffs. Sandy seafloor, patchy reef and seagrass fill the shallow bay floor.

Mabel Cove is located 300 m to the northeast and is another 150 m wide semicircular bay surrounded by calcarenite bluffs (Fig. 4.187). There are three beaches in the cove. Beach **RI 46** is located on the western point and consists of an 80 m long pocket of high tide sand, backed by the bluffs and fronted by a flat 50 m wide intertidal platform. The cove car park is located above the beach. Beach **RI 47** lies immediately to the east in the centre of the cove shore. It is a curving 150 m long reflective beach, fronted by a sandy bay floor, with waves averaging about 0.5 m. Beach **RI 48** lies 50 m to the north on the northern side of the bay. It is an 80 m long pocket of high tide sand, with two layers of beachrock, the lower a 30 m wide intertidal platform. Only the central beach is relatively free of rocks and more suitable for swimming.

Figure 4.187 Mabel Cove is a semi-circular cove containing three sheltered beaches.

RI 49-54 **MARJORIE-STARK BAYS**

No.	Beach	Rating	Type	Length
RI 49	Marjorie Bay	2	R+reef	600 m
RI 50	Rocky Bay (1)	3	R+rocks/reef	100 m
RI 51	Rocky Bay (2)	3	R+rocks/reef	80 m
RI 52	Lady Edeline Beach (W)	3	R+rocks/reef	600 m
RI 53	Lady Edeline Beach (E)	3	R+rocks/reef	650 m
RI 54	Stark Bay	3	R+rocks/reef	800 m
RI 55	Stark Bay (N 1)	3	R+platform	60 m
RI 56	Stark Bay (N 2)	3	R+reef	100 m
RI 57	Crayfish Pt	3	R+reef	300 m
Spring & neap tidal range = 0.4 & 0.0 m				

At Celia Rocks the shoreline turns and trends to the east, then northeast for 3 km to Crayfish Rocks. In between is a generally north-facing, low energy sandy shoreline containing nine beaches (RI 49-57). The low energy beaches are protected by their orientation, as well as extensive rocks and reefs lying off all the beaches. The Cape Vlamingh road runs behind the beaches with good access to the central three beaches (RI 52-54), and a 200-300 m walk to reach the eastern (RI 55-57) and western beaches (RI 49-51).

Marjorie Bay is a curving 800 m wide bay bordered by Celia Rocks to the west and Abraham Point in the east. In between is a 600 m long sandy beach (**RI 49**) bordered by the bluffs, rocks and reef of the point, but with a generally rock-free shoreline, with seagrass growing to the shore. However rocks and reefs partly fill the outer bay resulting in low wave to calm conditions at the shore. The beach is backed by an irregular foredune, with a walking track from the road reaching the centre of the beach.

Abraham Point marks the western boundary of **Rocky Bay**, a 1.2 km long north to northwest-facing curving bay, with rocks, reefs and seagrass meadows filling much of the bay floor. Four near continuous beaches occupy the bay shore with beaches RI 50 and 51 are located in lee of the point. Beach **RI 50** is a 100 m long beach bordered by the Rocky Point in the west and a low sand-draped bluff in the east. It has a patch of open water in the west with intertidal rocks and reef along the remainder. An unstable foredune rising to 30 m backs the western end of the beach. Beach **RI 51** continues on immediately to the east and is an 80 m long strip of high tide sand, fronted by continuous intertidal rocks and reefs, including some jagged calcarenite rocks forming the eastern boundary. Both beaches are usually calm.

Lady Edeline Beach commences immediately to the east and curves initially for 600 m to a small calcarenite outcrop. The western section (beach **RI 52**) faces north across inner rocks, reefs, and seagrass meadows. The eastern beach section (**RI 53**) continues to curve to the northeast for another 650 m to the centre of the sandy foreland that forms the boundary with Stark Bay. This section of the beach is clear of inshore rocks, with seagrass meadows dominating the shallow bay floor, and rocks and reefs further out. Crayfish pots dot the bay and boats sometimes anchor off the beach. Both beaches form the northern side of the 150 m wide isthmus that links with cape area with the island. The cape road reaches the shore at the boundary between the two sections and continues along right behind the western beach.

Stark Bay is a 650 m wide curving northwest-facing, embayment, containing a continuous 800 m long beach (**RI 54**). The southern sandy foreland has formed in lee of an inshore reef to which it is attached. The centre of the bay is more open with patchy seagrass growing close to shore, while reefs again lie off the northern side of the bay. The beach has an area of active dunes extending up to 300 m inland in lee of the foreland, with vegetated dunes behind the reminder. A spur off the island road leads to the northern end of the beach. During bigger swell a very heavy *left hand break* forms over the reefs in the bay.

Immediately north of Stark Bay is a 600 m long section of rock-dominated shore containing three beaches (RI 55-57). Beach **RI 55** is a 60 m long pocket of sand, backed and bordered by 5-10 m high calcarenite bluffs and fronted by a 50 m wide intertidal platform, containing a tidal pool. Beach **RI 56** lies 50 m to the north and is a 100 m long sandy beach, also backed and bordered by the bluffs, but with an open sandy seafloor bordered by reefs. These beaches lie 100 m north of the Stark Bay access track with no formal track to either beach.

Beach **RI 57** is a 300 m long west-facing beach that terminates in the north at Crayfish Point. It is backed by low dune-draped bluffs, and fronted by shallow reefs and some rocks towards the north. Waves are usually low to calm at the shore. A walking track off the island road reaches the northern end of the beach.

RI 58-63 RICEY BEACH-NORTH POINT

No.	Beach	Rating	Type	Length
RI 58	Ricey Beach	3	R+reef	600 m
RI 59	City of York Bay (1)	3	R+platform	80 m
RI 60	City of York Bay (2)	3	R+reef	250 m
RI 61	Catherine Beach	2	R+reef	600 m
RI 62	Little Armstrong Bay	2	R+reef	80 m
RI 63	North Pt (S)	5	R+reef	100 m
Spring & neap tidal range = 0.4 & 0.0 m				

The northwest section of Rottnest Island extends for 3 km from Crayfish Point to North Point the northern tip of the island. This is a northwest-facing relatively protected section of coast, dominated by calcarenite points and reefs, with bluffs increasing to the north. Six beaches (RI 58-63) lie along the shore with good access to four (RI 58, 60-62).

Ricey Beach (**RI 58**) is a crenulate 600 m long sandy beach that trends northeast from Crayfish Point (Fig. 4.188). The crenulations are induced by shallow reefs, which lie off the southern-central section and attach to the northern part of the beach. There is a walking track down to the beach from the road and a clear sandy central section for swimming, which is backed by a blowout extending about 50 m inland.

Figure 4.188 Ricey Beach has a crenulate shoreline induced by the adjoining reefs.

City of York Bay, site of an 1899 shipwreck, contains two pockets of sand bordered by small rocky points. The southern beach **RI 59** is a straight 80 m long strip of sand fronted by a 20 m wide intertidal platform. Immediately north is 250 m long beach **RI 60**, which curves between the two low boundary rocks and 30 m high Charlotte Point. The beach is steep and reflective beach with a generally sandy seafloor. Steep 40 m high vegetated bluffs back most of the beach, with a spur off the main road running down to the rear of the main beach.

Catherine Beach (**RI 61**) commences on the northern side of Charlotte Point and curves to the northeast for 600 m to a low northern foreland that terminates at a beachrock point (Fig. 4.189). A series of reefs lie 200 m off the beach resulting in low wave to calm conditions at the shore. Crayfish pots dot the bay and boats sometimes moor inside the reef. A low vegetated foredune backs the beach and the island road skirts the southern end of the beach providing good access.

Figure 4.189 Catherine Beach is a low energy beach sheltered by extensive inshore reefs.

Little Armstrong Bay (**RI 62**) lies 400 m to the east and consists of an 80 m long pocket of sand bordered by steep 30 m high calcarenite bluffs. Shallow reefs extend several hundred meters off the beach resulting in usually calm conditions at the shore. A circuit off the main road runs out to the top of the beach, which is a popular swimming spot.

Beach **RI 63** is a narrow, irregular 100 m long strip of high tide sand and located beneath 20 m high bluffs, 200 m southwest of North Point. The beach is fronted by a continuous platform that widens from 30 to 50 m towards the north, with reefs extending further out. There is no formal access to this beach, which can best be reached along the rocks from Little Armstrong Bay

PERTH BEACHES

Length of coast:	32 km	(2190-2222)
Number of beaches:	31	(WA 834-865)
Length of beaches:	28.9 km	

The Perth coastline extends essentially due north from Rous Head at the entrance to Fremantle Harbour and the Swan River to Mullaloo (Fig. 4.190). This is a near continuous section of sandy shoreline broken into 31 beaches by calcarenite outcrops and bluffs, together with several groynes and breakwaters.

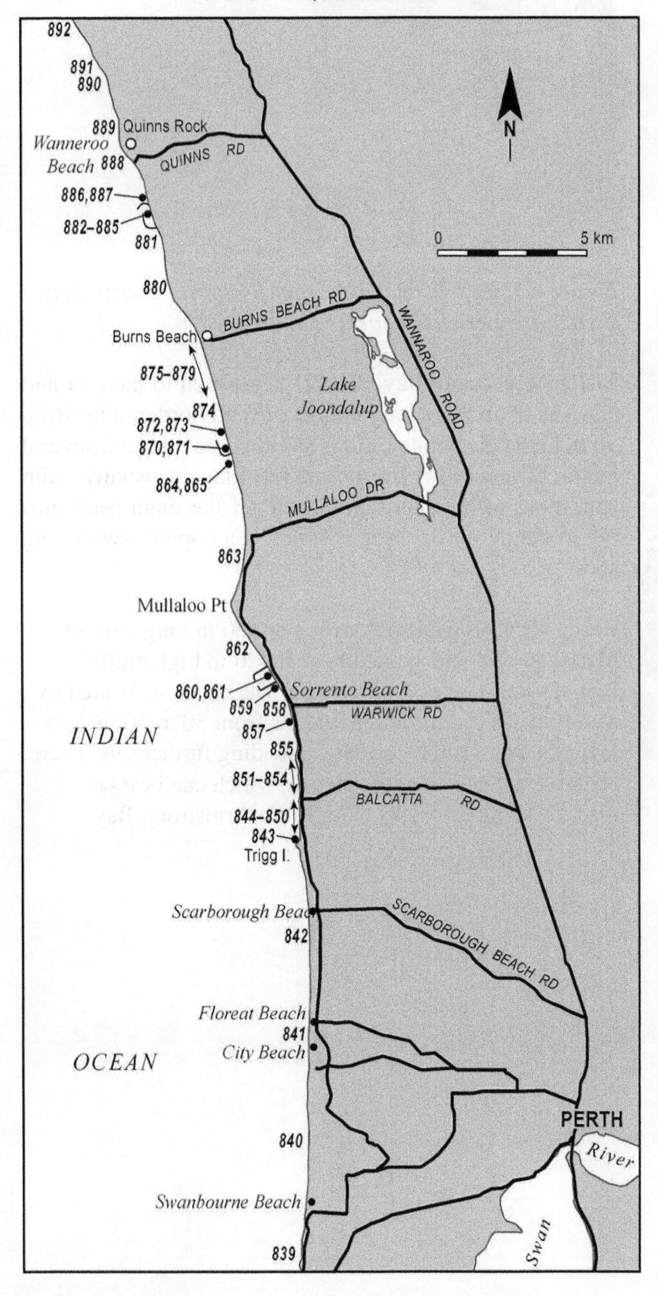

Figure 4.190 Region 4C, Perth to Quinns Rock, beaches WA 839-882

WA 834-838 NORTH FREMANTLE

No.	Beach	Rating	Type	Length
WA 834	Rous Head	4	R+seawalls	100 m
WA 835	Port Beach	4	LTT	1.25 km
WA 836	Leighton Beach	4	LTT	1.5 km
Fremantle Surf Life Saving Club				
Patrol period:	**Saturday**		**October-March**	
	Sunday		**November-March**	
WA 837	Cable Station	5	R+reef	700 m
WA 838	Dutch Inn	5	R+reef	800 m
Spring & neap tidal range = 0.7 & 0.5 m				

North Fremantle is a port suburb that lies wedged between the Indian Ocean and the Swan River mouth. Harbour entrance moles (training walls) and harbour facilities have modified the southern 1.5 km of what used to be a continuous sandy beach, including a new beach that has accumulated against the harbour walls (WA 834). A 2.5 km long beach now runs north from the end of the harbour walls to rocky bluffs below the Vlamingh Memorial. The southern end of the beach is called Port Beach (WA 835), the remainder Leighton and Mosman beaches (WA 836) (Fig. 4.189). Roads leading to the port facilities parallel the back of the beaches, and there are a number of car parks, which provide good access the length of both beaches. North of Leighton Beach calcarenite bluffs and reefs dominate the shore, with a narrow strip of sand divided by a central groyne into two narrow rocky beaches (WA 837 & 838).

Rous Head beach (**WA 834**) is a 100 m long wedge of sand that has accumulated between the two northern North Fremantle port structures. The beach is bordered by the protruding walls and backed by a straight seawall. The beach is located in the southern corner, with its length depending of the width of the beach. It is located within the port facilities and is not a public beach.

Port Beach (**WA 835**) commences on the northern side of the port facilities and gently curves to the north for 1.2 km to the centre of the sandy foreland (Fig. 4.191). The beach receives waves averaging about 1 m and occasionally higher and usually has an attached low tide bar, which is cut by rip during and following periods of higher waves. Higher waves also produce a beach break called *Sandtracks* towards the southern end of the beach.

This is a popular beach with good access and a large car park surrounding a central kiosk and restaurant, and a lifesaving watchtower.

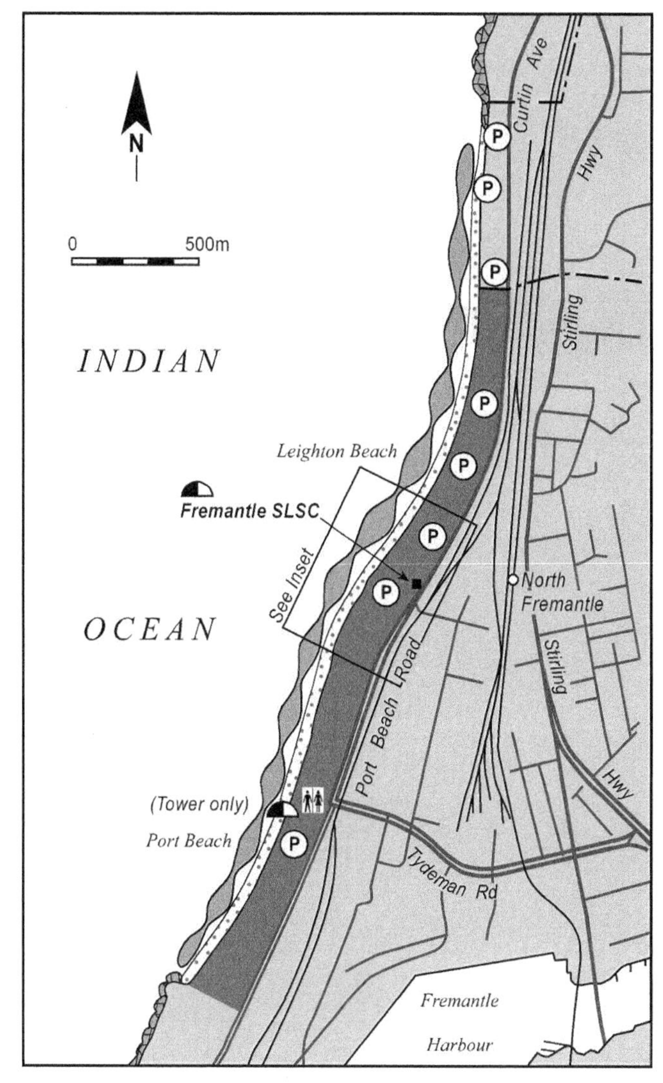

Figure 4.191 Port and Leighton are adjoining beaches, which receive sufficiently high waves to maintain a series of usually smaller rips. See Fig. 4.193 for insert.

Leighton Beach (WA 836) continues on from the centre of the foreland for another 1.5 km to the north. The northern section is also called **Mosman Beach** and terminates at the start of the bluff and rock outcrops. Like Port Beach waves average about 1 m and the low tide bar (Fig. 4.192) and occasional rips continue up to the rocks. There are beach breaks along here during higher swell. The **Fremantle Surf Life Saving Club** has been patrolling Leighton beach since 1935, and also maintains the annex and patrol tower overlooking Port Beach (Fig. 4.193). A separate Port Surf Life Saving Club operated at Port Beach between 1958 and 1976, when it was amalgamated with Fremantle SLSC, with the watchtower continuing to provide lifesaving services. Leighton Beach has a large car park around the surf club, which is flanked by a kiosk and changing rooms, with more car parks north of the surf club (Fig. 4.191).

Beaches WA 837 and 838 are two narrow beaches that extend for 1.5 km north of Leighton Beach. Beach **WA 837** is a 700 m long strip of high tide sand, wedged in between the base of 10 m high calcarenite bluffs and the fronting reefs that extend about 100-200 m off the beach. The reef produces some surf, and is the site of Australia's first surfing reef constructed in 2000, and known as *Cable Stations*. The beach terminates at the 50 m long rock groyne that separates it from beach **WA 838**. This is a similar 800 m long strip of sand, which continues north to the southern side of the Cottesloe groyne. The reef continues north, with breaks over the reefs known as *Seconds Left* and the *Dutch Inns*. Marine Parade runs along the top of the bluff providing good access to both beaches.

Swimming: Port and Leighton beaches are both patrolled and have a relatively low hazard rating under normal low wave conditions. Port with slightly lower waves is considered the least hazardous beach. However waves larger than 1 m will produce a number of rips along the beach and deeper rip channels.

Figure 4.192 Leighton beach and Fremantle SLSC, with typical narrow bar cut by small rip channels.

Surfing: Port and Leighton usually provide low beach breaks, with most surfers heading for the breaks over the *Cable Station* reef and off the northern *Dutch Inns* bluffs, below the Vlamingh memorial.

Summary: These two beaches bordered by port development, but have very good access, parking and facilities, and are usually less crowded, than some of the northern Perth beaches.

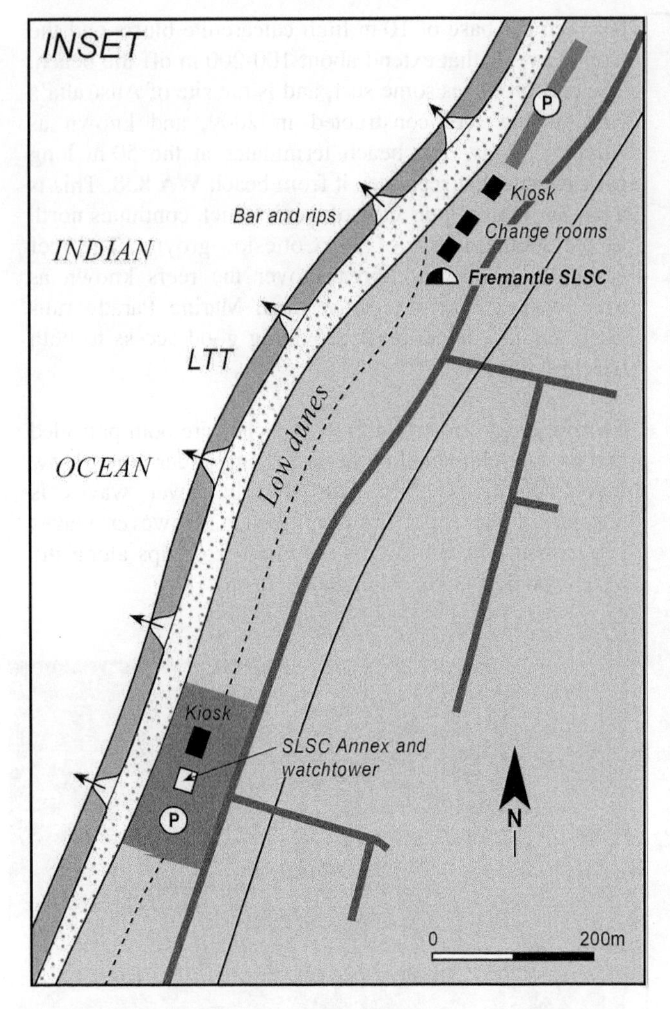

Figure 4.193 Fremantle SLSC patrols Leighton Beach.

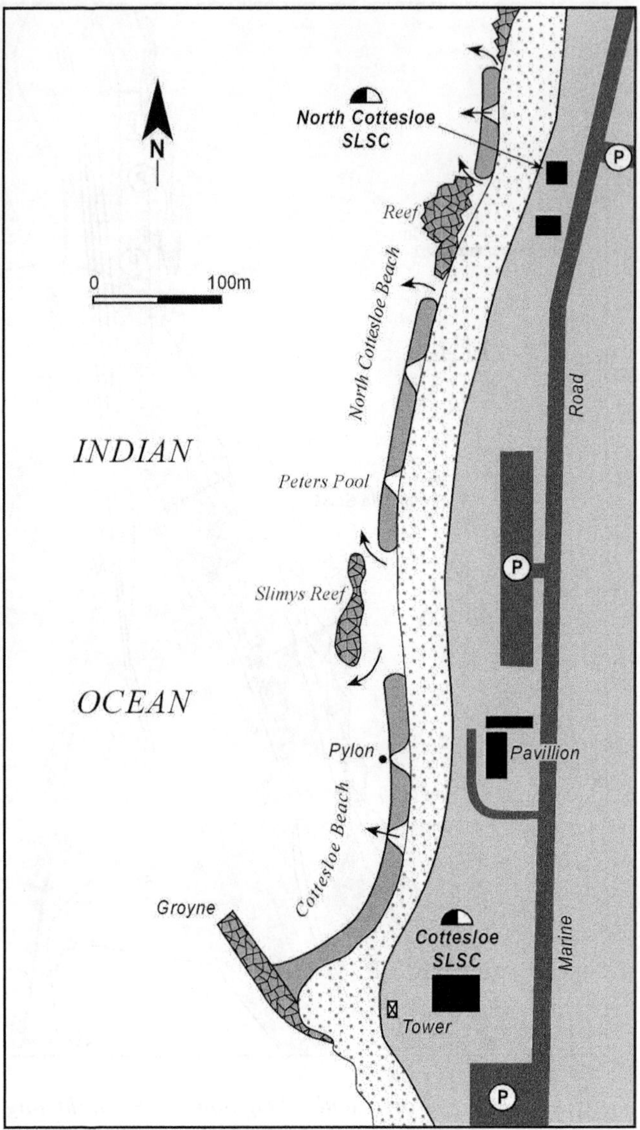

Figure 4.194 Cottesloe and North Cottesloe occupy a continuous sandy beach interrupted by some rock outcrops.

WA 835 COTTESLOE BEACH

No.	Beach	Rating	Type	Length
WA 835A	Cottesloe Beach	3	R/LTT	500 m
WA 835B	North Cottesloe	3	R/LTT	1 km

Cottesloe Surf Life Saving Club

Patrol period:	**Saturday**	**October-March**
	Sunday	**October-March**
Lifeguard:	**Mon-Fri**	**October-April**

North Cottesloe Surf Life Saving Club

Patrol period:	**Saturday**	**October-March**
	Sunday	**November-March**

Spring & neap tidal range = 0.7 & 0.5 m
Figure Cottesloe map 4.194

Cottesloe Beach is a 1.5 km stretch of west-facing sand and bluffs, which extends north from Mudurup Rocks to the southern rocks of Swanbourne Beach (Fig. 4.195). A 100 m long groyne has been built out across Mudrup Rocks, and defines the southern boundary. The beach was one of the first developed on the Perth coast. Following the opening of the Perth to Fremantle railway in 1880 people began walking the 1 km from Cottesloe railway station to the beach. As a result of its early popularity a 'caretaker' patrolled the beach as early as 1906. In 1909 is became the birth place of surfing and Surf Life Saving in Western Australia, with the establishment of the Cottesloe Surf Life Saving Club, with the North Cottesloe club following in 1912.

The entire beach is backed by Marine Parade. Between the Parade and the beach is the Cottesloe Surf Life Saving Club and a car park on the southern bluffs above the groyne, with an equipment building on the beach. Cottesloe patrols the southern half of the beach (**WA 839A**). Six hundred meters to the north is North Cottesloe Surf Life Saving Club right off the Parade, on the edge of the bluffs and overlooking the northern half of the beach (**WA 829B**). The Cottesloe end has a series of car parks, together with a grassy reserve between the parade and beach and a large beach pavilion. North Cottesloe has more limited parking and consequently is usually a less crowded beach.

Figure 4.195 View from Cottesloe to North Cottesloe, with typical low wave conditions.

While the beach runs continuously from Cottesloe to North Cottesloe it is interrupted by the groyne at the southern end and rocks and reefs north of the large the central car park and at either end of North Cottesloe beach. The more protected southern corner usually has lower waves and weak currents and is popular with families and children. Waves average about 1 m up the beach and usually maintain a steep beach with an attached 50 m wide bar. During winter and following higher waves rips usually commence about 100 m up the beach together with permanent rips against the reefs. The reefs in particular induce stronger currents and deeper rip channels and should be avoided by swimmers.

Swimming: Cottesloe is one of Perth's less hazardous beaches under normal low wave conditions. However conditions can rapidly turn hazardous with strong onshore winds and higher waves, which induce both a heavy shorebreak and strong rips.

Surfing: Cottesloe has usually narrow bar and a low beach break. During larger winter wave conditions the groyne has a right break on the south side called the *Cove*, and a larger left, called *North Cott Left*, running around the end of the groyne. Further up the beach is a left reef break called *Seconds*.

Summary: Colttesloe is Perth's oldest and still most popular surfing beach, and is patrolled by two surf life saving clubs.

WA 840-841 **SWANBOURNE-CITY**

No.	Beach	Rating	Type	Length
WA 840	Swanbourne Beach	3	R/LTT	5 km
WA 841	City Beach	4	LTT+groynes	500 m

Swanborne Surf Life Saving Club
Patrol period: Saturday November-March
Sunday October-March

City of Perth Surf Life Saving Club
Patrol period: Saturday October-March
Sunday October-March
Lifeguard: Mon-Fri December-February

Spring & neap tidal range = 0.7 & 0.5 m

The rocks at the southern end of Swanbourne Beach mark the beginning of a 12 km stretch of straight west-facing beach that terminates at Trigg Island. The continuous sandy beach is only interrupted in the centre by the two rock groynes at City Beach, dividing it into three beaches (WA 440-442). These three beaches contain five Surf Life Saving Clubs at Swanbourne, City, Floreat, Scarborough and Trigg beaches and represent the most heavily utilised section of the Perth coast (Fig. 4.196).

This section of the coast was originally backed by continuous sand dunes, extending in places a few kilometres inland, hence the endearing name 'sand gropers'. The presence of the dunes both restricted access to, and delayed the development of, the beaches. The gradual establishment of the Surf Life Saving Clubs indicates the growing popularity and development of this stretch of coast. City of Perth SLSC was the third city based club established in 1924, followed by Scarboro SLSC, which patrols Scarborough Beach, in 1928. Next was the southern Swanbourne-Nedlands in 1932, Floreat, originally called North City, in 1948, and finally Trigg Island in 1954.

Today much of the dune area behind the beaches is covered with commercial and residential development, with only the 2 km of dune area between Swanbourne and City beach, still in a relatively natural state, and occupied by the army's Campbell Barracks. Elsewhere the West Coast Highway backs the northern beaches between City Beach and Trigg Island, and there is good road access to Swanbourne.

The 12 km of beach runs relatively due north, with the only interruptions being the City Beach groyne. Waves average less than 1 m along the southern half, but increase in height to about 1 m along Scarborough and Trigg Island. These northern two beaches are the most hazardous on the Perth coast, accounting for 75% of all rescues in the Perth region.

Swanbourne Beach (**WA 840A**) begins at the rocks that separate it from North Cottesloe and runs due north to midway into 'No Mans Land' (WA 840B) the undeveloped dune area north of the clubhouse in front of

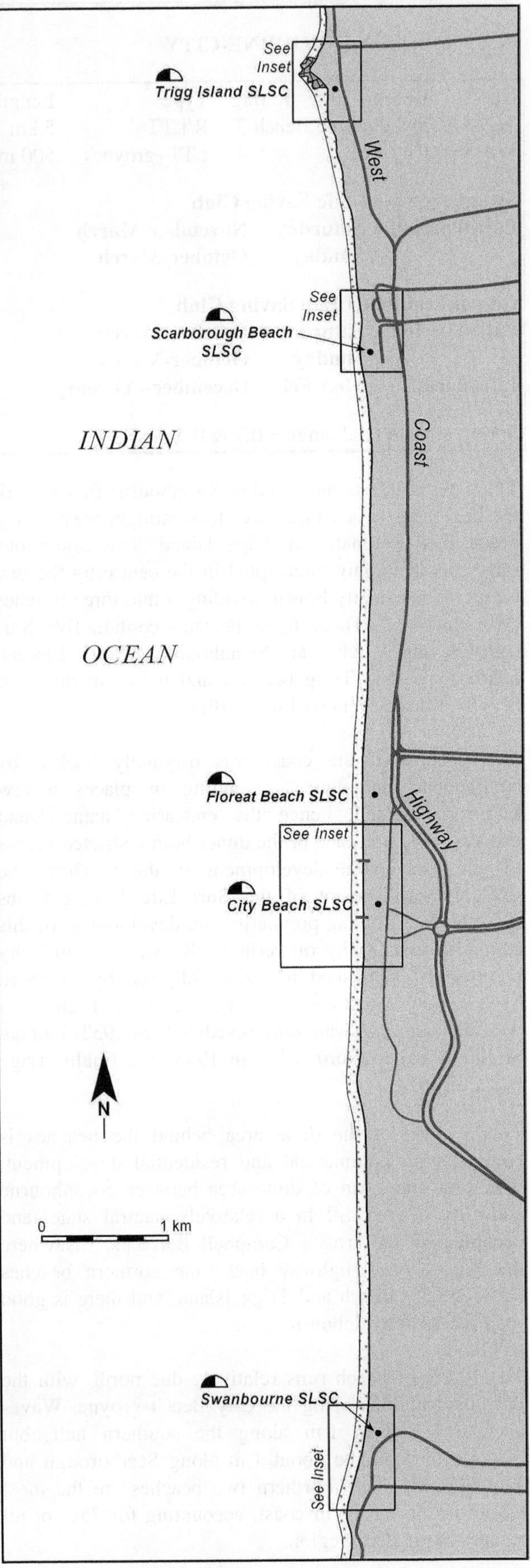

Campbell Barracks. The Swanbourne-Nedlands Surf Life Saving Club is located 500 m north of the rocks. The development of the beach followed the construction of a limestone road to the beach in 1930 with the Surf Club formed in 1932. Today the beach has a large surf club, car park and patrol tower (Fig. 4.197 and 4.198).

Figure 4.196 The straight 12 km long beach between Swanbourne and Trigg Island, also includes City, Floreat and Scarborough beaches and Surf Life Saving Clubs. See Figs. 4.197, 199, 202 and 203 for inserts.

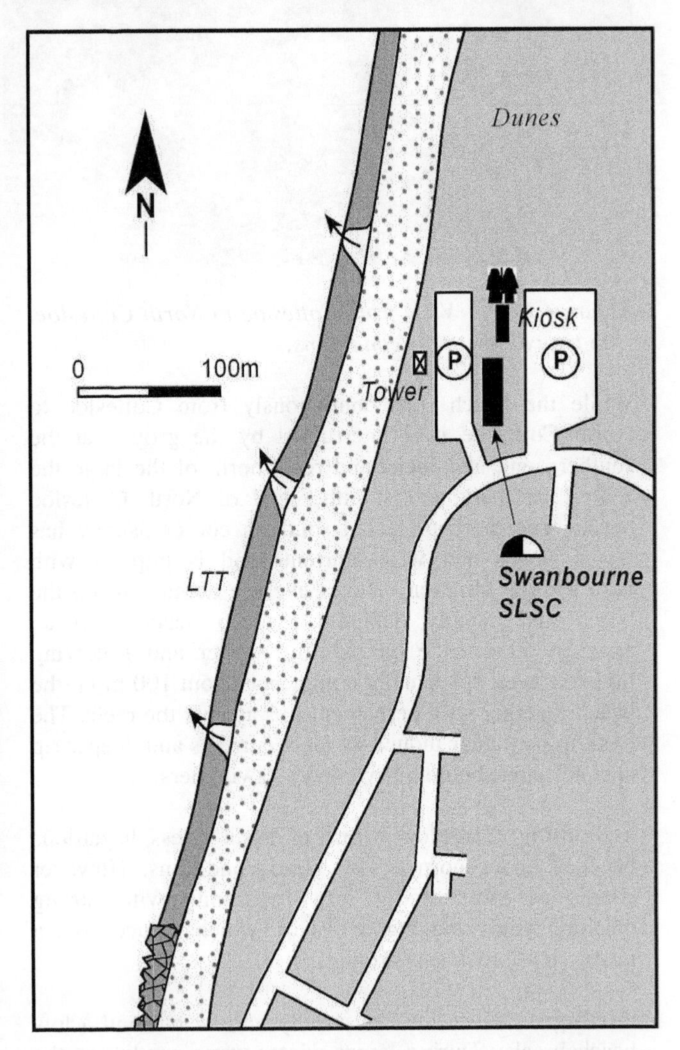

Figure 4.197 Swanbourne Beach is usually has an attached bar cut by occasional rips.

Figure 4.198 View of Swanbourne Surf Life Saving Club and beach.

The beach usually has low waves, averaging 0.5 to 1 m and a wide beach fronted by a steep swash zone and attached bar. During summer the bar is usually continuous with few rip holes, however during winter and following higher wave rip channels will cut across the bar every 100-200 m. The North Swanbourne 'No Man Lands' area of beach is backed by a 10-20 m high foredune containing several blowouts, then the Campbell Barracks.

City of Perth SLSC patrols a 2.5 km section of the beach from the northern end of 'No Mans Land' (**WA 840B**) up to the two groynes that lie either side of the club house and which demark the main 500 m long **City Beach (WA 841**) and patrol area (Fig. 4.199 and 4.200). A patrol tower is located at the tip of the southern groyne. When the Club was established in 1925 this section of dune backed beach was known as the 'Sahara desert'. A plank road was constructed across the dunes in 1918, followed by primitive shelters and toilets in 1921 and finally the surf club in 1925. Today the modern clubhouse overlooks the beach and is backed by extensive car parks and a large grassy foreshore reserve.

City Beach, like the adjoining beaches receives relatively low waves, averaging 0.5 to 1 m, and usually has a steep beach and a continuous bar. Rips only occur along the beach during higher waves, particularly in winter, however whenever waves are breaking across the bars, rips runs out against the groynes.

WA 842 **FLOREAT-SCARBOROUGH-TRIGG**

No.	Beach	Rating	Type	Length
WA 842A	Floreat Beach	3	R/LTT	2.5 km
WA 842B	Scarborough Beach	4	LTT	2.5 km
WA 842C	Trigg Beach	5	LTT/TBR+reef	1.4 km
			Total length	6.4 km

Floreat Surf Life Saving Club
Patrol period: Saturday November-March
Sunday October-March
Lifeguard: Mon-Fri December-February

Scarboro Surf Life Saving Club
Patrol period: Saturday October-March
Sunday October-March

Trigg Island Surf Life Saving Club
Patrol period: Saturday October-March
Sunday October-March

Spring & neap tidal range = 0.7 & 0.4 m

The northern City Beach groyne marks the beginning of a 6.4 km long continuous sand beach up to Trigg Island. Three surf life saving clubs patrol the beach, with Floreat on the south to the northern side of the groyne, then Scarboro 4 km up the beach, and Trigg Island just south of the low calcarenite of the island. The West Coast Highway backs the entire beach, with the best access at each of the three surf life saving clubs.

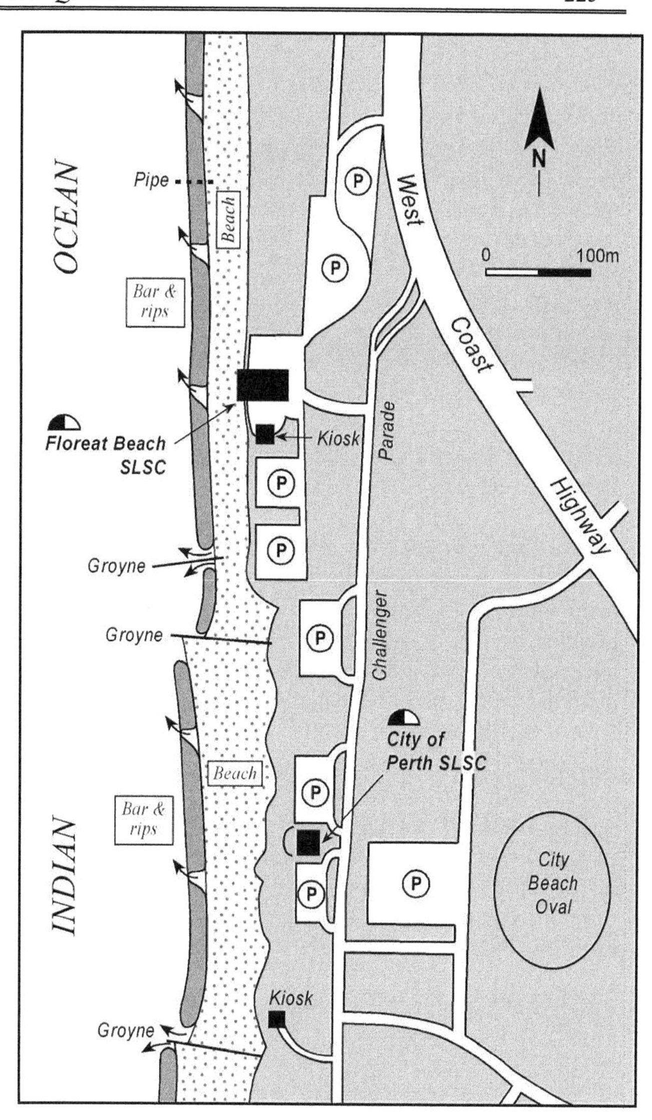

Figure 4.199 Adjoining City and Floreat beaches are patrolled by their respective Surf Life Saving Clubs. The beach usually has an attached bar cut by small rips.

Figure 4.200 View along City beach from the southern to the northern groyne.

Floreat Beach (WA 842A) begins on the northern side of the City Beach groyne. The Floreat Surf Life Saving Club is located 150 m north of the groyne and patrols the next 2.5 km of beach (Fig. 4.199). Patrols began at the beach following World War 2 when members from City Beach would patrol the area. This lead to the development of the North City Surf Life Saving Club in 1947, which changed its name to Floreat in 1954. The present club, kiosk, café, playground and volleyball courts are surrounded by car parks and reserves. The beach, apart from the groyne, is straight and relatively steep and reflective, with a continuous attached bar forming during higher waves. Rip channels tend to be absent in summer, however during winter storms and any period of higher waves they will cut channels across the bar every 100 to 200 m. The 3 km of beach between Floreat and Scarborough is backed by a 200 m wide dune reserve, with occasional car parks and the foredune rising to 10-15 m. This is a more difficult to assess section with less usage. The section north Floreat to just south of the Scarboro surf club is also known as **Brighton Beach**.

Scarborough Beach (WA 842B) has developed into one of Perth's more popular beaches. However, it and neighbouring Trigg Beach are the two more hazardous beaches on the Perth coast (Fig. 4.201). In fact the impetus to establish a surf club on the beach followed a mass drowning of six people at the Scarborough in 1916. Patrols began on the beach that year and continued until the establishment of the Scarboro Surf Life Saving Club in 1928. Since then the Scarboro lifesavers have rescued several thousand people, and performed almost 50% of all rescues in Western Australia, including a famous mass rescue in 1936. During this time Scarborough has developed from dune wasteland, into one of Perth's most exclusive beachfront suburbs.

Figure 4.201 View south from Trigg Island along Trigg and Scarborough beaches. Note the attached bar cut by rip channels.

The Scarboro Surf Club has extensive parking behind and a picnic area and foredune between it and the beach, with an elevated landscaped walkway extending north of the club (Fig. 4.202). A patrol tower in front of the club provides lifesavers with a good view of the surf. This 2.5 km section of beach is slightly more hazardous than its southern and northern neighbours owing to the occurrence of higher waves, coupled with its popularity. The higher waves both break more heavily on the bar, and are more likely to produce rips which in turn scour rip channels across the bar every 100-200 m. The car park at the end of Brighton Road is a popular spot to check for surf.

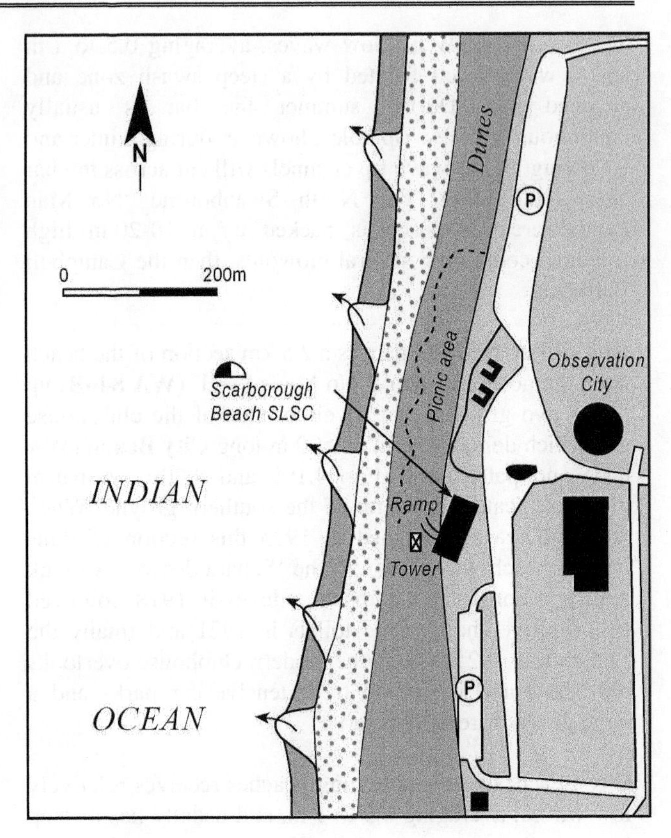

Figure 4.202 Scarborough Surf Life Saving Club patrols a beach dominated by a continuous bar usually cut by beach rips.

Trigg Island (WA 842C) is a low wave washed rocky islet that forms the northern boundary of the long Swanbourne to Trigg beaches. This northern part of the Perth coast was only developed in the 1950's, with the Trigg Island Surf Life Saving Club being established in 1954. In its first two years of operation it rescued 47 people from the more hazardous surf at this end of the beach. The large surf club and adjacent restaurant is located 100 m south of the island, and has extensive car parks to either side (Fig. 4.203). A watchtower also fronts the club, while the Western Australian Surfing has it headquarters immediately south of the clubhouse. The beach to the south of the club is called South Trigg.

The surf club patrols the northern 1.5 km of the beach including the infamous Blue Hole, a permanent rip that runs out on the south side of the island, and has been the scene of many rescues. Furthermore, like adjoining South Trigg and Scarborough beaches, waves are higher along Trigg Beach, and the continuous bar is more likely to be cut by rips and rip channels, making this the most hazardous beach in the Perth region.

Swimming: There are five Surf Life Saving Clubs and associated patrolled areas to swim at on this long beach. The patrolled area offers the safest swimming areas. However be a little more careful on Scarborough and Trigg as waves are usually bigger and rips more persistent and stronger. Always stay between the flags, on the bar and close inshore.

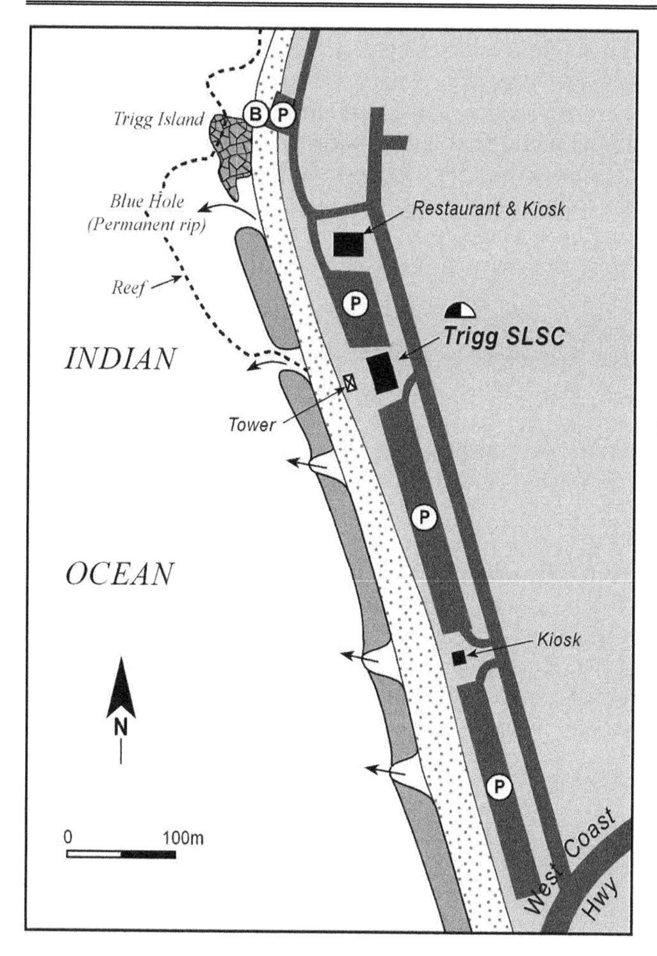

Figure 4.203 Trigg Surf Life Saving Club patrols the rip-dominated beach that extends south from Trigg Island,

Surfing: The best beach breaks are amongst the higher waves at the Scarborough-Trigg end of the beach. The main breaks are associated with the reefs and groynes. At City and Floreat the groynes attached most attention, while up at Trigg the *Third Car Park* is a popular location, particularly in winter and when rips cut the bar, as well as the reefs and permanent breaks around *Trigg Point*.

Summary: This is Perth's longest beach (WA 840-842), the site of five Surf Life Saving Clubs, and a beach with excellent access and parking facilities, extensive foreshore reserves and numerous facilities for beach goers and visitors including beachfront restaurants. However it is also the site over 75% of al surf rescues in Western Australia, so only swim at patrolled beaches between the flags.

MARMION MARINE PARK

Established:	1987 (first in WA)
Area:	10 500 ha
Coast length:	2208-2225 km (17 km)
Beaches:	WA 843-879 (36 beaches)

The Marmion Marine Park extends along the coast for 17 km from Trigg Island to Burns Beach and 5 km seaward encompassing an area of 10 500 ha of generally shallow seafloor between the coast and the numerous calcarenite reef systems that parallel the coast, as well as 36 beaches between Trigg and Burns.

WA 843-848 TRIGG ISLAND-HAMMERSLEY POOL

No.	Beach	Rating	Type	Length
WA 843	Trigg Island	5	R+platform/reef	150 m
WA 844	Benniou Beach	5	R+platform/reef	300 m
WA 845	Benniou (N)	5	R+platform/reef	80 m
WA 846	Bailey St	5	R+platform/reef	100 m
WA 847	Bailey St (N 1)	5	R+platform/reef	80 m
WA 848	Bailey St (N 2)	5	R+platform/reef	70 m
Spring & neap tidal range = 0.7 & 0.4				

Trigg Island marks the southern boundary of a 3.5 km long section of bluff-dominated shoreline that extends to Marmion Beach in the north and is backed by the suburbs of Trigg, North Beach, Waterman and Marmion. In between are 13 small beaches (WA 843-856) all backed by 10-15 m high calcarenite bluffs and dominated to varying degrees by calcarenite intertidal platforms and reefs. The West Coast Highway runs along the edge of the bluffs providing good access to all the beaches. The beaches receive relatively high waves for Perth averaging just over 1 m, with waves commonly breaking over the reefs. The first six beaches (WA 843-848) occupy the first 1 km of shore and front the suburb of Trigg.

Trigg Island beach (**WA 843**) is a 150 m long pocket of high tide sand located on the northern side of the island. It is fronted by a 50 m wide intertidal platform that runs the length of the beach. The beach has a boat ramp at the southern end in lee of the island, which can only be used at mid- to high-tide. A large car park backs the beach.

Benniou Beach (**WA 844**) lies immediately to the north and is a 300 m long strip of high tide sand backed by irregular vegetated 10 m high bluffs, with rocks outcropping along the beach and an irregular intertidal platform paralleling much of the beach. The main road runs along the top of the bluffs with a small car park behind the centre of the beach.

Beach **WA 845** occupies the next 80 m long gap in the bluffs. Sloping vegetated bluffs rise to 15 m behind the beach, which is bordered by protruding arms of calcarenite and fronted by a continuous 50 m wide platform. A walking track runs down the northern bluff to provide access to the beach. Its northern neighbour beach,

WA 846 is a similar 100 m long beach located at the end of Bailey St (Fig. 4.204). A car park occupies all the blufftop behind the beach, providing access via the southern boundary track.

Figure 4.204 The small rocky-bound beach at the end of Bailey Street is typical of this section of the Perth shoreline.

Bailey St beach (**WA 847**) is an 80 m long high tide beach, backed and bordered by 15 m high bluffs and fronted by the continuous intertidal platform, which is eroded in the centre to form a tidal pool. It can be accessed by the Bailey St car park, located immediately south of the beach. Beach **WA 848** lies immediately to the north and is a similar 70 m long pocket of high tide sand, fronted by the continuous platform. A walking track from the northern bluff runs down to the centre of the beach.

All six beaches usually receive waves averaging about 1 m, with lower waves at the shore owing to the reefs and platforms. Use care if swimming at any of these beaches as rocks and reefs abound, and during higher waves strong rips are present.

WA 849-852 NORTH BEACH

No.	Beach	Rating	Type	Length
WA 849	Meltams Pool	5	R+platform	500 m
WA 850	Centaur	5	R+platform	80 m
WA 851	Hammersley Pool	5	R+platform	350 m
WA 852	North Beach	4	R	150 m
Spring & neap tidal range = 0.7 & 0.4 m				

The suburb of **North Beach** has a 1.2 km long west-facing oceanfront consisting of 10-20 m high calcarenite bluffs, with four beaches (WA 849-852) located along their base. The West Coast Highway runs along the top of the bluffs providing good access to all the beaches.

Meltams Pool is located at the southern end of a crenulate 500 m long beach (**WA 849**) where it is afforded protection by calcarenite reefs. The beach is backed by the vegetated bluffs and fronted by an irregular 50-100 m wide platform-reef, which generates the irregularities in the shoreline. The road backs the bluffs, with a large car park cut into the bluffs behind the northern section of the beach. Beach **WA 850** lies 100 m to the north, below the Centaur monument. It is an 80 m long pocket of sand wedged between two protrusions of the bluffs and fronted by a continuous 20-50 m wide intertidal platform. A walking track, then the road, runs along the top of the backing bluffs.

Beach **WA 851** commences immediately to the north and is a 300 m long relatively narrow high tide beach, fronted by near continuous intertidal platform and reefs that enclose the Hammersley tidal pool. It is backed by the bluffs, with a car park located above the northern end of the beach, which provides access to its neighbour, North Beach.

North Beach (**WA 852**) is a 150 m long sandy reflective beach, free of reefs (Fig. 4.205). For this reason it is a popular swimming beach, with waves usually less than 1 m high surging up the moderately steep beach face. Car parks are located above either end of the beach, with a fenced fishing platform on the northern boundary rocks.

Figure 4.205 North Beach is one of the few beaches in this area free of rocks and reefs, and is consequently a popular swimming beach.

WA 853-856 WATERMAN

No.	Beach	Rating	Type	Length
WA 853	Hale St	5	R+reefs	200 m
WA 854	Margaret St	4	R	200 m
WA 855	Waterman Bay	4/5	R/LTT&R+reefs	500 m
WA 856	Lennard Pool	5	R+platform+reefs	300 m
Spring & neap tidal range = 0.7 & 0.4 m				

Waterman is a small beachfront suburb centred on Waterman Bay. It has 1.2 km of ocean shoreline consisting of the 20 m high bluffs with four beaches (WA 853-856) located along their base. The CSIRO Marine Laboratory is located on top of the bluffs between beaches WA 853 and 854.

Beach **WA 853** is a 200 m long strip of sand adjacent to Hale St. It commences in a pocket on the south side of the marine laboratory and continues along the base of the bluffs in front of the lab. It is a narrow beach, almost awash at high tide, with rocks and reefs paralleling the shoreline. Beach **WA 854** is located on the northern side of the lab in line with Margaret St. It starts as 100 m long open pocket of sand, free of rocks, then narrow to the north as a thin strip of sand between the bluffs and parallel reefs. A large car park is located along the rear of the northern half of the beach.

Waterman Bay beach (**WA 855**) is a 500 m long beach bordered by 20 m high bluffs (Fig. 4.206), with a lower central section and the West Coast Highway right behind. It is fronted by 50 m wide reefs in the south and along the narrower northern section, with a central 200 m long open water section, free of rocks and reefs, with usually a narrow attached bar. There is a viewing platform off the highway above the centre of the beach. Parking is restricted to the highway and backing streets. During high swell there are breaks over some of the reefs along Waterman, called *Gravers* or *Grabbers*.

Figure 4.206 Reefs fringe Waterman Beach, with an open central section used for swimming.

Troy Avenue terminates above beach **WA 856**, which is a continuation of Waterman beach, with a small section of protruding bluff separating the two. It is a 300 m long narrow high tide beach that occupies the next indentation in the bluffs and is fronted by a continuous intertidal platform, with reefs extending off the boundary bluffs and off the centre of the beach. The platform and reefs enclose the Lennard tidal pool. It is backed by vegetated dune-draped 15 m high bluffs, with a curving walking track down to the centre of the beach.

WA 857-859 **MARMION-SORRENTO**

No.	Beach	Rating	Type	Length
WA 857	Marmion Beach	4	LTT	800 km
WA 858	Sorrento Beach	4	LTT	600 m
WA 859	Sorrento (N)	4	LTT	300 m

Sorrento Surf Life Saving Club
Patrol period: Saturday October-March
Sunday November-March
Lifeguard: Mon-Fri December-March

Spring & neap tidal range = 0.7 & 0.4 m

Marmion Beach marks the end of the calcarenite bluffs and the beginning of what was once a continuous sand beach up to the protruding sandy foreland at Mullaloo Point, 4 km to the north. However since the 1990's the Sorrento groynes and the large Hillarys boat harbour have been constructed in the centre of the beach, dividing it into Marmion Beach (WA 857), the Sorrento beaches (WA 858 & 859), two beaches in the boat harbour (WA 860 & 861) and then the longer remaining beach up to Mullaloo Point (WA 863). The 1.5 km of shoreline between the end of the bluffs and the southern breakwater of the boat harbour contains three near continuous beaches (WA 857-859), only separated by the Sorrento groyne field.

Marimon Beach (WA 857) is a north-northwest trending 800 m long sandy beach, that commences at the end of the calcarenite bluffs and continues north as a continuous sandy beach with two patches of shore-attached reef along the southern section. The highway runs along the top of the bluffs and the Marmion Angling and Aquatics Club overlooks the southern end of the beach. A small car park is located towards the northern end.

Sorrento Beach (WA 858) is one of Perth's traditional northern beach suburbs and site of the Sorrento Surf Life Saving Club, which was established in 1958. It is part of the first continuous sandy beach north of Trigg Island. However, problems with beach erosion have lead to the construction of three 50 m long rock groynes across the 600 m long beach (Fig. 4.207). The groynes are located either side of the surf club (Figs. 4.208 & 4.209), while 500 m north of the club is the large curving southern breakwater of Hillarys Boat Harbour, which extend 1 km out to sea. Waves average about 1 m and maintain a relatively steep beach with a narrow, continuous attached bar. There is a large car park off the highway adjacent to the surf club. Today the surf club with its high watchtower sits between the two groynes, with a seawall fronting the club and car park.

Beach **WA 859** extends for 300 m from the northern groyne to the southern boat harbour wall, that extends several hundred metres seawards. This section is backed by a 50 m wide fenced foredune, then the highway, with a car park at the northern end adjacent to the boat harbour. It receives slightly higher waves and usually has an attached bar with occasional, rips, including a permanent rip that forms against the wall.

Swimming: Sorrento is one of Perth's least hazardous beaches, so long as you stay in the central patrol area and clear of the groynes.

Surfing: The highest waves tend to occur toward the northern end where they can produce low beach breaks. Only large southwest swell will produce a rideable wave.

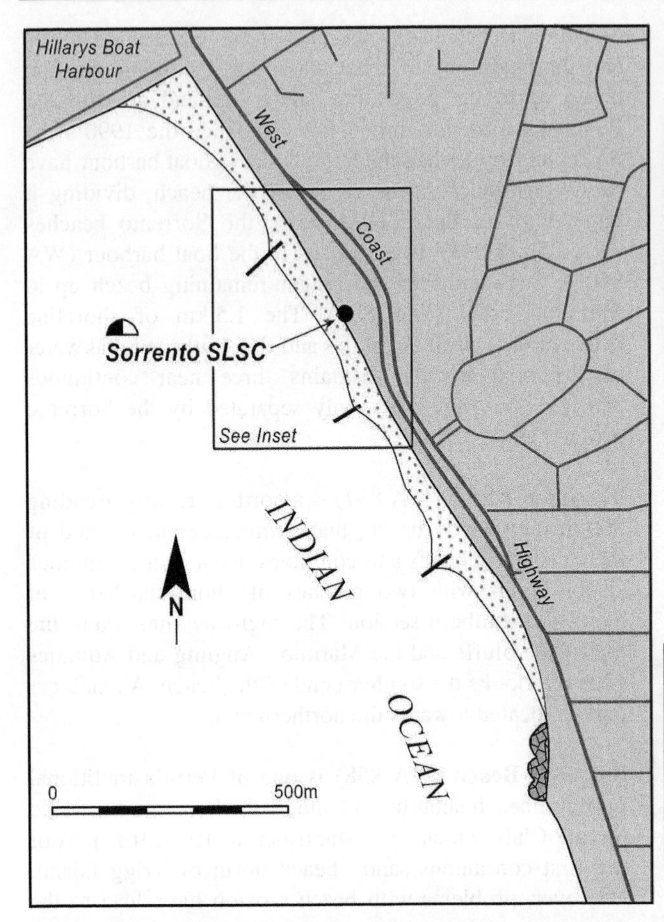

Figure 4.207 Sorrento Beach is located between the Marimon bluffs and Hillary Boat Harbour breakwater and is crossed by three rock groynes.

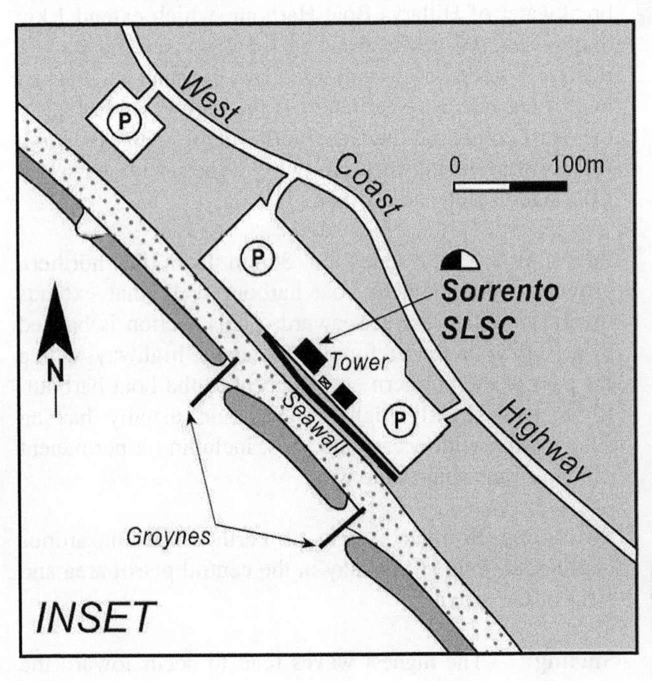

Figure 4.208 Sorrento Beach and Surf Life Saving Club.

Figure 4.209 View north along Sorrento Beach, with its three groynes.

WA 860-861 HILLARYS BOAT HARBOUR

No.	Beach	Rating	Type	Length
WA 860	Hillarys Boat Harbour (S)	1	R	200 m
WA 861	Hillarys Boat Harbour (N)	1	R	100 m

Lifeguard: Mon-Fri December-February

Spring & neap tidal range = 0.7 & 0.4 m

Hillarys Boat Harbour was constructed across the northern end of Sorrento beach during the 1980's. The harbour occupies 800 m of the former beach and is bounded by two attached breakwaters that extends up to 800 m seaward and enclosing a protected 25 ha small boat harbour, together with numerous boat, commercial and car parking facilities around the perimeter (Fig. 4.210). While the harbour is dominated by jetties and moored boats, two beaches (WA 860 & 861) were constructed, specifically as recreational beaches, at the eastern extremity of the harbour.

The southern beach **(WA 860)** extends for 200 m between the southern car park and a central 50 m long circular seawall that separates the two beaches. The beach is backed by commercial operations then a grassy reserve. The northern beach (**WA 861**) continues on the northern side of the seawall for another 100 m, and is backed by a continuous grassy reserve. The main boat ramp is located 50 m west of the beach. They are popular beaches with large car parks to either side, as well as commercial facilities nearby. They are usually calm and relatively safe, so long as you stay close to shore and clear of the boating activity. Some bathers get into trouble when they swim or drift into the deeper water off the beaches. As a consequence the beaches ares patrolled by lifeguards during December, January and February.

Figure 4.210 Hillarys Boat Harbour contains two sheltered beaches.

Figure 4.211 Mullaloo Point separated Hillarys from Whitfords beach.

WA 862-863 HILLARYS-WHITFORD-MULLALOO BEACHES

No.	Beach	Rating	Type	Length
WA 862	Hillarys Beach	4	R	1.9 km
WA 863	Whitfords-Mullaloo Beach	4→5	LTT-TBR	4 km

Mullaloo Surf Life Saving Club

Patrol period:	**Saturday**	**October-March**
	Sunday	**October-March**
Lifeguard:	**Mon-Fri**	**December-February**

WA 864	Mullaloo (N)	6	LTT+rocks/reef	150 m

Spring & neap tidal range = 0.7 & 0.4 m

To the north of Hillarys Boat Harbour is a continuous 6 km long stretch of coast containing two beach systems that lie to either side of Mullaloo (or Pinaroo) Point (Fig. 4.211). The point is a cuspate foreland formed in lee of the Little Island reef group which lies 2 km seaward. Wave refraction around the reefs has resulted in the shoreline prograding about 500 m seaward to form the central foreland. Waves are generally lower along the southern Hillarys Beach (WA 862), while they pick up to the north along Whitfords-Mullaloo Beach (WA 863). Both beaches are backed by a continuous recreational reserve with car parks and good access points.

Hillarys Beach (**WA 862**) extends for 1.9 km from the northern training wall of Hillarys Boat Harbour to the western tip of Mullaloo (Pinaroo) Point. Seagrass meadows and shallow reef commence 50-100 m off the beach. The beach is also sheltered by the boat harbour, which extends 600 m to the west and the offshore reefs, with waves averaging less than 1 m and reflective beach conditions usually prevailing. The beach is backed by a 5-10 m high foredune, then Hillarys Beach Park, which contains three car parks and a range of recreational and picnic facilities.

Whitfords Beach (**WA 863**) commences at the tip of the foreland and curves initially to the northeast for 1 km, then straightens and trends north as Mullaloo Beach for 3 km to the northern end of the foreland and the beginning of 10 m high calcarenite bluffs. Wave height increases up the beach with the highest waves around the Mullaloo surf club area. There is a large car park at the foreland, which is also the site of Whitfords Bay Sailing Club and a beach boat launching area, with water skiing permitted north of the point. An undeveloped 200-300 m wide reserve extends for nearly 2 km north of the point to the first of the Mullaloo access point.

The **Mullaloo Surf Life Saving Club** patrols the central-northern section of the beach (Fig. 4.212). The club, which was established in 1960, is located in a large grassy foreshore reserve between the road and the beach, with a large car park next to the clubhouse, and three car parks to the north. This is one of the higher energy Perth beaches with usually waves breaking across the attached bar and rips forming during and following periods of higher waves (Fig. 4.213 & 4.214). Surfers usually head to the northern end of Mullaloo where waves tend to be highest, with higher waves also producing a right-hand surf break along *Mullaloo Point.*

At the northern end of Mullaloo the calcarenite bluffs and reefs have isolated a 150 m long section of high tide beach (**WA 864**) bordered and fronted by a mixture of intertidal platforms and reefs, with a central 50 m long open section. Seagrass is at time piled up on the beach in lee of the reefs. Waves break over the reefs and during higher waves a rip flows out the central section.

Swimming: Mullaloo is a moderately safe beach when waves are low. However rips can occur, particularly north of the patrol area and care should be taken when they are present, especially against the northern boundary rocks and reefs.

Surfing: As waves tend to pick up, up the beach most surfers head for the northern car park and beach.

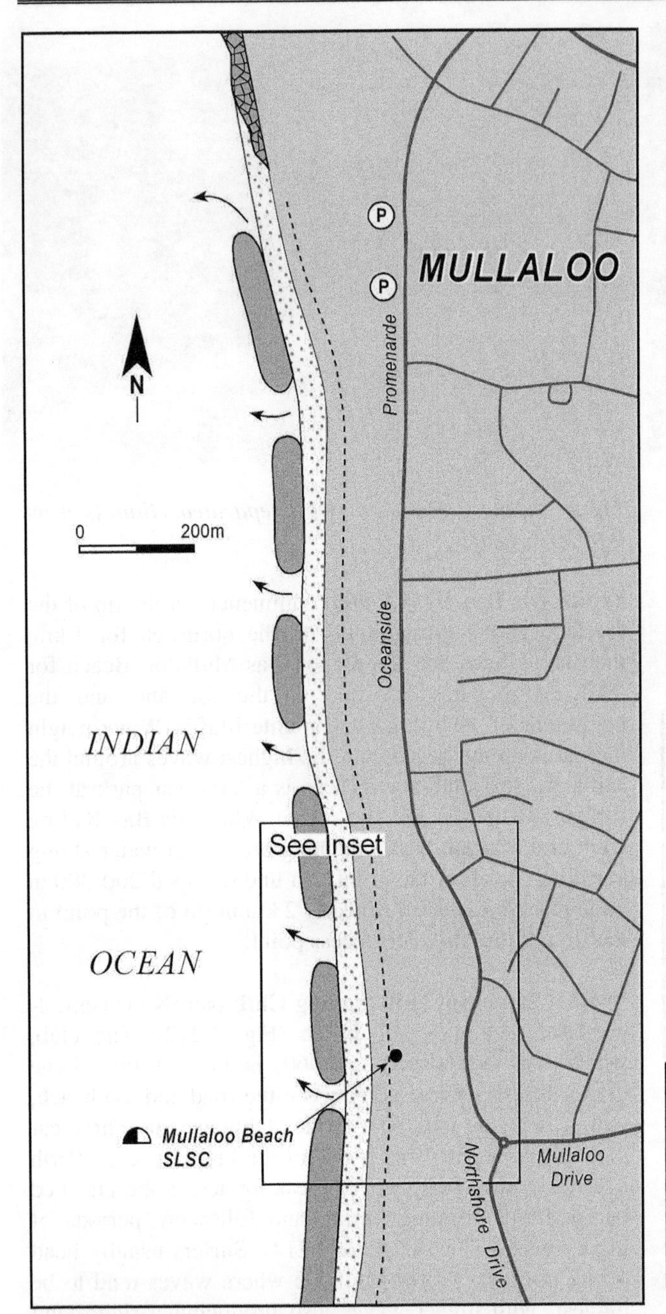

Figure 4.212 Mullaloo is Perth's northernmost beach and one of the more exposed, with rip channels usually cutting across the bar.

Figure 4.213 View north along Mullaloo Beach past the patrol section. Note the wider bar and rip channels.

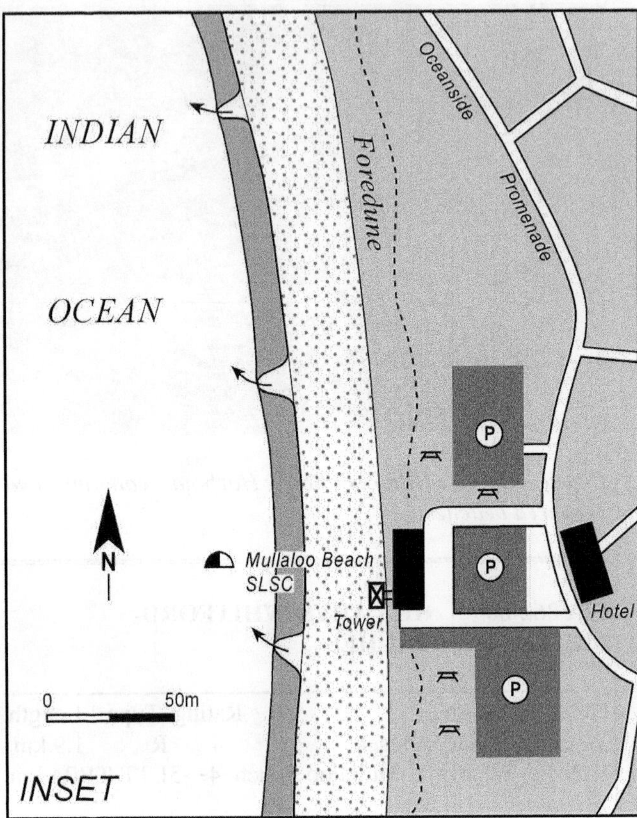

Figure 4.214 The central section of Mullaloo Beach is site of the Mullaloo Surf Life Saving Club.

WA 864-869 **OCEAN REEF (S)**

No.	Beach	Rating	Type	Length
WA 865	Mullaloo (N 1)	5	R+platform/reef	50 m
WA 866	Ocean Reef (S4)	5	R+platform/reef	40 m
WA 867	Ocean Reef (S3)	5	R+platform/reef	100 m
WA 868	Ocean Reef (S2)	5	R+platform/reef	150 m
WA 869	Ocean Reef (S1)	6	R+rocks/reef	50 m
Spring & neap tidal range = 0.7 & 0.4 m				

Mullaloo Beach terminates at the beginning of a calcarenite-dominated section of shore that extends 3 km north to Burns Beach. One kilometer north of Mullaloo the Ocean Reef Boat Harbour has been constructed out from this bluff section. Between the south wall of the boat harbour and Mullaloo are five pockets of sand (WA 865-869), each backed and bordered by 10 m high calcarenite bluffs, and fronted by a mixture of platforms, reefs and rocks. All the beaches are backed by a 500 m wide nature reserve, with a bike and walking track along the top of the bluff proving foot access to the beaches. While the beaches are accessible, they are generally unsuitable for swimming owing to the predominance of rocks and reefs.

Beach **WA 865** is located 300 north of the northern Mullaloo beach. It consists of a 50 m long wedge of high tide sand fronted by a 50 m wide intertidal platform, with some tidal pools. A righthand surf break called *Mossies*, is located adjacent to this beach.

Beaches **WA 866** commences 400 m to the north and is located 400 m south of the boat harbour. The beach consists of a 40 m long pocket of sand fronted by a platform and reefs. Beach **WA 867** lies immediately to the north and is a 100 m long beach connecting pockets of sand fronted by shallow reefs and a small central open section. Beach **WA 868** is a similar 150 m long beach containing three connected pockets of sand which decrease in width to the north, each fronted by near continuous shallow reefs. Beach **WA 869** lies adjacent to the southern wall of the boat harbour. It is a 50 m long strip of high tide sand, with large rocks littering the beach and shallow reefs extending 100 m offshore. The large boat harbour car park is located 100 m to the north.

WA 870-873 OCEAN REEF BOAT HARBOUR

No.	Beach	Rating	Type	Length
WA 870	Ocean Reef Boat Hbr (S)	2	R	80 m
WA 871	Ocean Reef Boat Hbr (N1)	2	R	80 m
WA 872	Ocean Reef Boat Hbr (N2)	2	R	70 m
WA 873	Ocean Reef (N)	5	R+reef	250 m
Spring & neap tidal range = 0.7 & 0.4 m				

Ocean Reef Boat Harbour was constructed in the 1990's along a section of predominately calcarenite shore. The harbour extends for 500 m along the shore and consists of a 400 m long attached breakwater, with an opening to the north, as well as three northern groynes. The harbour is solely used for boat launching and has five adjacent boat ramps backed by a large car park (Fig. 4.215). In addition a beach has been constructed immediately south of the ramps (WA 870), and two beaches are located between the groynes (WA 871 & 872), with beach WA 873 immediately north of the harbour. The entire development is backed by a 500 m wide reserve, then the Ocean Reef subdivision.

Figure 4.215 Ocean Reef boat harbour.

Beach **WA 870** is a curving 80 m long beach located at the southern end of the boat harbour. It is backed by a playground and has been designed to provide a calm swimming location within the harbour.

Beaches WA 871 and 872 are located north of the boat ramps and are bordered by the two northern groynes. Beach **WA 871** is an 80 m long reflective beach located between the main entrance groyne and the ramp and is usually calm. It is backed by a reserve. Any sand moving into the harbour accumulates at the northern end of this beach and is periodically removed. Beach **WA 872** is located immediately to the north between the main groyne and a second smaller groyne, behind which is located the Sea Sports Club and Sea Rescue Group. The beach is a 70 m long pocket of sand which receives low surging waves. A navigation marker is located behind the centre of the beach.

Beach **WA 873** extends from the small groyne north for 250 m to the beginning of the calcarenite bluffs. It has a central attached reef resulting in a crenulate beach, with waves averaging about 1 m surging up the steep beach face. Additional reefs lie off the beach, which is backed by vegetated dune-draped 10 m high bluffs, then the reserve.

WA 874-879 BEAUMARIS-BURNS BEACH

No.	Beach	Rating	Type	Length
WA 874	Beaumaris (S)	5	R+rocks/reef	500 m
WA 875	Beaumaris	5	R+rocks/reef	300 m
WA 876	Burns Beach (S3)	5	R+rocks/reef	350 m
WA 877	Burns Beach (S2)	5	R+rocks/reef	100 m
WA 878	Burns Beach (S1)	5	R+rocks/reef	400 m
WA 879	Burns Beach	4	R+LTT	200 m
Spring & neap tidal range = 0.7 & 0.4 m				

Beaumaris Beach is located at the end of Shenton Road. Between the southern end of the beach and Burns Beach is 2.5 km of calcarenite-dominated shore containing six beaches (WA 874-879) all located at the base of 10-20 m high calcarenite bluffs. Waves average about 1 m along the shore having been reduced by shallow calcarenite reefs, which extends 1-2 km offshore. There is access via a car park at the end of Shenton Drove to beach WA 875, and at Burns Beach to beaches WA 878 and 879, with a bike and walking track along the top of the bluff providing foot access to the intervening beaches (WA 876 & 877). All the beaches are backed by a 200-300 m wide reserve then Ocean Reef Road, which connects in the north to Burns Beach Road. To the east of the road are the Ocean Reef and Iluka subdivisions.

The southern Beaumaris Beach (**WA 874**) is a straight, west-facing 500 m long beach backed by vegetated slopes rising to 15 m and fronted by a fringe of calcarenite reef that widens to about 50 m in the north. A protruding section of bluff separates it from the main beach (**WA 875**), which continues north for 300 m to the next section of protruding bluff. The car park is located behind the bluff with fenced access down to the beach. The moderately steep sandy beach is backed by dune-draped bluffs, and fronted by a 50-80 m wide section of discontinuous calcarenite reefs, with some rocks along the base of the beach.

Beach **WA 876** commences on the northern side of the 100 m long bluff and is a narrow 350 m long high tide beach, backed by steep 10-25 m high bluffs and fronted by a near continuous intertidal platform/reef. The irregular bluff almost cuts the beach into three with some rocks debris scattered along the beach. Beach **WA 877** commences immediately to the north and is a similar 100 m long strip of high tide sand, wedged between the backing bluff and fronting reefs.

Beach **WA 878** commences 100 m to the north and is a 400 m strip of high tide sand, also located below the continuous 10-15 m high bluffs, and paralleled by the near continuous intertidal rocks and reef. It terminates 200 m south of the Burns Beach Road.

Burns Beach (**WA 879**) lies at the end of Burns beach Road in front of the caravan park, with a car park along the top of the backing 10 m high bluffs. The 200 m long beach is bordered by calcarenite bluffs and a reef in the south, while it opens up along the north half, to become relatively free of rocks. It terminates at a small rock groyne that has been built across the beach.

WA 880-881 **BURNS BEACH-MINDARIE**

No.	Beach	Rating	Type	Length
WA 880	Burns Beach (N)	4	R/LTT	3.8 km
WA 881	Mindarie (S)	5	R+rocks/reef	300 m

Spring & neap tidal range = 0.7& 0.4 m

To the north of Burns Beach is a continuous 4 km stretch of sand that extends up to the training walls of Mindarie Keys, and which contains two beaches (WA 880 & 881). Beach **WA 880** is a 3.8 km long reflective sand beach which forms a slight west-facing foreland to the lee of Burns Rock and other reefs which lie 500-1000 m offshore. The reefs lower waves to about 1 m at the shore. It is backed by dune-draped 10-20 m high calcarenite bluffs, and in places active and vegetated dunes that have transgressed up to 2.5 km inland and to heights of 50 m. While in 2001 most of the backing dunes were undeveloped, they are being encroached upon by the expanding Burns Beach settlement, which borders the southern end and the newer Mindarie subdivision at the northern end.

Beach **WA 881** extends south of the southern Mindarie Keys training wall for 300 m to a small bluff (Fig. 4.216), which separates it from the longer Burns Beach. The beach is fronted by near continuous 50-80 m wide intertidal rocks and reef and backed by dune-draped 20 m high bluffs. It can be accessed in the south from a bluff top car park, off Long Beach Promenade, and in the north from the southern Mindarie Keys car park. There is vehicle access along the Mindarie training wall, which includes a fishing platform for the disabled.

Figure 4.216 Beach WA 881 extends south of the Mindarie Boat Harbour southern wall. It has a mixture of sand and rocks in the surf zone.

WA 882-887 **MINDARIE KEYS**

No.	Beach	Rating	Type	Length
WA 882	Mindarie Keys (N)	1	R	100 m
WA 883	Mindarie Keys groyne (S)	3	R+platform	60 m
WA 884	Mindarie Keys groyne (N)	3	R+platform	120 m
WA 885	Mindarie Keys (N1)	4	R+platform	250 m
WA 886	Mindarie Keys (N2)	4	R+platform	50 m
WA 887	Mindarie Keys (N3)	4	R+platform	100 m

Spring & neap tidal range =0.7 & 0.4 m

Mindarie Keys is a 1990's 'keys' development located 2 km south of Quinns Rocks (Fig. 4.217) and cut into the sloping calcarenite that rises to 30 m. The development includes the excavated 19 ha keys and associated boat harbour with marina and boat launching facilities, all of which are backed by residential development on the surrounding calcarenite slopes, formerly known as Mindarie. The harbour entrance is located between the end of a 800 m long attached breakwater and a 100 m long entrance groyne (Fig. 4.218). One beach has been constructed deep inside the harbour (WA 822), with two accumulating either side of the entrance groyne (WA 883 & 884), the remainder (WA 884-887) located along the base of the calcarenite buffs immediately north of the harbour entrance.

Beach **WA 882** is located at the northern extremity of the inner harbour, with Rosslare Promenade running around the rear of the beach. The beach is 100 m long and faces south towards the main harbour area, with residential development lining each side and a grassy recreation reserve behind. It is calm apart from small wind waves, however deeper water lies off the beach.

Beach **WA 883** is located on the southern side of the entrance groyne. It consists of a 60 m long pocket of southwest-facing sand, wedged in between the groyne and sloping 15 m high bluffs, with houses on the crest of the bluffs. The end of the harbour seawall forms the southern boundary and an intertidal platform fronts the beach which is usually calm apart from boat wakes.

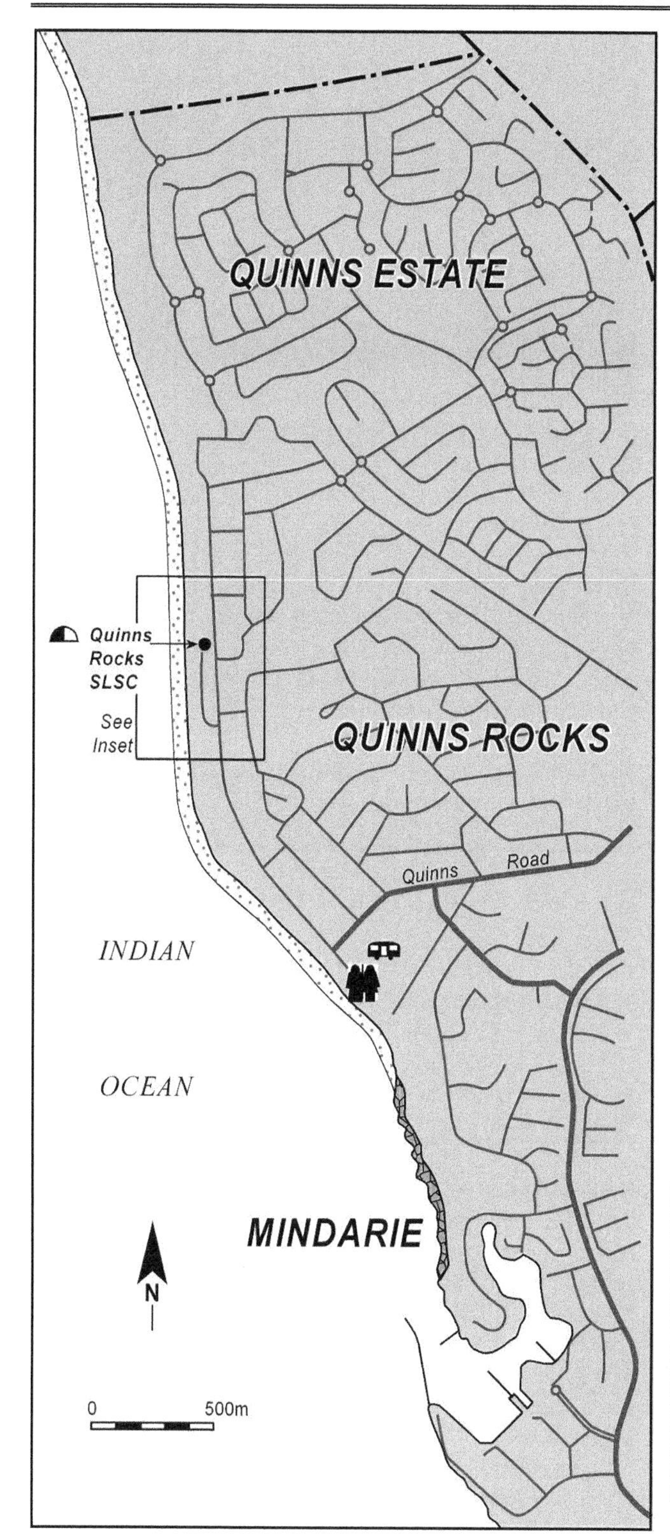

Figure 4.217 Mindarie Key boat harbour is the southern boundary of a continuous coastal settlement that extends north to Quinns Rocks and Quinns Estate. See Fig. 2.19 for insert.

Beach **WA 884** is located on the northern side of the 100 m long groyne, immediately outside the harbour entrance, through still partly protected by the end of the western breakwater. The 120 m long beach is backed by 15 m high vegetated bluffs, then Clarecastle Road. It is a high tide sandy beach with considerable calcarenite along the shore and a 50 100 m wide intertidal platform paralleling the shore.

Figure 4.218 Mindarie Keys Marine and community.

Beaches **WA 885** is located 100 m to the north and consists of a 250 m long west-facing beach, which is clear of the entrance breakwater and exposed to normal wave activity averaging about 1 m. The beach is backed by lower 5 m high vegetated bluffs then the northern end of Clarecastle Road. It is fronted by a 50 m wide intertidal platform, with rocks littering the northern end of the beach.

Beaches WA 886 and 887 lie immediately to the north. Beach **WA 886** is a 50 m long pocket of high tide sand bordered by low protruding bluffs, with rocks on the lower beach and a 50 m wide platform. Beach **WA 887** is a similar 100 m long high tide beach. Both beaches are backed by a sloping vegetated 150 m wide reserve then houses.

WA 888-889 QUINNS ROCKS

No.	Beach	Rating	Type	Length
WA 888	Quinns Rocks Beach	4	R/LTT	950 m
WA 889	Wanneroo Beach	4	R/LTT	3.2 km

Quinns/Mindarie Surf Life Saving Club
Patrol period: Sunday October-March

Lifeguard: Mon-Sat December-February

Spring & neap tidal range = 0.7 & 0.4 m

Quinns Rocks in the name of an area of shallow reefs located 1.5 km off the coast, 30 km north of Perth. The lower waves in lee of the rocks have resulted in the progradation of a 300 m wide sandy foreland. Today this foreland and its backing dunes are occupied by the coastal settlement of Quinns Rocks and the new northern Quinns Estate (Fig. 4.217). The settlement has a 2 km beachfront consisting of the Quinns Rocks Beach (WA 888) on the southern side of the foreland and Wanneroo Beach (WA 889) to the north. Both beaches are backed by a wide foreshore reserve occupied by 10 m high dunes, all backed by Ocean Drive together with a ramp for beach boat launching and a picnic area. At the southern end of the drive is a beach side caravan park, beyond which is the Mindarie Keys boat harbour and expanding residential development.

Quinns Rocks Beach (**WA 888**) commences at the end of the 10 m high calcarenite bluffs that extend 500 m north of the Mindarie beaches. The beach continues north for 950 m to the tip of the foreland where a low 100 m long attached rock breakwater has been constructed. The beach faces southwest but is protected from high waves by the offshore reefs, with waves averaging 0.5-1 m at the shore. These produce a relatively narrow steep reflective beach, with a narrow bar and rips occurring when waves exceed 1 m.

Wanneroo Beach (**WA 889**) commences on the northern side of the breakwater and continues north for 3.2 km to the next bluffs. The **Quinns-Mindarie Surf Life Saving Club**, founded in 1982, is located in the centre of the drive on the north side of the foreland, and just north of the post office and shops. It has a large car park adjacent to the small clubhouse (Fig. 4.219 & 4.220). The reserve then houses continue for another 2 km up the beach. Low swell and reflective beach conditions usually prevail, with a bar and rips occurring only during and following periods of higher waves.

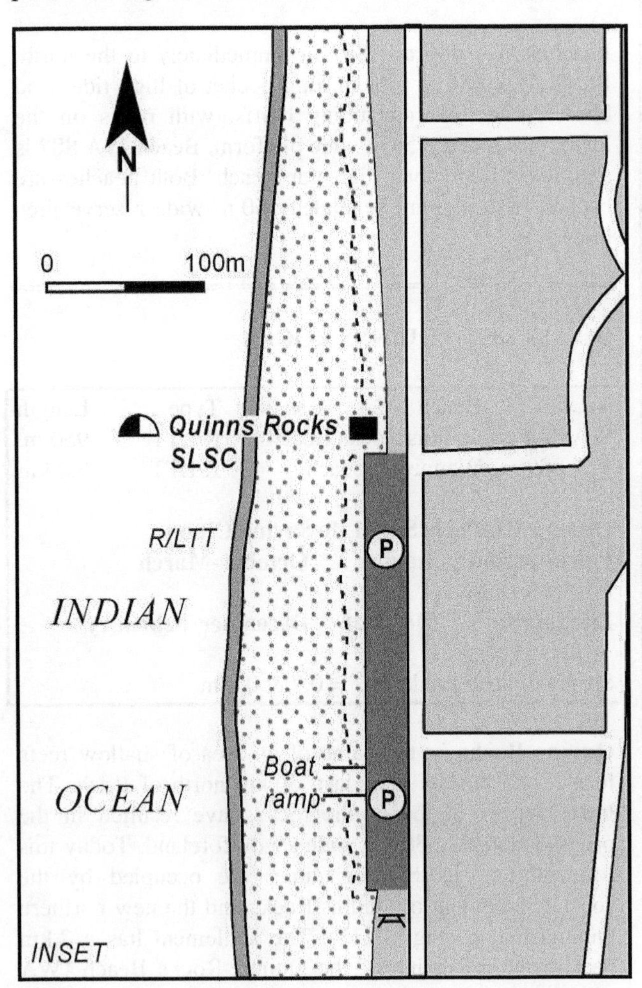

Figure 4.219 Quinns Rocks Surf Life Saving Club patrols a straight section of beach with usually low waves.

Figure 4.220 Quinns Rocks Surf Life Saving Club and the patrol area.

Swimming: Quinns Rocks-Wanneroo beaches have a relatively low hazard rating with usually low waves and few rips. However the beach can be steep, with deep water just off the beach, so care is required with young children and non-swimmers.

Surfing: Waves are usually too low and the surf zone too narrow to produce rideable waves.

WA 890-892 PAMELA SHOAL-YANCHEP

No.	Beach	Rating	Type	Length
WA 890	Eglington Hill (1)	5	R+platform	250 m
WA 891	Eglington Hill (2)	5	R+platform	100 m
WA 892	Pamela Shoal	4	R/LTT	3 km
WA 893	Alkimos-Pipidinny-Yanchep (S)	4	R/LTT	8.3 km
Spring & neap tidal range = 0.7 & 0.4 m				

Pamela Shoal is located 1-1.5 km offshore and is part of the near continuous drowned Pleistocene barrier that forms a series of shore-parallel reefs and calcarenite seabed. The shoaler reefs like Quinns Rocks and Pamela Shoals induce additional wave attenuation and refraction resulting in the formation of sandy forelands to their lee. The shoals are located 6 km north of Quinns Rocks, with three beaches (WA 890-892) located between the northern end of Wanneroo Beach and the foreland. North of the foreland is a continuous beach which extends to the northwest for 8 km past the Alkimos wreck and Pipidinny beach to the southern side of Yanchep.

Beach WA 890 and 891 are adjoining bluff-bound beaches located immediately north of the end of Wanneroo Beach. Beach **WA 890** is a 250 m long high tide beach fronted by a continuous 50-70 m wide intertidal platform. It is backed by a 14 m high foredune and then undeveloped (in 2001) vegetated dunes extending 1 km inland. Beach **WA 891** lies immediately to the north and is a 100 m long pocket of sand bounded by two protruding sections of calcarenite and backed by the vegetated dunes, which rise to 57 m at Eglington Hill. Vehicle tracks reach the back of both beaches.

Beach **WA 892** commences on the northern side of the bluff and continues north-northwest for 3 km past the slight Pamela Shoals foreland to the tip of the sandy foreland formed in lee of the Eglington Shoals. The beach receives waves averaging less than 1 m, which maintain a usually moderately steep reflective beach. A vegetated 10-15 m high foredune, then vegetated transgressive dunes extend up to 2 km inland. Vehicle tracks parallel the rear of the beach.

Beach **WA 893** commences on the northern side of the sandy foreland, initially curving to the north for 1.5 km to the lee of the prominent Alkimos wreck, then continue north-northwest for another 6.5 km past Pipidinny beach and the southern side of Yanchep to the beginning of the Yanchep beachrock. The beach totals 8.3 km in length, with vehicle access at Pipidinny and Yanchep, as well as 4 WD tracks backing most of the beach. For most of its length it is backed by a 10-20 m high foredune with a few blowouts, then vegetated undeveloped (in 2001) transgressive dunes extending 2-3 km inland. The northern 1.5 km is backed by the expanding settlement of Yanchep. Pleistocene calcarenite reef continue to parallel the shore lowering waves to less than 1 m, which maintain a moderately steep reflective beach face. The beach is sandy apart from a section of partly exposed beachrock along Pipidinny beach. During moderate to high swell there are reef breaks located about 1 km offshore at *Alkimos*, including an inner and outer break, which requires a boat to reach.

Region 4E CENTRAL WEST: Yanchep – Leeman

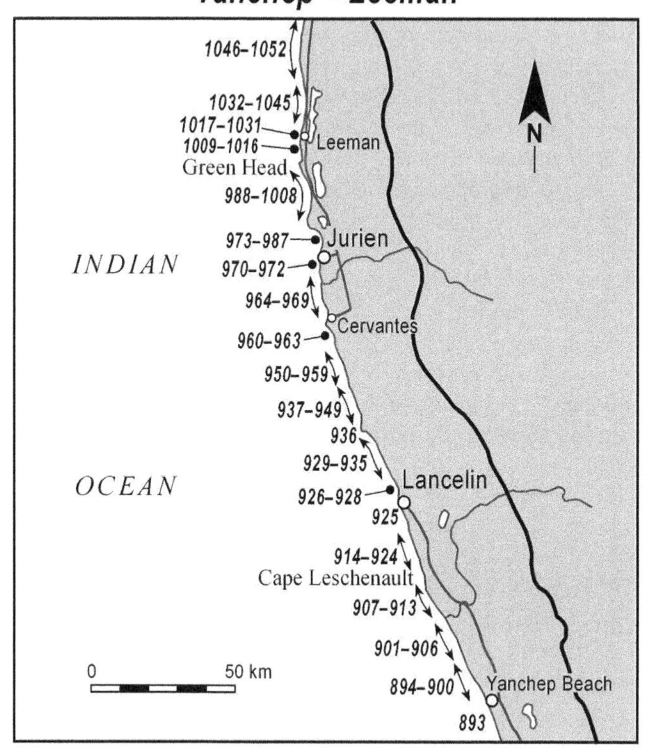

Figure 4.221 Region 4E Yanchep to Leeman, beaches WA 893-1052

WA 894-896 YANCHEP

No.	Beach	Rating	Type	Length
WA 894	Yanchep Lagoon (S)	4	R+reef	250 m
WA 895	Yanchep Lagoon	4	R+reef	300 m
WA 896	Yanchep Beach	4	LTT/TBR	1 km

Yanchep Surf Life Saving Club
Patrol period: Sunday November-March

Lifeguard: Mon-Fri December-February

Spring & neap tidal range = 0.5 & 0.0 m

Yanchep is a growing coastal settlement of several hundred located 50 km north of Perth, and 6 km off the Lancelin Road. The settlement spreads for 2 km along the coast, with most of the houses built on the backing undulating dune terrain. There is a caravan park 1 km north of the town, while a foreshore drive runs along the top of the 20 m high coastal dune, providing a good view of the beaches below. A large car park and kiosk overlook the southern end of the beach. Four beaches border the town - the long Yanchep-Alkimos to the south (WA 893), the smaller beachrock enclosed 'lagoonal' beaches in the centre (WA 894 & 995), and the patrolled open northern beach (WA 896) (Fig. 4.222).

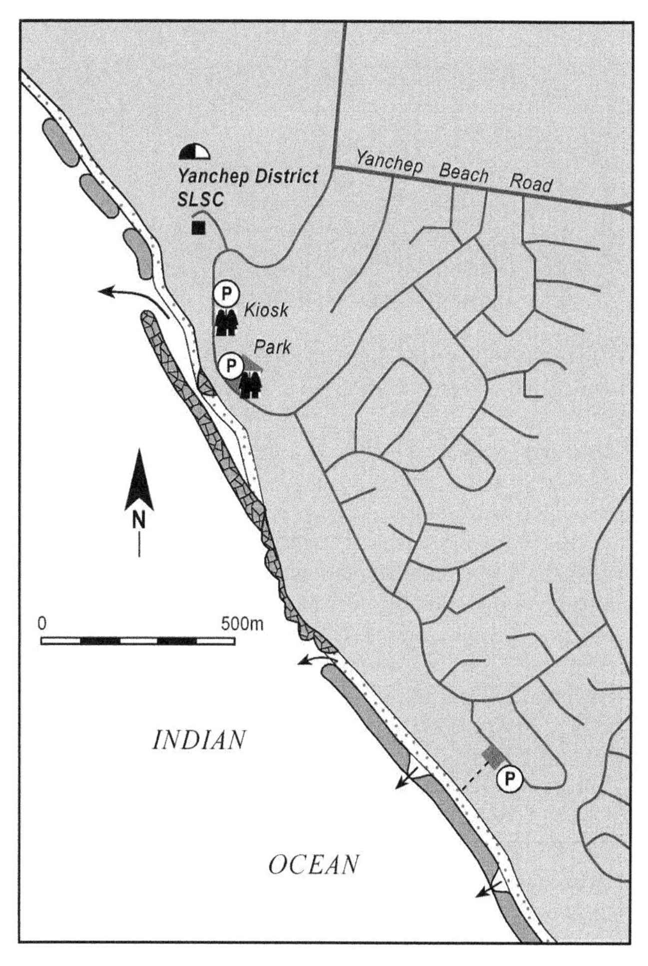

Figure 4.222 Yanchep is centred on a beachrock outcrop and backing bluffs with beaches to either side.

Yanchep lagoon is a 600 m long section of shore, paralleled by a beachrock reef, which is attached to the shore in the south with an opening to the north. Two beaches are located behind the reef. The southern beach (**WA 894**) is completely enclosed by the reef, which connects with calcarenite points to either end of the 250 m long beach. It is backed by vegetated bluffs rising to 15 m with a blufftop car park above the centre and steps down to the beach. The beach is relatively narrow and steep, with a usually calm lagoon, particularly at low tide, and waves averaging 1 m breaking over the reef located 50-80 m off the beach. Water depth varies in the lagoon, with both shallow sandy sections and some deeper holes.

The main **Yanchep Lagoon** beach (**WA 895**) extends north from the northern bluff for 300 m. The reef is attached at the southern end, with the beach curving to the east causing the lagoon to widen to the north. While waves are usually low in the lagoon, water rushing over the reef flows northward along the lagoon and out of the deep 50 m wide channel as a permanent rip (Fig. 4.223). Beware of this current if swimming in the channel. The beach is backed by vegetated 15 m high bluffs, a large blufftop car park and kiosk.

Figure 4.223 Yanchep Lagoon beach (right) and the Yanchep Surf Life Saving Club and patrolled beach (left)

Yanchep Beach (**WA 896**) begins on the northern side of the reef and continues north for 1 km to the Capricorn beach groyne, which is located in lee of a small reef. Waves average just over 1 m along the beach and usually maintain a low tide terrace, which is cut by beach rips during and following periods of higher waves. The blufftop **Yanchep Surf Life Saving Club**, founded in 1981, overlooks the southern end of the beach and patrols both the lagoon and main beaches.

Swimming: The safest swimming is at the southern lagoon beach (WA 894) and at the southern end of the main lagoon (WA 895). Be very careful near the lagoon opening as a strong current flows out into deep water. If swimming on the open beach stay clear of the rips and on the attached portion of the bar.

Surfing: The beachrock reef at *Yanchep Lagoon* can provide some good rights in a moderate to high swell. Otherwise there are usually some beach breaks north of the lagoon.

WA 897-900 CAPIRCORN-WRECK PT

No.	Beach	Rating	Type	Length
WA 897	Capricorn	5	LTT/TBR	1.25 km
WA 898	The Spot (S)	5	LTT/TBR	1.1 km
WA 899	The Spot	5	LTT/TBR	3.2 km
WA 900	Wreck Pt	5	LTT/TBR	300 m

Spring & neap tidal range = 0.5 & 0.0 m

To the north of Yanchep Beach the shoreline continues north-northwest for 6 km to Wreck Point. In between are four near continuous sandy beaches (WA 897-900), backed by the Capricorn Resort in the south and bordered by Two Rocks Marine in the north.

Capricorn beach (**WA 897**) fronts the Capricorn Resort and extends for 1.25 km from the groyne in front of the beach access point to a foreland attached calcarenite reef. A car park behind the groyne provides access to the northern end of Yanchep Beach (WA 896) as well as the Capricorn beach. Waves average about 1 m and maintain a reflective to low tide terrace beach. The resort backs the southern end of the beach, with vegetated transgressive dunes extending up to 5 km inland behind the remainder. A stabilising blowout backs the northern end of the beach.

Beach **WA 898** extends north of the small foreland for 1.1 km to the next reef-attached foreland known as **The Spot**. Wave height increases up the beach with a reflective southern end and a low tide terrace in amongst the reefs at the northern end. The beach is backed by vegetated dunes that rise rapidly to 20 m and extends up to 5 km inland, with a 20 ha blowout located behind the northern end. The Two Rocks Road crosses the dunes 1 km inland and a 4 WD track reaches the centre of the beach.

The Spot (**WA 899**) is a popular surfing location with slightly higher waves and a lefthand break along the southern calcarenite reef. A vehicle track from the Two Rock Road runs for 500 m straight out to the surfing break. The beach extends for 3.2 km to the north terminating at the cuspate Wreck Point. Waves average just over 1 m along the beach and usually maintain a low tide terrace, with rips cutting across the bar during periods of higher waves.

Wreck Point is tied to a calcarenite reef capped by a sea stack. It is a low sandy foreland that forms the boundary of beaches WA 899 and 900. Beach **WA 900** curves round north of the point for 300 m to the southern breakwater of Two Rocks marina and a second sea stack. The beach is moderately sheltered by the reefs, with lower waves breaking across the low gradient beach. Seagrass debris sometimes accumulates along the beach. There is a large car park in the adjacent marine and a low foredune and reserve behind the beach, which includes the Leeman Landing Memorial.

WA 901-906 TWO ROCKS-MOORE RIVER

No.	Beach	Rating	Type	Length
WA 901	Two Rocks (N)	5	R/LTT+reefs	3.6 km
WA 902	Two Rocks (N1)	4	R+reefs	400 m
WA 903	Two Rocks (N2)	5	LTR+reefs	2.2 km
WA 904	Tuart Rd	5	LTT+reefs	350 m
WA 905	Tuart Rd (N)	5	R/LTT+reefs	1.2 km
WA 906	Moore River (S)	4	R	9.8 km
Spring & neap tidal range =0.5 & 0.0 m				

Two Rocks refers to the two sea stacks on Wreck Point, which demarks either end of beach WA 900. Immediately north of the northern stack the Two Rocks marina was constructed in the 1970's. It occupies 800 m of the formerly continuous sandy shore and today is backed by the growing settlement of Two Rocks. The boat harbour consists of two attached breakwaters, with the southern overlapping the northern to provide a protected north-facing entrance. A boat ramp, several jetties and marina facilities are located in the 12 ha harbour (Fig. 4.224). On the backing slopes is a large car park, shopping centre and numerous facilities. Residential development extends 1 km north of the marina, with a reserve between the houses and the beach. To the north of the marina is a near continuous stretch of six sandy beaches (WA 901-906), interrupted only by reef-tied forelands. They extend for a total of 18 km to the mouth of the Moore River.

Figure 4.224 Two Rocks boat harbour.

Beach **WA 901** commences on the northern side of the north breakwater and trends north-northwest for 3.6 km to a low reef-tied sandy foreland. The southern corner of the beach has experienced erosion and is partly backed by a makeshift seawall to protect the backing dunes. The beach receives waves averaging just over 1 m, which in the south maintain a low tide terrace to at time a transverse bar rip system, with rips spaced every 200 m during and following periods of higher waves. The beach is interrupted by small patches of beachrock, with a continuous 500 m long section towards the northern end, which also encloses a section of 'lagoon' near the northern end of the beach. It is backed by a partially active foredune with several small blowouts, then vegetated transgressive dunes that extend up to 5 km inland. Several 4 WD track reach the back of the beach.

Beach **WA 902** is a 400 m long sandy beach located between two reef-tied sandy forelands. The boundary reefs extend a few hundred metres seawards and lower waves to less than 1 m at the shore, with reflective conditions usually prevailing. An active blowout has moved 400 m in from the beach, with vegetated dunes extending another 4 km inland. A four wheel drive track reaches the beach.

Beach **WA 903** continues north for another 2.2 km as a slightly curving west-facing beach, tied to reef-forelands at either end, and with many reefs located along and just off the shore. As a consequence the shoreline is highly variable depending on the presence and location of the reefs, with permanent rips forming against gaps in the reef when waves exceed 1 m. The beach is backed by generally vegetated, 30-40 m high transgressive dunes.

Beach WA **904** commences at the end of the straight west-trending track off the Tuart Road. It is a 350 m long section of sandy beach fronted by a discontinuous section of intertidal calcarenite, together with some reefs off the beach. Gaps in the reef develop into permanent rips when waves exceed 1 m. Vegetated dunes and 4 WD tracks back the beach. Beach **WA 905** continues north for 1.2 km to a more westerly inflection in the shore. Calcarenite continues to outcrop on and off the beach resulting in crenulations along the shore and permanent rips adjacent to the inner reef during period of higher waves. Vegetated dunes rising to 40 m and extending up to 3 km inland back the beach, together with several 4 WD tracks.

Beach **WA 906** commences at the inflection and trends north-northwest for 9.8 km to the usually blocked mouth of Moore River. The beach is relatively straight and free of reefs. Outer reefs however lower waves to less than 1 m, which usually maintain reflective beach conditions the length of the long beach. The entire beach is backed by predominately vegetated transgressive dunes rising in places to 50 m and extending 3-4 km inland. A few active blowout-parabolics back the central section of the beach and at least three 4 WD tracks reach the beach.

WA 907-910 GUILDERTON-SEABIRD

No.	Beach	Rating	Type	Length
WA 907	Moore River (N)	4	R	1.3 km
WA 908	Guilderton	4	R	700 m
WA 909	Guilderton (N1)	4	R	1.5 km
WA 910	Guilderton (N2)	4	R	1.8 km
WA 911	Cape Leschenault (S2)	4	R+reef	300 m
WA 912	Cape Leschenault (S1)	4	R	1.5 km
WA 913	Cape Leschenault	4	R	250 m
WA 914	Seabird	4	R	7.2 km
Spring & neap tidal range = 0.5 & 0.0 m				

Guilderton is a small settlement of about 500 people located on the northern side of the Moore River mouth (Fig. 4.425), 70 km north of Perth. It is named after the Dutch ship the Guilt Dragon, wrecked off the coast in 1656. Dune-draped calcarenite slopes rising to 50 m border the river, which has maintained its channel through multiple dune transgressions. The mouth is usually blocked only breaking open during winter rains. Much of Guilderton is located on the northern slopes, amongst the 20 to 30 m high undulating vegetated dunes, with a wide reserve adjacent to the beach. The only commercial activity is a caravan park and recreational area on flat land just inside the river mouth, together with a store. There is large car park on the 40 m high slopes above the river entrance, and a second vehicle access at the groyne, 1 km to the north. From the Moore River the coast trends northwest past Cape Leschenault, for 9 km to Seabird, where the coast continues to the north for another 5 km to the next inflection. In between are eight near continuous sandy beaches. All are protected by shore parallel calcarenite reefs and shoals which lie 1-2 km offshore, resulting in waves averaging 1 m or less at the shore and reflective conditions usually prevailing (Fig. 4.226). The only public access is at Guilderton and Seabird, the remainder requiring 4 WD to access through the backing vegetated dunes.

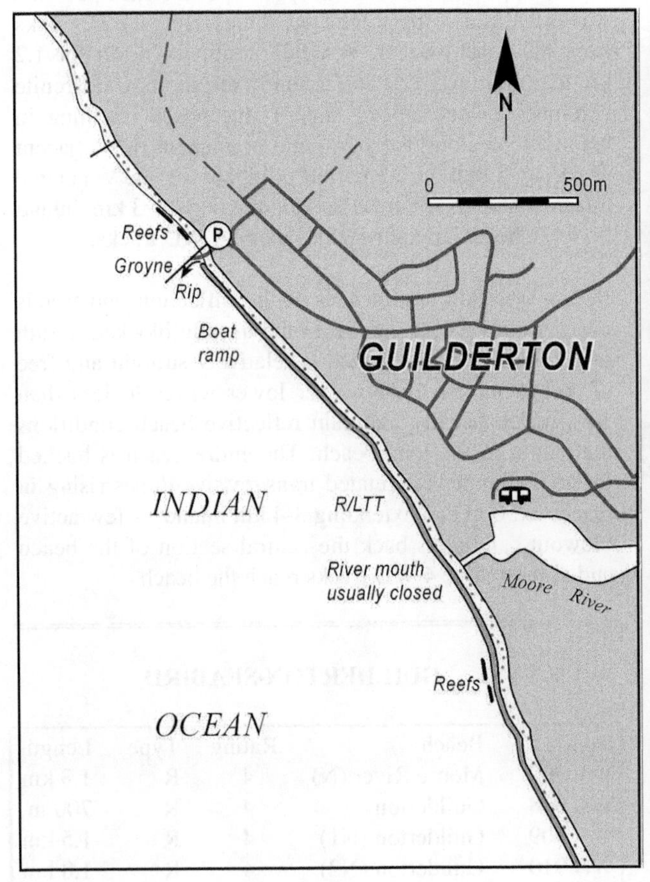

Figure 4.425 Guilderton is a small settlement at the mouth of the Moore River.

Beach **WA 907** faces southwest and extends from the river mouth for 1.3 km north to a 100 m long section of calcarenite that terminates at a groyne and beachside car park. The groyne was built to assist boat launching across the soft beach. Two lines of offshore reefs lower the ocean waves to about 0.5-1 m at the shoreline. The low waves produce a steep beach, with a narrow continuous bar, which is cut by rips during periods of larger waves. A permanent rip runs out against the groyne, and when the river does open it deposits a river mouth bar, which may induce additional rips. Dunes rising to 40 m back the beach, then a reserve and residential development.

Beach **WA 908** commences on the northern side of the groyne with good access via the car park, as well as 4 WD access onto the beach. It trends to the north for 700 m to a slight inflection at a patch of intertidal calcarenite with calcarenite also outcropping along the beach. Waves average about 1 m and can form rips against the groyne and amongst the rocks, particularity during higher wave conditions. At other times large seagrass berms can accumulate either side of the groyne. Vegetated dunes rising steeply to 20 m back the beach.

Swimming: Guilderton offers both usually calm river conditions, and usually low waves and deep water off the beach. During higher waves a heavy shorebreak and rips prevail and rocks dominates the beach north of the groyne. As the beach is not patrolled be careful if swimming on the ocean side.

Figure 4.226 The narrow mouth of the Moore River, with Guilderton behind and the lower energy beaches typical of this section of coast.

Beach **WA 909** extends north of the infection point for 1.5 km to a 500 m long section of calcarenite. The beach alternates between sections of intertidal calcarenite and some sandy areas free of rock. Waves average about 1 m and either break across the rocks or surge up a reflective beach face. Ten to twenty metre high vegetated dunes back the beach, with two areas of active blowouts in the south and centre. Vegetated dunes then extend up to 2.5 km inland.

Beach **WA 910** commences on the northern side of the calcarenite section and continues straight for 1.8 km to the next 100 m long section of calcarenite. Reflective beach conditions usually prevail, with a narrow bar forming during higher wave conditions. This is an undeveloped beach backed by generally well vegetated transgressive dunes, with access along the beach by 4 WD.

Beach **WA 911** is a 300 m long strip of high tide sand fronted by continuous irregular intertidal calcarenite platforms and reefs. It is bordered by two small protrusions in the backing dune-draped calcarenite bluffs. There is 4 WD access to the rear of the southern bluff.

Beach **WA 912** continues north for 1.5 km to **Cape Leschenault**, a 20 m high vegetated calcarenite bluff. The beach is backed by vegetated dunes and bluffs rising to 20 m and one blowout. The main 4 WD access track reaches the southern end of the beach and continues north to Seabird with several access points to the beach. Intertidal calcarenite dominates the shoreline with waves averaging about 1 m.

Beach **WA 913** is a 250 m long strip of high tide sand located between the two bluffs that form Cape Leschenault. The beach is fronted by 50-80 m wide intertidal reefs, with only a small open section at the northern end. It is bordered by vegetated 20 m high calcarenite bluffs and backed by vegetated dune surfaces, with a vehicle track reaching the centre of the beach.

Seabird beach (**WA 914**) commences on the northern side of Cape Leschenault, and trends 2 km northwest to Seabird, where the shore turns more to the north and continues on for another 5.2 km to the next calcarenite-induced inflection in the shore, called Second Bluff. This is a more continuous sandy beach (Fig. 4.227), with waves averaging just under 1 m along most of the beach, through dropping at Seabird in lee of inshore reefs. The lower waves permit fishing boats to anchor off Seabird and small boats are launched from the beach, though the Seabird jetty was destroyed by high waves. Seabird has a population of about 100 and consists of a row of beachfront houses, including a tavern and store, a large beachfront caravan park, and slowly expanding residential development extending up to 500 m inland. Either side of the settlement vegetated Holocene transgressive dunes extend a few hundred metres inland over Pleistocene calcarenite. Vehicle tracks run through the dunes to the north of Seabird with solitary houses located 2 and 5 km north of the settlement.

Figure 4.227 Seabird is located on a slight shoreline protrusion with usually low waves along the beach.

WA 915-920 **BRETON BAY-LEDGE PT**

No.	Beach	Rating	Type	Length
WA 915	Second Bluff	4	R	2.5 km
WA 916	Breton Bay	4	R	3.2 km
WA 917	First Bluff	5	R+reef	400 m
WA 918	First Bluff (N)	5	R+reef	1.3 km
WA 919	Manakoora	4	R	1.5 km
WA 920	Ledge Pt (S 1)	5	R+reef	1 km
Spring & neap tidal range = 0.6 & 0.4 m				

Seabird beach continues north of the settlement for 5 km to a section of outcropping calcarenite that terminates at a 20 m high calcarenite cliff and generates an inflection in the shore. To the north of the cliff is 14 km of essentially north-northwest trending continuous sandy beaches up to Ledge Point. The entire shoreline is paralleled by drowned Pleistocene barriers, which from lines of reefs and shoals located 1-2 km seaward and lower waves to 1 m or less at the shore. It is backed by generally vegetated, and in places cleared, Holocene transgressive dunes extending 2-3 km inland. The only development south of Ledge Point is the farmland and a few beachfront shacks, most accessed by private tracks. There is little public access to the shore.
Second Bluff marks the beginning of a slight indention on the shore. The indentation is partially occupied by the straight sandy beach **WA 915**, which trends north of the cliff for 2.5 km, to a section where the beach narrows against calcarenite outcrops. Waves average less than 1 m and reflective conditions usually prevail. There is a solitary house located 500 m north of the beginning of the beach, while vegetated dunes back the beach.

Beach **WA 916** is located in the slight indentation called Breton Bay, the end of the private Greenwood Road, with no public access. The beach trends straight north for 3.2 km, terminating at First Bluff, a slight protrusion in the shore marks the beginning of a calcarenite section. The beach is usually reflective and backed by a low foredune and generally cleared farmland, with a series of six beachfront shacks located midway along the beach.

Beaches 917 and 918 occupy the calcarenite section of shore. Beach **WA 917** is a 400 m long narrow high tide beach, backed by 5-10 m high calcarenite bluff, then vegetated transgressive dunes, with a large vegetated blowout at the northern end. It is fronted by shallow reefs extending 20-300 m off the shore, with usually calm conditions at the beach. Beach WA **918** consists of two scallops of sand totalling 1.3 km in length. The reefs continue off the beach, with low waves at the shore, while lower dunes and some calcarenite back the beach. A 4 WD track runs along the crest of the dunes behind both beaches.

Beach **WA 919** commences at the northern end of the calcarenite section and marks the beginning of a 1 km long section of active dunes. The beach is 1.5 km in length and backed by the active Manakoora sand dunes for most of its length, which extend up to 1 km inland.

Shallow reefs lie off the beach with waves averaging about 1 m and breaking on some of the inner reefs. It terminates at anther calcarenite section of shore containing beach WA 920. Beach **WA 920** is a 1 km long section of sandy shore fronted by shallow inshore reefs. Seagrass debris commonly accumulates in lee of the reef at the base of the low energy beach. Freshwater springs can be seen in amongst the intertidal reefs at low tide , while active dunes back much of the beach..

WA 921-924 **LEDGE POINT**

No.	Beach	Rating	Type	Length
WA 921	Ledge Pt (S)	4	R	3.3 km
WA 922	Ledge Pt groynes	4	R/LTT	200 m
WA 923	Ledge Pt (N)	4	LTT	2.9 km
WA 924	Ledge Pt (N1)	4/5	LTT/TBR	2.8 km
Spring & neap tidal range = 0.6 & 0.4 m				

Ledge Point is a small bit growing coastal settlement of about 300, located 100 km north of Perth and 4 km off the Lancelin Road. The ledge or reef off the point, has both lead to the development of the low sandy point (Fig. 4.228), as well as providing sufficient protection to permit mooring of about 20 cray boats in its lee. Ledge Point has traditionally been based around crayfishing, but is becoming increasingly popular as a holiday and retirement destination. The township occupies the low sandy foreland that has prograded a few hundred metres seaward in lee of a series of shallow reefs located up to 500 m off the shore. The fishermen's houses are located in a subdivision in lee of the beach, which also contains a small shopping area and a caravan park. At the beach is a large car park, toilet and picnic area. Beaches extends either side of the sandy point, south towards Seabird (WA 921) and north towards Lancelin (WA 922-924).

Beach **WA 921** commences on the northern side of the inshore reef and trends to the north-northwest, past one slight foreland for 3.3 km finally curving more to the west to terminate at the tip of Ledge Point. Waves average about 1 m and break across a narrow bar during higher waves with a reflective beach during more normal conditions. Generally vegetated Holocene transgressive dunes extend 2-3 km in from the beach, with several 4 WD tracks through the dunes to the beach. The northern 500 m of beach is backed by Ledge Point settlement, with a narrow foredune reserve then beachfront houses (Fig. 4.229). Dinghies are launched from the beach to access the fishing boats moor in lee of the reef that lies 500 m off the point.

Beach **WA 922** is located on the western side of the point and is bordered by two rock groynes located 200 apart. This is the designated swimming area and is backed by a low fenced foredune, large car park then houses. The beach is sheltered by the reef and waves are usually less than 0.5 m, with seagrass debris sometimes trapped against one or other of the groynes. There is surf on the reefs out off the point, all of which require a long paddle or boat to access.

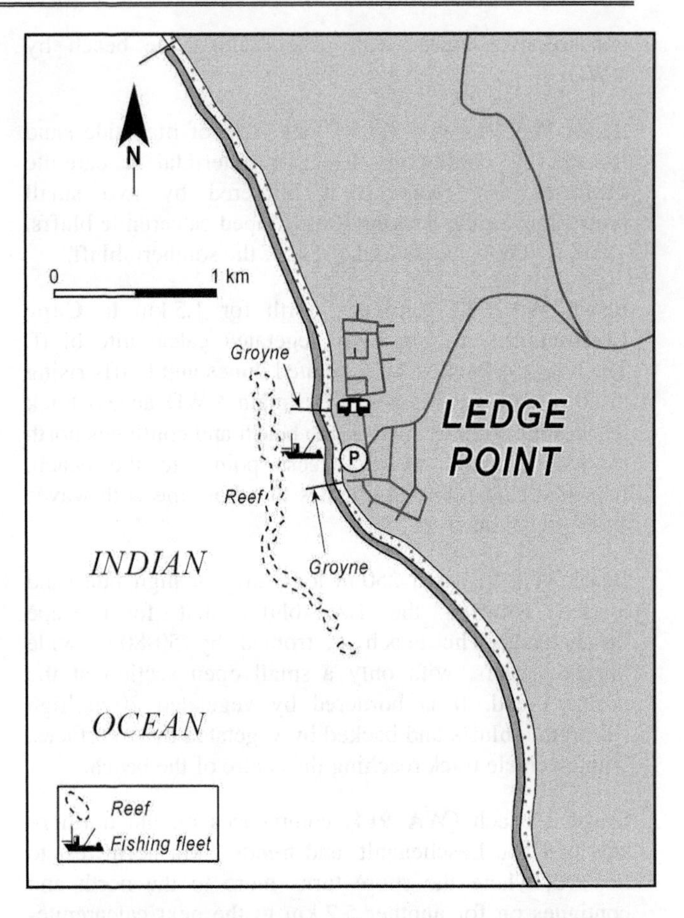

Figure 4.228 The small ledge Point community is located on a sandy foreland formed in lee of a shore parallel reef.

Figure 4.229 Ledge Point is located to the lee of a small calcarenite reef, with boats usually moored off the point.

Beach **WA 923** commences at the northern groyne and trends north-northwest for 2.9 km to the next small sandy foreland. Waves average less than 0.5 m at the point, but increase to about 1 m up the beach where the beach sand coarsens and the beach becomes narrow and steep, with often no bar. The wide foreshore reserve, caravan park, then houses back the southern 1 km of the beach, with a second beach access road and car park north of the caravan park. North of the houses generally vegetated transgressive dunes back the beach, with a 1 km wide area of active dunes inland of the northern end of the beach.

Beach **WA 924** commences on the northern side of the boundary protrusion and continues north-northwest for 2.9 km to the next slight sandy foreland. The beach is sheltered by reefs located off either foreland, with some slightly higher waves, up to 1 m reaching the more open central section of the beach, where they may form a narrow bar cut by occasional beach rips. Access is only by 4 WD along the beach or through the generally vegetated dunes, with some dune activity behind the southern 1 km of the beach.

Swimming: The Ledge Point beaches have a low hazard rating owing to their usually low waves and protected location. If you are swimming stay between the two groynes and clear of the boat launching area on the point, and watch for beach vehicle traffic. The best spot is to the south where there is a shallow bar off the beach. Rips are usually absent, but will occur when waves exceed 0.5 m.

WA 925-928 LANCELIN

No.	Beach	Rating	Type	Length
WA 925	Lancelin (S)	4→5	LTT→TBR	4.9 km
WA 926	Lancelin	3	R/LTT	1.8 km
WA 927	Lancelin (N)	2	R+reef	1.2 km
WA 928	Horseshoe Reef	2	R+reef	2.5 km
Spring & neap tide range:		0.6 & 0.4 m		

Lancelin is moderate size fishing and holiday town of 800 people located 110 km north of Perth, and at the end of the Lancelin Road. The road terminates here, with the coast to the north taken over by extensive bare coastal dunes and deflated surfaces. The town spreads for 5 km along the coast, occupying three sandy foreland formed in lee of Edwards Reef and Lancelin Island (Fig. 4.230). The sea forms the western boundary with active dunes bordering the eastern side of the settlement. The near continuous string of shallow reefs and islands, effectively block most of the ocean waves and permit the large fishing fleet to moor between the reefs and beach. Lancelin has a small shopping area, a tavern and beachfront caravan park. A 30 m high water tower dominates the horizon. There are also two jetties to service the fishing fleet. However most small boat launching is done from the beach. Like all the forelands along this section of coast sandy beaches extend to the south (WA 925) and north, with beaches WA 926-928 fronting the elongate township.

Beach **WA 925** commences at the sandy foreland it shares with its southern neighbour (WA 924) and trends north-northwest for 4.9 km finally curving to the west for 600 m on the southern side of the Lancelin foreland (Fig. 4.231). The beach is composed of finer sand, which together with waves averaging up to 1 m produces a wider, firm low gradient beach fronted by a continuous shallow bar. The highest waves occur 500 m to 2 km south of the point and usually maintain rips across the bar and surf on the reefs off the northern section of the beach.

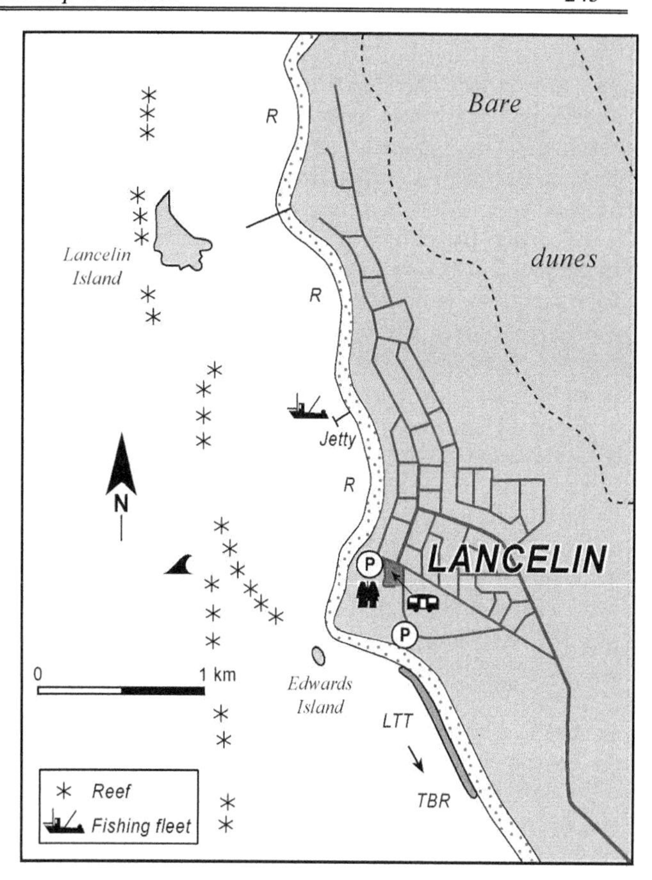

Figure 4.230 Lancelin occupies a low foreland and spreads north to a large active dune field.

Figure 4.231 Lancelin extends north from the prominent foreland, with the Indian Ocean to the west and active dunes to the east.

Lancelin beach (**WA 926**) commences at the tip of the foreland formed in lee of the small Edward Island and its larger surrounding reef. The beach curves round to the north for 1.8 km to the next foreland, where the main fishing jetty is located. The beach is sheltered by the reefs off each foreland, which provide only a small central opening for ocean waves. As a result conditions are often calm at the low gradient shoreline and numerous fishing boats moor off the northern end of the beach. The beach is backed by the main caravan park on the foreland, then a wide foreshore reserve and the township. Beach **WA 927** commences at the main jetty and curves to the north for 1.2 km to the next foreland, formed in lee of 20 m high Lancelin Island. A second jetty is located at the southern end of the beach. It is backed by a narrow reserve and beachfront houses. Wave remain low at the beach and low gradient reflective conditions usually prevail. Beach **WA 928** continues north of the island foreland for 2.5 km, terminating at the next foreland in lee of inner Horseshoe Reef. Houses extend along the southern kilometre of shore before they give way to a 1 km long section of active dunes, that extend up to 2 km inland and border the northern and much of the eastern side of the town. The beach remains sheltered by Lancelin Island and a near continuous chain of reefs that lie 500-1000 m offshore. A surf break called *Hole in the Wall* is located on reefs seaward of the main jetty.

Swimming: The safest swimming is in the main town beaches (WA 926-928) and harbour area, clear of the boat launching. The south beach (WA 925) has higher waves averaging 0.5 to 1 m and is more likely to have rips. The northern beach (WA 928) has low waves, but a steep beach and deep water close inshore.

Surfing: Lancelin picks up more swell than much of the coast to the south, and consequently has a number of spots on the south beach (WA 925) and outer reefs. The caravan park has a special beachfront area set aside of surfers. *Edward Island* in the south is a lefthand reef break, while straight out in front of the caravan park are two reef breaks located about 200 m offshore, known as *The Passage* and *Hole in the Wall.*

WA 929-935 HORSESHORE REEF-WEDGE PT

No.	Beach	Rating	Type	Length
WA 929	Horseshoe-Virgin Reefs	5→6	LTT-TBR	2.4 km
WA 930	Virgin Reef (N)	6	TBR+reef	1.6 km
WA 931	Dide Bay	4	LTT+reef	1.5 km
WA 932	Dide Bay-Narrow Neck	4	R/LTT	4.3 km
WA 933	Narrow Neck (N)	6	TBR	5.7 km
WA 934	Wedge Island (S 1)	6	TBR	5.2 km
WA 935	Wedge Island (S)	6→5	TBR-LTT	3.1 km
Spring & neap tidal range = 0.6 & 0.4 m				

North of Lancelin the shoreline remains relatively undeveloped for 60 km up to Cervantes. The former shack communities at Dide Bay and Narrow Neck were removed during the late 1990's, as the management of the coast became better organised, and the world famous Nambung National Park was better developed. The first 25 km of coast between Lancelin and Wedge Island is occupied by the three Lancelin beaches (WA 926-928), then seven near continuous sandy beaches, each bordered by sandy forelands formed in lee of the chains of calcarenite reefs and occasional small islands that parallels the coast, usually located a few hundred metres offshore off the beach. All the beaches face west-southwest towards the dominant waves, however wave energy at the shoreline depends on the nature of the reef, with higher waves (~ 1 m) only occurring where the significant gaps occur or the reef are deeper. All the beaches are accessible by 4 WD tracks that parallel the coast all the way to Cervantes. Both vegetated and active Holocene dunes extend up to 5 km inland, widening to 12 km in lee of the Wedge Island area. All the dunes are associated with northward migrating or orientated longwalled parabolics formed by the prevailing southerly winds.

The Nilgen Nature Reserve extends along the coast from 1 km north of Horseshore Reef to Dide Bay (beaches WA 929-931), followed by a Military Training area from Dide Bay to 4 km south of Wedge Island (beaches WA 931-934). The Wanagarren Nature Reserve brackets then crown land of Wedge Island and extends north to the southern boundary of Nambung National Park (beaches WA 934, 936-946).

Beach **WA 929** is located between the forelands in lee of Horseshoe and Virgin Reefs, both more prominent inner reefs, with the forelands attached to the inner parts of the reefs. The beach is 2.4 km long and while sheltered at either end, receives waves averaging about 1 m along the central section. These break both on inshore reefs and in places across a narrow attached bar, with up to eight partially reef-controlled rips forming during periods of higher waves. Partly active dunes extend up to 200 m inland, fronting generally stable Holocene transgressive dunes up to 5 km wide. There is surf along the beach and on the left- and right-hand breaks on *Virgin Reef.*

Beach **WA 930** extends for 1.6 km north of Virgin Reef to a small reef-tied foreland. Patchy reefs dominate the inshore, with the shoreline ranging from reef to sand bar, and rips forming adjacent to most of the reefs. Beach **WA 931** continues on the northern side of the foreland to Dide Bay point 1.5 km to the north. This beach is free of inshore reefs, but receives slightly lower waves owing to the presence of outer reefs. Reflective to low tide terrace conditions prevail along most of the beach, with the highest waves towards the less protected southern end. Both beaches are backed by generally vegetated transgressive dunes, apart from an area of active dunes in lee of Virgin Reef.

Beach **WA 932** commences at the prominent Dide Bay foreland, and extends north for 4.3 km to the Narrow Neck foreland. Inner and outer reefs parallel the beach resulting in lower wave at the shore. Waves are lower along the southern half where a low tide terrace is usually present (Fig. 1.2b), with slightly higher waves along the northern half occasionally forming beach rips. Vegetated dunes back the entire beach, while the area north of the bay is a Military Training Area, with access restricted during training exercises.

Beach **WA 933** extends from the Narrow Neck foreland for 5.7 km to the next foreland, which is tied to an inner reef-capped by a small sea stack. While the southern and northern corners of the beach are relatively sheltered, much of this longer beach receives waves averaging over 1 m which together with the fine beach sand maintain a 100 m wide low gradient surf zone, usually cut by rips every 200-300 m. Generally vegetated dunes back the beach, with two areas of northward migrating parabolics located in from the centre of the beach.

Beaches WA 934 and 935 occupy the 8 km long embayment formed between the northern end of beach WA 933 and the prominent Wedge Island foreland that extends 2 km west of the trend of the coast (Fig. 4.432). The beaches are separated by a smaller reef tied-foreland. Beach **WA 934** extends for 5.2 km between the southern forelands, and is relatively well exposed to waves averaging over 1 m. These maintain a low gradient 100 m wide surf zone, with rips spaced every 200-300 m. Vegetated dunes and two large active parabolics back the beach. Beach **WA 935** commences at the smaller foreland and trends to the north then west for 3.1 km to the tip of the Wedge Island foreland. The low Wedge Island lies just off the tip of the foreland. Waves are highest along the southern section, where they maintain a wide low gradient bar cut by 2-3 rips. Wave height decreases slightly towards the foreland and as the sand fines and the bar widens to 80 m. A large parabolic dune extends north of the beach and across the rear of the foreland to reach beach WA 936.

Figure 4.432 Beaches WA 935 & 936 are located to either side of the cuspate foreland formed in lee of Wedge Island.

WA 936-942 **WEDGE ISLAND (N)**

No.	Beach	Rating	Type	Length
WA 936	Wedge Is	4→5	LTT→TBR	4 km
WA 937	Wedge Is (N)	5	LTT/TBR	2.2 km
WA 938	Flat Rock reef	5	LTT/TBR	1.8 km
WA 939	Flat Rock reef (N)	5	LTT/TBR	2 km
Spring & neap tidal range = 0.6 & 0.4 m				

JURIEN BAY MARINE PARK

Wedge Island-Grey-Cervantes-Jurien-Green Head
Established: 2003
Coast length: 98 km (2338-2436 km)
Beaches: 79 (WA 935-1014)

Jurien Bay Marine Park extends for 88 km between south of Wedge Island to Green Head, and up to 12 km seaward, with a total area of approximately 90 000 ha. The park encompasses the intervening waters and islands, and is bordered along the shore by 79 near continuous sandy beaches, most sheltered by the numerous rocks, reefs, islets and island of the park.

The prominent Wedge Island foreland forms a major protrusion on the otherwise relative straight coastline between Lancelin and Cervantes. While the southern foreland beach (WA 935) faces south, the northern beach (WA 936) faces due west. The shoreline continues to the north-northwest with minor undulations induced by offshore calcarenite reefs, and past Flat Rock, by calcarenite outcrops along the shore. Between the tip of the foreland and the site of the old Flat Rock shack community is 12 km of coast containing eight beaches, the southern four (WA 936-939) are relatively free of rocks and reef, while the northern four (WA 940-943) are dominated by inshore reefs and backing calcarenite. The Lancelin-Cervantes 4 WD track runs along the beaches for 8 km to the lee of Flat Rock Reef, then follows the crest of the calcarenite section.

Wedge Island beach (**WA 936**) commences at the foreland and trends due north for 4 km to an intertidal outcrop of calcarenite. The continuous sandy beach is composed of fine carbonate sands and has a low gradient, with a low tide terrace to the lee of the southern reefs, while wave height increases sufficiently by the northern end to maintain a 50 m wide bar cut by rips during and following periods of higher waves. The Wedge Island shack community used to cover much of the foreland, with fishing boats moored off the southern end of the beach. The beach is backed initially by the vegetated old shack area, then by an active sand sheet that extends 3 km to the north and 1-2 km inland.

Beach **WA 937** commences on the northern side of the small reef and continues north for 2.2 km to the next reef-induced inflection in the shore. Waves average over 1 m and break across a 100 m wide low gradient bar, usually cut by 1-2 beach rips, with permanent rips flowing out against the boundary reefs. The active sand sheet continues to the rear of the beach. Beach **WA 938** extends for another 1.8 km to the north of the inflection, to a 300 m long section of intertidal calcarenite, located in lee of Flat Rock reef. Waves average just over 1 m and break across a central 100 m wide low gradient bar, usually cut by two beach rips. A 500 m wide stablising parabolic and vegetated dunes back the beach (Fig. 4.433).

Figure 4.433 An active parabolic dune is moving northward in lee of beach WA 938, which in turn is fronted by Flat Rock reef.

Beach **WA 939** commences at the centre of the boundary calcarenite and extends 2 km to the north terminating at the first outcrop of low calcarenite bluffs. The 4 WD track leaves the beach at the bluff and continues north through vegetated dunes to the rear of the beaches. Waves continue to maintain a wide low gradient bar cut by 2-3 beach rips, as well as breaking on the northern reefs.

WA 940-943 **FLAT ROCK**

No.	Beach	Rating	Type	Length
WA 940	Flat Rock (S3)	4	LTT+reef	400 m
WA 941	Flat Rock (S2)	3	R+reef	150 m
WA 942	Flat Rock (S1)	3	R+reef	200 m
WA 943	Flat Rock	3	R+reef	800 m

Spring & neap tidal range = 0.7 & 0.5 m

Flat Rock is the name of a reef located off beach WA 938, and of a former small shack community located to the lee of beach WA 943. The 4 WD track that leaves the beach at the end of beach WA 939 continues along the rear of the coast and connects the beaches to the north. Beaches WA 940 to 943 occupy 2.5 km of low calcarenite shoreline fronted by shallow inshore reefs extending up to 500 m offshore, with the main reef systems paralleling the shore 3 km seaward. Generally vegetated transgressive dunes extend 4-5 km inland.

Beach **WA 940** commences 50 m north of the end of beach WA 939. It is bordered and backed by low 3 m high calcarenite bluffs, with patchy reef to either end and offshore and only a 100 m long central sandy section, with a low gradient bar. Waves break on the reefs 100-300 m offshore, with usually calm conditions at the shore.

Beaches **WA 941** and **942** are located 200 m further north and consist of two narrow high tide sandy beaches 150 and 200 m long respectively, separated by a 100 m long section of low bluffs. Both beaches are bordered and backed by the low bluffs, and fronted by continuous intertidal calcarenite reefs. Waves break over outer reefs and then the inner reefs, with usually calm conditions against both beaches. The 4 WD track runs along the rear of both beaches.

Beach **WA 943** is located 700 m to the north and consists of a slightly seaward curving 800 m long sandy high tide beach, located to the lee of inner reefs, with patchy intertidal calcarenite paralleling most of the shore. The southern side of the foreland is known as Lawry Bay. Between the shore and the inner reefs is a 100 m wide open 'lagoon' with a generally sandy seafloor. Low waves to calm conditions commonly prevail at the beach. The former shack community was located along the rear of the southern half of the beach.

NAMBUNG NATIONAL PARK

Established:	1956	
Area:	approx. 17 500 ha	
Coast length:	21 km	(2357-2378 km)
Beaches:	16	(WA 947-963)

Nambung National Park is one of the better known Western Australian parks owing to the exposure of thousands 2-3 m high Pleistocene root casts known as 'the Pinnacles'. The stark pinnacles against the backdrop of deflated dune surfaces, with the coast behind, are a feature of many Western Australian tourist promotions. In addition the park offers a wide range of Holocene and Pleistocene vegetated and active dune environments, as well as 21 km of coastline containing 16 low energy beaches. Picnic facilities are provides at Hangover Bay (beach WA 968) and Kangaroo Point (beach WA 962).

WA 944 **FLAT ROCK (N)-GREY (S)**

No.	Beach	Rating	Type	Length
WA 944	Flat Rocks (N1)	3	R+reefs	300 m
WA 945	Flat Rocks (N2)	3	R+reefs	150 m
WA 946	Flat Rocks (N3)	3	R+reefs	750 m
WA 947	Grey (S2)	3	R+reefs	1 km
WA 948	Grey (S1)	3	R+reefs	1.2 km

Spring & neap tidal range = 0.7 & .5 m

Between the former Flat Rock shack area and Grey is a relatively straight 6 km of north-northwest trending shoreline, dominated by low calcarenite bluffs and inner and outer calcarenite reefs. To the north of Flat Rock beach (WA 943) are five beaches (WA 944-948) the northernmost terminating on the southern side of the small Grey headland. The 4 WD track runs along the rear of beaches WA 944-947, before turning 500 m inland to reemerge on the coast at Grey. All the beaches are backed by vegetated transgressive dunes rising to 30-40 m, with a large active sand sheet extending from 0.5-3 km east of Grey. The southern boundary of the Nambung National Park is located towards the southern end of beach WA 947.

Beach **944** is located below the 3-5 m high calcarenite bluffs that extend north of Flat Rock beach (WA 943). The 300 m long beach consists of a narrow strip of high tide sand fronted by a near continuous 50 m wide intertidal calcarenite reef, with inner reefs extending 500 m off the beach, and outer reefs 2-3 km offshore. Waves are low to calm at the shore. Beach WA **945** is located 100 to the north and is a 150 m long pocket of sand contained in a V-shaped gap in the bluffs, with the edge of the bluff surrounding the small beach, and a small grassy foredune located at the rear of the beach, below the bluffs. It is fronted by patchy shoreline reef, then the inner and outer reefs.

Beach **WA 946** is located 200 m to the north and is a relatively straight 750 m long high tide beach, for the most part fronted by irregular intertidal reef, with only the northern 50 m offering sandy access to the sea. It is backed by a 10-15 m high vegetated foredune, then the stable transgressive dunes extending a few kilometres inland. Two 4 WD tracks cross the foredune to reach the beach.

Beach **WA 947** commences 1.2 km to the north and is a 1 km long sandy high tide beach with a slight foreland in the centre and bordered by low calcarenite bluffs. The entire beach is fronted by a 100-150 m wide band of inner reefs, which result in usually low waves at the shore. It is backed by vegetated foredune and transgressive dunes, with a 4 WD track reaching the northern end of the beach.

Beach **WA 948** extends south of the low Grey headland for 1.2 km terminating 200 north of beach WA 947. The inner reefs continue along the entire beach, narrowing towards the northern end. Waves remain low at the shore, apart from the northern few hundred metres where the reef deepens and a narrow surf zone prevails. The transgressive dunes continue to the rear.

WA 949-951 **GREY**

No.	Beach	Rating	Type	Length
WA 949	Grey point	4	R+reef	100 m
WA 950	Grey	3	R	350 m
WA 951	Grey (N)	3	R	1.5 km
Spring & neap tidal range = 0.7 & 0.5 m				

Grey is a small fishing community consisting of about 50 shacks, located 20 km south of Cervantes. It occupies the rear of a low V-shaped calcarenite point and beaches WA 950 and 951, extending in total for 1 km alongshore and up to 300 m inland (Fig. 4.234). The area of freehold land is now surrounded by Nambung National Park. Two beaches border the point (WA 949 & 950) with beach WA 951 beginning the near continuous strip of sandy beaches that extend for 18 km to Cervantes.

Figure 4.234 The small community of Grey extends north of a rocky foreland, where it has a reef-sheltered low energy beach.

Beach **WA 949** is located along the southern side of the low point. It consists of a narrow 100 m long strip of high tide sand located below crumbling 10 m high bluffs and fronted by a 50 m wide irregular intertidal reef, with waves breaking over the edge of the reef. Outer reefs lie 2-3 km offshore.

Grey beach (**WA 950**) is the main beach and extends from the northern side of the point for 250 m to a small outcrop of calcarenite. It is bordered and almost enclosed by Longman Reef lying 100-200 m offshore, which forms a protected central sandy 'lagoon' and reflective shoreline. The lagoon is used as a mooring area by the fishing boats.

Beach **WA 951** commences 350 m north of the point and trends relatively straight and clear of inner reefs, for 1.5 km to the north-northwest, where fringing reefs then dominate the next 300 m long section of shore. Waves are lowered by the outer reef 2-3 km offshore and average about 0.5 m along the moderately steep reflective beach. The shack settlement extends along the southern 500 m of the beach, which is backed by 500 m of vegetated dunes, then the northern end of the active Grey sand sheet.

WA 952-956 **GREY (N)-SHAG ROCK**

No.	Beach	Rating	Type	Length
WA 952	Grey (N1)	3	R+reef	1.7 km
WA 953	Grey (N2)	3	R+reef	1 km
WA 954	Boggy Bay	3	R+reef	1.9 km
WA 955	Shag Rock (S)	3	R+reef	1.3 km
WA 956	Shag Rock	3	R+reef	300 m
Spring & neap tidal range = 0.7 & 0.5 m				

The northern end of beach WA 951 marks the beginning of a more irregular section of coast that continues for 16 km to Cervantes. The irregularities are a result of wave refraction over the patchy shore parallel reefs, which induce wave refraction and have lead to the formation of large sandy forelands and intervening bays.

The overall trend of the coast however remains to the north-northwest. Beaches WA 952 to 956 occupy a 6 km long section of coast dominated by both the inner and outer calcarenite reefs. It is backed by vegetated transgressive dunes extending 6-7 km inland, then an area of deflated Holocene and Pleistocene dunes that includes The Pinnacles, due west of beach WA 952. The coast 4 WD track runs a few hundred metres inland with spur tracks running out to most of the beaches.

Beach **WA 952** commences in lee of the attached calcarenite reef section and curves gently to the north for 1.7 km. For most of its length it is fronted by inner- to attached-patchy reef, with the outer reef located 2-3 km further out. Waves remain low at the shore and reflective conditions prevail at high tide. Beach **WA 953** commences on the northern side of a slight sandy foreland and curves to the north for 1 km to the next foreland. Inner reefs parallel the southern half of the each, with a shallow reef lying 100-200 m off the central-northern section of the beach, forming a 'lagoon' section between the reef and the relatively calm sandy shore. A 4 WD track reaches the centre of the beach.

Beach **WA 954** extends from the boundary sandy foreland for 1.9 km as an initially curving sandy beach, then along a blunt foreland to terminate at its northern tip. The beach is moderately open along its curving southern half and receives some low waves, which surge up the moderately steep beach face. Reefs fringe the northern half with low wave to calm conditions at the shore. Vegetated foredune and transgressive dune back the beach, which has an access track reaching the northern end of the southern section.

Beach **WA 955** occupies the next embayment and is a curving almost semicircular 1.3 km long sandy beach, bordered by prominent attached shallow reefs, but with a relatively reef-free central section (Fig. 4.235). This permits seagrass to gown on the sandy seafloor and low waves to reach the shore. A 200 m wide active parabolic dune extends for 500 m due north of the centre of the beach. There is a vehicle access track via the northern foreland to the beach. Beach **WA 956** occupies the northern west-facing foreland in lee of Shag Rock reef. The relatively straight 300 m long beach is fringed by continuous calcarenite reef that extends 200-300 m seaward, resulting in usually calm conditions at the shore.

WA 957-961 **HANGOVER BAY-KANGAROO POINT**

No.	Beach	Rating	Type	Length
WA 957	Hangover Bay (1)	2	R	550
WA 958	Hangover Bay (2)	2	R	400 m
WA 959	Hangover Bay (3)	2	R/LTT	1.4 km
WA 960	Kangaroo Pt (S)	2	R+reef	400 m
WA 961	Kangaroo Pt	2	R+reef	1.7 km
WA 962	Nambung Bay	2	R+reef	4.6 km
Spring & neap tidal range = 0.7 & 0.5 m				

Figure 4.235 View along beach WA 966 to the Shag Rock sandy foreland.

Hangover Bay is a 2.5 km wide southwest-facing bay located 10 km south of Cervantes. The bay is bordered by substantial cuspate forelands to either side, including the 2 km long Kangaroo Point to the north, followed by 4 km long Nambung Bay. Both the bay and point are the focus of coastal recreation within the Nambung National Park, with access and facilities at both locations. There are three continuous beaches (WA 957-959) within the bay, two (WA 960-961) on the point and then the longer Nambung Bay beach (WA 962).

The southernmost Hangover Bay beach (**WA 957**) is a curving 550 m long west-facing sandy beach tied to the extensive Shard Rock reef in the south and a second intertidal reef to the north, both extending 200-300 m off the beach, with a central more sandy seafloor. Waves are usually low to calm. A car park is located at the northern end of the beach in lee of the reef. Its neighbour (**WA 958**) extends from the northern foreland for another 400 m to the next more subdued foreland. It also has boundary reefs and a sandy centre narrowing to a central channel. Waves are usually low to calm at the shore. The main car park and a range of facilities including picnic area, bar-b-ques, toilets and beach boat launching facilities are provided in lee of the more open sandy section. Both beaches are backed by vegetated transgressive dunes.

The northern bay beach (**WA 959**) continues north of the second foreland for 1.4 km to the southern side of the Kangaroo Point foreland. The beach faces southwest, which combined with a more open central section permits low waves to usually reach the beach at time forming a narrow low tide terrace. The beach has prograded about 200 m into the bay and is backed by seven vegetated foredune ridges, with transgressive dunes behind. There is no formal vehicle access except along the beach.

Kangaroo Point extends north for 2 km and contains two beaches. Beach **WA 960** is located on the southern tip and is a 400 m long southwest-facing sand beach fronted by continuous shallow calcarenite reefs and seagrass meadows, with seagrass debris commonly piled up on the beach. It has also prograded about 100 m seaward and is backed by four recurved foredune ridges. The main point beach (**WA 961**) continues due north for 1.7 km to the southern end of Nambung Bay. It is a crenulate beach responding to the shallow reefs that extend along the shore and result in low wave to calm conditions at the shore. A series of crenulate foredune ridges back the beach extending up to 400 m to the east. The have been transgressed in the south by now vegetated parabolic dunes (Fig. 2.24a).

Nambung Bay (**WA 962**) is a curving 4.6 km long west-southwest-facing sandy beach tied to Kangaroo Point in the south and a small foreland to the north. A line of reefs lying 1-3 km offshore lower waves sufficiently to permit extensive seagrass meadows to gown almost to the shore the length of the bay resulting usually low waves to calm conditions at the shore. The Kangaroo Point recreational area is located at the southern end of the bay and contains a picnic area, bar-b-que facilities and toilets. Most of the bay shore is backed by vegetated transgressive dunes, apart from a parabolic that has blown from the centre of the beach up to 1.5 km to the north.

WA 963-966 CERVANTES

No.	Beach	Rating	Type	Length
WA 963	Hansen Bay	3	R	2.6 km
WA 964	Thirsty Point	2	R+sand flats	700 m
WA 965	Cervantes	1	R+seagrass	1.4 km
WA 966	Ronsard Bay	2	R	4.4 km
Spring & neap tidal range = 0.7 & 0.5 m				

Cervantes is a growing coastal town of several hundred, located 180 km north of Perth and 40 km in from the highway. It has gained increasing prominence in recent years as the gateway to Nambung National Park, site of the world famous 'pinnacles' formations. It is possible to travel up the coast from Lancelin to Cervantes by 4WD, however most people prefer the longer sealed road access. The town is located on a prominent sand foreland, called Thirsty Point, that protrudes over 1 km seaward in lee of shallow offshore reefs. To the south of the point is 2 km wide Hansen Bay (WA 963), and to the north 6 km long Ronsard Bay (WA 964-968), the southern portion of which contains the safest anchorage and houses the Cervantes fishing fleet and the town area (Fig. 4.236). The town has all basic facilities including a caravan park and licensed fishing club. There are three jetties to service the fishing boats, with small boat launching off the firm beach. Between Cervantes and Jurien, 22 km to the north, is a near continuous crenulate sandy shoreline containing nine beaches (WA 964-972).

Hansen Bay (**WA 963**) occupies the southern side of the Cervantes foreland, which terminates at Thirsty Point. The bay curves to the southeast for 2.6 km to the foreland it shares with Nambung Bay. Its orientation and the fact that the calcarenite reefs tend to parallel each side permit slightly higher waves to reach the northern shore. Seagrass and calm conditions dominate the southern half of the beach with low waves along the northern half out to the point. There are two roads out from Cervantes to car parks and beach access at the point and about 500 m and 1 km east of the point, the latter suitable for boat launching. Ruins of a jetty are located in the middle of the beach, with vehicle access also to the southern end. The northern end of the national park crosses the centre of the beach.

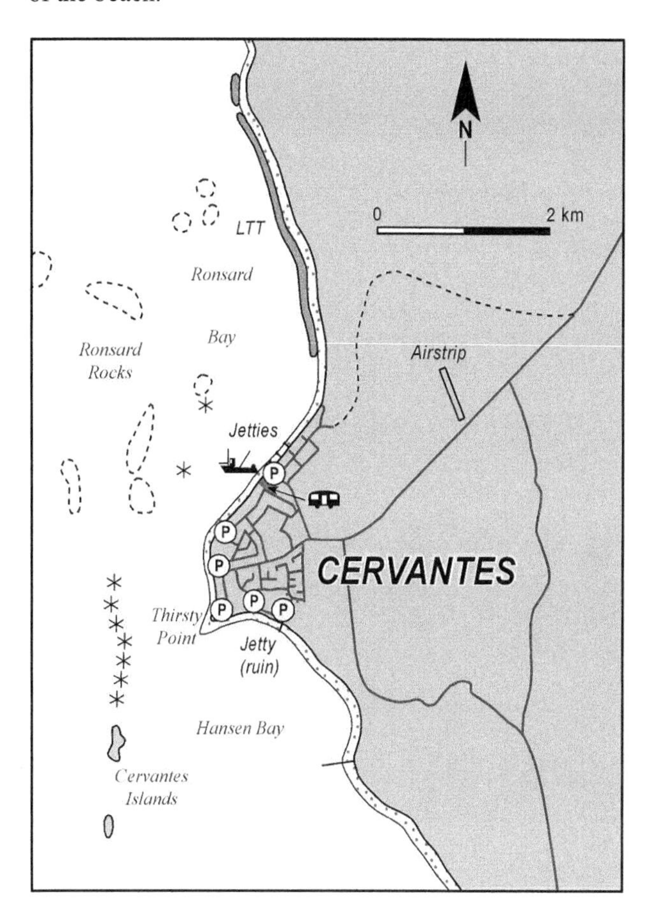

Figure 4.236 Cervantes occupies the Thirsty Point foreland, with the Rosnard Rocks (reefs) providing a relatively sheltered shoreline.

Thirsty Point beach (**WA 964**) extends from the southern tip of the foreland north for 700 m to a sandy inflection (Fig. 4.237). The beach faces west across a shallow sandy seabed, with reefs located several hundred metres offshore, which result in low wave to calm conditions at the shore. The beach is backed by a wide reserve containing a series of low vegetated foredune ridges, with car parks located at either end of the beach. This is one of the two main swimming beaches for the town.

Cervantes beach (**WA 965**) commences at the inflection and trends to the northeast for 1.4 km past three jetties to a 50 m long rock groyne. The beach is usually calm and seagrass meadows grow to the shore (Fig. 4.238). The beach is divided into four sections. The southern section is backed by two car parks and a dune reserve and is designated as a swimming area. Next is a boat launching area at the end of Weston St, followed by another swimming area in front of the large caravan park, with a dune reserve between the park and beach. Finally the northern section of beach, along the jetties, services the

Figure 4.237 The prominent Thirsty Point separates Hansen Bay (top) from Cervantes (left).

fishing fleet that anchors off the beach. Fish processing facilities back this section of the beach.

Figure 4.238 Cervantes township and beach are spread along the relatively calm shores of Ronsard Bay.

North of the groyne beach **WA 966** curves to the north for 4.4 km to the southern side of **Black Point**, a 200 m long 15 m high outcrop of calcarenite. Wave height increases slightly up the beach and the seagrass meadows move a few tens of meters off the shore, and are absent along the northern half of the beach. The sailing club is located 200 m north of the groyne with this section of beach used for swimming and launching sailing boats. North of the club the beach is undeveloped and backed by vegetated foredune ridges in the south grading to vegetated and some active parabolic dunes to the north (Fig. 2.24b), with a second smaller active dune on Black Point. A 4 WD track follows the rear of the beach to Black Point. The northern half of the beach and backing dunes are the southern most part of the Beekeepers Nature Reserve.

Swimming: The safest swimming is in the three designated swimming areas on the Cervantes beachfront. Elsewhere beaches are relatively safe when waves are low, however beware of boat traffic away from the swimming areas.

Surfing: The only surf is on some of the outside reefs, which require a boat and local knowledge to reach.

INDIAN OCEAN DRIVE is the name of the 100 km long coastal drive from Cervantes north to Jurien Bay, Green Head and Coolimba to connect with the Brand Highway, north of Cliff Head. The drive was completed in 2001 with the construction of the new Cervantes to Jurien Bay and Juiren Bay to Green Head sections. The through road has greatly improved access to these communities and provided a more direct and scenic coastal route to the north.

WA 967-972 BLACK PT-ISLAND PT

No.	Beach	Rating	Type	Length
WA 967	Black Pt (N)	3	R/LTT	800 m
WA 968	Black Pt (N1)	2	R	3 km
WA 969	Hill River (S)	2	R	3.8 km
WA 970	Hill River (N)	2	R	2 km
WA 971	Island Pt (S1)	2	R	2.3 km
WA 972	Island Pt (S)	2	R/LTT	4 km
Spring & neap tidal range = 0.7 & 0.5 m				

Black Point is one of the few calcarenite outcrops along this section of coast. The small outcrop marks the beginning of a continuous 16 km long section of sandy coast that terminates at Island Point the next major sand foreland, on the northern side of which is the town of Jurien. Between the two points are six beaches, separated by slight forelands and the Hill River. There is 4 WD access to Black Point and Hill River from the south, and to Hill River from the north, with most of the coastline undeveloped and backed by both active and vegetated transgressive dunes.

Beach **WA 967** commences on the northern side of 15 m high Black Point and extends for 800 m to the north to a 100 m long outcrop of calcarenite. A usually blocked elongate creek-lagoon backs the beach, with a few vegetated foredune ridges between the beach and creek. When the creek does break open is runs out against the rocks of Black Point. The beach receives waves averaging 0.5-1 m, which maintain a reflective to at time low tide terrace shoreline, usually free of rips.

Beach **WA 968** commences on the northern side of the second headland and continues north for 3 km to the centre of a slight sandy foreland formed in lee of some inshore reefs. It is backed by three active parabolic dunes, the largest extending from the southern headland for 2 km to the north. Waves average less than 1 m and maintain reflective to low tide terrace conditions. Beach **WA 969** commences on the foreland and continuous for another 3.8 km to the usually blocked mouth of the small Hill River, which is backed by a 10 ha lagoon. The beach is backed by a few vegetated foredune ridges, a large vegetated parabolic dune and a small area of dune activity extending a few hundred metres south of the river mouth. Waves continue to average less than 1 m and combined with coarser sand usually maintain a steep reflective beach.

Beach **WA 970** commences on the northern side of the river mouth and trends north for 2 km to a reef-induced inflection in the shore. Stabilising parabolic dunes back most of the beach. Waves increase slightly along the beach with a low tide terrace tending to form along the northern section of the shore, with waves also breaking over the shallow boundary reef. Beach **WA 971** extends north from the reef for 2.3 km to the next reef-induced small foreland. This is a more crenulate beach owing to inshore reefs and seagrass meadows with low waves and reflective beach conditions. It is backed by three areas of now stabilising dune activity.

Beach **WA 972** commences on the northern side of the foreland and curves to the north then northwest for 4 km to the tip of Island Point (Fig. 4.239). A 2 km long parabolic dune commences at the southern foreland, with the small shack settlement of Booka Valley, located between the active dune and the beach 1 km north of the foreland. The remainder of the beach is backed by a series of up to 30 curving foredune ridges that are part of the 6 km long foreland complex. As the beach turns to faces the southwest wave height increases slightly, with reflective to low tide terrace conditions along the northern foreland section. A gravel road from Jurien reaches the tip of the foreland providing access to the northern end.

Figure 4.239 View south from Island Point along beach WA 972.

WA 973-980 **JURIEN**

No.	Beach	Rating	Type	Length
WA 973	Island Pt (N)	1	R+sand flats	900 m
WA 974	Jurien	2	R	3 km
WA 975	Jurien groyne (1)	3	R/LTT	200 m
WA 976	Jurien groyne (1)	3	R/LTT	250 m
WA 977	Jurien harbour (S1)	3	R	150 m
WA 978	Jurien harbour (S2)	3	R	150 m
WA 979	Jurien harbour (N)	3	R	50 m
WA 980	Middle Head (S)	4	LTT	1.7 km
Spring & neap tidal range = 0.5 & 0.1 m				

Jurien Bay is the largest town on this section of coast with a population of several hundred. It is located 266 km north of Perth and 38 km in from the highway. The town is the centre of a large fishing fleet, which traditionally moored off the beach in lee of the extensive reefs, but now has a boat harbour excavated on the north side of the town (Fig. 4.240). The town has a full range of facilities and accommodation, all adjacent to the beach. Like most of the settlements along this section of coast, the town is located on a protruding cuspate foreland, called Island Point, that extends 3 km seaward in lee of Boullanger Island and contains over 30 converging foredune ridges. The point forms curving bays to either side. To the south is an open, low energy bay, very protected by offshore reef, with fishing shacks located 3 km south of the point. To the north is the protected Jurien ocean anchorage and the township spreading for 3 km along the beachfront, with the breakwater of the newer boat harbour now defining the northern end of the beach. A new section of the Indian Ocean Drive linking to Green Head was completed in 2001 providing a shorter, more scenic and direct route to the north.

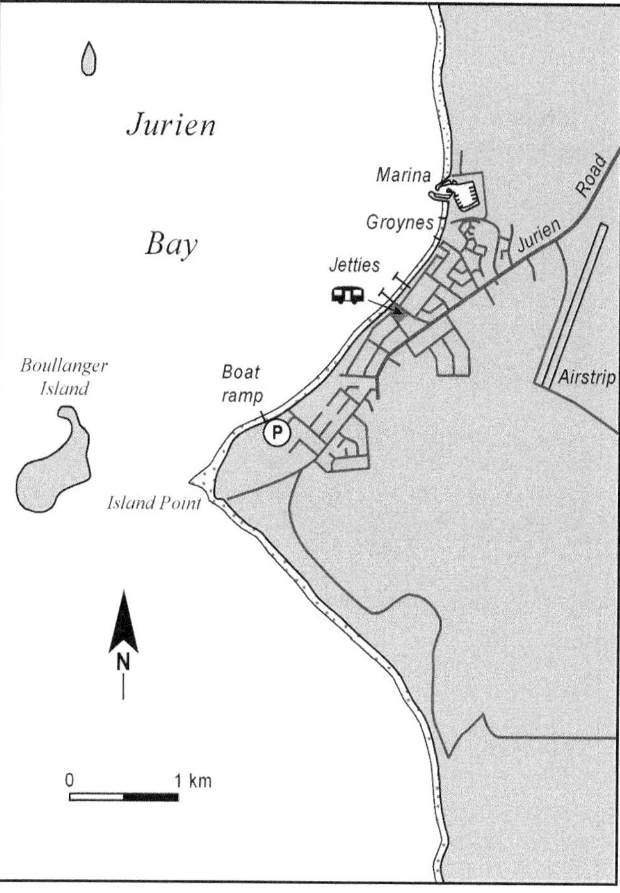

Figure 4.240 Jurien is located on the north side of Island Point, and has a relatively sheltered shoreline.

Island Point forms the southern boundary of 10 m wide west-facing Jurien Bay, which is bordered to the north by North Head. The foreland occupies 6 km of the shoreline, with beach WA 972 to the south, beach WA 973 on the tip of the foreland, and beach WA 973 the beginning of a once continuous beach to Middle Head. This beach is now broken by the harbour into six beaches (WA 974-980). Beaches WA 974 and 975 are bordered by groynes, beaches WA 976-979 are located within the new boat harbour, with beach WA 980 the northern extension of the original beach.

Island Point beach (**WA 973**) is a 900 m long northwest-facing beach located to the lee of 1 km long Boullanger Island. The protection afforded by the island has permitted considerable sand to accumulate in its lee, which extends to the shore as a shallow sandy seafloor and sand flats, crossed by shore transverse sandy ridges. Low vegetated foredune ridges back the beach, with vehicle access and car parks to either end. The beach is a designated fishing, swimming and sailing area.

Jurien beach (**WA 974**) commences on the northern side of the foreland and curves to the northwest then north for 3 km, to the first of the groynes. The beach receives waves averaging less than 0.5 m, which combine with the fine beach sand to maintain a wide low gradient beach and inshore (Fig. 4.241), seaward of which are seagrass meadows. A few jetties cross the central and northern end of the beach and boats are launched off the beach. The beach has the central jetties and boating area, with swimming areas to either side. The entire beach is backed by a low foredune reserve, containing numerous recreational and picnic facilities, then the township, with the caravan park centrally located near the main jetty.

Figure 4.241 Island Point, Jurien Beach and southern section of the township.

Swimming: Both the point and main beach are relatively safe under normal low waves, with the best swimming in the two designated areas and in the boat harbour swimming area. Rips may however cut across the bar following higher winter waves.

In the late 1980's the 10 ha Jurien Harbour was excavated and constructed on the northern side of the township, together with two small rock groyne on the southern side of the harbour. Beach **WA 975** is located between the first two groynes and consists of a 200 m long section of the main beach. Beach **WA 976** is located between the second groyne and the southern harbour breakwater that extends 200 m seaward. Both these beaches receive slightly higher waves and at times have a low tide terrace, which is cut by small rips during and following periods of higher waves. They are both backed by a wide dune reserve, then one of the newer residential developments.

Jurien Harbour consists of two main attached breakwaters, two smaller inner groynes designed to prevent sand moving into the harbour (Fig. 4.242) and the rectangular-shaped harbour which contains jetties and fish processing facilities. Beach **WA 977** is a protected 150 m long beach that has formed between the southern breakwater and the inner groyne. A low dune and car park back the beach. Beach **WA 978** is located inside the harbour and forms the southern side of the harbour between boundary seawalls. It is 150 m long and faces north up the harbour. It is backed by a low vegetated reserve, with a car park at its western end. Beach **WA 979** is a 50 m long pocket of south-facing sand located between the northern breakwater and inner groyne. It is backed by a low dune, then the port commercial area, with the port road terminating behind the beach. All three beaches are sheltered from ocean waves and are calm apart from boat wakes.

Figure 4.242 Jurien boat harbour.

Beach **WA 980** commences at the northern breakwater and continues north for 1.7 km to Middle Head, a 200 m long section of dune-draped calcarenite bluffs rising 10-15 m. Waves average up to 1 m along this section and often maintain a 50 m wide attached bar, which during strong sea breeze conditions may be cut by small 50 m beach rips spaced about 50 m apart. A foredune, then deflated area associated with a large parabolic, back the beach

WA 981-987 MIDDLE HD-PUMPKIN HOLLOW-NORTH HD

No.	Beach	Rating	Type	Length
WA 981	Middle Hd (N)	3	R/LTT	3.5 km
WA 982	Pumpkin Hollow (1)	3	R+reefs	150 m
WA 983	Pumpkin Hollow (2)	3	R+reefs	100 m
WA 984	Pumpkin Hollow (3)	3	R+reefs	350 m
WA 985	Pumpkin Hollow (4)	3	R+reefs	250 m
WA 986	Pumpkin Hollow (5)	3	R+reefs	80 m
WA 987	North Head (E)	3	R+reefs	1.6 km
Spring & neap tidal range = 0.5 & 0.1 m				

The northern half of Jurien Bay trends north of Middle Head then begins to curve to the west along Pumpkin Hollow to finally terminate at North Head. In between are two longer boundary beaches (WA 981 & 987) and five smaller beaches (WA 982-986) associated with the calcarenite bluffs of Pumpkin Hollow, a former shack community. The shacks were removed during 2001 when the Jurien to Green Head section of the Indian Ocean Drive was constructed. All the beaches were accessible by 4 WD tracks in 2001, however some of the tracks have since been closed.

Beach **WA 981** commences on the northern side of Middle Head and gently curves to the north then northwest for 3.5 km to the first calcarenite rocks of Pumpkin Hollow. The beach and all the northern bay becomes increasingly protected by shallow inner reefs and seagrass meadows, resulting in low waves at the shore and seagrass debris littering the beach. It is backed by a large southern and smaller northern parabolic dune, together with vegetated transgressive dunes. It is accessible via 4 WD tracks from either end.

Pumpkin Hollow consists of two sets of adjoining bluff-controlled beaches. Beaches WA 982-984 occupy a 600 m long section of the shore. An elongate finger of calcarenite forms the boundary between the Middle Head beach and beach **WA 982**. The beach continues north of the finger for 150 m to a 15 km high bluff. It is bordered by reefs to either end with a 60 m long central open sand section, while two shacks sit ontop of the dune-draped bluffs at the southern of the beach. Beach **WA 983** is located immediately to the north and consists of a narrow 100m long high tide beach located at the base of 10 m high bluffs, while it is fronted by patchy shallow reefs. A solitary shack backs the beach. Beach **WA 984** curves for 350 m to the west to a 15 km high bluff with a shack on its crest. The beach is fringed by shallow reefs with low waves at the shore. It is backed by dune-draped 10 m high bluffs with three shacks towards the southern end.

Beaches **WA 985** and **986** are adjoining 250 m and 80 m long, narrow beaches that extend northwest of the bluff point. They are backed by sloping dune-draped 10 m high bluffs, with a protrusion in the bluffs separating the two beaches. Patchy inter- and sub-tidal reefs extending up to 150 m offshore of the beaches resulting in usually calm conditions at the shore.

Beach WA **987** commences at the northern end of the bluffs where they is a blufftop shack and curves to the west and finally southwest for 1.6 km terminating in lee of **North Head**. Patchy reef lies off the beach and waves are low along the shore. It is backed by a 5-10 m high foredune and both vegetated and some active parabolic dunes.

WA 988-994 NORTH HEAD-SANDY PT

No.	Beach	Rating	Type	Length
WA 988	North Hd	2	R+reef	300 m
WA 989	North Hd (N1)	2	R+reef	200 m
WA 990	North Hd (N2)	2	R+reef	100 m
WA 991	North Hd (N3)	2	R+reef	900 m
WA 992	Sandland (S)	3	R/LTT	1.2 km
WA 993	Sandland (N)	3	R/LTT	1.9 km
WA 994	Sandy Pt (S)	3	R/LTT	700 m
Spring & neap tidal range = 0.5 & 0.1 m				

North Head marks the northern boundary of Jurien Bay and the beginning of a 6 km long section of coast that trends north to Sandy Point (Fig. 4.243). It is sheltered by inshore calcarenite reefs and Sandland Island. In between are eight near continuous sandy beaches (WA 988-995). Shack communities were located in lee of beaches WA 988, 990, 992 and 995, however these were all removed by 2001, and the myriad of 4 WD tracks rationalised when the Jurien-Green Head section of the Indian Ocean Drive was completed in the same year.

Figure 4.243 North Head, with beach WA 987 (right) and the smaller headland beaches (foregroud).

North Head beach (**WA 988**) is a curving, west-facing 300 m long sheltered beach located between the northern side of the head and a low sandy foreland attached to a calcarenite outcrop. Shallow reefs and seagrass dominate the small embayment, with waves usually low at the shore. Calcarenite outcrops along the northern section of the beach, while vegetated dunes overlying calcarenite back the beach. Beach **WA 989** is located on the northern side of the foreland and trends northeast for 200 m to a small 10 m high calcarenite point. It has an attached reef in the north, which moves offshore toward the southern end, with usually low waves at the beach.

Beach **WA 990** is a 100 m long pocket of sand located on the northern side of the point and bordered by a second low calcarenite outcrop to the north. Shallow reefs link the two outcrops, with low waves at the shore. Several shacks were located to the lee of this beach. Beach **WA 991** commences at the low outcrop and trends north for

900 m to a 200 m long section of calcarenite. The beach is backed by 10 m high foredune then vegetated transgressive dunes. Waves average about 0.5m and maintain a low gradient reflective beach in lee of the inshore reefs.

Beach **WA 992** commences on the northern side of the outcrop and curves to the north for 1.2 km to the low sandy foreland formed in lee of Sandland Island, part of a low calcarenite outcrop located 300 m offshore. The beach is initially well protected by attached reefs in the south, then opens up to a sandy beach with seagrass and reefs located 50-100 m off the beach. A major settlement of about 40 shacks located behind the southern half of this beach was removed in 2001. Beach **WA 993** extends from the northern side of the foreland for 1.9 km, as a curving west-facing beach to a low calcarenite outcrop that forms the next foreland. The beach is clear of inshore reefs and usually has a moderately steep reflective beach grading to a low tide terrace during periods of higher waves. About 20 shacks were located behind the centre of the beach prior to 2001, in amongst the vegetated transgressive dunes.

Beach **WA 994** continues north of the foreland for 700 m to southern rocks of Sandy Point, a dune-draped calcarenite point. The beach faces west and receives waves averaging over 0.5 m, which break across shore-attached central and northern reefs, as well as maintaining a reflective to low tide terrace beach. An unstable foredune, then active dunes extending 200-300 m inland, back the beach.

WA 995-105 **SANDY PT (N)**

No.	Beach	Rating	Type	Length
WA 995	Sandy Pt (N)	2	R	1 km
WA 996	Sandy Pt (N1)	3	R+platform/reef	100 m
WA 997	Sandy Pt (N2)	3	R+reef	800 m
WA 998	Sandy Pt (N3)	3	R+reef	200 m
WA 999	Sandy Pt (N4)	3	R+platform/reef	200 m
WA 1000	Sandy Pt (N5)	3	R+platform/reef	150 m
WA 1001	Sandy Pt (N6)	3	R+reef	100 m
WA 1002	Sandy Pt (N7)	3	R+reef	250 m
WA 1003	Sandy Pt (N8)	3	R+reef	400 m
WA 1004	Sandy Pt (N9)	3	R+reef	600 m
WA 1005	Sandy Pt (N10)	3	R+reef	200 m
Spring & neap tidal range = 05 & 0.1 m				

On the north side of Sandy Point is the beginning of a 5 km long section of shoreline dominated by inshore calcarenite reefs and dune-draped 10-20 m high calcarenite shoreline. It contains 11 generally small beaches, each bordered by calcarenite reefs and/or bluffs, and fronted by generally shallow reefs, then the shore parallel Fisherman Islands reef located 5-6 km offshore, resulting in very low waves at the shore. The beaches are backed by vegetated transgressive dunes, which extend for 2-3 km inland with the newer section of the Indian Ocean Drive generally following the rear of the dunes. Shacks were located at the southern ends of beaches WA 995 and 997 until 2001, otherwise the coast is undeveloped, with limited access.

Beach **WA 995** commences on the northern side of Sandy Point and curves to the north for 1 km (Fig. 4.244), terminates at a 20 m high calcarenite headland surrounded by shallow reefs. The inshore is free of reefs, however reefs in line with the point lower waves to usually less than 0.5 m maintaining reflective beach conditions. There is a sand blowout at the southern end of the beach and a larger northward migrating parabolic located 0.5-1 km inland, with the former shack settlement located along the southern half of the beach. Beach **WA 996** is a 100 m long strip of high tide sand located along the base of the headland. It is backed by steep 20 m high bluffs and fronted by irregular intertidal reefs extending 50-100 m offshore.

Figure 4.244 Beach WA 995 curves to the north of Sandy Point.

Beach **WA 997** commences on the northern side of the headland and trends north for 800 m. It is initially fronted by shallow reefs, with a seagrass-fringed central section and boundary reefs to the north. It was backed by a few shacks in the south. The southern bluff preserves some well developed lithocasts – tree trunks and roots encased in calcarenite.

Beach WA 998 commences a 2 km long section of shoreline backed by near continuous 10-20 m high calcarenite bluffs, all draped with vegetated transgressive dunes. Beach **WA 998** is a narrow 200 m long beach wedged in between the backing bluffs and scattered inshore reefs and seagrass meadows. It is bordered by protrusions in the bluffs, with beach **WA 999** continuing on the northern side as a curving 200 m long sandy beach. It is backed by vegetated bluffs and fronted by continuous, though irregular, 10-50 m wide calcarenite, with sand then seagrass meadows 50-100 m off the beach.

Beach **WA 1000** commences on the northern side of an 80 m long section of protruding bluffs and curves to the north for 150 m, with the bluffs decreasing in height to the north, and a 4 WD tracks reaching the top of the bluffs. It has an open southern section with seagrass meadows 50 m off the shore, while calcarenite increases along the shore to the north. Beach **WA 1001** is located immediately to the north and is a protruding 100 m long patch of sand in lee of a central reef, with shallow reef extending the length of the beach. It is backed by low dune-draped bluffs.

Beach **WA 1002** is a narrow 250 m long beach located along the base of steep bluffs, which rise from 5 m in the south to 20 m in the north. It is fronted by continuous irregular inter- and subtidal reefs. Beach **WA 1003** is a curving 400 long beach backed by continuous dune-draped low bluffs, with the dunes increasing to 20 m in the north. The beach narrows to the north as the bluffs encroach on the beach. Several shacks are located on the northern dune section. The shoreline is sandy with seagrass meadows 20-50 m off the beach then inshore reefs.

Beach **WA 1004** commences 150 m to the north and is a 600 m long narrow high tide beach. It has a 200 m long southern section backed by a foredune, while the central-northern section has a narrow beach awash at high tide and backed by eroding bluffs. Four wheel drive tracks reach the southern end of the beach. Vegetated transgressive dunes originating from the sandy section cover the rear of the northern bluffs. Beach **WA 1005** is located around a 200 m long calcarenite point and consists of a 200 m long strip of northwest-facing high tide sand wedged between the base of 10-20 m high bluffs and scattered inshore rocks and reefs.

WA 1006-1008 **SOUTH BAY**

No.	Beach	Rating	Type	Length
WA 1006	South Bay (S2)	3	R+reef	1.2 km
WA 1007	South Bay (S1)	3	R+reef	1.1 km
WA 1008	South Bay	3	R/LTT	6.5 km
Spring & neap tidal range = 0.7 & 0.4 m				

South Bay is the southern settlement of Green Head and also the name of the open bay that curves to the south towards Sandy Point. The first 9 km of shoreline contains three near continuous sandy beaches (WA 1006-1008), which terminate at the beginning of the calcarenite-dominated section.

Beach **WA 1006** begins against the 20 m high vegetated calcarenite bluffs its shares with beach WA 1005. As the bluffs trend inland the beach continues due north and is backed by a well developed 10-20 m high vegetated foredune, grading into vegetated transgressive dunes at the northern end of the 1.2 km long beach. The beach narrows to the north as shallow reefs impinge upon the shoreline. The reefs and associated seagrass meadows front the entire beach, with seagrass debris commonly piled up on the beach.

Beach **WA 1007** commences just to the north and continues north for 1.1 km past a central bluff-backed section to a 300 m long section of 20 m high calcarenite bluffs which forms its northern boundary. Shallow reefs and seagrass extend to the shoreline, with usually calm conditions and seagrass debris deposited on the beach. It is backed by vegetated transgressive dune and some deflated surfaces.

South Bay beach (**WA 1008**) commences on the northern side of the bluffs and trends due north and finally curves to the west in lee of Green Head, forming the 'south bay' of the head. Wave energy increases slightly along the beach and the seagrass meadows move about 50 m off the shore. The wind and waves push the seagrass debris to the north and often large berm of debris accumulates in the northern corner, where boats are launched off the beach. Most of the beach is backed by a foredune, then two areas of dune transgression (Fig. 4.245). The southern parabolic has deflated the area behind beach WA 1007 and is moving northward on a 2 km wide front. The second is a central 1 km long 500 m wide parabolic. The northern corner of the bay is backed by a small reserve and beach access, then the houses of South Bay and Green Head.

Figure 4.245 South Bay has a reflective beach backed by extensive vegetated and active dunes.

WA 1009-1016 **GREEN HEAD**

No.	Beach	Rating	Type	Length
WA 1009	Green Head (1)	3	R+reef	100 m
WA 1010	Green Head (2)	3	R+reef	150 m
WA 1011	Green Head (3)	3	R+reef	100 m
WA 1012	Green Head (4)	3	R+reef	100 m
WA 1013	Green Head (5)	3	R+reef	100 m
WA 1014	Dynamite Bay	2	R	250 m
WA 1015	Cambawarra Hd	2	R+reef	250 m
WA 1016	Anchorage Bay	2	R+reef	2.2 km
Spring & neap tidal range = 0.7 & 0.4 m				

Green Head is a small but growing coastal community of a few hundred located 288 km north of Perth. It was surveyed in 1959 and the township gazetted in 1966. The new link road with Jurien and the removal of many coastal shacks to either side of the town, will encourage growth of the town. The town is located to the lee of 30 m high Green Head, with South Bay (WA 1008) curving to the south, five small pockets of sand located around the head (WA 1009-1013), and Dynamite Bay (WA 1014), Cambawarra Head (WA 1015) and the longer Anchorage Bay (WA 1016) forming the western boundary.

The five Green Head beaches (WA 1009-1013) are all small pockets of sand located below the 10-20 m high calcarenite bluffs of the head, and surrounded by rock platforms, shallow reefs and seagrass meadows. Beach **WA 1009** is a narrow, curving 100 m long strip of sand located at the southern tip of the head, with the headland car park located to the rear of the backing bluff. Beach **WA 1010** is a 200 m long cuspate-shaped beach tied to a small central reef, with seagrass growing to within 20 m of the shore, and reefs further out. Beach **WA 1011** is a crenulate 100 m long high tide beach, with a sand blow at the southern end. It is backed by the bluffs and fronted by 200 m wide shallow reefs. Beach **WA 1012** is a narrow strip of high tide sand on the southwestern tip of the point, with steep bluffs behind, platforms to either end and shallow reef off the beach. Beach **WA 1013** lies at the western tip of the head, and is a curving 100 m high tide beach, with 100 m wide platforms to either side, and a small central patch of sand.

Dynamite Bay (WA 1014) is located on the northern side of the head, with Cambawarra Head fronting the northern point of the semi-circular shaped bay (Fig. 4.246). The 200 m long curving beach occupies the eastern end of the bay, with only low waves reaching the shore. Apart from some occasional seagrass debris the beach and bay floor is sandy. This is the main recreational beach for Green Head and has a large central car park and toilet block, and is located just 200 m from the town's small commercial area.

Figure 4.246 Dynamite Bay with Green Head to the right and the township in the background.

Beach **WA 1015** is located on the northern side of Cambawarra Head and consists of a narrow curving 250 m long beach, backed by an elevated beachrock platform, then dune-draped bluffs. Shallow reefs extend 200-300 m off the beach and conditions are usually calm at the shore. The Green Head caravan park is located to the lee of the beach.

Anchorage Bay (WA 1016) is the site of Green Head jetty and the main anchorage for the fishing boats, with part of the township to the lee of the southern half of the beach. The jetty and boats are located towards the southern end of the 2.2 km long beach in lee of the shallow reefs that extends north from Cambawarra Head. The reefs and seagrass fill much of the beach and seagrass debris is usually piled up along the southern half of the beach. Only the northern third is relatively free of seagrass meadows. There is vehicle access to either end of the beach. An elongate active dune extends north of the bay to the lee of Billy Goat Bay.

WA 1017-1024 **BILLY GOAT BAY**

No.	Beach	Rating	Type	Length
WA 1017	Point Louise	2	R+platform/reef	70 m
WA 1018	Point Louise (N)	2	R+reef	80 m
WA 1019	Point Break	2	R+reef	250 m
WA 1020	Billy Goat Bay	2	R+reef	1 km
WA 1021	Lipfert Island	2	R+reef	300 m
WA 1022	Tea Tree	2	R+reef	500 m
WA 1023	Little Anchorage	2	R+reef	1.2 km
WA 1024	Shell Beach	2	R+reef	2.5 km
Spring & neap tidal range = 0.7 & 0.4 m				

To the north of Point Louise is 6 km long **Anchorage Bay** foreshore reserve. Until the late 1990's numerous shacks occupied this section of shore. These were removed and the reserve laid out in 2001. There is access in the south and towards the northern end with a gravel road running from Point Louise to the southern end of Shell Bay, with the active dune backing the southern half of the bay (Fig. 4.247). Eight very low energy beaches (WA 1017-1024) occupy the shore. They are all fronted by wide shallow calcarenite reefs and some low islands, with outer reefs extending up to 4 km offshore.

Figure 4.247 Point Louise with Anchorage bay to the right and the Billy Goat Bay to the left.

Point Louise is a 400 m long, 10-20 m high section of calcarenite that terminates at a narrow finger of calcarenite, below which is the narrow 70 m long beach **WA 1017**. The Point Louise car park is located above the northern end of the beach, which is fronted by a continuous 20-30 m wide platform, with reefs further out. Beach **WA 1018** is located 300 m to the north and consists of an 80 m long strip of high tide sand backed by 10 m high bluffs and fronted by extensive shallow reefs. It shares a car park with beach **WA 1019**, a 250 m long narrow beach, backed by a dune-draped 10 m high bluffs, with seagrass meadows and shallow reefs extending seaward. The Surfers Point Break car park is located above these beaches and provide a view of and access to reef breaks that are located 100-200 m off the point. These breaks only work when there is a sizable outside swell. The beaches are all usually calm and partly covered with seagrass debris.

Billy Goat Bay (WA 1020) commences in lee of the northern tip of the point. The car park is located at the northern north-facing end of the beach, with the only toilet facilities in the reserve also located here. A shallow reef lies off the usually calm beach, which curves and trends north for 1 km, past Lipfert Island and Orton Rock, to a small calcarenite headland. The main southern access road reaches the rear of the northern end of the beach. Shallow reefs and seagrass meadows dominate the shore and calm conditions and seagrass debris dominates the beach. Beach **WA 2021** commences on the northern side of the 100 m long point and extends north for another 300 m to the next calcarenite outcrop. The beach is fronted by the shallow reefs with calm conditions at the shore.

Tea Tree beach **(WA 1022)** is located 200 m to the north and is a 500 m long beach, bordered in the north by a prominent rock, with reefs off the beach and seagrass usually piled along the shore. **Little Anchorage** beach **(WA 1023)** commences 200 m to the north and is a 1.2 km long beach that curves to the north to terminate at a sandy foreland. A boat launching area is located at the southern end. A small island and shallow reef lie off the southern end, with seagrass and reefs running the length of the beach, resulting in calm conditions along the shore.

Shell Bay is a curving 2.5 km long beach that extends north of the foreland to a straight section of calcarenite that forms a more prominent foreland. The bay is occupied by **Shell Beach (WA 1024)** another reef-dominated low energy beach (Fig. 4.248), with seagrass growing close to shore and littering the beach.

WA 1025-1030 **WEBB ISLAND**

No.	Beach	Rating	Type	Length
WA 1025	Webb Island (S)	2	R+reefs/seagrass	1.1 km
WA 1026	Webb Island	2	R+reefs/seagrass	1.2 km
WA 1027	Webb Island (N1)	2	R+reefs/seagrass	500 m
WA 1028	Webb Island (N2)	2	R+reefs/seagrass	600 m
WA 1029	Bill Wilson Reef (1)	2	R+reefs/seagrass	400 m
WA 1030	Bill Wilson Reef (2)	2	R+reefs/seagrass	300 m
Spring & neap tidal range = 0.7 & 0.4 m				

Figure 4.248 Shell Beach is a low energy beach fringed by seagrass.

Webb Island is an outcrop of calcarenite located 200 m offshore and 3 km south of Leeman. Between the northern end of Shell Bay and the Leeman beaches is 4.5 km of very low energy shoreline containing six beaches (WA 1025-1030) roughly located either side of the island. The island is part of the shallow calcarenite reefs that extends up to 6 km offshore, effective blocking ocean waves from the shore. The Green Head-Leeman section of the Indian Ocean Drive parallels the coast approximately 1 km inland, however apart from 4 WD tracks there is no development along this section of shore.

Beach **WA 1025** commences on the northern side of the 300 m long straight section of low calcarenite bluffs, which separates it from Shell Bay (WA 1024). The beach trends to the north for 1.1 km with a slight central foreland. Shallow reefs and seagrass meadows dominate the inshore with usually calm conditions at the shore, and seagrass debris piled along the beach. A vehicle track from the main road reaches the centre of the beach and runs south to the boundary bluffs.

A 300 m long calcarenite bluff forms the southern boundary of beach **WA 1026**, which continues north for 1.2 km to a dune-capped calcarenite point. The beach is backed by vegetated dunes and fronted by the reefs including Webb Island located just off the northern section of the beach. Low waves and seagrass dominate the inshore and shoreline. Beach **WA 1027** is located on the northern side of the small point and initially curves to the east before trending north for 500 m to the next low calcarenite point. Both beaches are accessible by 4 WD tracks that parallel the coast.

Beach **WA 1028** extends north of the calcarenite point as a crenulate 600 m long beach that narrows to the north as low calcarenite bluffs impinge upon the beach. Low waves and seagrass debris litters the shoreline. Beach **WA 1029** occupies the next slight embayment in lee of Bill Wilson Reef. It is a narrow strip of sand bordered by seagrass growing to the shore and backed by low calcarenite bluffs. Beach **WA 1030** lies 500 m to the north, past a series of more irregular calcarenite bluffs. It is a narrow 300 m long strip of high tide sand bordered by low bluffs to either end and with the reefs and seagrass extending to the shore. The township of Leeman commences immediately north of the beach.

WA 1031-1032 **LEEMAN**

No.	Beach	Rating	Type	Length
WA 1031	Leeman (S)	2	R+reefs/seagrass	600 m
WA 1032	Leeman	2	R+reefs/seagrass	1.1 km
Spring & neap tidal range = 0.7 & 0.4 m				

Leeman is a small fishing and holiday town of several hundred, located 295 km north of Perth. It has developed around the sheltered anchorage in lee of Drummond Rock and has expanded both north and south of the cove. It is located on low ground between a low calcarenite shoreline and a series of 1-3 km wide salt lakes that lies 1-2 km east of the town and parallel the coast for 18 km. Two beaches front the township (WA 1031 & 1032).

Beach **WA 1031** extends south of Drummond Rock for 600 m to the northern end of the low bluffs. It is a narrow moderately steep sandy beach backed by a low vegetated dune reserve, then the southern houses of Leeman with the caravan park behind the southern end. A small car park is located behind the centre of the beach. Seagrass meadows commence about 50 m offshore making it more suitable for swimming, with shallow reefs extending seaward.

Leeman beach (**WA 1032**) commences at a curves in the shoreline in lee of Drummond Rock (Fig. 4.249). It then trends north for 1.1 km to a series of low bluffs. It is a sheltered very low energy beach with two jetties and a boat ramp located towards the southern end and fishing boats moored off the beach. A foreshore reserve runs the length of the beach, with the town centre backing the southern end of the beach.

Figure 4.249 Leeman and the low energy Leeman beach extend north of Drummond Rock.

BEEKEEPERS NATURE RESERVE

Established: 1991
Area: 68 063 ha
Coast length: 42 km (2384-2387, 2452-2457, 2463-2477, 2502-2522 km)
Beaches: 18 (WA 966, 1033-1038, 1043-1050, 1059-1060)

Beekeepers Nature Reserve is relatively narrow, discontinuous reserve that commences in the south just above Cervantes and continues north, generally paralleling the coast for 135 km to just south of Port Denison. The reserve encompasses 18 generally low energy beaches, along 42 km of the coast, as well as large areas of active and vegetated Holocene longwalled parabolic dunes, and the backing vegetated Pleistocene barrier systems. The Indian Ocean Drive provides access too much of the reserve.

WA 1033-1042 **LEEMAN-COOLIMBA**

No.	Beach	Rating	Type	Length
WA 1033	Taylor Bay (S)	2	R+seagrass/reefs	150 m
WA 1034	Taylor Bay	2	R+seagrass/reefs	100 m
WA 1035	Taylor Bay (N)	2	R+seagrass/reefs	300 m
WA 1036	Dumper Bay	2	R+seagrass/reefs	500 m
WA 1037	Dumper Bay (N)	2	R+seagrass/reefs	200 m
WA 1038	Mace Beach (S)	2	R+seagrass/reefs	250 m
WA 1039	Mace Beach	2	R+seagrass/reefs	1 km
WA 1040	Coolimba (S 2)	2	R+seagrass/reefs	150 m
WA 1041	Coolimba (S 1)	2	R+seagrass/reefs	100 m
Spring & neap tidal range = 0.7 & 0.4 m				

To the north of Leeman the coast trends essentially due north for 70 km to Port Denison. The entire stretch is fronted by extensive shallow calcarenite reefs extending a few kilometres offshore, with seagrass meadows dominating the shallow water and in most locations growing right to the shore. The Indian Ocean Drive parallels the rear of the coast providing numerous access points to shacks and the small communities of Coolimba, 10 km north of Leeman, and Illawong, 27 km to the north, before joining the Brand Highway 52 km to the north. Apart from the fishing shacks and these two small communities, neither with any commercial establishments, there is no other development along this section of coast. The first 10 km north to Coolimba contains nine generally small beaches (WA 1033-1041), most backed and/or bordered by low calcarenite bluffs. Beaches WA 1033-1038 are part of Beekeepers Nature Reserve.

Beach **WA 1033** commences 2 km north of Leeman and is a 150 m long strip of sand bordered and backed by dune-draped 10 m high bluffs, rising to a 15 m high point in the north, with seagrass growing to within 20 m of the shore. Three shacks back the beach, which lies just 100 m west of the Indian Ocean Drive. Beach **WA 1034** is located in the small Taylor Bay on the northern side of the 200 m long calcarenite point, and is a 100 m long strip of sand located at the base of steep, eroding 10 m high bluffs, with rock debris littering the rear of the narrow beach. Seagrass lies 10-20 off the beach, with reefs further out.

Beach **WA 1035** is located 500 m to the north, with the Indian Ocean Drive clipping the southern end of the 300 m long beach. It consists of a narrow high tide beach, with seagrass growing to the shore and usually covered with seagrass debris. It is backed by dune-draped 10 m high bluffs. Beach **WA 1036** occupies a slight indention in the shoreline called Dumper Bay and curves to the north for 500 m. About 15 shacks are located around the southern corner of the beach, which is lies 200 m west of the road. Seagrass grows to the shoreline, with the shallow reefs extending several kilometre seaward. An active parabolic commences on the northern side of the beach and trend north for 500 m.

Beach **WA 1037** lies 1 km to the north and is a 200 m long sandy beach backed by a 5-10 m high foredune, with low calcarenite bluffs to either end and seagrass growing to the shore. A solitary shack is located on the southern bluffs, with a vehicle track running 500 m east to the main road. Beach **WA 1038** is located 100 m to the north and is a 250 m long sandy beach, backed by low vegetated dunes and three southern and one northern shack, then ridges of north-trending vegetated longwalled parabolic dunes. Because of the dune ridges the access track connects with the southern vehicle access.

Mace Beach **WA 1039** is curving 1 km long beach, which has a low southern section where three shacks are located, with higher vegetated dune ridges backing the central-northern section of the beach. Seagrass meadows grow almost to the shore and its debris commonly litters the beach. Beach **WA 1040** is a 150 m long strip of sand bordered by 10 m high calcarenite bluffs. A vehicle a track runs along the rear of the beach providing access to 12 shacks that back the beach. Seagrass debris litters the beach with meadows growing to the shore.

Beach **WA 1041** lies 1 km further north and 1 km south of Coolimba. It is a 100 m long pocket of sand backed by vegetated dune-draped 10 m high bluffs, with an eroding 10-20 m high calcarenite point to the north. It is backed by vegetated dune ridges with a vehicle track across the ridges to the beach.

WA 1042-1045 **COOLIMBA**

No.	Beach	Rating	Type	Length
WA 1042	Coolimba (S)	2	R+seagrass/reefs	600 m
WA 1043	Coolimba	2	R+seagrass/reefs	2.5 km
WA 1044	Coolimba (N1)	2	R+seagrass/reefs	300 m
WA 1045	Coolimba (N2)	2	R+seagrass/reefs	2.9 km
Spring & neap tidal range = 0.7 & 0.4 m				

Coolimba is small fishing community consisting of an irregular collection of about 30 shacks. It is also known Desperate Bay, and site of the Desperate Bay Receival Depot for the small fishing fleet. It is located 2 km off the main road, with the access road terminating at the beach (Fig. 2.250) between the small jetty and the shacks that extend for 1 km to the south.

Figure 4.250 The small Coolimba fishing community and anchorage.

Beach **WA 1042** extends for 600 m south of the 10 m high calcarenite bluffs that mark the southern end of Desperate Bay. The beach is an irregular, narrow high tide strip of sand wedged in between the backing bluffs. The bluffs alternate between vegetated and eroding, while shallow reefs and seagrass meadows extends several kilometres seaward. Behind the bluffs are the scattered shacks of Coolimba, set in amongst the vegetated deflation hollow of a longwalled parabolic.

Beach **WA 1043** is the main Coolimba Beach and site of the small jetty with the fishing boats moored just off the beach. The beach commences in lee of the bluffs and trends north for 2.5 km to the next set of bluffs. It is sheltered by the wide shallow reefs and seagrass meadows with usually low wind waves or calm conditions at the shore. It is backed by a central 2 km long active parabolic and vegetated dunes. The fish depot and parking area is located behind the jetty, with shacks extending to the south.

Beach **WA 1044** is a low energy 300 m long pocket of sand located between two sets of 10 m high eroding calcarenite bluffs. The beach is widest in the centre where the bluffs decrease in height and a 4 WD track reaches the shore, narrowing to either end below the bluffs. Beach **WA 1045** commences on the northern side of the bluffs and continues north for 2.9 km to the next more northerly inflection in the short. A foredune, then vegetated transgressive dunes extend 2 km inland, with a 4 WD access track reaching the foreland at the northern end of the beach.

WA 1046-1048 **GUM TREE BAY**

No.	Beach	Rating	Type	Length
WA 1046	Gum Tree Bay (S)	2	R+seagrass/reefs	2 km
WA 1047	Gum Tree Bay	2	R+seagrass/reefs	2.8 km
WA 1048	Gum Tree Bay (N)	2	R+seagrass/reefs	3.3 km
Spring & neap tidal range = 0/7 & 0.4 m				

Gum Tree Bay is a collection of about 50 fishing shacks located along 1 km of shoreline in the slight indentation in the coast called Gum Tree Bay. The shacks are located within Beekeepers Nature Reserve and will eventually be removed. The bay is located 500 m west of the main road and readily accessible via two tracks. To the north and south of the beach is a near continuous slightly crenulate shoreline containing a southern beach (WA 1046), the bay beach (WA 1047) and longer northern beach (WA 1048) that terminates at the southern Illawong bluffs.

Beach **WA 1046** commences at the slight foreland boundary with beach WA 1045, and trends north as a continuous sandy beach for 2 km to the southern foreland boundary of Gum Tree Bay. It is backed by hummocky vegetated parabolic dunes, with vehicle access to either end. Shallow reefs and seagrass meadows extend well seaward of the beach.

Gum Tree Bay beach (**WA 1047**) is a curving 2.8 km long slight indentation in the shoreline bordered by sandy forelands. The Gum Tree Bay shacks occupy the southern half of the bay, with some shacks located on the edge of the beach and in danger of erosion (Fig. 4.251). Seagrass meadows and shallow reefs extend well seaward and seagrass debris commonly litters the high tide beach. A 2 km long blowout extends north of the northern half of the beach, with vegetated dunes backing the remainder, including a small stand of gum trees just behind the centre of the beach, which lends its name to the bay.

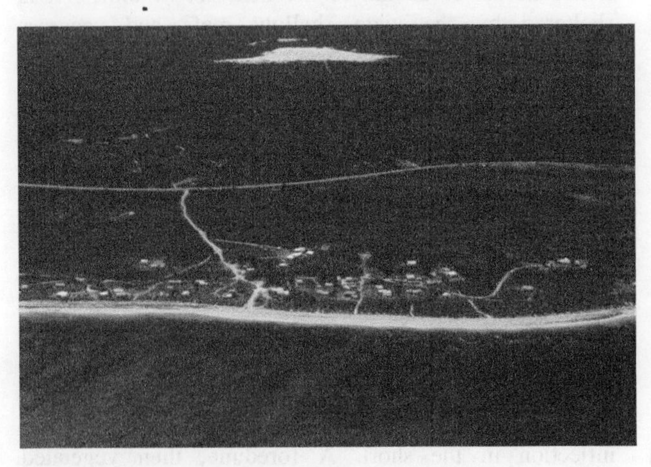

Figure 4.251 Gum Tree Bay has a low energy beach with seagrass growing to the shore.

Beach **WA 1048** commences at the northern bay foreland and continues north as a slightly crenulate sandy beach for 3.3 km to a small calcarenite bluff that forms the boundary with beach WA 1049. The beach is backed by a 2 km long blowout along its northern half, with vegetated dunes backing the rest. Seagrass meadows and shallow reefs continue alongshore, with usually low wave to calm conditions at the beach.

WA 1049-1052 **ILLAWONG**

No.	Beach	Rating	Type	Length
WA 1049	Illawong (S3)	2	R+seagrass/reefs	1.3 km
WA 1050	Illawong (S2)	2	R+seagrass/reefs	900 m
WA 1051	Illawong (S1)	2	R+seagrass/reefs	1.2 km
WA 1052	Illawong	2	R+seagrass/reefs	600 m
WA 1053	Illawong (N)	2	R+seagrass/reefs	4.2 km
Spring & neap tidal range = 0.7 & 0.4 m				

Illawong is small fishing community located 10 km north of Gum Tree Bay and just a few hundred metres off the Indian Ocean Drive. The townsite, which was previously known as Sandy Bay, was gazetted in 1972. Today is consists of a collection of about 30 shacks located either side of a small calcarenite point and nestled in vegetated dune hollows and along the shore for a distance of about 500 m. Three bluff-separated beaches with shacks extends south of the main settlement (WA 1049-1051), with the longer beach (WA 1053) extending north to Knobby Head.

Beach **WA 1049** commences on the northern side of a 50 m long section of low calcarenite that forms a slight inflection in the shore. The beach trends north for 1.3 km to a 100 m long dune-draped calcarenite bluff. The northern end of the Gum Tree Bay parabolic backs the southern end of the beach, with several beachfront shacks between the shore and active dunes. Low waves, seagrass and reefs continue along the shore. Beach **WA 1050** commences in lee of the bluffs and continues north for 900 m to a 200 m long section of low dune-covered bluffs. One shack is located on the crest of the southern bluffs with two more just up the beach. Vegetated parabolic dunes extend from the shore 1.5 km east to the main road. Seagrass grows to the shore with shallow reefs extending several kilometres offshore.

Beach **WA 1051** commences on the northern side of a 20 m high section of calcarenite bluffs that forms a more substantial indentation in the otherwise relatively straight shoreline. The 1.2 km long beach continues north along the base of 10-20 m high dune-covered bluffs, to a 200 m long cliffed section. Vegetated parabolic dunes extend east to the Indian Ocean Drive, which lies 200 m east of the northern bluffs. **Illawong** beach (**WA 1052**) extends from the northern side of the bluffs for 600 m to a 10 m high, 50 m long vegetated headland. A total of about 20 shacks grouped into several small clusters back the beach, including one on the boundary headland, with the main road running just 200-300 m inland. Another 10 shacks are located on the northern side of the headland. Seagrass meadows front the southern section of the beach with a more sandy seafloor along the central-northern section.

Beach **WA 1053** commences on the northern side of the small headland and trends due north for 4.2 km to a subtle sandy foreland called Knobby Head. Seagrass returns to the shoreline and continues the length of the low energy beach. About 15 shacks are scattered along the beachfront, with the main road running ontop of a calcarenite bluff, 200-300 m to the east.

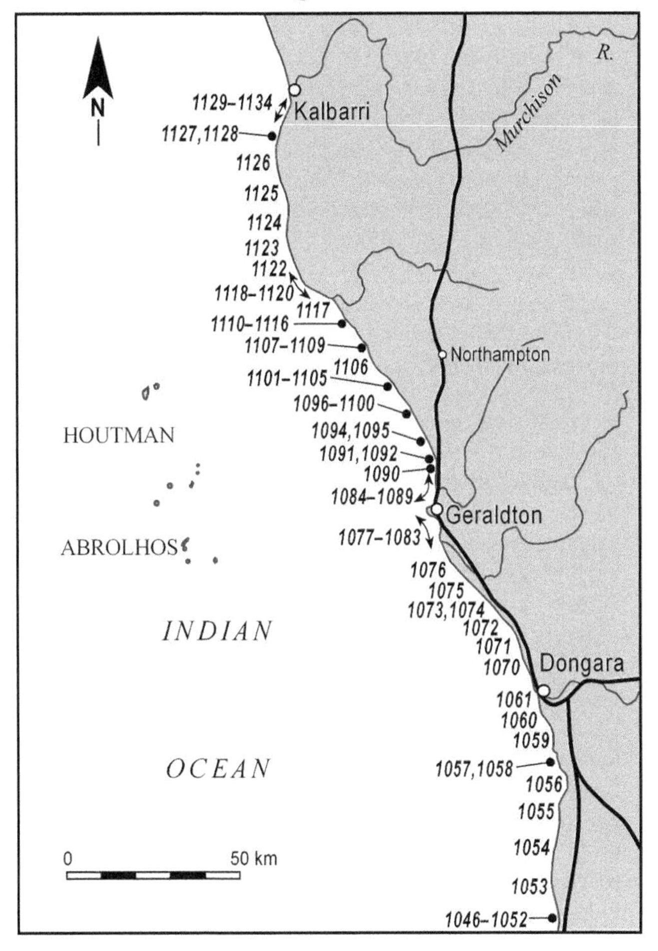

Figure 4.252 Region 4F, beaches WA 1053-1134

WA 1054-1058 **KNOBBY HEAD-CLIFF HEAD**

No.	Beach	Rating	Type	Length
WA 1054	Knobby Head	2	R+seagrass/reefs	6.6 km
WA 1055	Freshwater Pt	2	R+seagrass/reefs	1.4 km
WA 1056	Cliff Head (S)	2	R+seagrass/reefs	6.7 km
WA 1057	Cliff Head	2	R+seagrass/reefs	1.3 km
WA 1058	Cliff Head (N)	2	R+seagrass/reefs	1.6 km
Spring & neap tidal range = 0.7 & 0.4 m				

Knobby Head and Cliff Head are more prominent parts of a 20 km long, 10-20 km high calcarenite bluff that trends north of Illawong to just beyond Cliff Head. The Indian Ocean Drive for the most part runs along the crest of the bluff. Between the bluff and the shore is a narrow zone of Holocene beach ridges and low foredunes. Shallow reefs and seagrass meadows extend several kilometres seaward of the shore ensuring zero deepwater waves energy, with only low wind waves or calms prevailing. The small fishing communities of Knobby Head, Freshwater Point and Cliff Head are located along the shore together with scattered shacks, all located within a few hundred meters of the main road.

Knobby Head beach (**WA 1054**) commences at the sandy inflection of the same name and trends north-northeast for 1 km to the main Knobby Head shack area, then due north for another 5.6 km to the 10 m high calcarenite of Freshwater Point. Seagrass grows to the shore the length of the beach and is commonly piled up on the shore. The bluffs rise to 30 m at the shack settlement forming a 300 m long relic seacliff, now fronted by 100 m of shoreline. The shacks extend either side of the 'head', as well as a few scattered clusters along the beach.

Freshwater Point forms a prominent headland boundary between beaches WA 1054 and 1055. Freshwater Point beach (**WA 1055**) commences at the base of the 20 m high point and continues north for 1.4 km (Fig. 4.253) to two small adjoining bluffs. Seagrass grows to the shore and is usually piled up in the southern corner of the beach. A fishing and refueling depot and landing site is located in the southern corner between the backing bluffs and the shore. A radio tower and navigation light are located on the point.

Figure 4.253 Freshwater Point beach is a crenulate, low energy beach with seagrass growing to the shore.

Beach **WA 1056** commences on the northern side of the two bluffs and continue north for 6.7 km to Cliff Head. This is a crenulate very low energy beach with seagrass growing to the shore. Holocene beach ridges and foredunes have prograded up to a few hundred metres west of the backing bluffline, out across the shallow calcarenite seafloor. There are a few shacks located just north of the southern bluffline, otherwise apart from 4 WD tracks there is little development along the beach.

Cliff Head beach (**WA 1057**) is located in a small indentation in the bluffline, with the beach prograding about 100 m seaward in the centre. Ten metre high calcarenite bluffs border the 1.3 km long low energy beach. About 20 fishing shacks have been built on the slopes of the bluff overlooking the beach. Seagrass debris is often piled up on the beach with the meadows and shallow reefs extending a few kilometres seaward. The northern Cliff Head settlement is located 1 km to the north along the northern end of the bluffline and along beach (**WA 1058**). The beach is initially backed by a 50-100 m wide plain and vegetated 20 m high bluffs, with about 30 shacks occupying the 500 m long section. Then as the bluffs diminish the beach continues for another 1 km to a low sandy foreland that marks the boundary with the long beach WA 1059. Fishing shacks continue north of the bluffs, with a seagrass berm often extending the length of the beach. The main road parallels the shoreline 200-300 m inland.

WA 1059-1061 **CARSONS-TEN MILE-SOUTH BEACH**

No.	Beach	Rating	Type	Length
WA 1059	Carsons Beach	2→4	R→LTT	12.8 km
WA 1060	Ten Mile Beach	6	TBR	10.5 km
WA 1061	South Beach	7→6	D→TBR	3.8 km

Spring & neap tidal range = 0.7 & 0.4 m

Two kilometres to the north of Cliff Head the shoreline trends to the north-northwest and wave height gradual increases as offshore reefs lessen and deepen. As a consequence of the higher waves and slight orientation into the dominant southerly winds, a major accumulation of beach and dune sand has occurred between Cliff Head and Port Denison, 24 km to the north. Three beaches occupy this section, Carsons (WA 1059), Ten Mile (WA 1060) and South Beach (WA 1061). Massive longwalled parabolic dunes back all three beaches (Fig. 4.254) restricting public access to the south and north, with 4 WD tracks elsewhere. Most of Carsons and Ten Mile beaches and their backing dunes are the northernmost part of the long Beekeepers Nature Reserve.

Figure 4.254 Longwalled parabolic dunes back the lower energy Carsons Beach.

Carsons Beach (WA 1059) commences at the slight foreland 2 km north of Cliff Head and trends to the north-northwest for 12.8 km to the large cuspate foreland called White Point. The foreland has resulted from wave refraction around Leander Reef, located 12 km to the west. The beach receives low waves in the south, with seagrass growing to within 100 m of the shore and with its debris commonly littering the beach. Wave height increases to about 1 m by White Point and a low tide terrace dominates the northern several kilometres of beach. Massive 5-6 km wide northward trending longwalled parabolics back the entire beach. They are generally vegetated in the south, with three large active dunes in the north moving to the lee of White Point. The Dongera-Eneabba railway line runs 1-4 km in from the beach, with the Brand Highway 3-5 km inland, with 4 WD tracks reaching the beach in several locations.

Ten Mile Beach (WA 1060) commences on the northern side of White Point and curves to the north and finally northwest for 10.5 km to the next sandy foreland, formed in lee of Jack Reef. This beach is exposed to waves averaging over 1 m which break across a low gradient 150 m wide, rip dominated surf zone, with rips usually spaced about every 300 m. The beach is backed in the south by a series of 15 foredune ridges, then earlier parabolics, while active and some vegetated parabolics dominate the centre and northern end. The only access to the beach is by 4 WD.

South Beach (WA 1061) extends north of the Jacks Reef foreland for 3.8 km to the calcarenite rocks and platform of Leander Point. The beach faces west and receives waves averaging over 1 m which, combine with the fine calcareous sand to maintain a low gradient 100-150 wide bar, which tends to be dissipative in the south (Fig. 4.255), grading to rips in the north, the rips partly induced by two small calcarenite reefs in the surf and the northern platform. The dissipative section may have up to three bars and lines of breakers, with rips dominating the inner bar. This a popular surfing area with breaks along the beach as well as a lefthand break on the reefs off Leander Point. The beach is backed by a series of generally vegetated longwalled parabolic dunes extending up to 5 km northward. These are bordered by the railway line to the east, the Kallis Road in the north, with the Dongara landing ground behind the northern end, and finally the southern area of Port Denison. There is a car park at the north end, which also provides 4 WD access to the usually firm low gradient beach.

Figure 4.255 The more exposed dissipative section of South Beach with wave breaking across the three bars.

WA 1062-1068 **PORT DENISON-DONGARA**

No.	Beach	Rating	Type	Length
WA 1062	Port Denison Harbour	1	R	700 m
WA 1063	Arurine Bay (Surf Beach)	3-4	R→LTT	1.1 km
Dongara/Denison Surf Life Saving Club				
Patrol period: Sunday	**November-March**			
WA 1064	Irwin River mouth (S)	3	R+reef	400 m
WA 1065	Irwin River mouth (N)	3	R+reef	300 m
WA 1066	Dongara	4	R/LTT	800 m
WA 1067	King Bay	3	R+reef	4.1 km
WA 1068	Five Mile Beach	3	R+reef	300 m
Spring & neap tidal range = 0.7 & 0.4 m				

Dongara is an older town of about 2000 located at the mouth of the Irwin River (Fig. 2.356). The river is not navigable and an open port and jetty was established in the 1860's 2 km to the south in Arurine Bay in the lee of Leander Point. In the 1980's the port was partly enclosed with two 500 m long attached breakwaters, which provide a calm 20 ha harbour. Today Dongara-Port Denison is a growing fishing, holiday and retirement town. It is also the first coastal town on the Brand Highway north of Perth, 360 km to the south. The harbour construction divided the Arurine Bay beach into two beaches (WA 1062 & 1063), while beaches WA 1064 and 1065 straddled the usually closed river mouth. Beaches WA 1066-1068 begin a near continuous chain of beaches that trend north to Greenough. All the beaches are moderately protected by scattered inshore calcarenite reefs, with generally lower waves at the shore.

Port Denison harbour beach (**WA 1062**) occupies the 700 m long southeastern shore of the harbour and faces northwest towards the narrow harbour entrance (Fig. 4.257). Conditions are usually calm apart from low wind waves. Shallow reefs outcrop along the southern corner of the beach, with a boat ramp located in the centre, then the main 150 m long jetty. A second large boat ramp and commercial jetty occupies the northern end of the beach adjacent to the northern breakwater. The entire beach is backed by a foreshore reserve, which provides various recreational facilities including car parks, picnic-barbeque areas and playgrounds, then the shops and houses of Port Denison.

Beaches WA 1064 and 1065 lie either side of the **Irwin River** mouth and Thungarra estuary (Fig. 4.258). The nature and length of each beach depends on whether the river is open, and where it breaks out when open. The southern beach **WA 1064** is approximately 400 m long extending from the southern boundary reefs and rocks, as a double crenulate beach in lee of a detached and attached reef, to the usually blocked river mouth. Beach **WA 1065** curves on the northern side for another 300 m to a sandy foreland formed in lee of a small reef. Both beaches usually receive wave averaging less than 1 m, which surge up steep coarse reflective beach. They are backed by an unstable backbeach consisting of washover deposits and small sand dunes, then the estuary.

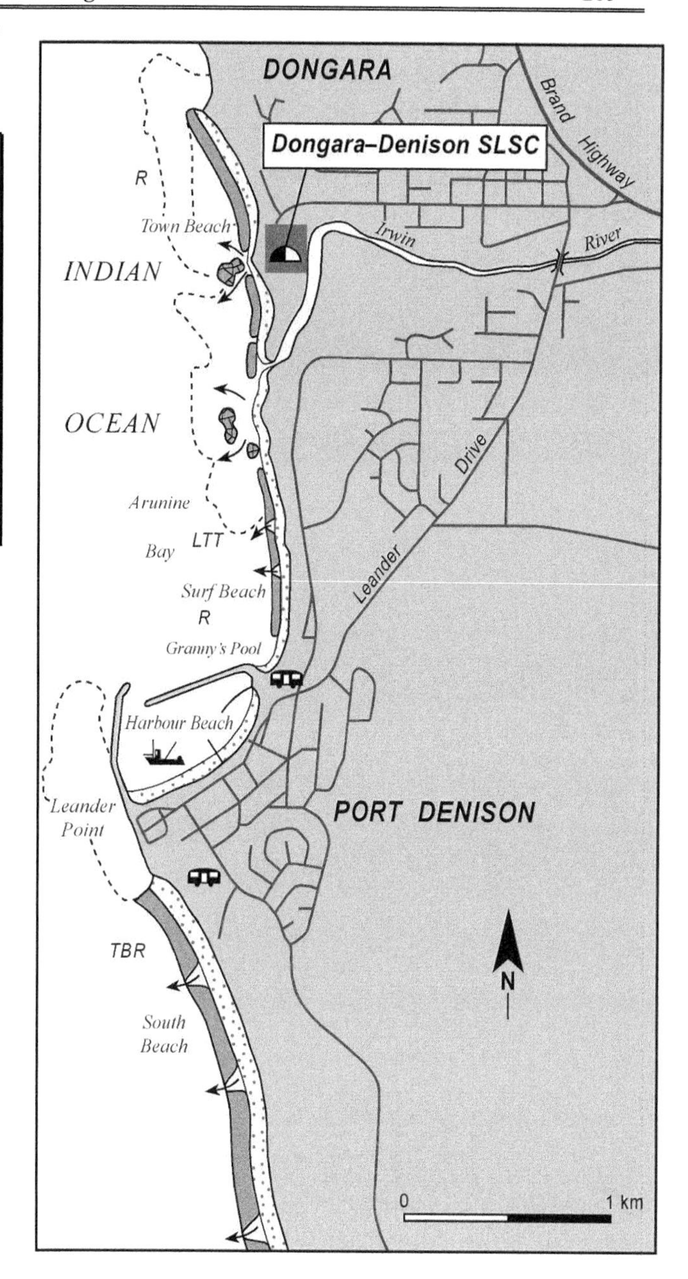

Figure 4.256 The South Beach-Port Denison-Dongara area has a range of beaches from sheltered in the harbour, to rip-dominated to the south and reef-dominated to the north.

Arurine Bay beach (**WA 1063**, also known as **Surf Beach**) extends north of the breakwater for 1.1 km to an attached calcarenite reef, which forms a small foreland. The beach receives slight protection in the southern corner with waves increasing up the beach to about 1 m along the northern section. These maintain a moderately step reflective beach in the south grading into a low tide terrace to the north, with reefs fringing the northern section. The Port Denison caravan park backs the southern corner of the each, then a grassy foreshore reserve between the beach and Ocean Drive.

Figure 4.257 Port Denison is sheltered by two attached breakwaters.

Figure 4.258 Mouth of the Irwin River at Dongara.

Dongara beach (**WA 1066**) is the main ocean beach for Dongara and located at the end of the town's main street. It is also known as **Ocean** or **Town** beach. Dongara/Denison Surf Life Saving Club was founded in 2004 and patrols Town Beach. The beach faces west and curves for 800 m between reef-attached sand forelands. The boundary and offshore reefs lower waves to lees than 1 m, which usually maintain a steep reflective beach to each end, with at times a central low tide terrace where the waves are highest. A dune reserve runs the length of the beach with the Dongara caravan park at the southern end next to the main car park and beach access. A crayfish factory is located behind the centre, with houses backing the northern half.

Swimming: The safest swimming in Dongara is at the harbour beach clear of the boat launching area. It also has the best facilities. South Beach has the best surf, though rips are present, while Surf and Dongara beaches, while they have lower waves can have a heavy shorebreak and deep water right off the beach.

Surfing: South Beach has the most consistent surf and the best beach and some reef breaks, with good access if you have a 4WD vehicle. The Point itself, known as *Dension*, also has some heavier reef breaks, while conditions at Surf and Dongara beaches can range from calm to reasonable beach and reef breaks.

Beach **WA 1067** extends north of Dongara for 4.1 km, along a relatively open section of coast known as King Bay. This is a crenulate north-trending beach, which is paralleled by lines of scattered reefs extending a few hundred meters offshore with only low waves reaching the shore to maintain a steep reflective beach. The beach is backed by vegetated transgressive dunes, with no development apart form 4 WD tracks. The 50 ha Dongara Nature Reserve is located in the dunes behind the centre of the beach.

Five Mile Beach (**WA 1068**) occupies a curving 300 m long embayment at the northern end of King Bay. The beach is bordered by reef-tied forelands, with an open central section. Waves remain low and reflective conditions prevail. A 4 WD track from the highway winds through the backing vegetated dunes to the beach.

WA 1069-1071 FIVE-SEVEN-NINE MILE BEACHES

No.	Beach	Rating	Type	Length
WA 1069	Five-Seven Mile Beach	3	R+reefs	4.9 km
WA 1070	Seven-Nine Mile Beach	3	R+reefs	4.8 km
WA 1071	Nine Mile Beach	5	R/LTT+reefs	1.4 km
Spring & neap tidal range = 0.7 & 0.4 m				

Between Five and Nine Mile beaches is 11 km of continuous crenulate sandy beaches that trend essentially straight to the north-northwest. The beaches are paralleled by scattered inshore calcarenite reefs extending up to 1 km offshore, with deeper reefs further out. The reefs induce the shoreline crenulations and in places attach to the shore. The only public access on this section is the road out to Seven Mile Beach, with a few private houses elsewhere and some 4WD tracks.

Beach **WA 1069** commences on the northern side of the Five Mile Beach foreland and trends to the north for 4.9 km to the more prominent Seven Mile Beach foreland (Fig. 4.259). The beach in between is dominated by the shore parallel reefs, with waves breaking for a few hundred metres across the reefs, resulting in a calmer inner 'lagoon' with waves averaging about 0.5 m at the shore. The waves maintain a moderately steep reflective beach, with a few scattered reefs at and close to the shore. The beach is backed by a mixture of vegetated foredune ridges and some transgressive dunes extending 1-2 km inland, together with one private residence.

Seven Mile Beach (**WA 1070**) is readily accessible via a sealed road from the highway, 2 km to the east. There is a large car park and area for beach boat launching at the shore and small fishing boats are often moored off the beach. The beach trends north for 4.8 km to the next major foreland at Nine Mile Beach. The reefs continue to parallel the coast and waves average about 0.5 m along the reflective beach. The reefs narrow to the north, with some of the reef-induced surf, almost reaching the shore and two areas of reef attachment. Seagrass is commonly washed up on the southern section of beach. Vegetated transgressive dune extending up to 2 km inland back the beach, with a 4 WD track running though the dunes the length of the beach.

Figure 4.259 View along Five and Seven Mile beaches, both located to the lee of continuous outer reefs.

Nine Mile Beach (**WA 1071**) marks a more northerly shift in the orientation of the shore and the northern end of the reef-dominated beaches. The beach is 1.4 km long and has several extensive calcarenite outcrops along the shore, as well as scattered reefs extending 200-300 offshore. Wave height increases along the beach, with a low tide terrace in amongst the northern attached reefs. A 4 WD track runs through the vegetated dunes the length of the beach.

WA 1072-1074 NINE MILE-FLAT ROCK-CAPE BURNEY

No.	Beach	Rating inner	Type inner	outer bar	Length
WA 1072	Nine Mile (N)	7	TBR	RBB	11 km
WA 1073	Flat Rock (S)	5	R/LTT		3.6 km
WA 1074	Flat Rock	4	R+reef		1.9 km
WA 1075	S Bend	6	R/LTT+reef		4.6 km
WA 1076	Lucys Beach	8	R+platform		15 km
Spring & neap tidal range = 0.7 & 0.4 m					

At Nine Mile beach the shoreline turns and trends to the northwest for 40 km to Cape Burney and the mouth of the Greenough River. In between is an exposed high energy shoreline with deepwater waves averaging over 1.5 m. The shoreline alternated between exposed sections with wide rip dominated surf, and reef-bound sections with wide reef-surf zone fronting lower energy beaches. All the beaches are backed by 20-60 m high active and vegetated dunes extending 2-3 km inland, then the fertile Greenough valley. The first 17 km between Nine Mile and Flat Rock contains three beaches (WA 1072-1074), with public access only at Flat Rock. These are followed by S Bend beach (WA 1075), then the long Lucys Beach (WA 1076) with access in the south at McCartney's Road and in the north via Cape Burney.

Beach **WA 1072** is one of the longer and higher energy beaches on the west coast. The 11 km long beach extends from the inflection at the northern end of Nine Mile beach to the next prominent foreland, induced by the reemergence of the inshore reefs. In between is a high energy double bar surf zone averaging 300-400 m in width. The inner bar is dominated by large beach rips spaced about every 500 m (Fig. 4.260), with widely space rips on the outer bar. Vegetated and some active parabolic dunes extend 2-3 km inland of the beach. The only access is via private farm tracks and to a few houses nestled in the dunes.

Figure 4.260 Well developed rips (arrows) persist along Nine Mile beach, as higher waves reach the shore.

Beach **WA 1073** is a reef-protected beach that extends for 3.6 km to the southern side of Flat Rock. The reefs lie about 500 m offshore and parallel the beach. A heavy wave breaking occurs across the reefs, with a relatively calmer lagoon between the reefs and shore with lower waves at the shore resulting in a reflective to low tide terrace beach. The only access is via private tracks across the 2 km wide dunes from the backing farmland or along the beach from Flat Rock.

Flat Rock is accessible via a 2.5m long gravel road off the Brand Highway. This is a popular surfing location with breaks on the reefs off the beach. The 'flat rock' is a 50-100 m wide intertidal calcarenite rock platform that is located directly in front of the car park. A break in the outer reefs permits higher waves to reach the platform and produce the surfing breaks relatively close to shore. The beach (**WA 1074**) extends 1.9 km north of the car park as a narrow, crenulate steep reflective high tide beach partly fronted by beachrock and platforms

S Bend beach (**WA 1075**) is located west of the prominent S bend in the Brand Highway. A closed track leads from the rear of the S Bend caravan park across the dunes, otherwise there is public access in the north via Lucys Beach access. The beach extends straight for 4.6 km between the northern end of Flat Rock and a prominent sandy foreland attached to a 150 m long section of shore parallel calcarenite. The entire beach is paralleled by a line of calcarenite reefs located 200-300 m offshore. These lower waves to less than 1 m at the shore, where a combination of reflective to low tide terrace conditions (Fig. 4.261), together with patchy outcrops of calcarenite dominate. The waves and reefs induce a number of permanent rips along the shore.

Figure 4.261 View south along the beach at S Bend, with wave breaking on the outer reefs and lower energy conditions at the shore.

Lucys Beach (**WA 1076**) commences on the northern side of the foreland and continues to the northwest for 15 km to Cape Burney. The calcarenite merges with the shoreline along this beach permitting higher waves to reach the shore, with calcarenite rocks, platforms and reefs dominating the shoreline and narrow inner surf zone. As a consequence this is a highly crenulate and hazardous beach. The northern few kilometers consists of a high tide sand beach fronted by a continuous 20-50 m wide intertidal calcarenite platform, with scattered reefs off the platform. The reefs result in several permanent rips forming along the beach. The beach is backed by the continuous partly active dune field, with public access at either end, and three tracks from farmland also crossing the dunes. In addition sandy 4 WD tracks parallel much of the beach. The tracks and beach are primarily used by beach fishers.

WA 1077-1080 **GREENOUGH RIVER MOUTH**

No.	Beach	Rating	Type	Length
WA 1077	Greenough R mouth (S)	4	LTT	300 m
WA 1078	Greenough R mouth (N)	6	TBR	1.3 km
WA 1079	Southgate Dunes (S)	5	R+platform	1.5 km
WA 1080	Southgate Dunes (N)	4	R+reef	1 km
Spring & neap tidal range = 0.5 & 0.1 m				

The **Greenough River** has been deflected 17 km to the north and runs along behind the dunes of beach WA 1076 to enter the sea in lee of the low Cape Burney. The river mouth is blocked during the dry summer months only breaking open during winter rains. The river, river mouth and adjacent beaches and dunes have long been a focus of a range of recreational activities and were patrolled by Surf Life Savers from Geraldton as early as 1934. It is located just 2 km off the Brand Highway and has a large caravan park and store behind the dunes and two large car parks at the beach. The river mouth is popular for both boating activities on the river as well as surfing and fishing at the beach. Two adjoining beaches (WA 1077 & 1078) block the river mouth, while beaches WA 1079 and 1080 bordered the western side of the large active Southgate Dunes which have originated from the northern river mouth beach.

Beach **WA 1077** is a curving 300 m long west-facing beach that is tied to the northern calcarenite rocks of Cape Burney in the south and a cuspate foreland formed to the lee of a small reef to the north (Fig. 4.262). The river usually breaks out across this beach. When the river is closed a wide berm and overwash flats back the beach. Waves are reduced slightly by Cape Burney and the reef and average about 1 m and break across a low gradient fine sand beach to maintain a 50 m wide low tide terrace. The beach has a moderate hazard rating when closed, however when the river is open a deep channel and strong currents flow across the beach resulting in more dangerous conditions. Vehicles cross the back of the beach, when closed, to reach the 4 WD tracks on Cape Burney and proceed south to beach WA 1076.

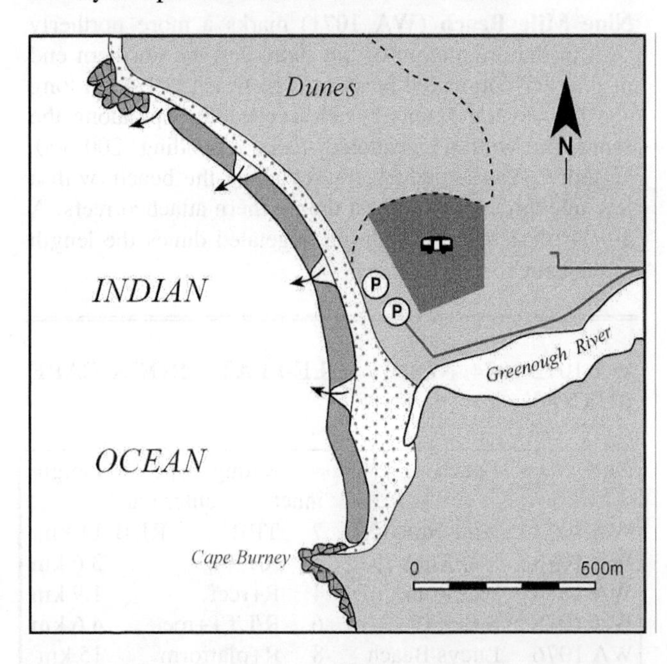

Figure 4.262 Greenough Beach is located across the mouth of the Greenough River (see Fig. 4.260 for location).

The northern river mouth beach (**WA 1078**) extends from the foreland to the north, then northwest as a curving 1.3 km long beach (Fig. 4.263). The southern half is exposed to wave averaging about 1.5 m which maintain a 100 wide surf zone usually cut by two beach rips and a third permanent rip against the northern reefs. A 50 m wide reef fringes the northern third of the beach resulting in unsuitable swimming conditions. There is a car park behind the southern end of the beach, backed by a 500 m wide dune section, then the caravan park and settlement.

The **Southgate Dunes** have originated from sand blowing north from beach WA 1078. The largest active dunes extend 3.5 km to the north and cover an area of about 300 ha. Beach **WA 1079** borders the southwestern side of the dunes from the northern end of beach WA 1078 to the westernmost tip of the dune field, a distance of 1.5 km. The beach consists of a strip of high tide sand fringed by continuous 50-100 m wide intertidal calcarenite platform with reef further out. Higher waves break over the reefs, with low waves at the shore. It is backed by a generally vegetated foredune, then the dune field. Beach **WA 1080** continues from the western tip, north-northeast for 1 km to a small reef attached foreland. This beach is fronted initially by a 50 m wide intertidal platform, which gives way to an sandy section then the boundary reef. It is protected by the reefs with usually low waves at the shore and largely backed by active dunes that are blowing north and sliding down onto the beach..

Figure 4.263 The Greenough River mouth receives moderate waves with rips and reefs persisting along the northern beach.

WA 1081 TARCOOLA/BACK/MAHOMETS BEACH (GERALDTON SLSC)

No.	Beach		
WA 1081	Tarcoola/Back/Mahomets		
	Rating	Type	Length
	3→6→4	LTT→TBR→LTT	6.3 km

GERALDTON (SLSC)
Geraldton Surf Life Saving Club

Patrol period:	**Saturday**	**December-March**
	Sunday	**November-March**
Lifeguard:	**Mon-Fri**	**mid Dec-January**

Spring & neap tidal range = 0.5 & 0.1 m

Geraldton has a population of about 30 000 and is the largest town north of Perth. It is situated either side of Point Moore, which protrudes 4 km seaward, thereby providing the more sheltered anchorage in Champion Bay that lead to it use as a port from 1840 and the gradual development as a major port and city. Today major harbour construction along Champion Bay provides safe anchorage for vessels including large grain carriers. The harbour also includes the Fisherman's Wharf area, boat ramps, yacht club and a two small boat harbour and the newer Batavia Coast Marina and residential development.

The Geraldton shoreline is for the most part sandy beaches with a total of nine beaches (WA 1081-1090) (Fig. 4.264). To the south of the city are Greenough River mouth beaches (WA 1077 & 1078), Tarcoola-Back beach site of the Geraldton Surf Life Saving Club (WA 1081) and Grey beach (WA 1982) between Separation Point and Point Moore. These are all exposed to moderate southerly swell and have reasonable surf. To the north of the harbour development are more protected St George and Sunset beaches.

Tarcoola-Back Beach (**WA 1081**) is the site of the Geraldton Surf Life Saving Club (Fig. 4.265). The Club was founded in 1930, following the development of a swimming club in 1918 and Swimming and Surf Life Saving Club in 1926. The original clubhouse was lost to beach erosion and demolished in 1967. The present club sits in a large foreshore reserve on Mahomets Flats between Tarcoola and Back beaches. It has a patrol tower located on the foredune, which provides a good view along both sections of the beach. The beach begins in lee of the Southgate Dunes and runs northwest as Tarcoola Beach for 4 km, before it swings round to face south by Separation Point, with the last 2 km called Back Beach. The beach is exposed to southerly ocean swells with waves averaging over 1 m, and occasionally reaching 2 to 3 m on the more open Tarcoola section where they maintain a continuous bar cut by rips every few hundred meters (Fig. 4.266). The southern and Back beaches sections are a little more protected with a usually continuous bar free of rips with two small patches of reefs also along Back Beach. The beach is backed by a 300 m wide low foredune ridge plain, which now contains a continuous foreshore reserve then the houses of Tarcoola in the south and Mahomets Flats in the north. The Brand Highway parallels the beach 400 m inland, with several access points and car parks along Glendinning and Willcock drives which back the reserve.

Swimming: This is an exposed and at time rip dominated beach so be careful if swimming here. Swim at the surf club and in the patrolled area.

Surfing: There are numerous beach breaks the length of the beach with the larger waves along the northern part of Tarcoola Beach.

WA 1082-1085 POINT MOORE (GERALDTON)

No.	Beach	Rating	Type	Length
WA 1082	Greys Beach	4	LTT	2.5 km
WA 1083	Point Moore	4	R/LTT+reef	800 m
WA 1085	Point Moore (N)	3	R/LTT	600 m
WA 1085	Pages Beac h	3	R/LTT	400 m
Spring & neap tidal range = 0/5 & 0.1 m				

Point Moore is a prominent sandy foreland formed in lee of the Point Moore Reefs, which forms the western boundary of Geraldton and the southern side of Champion Bay, site of Geraldton Harbour. The point

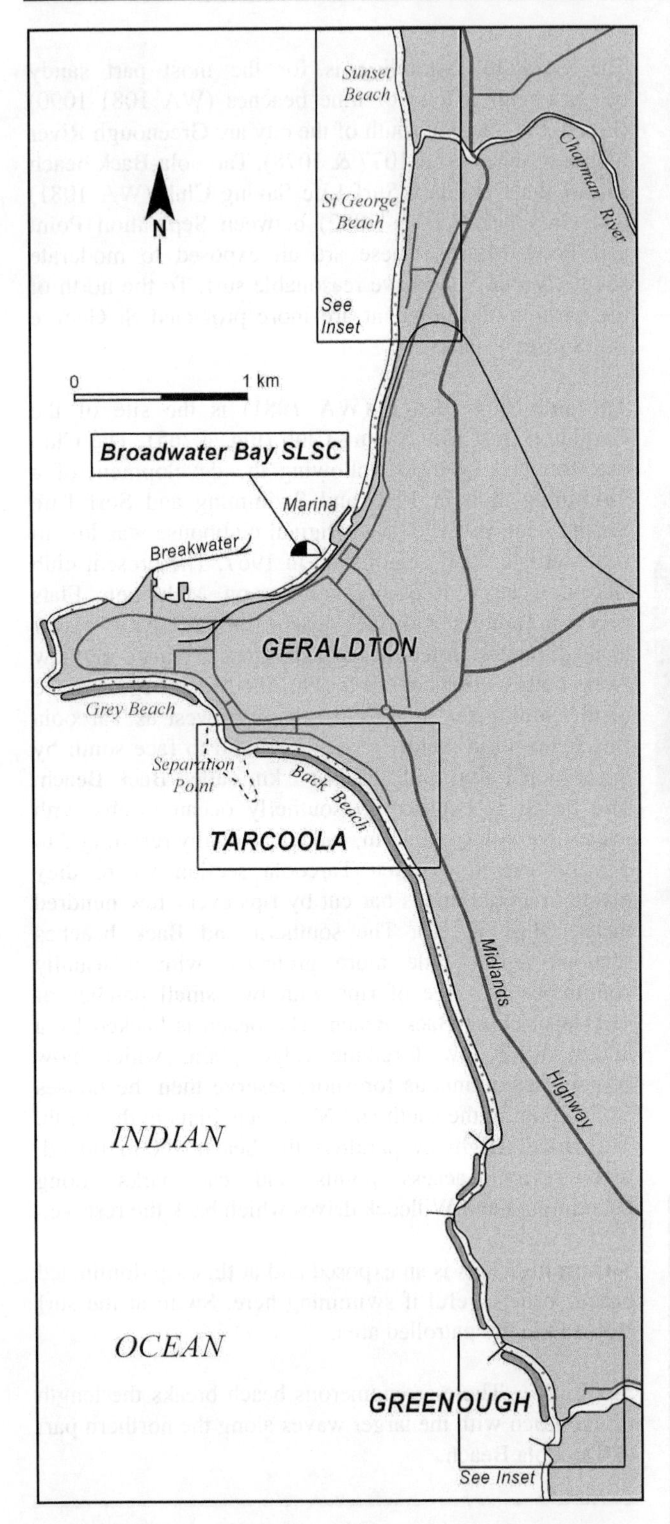

Figure 4.264 Geraldton is located to the lee of the prominent Point Moore, with beaches occupying most of the shoreline on the point and to the north and south. See Fig. 4.262, 265 & 269 for inserts.

protrudes 4 km west of the north-south trend of the coast, however the reefs provide moderate shelter from the deepwater waves with wave averaging 1 m or less at the shore. Willcock Drive runs right around the low point area and provides good access to all the beaches (WA 1082-1085). The reefs off the Point are very popular for windsurfing.

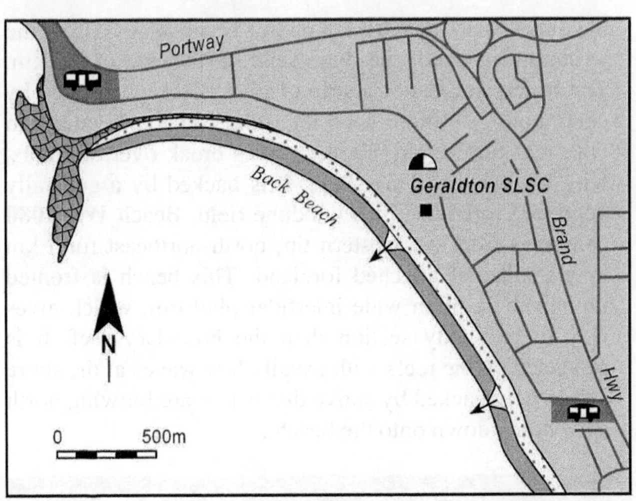

Figure 4.265 Geraldton Surf Life Saving Club is located on the more exposed, rip-dominated Tarcoola beach.

Figure 4.266 Tarcoola Beach with the Geraldton Surf Life Saving Club (centre) and rip-dominated surf zone.

Greys Beach (WA 1082) forms the southern side of the point and curves to the north, then west for 2.5 km between Separation Point and Point Moore. Both points are low sandy forelands formed in lee of attached intertidal calcarenite reefs, with additional reefs extending up to 1 km westward of Point Moore. Greys Beach faces squarely into the dominant waves and winds. The waves are reduced by the reef and average about 1 m and usually maintaining a 50 m wide low tide terrace, which may be cut by rips following periods of higher waves. The beach is backed by a 10 m high scarped foredune, a 100 m wide dune reserve then the drive, with car parks located to either end. A caravan park is located to the lee of Separation Point.

At **Point Moore** the shoreline turns and curves to the north for 800 m as beach **WA 1083**. Its northern boundary is a prominent sandy foreland tied to a large intertidal reef, which extends 200 m to the west (Fig. 4.267). This boundary and outer reefs lower waves to less than 1 m, which combine with the fine sand to maintain a lower energy low tide terrace along the southern half, with the reef platform fronting the northern half of the beach. There are surfable waves on some of the reefs off the point including one about 1 km out called Hells Gate, the gate referring to a break in the reef used by the fishing boats. A 200 m wide dune reserve backs the beach with a large southern car park, then houses east of Willcock Drive, and the Point Moore lighthouse opposite the car park.

Figure 4.267 Point Moore is a prominent foreland surrounded by sandy beaches.

Beach **WA 1084** commences at the northern foreland and trends east for 600 m to a shore parallel groyne that is attached to a small reef. The beach is sheltered from most waves by the western boundary reef and its northerly orientation, with usually low waves to calm conditions at the shore. It is backed by a 300 m wide grassy dune reserve, with a car park on the boundary western foreland and at the western end.

Pages Beach (**WA 1085**) is located between the western groyne and the western side of the 1.5 km long attached breakwater that forms the main protection for Geraldton Harbour. The beach has prograded a few hundred meters north at its eastern end causing it to curve and faces the northwest. The western end is very sheltered by the groyne and outer reefs, with slightly higher waves towards to eastern end, as a consequence the waves maintain a reflective beach in the east grading into a narrow attached bar to the west. Most of the beach is backed by a narrow dune reserve, then a large recreational reserve with a range of picnic facilities, as well as a central boat launching area. This is a popular low energy beach where the prevailing southerly winds blow offshore.

WA 1086-1087 **TOWN BEACH (GERALDTON) BROADWATER BAY SLSC**

No.	Beach	Rating	Type	Length
WA 1086	Town Beach (1)	3	R/LTT	200 m
WA 1087	Town Beach (2)	3	R/LTT	200 m
WA 1088	Broadwater Bay	3	R/LTT	150 m

Broadwater Bay Surf Life Saving Club
Patrol period: Sunday November-March

Spring & neap tidal range = 0.5 & 0.1 m

Town Beach is located at the base of downtown Geraldton, with the main commercial district along Marine Terrace paralleling the beach 200 m to the south. The beach forms part of the once continuous sandy shoreline of Champion Bay, that extended east from the northern side of Point Moore then curved to the north for 8 km to Chapman River mouth. Today major port development has replaced 3 km of the southern sandy shore with harbour, port and marina facilities and divided the once continuous beach into six remnants, two west of the port (WA 1084 & 1085), three wedged between the major port and the newer Batavia Coast Marina (WA 1086-1088), and the longer Champion Bay beach (WA 1089) left to continue north to the river mouth.

The Town beaches are part of a 550 m long section of remnant shoreline divided into three beaches by two groynes and bordered by the attached harbour breakwaters. All the beaches receive low refracted waves averaging about 0.5 m and increasing slightly to the north. They break across a fine sand low gradient beach and narrow surf zone. Beach **WA 1086** commences at the 50 m long groyne that forms the eastern side of the yacht marina and large boat ramp and trends to the east for 200 m to a forked 80 m long groyne and jetty. A seawall, yacht club, and large car park back the beach with an elevated boat ramp crossing the western end. Beach **WA 1087** extends for 200 m between two rock groynes. It is backed by a continuous rock seawall, then Foreshore Drive, with a small patch of grass in the western corner where the beach has prograded slightly. Beach **WA 1088** continues northeast of the groyne for another 150 m to the southern boundary of the Batavia Marina development, which occupies then next 1 km of the shore (Fig. 4.268). The beach is patrolled by **Broadwater Bay Surf Life Saving Club**, which was founded in 2003. It usually has am moderately steep swash zone and low gradient attached bar. It is backed by a seawall and the road, with a large car park behind the northern Batavia boundary seawall.

Figure 4.268 The Batavia Marina, with Broadwater Bay beach to the right.

WA 1089-190 CHAMPION BAY-SUNSET BEACH (GERALDTON)

No.	Beach	Rating	Type	Length
WA 1089	Champion Bay/St Georges	4	R+reef/seagrass	3.8 km
WA 1090	Sunset Beach	4→6	LTT→TBR	5 km

Spring & neap tidal range = 0.5 & 0.1 m

Champion Bay-St Georges-Bluff Point beach (**WA 1089**) is the 3.8 km long remnant of the beach that once extended for 8 km south and west to Point Moore. It now commences on the northern side of the Batavia marina where it is called St Georges Beach and trends north past low Bluff Point (Bluff Point Beach), to the usually blocked 100 m wide mouth of the Chapman River (Fig. 4.269). Shallow seagrass-covered calcarenite reef parallels most of the beach extending up to 1 km offshore and result in very low wave conditions at the shore and seagrass commonly piled up on the beach. A foreshore reserve, then Chapman Drive and Kempton Street parallel the rear beach with good access the length of the beach. The very southern corner had prograded in lee of the marina breakwater and faces north, however a few hundred metres up the beach erosion has cut the beach back to the road and a seawall backs the beach. At the beginning of Kempton Street is a park, car park and boat ramp.

Sunset Beach (**WA 1090**) commences at the river mouth (Fig. 4.270) and trends to the north-northwest for 5 km to the southern reefs of Drummond Cove. The beach is initially sheltered by the reefs with reflective conditions at the river mouth. The southern 1 km of beach is backed by the Sunset Beach a caravan park and housing estate. As the reefs diminish to the north the waves pick up and maintain a low tide terrace extending 2 km north of the river, beyond which waves average 1.5 m and maintain a well develop transverse bar and rip system, with rips spaced about every 300 m along the northern 3 km of the beach. The higher waves and slight southerly orientation of the beach have permitted a series of nested parabolics to develop in lee of the beach. These have migrated up to 4 km northward to reach Drummond Cove and attain an elevation of 30 m at Sandy Hill. The central-northern section of the beach is only accessible by 4 WD.

WA 1091-1094 DRUMMOND COVE

No.	Beach	Rating	Type	Length
WA 1091	Drummond Cove (S)	4	R+reef	1 km
WA 1092	Drummond Cove	3	R	1.3 km
WA 1093	Drummond Cove (N)	5	R/LTT+reef	2.5 km
WA 1094	Drummond Cove (N1)	4	R	900 m

Spring & neap tidal range = 0.5 & 0.1 m

Drummond Cove is part of an open, west-facing, 3 km long embayment bordered by reef-tied forelands. The southern end of the embayment, to the lee of beach WA 1092, has been developed for housing in amongst the dunes that originated from Sunset Beach to the south. The community is essentially the northern-most beach suburb of Geraldton. Either side of the 1 km long settlement are five beach systems (WA 1091-1094).

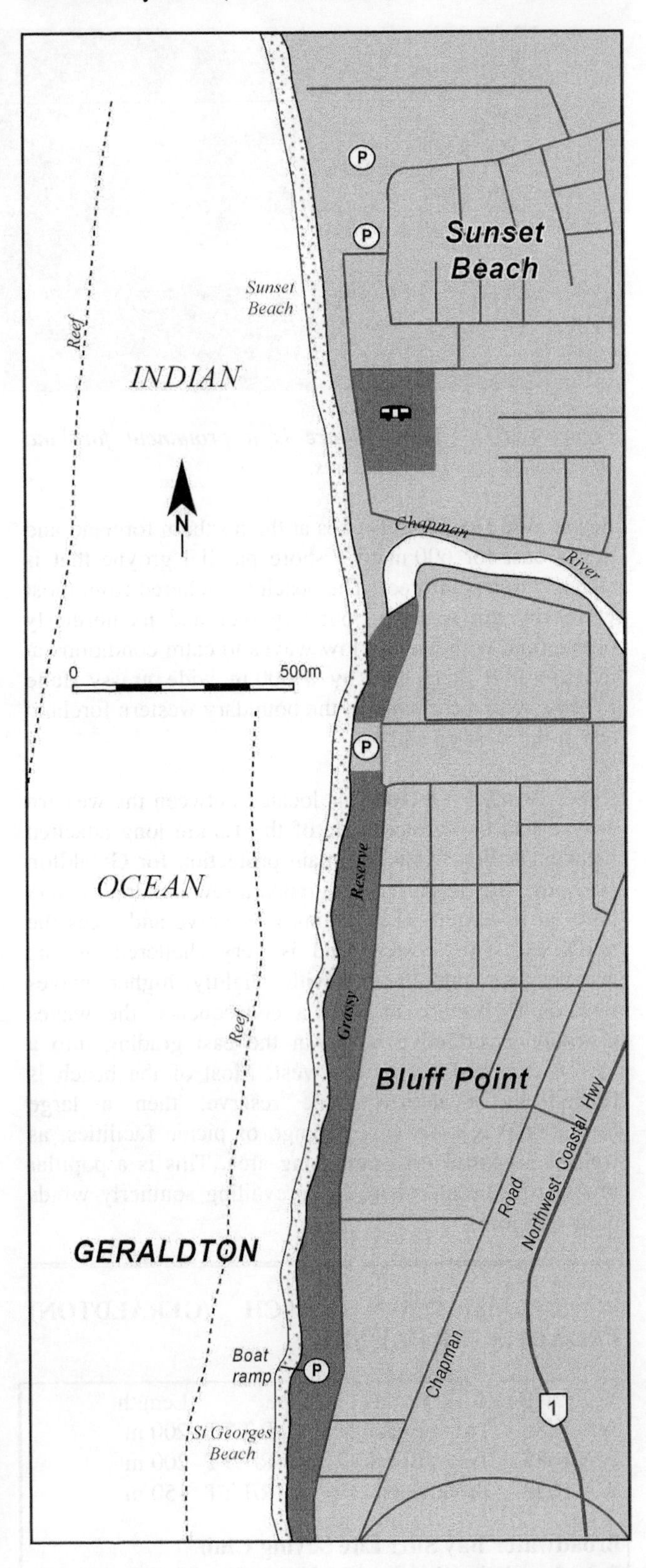

Figure 4.269 St George and Sunset beaches extend north of Geraldton across the mouth of the usually blocked Chapman River.

Figure 4.270 The Chapman River mouth separates St George and Sunset beaches. Shallow reefs dominate the inshore.

Beach **WA 1091** is a crenulate 1 km long high tide beach fringed by continuous, though irregular 200-300 m wide calcarenite reefs and platforms. Waves break over the outer edge of the reefs with generally low waves at the shore. There is some surf along the edge of channels in the reef with easy access from the shore. A 4 WD track from Drummond Cove winds around the sandy foreland, which is backed by vegetated transgressive dunes rising to 20 m.

Drummond Cove beach (**WA 1092**) is a curving west-facing 1.3 km long low energy beach sheltered by a chain of patchy reefs, which extend a few hundred metres offshore (Fig. 4.271). The beach is narrow, crenulate and backed by the northern end of the Sunset Beach transgressive dunes. The expanding Drummond Cove housing estate extends for about 1 km in amongst the backing dunefield, with the northern houses separated from the beach by a narrow foreshore reserve. There are six car parks and two areas for beach boat launching, along the usually low energy beach. Vehicles are permitted to access the beach from the southern car park to drive out to beach WA 1091.

Figure 4.271 Drummond Cove beach is located to the lee of extensive shallow reefs, resulting in low waves at the shore.

Beach **WA 1093** commences towards the northern end of the settlement about 500 m south of where the access road reaches the shore. The beach curves to the north then north-northwest for 2.5 km and terminates at the next reef-tied sandy foreland. This beach has less reef and receives waves averaging about 1 m, which form a low tide terrace in the south. To the north reefs increase with waves breaking over inner reefs and the beach a mixture of reflective to low tide terrace conditions. During periods of higher waves rips form along the beach and amongst the inner reefs. The beach is backed by a 50 ha active parabolic dune which extends 1 km to the north to reach the Buller River. Access north of the road is by 4 WD or foot.

Beach **WA 1094** occupies the foreland and is a 900 m long crenulate beach located in lee of two section of shore parallel reef, with a foreland to the lee of each. Waves break over the reefs with generally low waves and reflective conditions at the shore, as well as areas of reef attached to the shore. The active Drummond Cove dunes pass the rear of the beach, with vegetated transgressive dunes between the sand sheet and the shore. A 4 WD track runs along the rear of the beach.

WA 1095-1097 **BULLER RIVER**

No.	Beach	Rating	Type	Length
WA 1095	Buller R	3	R/LTT	800 m
WA 1096	Buller R (N1)	4	LTT	1.4 km
WA 1097	Buller R (N2)	3	R/LTT	4 km
Spring & neap tidal range = 0.5 & 0.1 m				

Buller River is a usually dry stream that reaches the shore 4 km north of Drummond Cove and 2 km west of the highway. The drowned river valley forms a slight embayment occupied by beach **WA 1095**, which curves for 800 m between boundary sandy forelands. Shore parallel reefs attach to the southern foreland and continue across the embayment resulting in low waves at the shore. The low waves and fine sand combine to maintain a reflective to low tide terrace beach. The small stream reaches the centre of the beach, with a vegetated dune to the south and an unstable dune to the north, which develops into a 1.5 km long northward migrating parabolic dune. A farmhouse is located on the northern side of the river mouth behind the foredune, with 4 WD tracks running though the dunes on both sides of the river.

Beach **WA 1096** commences at the slight northern foreland and continues north for 1.4 km to a slight westerly protrusion in the shore at the next sandy foreland. The beach remains sheltered by scattered shore parallel reefs that extend a few hundred metres offshore. Waves average less than 1 m and maintain a narrow low tide terrace, which may be cut by beach rips during and following periods of higher waves. The beach is backed by the active dune sheet that originates from the river mouth, with only the northern foreland section vegetated.

Beach **WA 1097** continues to the north-northwest for the foreland for 4 km to the southern foreland of the Oakajee

River mouth. This is a crenulate sandy beach with shore attached as well as a 500 m wide band of inshore and offshore reefs. Some of the inshore reefs appear to be coral formations Waves break across the scattered outer reefs and average about 0.5 m at the shore with reflective conditions dominating. The southern half of the beach is backed by a 1 km wide vegetated band of transgressive dunes, while an active parabolic originates from the northern half and extends northward for up to 1.5 km, with older vegetated dunes to the east and north. A few 4 WD track reach the shore and parallel the rear of the beach.

WA 1098-1101 **OAKAJEE R-CORONATION**

No.	Beach	Rating	Type	Length
WA 1098	Oakajee River	6	LTT/TBR	1.4 km
WA 1099	Oakajee R (N)	5	R/LTT+reefs	2.8 km
WA 1100	Coronation Beach	3	R+outer reefs	2 km
Spring & neap tidal range = 0.5 & 0.1 m				

The **Oakajee River** reaches the coast 7 km north of the Buller River and 5 km west of the highway. It flows though an incised river valley and reaches the shore in a slight embayment, with the river usually dry and blocked at the shore. Beach WA 1098 occupies the river mouth, with beaches WA 1099-1101 extending 5 km north to Oakabella Creek. The only public access to the coast is at Coronation Beach (WA 1100).

The Oakajee River mouth beach (**WA 1098**) is a curving 1.4 km sandy beach bordered by reef-induced sandy forelands. The reef deepens across the slight embayment with a 500 m wide sandy channel extending seaward of the river mouth. Waves break over the boundary reefs arriving unbroken along the central beach section, where they average about 1 m. The waves combine with the coarser sand to maintain a reflective to low tide terrace beach, with rips forming during periods of higher waves. The beach is backed by vegetated transgressive dunes to either side of the river, with farmland following the river valley to the rear of the foredune. A 4 WD track runs out to the beach.

Beach **WA 1099** commences at the northern reef-induced sandy foreland and trends to the north-northwest for 2.8 km to the southern foreland of Coronation Beach. This is a relative straight beach paralleled by a near continuous band of beachrock, which is attached to the southern half of the beach and parallels the northern half about 50-100 m offshore, with patchy reefs extending further seawards. Waves break across the outer and inner reefs resulting in low waves to calm conditions at the shore and a calm lagoon along the northern half between the beach and inner beachrock reef. The beach is backed by vegetated Holocene transgressive dunes then farmland, with a 4 WD track from Coronation Beach running the length of the beach.

Coronation Beach (**WA 1100**) is a curving then straight 2 km long sheltered beach with a 100 m wide intertidal platform attached to the southern foreland and the shore parallel reefs continuing north 300-400 m offshore. The reefs result in a relatively calm 'lagoon' between the reef and shore (Fig. 4.272). The beach is accessible via 7 km long gravel road from the highway. The relatively steep beach is used to launch boats in the 'lagoon' and some boats are moored off the beach. These is an informal camping area at the southern end and 4 WD access to the beach. This is a popular surfing and fishing location with a range of breaks out on the southern reef. The northern half of the beach continues straight in lee of the reef to the southern foreland of the Oakabella Creek mouth.

Figure 4.272 Coronation Beach with the southern sheltered 'lagoon' area and surf out on the reefs.

WA 1101-1102 **OAKABELLA CREEK**

No.	Beach	Rating	Type	Length
WA 1101	Oakabella Creek	6	TBR	1.4 km
WA 1102	Oakabella Ck (N)	3	R+reef	3.5 km
Spring & neap tidal range = 0.5 & 0.1 km				

Oakabella Creek is a usually dry creek, which meanders through a 2 km long valley, cut though the bordering 100 m high Pleistocene calcarenite, to reach the shore 5 km north of Oakajee River. The creek is surrounded by farmland with no public access to the shore. The creek mouth beach (**WA 1101**) occupies a slight indentation caused by the valley and consists of a curving 1.4 km long beach, with boundary reef-tied forelands to either side. The southern reefs partly protect the beach while elsewhere waves averaging over 1 m reach the shore and maintain a 50 m wide bar usually cut by four beach rips and a permanent rip against the northern boundary reef. A 4 WD track emerges from the usually dry creek bed and connects with tracks running to the north and south along the rear of the beach.

Beach **WA 1102** commences at the slight northern foreland and trends relatively straight to the north-northwest for 3.5 km to the southern side of Woolawar Gully creek mouth. The entire beach is paralleled by calcarenite reefs. They are attached at either end forming a 50 m wide intertidal platform and lie 100 m offshore along the centre forming a calmer 'lagoon' in their lee.

The beach is backed by a low foredune and 200 m wide vegetated sandy plain abutting vegetated Pleistocene calcarenite that rises to 80 m. A solitary shack is located behind the centre of the beach with 4 WD tracks running the length of the beach.

WA 1103-104 **WOOLAWAR GULLY**

No.	Beach	Rating	Type	Length
WA 1003	Woolawar Gully	6	TBR	1.2 km
WA 1004	Teakle Hill	8	R+platform/reef	11.3 km
Spring & neap tidal range = 0.6 & 0.5 m				

Woolawar Gully is a deeply incised stream that cuts a 100 m deep gully though the Pleistocene calcarenite to reach the shore 5 km north of Oakabella Creek. Farmland extends down the sides of small usually dry creek, which enters the sea in the centre of beach **WA 1103**. This is a curving 1.2 km long beach bordered by shore-attached and offshore reefs to either side. A moderately open gap between the reefs permit waves averaging 1-1.5 km to reach the shore. The waves maintain a 50 m wide bar usually cut by three beach rips, with permanent rips against the boundary reefs. There is a fishing shack behind the southern side of the creek mouth and 4 WD access to the beach.

Beach **WA 1004** commences on the northern side of the creek mouth and trends straight to the northwest for 11.3 km to the southern side of the Bowes River mouth. The beach is fronted by a continuous intertidal calcarenite platform and a shore parallel reef located about 150 m off the beach. Waves break over the outer reef and again over the platform to finally reach the high tide sandy beach. The waves, platform and reef result in hazardous beach conditions. The beach is backed by an unstable 50 m wide foredune with a few small blowouts, and then slopes rising to a 100 m high calcarenite ridge called Teakle Hill. Four wheel drive tracks parallel the rear of the beach.

WA 1105-1106 **BOWES RIVER**

No.	Beach	Rating	Type	Length
WA 1105	Bowes River mouth	6	TBR	400 m
WA 1106	Bowes R (N)	8	R+platform	3.3 km
Spring & neap tidal range = 0.6 & 0.5 m				

The **Bowes River** drains a deeply incised hinterland including an area of Precambrian granites resulting in predominately coarse quartz sand on the river mouth beach. The river mouth is accessible along the 2 km long Bowes River Road, which follows the usually blocked river and floodplain to the coast. The steep beach (**WA 1105**) consists of an exposed 400 m long section of sand bordered by the platform-fronted beaches to either side together with some calcarenite outcrops along and off the beach. Waves average about 1.5 m resulting in a 50 m wide surf zone dominated by two permanent rips with some good beach and reef breaks to either side of the river mouth.

Beach **WA 1106** commences on the northern side of the river mouth and trends to the northwest for 3.3 km to the beginning of Horrocks Lagoon. The beach is fronted by a continuous 50 m wide intertidal platform with waves breaking heavily on the outer edge of the platform to reach the sandy high tide beach. Conditions are hazardous the length of the beach. An unstable foredune and active dunes back the beach, with a series of parabolics expanding northward along the rear of the beach and extending up to 1 km inland. The beach is accessible by 4 WD tracks from the river mouth and through the dunes from Horrocks.

WA 1107-1108 **HORROCKS**

No.	Beach	Rating	Type	Length
WA 1107	Horrocks	3	R	2.5 km
WA 1108	Three Mile Beach	6	R+platform	2.6 km
Spring & neap tidal range = 0.6 & 0.5 m				

Horrocks is a small coastal settlement located on the shore of Horrocks Lagoon, 20 km west of Northampton. It consists of a few rows of houses, a large oval, caravan park, boat storage area, jetty, boat ramp and small fishing fleet which moors in the lagoon to the lee of a shore-parallel calcarenite reef. The main beach (WA 1107) fronts the settlement with Three Mile Beach to the north.

Horrocks beach (**WA 1107**) is sheltered curving 2.5 km long beach which has formed to the lee of a 5 km long attached northwest-trending reef, with the beach curving to the north to form a lagoon which widens to 500 m (Fig. 4.273). Waves break heavily on the reef where they produce a heavy lefthand surf break, with low wave to calm conditions at the shore and a reflective beach. A gap in the reef towards the northern end of the beach is used by the fishing boats and also permits low waves to reach the shore along the northern section. The town backs the southern end of the beach, with vegetated foredunes to the north grading into active transgressive dunes.

Figure 4.273 Horrocks beach, lagoon and protecting beachrock reef.

Three Mile Beach (**WA 1108**) begins as the shoreline curves to the northwest for 2.6 km. The gap between the shore and reef gradually narrows to where it becomes a discontinuous shore-attached reef, with one 200 m long reef-enclosed lagoon, and a at 100 m wide intertidal platform at the northern end of the beach. Waves break heavily on the reef and platform with low wave to calm conditions at the shore. A large parabolic dune originates at the northern end of Horrocks and southern end of Three Mile and extends 3 km to the north and up to 1 km inland. The beach is backed by vegetated foredune, with vehicle tracks the length of the beach, and the active parabolic further inland.

WA 1109-1115 **WHALEBOAT COVE-MENAI HILLS**

No.	Beach	Rating	Type	Length
WA 1109	Whaleboat Cove	3→6	R→TBR	4.6 km
WA 1110	Deep Gorge	7	LTT/TBR+reefs	1.6 km
WA 1111	White Cliffs	5	LTT+cliffs	200 m
WA 1112	White Cliffs (N)	7	TBR+reefs	500 m
WA 1113	Menai Hills (1)	6	R/LTT+reefs	200 m
WA 1114	Menai Hills (2)	8	R+platform	300 m
WA 1115	Menai Hills (2)	6	R+ platform	2.5 km
Spring & neap tidal range = 0.6 & 0.5 m				

The northern end of Three Mile beach marks the beginning of a 15 km long section dominated by northwest-trending calcarenite cliffs rising to 130 m along the northern Menai Hills. Beaches WA 1109-1115 occupy the first 13 km of the shore. They are wedged in along the base of the cliffs and are either exposed directly to high waves or fronted by platforms and reefs. There is no formal access to this section of shore, with only 4 WD tracks through farmland and backing scrub, reaching the top of the cliffs and in places descending down to some of the beaches.

Whalebone Cove (**WA 1109**) is located in lee of the large calcarenite platforms that forms its boundary with Three Mile Beach (Fig. 4.274). The reef extends 300 m north of the boundary providing a sheltered cove and anchorage for fishing boats. The 4.6 km long beach trends to the north in lee of the reef as a low energy reflective beach as far as the cliffs, it then curves to the northwest becoming more exposed to the north. The beach accordingly grades to a low tide terrace and finally a transverse bar and rip along the northern few hundred metres. The cliffed section has a narrow (50-100 m) wide foredune plain backing the beach, then steep sparsely vegetated slopes rising 80 m to a flat-topped surface. There is 4 WD access from Three Mile Beach in the south, with tracks along the crest of the cliffs, but no northern vehicle access. There are left-hand surf breaks out on the southern reef with access across the platform to the waters edge.

Deep Gorge dissects the 100 m high calcarenite slopes at the northern end of beach WA 1109 and marks the beginning of beach **WA 1110**. This is a protruding and crenulate 1.6 km northwest-trending beach that is backed by vegetated slopes rising steeply to 110 m at Gill Hill, and is fronted by three sections of attached reef, with shore parallel reefs extending 200 m offshore. Waves averaging 1.5 m break over the outer reefs and again on the platforms and across the intervening pockets of sand, each of which has a 50 wide bar and is drained by a permanent rip. A 4 WD track descends a dry gully most way down the northern slopes to a circular parking area, with a walking track down to the beach. This beach is used by the occasional rock fishers and surfers looking for waves over the numerous reefs.

Figure 4.274 Whalebone Cove (left) commences at the northern end of Three Mile Beach (right).

White Cliffs are located in a northerly indentation in the cliffs 300 m north of the northern end of Deep Gorge beach. The 120 m high calcarenite cliffs are scarped and eroding in the southern corner, exposing the fresh calcarenite and the 'white' appearance. Beach **WA 1111** is a narrow 200 m long beach located at the base of the cliffs. It has rock debris scattered over the beach and in the narrow surf zone. Waves are lowered by inshore and offshore reefs and average about 1 m at the shore. There is no vehicle or foot access to the beach. Beach **WA 1112** is located 500 m to the north and consists of a southern 100 m long sandy section backed by 100 m high cliffs, a 100 m long central platform-fringed protrusion backed by a vegetated foredune, then the slopes, and an 300 m long northern sandy section. Waves average up to 1.5 m and maintain an 80 m wide rip-dominated surf zone on the sandy section, while waves break over the reefs just off the platform. A 4 WD track across farmland reaches the top of the bluffs with a foot track down to the centre of the platform section.

Beach **WA 1113** commences at the next platform and reef-induced protrusion. The narrow 200 m long high tide beach is wedged in between the base of steep vegetated slopes and a protruding 50 m wide intertidal platform. It is backed by dune-draped slopes and a gully, with a 4 WD track descending through the gully from the crest of the 120m high Menai Hill ridge. Beach **WA 1114** commences 100 m to the north and is a similar 300 m long narrow crenulate platform beach with waves breaking over the 50 m wide platform to reach the high tide beach. It is backed by dune-draped slopes, which ultimately rise to 120 m. A 4 WD track from the beach WA 1113 winds its way along the slopes the length of the beach and on to beach WA 1114. There are various surf breaks over the reefs and in some gaps in the platforms on both beaches.

Beach **WA 1115** commences 300 m to the north and is a 2.5 km long high tide beach fronted by a continuous 50 m wide intertidal calcarenite platform, apart from a 100 m long open sandy pocket at the very southern end. Waves averaging about 1 m break across a 50 wide bar on the sandy section and across the platform the remainder of the beach, only reaching the beach at high tide. The beach is backed by dune-draped slopes rising to 100 m, with a few blowouts and deflation surfaces in the south, and an active 500 m long blowout towards the northern end, which overlaps with beach WA 1116. There is no vehicle access to the shore.

Beach WA 1115 terminates at an indentation in the shore occupied by a 200 m long sandy beach that marks the beginning of 1.7 km long beach **WA 1116**. Waves averaging about 1 m break across a 50 m wide bar drained by a permanent northern rip. The remainder is fronted by a 50 m wide intertidal platform that protrudes seaward, then curves back into a small reef-filled embayment. The beach is backed by the northern end of the dune from beach WA 1115, then a central stabilising blowout, and a northern blowout originating from the northern embayment. All three are draped over the steep backing calcarenite slopes, which rise to 120 m. The slopes and backing surface is covered by dense scrubs, with no vehicle access to the beach.

WA 1117-1119 **BROKEN ANCHOR BAY**

No.	Beach	Rating	Type	Length
WA 1117	Broken Anchor Bay	6	TBR-LTT	3 km
WA 1118	Broken Anchor Bay (N)	6	R/LTT-TBR	3.3 km
WA 1119	Port Gregory (S)	5	R+reefs	200 m
Spring & neap tidal range = 0.6 & 0.5 m				

Broken Anchor Bay is an open southwest-facing curve in the shoreline, which is centred on the usually blocked mouth of the Hutt River in lee of Archdeacon Ledge (Fig. 4.275). The curvature has resulted from the formation of the large 30 km long Hutt Lagoon barrier that commences at the river mouth, extending for 7 km to the northwest past Gregory before turning to the north-northwest. Beach WA 1117 forms the southern side of the bay, while beach WA 1118 extends to the north and beach WA 1119 is a small hinge beach at the base of Port Gregory lagoon.

Beach **WA 1117** commences at the northern end of beach WA 1116, where the shoreline turns and trends northwest, while the backing cliffline continues to the north-northwest, forming an escarpment that becomes the eastern shore of Hutt Lagoon 3 km to the north. The beach extends for 3 km to the river mouth. The continuous sandy beach is exposed to waves averaging 1.5 m along the central-southern section, which decrease slightly towards the river mouth. The higher waves maintain a 50 m wide bar cut by rips every few hundred metres. It becomes a more continuous bar as the sediment coarsens and the beach steepens to the north. The entire beach faces more into the southerly winds and is backed by active parabolics, which climb the backing slopes to elevations of 60 m and terminate on the slopes and at the Hutt River channel. The river drains an area of heavily weathered and dissected sandstones, whose eroded material forms the steep coarse quartz-rich beach. The river occasionally breakout in winter and induces additional instability in the beach and backing foredunes. A short track off the Port Gregory road leads to the northern side of the river mouth.

Figure 4.275 The Hutt River mouth at Archdeacon Ledge. In this view the river flows several hundred metres to the south before crossing the steep reflective berm crest.

Beach **WA 1118** commences on the northern side of the river mouth and trends to the northwest for 3.3 km, forming the northern arm of the bay. The southern section of the beach is partly sheltered by Archdeacon Ledge, a reef located 2 km west of the river mouth. The combination of coarser sands and waves averaging about 1 m result in a steep reflective beach, grading into a low tide terrace to transverse bar rip along the northern few hundred metres. A series of active and vegetated parabolic dunes back the beach and extend up to 1.5 km inland to the southern shores of Hutt Lagoon and reach 35 m in elevation. The Port Gregory road runs between the eastern edge of the dunes and lagoon shore.

Beach **WA 1119** is located to the lee of the southern tip of the shore parallel calcarenite reef that extends 5 km to the northwest to enclose Port Gregory lagoon. The 200 m long beach is located to the lee of a small gap in the reef. It has prominent boundary sandy foreland and a curving centre. Wave break over the reef with a permanent rip returning the water out though the gap. A 4 WD track winds through the backing vegetated dunes to the beach and then runs along beach WA 1118.

WA 1120 PORT GREGORY

No.	Beach	Rating	Type	Length
WA 1120	Port Gregory	2	R+reef	3.8 km
Spring & neap tidal range = 0.6 & 0.5 km				

Port Gregory is the site of the small town of Gregory, which was established as a port in lee of the 5 km long beachrock reef in 1853. The town now has a fishing fleet with a jetty and boat ramp, and is a holiday destination, with a caravan park and store. The houses are located in amongst the vegetated dunes behind the beach, with a central beach access point at the jetty.

Port Gregory beach (**WA 1120**) faces southwest into the prevailing winds, but is sheltered from waves by the continuous beachrock reef and backing shoals that extends the length of the 3.8 km long beach (Fig. 4.276). Water flowing across the reef returns via a deep channel that runs parallel to the shore and back out to sea through Leander Passage. Waves are usually calm along the beach, particularly at low tide, only picking up close to the prominent sandy foreland that forms the northern boundary in lee of the smaller Gold Digger Passage. This is a sheltered usually calm 'lagoon' beach with the main hazards being the deeper water in the channel and closer to Leander Passage the permanent current that returns the lagoon water seaward. The town is located midway along the beach. The remainder of the beach is backed by generally vegetated north-trending parabolic dunes that have formed a 1 km wide barrier that blocks the southern third of 3000 ha Hutt Lagoon, and enclosed salt lake.

Figure 4.276 Port Gregory is located to the lee of the prominent 5 km long beachrock reef.

Today the **Hutt Lagoon** is totally isolated from the sea. This was not always the case as when it was first flooded by the sea level rise approximately 6 000 years ago, there was an entrance 1-2 km southeast of Shoal Point. At this time waves moved through the entrance and built 1 km wide series of about 20 low foredune ridges, now located 2-3 km east of the point on the northeast shore of the lagoon. Since the lagoon closed to the sea all rainwater flowing into the lagoon evaporates and has over time deposited a thick deposit of salt and gypsum on the lagoon floor.

WA 1121-1125 SHOAL PT-LUCKY BAY-RED BLUFF

No.	Beach	Rating	Type inner	Type outer bar	Length
WA 1121	Leander Passage-Shoal Pt	7	R→LTT	LBT	10.6 km
WA 1122	Shoal Pt (N)	7	TBR+reef		5 km
WA 1123	Sandlewood Bay	4	R+reef		1.9 km
WA 1124	Halfway Bay	6	R+reef		1.9 km
WA 1125	Lucky Bay	4	R+reef		4.9 km
WA 1126	Wagoe	8	R+platform		15.5 km
Spring & neap tidal range = 0.4 & 0.1 m					

At Leander Passage the shoreline turns and trends northwest for 5 km to Shoal Point, where it turns again to the north-northwest and continues straight towards Bluff Point 29 km to the north. In between is an exposed undeveloped section of coast backed by extensive unstable dune systems. The first 22 km up to Lucky Bay contains five beaches (WA 1121-1125), with Wagoe beach (WA 1126) extending the remaining 15.5 km to the southern slopes of Bluff Point.

Beach **WA 1121** commences at the sandy foreland to the lee of Gold Digger Passage and trends to the northwest and finally west for 10.6 km to the sandy Shoal Point. The beach faces into the prevailing winds and waves, with waves averaging over 1.5 m. They combine with the fine beach sand to maintain a 200 m wide double bar system (Fig. 2.6b), with a continuous longshore bar and an inner beach which grades from reflective in the south to a more low tide terrace and at time transverse bar and rip in the north. Only the very southern end in lee of the northern end of the reef has a single bar. Waves break heavily on the outer bar, reform in the longshore trough and break again at the shore, with inner beach rips common along the northern half, while larger scale rips drain the tough across the outer bar. The beach is backed by transgressive dunes which are moderately active in the south and centre and which form a 1 km wide barrier, which is the western boundary of much of Hutt Lagoon. Access is via 4 WD tracks through the dune and along the beach from Port Gregory.

At **Shoal Point** the shoreline turns and trends to the north-northwest to Bluff Point paralleling a straight Pleistocene beachrock reef system. The reef is initially submerged between Shoal Point and Sandlewood Bay, then emerges and parallels the beaches north to Lucky Bay, then joins the coast up to Bluff Point, a total distance of 27 km. Outcrops and sandy forelands induced by the reef divide the first two sections into four beaches (WA 1122-1125). Waves remain high however, with the presence and depth of the reef determining the level of wave energy at the shore.

Beach **WA 1122** extends north-northwest from Shoal Point for 5 km. It is paralleled by the submerged reef, which emerges at Sandlewood Bay. The reef lies 400 m offshore permitting higher waves to reach the shore and break across a 100 m wide rip-dominated surf zone. The reef emerges to the north and some shallow reefs and reef-induce rips occupy the northern few hundred metres of the surf. The entire beach is backed by vegetated, generally deflated dune surfaces extending 2-3 km inland, with 4 WD access via Sandlewood Bay.

Sandlewood Bay beach (**WA 1123**) commences on the northern side of a prominent sandy foreland that at times attaches to the first of the emerged beachrock reef, that continue to the north. The shallow reef parallels the shore approximately 100 m offshore and causes wave breaking with relatively low wave to calm reflective conditions in the backing 'lagoon'. A 4 WD track reaches the southern end of the beach and lagoon, adjacent to a gap in the reef through which the lagoon is drained by a permanent current. A second gap 1 km to the north forms another permanent rip. Conditions are best for swimming away from these gaps in the area well protected by the reefs. The gap in the reefs results in deepwater, high waves and rip currents and should be avoided. The entire 1.9 km long beach has a total of seven gaps and associated currents. Vegetated deflated dune surfaces back the beach.

Halfway Bay beach (**WA 1124**) begins on the northern side of a 150 m long section of reef that is attached to the shore for much of the length of the 1.9 km long beach and forms prominent bays and permanent rips to either end. In the south there is a large gap in the reef which permits higher wave to reach the shore resulting in a 100 m wide surf zone dominated by strong beach-reef controlled rips. The reef then reestablishes itself with a more continuous reef and backing 'lagoon' cut by two central gaps and permanent rips. A sandy 4 WD track runs out to a small collection of fishing shacks located behind the southern more exposed section of the beach. The beach is backed by vegetated dunes grading into more active dunes along the northern half.

Lucky Bay beach (**WA 1125**) is a 4.9 km long sandy beach backed by bare active sand dunes that extend the length of the beach and further north and up to 1.5 km inland. The beachrock reef continue to parallel the beach about 100 m offshore, with a few gaps resulting in five cuspate beaches-forelands along the shore, and a continuous 'lagoon' with none of the forelands reaching the reef (Fig. 4.277). Waves average about 0.5 m at high tide resulting in reflective conditions the entire beach, and calm conditions at low tide. At the northern end the beach rejoins the beachrock.

Beach **WA 1126** is a straight 15.5 km long strip of beachrock and sand that extends from the northern end of Lucky Bay straight to the north-northwest to within 1 km of 80 m high **Bluff Point**. The beach is a continuation of the beachrock reefs that begin at Shoal Point. Along this section it forms an inter to supra-tidal platform in front of the beach along this section. Waves averaging up to 1.5 m break heavily on the outer edge of the 50 m wide platform only reaching the beach at high tide. In places the reef has been broken up and tidal pools and holes front and parallel the beach. The beach is backed by extensive northward migrating dunes, which extends up to 2 km inland. They average 20 m high and behind the northern 2 km of beach climb the southern 80 m high slopes of Bluff Point and continue north of the beach for 1.5 km as clifftop dunes. There is 4 WD access to be beach from the Balline Road, with Wagoe Well and Wagoe farm located behind the centre and northern section of beach respectively.

Figure 4.277 Lucky Bay is an elongate lagoon formed to the lee of a continuous beachrock reef.

KALBARRI NATIONAL PARK

Area:	183 096 ha	
Coast length:	13 km	(2717-2730 km)
Beaches:	2	(WA 1127 & 1128)

Kalbarri National Park is one of Australia's more spectacular national parks dominated by the bright red Ordovician sandstone along the Murchison River and coastal cliffs. The deeply carved and exposed sandstone results in brilliant scenery, which can be viewed, from the crest of the red plateau along the river channel and the coastal cliffs.

WA 1127-1128 **POT ALLEY GORGE**

No.	Beach	Rating	Type	Length
WA 1127	Pot Alley Gorge	5	R+rocks	70 m
WA 1128	Rainbow Valley	6	LTT+rocks	40 m
Spring & neap tidal range = 0.4 & 0.1 m				

The Kalbarri coast commences at Red Bluff (Fig. 4.278) and trends to the north for 17 km to the Murchison river mouth at Kalbarri. The first 13 km of high red cliffs are part of the national park and contain a number of scenic viewpoints of the cliffs, sea stacks, gullies and tidal pools. Two small beaches (WA 1127 & 1128) are located 10 km north of the bluffs in Pot Alley Gorge and adjacent Rainbow Valley

Figure 4.278 Red Bluff (right) marks the boundary between the lower sandy coast to the south, and the rocky Kalbarri coast to the north.

Beach **WA 1127** is a 70 m long pocket of sand located at the mouth of Pot Alley Gorge, a small but deep gorge carved in the 80 m high sandstone plateau by a 2 km long usually dry creek. The beach faces northwest and receives waves averaging about 1 m, which break across the bordering 50 m wide rock platforms and surge up a usually steep reflective beach. The beach is backed by 20 m high cliffs then the steep slopes of the gorge, with the creek bed entering on the southern side. An access road and car park are located above the southern side of the gorge.

Beach **WA 1128** is located 200 m to the north and lies at the convergence of three small but deep gorges, the northern called Rainbow Valley. The beach is just 40 m long, bordered by narrow rock platforms, with rocks also littering the narrow surf zone. A small but permanent rip runs out the narrow entrance of the gorge. A sealed road runs down Rainbow Valley to the rear of the small beach.

WA 1129-1132 **KALBARRI BEACHES**

No.	Beach	Rating	Type	Length
WA 1129	Red Bluff	6	R+reef	700 m
WA 1130	Jakes (S)	5	R+reef	100 m
WA 1131	Jakes	6	R+platform	1 km
WA 1132	Blue Holes	6	R+platform	1 km
WA 1133	Kalbarri	2	R	300 m
WA 1134	Kalbarri (N)	3	R+reef	300 m
Spring & neap tidal range = 0.4 & 0.1 m				

Kalbarri is a growing coastal town of 3 000 people located at the mouth of the Murchison River (Fig. 4.279). It lies 598 km north of Perth and 66 km in from the highway, with only a gravel road access from the south and no access to the north. The natural beauty of the Kalbarri region attracts a growing number of national and international visitors. It is also a place where surfers head, as it is one of the first readily accessible places north of Perth where the full force of the Indian Ocean swell reaches the coast producing some excellent surf. Much of the cliffed coast to the south is included in the Kalbarri National Park, which contains both spectacular coastal cliffs, and well as the gorges of the Murchison River.

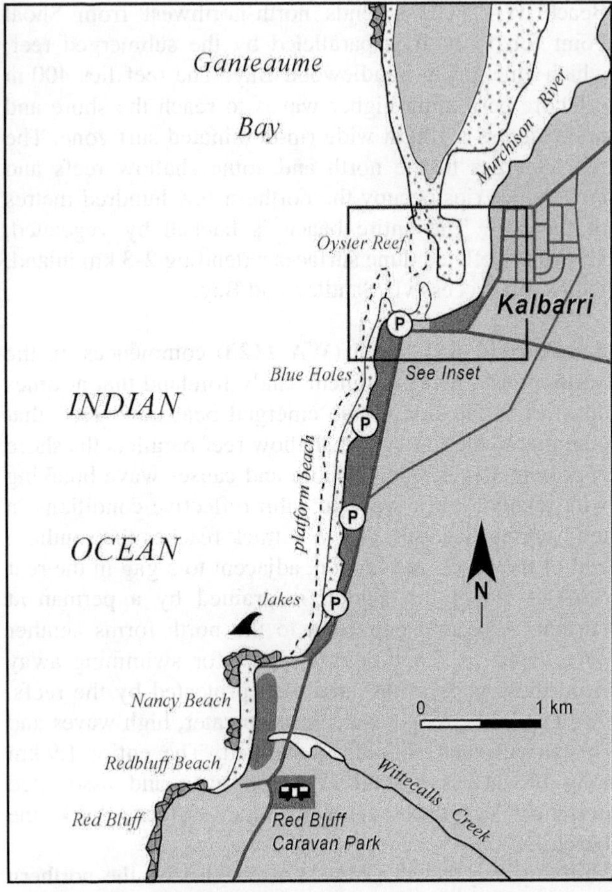

Figure 4.279 Kalbarri is located at the mouth of the Murchison River, with exposed beaches extending to the south. See Fig. 2.81 for insert.

The town is located on the southern banks of the Murchison River estuary. A road parallels the river with a wide riverbank reserve fronting the river. The mouth of the river is protected by two parallel reefs, the outer an extension of beachrock reefs that parallel the beach all the way from Jakes, and the inner Oyster Reef an old lithified beach. Fishing boats can negotiate the entrance in low swell, through this becomes a hazardous trip in moderate to high swell. There are five beaches in and near Kalbarri, the southern Red Bluff and Jakes beaches (WA 1129 & 1130), the rocky Jakes and Blue Holes (WA 1131 & 1132) and the protected Kalbarri Beach (WA 1133) in lee of Oyster Reef and site of Kalbarri Surf Life Saving Club.

Red Buff is a prominent 40 m high red sandstone bluff and the northern coastal boundary of the Kalbarri National Park. It is located 3 km south of Kalbarri and there is a caravan park and store just behind the beach. There is also a large car park and boat ramp at the southern end. **Nancy Beach** (**WA 1129**) faces west and runs 700 m north of the bluff to the low rocks and bluff section that separate it from Jakes Beach. It is composed of coarser sand and in interrupted by a number of rocks and reefs immediately off the beach. Waves average 1-1.5 m and tend to break heavily on the reefs or as a heavy shorebreak on the steep, soft beach. Wittacarra Creek is located towards the northern end of the beach and usually blocked by a steep berm. A tracks runs along the southern side of the creek to the beach providing 4WD access for beach fishers and during calm conditions beach boat launching. A small plaque at the creek commemorates this site as the spot of the first permanent white 'settlement' in Australia. In 1629, two of the mutinous crew of the *Batavia* wreck were permanently castaway at this location.

Jakes is the name of Kalbarri's best surf break located on Jacques Point and the adjoining beach. The point consists of a wide red sandstone rock platforms, with the left break running along the northern edge of the rocks. There are two car parks on sandy bluffs overlooking the point and beach and a third on the foredune, but no facilities. Beach **WA 1130** is located in the southern corner and consists of a 50 m pocket of sand bordered by sandstone platforms and backed by low vegetated bluffs then the road. The beach consists of a high tide strip of steep pink sand separated from the water by flat sandstone and beachrock that is exposed at low tide. The main Jakes beach (**WA 1131**) extends for 1 km to the north and is a continuous slightly crenulate high tide platform beach, fronted by a continuous, though irregular intertidal sandstone platform, and backed by low dune-draped bluffs, with the road 100 m inland. Car parks are located at either end of the beach, with numerous assess tracks from the main road.

Blue Holes beach (**WA 1132**) is a sinuous 1 km long, west-facing platform beach that extends north from Jakes beach to Chinaman's Rocks at the mouth of the Murchison River (Fig. 4.280). The beach is composed of medium sand, pink in places, which produces a relatively steep, soft beach. However it is fronted by a near continuous low tide rock flat waves breaking heavily on the rocks. There are breaks in the rocks producing the 'blue holes' or tidal pools, with strong rips flowing out of the holes. The road to Red Bluff runs along the top of the 20 m high dune-covered bluffs 100 m behind the beach and connects to three car parks and a lookout. The car parks provide access either side of Blue Holes and to the southern hole in the platform.

Figure 4.280 View along Blue Hole beach to Kalbarri and the Murchison River mouth.

Kalbarri Beach (**WA 1133**) is the main swimming beach for Kalbarri. It is also known as Chinaman's beach, and is located just inside the river mouth on the 300 m long spit of bare sand that extend north in lee of Oyster Reef (Fig. 4.281). This beach is in full view of the backing town and has a large car park just off the Red Bluff Road. Waves break over Oyster Reef and a second northern reef and average less than 0.5 m at the steep, barless beach. There are a few reefs off the southern end while strong tidal currents flow in the channel parallel to the beach and around the tip of the spit and into the river.

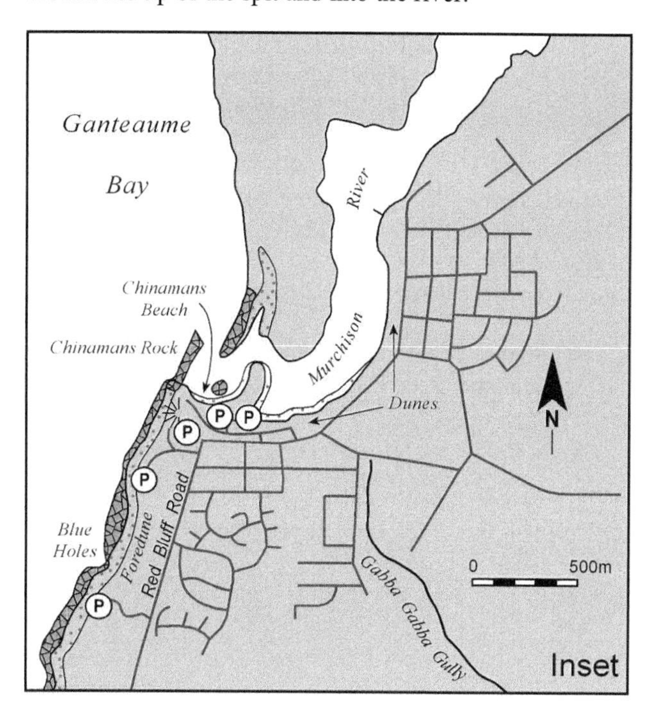

Figure 4.281 The Murchinson River enters the sea between beachrock reefs, with the sheltered Chinamans Beach inside the reef and exposed rocky beaches outside.

Beach **WA 1134** is located directly opposite the main beach on the northern side of the tidal channel in lee of the northern beachrock reef. This beach is even more sheltered than the main beach with low waves only reaching the beach at high tide. The reef and beach link at the northern end to form a shallow, narrowing 'lagoon', while the southern end borders the deep tidal channel with its strong currents. It is backed by a low unstable dune-covered section of the tidal delta, with no direct access from Kalbarri, other than by boat.

Swimming: The Kalbarri spit is the least hazardous beach in the area and is patrolled, however even here care must be taken to avoid the tidal current in the deeper tidal channel, which runs just off spit. The ocean beaches are hazardous owing to the usually higher swell and numerous rocks, reefs and strong rips. They are only suitable for swimming when calm, with the best being at Red Bluff and Nancy Beach.

Surfing: Kalbarri is famous for its surf and many Perth based surfers are prepared to drive the 600 km for a weekend of surf. The main spot is the left on Jacques Point, called *Jakes*. It is a lefthander breaking over reef, which can be surfed to 5 m. There is also a left at *Red Bluff*, inner reef/beach breaks at Red Bluff beach, and reef breaks at the car parks along Jakes and *Blue Holes* beaches.

Region 5 Carnarvon Coast: Murchison River to Cape Preston

Length of coast: 2036 km (2734-4770 km)
Number of beaches: 574 (WA 1135-1708)

Regional maps:

National Parks:	km	length	beaches	number
Shark Bay World Heritage Area	2788-3837	1049 km	266	WA 1161-1427
Ningaloo Marine Park	4018-4276	258 km	138	WA 1478-1606

(includes several national parks and reserves - see page 284)

Coastal towns: (Kalbarri), Denham, Carnarvon, Coral Bay, Exmouth, Onslow
Other locations: Steep Point, Useless Loop, Nanga, Monkey Mia, Gladstone, Garnaloo,

Region 5A CARNARVON: Kalbarri – Steep Point

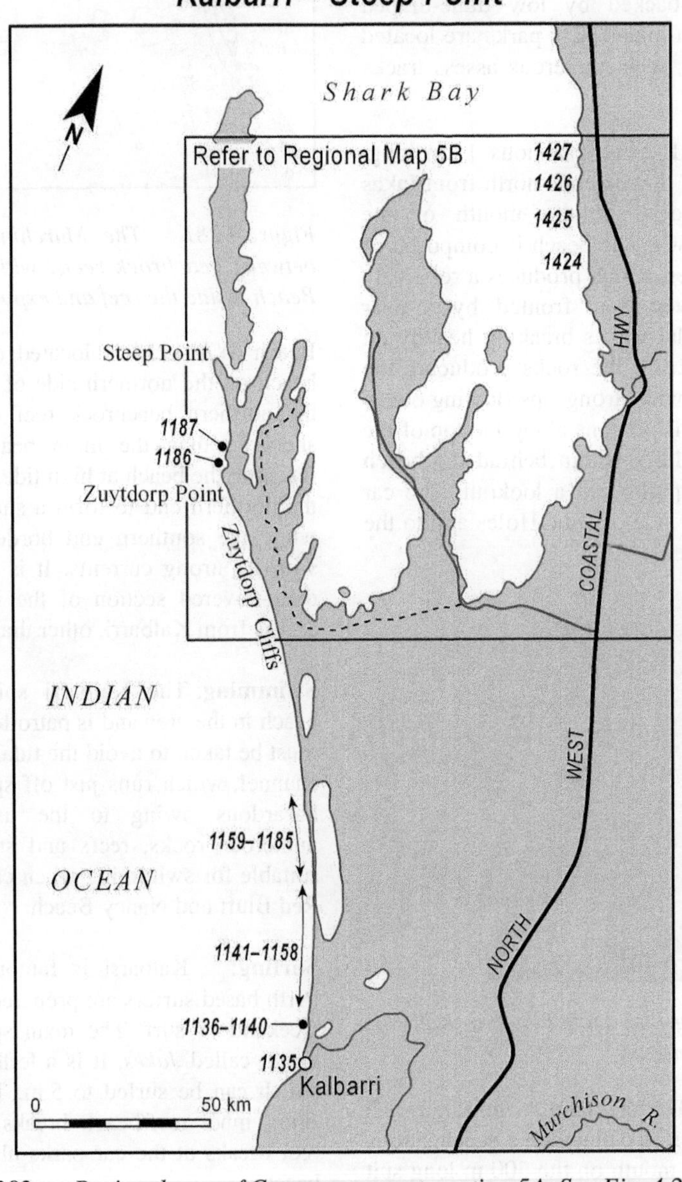

Figure 4.282 Regional map of Carnarvon coast, section 5A. See Fig. 4.292 for section 5B.

The **Carnarvon coastal region** extends from the Murchison River at Kalbarri to Cape Preston, a linear distance of 800 km, but with a shoreline length of 2036 km, which includes Shark Bay and Exmouth Gulf. This coastal region is centered ion the Tropic of Capricorn and contains Australia's most arid section of coast, with all having an annual rainfall less than 400 mm dropping to less than 200 mm in parts of Shark Bay, and all being classified as desert. The coast is a mix of exposed relatively high wave energy and micro tides section south of North West Cape, and the sheltered Shark Bay and Exmouth Gulf, together with the most sheltered northern coast between the gulf and Cape Preston, all with tide increasing northward from 2-4 m. Despite its aridity the coast is bordered by the growing Kalbarri to the south and Karratha to the north, and contains the towns of Denham, Carnarvon, Exmouth and Onslow. In addition its possesses the Shark Bay World Heritage Area and Ningaloo Marine Park, the most poleward coastal coral reef system in the world.

WA 1135-1141 **GANTHEAUME BAY-SECOND FENCE**

No.	Beach	Rating	Type	Length
WA 1135	Gantheaume Bay	8	LTT/TBR+reef	13.5 km
WA 1136	First Fence	8	LTT+reef	200 m
WA 1137	First Fence (N 1)	8	LTT+reef	800 m
WA 1138	First Fence (N 2)	8	LTT/TBR+reef	2.3 km
WA 1139	Second Fence	8	LTT+reef	300 m
WA 1140	Second Fence (N1)	8	LTT+reef	6.4 km
WA 1141	Second Fence (N2)	8	LTT+reef	600 m
Spring & neap tidal range = 0.4 & 0.1 km				

The Murchison River mouth at Kalbarri marks a major change in the geology, orientation and nature of the coast. The Silurian red sandstones that dominate the hinterland and coast to the south as far as Bluff Point, are replaced by one of the most massive Pleistocene dune calcarenite deposits in the world. It commences at the river mouth and extends to the northwest initially a steep 100-200m high cliffs, for 176 km to Zuytdrop Point, part of the way as the Zuytdorp Cliffs, then turns and trends more northerly for another 34 km to Steep Point, the westernmost tip of the Australian continent, before breaking into a series of calcarenite islands that continue north for another 155 km, finally terminating at Cape Ronsard on Bernier Island. In all the deposits extend for 365 km. The cliffs represent successive layers of dune sands that have accumulated during each high stand of sea level over the past 2 million years, the last rise adding the newer active sand dunes along the coast. During period of lower sea level the dunes are partially lithified (cemented) into aeolian calcarenite or dunerock. The successive layers have been built up in places to an elevation of over 200 m.

The entire coastline along this part of the coast is remote, cliffed and difficult to access, and backed by grazing land. The highway lies 50-100 m to the east, with the only public access at Kalbarri in the south and Shark Bay in the north, and no public access in between.

The coast, while dominated by cliffs, does have a numerous beaches along their base, most impacted to significant degrees by the presence of calcarenite bluffs, rocks, platforms and reefs. The first section of the coast between the river mouth and the Second Fence, 26 km to the northwest, contains seven beaches (WA 1135-1141), which occupy 24 km of the shore.

Beach **WA 1135** commences on the northern side of the river mouth in the open west-facing Gantheaume Bay. It is a high tide sandy beach fronted by a straight beachrock intertidal reef/platform. It continues essentially uninterrupted to the north for 13.5 km terminating about 2 km north of First Fence. The beach receives waves averaging about 1.5 m, which break heavily on and at time seaward of the reef, only reaching the beach at high tide. In some places the reef has been breached and partially eroded and surf and strong rips are generated in the gaps. The beach is backed by dune-draped calcarenite that rises from 90 m in the south to 160 m in the north (Fig. 4.283). Four wheel drive tracks run along the tops of the cliffs with occasional access down to the beach.

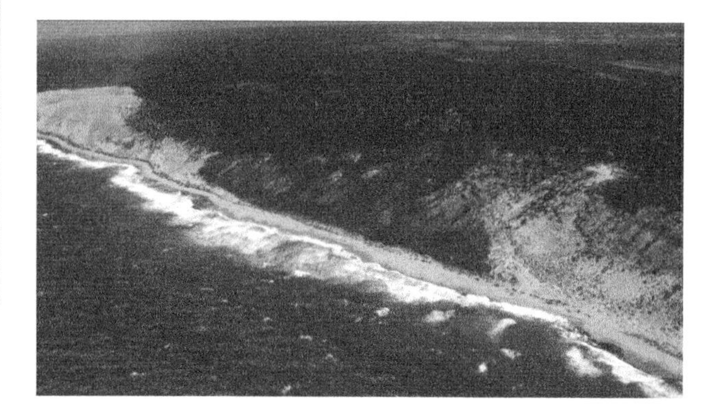

Figure 4.283 The exposed beach WA 1135 is backed by dunes climbing the 90-160 m high calcarenite cliffs.

Beach **WA 1136** is a 200 m long pocket of sand located 500 m north of beach WA 1135 and 500 m south of beach WA 1137. It occupies a slight indentation in the backing 140 m high bluffs. It is a inter to high-tide beach, with a continuous reef along the base of the beach and a small dune climbing up the slopes on the northern side to reach the crest. Vehicle tracks run along the top of the backing slopes.

Beach **WA 1137** is an 800 m long beach bordered by the second two cliff protrusions and beachrock reefs with scattered reefs along the beach and a sandy tidal pool at the northern end (Fig. 4.284). It has active dunes climbing the backing slopes and a 4 WD track that reaches the southern end of the beach. Beach **WA 1138** commences on the northern side of the third protrusion and continues northwest for 2.3 km with an irregular shore-attached beachrock reef running the length of the beach. It is backed by steep sparsely vegetated slopes rising to 80 m, which are scarped along their base. A 4 WD track reaches the southern end of the beach.

Figure 4.284 Beach WA 1137 is typical of the small pockets of exposed sand scattered along the base of the southern Zuytdrop Cliffs.

Beach **WA 1139** lies at the base of the **Second Fence** that terminates at the crest of an 80 m high cliff. The beach is 300 m long and consists of a north and southern high tide pink sandy section and central eroded section that forms a tidal pool, all fronted by a continuous beachrock reef/platform. A foot track from the fence line descends the steep slopes to the northern end of the beach.

Beach **WA 1140** commences 300 m to the north and is a continuous straight 6.4 km long high tide beach fronted by a continuous intertidal beachrock reef. It is backed by steep slopes rising to a 150 m high cliffline, which becomes increasingly gullied to the north. Waves averaging 1.5 m break across a 50 m wide reef-dominated surf zone. Beach **WA 1141** is a 600 m long sandy beach backed by steep 100 m high slopes. It is fronted by continuous beachrock reef, with a central gap maintaining a permanent rip. A small patchy foredune is located at the base of the backing slopes.

WA 1141-1151 **THIRD-FOURTH FENCE**

No.	Beach	Rating	Type	Length
WA 1142	Third Fence	8	LTT+reef	1.9 km
WA 1143	Third Fence (N1)	8	LTT+reef	4.7 km
WA 1144	Third Fence (N2)	8	LTT+reef	150 m
WA 1145	Fourth Fence	8	LTT+reef	200 m
WA 1146	Fourth Fence (N1)	8	LTT+reef	200 m
WA 1147	Fourth Fence (N2)	8	LTT+reef	500 m
WA 1148	Fourth Fence (N3)	8	LTT+reef	150 m
WA 1149	Fourth Fence (N4)	8	LTT+reef	300 m
WA 1150	Fourth Fence (N5)	8	LTT+reef	150 m
WA 1151	Fourth Fence (N6)	8	LTT+reef	500 m
Spring & neap tidal range = 0.4 & 0.1 m				

To the north of Third Fence the cliffed coastline continues on straight to the northwest past Fourth Fence. The next 16 km of coast contains 10 beaches (WA 1142-1151) of varying length, which occupy 9 km of the shore, the remainder dominated by exposed rocks at the base of the cliff. All the beaches are exposed to waves averaging about 1.5 m and all are dominated by irregular, through continuous intertidal calcarenite reefs and platforms. All the beaches are backed by sparsely vegetated slopes rising to 140 m with vehicle access only to above beaches WA 1143 and 1148.

Beach **WA 1142** commences just to the north of Third Fence and trends straight for 1.9 km. The coarse high tide beach contains pink sands and is wedged in between the base of the shore and the irregular platform and reef, with waves breaking across a 50 m wide surf zone. Beach **WA 1143** commences 200 m to the north and is a similar 4.7 km long beach, with a small foredune in places located at the base of the slopes. The Fourth Fence terminates on the slopes above the northern end, with a vehicle track winding down to the bluffs above the northern end of the beach.

Beach **WA 1144** is a 150 m long pocket of high tide sand located directly below the end of the Fourth Fence and immediately north of beach WE 1143. It is fronted by a 50 m wide platform, with a 150 m long tidal pool backing the platform on the northern side of the beach. Beach **WA 1145** lies 300 m to the north and is a similar 200 m long strip of high tide sand, with a slightly protruding platform and small central tidal pool. A sand shute at the northern end of the beach is feeding sand into a climbing dune that reaches part way up the backing slopes.

Beach **WA 1146** is located 200 m to the north and is a 200 m long strip of high tide sand fronted by an irregular beachrock reef. It is backed by sparsely vegetated slopes rising to 140 m, which are scarped along their base. Beach **WA 1147** lies 500 m further north and is a 500 m long beach located along the base of steep slopes rising to 140 m, with rock debris littering the rear of the beach. It has a continuous intertidal calcarenite platform with a permanent rip flowing out of a central break in the reef.

One kilometer of rocky shore separates beaches WA 1147 and **1148**, with the latter a 150 m long pocket of high tide sand. It is backed by scarped then steep slopes and fronted by a continuous irregular calcarenite reef. A vehicle track terminates on the slopes at the southern end of the beach. Beach **WA 1149** commences 500 m to the north and is a 300 m long high tide beach wedged in between scarped 10-20 m high toe of the slopes and an irregular calcarenite reef.

Beach **WA 1150** lies immediately to the north, with a protruding section of the bluffs separating the two beaches. It consists of a 150 m long strip of high tide sand backed by steep vegetated slopes that descend to the rear of the beach, with the beachrock reef dominating the 50 m wide surf zone. It terminates at a narrow protrusion in the boundary bluffs with a small gully behind. A 4 WD track terminates on the slopes on the northern side of the gully providing access to the beaches either side. Beach **WA 1151** continues on the northern side for 500 m to the next more substantial protrusion. It is similarly backed by vegetated slopes, with the shore-attached reef continuing the length of the beach.

WA 1152-1160 **FIFTH FENCE**

No.	Beach	Rating	Type	Length
WA 1152	Fifth Fence	8	R/LTT+reef	500 m
WA 1153	Fifth Fence (N1)	8	R/LTT+reef	300 m
WA 1154	Fifth Fence (N2)	8	R/LTT+reef	1.6 km
WA 1155	Fifth Fence (N3)	8	R/LTT+reef	400 m
WA 1156	Fifth Fence (N4)	8	R/LTT+reef	800 m
WA 1157	Fifth Fence (N5)	8	TBR+reef	200 m
WA 1158	Fifth Fence (N6)	8	TBR+reef	500 m
WA 1159	Fifth Fence (N7)	8	R/LTT+reef	150 m
WA 1160	Fifth Fence (N8)	8	TBR+reef	2.3 km
Spring & neap tidal range = 0.4 & 0.1 m				

The **Fifth Fence** reaches the coast 45 km north of Kalbarri with the steep cliffline extending to the north and south. Between the fence and a major inflection in the shore 21 km to the north are nine beaches (WA 1152-1160). Like the beach to the south they are all dominated by the bordering calcarenite bluffs, slopes and shore attached reefs, with waves averaging about 1.5 m and breaking across a 50 m wide surf zone. The only vehicle access is to the slopes above beaches WA 1153, 1154 and 1159.

Beach **WA 1152** is a 500 m long high tide sand beach backed by slopes rising to 160 m, and paralleled by the fringing calcarenite reef (Fig. 4.285). A dune originating along the northern end of the beach has partly climbed the backing slopes and migrated along the side of the slopes for 500 m to the north, to reach beach WA 1153. Beach **WA 1153** is a 300 m long beach with the sand dunes draping the backing southern slopes. The base of the slopes are scarped, which increases in height to the northern end. Rock platforms border each end with broken reef in between. A 4 WD track partly descends the slopes between beaches WA 1153 and 1154, with foot access down to the shore.

Figure. 4.285 Beaches WA 1152 and 1153 are both dominated by rock outcrops and backed by calcarenite slopes rising to 160 m.

Beach **WA 1154** is a crenulate, though straight, 1.6 km long high tide beach with several calcarenite outcrops along the beach, some of which induce permanent rips through the 50-80 m wide reef-dominated surf zone. Two areas of dune activity back the beach with the dunes transgressing northward along the lower side of the backing 160 m high slopes.

Beach **WA 1155** is located immediately to the north and is a 400 m narrow strip of sand wedged in at the base of steep scarped slopes, with the reef dominating the 50 m wide surf zone. A few climbing blowouts are located on the northern slopes. Beach **WA 1156** commences 150 m to the north and is an 800 m long crenulate beach with two major rock outcrops on the lower beach, as well as reef in the surf zone causing three permanent rips to drain part of the surf. The entire beach is backed by climbing dunes, which reach the top of the initial 60-70 m high slops and continue to the north behind beach WA 1157.

Beach **WA 1157** curves for 200 m between two narrow calcarenite points. It is backed by the dune-covered slopes, and fronted by an 80 m wide surf zone drained by a permanent northern rip. Beach WA **1158** commences immediately to the north and continues for 500 m to a more prominent northern point. It is backed by vegetated slopes and fronted by an 80 m wide rip and reef dominated surf zone. Beach **WA 1159** lies on the northern side of the point and is a 150 m long strip of high tide sand wedged in at the base of steep vegetated slopes, with a vehicle track terminating above the beach. Shallow reefs dominate the 70 m wide surf zone.

Beach **WA 1160** is one of the more prominent beaches along this section. It is 2.3 km in length and is relatively free of reef in the surf, resulting in an 80 m wide rip-dominated surf zone drained by four permanent reef-induced rips. The entire beach is feeding an active dune that as climbed 80 m up the backing 160 m high slopes and migrated 1 km past the northern end of the beach (Fig. 4.286). A vehicle track reaches the southern end of the dunes.

Figure 4.286 Beach WA 1160 has a rip-dominated surf zone and active dunes climbing north along the 160 m high calcarenite slopes,

SHARK BAY WORLD HERITAGE AREA

Established: 1991
Area: 2 300 000 ha
Coast length: 1049 km (2788-3837 km)
No. beaches 266 (WA 1161-1427)

Includes:		
	Shark Bay Marine Park	881 000 ha
	Hamlin Pool Marine NR	127 000 ha
	Zuytdorp Nature Reserve	59 000 ha
	Bernier & Dorre Islands NR	9720 ha
	Nanga Nature Reserve (proposed)	
	Francis Peron National Park	52 500 ha
	Dirk Hartog Island National Park (proposed)	

Shark Bay is a large shallow epicontinental sea bordered on the west by the northern extension of the Zuytdorp Cliffs between Edel Land to Steep Point, and Dirk Hartog, Dorre and Bernier islands, with Peron Peninsula dividing the southern bay in two. The cliffs, island and peninsula are all composed Tertiary and Quaternary marine sands deposited as beaches and massive dune systems. The bay has therefore been forming over the past 10's millions of years, with its shallow seafloor exposed whenever sea level has fallen, then flooding during high stands of the sea, like at present. The combination of large shallow (< 20 m deep) bay, with a low tide range, negligible freshwater inflow and high evaporation rates has produce a system whereby the salinity increased southward into the bay reaching as high as 70 $^o/_{oo}$ in Hamelin Pool, twice that of sea water. The clear, clean waters, abundant sea grass meadows and range of salinity has resulted in an abundant, diverse and in places unique marine life, one of the main attractions of the bay.

ZUYTDORP NATURE RESERVE

Established: 1992
Area: 59 000 ha
Coast length: 5 km (2800-2805 km)
Beaches: none

The Zuytdorp Nature Reserve is designed to protect the land surrounding the *Zuytdorp* wreck site including the inland soaks some of the survivors may have used following the wreck. Access to the reserve is restricted, while access to the wreck site is also very hazardous.

WA 1161-1169 **ZUYTDORP WRECK (S)**

No.	Beach	Rating	Type	Length
WA 1161	Zuytdorp Wreck (S9)	8	LTT+reef	300 m
WA 1162	Zuytdorp Wreck (S8)	8	LTT+reef	1.7 km
WA 1163	Zuytdorp Wreck (S7)	8	LTT+reef	1 km
WA 1164	Zuytdorp Wreck (S6)	8	LTT+reef	300 m
WA 1165	Zuytdorp Wreck (S5)	8	LTT+reef	600 m
WA 1166	Zuytdorp Wreck (S4)	8	LTT+reef	500 m
WA 1167	Zuytdorp Wreck (S3)	8	LTT+reef	200 m
WA 1168	Zuytdorp Wreck (S 2)	8	LTT+reef	500 m
WA 1169	Zuytdorp Wreck (S)	8	LTT+reef	100 m
Spring & neap tidal range = 0.4 & 0.1 m				

The Dutch ship the *Zuytdorp* was wrecked at the base of 100 m high eroding cliffs (Fig. 4.287) approximately 70 km north of Kalbarri in 1712. There is evidence to suggest some survived the wreck and erected a makeshift sheltered to the lee of the backing cliffs. Debris from the wreck lodged in crevasses at the base of the cliffs and was known to the local aboriginals and with their assistance the debris was located in 1927 by a local stockman Tom Pepper. Expeditions commenced in 1941 to locate the actual wreck, but it was not until the use of SCUBA equipment that this was achieved in 1964. Today apart from rough vehicle tracks little has physically change along this section of coast since the wreck.

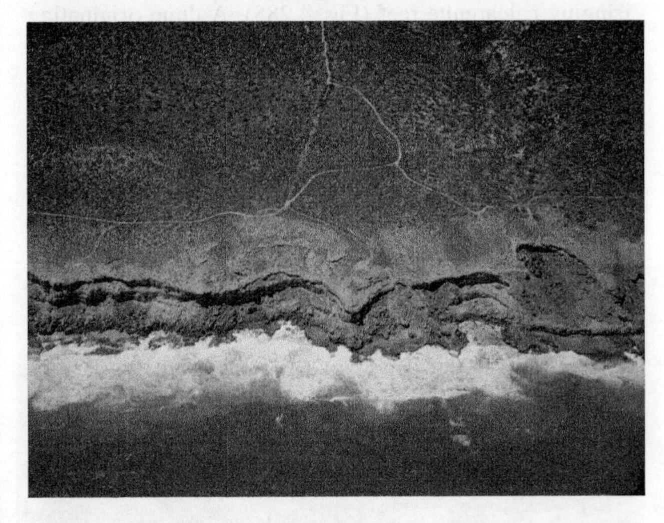

Figure 4.287 Site of the wreck of the 'Zuytdorp' in 1712.

Beaches **WA 1161** is located approximately 14 km south of the wreck site and between it and the wreck are nine beaches located at the base of the cliffs, now named after the Zuytdorp. They are all fronted by the near continuous intertidal reefs, rocks and platforms. Waves average about 1.5 m and persist much of the year making this an exposed, rocky and dangerous coast.

Beach **WA 1161** is located immediately below the end of the State (dog) Barrier Fence designed to keep dingo out of the south. The beach curves slightly for 300 m along the base of a 20 m high bluff, which rise inland to reach 150 m, 2 km form the shore. The beach is bordered by protruding platforms with a continuous platform fronting the beach. Beach **WA 1162** commences on the northern side of the boundary protrusion and continues north for 1.7 m along the base of the bluffs and slopes and fronted by a near continuous intertidal platform and reefs.

Beach **WA 1163** is located 1 km to the north and is a 1 km long beach that extends between the bluffs/slopes and intertidal reef, terminating at the mouth of a winding narrow gully, with a vehicle track running along its northern side. Beach **WA 1164** lies another 1 km to the north and is a double crenulate 300 m long high tide beach fronted by a continuous platform. A steep gully cuts through the backing 60 m high cliffs to reach the centre of the beach. A 100 m long protrusion in the cliffs separates it from beach **WA 1165**, a narrow 600 m long high tide beach located at the base of the 60 m high cliffs, with rock debris littering the rear of the beach and a continuous, though irregular platform running the length of the beach.

Beach **WA 1166** is located 100 m to the north and is a similar 500 m long narrow high tide beach, with three narrow gullies cutting through the backing cliffs and a continuous platform fronting the beach. Beach **WA 1167** is a 200 m long beach consisting of two pockets of sand partly separated by a central protrusion and rock falls, with much rock debris in the surf. The platform is partly eroded and wave break across a 60-80 m wide rock strewn surf zone with a permanent rip existing against the northern cliffs. A vehicle track and solitary hut is located on the cliffs 200 m north of the beach.

Beach **WA 1168** is a narrow 500 m long strip of high tide sand wedged in along the base of 60 m high cliffs, with rock debris littering the back of the beach and an irregular platform running the length of the beach. A vehicle track reaches the bluffs above the northern end of the beach, with an adjacent gully providing foot access to the shore. Beach **WA 1169** is a 100 m long pocket of sand located at the mouth of a small gully. It lies in lee of a rocky gap in the platform, with waves breaking over a 50 m wide platform and rock strewn surf zone.

WA 1170-1185 **ZUYTDORP WRECK (N)**

No.	Beach	Rating	Type	Length
WA 1170	Zuytdorp Wreck (N1)	9	LTT+rocks	50 m
WA 1171	Zuytdorp Wreck (N2)	8	LTT+rocks	70 m
WA 1172	Zuytdorp Wreck (N3)	8	LTT+rocks	60 m
WA 1173	Zuytdorp Wreck (N4)	8	LTT+rocks	200 m
WA 1174	Zuytdorp Wreck (N5)	8	LTT+reef	250 m
WA 1175	Zuytdorp Wreck (N6)	8	LTT+reef	300 m
WA 1176	Zuytdorp Wreck (N7)	8	LTT+reef	1.2 km
WA 1177	Zuytdorp Wreck (N8)	8	LTT+reef	300 m
WA 1178	Zuytdorp Wreck (N9)	8	LTT+reef	150 m
WA 1179	Zuytdorp Wreck (N10)	8	LTT+reef	300 m
WA 1180	Zuytdorp Wreck (N11)	8	LTT+reef	300 m
WA 1181	Zuytdorp Wreck (N12)	8	LTT+reef	800 m
WA 1182	Zuytdorp Wreck (N13)	8	LTT+reef	600 m
WA 1183	Zuytdorp Wreck (N14)	8	LTT+reef	300 m
WA 1184	Zuytdorp Wreck (N15)	8	LTT+reef	150 m
WA 1185	Zuytdorp Wreck (N16)	8	LTT+reef	250 m

Spring & neap tidal range = 0.4 & 0.1 m

The Zuytdorp wreck is located along a 5 km long section of the cliffs with no high tide beaches. These commence about 3 km north of the wreck site at beach WA 1170 and continue north for 17 km to the first significant inflection in the cliffs since Kalbarri. In between are 16 beaches (WA 1170-1185), all located at the base of the 100 m high cliffs and all dominated by calcarenite rocks, reefs and platforms and exposed wave waves averaging about 1.5 m (Fig. 4.288). There is vehicle access to the cliffs above a few of the beaches, with access to most beaches down steep, narrow gullies that dissect the cliffs.

Figure 4.288 Beaches 170 and 171 are located in narrow gullies at the base of the 100 m high calcarenite cliffs.

Beach **WA 1170** is a 50 m long pocket of sand located 100 m inside the mouth of a steep gully, whose slides extend seaward of the beach forming a small rock embayment. Large rock debris surrounds the beach and also litters the 80 m wide surf zone. Beach **WA 1171** lies 500 m to the north in the mouth of the next gully. It is a 70 m long pocket of sand bordered by cliffs rising to 100 m, with rocks littering the 60 m wide surf zone.

Beach **WA 1172** is located in the next gully 1 km to the north and consists of a 60 m long pocket of sand located inside the mouth of a small embayed gully. A rim of dense vegetation borders the beach, and the usually dry creek bed dissects the beach. Sheer cliffs rising to 100 m border the beach and a small but strong permanent rip drains the small embayment. A vehicle track terminates on the northern side of the gully with foot access down the gully to the beach.

Beach **WA 1173** is a 400 m long high tide beach backed by steep scarped slopes and gullies and paralleled by an irregular 50-70 m wide platform and a few rocks extending up to 100 m offshore. Beach **WA 1174** lies immediately to the north and is a 300 m long beach, wider in the south and narrowing to the north, in lee of an irregular intertidal calcarenite platform. A gully backs the southern end with steep slopes behind the northern section.

Beach **WA 1175** is an irregular 800 m long beach consisting of a narrow southern section and a wider northern pocket located at the mouth of a small steep gully. The entire beach is fronted by a continuous 50-70 m wide intertidal calcarenite platform, with waves breaking just seaward of the edge of the platform. Beach **WA 1176** commences 200 m to the north and is a 250 m long sandy beach backed by 20-30 m high scarped slopes and fronted by three sections of reef, with rocks in between, resulting in an irregular reef- and rock-dominated surf zone. A 4 WD track terminates on the slopes between the two beaches providing steep foot access to both.

Beach **WA 1177** is a slightly curving 300 m long beach that narrows to the north as the fronting platform narrows and disintegrates into a series of rocks within the surf zone. It is backed by dune-dusted, then scarped slopes that rise to 80 m. Beach **WA 1178** is a 150 m long beach that commences as a narrow strip of sand between an 80 m wide platform and 20 m high scarped slopes, widening into a northern gully. The gully has acted as a shute and dune sand dusts the backing slopes. Beach **WA 1179** is a 300 m long sandy beach located along the base of scarped irregular 80 m high cliffs, with a broken platform paralleling the beach. A 4 WD track descends to the headland separating the two beaches, with steep foot access down to the shore.

Beach **WA 1180** is a straight 300 m long sandy beach with a calcarenite platform bordering each end and a more open 100 m long central section. The beach widens behind the southern platform with a small foredune at the rear of the beach. It is backed by irregular 30-40 m high scarped slopes. A curving section of beachrock separates it from beach **WA 1181** which continues north for 800 km as a relatively continuous high tide sand beach with irregular intertidal platforms and rocks (Fig. 4.289). Steep slopes rising to 60 m back the beach. Along the northern half of the beach a mixture of vegetated slope debris and a climbing foredune ramps up against the slopes. There are two vehicle tracks that terminate behind both beaches. A steep gully separates it from beach **WA 1182** that continues north for 600 m past a second gully to terminate at a third hanging gully. A 70 m wide platform parallels the beach with two areas of tidal pools towards the southern end, and a patchy foredune abutting the base of the 80 m high cliffs. Beach **WA 1183** continues on the northern side of the boundary gully for 300 m. It is backed by 100 m high cliffs and fronted by a 70 m wide intertidal platform. A permanent rip exits the deeper surf zone in front of the boundary gully.

Beach **WA 1184** is a protruding 150 m long strip of sand located below a steep hanging gully with 60 m high cliffs to either side. It protrudes slightly in lee of a 100 m wide intertidal platform and reef. Beach **WA 1185** lies 100 m to the north and is a 250 m long strip of high tide sand with an irregular 70 m wide platform. It is backed by 50 m high cliffs. The slopes behind both beaches have a veneer of active dune sand located about 500 m inland.

Figure 4.289 Beach WA 1181 (right) is wedged in between the cliffs and continuous rock platform.

The **Zuytdorp Cliffs** commence at beach WA 1161 and trend essentially straight to the northwest for 124 km to Zuytdorp Point, the southern boundary of False Entrance or Dulverton Bay. In between beaches WA 1161-1185 occupy the first 30 km of the cliff base, with the remaining 94 km composed of steep cliffs and devoid of any sand or beach deposits. The cliffs range in height from 80 to 240 m (Fig. 2.22a). Despite the absence of beaches Holocene dunes have in the past ramped up the cliff and transgressed across the backing calcarenite plateau, as longwalled parabolic for distances up to 6 km.

At **Zuytdorp Point** occasional high southerly waves break over the base of the cliffs and deposit spray and swash on top of the 30 m high calcarenite cliffs. The swash drains northward along a sloping 500 m long overwash shute into False Entrance. During extreme wave conditions boulder up to a few tonne in weight are thrown up on to the top of the cliffs. Blowholes are also located along the exposed side of the point.

WA 1186-1187 FALSE ENTRANCE-CATFISH BAY

No.	Beach	Rating	Type	Length
WA 1186	False Entrance	7	TBR+reefs	2.8 km
WA 1187	Catfish Bay	8	R+reef	3 km
Spring & neap tidal range = 0.4 & 0.1 m				

The straight section of largely cliffed coast that commences at Murchison River mouth at Kalbarri finally terminates at Zuytdorp Point, to the lee of which is False Entrance, also known as Dulverton Bay. It is the 'false entrance' to Shark Bay, the actual entrance is located in lee of Steep Point 32 km to the north. This bay and Catfish Bay 5 km to the north contain the only two beach systems (WA 1186 & 1187) on this section of coast.

False Entrance bay has a curving 2.8 km long beach (**WA 1186**) that contains segments of shore parallels beachrock and inshore reef, as well as a few open sections backed by a low gradient beach (Fig. 4.290). Waves average about 1.5 m and break across a surf zone that widens from 50 m in the south to 150 m in the north and is dominated by six permanent reef-controlled rips. There is good surf across the reefs, particularly towards the more protected southern end. A sandy 4 WD track off the Steep Point track runs out to the bay, to an informal camping area behind the high foredune, then onto Zuytdorp Point, with another spur tracking north along the beach and coast to Crayfish Bay. The beach supplies one of the longest active dunes in Australia. Sand blows out of the southwest-facing, central-northern section of beach and is actively transgressing up to 20 km to the north along Bellefin Prong, with older dunes continuing on for another 18 km, 38 km in all.

Figure 4.290 View along Crayfish Bay beach with waves breaking across the reef-littered surf zone.

Crayfish Bay (also known as Epineux Bay) contains the curving 3 km long beach **WA 1187**. The beach grades from a steep boulder beach in the south to a steep sand beach in the north, with both sections fronted by a continuous intertidal calcarenite platform. Waves break seaward of and then heavily on the outer edge of the platform only reaching the beach at high tide. A permanent rip runs out against the northern point, while there is a *lefthand break* at the southern end of the beach. The boulders are composed of eroded slabs of beachrock and form a 5-6 km high steep boulder berm. Midway along the boulder berm is a low walled dugout, which is has been suggested may have been made by shipwrecked sailors. The northern sandy end of the beach is also backed by a stabilising parabolic that in the past has linked with the main False Entrance parabolic. The 4 WD track runs along the back of the beach and on to Steep Point.

Five kilometres north of Catfish Bay is **Thunder Bay** and cliffed rocky embayment containing a number of blowholes and an elevated boulder beach, the boulders deposited during high wave events. A spur off the coast track runs out to the top of the bay where there is an informal camp site.

DIRK HARTOG, DOORE & BERNIER ISLANDS

Dirk Hartog, Doore and Bernier islands are a 160 km long continuation of the calcarenite cliffs that commence at Gantheaume Bay. They have been separated from the mainland by the sea level rise. Today a 1.5 km wide channel through South Passage separates Dirk Hartog Island from the Steep Point region, while a 25 km wide opening separates Dirk Hartog and Doore, while only a few hundred metres lie between Doore and Bernier. Dirk Hartog is the largest island with an area of 62 000 ha, it is 77 km long, between 3-11 km wide with 200 km of shoreline. Doore Island is 31 km long and Bernier 25 km, both having a maximum width of 3 km. The three islands are dominated by exposed steep cliffs rising 100-200 m along their exposed western sides, usually capped by both vegetated and active Holocene sand dunes (Fig. 4.291). On the eastern sides the quieter waters of Shark Bay result in low energy conditions, sand flats and seagrass meadows. There are a few high energy exposed beaches on the northern end of Dirk Hartog Island, where there is apparently some good surf, with near continuous low energy beaches along its eastern shore. Doore and Bernier islands are nature reserves, while a national park is proposed for Dirk Hartog Island.

Figure 4.291 The western shore of Dirk Hartog Island consists of 100-200 m high calcarenite cliffs capped by longwalled parabolic dunes.

SHARK BAY

Shark Bay is a 1 300 000 ha shallow, sheltered bay that extends for 250 km from north to south and is up to 110 km wide. The bay has formed to the lee of massive Pleistocene barrier systems, including the 120 km long Peron Peninsula, which dissected the southern half of the bay, and Dirk Hartog, Dorre and Bernier islands, which extend for 150 m along the western boundary (Fig. 4.292). Because of these and other smaller elongate peninsulas, the bay has a mainland shoreline of 900 km, all consisting of low to very low energy beaches and sand flats.

WA 1188-1196 STEEP PT-SOUTH PASSAGE

No.	Beach	Rating	Type	Length
WA 1188	Steep Pt (E)	4	R+rock flats	300 m
WA 1189	Steep Pt (E 1)	4	R+rock reef	100 m
WA 1190	Steep Pt (E 2)	3	R+rocks	700 m
WA 1191	Cape Ransonnet (W)	2	R+sand flats	300 m
WA 1192	Cape Ransonnet (E)	2	R+sand flats	1.8 km
WA 1993	Cape Ransonnet (E 1)	2	R+sand flats	150 m
WA 1194	Cape Ransonnet (E 2)	2	R+sand flats	50 m
WA 1995	Cape Ransonnet (E 3)	2	R+sand flats	80 m
WA 1196	Shoal Flats	2	R+sand flats	1.3 km
Spring & neap tidal range = 0.6 & 0.2 m				

Steep Point is the western most tip of the Australian continent (Fig. 4.293). To reach the point from the highway requires a 50 km drive to Hamelin Pool, then a 130 km long 4 WD drive through Edel Land and across massive dunes to finally reach the shore at South Passage, then a 10 km long drive along the north-facing shore to the point. The southern shore of Dirk Hartog Island forms the northern shore of the passage. Despite the distances and terrain many people make the trip to see this geographical site and to camp along South Passage and fish the area. The point is a 40 m high calcarenite bluff on the crest of which people have built scores of small rock cairns to mark their arrival. There is a rough exposed camp site for fishers immediately east of the point, with better camping areas along the shore east of Cape Ransonnet. Camping is by permit obtainable from the ranger station located in lee of beach WA 1194. Between the point and Shoal Flats 10 km to the southeast are nine beaches (WA 1188-1196) all accessible along the track. The beaches are located along a calcarenite shoreline backed by vegetated parabolic arms, the parabolic originating from the cliffs to the south. The beaches face north and wave energy decreases into the passage.

Beach **WA 1188** is the westernmost beach on the continent. It is located 700 m east of Steep Point at the entrance to South Passage. It faces north across the passage to Surf Point 3 km to the north, on Dirk Hartog Island. The beach consists of a 300 m long convex accumulation of coarse carbonate material, located to the lee of a 100 m wide intertidal calcarenite platform, with sloping vegetated bluff rising to 20 m behind. Wave refract around the point and average about 1 m on the edge of the platform. The beach faces towards the eastern side of the point where there is a lefthand surf break over the over edge of the platform (Fig. 4.294). There is also surf out on *Surf Point*, which can only be reached by boat.

Beach **WA 1189** is located 200 m further east and is a 100 m long strip of sand wedged in between low eroding calcarenite bluffs and vegetated slopes and the shallow sand and calcarenite flats. Waves average less than 1 m and surge up the steep beach at high tide.

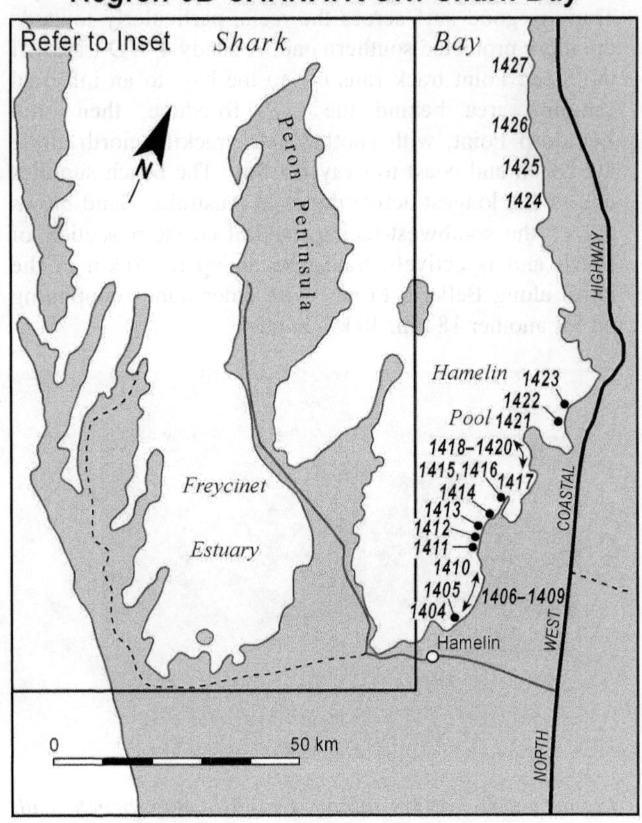

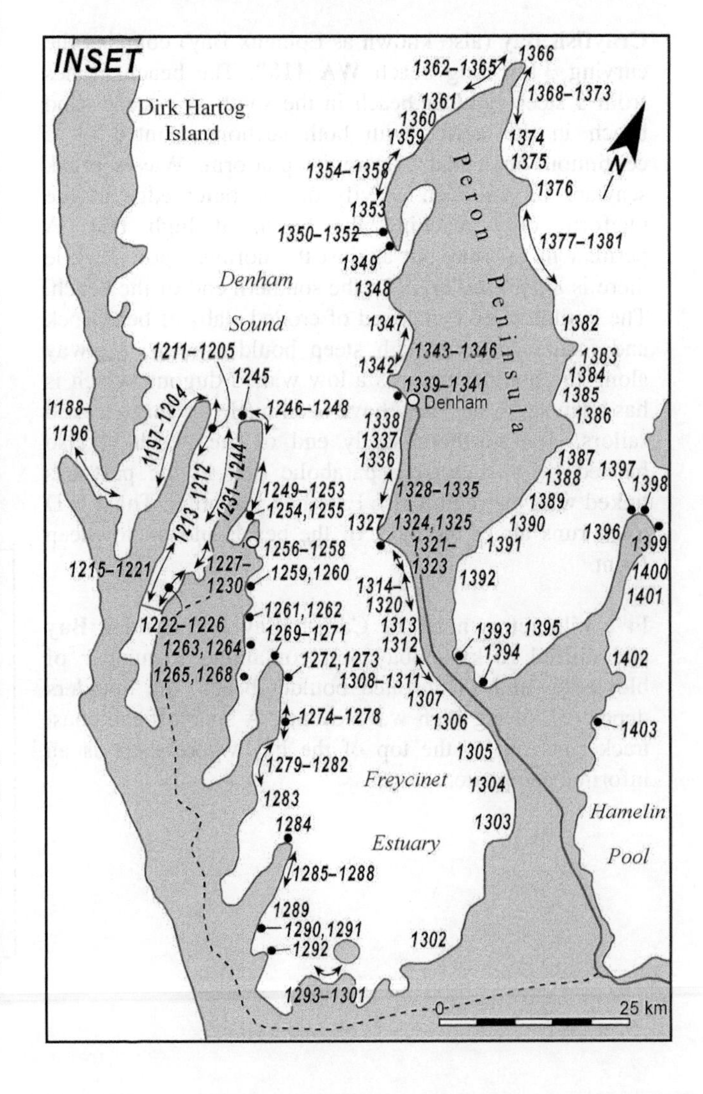

Figure 4.292 (right) Regional map of Shark Bay (top right). The southern section of Shark Bay (lower), is bordered by Dirk Hartog Island and divided in two by the long Peron Peninsula.

Figure 4.293 Steep Point (left) is the western most tip of the continent and forms the southern headland of the entrance to Shark Bay.

Figure 4.294 Beach WA 1188 the westernmost on the continent, with the western most surf break.

Beach **WA 1190** is a narrow 700 m long strip of high tide sand located along the northern shore of the second point east of Steep Point. It is backed by vegetated slopes rising 20-30 m, and fronted by a near continuous intertidal calcarenite platform, with shallow sand and reef flats off the beach.

Cape Ransonnett lies 4 km inside the passage with beaches to either side of the low sandy cape. Beach **WA 1191** occupies the western side and is a 300 m long sand beach bordered in the west by low dune-capped calcarenite bluffs, and terminates at the sandy foreland of the cape. The foreland has prograded 300 m into the passage, with shallow sand flats to either side. Because of the access to deep water off the foreland it is used to launch small fishing boats. Beach **WA 1192** commences at the foreland and curves round to the south then east for 1.8 km. Shallow sands flats fill the bay in between with usually calm conditions at the shore. This is the main camping area for the area, with campers usually spread along the low shoreline.

Beaches WA 1193, 1194 and 1195 occupy the eastern ends of the next three small indentations in the next 1.5 km of low calcarenite shoreline, with sand flats widening to the east in front of the beaches. Beach **WA 1193** is a 150 m long strip of sand wedged between the sand flats and the backing calcarenite and slopes. The ranger station is located behind beach **WA 1194** a 50 m long pocket of sand bordered by the low calcarenite bluffs. Beach **WA 1195** is an 80 m long strip of sand abutting the western 5 m high calcarenite point. It is used as a fishing camp.

Beach **WA 1196** is the first beach reached when driving out to the point area (Fig. 4.296). It is a very low energy 1.3 km long beach fronted by the 1 km wide Shoal (sand) Flats that widen to 2 km in the east. It faces north into Blind Strait between Dirk Hartog Island and Bellefin Prong. The main access track runs along the beach, which is backed by 10 m high vegetated dunes.

Figure 4.295 Low energy beach WA 1196 forms part of the access track to Steep Point.

WA 1197-1204 **BELLEFIN PRONG (W)**

No.	Beach	Rating	Type	Length
WA 1197	Bellefin Prong (W1)	1	R+sand flats	900 m
WA 1198	Bellefin Prong (W2)	1	R+sand flats	2.6 km
WA 1199	Bellefin Prong (W3)	1	R+sand flats	1.8 km
WA 1200	Bellefin Prong (W4)	1	R+sand flats	2.8 km
WA 1201	Bellefin Prong (W5)	1	R+sand flats	600 m
WA 1202	Bellefin Prong (W6)	1	R+sand flats	2.8 km
WA 1203	Bellefin Prong (W7)	1	R+sand flats	2.6 km
WA 1204	Cape Bellefin (W)	1	R+sand flats	3.5 km
Spring & neap tidal range = 0.6 & 0.2 m				

Bellefin Prong is a 30 m long peninsula that trends due north to Cape Bellefin. It is 5 km wide at its base narrowing to 1 km at the cape and average 20-30 m in elevation, but reach a maximum of 65 m. The peninsula separates Blind Strait on the west from Useless Loop on the east. The peninsula has originated from the accumulation of dune sand that has blown north from the False Entrance region during both Pleistocene and Holocene high stands in the sea. The present dune field extends north from False Entrance beach as an initial 20 km long active dune, halfway up the peninsula, with older vegetated and active dunes extending all the way to the cape. The whole surface of the peninsula is composed of trailing arms and elongate deflation hollows of the parabolics. The entire peninsula is surrounded by low energy beaches exposed only to wind waves, with sand flats extending 1-3 km into the bays and mangroves along the southeastern side. Mangroves are precluded from Useless Loop owing to the high salinity and now the salt evaporators.

Beaches WA 1197-1204 occupy the northern 18 km of the eastern side of the prong. All faces west across Blind Strait towards the southeastern shores of Dirk Hartog Island 5-8 km across the strait. There is no formal vehicle access to the peninsula. All beaches can however be accessed by small boat at high tide.

Beach **WA 1197** commences as the mangroves diminish and trends north for 900 m to a small projecting calcarenite point. Sand flats extend 2 km to the east. Beach **WA 1198** continues on the northern side of the point for 2.6 km as a north-trending crenulate beach, which terminates at a more prominent sandy foreland formed in lee of a small calcarenite reef. The sand flats widen to 3 km in the north, while vegetated dunes back both beaches.

Beach **WA 1199** is located in a curving 1.8 km long embayment between the southern foreland and a second northern foreland. The bay is filled with shallow sand flats, with deeper flats extending up to 3 km into the strait. Beach **WA 1200** continues north of the foreland as a crenulate 2.8 km long beach, the crenulations induced by vegetated northern trending dune lobes. It terminates at a small recurved spit, which borders the southern side of a small tidal creek. Beach **WA 1201** extends from the western side of the creek mouth for 600 m west to a sandy foreland, exposing the beach to the southerly winds. The winds have generated a couple of now vegetated blowouts extending up to 200 m inland.

Beach **WA 1202** commences at the foreland and trends straight almost due north for 2.8 km to the next embayment. The beach is backed by three shore parallel vegetated dune ridges, each part of the trailing arms of the longwalled parabolic dunes. It is fronted by 2 km wide sand flats. Beach **WA 1203** occupies the curving 2.6 km long west-facing embayment, with sand flats extending 300 m into the strait. It curves round towards the northwest along its northern shore, where three foredune ridges have built out to form the northern foreland. Beach **WA 1204** commences at the tip of the foreland and trends to the north-northeast for 3.5 km to Cape Bellefin, which narrows to a V-shaped tip. It is backed by truncated dune ridges, which form crenulations in the shore, with sand flats narrowing to 1 km at the cape.

WA 1205-1221 **BELLEFIN PRONG (E)**

No.	Beach	Rating	Type	Length
WA 1205	Cape Bellefin (E)	1	R+sand flats	1.3 km
WA 1206	Bellefin Prong (E1)	1	R+sand flats	400 m
WA 1207	Bellefin Prong (E2)	1	R+sand flats	1.7 km
WA 1207	Bellefin Prong (E3)	1	R+sand flats	700 m
WA 1209	Bellefin Prong (E4)	1	R+sand flats	1.2 km
WA 1210	Bellefin Prong (E5)	1	R+sand flats	900 m
WA 1211	Bellefin Prong (E6)	1	R+sand flats	1.1 km
WA 1212	Bellefin Prong (E7)	1	R+sand flats	3.3 km
WA 1213	Bellefin Prong (E8)	1	R+sand flats	4 km
WA 1214	Bellefin Prong (E9)	1	R+sand flats	3.1 km
WA 1215	Bellefin Prong (E10)	1	R+sand flats	2.3 km
WA 1216	Bellefin Prong (E11)	1	R+sand flats	400 m
WA 1217	Bellefin Prong (E12)	1	R+sand flats	1.5 km
WA 1218	Bellefin Prong (E13)	1	R+sand flats	1.3 km
WA 1219	Bellefin Prong (E14)	1	R+sand flats	1.8 km
WA 1220	Bellefin Prong (E15)	1	R+sand flats	500 m
WA 1221	Bellefin Prong (E16)	1	R+sand flats	200 m

Spring & neap tidal range = 0.6 & 0.2 m

The eastern side of the prong extends almost due south from Cape Bellefin for 27 km to the seawall that encloses the Useless Inlet salt ponds. The shoreline continues south inside the evaporation ponds for another 16 km, inline with Zuytdorp Point on the open coast. Beaches WA 1205-1221 are located along the eastern side of the prong between the cape and the seawall. All the beaches are backed by the generally vegetated transgressive dunes, with two areas of activity in lee of beaches WA 1212 and 1213 and beach WA 1219. There is vehicle access to beach WA 1221 at the seawall, and a vehicle track from here north to beaches WA 1220 and 1219. The remainder have no formal land access, with best access by boat. All the beaches receive only low wind waves and are fronted by 1-2 km wide sand flats.

Beach **WA 1205** forms the eastern side of the pointed cape and curves for 1.3 km southeast from the cape to a low sandy foreland. Vegetated dunes back the beach, with sand flats extending 1 km to the east. Beach **WA 1206** commences at the foreland and curves to the south for 400 m to a second small foreland. The beach faces slightly south and is backed by some minor dune transgression, then the vegetated transgressive dunes. Beach **WA 1207** continues south of the second foreland for 1.7 km in a curving east-facing open embayment fronted by 1 km wide sand flats. Two transgressive dune ridges intersect the shoreline.

Beach **WA 1208** is a slightly convex and crenulate 700 m long narrow beach backed by a foredune, then a vegetated deflation basin. Beach **WA 1209** is a crenulate slightly embayed continuation that extends for 1.2 km to the south. Both beaches front 1 km wide sand flats. Beach **WA 1210** is a curving 900 m long slightly embayed east-facing beach backed by a northern foredune and fronted by 1.5 km wide sands flats. Beach **WA 1211** extends almost due south for another 1.1 km to a sand flats filled 1.5 km long embayment, with a few mangroves in the northern corner. Some now vegetated Holocene blowouts and parabolics up to 1.5 km long have blown out of the northern side of the embayment.

Beach **WA 1212** is commences on the southern side of the embayment and trends to the south then south-southeast for 3.3 km. The northern depositional lobe of a 6 km long active parabolic is spilling onto the southern half of the beach causing it to prograde into the inlet, producing a shallow sandy inner section on the 1 km wide sand flats. Beach **WA 1213** commences on the southern side of the dune-induced protrusion and continues due south for 4 km. The beach is paralleled by the transgressive dune, with a strip of vegetated dunes between the shore and active dune. Sand flats widen to 2 km off the beach. The beach terminates at small sandy foreland, with beach **WA 1214** continuing on the southern side for another 3.1 km. It is backed by a vegetated longitudinal dune ridge and fronted by 1.5 km wide sand flats. A sandy foreland forms the southern boundary, with beach **WA 1215** curving slightly to the south for 2.3 km to a more prominent foreland, with a few mangroves on the northern side of the foreland.

Beach **WA 1216** is located immediately south of the foreland on a 1 km long protruding section of the dunes. It occupies the northern 400 m of the protrusion. Beach **WA 1217** curves around in an east-facing embayment on the southern side of the protrusion. Two now vegetated parabolics have blown out of the northern shore of the beach and across the protrusion to the southern shore of beach WA 1215.

Beach **WA 1218** occupies the north side of a crenulate 1.5 km long protrusion that appears to be composed of a series of low beach ridges. The beach is 1.3 km long curving round in two continuous crenulations.

Beach **WA 1219** commences 1 km to the south and is a curving 1.8 km long beach whose western most shore is being transgressive by the long False Entrance dune field, which commences 15 km to the south. The dunes are depositing sand along the centre of the beach partly filling the shallow embayment.

Beach **WA 1220** lies immediately to the south and is a double crenulate 500 m long low energy beach, with a few small mangroves in the centre and low calcarenite bluffs to either end. Beach **WA 1221** commences on the southern side of the bluffs and trends south for 200 m (Fig. 4.296) to the salt ponds seawall, which extends for 3.5 km across the inlet to the eastern shore. The beach has a couple of mangroves at the base of the bordering calcarenite and curves round to the side of the wall. It is the only beach readily accessible by vehicle. This site was chosen to build the wall, as to the south of this point the sand flats link across the 3 km wide inlet providing a shallow seafloor for the evaporation ponds.

Figure 4.296 Beach WA 1221 is located deep in Bellefin Prong, adjacent to the salt pond walls.

WA 1222-1244 **HEIRISSON PRONG (E)**

No.	Beach	Rating	Type	Length
WA 1222	Heirisson Prong (E1)	1	R+sand flats	300 m
WA 1223	Heirisson Prong (E2)	1	R+sand flats	1.2 km
WA 1224	Heirisson Prong (E3)	1	R+sand flats	600 m
WA 1225	Heirisson Prong (E4)	1	R+sand flats	300 m
WA 1226	Heirisson Prong (E5)	1	R+sand flats	900 m
WA 1227	Heirisson Prong (E6)	1	R+tidal flats	700 m
WA 1228	Heirisson Prong (E7)	1	R+tidal flats	1.5 km
WA 1229	Heirisson Prong (E8)	1	R+sand flats	200 m
WA 1230	Heirisson Prong (E9)	1	R+sand flats	1.3 km
WA 1231	Heirisson Prong (E10)	1	R+sand flats	900 m
WA 1232	Heirisson Prong (E11)	1	R+sand flats	1.1 km
WA 1233	Heirisson Prong (E12)	1	R+sand flats	1.4 km
WA 1234	Heirisson Prong (E13)	1	R+sand flats	700 m
WA 1235	Heirisson Prong (E14)	1	R+sand flats	1.2 km
WA 1236	Heirisson Prong (E15)	1	R+seagrass	3.5 km
WA 1237	Heirisson Prong (E16)	1	R+sand flats	400 m
WA 1238	Heirisson Prong (E17)	1	R+sand flats	300 m
WA 1239	Heirisson Prong (E18)	1	R+sand flats	300 m
WA 1240	Heirisson Prong (E19)	1	R+seagrass	1.5 km
WA 1241	Heirisson Prong (E20)	1	R+seagrass	1.4 km
WA 1242	Heirisson Prong (E21)	1	R+sand flats	500 m
WA 1243	Heirisson Prong (E22)	1	R+sand flats	300 m
WA 1244	Heirisson Prong (E23)	1	R+sand flats	300 m
WA 1245	Cape Heirisson	1	R+sand flats	600 m
Spring & neap tidal range = 0.6 & 0.2 m				

Heirisson Prong is a 1-2 km wide, 17 km long, north-trending peninsula that forms the northern side of Useless Inlet and in the south form the western shore of Useless Loop on its eastern base. Beaches WA 1222-1244 extend from 1.5 km north of the Useless Inlet salt ponds to Cape Heirisson 25 km due north. In between are the 22 usually crenulate or embayed beaches, most backed by vegetated calcarenite bluffs, with some vegetated Holocene dunes blanketing the backing calcarenite peninsula. All the beaches receive only low wind waves and are fronted by sand flats averaging 1 km in width. The salt town of Useless Loop is located on the eastern side of the prong, and numerous vehicle tracks criss-cross the peninsula providing vehicle access to many of the beaches

Beach **WA 1222** is located 1.5 km north of the salt pond wall. It is a curving, northwest-facing 300 m long pocket of sand bordered by vegetated calcarenite bluffs. Sand flats extends 1 km into the inlet, and a vehicle track reaches the eastern end of the beach. Beach **WA 1223** is located 100 m to the east and is a curving 1.2 km long beach, bordered by small calcarenite points. A vehicle track runs along the rear of the beach.

Beach **WA 1224** is located 1.5 km to the north and is a narrow, 600 m long west-facing strip of sand located along the base of vegetated bluffs.

Beaches WA 1225-1227 are part of a sedimentary cell that has been slowly infilling a 40 ha tidal creek. Beach **WA 1225** commences at an easterly inflection in the shore and is a 300 m long northwest-facing sand beach located between a western calcarenite bluff and a small reef that separates it from beach WA 1226. Beach **WA 1226** continues east for 900 m to the former mouth of the creek. It faces north across 1 km wide sand flats. Beach **WA 1227** is the outer most in a series of four recurved spits that extends up to 700 m into the tidal creek. It is a crenulate southeast-trending spit that terminates at the creek channel, which drains an area of tidal and salt flats.

Beach **WA 1228** commences on the eastern side of the inlet and trends north for 1.5 km as a narrow crenulate beach. It is initially fronted by the tidal delta, then by 1 km wide sand flats. A vehicle track runs out to the creek mouth at the southern end of the beach. Beach **WA 1229** lies 100 m to the north and is located in a north-facing indentation in the shore, and consists of a 200 m long strip of sand backed and bordered by vegetated calcarenite bluffs. Beach **WA 1230** continues to the north as a 1.3 km long series small crenulate embayments, with the beach varying in width while generally backed by calcarenite bluffs. It terminates at the next easterly inflection in the shore, with beach **WA 1231** continuing on the northern side. This beach is highly crenulate with a series of small embayments linked by a near continuous strip of high tide sand. A vehicle track commences at beach WA 1229 and runs north behind the beaches.

Beach **WA 1332** is a 1.1 km long beach that occupies the shoreline of the next west-facing embayment. The beach is widest in the south where sand moves around the southern point, narrowing to the north below the backing vegetated bluffs. Beach **WA 1333** is located in the next embayment. It is 1.4 km long and curves around a crenulate embayment. It is widest to either end, narrowing along the central section where some calcarenite bluffs protrude onto the beach. A foredune has formed behind the north, southwest-facing section of the beach.

Beaches WA 1134 and 1135 are located in two embayments 3 km west of Useless Loop township. Beach **WA 1134** spirals north for 700 m between low calcarenite bluffs. It has prograded up to 100 m at its northern end. Beach **WA 1135** occupies the next embayment and also curves to the north for 1.2 km with a slightly protruding bluff in the southern corner that divides the beach into two sections. A road from Useless Loop runs along the back of both beaches.

Beach **WA 1236** occupies a 2.5 km wide west-facing embayment. The beach curves around the shoreline for 3.5 km. It receives sediment around the southern point, which shoals the sand flats, then it narrows along the centre and widens again along the northern section which faces into the southerly winds. There has been some minor Holocene dune activity towards the northern end. The sand flats also narrow to the north permitting seagrass to grow to the shore. Beach **WA 1237** extends from the northern point as a recurved spit extending 400 m into the next embayment and forming a shallow, sheltered southern corner, with some mangroves growing in lee of the spit. Beach **WA 1238** occupies the northern end of the same embayment and forms the western side of a low vegetated sandy foreland.

Beach **WA 1239** is tucked-in on the protected northern side of the foreland. It is a curving 300 m long beach backed by a 50 m wide series of low beach ridges. It is bordered to the north by a 10 m high calcarenite bluff, with beaches WA **1240** continuing on the northern side for another 1.5 km. It is a relative straight southwest-facing beach backed by two foredune ridges and fronted by seagrass meadows, which grows in the deeper water just off the shoreline.

Beach **WA 1241** occupies the next embayment and is a 1.4 km long southwest-facing beach also fronted by deeper water with seagrass growing to the shore. It is backed by some minor prior dune transgression, with a vehicle track running out to the northern point. Beach **WA 1242** commences at the point and trends to the north as an active sand spit that may reach 500 m in length. It is backed by a shallow protected sandy area occupied by a few mangroves. Beach **WA 1143** lies 600 m to the northwest and is a narrow curving 300 m long strip of sand, backed and bounded by 10 m high calcarenite bluffs. The sand flats of Cape Heirisson extend 3 km to the west. Beach **WA 1244** is located on the northern side of the next calcarenite point, and consists of a narrow 600 m long strip of sand running along the base of the bluffs and fronted by the wide sand flats.

Cape Heirisson is a 25 m high vegetated calcarenite bluff, with beach **WA 1455** running for 1.5 km along the base of its northern side. The beach faces into Shark Bay across 1 km wide sand flats (Fig. 4.297). It is backed by steep, vegetated bluffs, which rise to 25 m in the centre, where a vehicle track reaches the top of the bluffs.

Figure 4.297 Cape Heirsson and beach WA 1455 lie at the tip of the 17 km long Heirsson Prong. Sand flats and seagrass meadows surround the peninsula.

WA 1246-1252 **HEIRISSON PRONG (E)**

No.	Beach	Rating	Type	Length
WA 1246	Cape Heirisson (E)	1	R+sand flats	200 m
WA 1247	Cape Heirisson (S1)	1	R+sand flats	100 m
WA 1248	Cape Heirisson (S2)	1	R+sand flats	250 m
WA 1249	Cape Heirisson (S3)	1	R+sand flats	500 m
WA 1250	Cape Heirisson (S4)	1	R+sand flats	600 m
WA 1251	Cape Heirisson (S5)	1	R+sand flats	700 m
WA 1252	Smith Rock (1)	1	R+sand flats	900 m
Spring & neap tidal range = 0.6 & 0.2 m				

Cape Heirisson forms the northern tip of Heirisson Prong and one of the western boundaries of 15 km wide Freycinet Reach. The reach connects to Henri Freycinet Harbour, which extends for 80 m southeast of the cape, and whose eastern boundary is the long Peron Peninsula. The first 9 km south of the cape contains seven beaches (WA 1246-1252), all low energy east-facing strips of sand fronted by sand flats ranging from a few hundred metres to 1.5 km wide, with seagrass growing in the deeper waters beyond the flats. A vehicle track from Useless Loop runs up the narrow prong and provides access to most of the beaches.

Beaches WA 1246-1248 occupy three small indentations on the eastern side of the cape. Beach **WA 1246** is a 200 m long narrow strip of high tide sand located at the base of 10-15 m high calcarenite bluffs. Beach **WA 1247** lies 200 m to the south in the next indentation and is a similar 100 m long strip of wider sand, backed by bluffs. Beach **WA 1248** lies another 100 m further south and is a 250 m long beach with backing bluffs. It also has a small foredune along the central-northern section, which faces slightly into the south. All three beaches are fronted by 200-300 m wide sand flats, then seagrass meadows.

Beaches WA 1249 and 1250 form the sides of a low sandy foreland formed (Fig. 4.298) in lee of a low calcarenite islet located 1 km east of the shore. Beach **WA 1249** is a curving 500 m long sand beach that forms the northern side of the foreland and connects in the north to the bluffs of the prong. Sand flats are 100 m wide at the bluffs widening to 1.5 km at the foreland. Beach **WA 1250** commences at the foreland and trends to the south for 600 m as a crenulate beach which initially attaches the foreland to the bluffs, then runs along the base of the bluffs for 300 m. The foreland is composed of a series of low foredunes, which have prograded eastward in lee of the islet. A vehicle track crosses the ridges.

Figure 4.298 Beaches WA 1249 (right) and 1250 (left) occupy a protruding low sandy foreland.

Beach **WA 1251** is located 1 km to the south. It is a curving 700 m long east-facing beach, backed and bordered by 10 m high calcarenite bluffs, and fronted by 1 km wide sand flats. Beach **WA 1252** occupies the northern half of the next curving embayment. It extends south for 900 m to a section of bluffs. It is backed by a foredune, then the bluffs, and fronted by sand flats that widen to 1.5 km in the south.

WA 1253-1258 **USELESS LOOP**

No.	Beach	Rating	Type	Length
WA 1253	Smith Rock (2)	1	R+sand flats	1.1 km
WA 1254	Spit (N)	1	R+sand flats	200 m
WA 1255	Spit (S)	1	R+sand flats	100 m
WA 1256	Useless Loop (1)	1	R+sand flats	400 m
WA 1257	Useless Loop (2)	1	R+sand flats	100 m
WA 1258	Useless Loop (3)	1	R+sand flats	1 km
Spring & neap tidal range = 0.6 & 0.2 m				

Useless Loop is the name of a 6 km deep, 1-2 km wide embayment that has since the 1960's been dammed and divided for salt evaporation ponds. It is also the name of the small salt town located on the eastern side of the entrance to the former loop. Beaches WA 1253 to 1258 extend either side of the loop and the township.

Beach **WA 1253** is located to the lee of the small Smith Rocks and is a 1.1 km long northeast-facing low energy beach fronted by 1.5 km wide sand flats. It is bordered by a section of low calcarenite bluffs in the north, while the southern end is crossed by the 2.5 km long jetty for the salt works. The jetty ties the small Slope Island to the mainland. The beach is backed by a few low foredunes, then the backing bluffs. The base of the jetty separates it from the small beach **WA 1254**, which continues east of the jetty for 200 m to the northern entrance to the loop. The beach is backed by the building for the salt works. Beach **WA 1255** is located just inside the loop and looks east across the seawall for the salt ponds. It is a crenulate 300 m long beach formed from sand moving around the point and into the loop as a crenulate spit. Sand flats extend 500 m off the beach. It is backed by low vegetated ridges and then a road for the salt works.

Beach **WA 1256** is located on the eastern side of the loop and commences against the eastern end of the seawall for the salt ponds (Fig. 4.299). The beach is attached to the wall at its western end and faces north. It extends for 400 m curving round a low sandy point along the eastern half and terminating against a 200 m long L-shaped rock jetty and boat ramp. Sand flats narrow to 100 m at the jetty. Beach **WA 1257** commences on the eastern side of the jetty and extends for 100 m to steep 20 m high calcarenite bluffs. A road to the jetty crosses between the two beaches. Beach **WA 1258** commences 500 m to the east at the end of the bluffs and continues to curve to the southeast for 900 m to a protruding sandy point. Sand flats averaging 200 m wide front the beach. The small community of Useless Loop is located on the bluffs to the lee of beaches WA 1256 and 1257, and western end of WA 1258.

Figure 4.299 Useless Loop township, with the salt ponds to the right and low energy beaches along the shoreline.

WA 1259-1268 USELESS LOOP-BRIGGS ROCK

No.	Beach	Rating	Type	Length
WA 1259	Useless Loop (S1)	1	R+sand flats	1.2 km
WA 1260	Useless Loop (S2)	1	R+sand flats	2.2 km
WA 1261	Useless Loop (S3)	1	R+sand flats	500 m
WA 1262	Useless Loop (S4)	1	R+sand flats	800 m
WA 1263	Useless Loop (S5)	1	R+sand flats	1.1 km
WA 1264	Useless Loop (S6)	1	R+sand flats	900 m
WA 1265	Briggs Rock (N3)	1	R+sand flats	200 m
WA 1266	Briggs Rock (N2)	1	R+sand flats	250 m
WA 1267	Briggs Rock (N1)	1	R+sand flats	1.8 km
WA 1268	Briggs Rock	1	R+sand flats	700 m
Spring & neap tidal range = 0.6 & 0.2 m				

The shoreline to the south of Useless Loop trends south-southeast towards Boat Harbour Loop, a protected 20 km long south trending loop. Between beach WA 1259 and the entrance of the loop at Briggs Rock is 16 km of crenulate shore containing 10 beaches (WA 1259-1268). The northern few beaches are accessible by vehicle tracks from Useless Loop, while the southern beaches have no formal vehicle access, with the best access by boat at high tide. All the beaches are very low energy, face east and are fronted by sand flats averaging 1-2 km in width and backed by 5-10 m high, often dune-draped calcarenite bluffs.

Beach **WA 1259** commences on the south side of the eastern drain for Useless Loop and curves to the south for 1.2 km to a section of low calcarenite bluffs that separates it from beach WA 1260. Low vegetated terrain and a salt lake back it, while it is fronted by 300 m wide sand flats, then seagrass meadows. Beach **WA 1260** continues to curve to the east for 2.2 km to a narrow foreland attached to a small bluffed calcarenite point (Fig. 4.300). The beach has prograded up to 200 m towards the foreland. A vehicle track winds out through the scrub to the foreland. Sand flats are 300 m wide at the foreland narrowing to 100 m at the western end of the beach. Beach **WA 1261** commences on the eastern side of the foreland and curves to the south for 500 m to a low protruding calcarenite bluff. Beach **WA 1262** extends south of the bluff for 800 m as a narrow beach backed by a continuous vegetated 10 m high calcarenite bluff. Both beaches are fronted by sand flats up to 1 km wide, then scattered seagrass meadows.

Figure 4.300 Beaches WA 1260 (right) and 1261 (centre) fronted by sand flats and seagrass meadows.

Beach **WA 1263** is located 3 km to the south. It is a curving 1.1 km long beach backed by 10 m high bluffs, with a narrow foredune behind the southern half. A 200 m long section of bluff separates it from beach **WA 1264** that continues southeast for 900 m to a calcarenite-tipped southern inflection in the shore. Sand flats extend 1-1.5 km off both beaches.

Beaches WA 1265 to 1268 occupy the 4.5 km of shoreline that extends north from Briggs Rock, to the entrance to Boat Haven Loop. Beach **WA 1265** is a

200 m long strip of sand bordered and backed by 10 m high calcarenite bluffs. Beach **WA 1266** is a similar 250 m long beach, with sand flats extending 1.5 km off both beaches. Beach **WA 1267** is a 1.8 km long beach located below the bluffs, with the beach widening to the south where it is backed by a 50 m wide foredune. A 300 m long section of bluffs separates it from beach **WA 1268**, a 700 m long beach, backed by 50-100 m wide foredune ridges, then the bluffs. It terminates at Briggs Rock where the shoreline turns and trends due south into the loop. The northern sand flats narrow to 1 km at the rocks while they extend 1.5 km east to the 200 m wide loop tidal channel.

WA 1269-1283 **CARARANG PENINSULA**

No.	Beach	Rating	Type	Length
WA 1269	Cararang Peninsula (N1)	1	R+sand flats	200 m
WA 1270	Cararang Peninsula (N1)	1	R+sand flats	400 m
WA 1271	Cararang Peninsula (N1)	1	R+sand flats	1.9 km
WA 1272	Cararang Peninsula (S1)	1	R+sand flats	250 m
WA 1273	Cararang Peninsula (S2)	1	R+sand flats	150 m
WA 1274	Cararang Peninsula (S3)	1	R+sand flats	500 m
WA 1275	Cararang Peninsula (S4)	1	R+sand flats	1.1 km
WA 1276	Cararang Peninsula (S5)	1	R+sand flats	600 m
WA 1277	Cararang Peninsula (S6)	1	R+sand flats	150 m
WA 1278	Cararang Peninsula (S7)	1	R+sand flats	900 m
WA 1279	Cararang Peninsula (S8)	1	R+sand flats	300 m
WA 1280	Cararang Peninsula (S9)	1	R+sand flats	1.4 km
WA 1281	Cararang Peninsula (S10)	1	R+sand flats	200 m
WA 1282	Cararang Peninsula (S11)	1	R+sand flats	500 m
WA 1283	Cararang Peninsula (S12)	1	R+sand flats	600 m
Spring & neap tidal range = 0.6 & 0.2 m				

Cararang Peninsula a 36 km long, 2-6 km wide north-trending calcarenite peninsula, that is bordered by the long elongate Boat Haven Loop on the west and the open Freycinet Estuary, then the 3 km wide Depuch Loop on its eastern side. Much of the peninsula has very low energy meso-saline shores. Low energy beaches only occur along the slightly more exposed northeastern section that faces into the greater fetch of Freycinet Estuary. Between the tip of the peninsula and the beginning of Depuch Loop 16 km to the south are 14 beaches (WA 1269-1283). They are only exposed to low wind waves and all are fronted by generally wide sand flats, and seagrass meadows, which diminish to the south, as salinity increases. The entire peninsula is part of Cararang Station and not open to the public. Station tracks lead to the southern beach WA 1283, with the remainder best accessed by boat. The peninsula is rimmed by 10-20 m high vegetated calcarenite bluffs and rises gently to a central 40-70 m high Pleistocene ridge.

Beaches WA 1269-1271 occupy the northern 3 km of the peninsula (Fig. 4.301). Beach **WA 1269** is located at the northwestern tip of the peninsula and forms the eastern boundary of Boat Haven Loop. It is a convex 200 m long low sandy beach fronted by sand flats extending 2 km across the entrance to the loop tidal channel. Beach **WA 1270** is located immediately to the east and is a 400 m long north-facing strip of low energy high tide sand fronted by 1 km wide sand flats which are rimmed by seagrass meadows. Beach **WA 1271** curves to the east for 1.9 km to the northeastern tip of the peninsula. It is fronted by 100-200 m wide sand flats, then a deeper 'lagoon' impounded by an intertidal to submerged narrow 2.5 km long spit that originates from the eastern tip of the beach. This spit represents the terminus for the sand being moved by the southerly winds and waves up the western side of the peninsula.

Figure 4.301 The northern end of Cararang Peninsula includes beaches WA 1269-1271 fronted by wide sand flats.

Beach **WA 1272** lies at the northern tip of the eastern side of the peninsula with sand moving slowly along the beach to supply the spit. It is a 250 m long sandy beach backed by a low foredune, with low bluffs to the south and the beginning of the spit to the north. It faces east across 1.5 km wide sand flats. Beach **WA 1273** is located in the next small indentation in the bluffs just to the south. It is a 150 m long pocket of sand bordered and backed by 10 m high calcarenite bluffs, with sand flats extending 1 km east to extensive seagrass meadows.

Beach **WA 1274** lies 1.5 km to the south. It is a curving 500 m long beach backed by low vegetated bluffs and fronted by a 200 m wide band of shallow sand flats, then deeper flats extending another 1 km into the bay, with seagrass fringing the flats. Beach **WA 1275** is located immediately to the south in the next curving calcarenite-rimed embayment. The beach is narrow, 1.1 km in length, with a slight southern protrusion cutting across the high tide beach. Beach **WA 1276** commences on the southern side of a 100 m wide boundary bluff. It is a similar 600 m long beach, also with a slight central protrusion in the bluffs, Both beaches share sand flats which narrow to 500 m in the south and which are fringed by seagrass meadows.

Beach **WA 1277** lies 100 m further south and is a 150 m long beach tucked in the northern corner of the next open calcarenite-rimmed embayment. It is backed by 15 m high vegetated bluffs and faces southeast across 500 m wide sand flats. One kilometer of bluffs separate it from beach **WA 1278** which occupies the southern end of the same open embayment. The beach extends for 900 m below bluffs, which rise to 20 m in the centre. It has prograded about 100 m onto the flats.

Beach WA **1279** is located to the lee of Charlie Island, 1.5 km to the south. It is a curving 300 m long beach backed by vegetated 10 m high bluffs and fronted by sand flats that extend 2 km east to the low island. Beach **WA 1280** occupies the southern part of the next embayment, 1 km further south. It is a 1.4 km long beach with a central protruding bluff fronted by continuous sand flats, which narrow to 300 m in the south (Fig. 4.302). The northern part of the beach has prograded about 50 m. Beach **WA 1281** lies 200 m to the south and consists of a 200 m long pocket of sand fronted by sand flats which narrow to 200 m then a narrow fringe of seagrass. Beach **WA 1282** is located another 300 m to the south and is a curving 500 m long beach with sand flats widening to 300 m at its southern end.

Figure 4.302 Low energy beaches WA 1280-1282 occupy past of the eastern shore of Cararang Peninsula.

Beach WA **1283** is located 3 km to the south at the end of a road from Cararang Station, 7 km to the southwest. The beach faces east and curves round for 600 m between low calcarenite points. It is backed by several beach ridges, which have prograded the centre of the beach about 200 m, while sand flats extend 500 m into the bay.

WA 1284-1291 **GIRAUD POINT**

No.	Beach	Rating	Type	Length
WA 1284	Giraud Pt	1	R+sand flats	700 m
WA 1285	Giraud Pt (S1)	1	R+sand flats	300 m
WA 1286	Giraud Pt (S2)	1	R+sand flats	500 m
WA 1287	Giraud Pt (S3)	1	R+sand flats	300 m
WA 1288	Giraud Pt (S4)	1	R+sand flats	500 m
WA 1289	Giraud Pt (S5)	1	R+sand flats	700 m
WA 1290	Giraud Pt (S6)	1	R+sand flats	500 m
WA 1291	Giraud Pt (S7)	1	R+sand flats	700 m
Spring & neap tidal range = 0.6& 0.2 m				

Giraud Point is located at the northern tip of a 16 km long, north-trending calcarenite peninsula that is bordered by the protected Depuch Loop to the west and the more open lower reaches of Freycinet Estuary to the east. The low undulating peninsula rises to a central height of 40 m in the south and 30 m in the north. Beaches WA 1284-1291 are spread along the eastern side of the peninsula between the point and 16 km to the south. All are low energy beaches fronted by wide sand flats, fringed by seagrass meadows. The remainder of the shore consists of low calcarenite bluffs, also fronted by the continuous sand flats. A vehicle track runs up the centre of the peninsula which is 4 km wide at it base narrowing to 1 km towards the point. One spur reaches the shore of the loop, the other the point, with sidetracks off to most of the beaches located along the eastern side of the point.

Beach **WA 1284** is located along the northeast-facing side of 10 m high Giraud Point, with a vehicle track reaching the northern end of the beach. It is bordered by vegetated calcarenite bluffs and backed by gently sloping vegetated terrain, together with two small salt lakes. Sand flats, fringed by seagrass meadows, extend 1 km off the beach, widening to the east.

Beach **WA 1285** is located 1 km to the south and consists of a 300 m long strip of high tide sand backed by low vegetated bluffs, with a central rock outcrop. A 200 m long low bluff separates it from beach **WA 1286**, which continues to the south for another 500 m to a slightly protruding calcarenite point. Both beaches are fronted by continuous sand flats widening from 1 km in the north to 3 km in the south.

Beach **WA 1287** is located on the southern side of the 300 m long rocky point and trends relatively straight for another 300 m to the next rocky outcrop. A vehicle track reaches the southern end of the beach. Beach **WA 1288** lies on the southern side of the 200 m long outcrop and extends south for 500 m, wrapping round a slight foreland at its southern end. Both beaches are backed by low vegetated bluffs, and fronted by 2 km wide sand flats, which link to Baudin Island 1.5 km southeast of beach WA 1288.

Beach **WA 1289** lies 4 km further down the peninsula and is a 700 m long sandy beach that commences below 15 m high scarped bluffs, then widens to the south as a slight sandy foreland that has prograded about 100 m to the east. It terminates against a low rocky point, with a vehicle track reaching the southern corner of the beach.

Beaches WA 1290 and 1291 occupy a curving semi-circular shaped, east-facing embayment located 1 km south of Three Bears Island. Beach **WA 1290** occupies 500 m of the northern shoreline and faces southeast. Because of its southern orientation it is backed by some active dunes extending 150 m to the north and some older vegetated parabolics that extend 500 m to the north. Beach **WA 1291** curves along the centre and southern shore of the bay and faces northeast. Sand flats fill the bay and extend 1 km east of the beaches. The main peninsula vehicle track runs along the back of the bay.

WA 1292-1301 **BABA HEAD**

No.	Beach	Rating	Type	Length
WA 1292	Disappointment Loop (W)	1	R+sand flats	700 m
WA 1293	Baba Head	1	R+sand flats	150 m
WA 1294	Baba Head (E1)	1	R+sand flats	1 km
WA 1295	Baba Head (E2)	1	R+sand flats	100 m
WA 1296	Baba Head (E3)	1	R+sand flats	800 m
WA 1297	Baba Head (E4)	1	R+sand flats	1.6 km
WA 1298	Baba Head (E5)	1	R+sand flats	700 m
WA 1299	Baba Head (E6)	1	R+sand flats	300 m
WA 1300	Baba Head (E7)	1	R+sand flats	400 m
WA 1301	Baba Head (E8)	1	R+sand flats	600 m
Spring & neap tidal range = 0.6 & 0.2 m				

The southern base of Freycinet Estuary between the base of the Giraud Point peninsula and Fording Point is 18 km wide and contains 35 km of crenulate shoreline including the shallow 9 km long Disappointment Loop. Baba Head is located at the eastern entrance to the loop and roughly in the centre of the base of the harbour. Beach WA 1292 lies on the western entrance to the loop, while beaches WA 1293-1301 extend for 9 km to the east. This is a very low energy meso-saline environment, with the low beaches all fronted by wide sand flats. Tamala station is located 10 km south of the head and several vehicle tracks reach some of the beaches.

Beach **WA 1292** is an isolated beach located at the tip of a 3 km long dune-capped calcarenite peninsula. The 700 m long beach occupies the northern side of the point and faces due north across 1.5 km wide sand flats and up the harbour. It is backed by a series of north-south trailing arms of longwalled parabolics that where initiated in the early-mid Holocene on the Zuytdorp Cliffs 15 km to the south and have blown across the intervening calcarenite plain of the peninsula.

Baba Head is located 3 km to the east with the shallow Disappointment Loop in between. The head protrudes into relatively deep water, with the sand flats narrowing to about 200 m. The small 100 m long beach **WA 1293** is located along the tip of the head, with low calcarenite points bordering the beach. A vehicle track terminates at its western end.

Beach **WA 1294** is located 1 km to the southeast and is a very low energy 1 km long north-facing beach fronted by 1 km wide ridged sand flats containing up to 20 m low subdued sand ridges. The eastern end of the beach terminates at tidal flats that border a small tidal creek.

Beach WA **1295** is a 100 m long pocket beach wedged between two low calcarenite points, 3 km east of Baba Head. A vehicle track runs due north from Tamala Station across the salt flats to the beach where there is a solitary hut. The salt flats narrow to 100 m off the beach and provides the best land access to deep water in the southern harbour. For this reason it has been used as a landing for the station.

Beach **WA 1296** lies 1 km to the east and is an 800 m long beach bordered by low calcarenite points, backed by some minor shoreline progradation, and fronted by 1.5 km wide sand flats, which contain a few faint sand ridges. A small tidal creek enters a small embayment immediately east of the beach.

Beaches WA 1297-1301 are part of a continuous series of five beaches each separated by small calcarenite points that occupy 4 km of the shore and all fronted by continuous 1 km wide sand flats. Beach **WA 1297** commences on the eastern side of the tidal creek and curves to the east then north for 1.6 km terminating at a low point. Sand flats extend 2 km north of the shore, with some sand ridges on the outer flats. Beach **WA 1298** continues to the northeast for 700 m to the next point, with beach **WA 1299** extending another 300 m to a low 100 m long vegetated point. Beach **WA 1300** is a 400 m long beach bounded by two low points, with beach **WA 1301** commencing on the eastern side of the northernmost point and curving round to the east for 600 m as a subdued convex foreland. To the east it merges with wide sandy tidal flats of a 3 km wide embayment, with low Fording Point located on the eastern side of the bay (Fig. 4.303). Sand flats extend 2 km north of beach WA 1301 and fill much of the embayment. Vehicle tracks and a fence line reach the boundary between beaches WA 1297 and 1298, and a vehicle track reaches the northern point towards the end of beach WA 1301.

Figure 4.303 Beach WA 1301 (right) faces across a very low energy shallow bay to Fording Point (left)

WA 1302-1305 **GARDEN PT-GNUNGAH WELL**

No.	Beach	Rating	Type	Length
WA 1302	Garden Pt (S)	1	R+sand flats	14.2 km
WA 1303	Garden Pt (N)	1	R+sand flats	5.5 km
WA 1304	Gnungah Well	1	R+sand flats	4.7 km
WA 1305	Gnungah Well (N)	1	R+sand flats	3.5 km
Spring & neap tidal range = 0.6 & 0.2 m				

The southeastern shores of Freycinet Harbour either side of Fording Point contain very long energy, low gradient meso-saline sandy tidal flats, backed in places by extensive salt flats. Nine kilometres northeast of Fording Point low waves again reach the shore and beach WA

1302 commences. Beaches WA 1302-1305 occupy the next 28 km of north-trending shore, with all beaches exposed to low wind waves and fronted by sand flats up to 5 km wide.

Beach **WA 1302** commences at a slight indignation in the shore partially filled by recurved spits. It continues northeast, then north for 14.2 km to low sandy Garden Point. This is a very low energy high tide beach fronted by 2-3 km wide sand flats including one section 5 km south of the point, containing two recurved spits located 1 km off the beach. The beach is low and backed by a few beach ridges and low gradient vegetated terrain. Apart from a solitary fence line midway along the beach there is no vehicle access. It terminates at Garden Point, beyond which is a 7 km long series of elongate recurved spits, with beach WA 1303 to their lee. The spit attest to the long term northward transport of sand along this shore by the southerly wind waves. The spit is low, largely unvegetated and awash during storms. Beach **WA 1303** curves round to the north in lee of the spit for 5.5 km, as a very low energy, protected beach, terminating at an easterly indentation in the shore.

Beach **WA 1304** commences on the northern side of the indentation and continuous to the north-northwest for 4.7 km. It is initially protected by the northern 2 km of the spit, and then by 4 km wide sand flats. Very low energy conditions continue north to a section of 20 m high red bluffs. Fence lines reach both ends of the beach. Beach **WA 1305** starts below the bluffs and continue to the northwest for 3.5 km, terminating at a subdued sandy foreland backed by an 80 ha salt lake. Sand flats narrow from 5 km to 2 km at the northern end of the beach, with dense seagrass meadows beyond the flats.

WA 1306 NANGA BAY

No.	Beach	Rating	Type	Length
WA 1306	Nanga	1→2	R+sand flats→R	11 km
Spring & neap tidal range = 0.6 & 0.2 m				

Nanga Bay is an open slightly curving embayment that occupies part of the northeastern shore of Freycinet Harbour. The bay is centred on Nanga Station, an historic sheep station, and since the 1980's a centre for tourism. Nanga now has a holiday resort and caravan park and is located just 3 km off the main road to Denham (Fig. 4.304).

Nanga beach (**WA 1306**) extends either side of the station, with a total length of 11 km. It faces southwest into the 40 km long fetch of the harbour and during strong southerlies can receive low wind waves. As a consequence the sand flats which are 2 km wide in the south, narrow to 500 m at the station, 50 m wide at the feral fence and are non-existent at the northern end against the low northern bluffs. As a consequence of the deeper water (1-2 m) close to shore dense seagrass grows off the edge of the flats. The beach is heavily utilized and has and area of bare sand behind the low, narrow swash zone. It is used for recreation, swimming and boat launching and has vehicle access and shade shelters, while the resort offering a range of accommodation and facilities as well as a store and kiosk.

Figure 4.304 Nanga Bay station and tourist resort, and adjacent low energy beach and seagrass meadows.

WA 1307-1313 NANGA BAY-GOULET BLUFF

No.	Beach	Rating	Type	Length
WA 1307	Nanga Bay (N1)	2	R+seagrass	1.4 km
WA 1308	Nanga Bay (N2)	2	R+seagrass	200 m
WA 1309	Nanga Bay (N3)	2	R+seagrass	200 m
WA 1310	Goulet Bluff (S2)	1	R+sand flats	700 m
WA 1311	Goulet Bluff (S1)	1	R+sand flats	1.1 km
WA 1312	Goulet Bluff (N1)	1	R+sand flats	2.2 km
WA 1313	Goulet Bluff (N2)	1	R+sand flats	1.4 km
WA 1314	Goulet Bluff (N3)	1	R+sand flats	2.6 km
Spring & neap tidal range = 0.6 & 0.2 m				

The northern end of Nanga Bay curves round to finally trend west at Goulet Bluff. The northern bay and bluff area are the southern side of the narrow Taillefer Isthmus which connects Peron Peninsula with the mainland. The isthmus narrow to 1.5 km to the lee of the bluffs at beach WA 1312.

The five northern bay beaches (WA 1307-1311) face south into the longer fetch of Freycinet Harbour and receive wind waves generated across the bay. The three northern bluff beaches (WA 1312-1314) faces west and are fronted by wide shallow sand flats built from sand moving northward around the bluff.

Beach **WA 1307** commences 200 m north of Nanga beach with low 5 m high eroding calcarenite bluffs separating the two. The 1.4 km long beach continues west-northwest beneath the bluffs, as a narrow high tide beach, with seagrass growing to the shore. The low bluffs continue to the west with beaches **WA 1308** and **1309** located 1 km along the bluffs. They are both 200 m long, narrow high tide beaches, fronted by the seagrass meadows. All three beaches are accessible via 4 WD track from the main Denham Road, which parallels the coast 2 km to the north.

Beach **WA 1310** commences at the northern end of the low bluffs and has formed across the mouth of a now infilled embayment between the bluffs and the southern 40 m high Goulet Bluff. The 700 m long beach connects the two bluffs and is backed by vegetated transgressive dunes that extend up to 500 m inland across a usually dry creek bed, which breeches the beach during very infrequent heavy rain. The dry creek and salt flats connect with the southern shores of Lharidon Bight and link the mainland with the peninsula. The salt flats are presently used as a landing ground.

Beach **WA 1311** is located on the northern side of the narrow southern spur and continues west for 1.1 km along the base of the main 40 m high bluff to the sandy point that surrounds the northern end of the bluff. The southern half of the beach has filled a small 150 m deep embayment, while the northern half runs along the base of the bluffs to the point. Both beaches are fronted by sand flats that widen to 200 m at the point, then seagrass meadows, with seagrass debris usually littering the beaches. A gravel road runs out from the Denham road to the bluffs, with spurs off the to the three beaches (WA 1310-1312).

The southerly wind waves move sand along the northern end of Nanga Bay and around Goulet Bluff to beaches WA 1312 and 1313. Beach **WA 1312** is a 2.2 km long north-trending beach, with long sand spits episodically moving along the front of the beach at times enclosing a series of small elongate lagoon/s between the spit/s and beach (Fig. 4.305). The spits eventually attach to the beach and the sand continues along the narrow northern half below 30-50 m high scarped bluffs. Sand flats, deposited by the longshore sand transport, extend up to 3 km west of the shoreline.

Figure 4.305 Sand moves north around Goulet Bluff to episodically widen beach WA 1312 (left).

Beach **WA 1313** is located on the northern side of the bluffs and is more sheltered from the southern waves. The waves do however usually maintain a small migrating spit at the southern end of the 1.4 km long beach, with a narrow lagoon between the spit and beach. A 200 m long gravel road runs out on to a car park on the 10 m high bluffs above the centre of the beach.

Beach **WA 1314** commences 300 m to the northwest and consists of a 2.6 km long narrow high tide beach located at the base of 10-20 m high bluffs that continue north to the next major inflection in the shore. The sand flats are 3 km wide in the south narrowing to 1 km off the northern boundary. The beach faces southwest into the prevailing winds and there has been some minor dune transgression up onto the backing bluffs.

WA 1315-1323 **WHALEBONE ROAD**

No.	Beach	Rating	Type	Length
WA 1315	Whalebone Rd (S4)	1	R+sand flats	400 m
WA 1316	Whalebone Rd (S3)	1	R+sand flats	200 m
WA 1317	Whalebone Rd (S2)	1	R+sand flats	100 m
WA 1318	Whalebone Rd (S1)	1	R+sand flats	250 m
WA 1319	Whalebone Rd	1	R+sand flats	200 m
WA 1320	Whalebone Rd (N1)	1	R+sand flats	400 m
WA 1321	Whalebone Rd (N2)	1	R+sand flats	200 m
WA 1322	Whalebone Rd (N3)	2	R	200 m
WA 1323	Whalebone Rd (N5)	2	R	200 m
Spring & neap tidal range = 0.6 & 0.2 m				

On the northern side of the Taillefeer Isthmus the Peron Peninsula begins to widen to several kilometres. Along the western shore the shoreline north of the Goulet Bluff beaches trends northwest towards Eagle Bluff. The first coastal vehicle assess north of Goulet Bluff is the Whalebone Road which runs straight for 1 km to the bluffs above the southern end of beach WA 1319. Beaches WA 1315-1318 are all pocket beaches located in slight embayments extending 5 km to the southeast, while beaches WA 1320-1323 extend 3 km to the northwest towards Fowlers Camp, the next vehicle access point. All these beaches are part of a littoral sediment cell with sediment slowly moving northeast along the coast, with generally narrow to non-existent beaches forming on the south-facing bluffed section, some backed by minor dune transgression, while more protected sand spits and wider pocket beaches accumulating in each of the small northerly inflections in the bluffs.

Beach **WA 1315** is located in the first of the series of inflections in the 20 m high bluffs, each of which form small more northerly facing embayments. The beach curves for 400 m and is backed by 50 m wide foredune plain in the centre, narrowing to either end. Episodic sand pulses in the forms of recurved spits move around the bluff widening the beach and at time forming a small lagoon between the beach and spit. The width of the beach therefore varies considerably depending of the presence and location of sand spits. A vehicle track winds across the backing plain to the northern end of the beach, then down the bluff and along the rear of the dunes to the southern end. Sand flats extend 1 km of the beach.

Beach **WA 1316** is located 1 km to the north and is a similar though smaller 200 m long pocket of sand also located in a small embayment formed by an inflection in the 20 m high bluffs. Sand pulses from beach WA 1315 move along the bluffs as sand waves then from a small spit across the beach, widening the beach in the process.

As the spits move on the beach narrows. The beach is backed by a narrow foredune, then the bluffs, with a vehicle track from the main road running 1 km out to the bluff and down onto the southern end of the beach. Sand flats and patchy seagrass meadows average about 1 km in width.

Beach **WA 1317** lies another 300 m to the north and is a 100 m long pocket of sand located in lee of the next inflection in the bluffs. Sand waves continue to move along and through this small beach, then northward along the base of the 20 m high bluffs. A vehicle track runs straight out from the main road to the bluffs above the beach. The sand flats and seagrass meadows continue along the beach. Two vegetated transgressive dune that originated from the southern end of the beach sit atop the backing bluffs.

Beach **WA 1318** is located 1 km further north and 500 m south of the Whalebone Road, with a spur off the end of the road to the bluffs above the northern end of the beach. The beach is located in lee of an inflection in the bluffs, which reach 30 m in height at the southern point. The beach curves round to the north, then northwest for 250 m with sand flats filling the small embayment and extending 500 m off the beach (Fig. 4.306). It is backed by a 10 m high dune-draped bluffs, with a small clifftop dune transgressing the northern bluffs. The Whalebone Road beach (**WA 1319**) is located at the end of the road with the blufftop car park providing a view up the curving 200 m long west-facing beach. The beach is backed by a 50 m wide foredune plain, from which a dune has transgressed 500 m north to link with the dunes to the lee of beach WA 1320. A vehicle track from the Whalebone road runs along the rear, then across this dune to reach the southern boundary headland of the next beach.

Figure 4.306 Beach WA 1318 is a low energy beach and sand flats curving between calcarenite bluffs.

Beach **WA 1320** is located in lee of a rounded 40 m high calcarenite headland, with the 400 m long west-facing beach curving round to the lee of the point to the northern bluffs. It is backed by a 100 m wide dune field including the northern end of the dune from beach WA 1319. Two vehicle tracks reach the centre of the beach and the southern headland. Sand pulses move round the headland and vary the width of the beach, as well as contributing to the sand flats that fill the small sandy embayment and extends 500 m out to the seagrass meadows.

Beaches WA 1321-1323 are located along the base of the 10-20 high bluffs that extends for 3 km west of beach WA 1320. Beach **WA 1321** is a 200 m long strip of high tide sand bounded and backed by 10 m high bluffs while sand flats narrow to 100 m at its northern end. Beach **WA 1322** is a similar 200 m long pocket of sand and narrow foredune, wedged in between the bluffs with the narrower sand flats then dense seagrass meadows off the beach. Beach **WA 1323** is located immediately east of the boundary point and consists of two adjoining pockets of sand linked by 100 m wide sand flats. A vehicle track descends the backing bluffs and blanketing foredune onto the eastern part of the 200 m long beach.

WA 1324-1326 **FOWLERS CAMP**

No.	Beach	Rating	Type	Length
WA 1324	Fowlers Camp	1	R+sand flats	600 m
WA 1325	Fowlers Camp (W)	2	LTT	2.8 km
WA 1326	Eagle Bluff (E)	2	LTT	400 m
Spring & neap tidal range = 0.6 & 0.2 m				

Fowlers Camp is located to the lee of a semicircular 1 km wide west-facing embayment that is filled with shallow sand flats. It faces towards Eagle Bluff 4 km to the west. Beach WA 1324 is located in the protected embayment, while beaches WA 1325 and 1326 lie along the base of the bluffs between the embayment and at Eagle Bluff.

Beach **WA 1324** is a curving 600 m long very low energy shelly beach, with sand moving north past beaches WA 1321-1323 and being deposited as a recurved spit attached to the southern end of the beach. The sand also fills the bay with sand. Because of the protection afforded by its orientation and the spit, a line of mangroves grows along the southern shore. The beach curves round to the north to terminate against a 100 m long 20 m high bluff. A gravel road runs straight out 1.5 km from the main road to the southern end of the beach and also providing access to the beach WA 1323.

Beach **WA 1325** commences on the western side of the bluff and continues essentially due west for 2.8 km to a small finger of calcarenite that separates it from beaches WA 1326. The wide sand flats of the embayment rapidly narrow along the beach, which towards its western end consists of a high tide beach, 50m wide low tide terrace then the seagrass meadows about 100 m offshore. The beach faces south exposing it to prevailing winds and higher wind waves. As a result some older vegetated and more recent active dune transgression covers the backing bluffs for 200-300 m inland. Beach **WA 1326** continues on the western side of the point for another 400 m to the 40 m high cliffs of Eagle Bluff. A narrow foredune is wedged between the beach and bluffs, while the beach is fronted by a 50 m wide low tide terrace then the seagrass meadows (Fig. 1.2c). The Eagle Bluff gravel road leads to a car park on the bluffs above the beach with walking tracks down to the beach and its eastern neighbour.

WA 1328-1336 **EAGLE BLUFF (N)**

No.	Beach	Rating	Type	Length
WA 1327	Eagle Bluff (N1)	1	R+sand flats	2 km
WA 1328	Eagle Bluff (N2)	1	R+sand flats	1 km
WA 1329	Eagle Bluff (N3)	1	R+sand flats	700 m
WA 1330	Eagle Bluff (N4)	1	R+sand flats	600 m
WA 1331	Eagle Bluff (N5)	1	R+sand flats	800 m
WA 1332	Eagle Bluff (N6)	1	R+sand flats	600 m
WA 1333	Eagle Bluff (N7)	1	R+sand flats	700 m
WA 1334	Eagle Bluff (N8)	1	R+sand flats	400 m
WA 1335	Eagle Bluff (N9)	1	R+sand flats	800 m
WA 1336	Eagle Bluff (N10)	1	R+sand flats	3.3 km
WA 1337	Denham Lookout (S)	1	R+sand flats	2 km
Spring & neap tidal range = 0.8 & 0.2 m				

At the prominent 40 m high **Eagle Bluff** the shoreline turns and trends north towards Denham 20 km to the north. The coast in between is dominated by crenulate 20-40 m high calcarenite bluffs, with beaches WA 1328-1338 located in the embayments between the calcarenite points, with continuous 1.5-2 km wide sand flats paralleling the shore (Fig. 4.307). Sediment moves northward along the shore producing sand pluses moving slowly past the beaches and bluffs with small recurved spits forming at the southern ends of the embayed beaches. There is gravel road access to beaches WA 1327, 1336 and 1338, while the others are only accessible by rough 4 WD tracks.

Figure 4.307 View north along beaches WA 1328 to 1336, all fronted by continuous 1.5-2 km wide sand flats.

Beach **WA 1327** is located in the first embayment on the northern side of Eagle Bluffs. The beach is protected by the bluffs and consists of a curving 2 km long west-facing low energy beach, with a few scattered mangroves along the southern shore and flats extending 1.5 km offshore to the seagrass meadows. A 15 ha salt lake backs the centre of the beach with a small tidal creek linking its to the shore and maintaining a projecting ebb tide delta. A gravel road off the Eagle Bluff road terminates at the creek mouth. The beach is backed by a 100 m wide band of foredunes, which along the southwest-facing northern section of the beach have been blown out to form a 500 m long vegetated parabolic dune.

Beach **WA 1328** is located immediately to the north in the next bluff-bound embayment. It is a 1 km long beach that curves round to the northwest, orientating it more into the southerly winds and waves. As a result two parabolic dunes have blown in from the beach and transgressed 200-300 m to the north. Sand flats covered by subdued sand waves extend 1.5 km to the west.

A 45 km high calcarenite point separates beaches WA 1328 and 1329, with beach **WA 1329** curving round as a narrow beach below the high bluffs for 700 m. Because of the steep bluffs there is no foot access to the beach. Beach **WA 1330** is located in the next embayment and is a similar curving, 600 m long narrow high tide beach. The bluffs decrease to about 20 m in the north, where they have been climbed by a dune ramp with dunes extending 100 m across the crest of the bluffs. A vehicle track runs along the top of the bluffs, but with no safe access down to either beach.

Beach **WA 1331** is located immediately to the north and extends from the southern 20 m high bluff point for 800 m to a low dune-capped calcarenite point. The northern half of the beach is backed by an unstable foredune with dunes climbing the bluffs and blanketing the backing point. Beach **WA 1332** is a slightly curving 600 km long beach with dunes from beach WA 1331 reaching the southern half of the beach, while sand also moves round the southern point. Shallow sand flats extend 1.5 km off the beach to the seagrass meadows.

Beach **WA 1333** is a curving 400 m long west-facing beach tied between two protruding sections of the backing 15 m high bluffs. Sand spits form at the southern point, while there is minor dune transgression over the northern point. A spur off the backing 4 WD track reaches the centre of the beach. Beach **WA 1334** is a curving 400 m long beach tied to an exposed low calcarenite point in the south and 20 m high vegetated bluffs to the north. Beach **WA 1335** continues on the northern side of the bluffs for 800 m as a curving bluff-backed beach. The beach narrows to the north and is backed by a foredune ramped up against the backing bluffs. The 4 WD track winds along the top of the bluffs with only steep foot access down to the beaches. Sand flats extends 1.5 km off all three beaches.

Beach **WA 1336** is a curving 3.3 km long beach located along the shore of an open west-facing embayment. The narrow beach commences at the base of 40 m high bluffs, which curve round to the north then northwest. The foredune has climbed the lower more southerly-facing northern bluffs with a few small active blufftop parabolics extending about 100 m inland over older, larger vegetated transgressive dunes. Sand flats fill the bay and extend 1.5 km offshore (Fig. 2.18b) narrowing to 1 km in the north. The 4 WD track reaches the southern bluffs and provides access to and a view up the beach, while a second track off the main road reaches the northern section of the beach.

At the northern end of beach WA1336 the shoreline turns and trends due north towards Denham. Beach **WA 1337** occupies the first 3.5 km of crenulate bluffed shoreline consisting of crenulate bluffs which are low in the south rising to 40 m in the north. The beach consists of a narrow strip of high tide sand that widens into some of the small embayments, and is replaced by 1 km of steep, 40 m high bluffs past the northern end. There are some minor climbing dunes along the northern side of the first three embayments, while sand flats extend 1.5 km offshore. An abandoned landing ground is located adjacent to the main road 1 km east of the beach, with a 4 WD track out via the landing ground to the top of the bluffs.

WA 1338-1341 **DENHAM**

No.	Beach	Rating	Type	Length
WA 1338	Denham Lookout	1	R+sand flats	2 km
WA 1339	Denham	1	R+sand flats	1.5 km
WA 1340	Denham (N)	1	R+sand flats	500 m
WA 1341	Lagoon Pt (S)	1	R+sand flats	1 km
Spring & neap tidal range = 0.8 & 0.2 m				

Denham is a growing town of over 1000 that has, over the past 20 years, transformed from a quiet fishing village into a major tourist destination. Today more than 160 000 visitors a year make the long trip out to Denham and usually onto Monkey Mia to see the dolphins, as well as some of the other features such as Shell Beach (WA 1394) the stromatolites at Hamelin Pool (WA 1403). Most however do not get to see the physical size and beauty of the Shark Bay region with its wide variety of coastal and bay environments, habitats, flora and fauna.

Denham is located on a more exposed southwest-facing shore (Fig. 4.308), which was chosen as a town site because the sand flats narrow to 500 m at the jetty, providing reasonable access to deeper water for shipping and boating (Fig. 4.309). Pearling commenced in the area in the 1850's with the town site gazetted in 1898. Today the town spreads along the shore behind beaches WA 1339 and 1340 and up onto the backing calcarenite slopes, with much of the newer development taking place on the backing slopes. Then main road, Knight Terrace, runs along the rear of the beaches, with a foreshore reserve between the road and shore. The main jetty and boating facilities are located in the centre of the town and divide the once continuous beach in two.

Beach **WA 1338** is located 2 km southeast of the town at the base of 50 m high bluffs. There is a car park on top of the bluffs called Denham Lookout, which provides a view of the beach and north to Denham, while a vehicle track runs down the bluffs to the northern end of the beach. This is a 2 km long low energy west-facing beach that has prograded about 200 m out onto the 1.5 km wide sand flats (Fig. 2.18a). The progradation has been in the form of a series of several recurved spits-beach ridges centred around a small tidal creek, which drains a small area of salt flats and has built an ebb tidal delta on the flats.

Denham beach (**WA 1339**) is a straight 1.5 km long beach that commences as the boundary bluffs trend slightly inland. The beach terminates on the southern side of the jetty and its associate structures include the dredged channel for boat access and the boat ramp. The beach is backed by a continuous rock seawall, together with a rock then wooden groyne towards the northern end, then the grassy reserve, road and main street of Denham. The main road and first row of houses are built on a low sandy plain that has prograded about 100 m onto the flats with the remainder of the town spread up onto the backing 30 m high slopes. The beach is fronted by sand flats, which narrow from 1 km in the south to 500 m at the jetty. Small dinghies are pulled up in the beach and larger boats moor off the end of the sand flats over the deeper seagrass meadows.

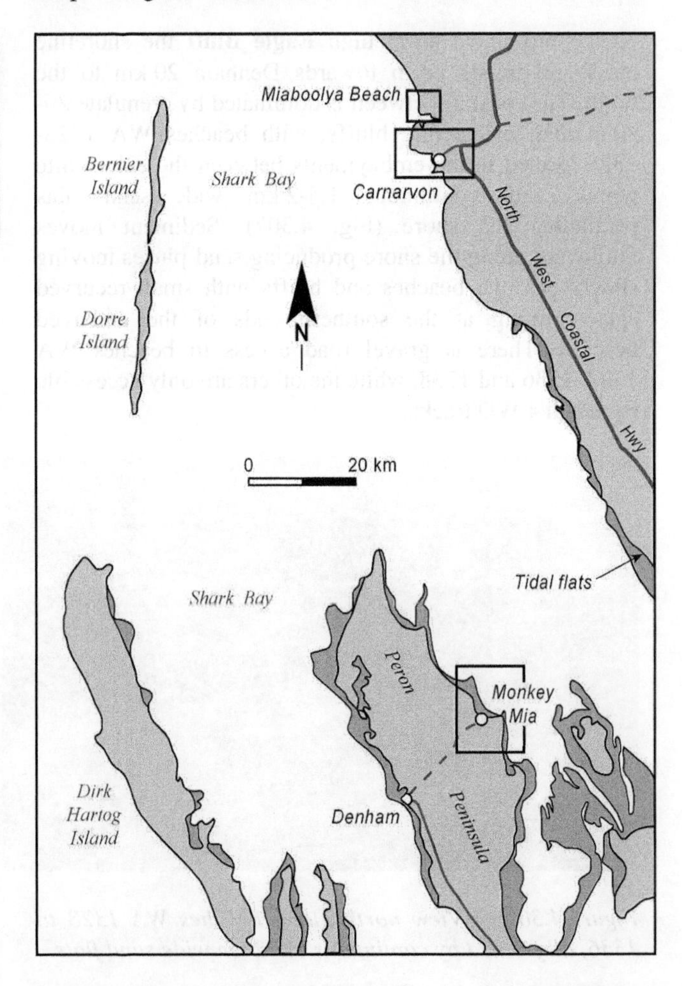

Figure 4.308 The northern section of Shark Bay, with the town of Denham located on Peron Peninsula. See Figs. 4.316 & 323 & 324 for inserts.

Figure 4.309 Denham is located on slopes to the lee of the town beach and boating facilities, including the channel dredged across the sand flats.

Beach **WA 1340** commences on the northern side of the jetty and continues north for 500 m to where the bluffs return to the shore. This beach has experienced erosion owing to the construction of the jetty, which has blocked the northerly movement of sand. As a consequence some sand is periodically dumped on to the beach to protect the foreshore. It is backed by the main road and houses and fronted by the 500 m wide sand flats.

Beach **WA 1341** extends from 500 m north of beach WA 1340 for 1 km to Lagoon Point, a 20 m high calcarenite bluff, where the shoreline turns 90° and trends north. A 4 WD track runs out to the bluffs above the point. The beach commences as a narrow strip of sand below 20 m high vegetated bluffs and widens to the north where its forms a recurved spit as sand moves around the point. The northern widening of the beach and the spit are an indication of the longshore sand transport. The blockage of sand by the jetty is also impacting upon this beach and those to the north by decreasing the sand supply and ultimately beach width.

FRANCOIS PERON NATIONAL PARK

Established: 1993
Area: 52 500 ha
Coast length: 92 km (3351-3443 km)
Beaches: 37 (WA 1345-1382)

Francois Peron National Park covers the northern end of the Peron Peninsula, an area of 40 000 ha including 92 km of low energy shoreline and 37 beaches. The focal point of the park is the old Peron Homestead, with 4 WD tracks then leading north to Cape Peron North and Herald Bight. The Monkey Mia road parallels the southern boundary of the park.

WA 1342-1348 **LAGOON PT-MIDDLE BLUFF**

No.	Beach	Rating	Type	Length
WA 1342	Lagoon Pt (N)	1	R+sand flats	1.6 km
WA 1343	Little Lagoon	1	R+sand flats	1.4 km
WA 1344	Little Lagoon (N1)	1	R+sand flats	1.8 km
WA 1345	Little Lagoon (N2)	1	R+sand flats	600 m
WA 1346	Little Lagoon (N3)	1	R+sand flats	600 m
WA 1347	Little Lagoon (N4)	1	R+sand flats	300 m
WA 1348	Middle Bluff	1	R+sand flats	1.3 km
Spring & neap tidal range = 0.8 & 0.2 m				

Lagoon Point marks the northern end of the Denham beaches and the beginning of an open 13 km wide west-southwest-facing embayment, which extends north to Middle Bluff. In between are seven beaches (WA 1342-1348) spread along and separated by 10-20 m high calcarenite bluffs. All the beaches are fronted by continuous 2-3 km wide sand flats. Francois Peron National Park commences at beach WA 1346 and incorporates the entire northern end of the peninsula. The only vehicle access to the beaches is via 4 WD tracks. The Lagoon Point walking track commences at the northern end of Knight Terrace and follows the buffs to the point and along to the lagoon inlet and then along the creek to Little Lagoon.

Beach **WA 1342** commences at Lagoon Point where the shoreline turns and trends to the north. Sand moving round the point has generated a series of migratory recurved spits which have over time prograded the northern end of the beach 100-200 m, as a series of beach ridge-spits, backed by 10 m high bluffs. The beach is crenulate and 1.6 km long terminating at the Little Lagoon tidal creek (Fig. 4.310). The lagoon is an oval-shaped 50 ha saline lake connected by a curving 1 km long creek to the entrance. Beach **WA 1343** continues on the northern side of the small tidal creek for another 1.4 km. It is also backed by beach ridges, which narrow to the north where a 10 m high bluff forms the northern boundary. There are scattered mangroves at the creek mouth, while sand flats extends 2 km bayward of the beaches. There are vehicle tracks to Lagoon Point and towards the northern end of beach WA 1343.

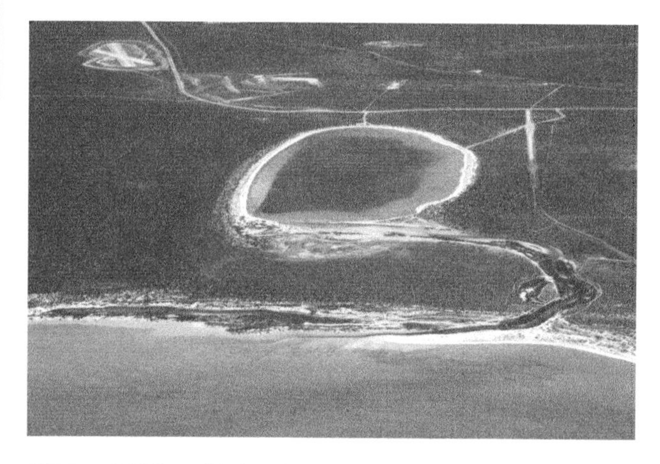

Figure 4.310 Little Lagoon with its tidal creek separating beaches WA 1342 and 1343.

Beach **WA 1344** occupies the next embayment. It curves for 1.8 km around the bluff-backed shoreline. The northern more southerly-facing section of the bay has experienced some minor blufftop dune transgression, extending up to 200 inland. The continuous sand flats narrow to 1.5 km off the northern end of the beach.

Beach **WA 1345** is located 1 km to the north in a smaller embayment, the beach curves for 600 m along the shore also facing slightly south along its northern section, where a 100 m wide area of dune transgression has covered the backing bluffs. Beach **WA 1346** occupies the southern half of the next embayment. It is a narrow 600 m long beach backed by 10 m high red bluffs, which are also blanketed by 50-100 m wide dune transgression along the central-northern section of the beach. Beach **WA 1347** occupies the northern part of the embayment and consists of a 300 m long narrow high tide beach located beneath 20 m high bluffs. A vehicle track reaches the top of the bluffs above the northern end of the beach. All three beaches are fronted by 2 km wide sand flats.

Middle Bluff forms the northern end of the larger embayment and the southern boundary of a 10 km long lagoonal system connected by a 200 m wide tidal inlet that flows parallel to beach **WA 1348**. The beach commences at the bluff and trends to the northeast as a 1.3 km long series of multiple overlapping recurved spits, with salt flats, and then mangroves occupying the inter-ridge depressions (Fig. 4.311). The nature, shape and length of the beach vary as new spits move slowly along the beach adding sand to the northern embayment, which has sand and tidal deposits extending 3 km offshore. The end of the spit reaches across the sand flats to the deeper tidal channel. In addition to the spit, there has been some dune transgression across the rear of the 300 m wide point, now indicated by vegetated walls of former parabolics.

Figure 4.311 Beach WA 1348 consists of a very low energy series of recurved spits that extend north of Middle Bluff.

WA 1349-1359 CAPE LESUEUR

No.	Beach	Rating	Type	Length
WA 1349	Cape Lesueur (S5)	1	R+sand flats	800 m
WA 1350	Cape Lesueur (S4)	1	R+sand flats	300 m
WA 1351	Cape Lesueur (S3)	1	R+sand flats	100 m
WA 1352	Cape Lesueur (S2)	1	R+sand flats	1.3 km
WA 1353	Cape Lesueur (S1)	1	R+sand flats	2 km
WA 1354	Cape Lesueur	1	R+sand flats	1 km
WA 1355	Cape Lesueur (N1)	1	R+sand flats	1.6 km
WA 1356	Cape Lesueur (N2)	1	R+sand flats	500 m
WA 1357	Cape Lesueur (N3)	1	R+sand flats	4.5 km
WA 1358	Cape Lesueur (N4)	1	R+sand flats	2.8 km
Spring & neap tidal range = 0.8 & 0.2 m				

The estuary at Middle Bluff is bordered on its northern and eastern side by a 10 km long peninsula that trends south-southeast from Cape Lesueur to the northern entrance to the estuary. In between the estuary and the cape are five beaches (WA 1349-1353) all backed by 20-40 m high bluffs and fronted by 2-3 km wide sand flats. North of the cape the shoreline trends north then northwest for 12 km to the next major inflection. Beaches WA 1354-1358 are located along this section. This is an undeveloped part of the national park with only 4 WD access along an old fence line to beach WA 1354.

Beach **WA 1349** is located on the northern side of the 1 km wide estuary mouth and 1.5 km north of Middle Bluff beach (WA 1348). Sand flats extend 1-1.5 km south of the beach to the tidal channel. The beach faces southwest and is backed by low vegetated pindan. Beaches WA 1350 and 1351 occupy the next two small embayments to the north. Beach **WA 1350** is a curving 300 m long strip of high tide sand backed by gently sloping vegetated terrain, while beach **WA 1351** is a 100 m long strip of sand that terminates against 15 m high red bluffs. Both beaches face west across 2 km wide sand flats.

Beach **WA 1352** lies 2 km further north and is a 1.3 km long narrow beach that commences in the southern part of a small embayment but for most of its length trends straight northwest below 20-40 m high scarped and vegetated bluffs. Sand flats extend 2 km west of the beach.

Beach **WA 1353** commences 2 km south of Cape Lesueur and trends relatively straight northwest to the cape. For the first 1.5 km it is wedged in at the base of 40 m high red bluffs, while the northern 500 m forms the western side of the sandy foreland that forms the cape. Sand moves northward along the beach and around the cape to supply beach WA 1354. In addition sand dunes have blown up the bluffs behind the northern end and onto the blufftop. Beach **WA 1354** forms the northern side of the cape foreland. This is a 1 km long, lower energy beach with sand flats extending 3 km offshore. In addition the sand that moves around the cape continues along the beach as recurved spits, which periodically move northwards enclosing a small 'lagoon' between the spit and the beach. The nature, width and length of the beach depend upon the presence of the spits. Vegetated bluff back the beach deposits and a vehicle track reaching the top of the bluffs above the cape, called the Peron Hills.

Beach **WA 1355** is located immediately north of the spit and consists of a curving 1.6 km long narrow beach located at the base of 30 m high red bluffs along its southern half, with lower sloping bluffs backing the northern half. Winds have ramped the foredune against the bluffs, in places almost reaching the crest. Beach **WA 1356** lies immediately to the north and occupies a small 500 m long curving embayment, the southern end of which is backed by a slumped 35 m high red bluffs. Sand from beach WA 1355 moves northward into the small embayment as a sand spit and continues north.

Beach **WA 1357** is a relatively straight north-northeast-trending 4.5 km beach that for its entire length is backed by red scarped 20-30 m high bluffs, with some dune sand ramped against the base of the bluffs. The beach is narrow and fronted by 2 km wide sand flats. It terminates at a small embayment on the northern side of which is a sand spit and the beginning of beach **WA 1358** which continues on straight for 2.8 km to the next major (unnamed) sand foreland (Fig. 4.312). The southern half of the beach lies below 30 m high bluffs, while the northern half forms the western side of the low sandy foreland. This foreland represents a major sink for the northward sand transport. It has an area of about 60 ha and is composed of a mixture of western foredunes and northern recurved spits together with some dune activity.

Figure 4.312 Beach WA 1358 (lower) forms the southern side of a prominent sandy foreland, with beach WA 1359 on the northern side.

WA 1359-1367 **CATTLE WELL-CAPE PERON NORTH**

No.	Beach	Rating	Type	Length
WA 1359	Cattle Well (S)	1	R+sand flats	3.3 km
WA 1360	Cattle Well	1	R+ridged sand flats	1.9 km
WA 1361	South Gregory	1	R+ridged sand flats	2.3 km
WA 1362	Gregory	1	R+sand flats	1 km
WA 1363	Gregory (N)	2	R	1.5 km
WA 1364	Bottle Bay	2	R	1.6 km
WA 1365	Bottle Bay (N)	2	R	1.8 km
WA 1366	Cape Peron N (W)	1	R+sand flats	500 m
WA 1367	Cape Peron N	1	R+sand flats	200 m
Spring & neap tidal range = 0.8 & 0.2 m				

The northwestern 15 km of the Peron Peninsula between Cattle Well and Cape Peron North is the most accessible section of shore in the national park, with marked 4 WD tracks leading to most of the beaches between Cattle Well and the cape (WA 1359-1367). There are also basic camping areas at a number of the beaches. The foreland at the southern end of Cattle Well beach and its sand flats that extend up to 5 km to the west of the foreland are a major sediment sink, with the result that sand flats are markedly narrower to the north averaging a few hundred metres or less. In addition the northwest cape is exposed to slightly higher wind waves together with refracted ocean swell through the 25 km wide gap between Dirk Hartog and Dorre islands, resulting in higher wind waves and low swell (up to 0.5 m) along this shore.

Beach **WA 1359** commences on the northern side of the large sand foreland it shares with beach WA 1360 and curves to the northeast then north for 3.3 km to the beachrock outcrop at Cattle Well. The sand flats narrow to a few hundred meters and waves averaging about 0.3 m reach the shore. The beach is backed by the 600 m wide foreland in the south, which extends for 1.5 km from where a low foredune and low vegetated bluffs back the beach. At the northern end there is a small sandy foreland capped by 20 m high partly active dunes. Seagrass eroded by the higher waves is often piled up on the beach.

Cattle Well beach (**WA 1360**) commences on the northern side of the foreland, which is tried to a beachrock reef and capped by active dunes rising to 20 m. The 2.3 km long beach initially curves round in lee of the reef then trends northeast to the next beachrock outcrop at South Gregory. The beach protrudes slightly bayward where a series of low foredunes has prograded about 100 m bayward. The foredunes are back by 10-20 m high red bluffs. A sandy track off the cape track runs out to the beach where these is a basic camping area.

South Gregory beach (**WA 1361**) commences at a beachrock outcrop. It is a crenulate beach backed by pindan bluffs, with a permanently blocked mouth of an old tidal creek cutting through the bluffs towards the southern end. The bluffs then increase to 20 m at the northern end. Ridged sand flats extend 300 m offshore narrowing to the north as wave energy increases (Fig. 2.17a), with waves averaging about 0.4 m. The 1 km long South Gregory track leads to the beach and a basic camping area. A fence line, which crosses the entire peninsula, terminates towards the northern end of the beach.

Gregory beach (**WA 1362**) is located to the lee of a prominent shore parallel beachrock reef that extends north of the point for 2 km enclosing the entire beach in a protected lagoon, as well as the southern part of beach WA 1363. Sand flats shoal within the lagoon, with a water exiting via a deeper channel through a southern break in the reef. The 1 km long beach is narrow and backed by a low vegetated bluff. The Gregory track connects to the beach where camping is permitted. Beach **WA 1363** commences at an easterly inflection in the shore and curves to the north for 1.5 km to the next easterly inflection. The beachrock reef lies 1 km offshore forming a wide deeper lagoon open to the north. The deeper water permits patchy seagrass to grow close to shore. There has been sight progradation at the southern end of the beach while 10 m high red bluff backs the northern end.

Bottle Bay beach (**WA 1364**) commences at a small beachrock point and trends to the northeast for 1.6 km to the next inflection. Deeper water lies off the reflective beach with seagrass growing close to shore. The track to the bay reaches the southern part of the beach, just above a 10 ha salt lake. Camping is permitted at the bay. Beach **WA 1365** commences at a beachrock reef, which continues as a submerged reef for the first 1 km of the 1.8 km long beach. The reef-protected section is backed by a reflective beach backed by 20 m high red bluffs. The beachrock merges with the shore along the northern section permitting higher waves to reach the shore. This section has experienced dune transgression up over the backing bluff and northward for 500 m to the base of beach WA 1366.

Beach **WA 1366** commences in a small embayment on the northern side of the dunes. Submerged beachrock parallels the 500 m long beach, terminating about 300 m offshore. Waves average about 0.5 m and the beach narrows to the north. Bright red 20 m high pindan bluffs back the beach. A vehicle track runs along the top of the bluffs providing access to and views of the beach. Beach **WA 1367** is located at the northwestern tip of the cape. It consists of a 200 m long sliver of sand located at the base of the 20 m high red bluffs, with sand flats extending 200 m offshore. The cape track terminates at the eastern end of the beach.

WA 1368-1375 **CAPE PERON NORTH-HERALD BIGHT**

No.	Beach	Rating	Type	Length
WA 1368	Cape Peron North (E)	1	R+sand flats	1 km
WA 1369	Cape Peron North (S1)	1	R+sand flats	500 m
WA 1370	Cape Peron North (S2)	1	R+sand flats	500 m
WA 1371	Cape Peron North (S3)	1	R+sand flats	400 m
WA 1372	Cape Peron North (S4)	1	R+sand flats	1.8 km
WA 1373	Cape Peron North (S 5)	1	R+sand flats	1.7 km
WA 1374	Herald Bight (N)	1	R+sand flats	3.9 km
WA 1375	Herald Bight	1	R+sand flats	6.2 km
Spring & neap tidal range = 1.2 & 0.5 m				

Cape Peron North marks the northern tip of Peron Peninsula (Fig. 4.313). To the east lies the eastern half of Shark Bay with Hopeless Reach paralleling the northeastern shore. At the cape the shoreline trends due south for 8 km then beings to curve round to the east into 6 km wide Herald Bight, which is bordered in the east by Guichenault Point. Between the cape and point is 22 km of low energy shoreline containing eight beaches (WA 1368-1375), all fronted by sand flats up to 1 km wide then seagrass meadows, while they are backed by the bright red pindan bluffs averaging 20 m in height. The only vehicle access is in the north at the cape and lighthouse, and at Herald Bight (WA 1375).

Figure 4.313 Cape Peron North has bright red bluffs, white sandy beaches surrounded by sand flats and seagrass meadows.

Beach **WA 1368** extends for 1 km to the southeast of the cape. It is a relatively straight beach, which has prograded about 100 m in the north, narrowing to the south to terminate against the protruding 30 m high red bluffs. Sand flats widen from 50 m at the cape to 400 m off the southern end. The cape track runs along to top of the bluffs with access down the vegetated bluffs to the beach. Beach **WA 1369** is located on the southern side of the point and commences beneath the cape lighthouse. It is a 500 m long curving beach, narrow in the north, widening to about 50 m in the south where it curves into a small north-facing embayment formed by the next point. Here the foredune has been ramped up against the bluffs. It can be viewed from the lighthouse, but there is no safe access down the steep bluffs to the beach. Sand flats extend about 300 m off the centre of the beach narrowing to each end.

Beach **WA 1370** commences on the southern side of the point and consists a narrow 400 m long beach wedged in at the base of vegetated 20 m high bluffs with no safe access down to the beach. There are two patches of blufftop dunes that have climbed up the bluffs. It terminates at a slight eastern inflection in the bluffs beyond which is 500 m long beach **WA 1371**. This is a similar narrow crenulate high tide beach that follows the base of the 20 m high bluffs, with no safe access down to the beach. Beach **WA 1372** commences on the southern side of an easterly inflection in the bluffs. It continues south for 1.8 km below the steep 20 m high bluffs, with one slight central inflection in the bluffs and shore. There is a well vegetated 1 km long clifftop parabolic behind the northern half of the beach, while the southern half is backed by a 100 m wide foredune ridges, and terminates against a 100 m easterly protrusion in the bluffs. Continuous 300 m wide sand flats then seagrass meadows parallel all three beaches. Beach **WA 1373** commences 200 m to the south and is a relatively straight south-trending 1.7 km long beach that curves round in the southern corner against a 200 m easterly protrusion in the bluffs. The beach is narrow for most of its length, widening in the southern corner to a 100 m wide series of low foredune ridges.

The protrusion marks the northwest boundary of Herald Bight. Beach **WA 1374** commences on the southern side of the protrusion and continues south-southeast for 3.9 km. The beach is relatively straight and has prograded about 150 m in the centre, while it narrows to either end, and is backed by continuous 20 m high bluffs. The sand flats widen to 1.5 km off the centre of the beach.

Beach **WA 1375** occupies the southern shore of Herald Bight. It commences at an easterly point and trends southeast, then east to the western mangrove-lined base of Guichenault Point. The only vehicle access in this area reaches the western side of the beach at an area where the sand flats narrow to 200 m (Fig. 4.314), providing a more suitable area for boat launching, as elsewhere the flats average 1 km in width. Apart from the access track there are no facilities, though camping is permitted. The beach is backed by a series of narrow foredune ridges, which widen in the centre where they are backed by a 1 km long salt lake, and in the east a 200 m wide series of ridges forms the western base of Guichenault Point. The sand flats transform to tidal flats in this corner, with some small mangroves growing along the base of the beach.

Figure 4.314 The Herald Bight beach sand flats narrow to 200 m at the access track and boat launching area (centre).

WA 1376-1382 **GUICHENAULT PT-MONKEY MIA**

No.	Beach	Rating	Type	Length
WA 1376	Guichenault Pt	1	R+sand flats	2.8 km
WA 1377	Herald Bluff	1	R+sand flats	1.2 km
WA 1378	Herald Bluff (S1)	1	R+ridged sand flats	400 m
WA 1379	Herald Bluff (S2)	1	R+ridged sand flats	200 m
WA 1380	Herald Bluff (S3)	1	R+ridged sand flats	900 m
WA 1381	Cape Rose	1	R+ridged sand flats	8.5 km
WA 1382	Monkey Mia	1	R+sand flats	9.2 km
WA 1383	Monkey Mia (S)	1	R+sand flats	700 m
Spring & neap tidal range = 1.2 & 0.5 m				

Guichenault Point represents a major terminus for northerly sand transport along the western shore of Peron Peninsula. The transport originates along the Eastern Bluffs and in Red Cliff Bay and has moved up to 30 km northward along the shore and shore parallel sands flats to the point and beyond as a submerged 6 km long sand shoal. The section of shore between the tip of the point and Monkey Mia, 24 km to the southeast contains seven beaches (WA 1376-1382), all backed by the southern extension of the Herald Bluffs, with beach WA 1383 located on the southern side of the Monkey Mia sand foreland.

The point consists of a 3 km long north-trending series of migratory multiple recurved spits, which have built the shoreline both northward into Herald Bight as well as 500 m to the west, as a low-lying area largely covered by mangroves (Fig. 4.315). The beach (**WA 1376**) forms the eastern shore. It is a 2.8 km long discontinuous, crenulate series of usually two to three slowly moving and merging low, narrow spits. Sand flats extend 1.5 km to the east and up to 6 km to the north, while the mangroves back the beach. Its base is attached to the low foredune ridges at the eastern end of beach WA 1375 and on its eastern side to the 30 m high steep red Herald Bluffs that dominate the shoreline down to Monkey Mia.

Figure 4.315 Guichenault Point is the terminus for northward moving sand.

At Herald Bluff the shoreline turns and trends to the south as a series of 20-40 m high slightly crenulate bluffs, with four beaches located in some of the indentations. All are backed by steep bluffs with no vehicle or safe foot access, boats being the best access. Beach **WA 1377** is located 2 km south along Herald Bluffs and is a relatively straight slightly protruding 1.2 km long beach. It is backed by steep 40 m high bluff cut by two central gullies. A dune has climbed the southern gully up onto the top of the bluffs. A 50 m wide foredune plain backs the southern protruding section, while sand flats and patchy seagrass extends 700 m offshore. Beach **WA 1378** lies 300 m to the south in the next slight indentation in the bluffs, which decrease to 20 m in height at the southern end of the 400 m long beach. The beach is narrow with a small foredune and fronted by 500 m wide ridged sand flats (Fig. 2.17b). Beach **WA 1379** is located 200 m to the south in the next indentation, and is a 200 m long strip of high tide sand backed by steep 20-30 m high red bluffs. Ridged sand flats extend 500 m off the beach.

Beach **WA 1380** is located 1 km further south in the next indentation in the bluffs. It is a straight 900 m long beach backed by a small foredune then steep bluffs that rise from 20 m in the north to 40 m in the south, while ridged sand flats extends 700 m offshore. The fence line and vehicle track reaches the top of the bluffs just to the north of the beach.

Beach **WA 1381** commences 200 m to the south as the bluffs curve round to trend south. It follows the base of the bluffs for 8.5 km as a curving southeast then more easterly trending beach. The bluffs reach 50 m in height, then decreases to 20 m at the southern end at Cape Rose. At the cape the shore turns and trends to the southeast. A 200 m wide series of low foredunes has prograded out from the backing 20 m high bluffs to form the cape. Sand flats parallel the beach widening from 300 m in the north to 2 km off the cape. A vehicle track reaches the bluffs above the centre of the beach.

Monkey Mia beach (WA 1382) commences at the slowly curving sandy Cape Rose and trends to the southeast and finally east to the sand foreland called Monkey Mia (Fig. 4.316). There are some crenulations along the beach where shore transverse sand waves attach to the shore, including at Monkey Mia. Most of the 9.5 km long beach is backed by 10-30 m high bluffs. A 4 WD track off the Monkey Mia road runs out to the centre of the beach. The foreland begins 2 km east of the tip as a series of low foredune ridges which form a 1 km wide sand plain at Monkey Mia. The sand flats narrow from 1 km at the cape to 50 m at the tip of Monkey Mia where the jetty is located (Fig. 4.317). The deeper water off the foreland is one reason why the dolphins can easily reach the shore, one of the few locations of this type in the entire bay. It is also the reason the Monkey Mia foreland was established early on as a pearling and fishing camp and boat launching area. Today it is dominated by the dolphin trade, with a large car park, caravan park, resort, national park facility including a dolphin information centre, jetty, boat ramp and managed sessions with the dolphins. Small boats are moored off the beach and pulled up on the shore. There is also a walking track that includes two lookouts on the backing 20 km high bluffs. It is serviced by a 25 km long sealed road from Denham.

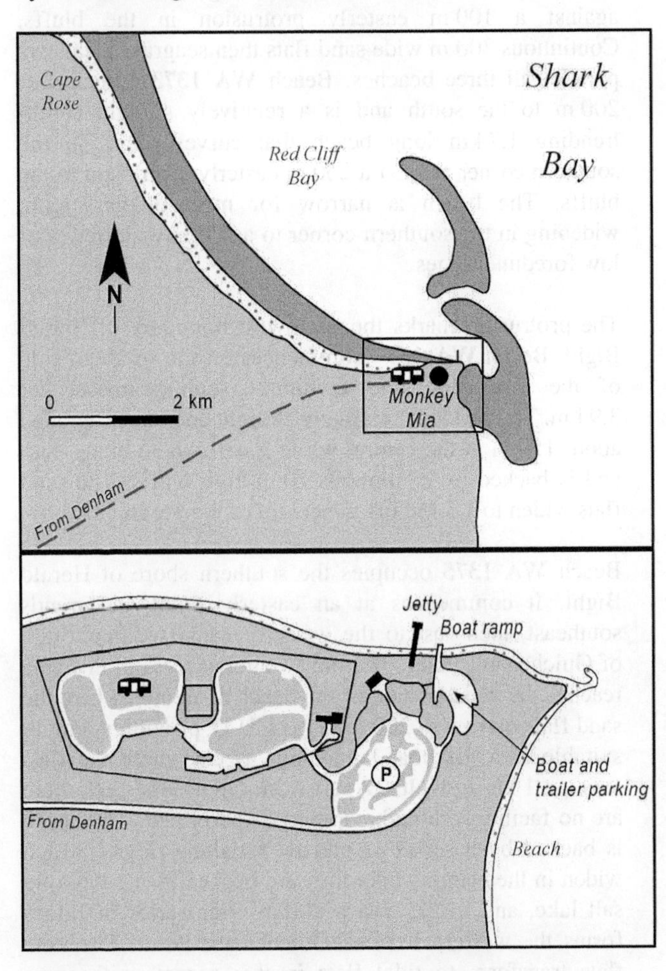

Figure 4.316 Monkey Mia is located at the tip of a sandy foreland. It has a large car park, caravan park and dolphin information centre (insert). See Figure 4.308 for location.

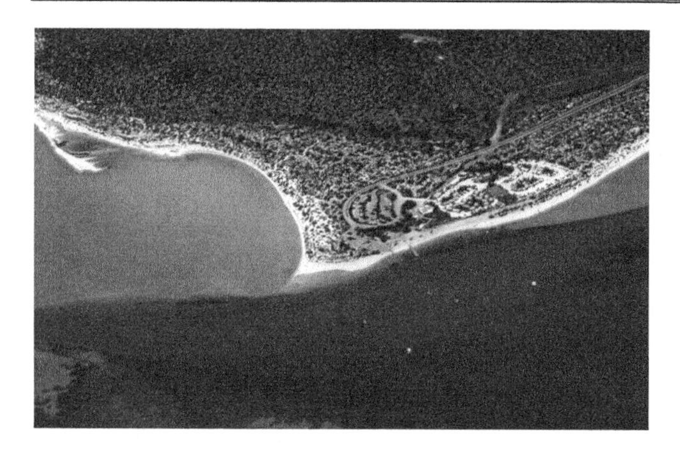

Figure 4.317 Monkey Mia foreland. Note the narrowing of the sands flats, which permits the dolphins to come in close to shore at the jetty.

Beach **WA 1383** is located on the southern side of the Monkey Mia foreland and curves to the south for 700 m to a sand foreland that has accumulated from sand moving northward along the next sediment cell. The foreland consists of several vegetated recurved spits that have prograded 200 m out from the backing red bluffs, with recurved spits episodically moving slowly round the point and on to Monkey Mia. The beach is backed by a 10 m high foredune, with the Monkey Mia car park backing the northern end of the beach.

WA 1384-1390 **EASTERN BLUFF-LHARIDON BIGHT**

No.	Beach	Rating	Type	Length
WA 1384	Eastern Bluff (N)	1	R+sand flats	2.7 km
WA 1385	Eastern Bluff (S)	1	R+sand flats	3.5 km
WA 1386	Dubaut Ck	1	R+sand flats	4 km
WA 1387	Dubaut Pt (S)	1	R+sand flats	5.5 km
WA 1388	Dubaut Pt (S1)	1	R+sand flats	2.5 km
WA 1389	Dubaut Pt (S2)	1	R+sand flats	2.4 km
WA 1390	Dubaut Pt (S3)	1	R+sand flats	6.5 km
Spring & neap tidal range = 1.2 & 0.5 m				

At Monkey Mia the shoreline turns and trends south towards Dubaut Point and then down into Lharidon Bight, which terminates 45 km to the south at Shell Beach (WA 1394). Beaches WA 1384-1390 occupy the first 25 km of shoreline and form the northwestern shoreline of the bight. The prevailing southerly winds and wind waves generated in the shallow bight slowly move sand northward along the eastern shore, forming substantial recurved spits along some of the beaches. There is no public vehicle access to these beaches south of Monkey Mia.

The **Eastern Bluff** is a 3 km long section of steep 20-30 m high red bluffs that backs parts of beaches WA 1384 and 1385. Beach **WA 1384** commences towards the northern end of the bluffs and trends to the north for 2.7 km as a crenulate sandy beach to the recurved spit that partly overlaps beach WA 1383. There are patches of vegetated sand plain up to 200 m wide between the bluffs and shoreline. Beach **WA 1385** extends south of the steep bluffed section for 3.5 km to the mouth of Dubaut Creek. For most of its length it is a narrow crenulate beach located at the base of the bluffs, which gradually decrease in height towards the creek. A 1 km long sand plain has prograded about 200 m on the northern side of the creek mouth, with mangroves lining the creek mouth as well as scattered along the adjacent beach. Sand flats averaging several hundred metres wide parallel both beaches.

At Dubaut Creek the shoreline turns and trend east for 4 km to **Dubaut Point**, a low sand point backed by vegetated pindan rising gradually to 30 m. The 50 m wide creek drains a 150 ha salt lake, that has been partly enclosed by a 3 km long sand barrier that has prograded across its northern side as part of beach WA 1386. Low energy north-facing beach **WA 1386** extends from the creek mouth to the point. It is fronted by 500 m wide sands flats, with scattered mangroves growing along the beach. Sand moving north along the eastern shore of Lharidon Bight has accumulated at Dubaut Point prograding the shoreline up to 1 km out from the backing pindan slopes. In addition sand flats associated with the sand transport moving into the bight extend up to 3 km into the bight to the deeper tidal channels.

Beaches WA 1387-1390 extend for 15 km south of the point and are all part of the northerly movement of sand along this shore. Beach **WA 1387** is a 5.5 km long east-facing crenulate sand beach backed by patchy narrow sand plain and then the red pindan rising to 20 m. Sand flats extends up to 2 km off the beach and a narrow, elongate sand spit up to 2 km long is slowing moving northward along the beach toward Dubaut Point. Beach **WA 1388** commences at a southwesterly inflection in the shore and curves to the southwest then south for 2.5 km to a curving section of low bluffs. The beach is backed by a few low foredunes and fronted 1.5 km wide sand flats. Beach **WA 1389** commences on the southern side of the low bluffs and continues south for 2.4 km to the northern side of a 1 km long eastward-pointing sandy foreland which is accessible via a vehicle track. Foredune ridges widen to either end of the beach, narrowing in the centre. Two kilometre wide sand flats extend east to the deeper tidal channel of the southern bight.

Beach **WA 1390** commences on the southern side of the foreland and trends to the south for 6.5 km as a series of low energy sandy crenulations, backed by sand plains up to a few hundred metres wide then terrain rising to 20-30 m. Towards the southern end of the beach a major 3 km long recurved sand spit extends east across the sand flats towards the end of the tidal channel. Elsewhere sand flats, ridged by tidal currents towards their outer edge, extend 2-3 m km out from the shore to the main tidal channel.

WA 1391-1396 **LHARIDON BIGHT/SHELL BEACH**

No.	Beach	Rating	Type	Length
WA 1391	Lharidon Bight (W1)	1	R+sand flats	9 km
WA 1392	Lharidon Bight (W2)	1	R+sand flats	11 km
WA 1393	Lharidon Bight (W3)	1	R+sand flats	2.5 km
WA 1394	Shell Beach	1	R+sand flats	6.5 km
WA 1395	Lharidon Bight (W1)	1	R+sand flats	11.5 km
WA 1396	Lharidon Bight (W2)	1	R+sand flats	8 km
Spring & neap tidal range = 1.2 & 0.5 m				

Lharidon Bight is an 18 km wide, 30 km deep embayment, bordered by Peron Peninsula to the west and Petit Point peninsula to the east. The 12 km wide mouth is largely filled with a shallow sand sill, while water depth in the bight averages less than 20 m. Because of the restricted circulation, low rainfall and high evaporation salinities reach 70°/$_{oo}$ in the south of the bight and precludes most marine species. One that can tolerate the salty water is the Cardiid Cockle (*Fragum erugatum),* a small white bivalve mollusk. Since the bight was flooded by the rising sea level about 6 500 years ago, the remains of this small shell has been slowly accumulating at the southern base of the bay, now knows as Shell Beach (WA 1394) (Fig. 4.318) and as a consequence one of the more famous beaches in Australia. Most of the bight consists of very low energy sandy-shelly shorelines, fronted by wide sand flats, with the southerly winds and waves generating northerly sand transport along both the western and eastern side of the bay, forming in places elongate recurved sand spits. The Dedham Road provides good access to the southern two beaches (WA 1393 & 1394) with only 4 WD access to the other beaches.

Figure 4.318 Shell Beach is composed entirely of the small white cockle (Fragum erugatum).

Beach **WA 1391** commences at the southern end of beach WA 1390, at a southwesterly inflection in the shore and the beginning of the central-southern bight region, south of the sill. The 9 km long beach contains both a protected sandy shoreline fronted by a series of multiple recurved spits located on the sand flats up to 2 km east of the shoreline, with a shallow, intertidal 'lagoon' in between. Sand slowly moves along the sand spits and bypasses the shoreline beach while sand flats continues for up to 3 km east of the spits. Beach **WA 1392** is a similar 11 km long beach that commences as a detached sand spit, then as it trends south the sand spits approach the shoreline gradually merging with the shoreline along the southern 4 km of the beach. The sand flats are 2 km wide in the south widening to 4 km in the north.

Beach **WA 1393** commences at the southern end of beach WA 1392 and trends to the southwest for 2.5 km as a low shelly spit, into a 2 km wide 3 km deep shallow bay that forms the southwestern end of the bight. A track off the Denham Road runs out to the tidal flats that fill the remainder of the bay. This bay backs onto the 2 km wide Taillefer Isthmus and a tidal creek at the southern end of the beach drains salt flats that link across to the rear of Goulet Bluff and beach WA 1310 thereby joining Peron Peninsula with the mainland.

Shell Beach (WA 1394) occupies a 3 km wide, 3.5 km deep bay that forms the southeast end of Lharidon Bight. The 6.5 km long curving, U-shaped bay faces north and is fringed by sand flats averaging about 1 km wide. Occasional northerly wind waves transport the cockle shells from the bay floor across the sand flats to the famous beach, and over time have built out the beach up to 200 m into the bay, with the former beaches marked by a series of shore parallel beach ridge crests (Fig. 4.316), which are gradually being degraded by pedestrian traffic. A gravel road runs out to the rear of the beach where there is a car park and toilets, but no other facilities. Besides being a major tourist attraction the shell grit is quarried for road base and the older cemented shells are cut out into slabs for building material.

Beach **WA 1395** commences on the eastern shore of Lharidon Bight and extends northeast of Shell Beach for 11.5 km terminating at a 3 km long recurved spit that extends up to 1.5 km out across the sands flats. The sand flats widens from 1 km the south to 3 km in the north and represent the long term accumulation of material slowly moving northward along the beach, and as smaller spits along the sand flats.

Beach **WA 1396** commences on the northern side of the large spit and continues to the north-northeast for another 8 km to the beginning of the eastern side of the bight sill. Beyond this point tidal flats dominate the shoreline up to Petit Point. The beach commences as a long energy sandy beach attached to the shore, then after 3 km begins to detach as a series of multiple receives spits that move along the sand flats up to 1.5 km off the shore. They finally merge with the shallow sill sediments, which extends west for 14 km across the mouth of the bight.

HAMELIN POOL MARINE NATURE RESERVE

Coast: 125 km (3545-3670 km)
Beaches: 21 (WA 1400-1421)

Hamelin Pool Marine Nature Reserve is part of the Shark Bay World Heritage Area, and occupies most of Hamelin Pool, the southeastern most section of the bay. It is also the most remote from ocean circulation with together with the shallow Faure Flats (sill) which spread right across the 26 km wide entrance to the pool, very low tide range (< 0.3 m), low rainfall and high evaporation, all contribute to the high salinities in the pool. As a consequent salinities reach 70°/$_{oo}$, twice that of seawater, and shallow bay floor the world's best example of algal stromatolities grow on the sand flats at the southern limit of the bay. The stromatolities can be viewed from an elevated walkway at Hamelin Pool.

WA 1397-1403 **PETIT PT-HAMELIN POOL (W)**

No.	Beach	Rating	Type	Length
WA 1397	Petit Pt	1	R+sand flats	2.5 km
WA 1398	Petit Pt (E)	1	R+sand flats	2.3 km
WA 1399	Petit Pt (E 1)	1	R+sand flats	2.9 km
WA 1400	Hamelin Pool (W1)	1	R+sand flats	6.8 km
WA 1401	Hamelin Pool (W1)	1	R+sand flats	7 km
WA 1402	Hamelin Pool (W1)	1	R+sand flats	6.5 km
WA 1403	Hamelin Pool (W1)	1	R+rock/tidal flats	2 km
Spring & neap tidal range = 1.2& 0.5 m				

Petit Point is a low sandy point located at the northern tip of the 30 km long sediment cell that commences on the eastern end of Shell Beach and gradually moves sediment and recurved spits up through beaches WA 1395 and 1396 towards the point. The point represents an accumulation of sand in the form of at least seven recurved spits, the tip of the curve forming the point. Beach **WA 1397** is a convex-shaped, north-facing 2.5 km long low outer sand spit that forms the point. It is attached in the west to a line of older red dunes, with the Holocene spits splaying out to the north and east. The shoreline is fronted by sand flats up to 7 km wide in the west narrowing to 1.5 km in the east. These flats parallel one of the main tidal channels for Hamelin Pool.

Beach **WA 1398** is commences 1 km east of Petit Point and forms the northern sink for sediment moving up the eastern side of the peninsula from Hamelin Pool. The beach is tied in the east to a westerly turn in the shoreline and splays out to the west for 2.3 km as a low recurved spit that has migrated across an older Holocene shoreline and beach ridge plain (Fig. 4.319), with 1 km wide sand flats between the original shoreline and the spit. The beach-spit parallels the main tidal channel with sand flats narrowing to 100 m off the eastern end of the beach.

Figure 4.319 Sediment moving north to Petit Point accumulated in the long low sand spit along beach WA 1398.

Beach **WA 1399** extends for 2.9 km due south of the inflection down into Hamelin Pool to a low dune-draped pindan point. The beach links the two points of the slight embayment and consists of a low overwashed prone shore, backed by wide overwash flats and shallow 'lagoon', with the original pindan shoreline several hundred metres to the rear.

Beach **WA 1400** commences at the southern point and curves to the south-southeast then south for 6.8 km to a line of red bluffs, which also mark the narrowest part of the Petit Point peninsula, with beach WA 1396 located 1.5 km to the west on the other side of the peninsula. The beach commences in the south as a narrow sand beach below the base of the 40 m high bluffs. It gradually widens to a few hundred metres in the north and as it begins to face slightly into the south active dunes back the beach. These dunes maintain a 3 km long dune sheet, which at the northern end of the beach have transgressed across the boundary point onto beach WA 1399. Both active and older vegetated transgressive dunes back the northern end and point. Beachrock is exposed along the eroded northern shoreline, then 1 km wide sands flats.

Beach **WA 1401** terminates at the southern end of beach WA 1400 at an inflection the backing red bluffs. It extends 7 km to the south as a sand beach backed by a narrow sand plain and 20-30 m high red bluffs. From the southern end it parallels the bluffs for 5 km widening gradually to the north. The shore parallel sand flats also widen from 500 m to 3 km in the north. As the bluffs turn to the west, the beach has deposited a 2.8 km long up to 1 km wide series of multiple recurved spit and beach ridges that in the north are gradually encroaching on beach WA 1400.

Beach **WA 1402** extends south of the inflection in the bluffs as a 6.5 km long beach that curves to the south, then east and finally northeast. The northern 3 km runs below steep 40 m high red bluffs, then as it turns to the east is broadens into a 2 km wide beach ridge plain, whose eastern point extends bayward as 4 km wide sand flats. Beach **WA 1403** commences on the eastern side of the beach ridge point and trends to the southeast, then south for 12 km. This beach represents a transition on the western shore of Hamelin Pool from a beach and sand flat to tidal flats. The beach is initially highly crenulate owing to the shape of the backing beach ridges and spits. It consists of a beach ridge plain gradually narrowing to the south, fronted by a highly crenulate sandy beach, then exposed intertidal rock-beachrock flats, grading into sand flats, with the rock outcrops inducing the crenulations in the shoreline. The beach ridge plain narrows to about 100 m below a central section of 20 m high red bluffs, then widens to 1 km in the south. A vehicle track reaches the southern section of beach, runs up the beach and across the northern beach ridge point to beach WA 1400.

WA 1404-1408 **FLINT CLIFF**

No.	Beach	Rating	Type	Length
WA 1404	Flint Cliff (W)	2	R+rock/sand flats	2.5 km
WA 1405	Flint Cliff (N1)	2	R+rock/sand flats	6.5 km
WA 1406	Flint Cliff (N2)	2	R+rock/sand flats	1.6 km
WA 1407	Flint Cliff (N3)	2	R+rock/sand flats	1.2 km
WA 1408	Flint Cliff (N4)	2	R+rock/sand flats	2.4 km
WA 1409	Carbla Pt (S)	2	R+rock/sand flats	4.8 km
Spring & neap tidal range = 0.3 & 0.1 m				

Flint Cliff is a 200 m long section of 10 m high bluffs located 3 km east of the Hamelin Pool stromatolite reserve. The reserve marks the boundary between the tidal flats to the southwest and the beginning of very low energy beach systems, fronted initially by rock and sand flats that extend to the northeast. The first beach east of the stromatolities is beach WA 1404.

Beach **WA 1404** is a northwest-facing 2.5 km long convex beach that terminates in the north at Flint Cliffs. It is a low energy beach fronted by calcarenite rocks and sands flats extending up to 1 km off the shore. The beach is composed of coarse shelly sand and only reached by occasional wind waves at high tide. It has formed from northeast sand migration that has developed a series of up to 15 beach ridge-spits, which have prograded the shoreline up to 700 m. The Hamelin Homestead caravan park, store and a shell block quarry are located behind the southern end of the beach.

To the north of Flint Cliff is an open 13 km wide west-facing embayment, bordered by Carbla Point to the north. In between are five low energy beaches (WA 1405-1409), all fronted by intertidal rocks then sand flats, and backed by beach ridge plains. All are accessible by vehicle tracks.

Beach **WA 1405** commences on the northern side of Flint Cliff and curves to the northeast then north for 6.5 km to a low reef-tied sand foreland. The beach is backed by a 600 m wide inner and outer series of beach ridge-foredunes that have blocked a drainage system, with a usually dry creek reaching the centre of the beach. It is fronted by irregular inner rock flats, then sand flats extending up to 1 km offshore. A vehicle track parallels the rear of the beach.

Beach **WA 1406** commences on the northern side of the foreland and trends to the northeast for 1.6 km to the next low rock-tied sandy foreland. The beach is backed by a 150 m wide beach ridge plain and a narrow salty back barrier depression. It is fronted by sand and rock flats, with the rock extending up to 3 km offshore. Beach **WA 1407** continues immediately to the north and is a similar 1.2 km long beach. It is backed by a central 200 m wide beach ridge plain, with rock and sand flats extending 1.5 km into the bay. Beach **WA 1408** curves to the north for 2.4 km to a more prominent sand foreland tied to low red bluffs. It is backed by a continuous 200 m wide beach ridge plain, then a blocked drainage system, which exits via the northern side of the foreland. Irregular 100 m wide rock flats, then 200 m wide sand flats front the beach. All three beaches are backed by a vehicle track.

Beach **WA 1409** commences on the northern side of the foreland and curves to the north for 4.8 km to low sandy Carbla Point, named after Carbla Homestead located 10 km to the northeast. A vehicle track from the homestead reaches the northern point. The beach is paralleled by a 100-200 m wide band of irregular rock flats, edged by sand while it is backed by a double beach ridge barrier. The outer widens to 400 m in the south, while the inner backs the centre of the beach. A usually dry creek crosses the southern end of the beach.

WA 1410-1416 **CARBLA PT-YARINGA PT**

No.	Beach	Rating	Type	Length
WA 1410	Carbla Pt (N1)	1	R+sand flats	2.8 km
WA 1411	Carbla Pt (N2)	2	R+rock/sand flats	1.2 km
WA 1412	Carbla Pt (N2)	2	R+rock/sand flats	4.4 km
WA 1413	Yaringa Pt (S)	2	R+rock/sand flats	1.7 km
WA 1414	Yaringa Pt	2	R+rock/sand flats	1.4 km
WA 1415	Yaringa Pt (N)	1	R+tidal flats	4 km
WA 1416	Hutchinson Is	1	R+tidal flats	3 km
Spring & neap tidal range = 0.3 & 0.1 m				

At **Carbla Point** the shoreline turns and trends roughly to the north with a series of small foreland-bounded embayments occupying the 11 km of shore up to Yaringa Point. In between are five beaches (WA 1410-1413). At Yaringa Point a long narrow sand spit extends to the north as beaches WA 1414 and 1415. All beaches are backed by a low vegetated coastal plain with only vehicle access to Carbla Point.

Beach **WA 1410** trends northwest of the point for 2.8 km to a reef-tied sand foreland. The beach faces southwest into the prevailing wind waves generates across Hamelin Pool and has a steep reflective beach fronted by deeper sand flats, with rock flats restricted to the bordering forelands. It is backed by some low active foredunes and a 50-100 m wide band of foredune ridges. Beach **WA 1411** commences on the northern side of the foreland and curves to the north for 1.2 km to the next reef-tied sandy foreland. Beachrock reefs outcrop along the shoreline fringed by a band of sand flats totalling about 100 m in width. A 100 m wide band of low foredunes, narrowing towards the points, back the beach, with an older inner barrier behind. Beach **WA 1412** occupies the next foreland-bound embayment, and is a more crenulate 4.4 km long beach fronted by 200-300 m wide irregular rock flats, covered in places with a veneer of sand. The equally irregular shoreline is backed by a crenulate 100-200 m wide series of foredunes, which are cut by a usually dry creek towards the centre.

Beach **WA 1413** continues to the north for 1.7 km between boundary rock-tied forelands. It is fronted by a 50-100 m wide band of beachrock, then deeper sand flats, and backed by foredune ridges, which narrow, from 100 m in the south to 50 m in the north. Beach **WA 1414** occupies the finally section of shore before Yaringa Point. It is a relatively straight north-trending 1.4 km long beach that terminates at the low sandy point. Irregular beachrock reef up to 80 m wide parallels the shoreline inducing shoreline crenulations. It is backed by a 100 m wide series of beach-foredune ridges.

Yaringa Point marks the southern boundary of a 15 km wide, 6 km deep embayment, and as a result has acted as a sink form sediment moving northward along this section of shore. The sediment has built two beaches (WA 1415 & 1416), as well a substantially filling the embayment (Fig. 4.320). Beach **WA 1415** commences at the point and initially trends to the northeast for 2.5 km as a narrow vegetated beach ridge, which the splays into a fan of older vegetated recurved spits, at well as curving to the north as a 1.5 km long active spit. The beach is fronted by 2 km wide tidal flats and backed by supratidal salt flats extending up to 5 km to the east. Five hundred meters north of the tip of the spit is Hutchinson Island (beach **WA 1416**), which is a low, narrow 3 km long barrier island, also fronted by tidal flats and backed by a 3 km wide embayment.

Figure 4.320 Yaringa Point with beach WA 1414 to the south and 1415 to the north (top).

WA 1417-1423 **KOPKE PT-GLADSTONE**

No.	Beach	Rating	Type	Length
WA 1417	Kopke Pt (S3)	1	R+tidal flats	1.8 km
WA 1418	Kopke Pt (S2)	1	R+tidal flats	1.6 km
WA 1419	Kopke Pt (S1)	1	R+tidal flats	1.1 km
WA 1420	Kopke Pt	1	R+tidal flats	3 km
WA 1421	Kopke Pt spit	1	R+tidal flats	2 km
WA 1422	Gladstone (S)	1	R+tidal flats	1.2 km
WA 1423	Gladstone Jetty	1	R+tidal flats	2.2 km
Spring & neap tidal range = 0.3 & 0.1 m				

Kopke Point is a low sandy point adjacent to the shallow 26 km wide Faure Sill which shoals the entrance to Hamelin Pool. As a consequence of the sill shallow waters and wide sand and tidal flats dominate the shoreline lowering waves at the shore. Four very low energy beaches (WA 1417-1420) extend 10 km south of the point, with beach WA 1421 prograding north of the point, and Gladstone Bay and its two sheltered beaches (WA 1422 & 1423) located in lee of the point. There is public access to the coast at Gladstone.

Beach **WA 1417** forms the shoreline of a 1.8 km long series of beach ridges, which are part of 1.5 km wide beach ridge-salt flat plain composed of three to four sets of crenulate beach ridges and recurved spits, located on a 2 km wide supra- to intertidal flats. The beach is a low west-facing, very low energy shoreline fronted by a mixture of irregular sandy tidal flats and some linear beachrock reef along its northern end. Land access requires transversing the salt flat.

Beach **WA 1418** commences 200 m to the north as a splaying series of 15 recurved spits which merge to the north into a 1.6 km long, 200 m wide beach ridge plain connected to backing coastal plain along its central-northern sector. It is fronted by 300 m wide tidal flats, then the sand of the sill. Beach **WA 1419** continues the beach ridges to the north for another 1.1 km, where they merges into a 300 m wide sand plain located to the lee of irregular intertidal rock flats. A straight vehicle track reaches the shore at the very northern end of the beach, which also marks the beginning of beach WA 1420.

Beach **WA 1420** is a 3 km long series of crenulate beach ridges, which splay north into a shore-attached inner and outer spit and are separated by 300 m wide salt flats. The outer ridge is fronted by tidal flats and then the shallow sill. The inner ridges reemerge at the shoreline along the southern side of Kopke Point. Beach **WA 1421** commences at a northerly inflection in the shore and continue north for 2 km as a convex-shaped curving very low energy shoreline, which also forms **Kopke Point**. It is backed by coastal plain in the south and centre, forming a small recurved spit at its northern terminus. Tidal flats extend west of the beach merging with the shallow sill.

Gladstone Bay is a 3 km wide, shallow north-facing, 100 ha embayment located to the lee of Kopke Point. The bay is bordered by the shallow waters of Kopke Point and Faure Sill to the west and the Wooramel river delta to the north, with a narrow tidal channel connecting the deeper wasters of Shark Bay to the bay. The deeper bay waters reach to within 100 m of a section of shore backed by coastal plain. At this point a woolstore, rock causeway and jetty were constructed in 1910 to permit lighters to access the shore and pick up wool from the surrounding stations. Today the ruins of the low 80 m long jetty still stand with beaches WA 1422 and 1423 located either side (Fig. 4.321).

Figure 4.321 Gladstone Bay, once the site of the wool jetty is now a popular camping area.

Beach **WA 1422** extends for 1.2 km southeast of the jetty as a series of degrading beach ridge-spits, that terminate among sandy tidal flats with a 4 WD track following the shoreline to the south. Beach **WA 1423** commences at the jetty and trends to the north for 2.2 km till it meets with the shallow tidal flats of the Wooramel delta. It is backed by an irregular 50-200 m wide beach-foredune ridge plain. The beach is a low, narrow (2-5 m) and composed of sandy-shelly beach with stunted vegetation growing almost to the shore, and intertidal vegetation on the tidal flats off the beach. The jetty and beach area is a camping reserve with numerous campers and caravans dotting the back of the beach and dunes, especially during winter. There are basic toilet facilities but no water.

The **WOORAMEL RIVER DELTA** reaches the coast 8 km north of Gladstone, with the delta front extending for another 24 km to the north. The highway crosses the usually dry river bed 11 km upstream of the delta shoreline, with the river spreading out to the west of the highway across the 200 km^2 delta plain, which is in turn fronted by a 200 km^2 inter to subtidal delta front, the latter extending 10 km into Shark Bay.

WA 1424-1427 **NEW BEACH-UENDOO CK**

No.	Beach	Rating	Type	Length
WA 1424	New Beach	1	R+tidal flats	2 km
WA 1425	Bush Bay	1	R+tidal flats	800 m
WA 1426	Grey Pt (S)	1	R+tidal flats	1 km
WA 1427	Uendoo Ck	1	R+tidal flats	1.5 km

Spring & neap tidal range = 0.9 & 0.5 m

To the north of Gladstone is a 90 km section of tidal flats, with mangroves increasing to the north as the high salinities of Hamelin Pool decrease towards normal ocean salinity and the spring tidal range picks up to 0.9 m. The tidal flats extend 2-5 km offshore resulting in very low waves at the shore. The first section of shore is dominated by the **Wooramel River** delta, which enters the northern side of Gladstone Bay and occupies about 10 km of the shore. To the north of the delta wide tidal and salt flats dominate for 50 km to Long Point. The point marks the first in a series six assemblages of protruding recurved spits, each backed by mangroves and salt flats. These occupy the next 30 km of shore to 5 km past New Beach. The entire shoreline is very low gradient and in sections consists of wide supratidal salt flats, with mangroves in the protected lee of the spits, and tidal flats extending offshore, and no beaches.

At the northern end of this system are four low energy beaches (WA 1424-1427) each accessible by 2 WD via a gravel road from the highway. They are spread along a 30 km long section of mangrove-dominated shoreline (Fig. 1.18), the beaches occupying the few gaps in the mangroves.

New Beach (**WA 1424**) is a low crenulate, narrow strip of high tide sand that extends for 2 km to the lee of 2 km wide tidal flats, with scattered mangroves on the flats. The beach is located 4 km off the highway. **Bush Bay** (**WA 1425**) is accessible via the same road and is located 7 km north of New Beach. The 'bay', is an 800 m long gap in the mangroves and consists of a slightly higher energy wider beach fronted by 3 km wide tidal flats, with mangroves to either end of the beach. Both beaches are reasonably popular camping and fishing spots, with however no facilities. Some fishers drive their boats out across the tidal flats to launch their boats at low tide.

Beach **WA 1426** lies 8 km to the north and is accessible via a shore parallel gravel road, which runs along the rear of the beach. It is a low, narrow 1 km long strip of high tide sand fronted by 2 km wide tidal flats, with mangroves to either end. One of the access roads from the highway reaches the coast track 1 km north of this beach.

Low mangrove-lined **Grey Point** separates beach WA 1425 from Uendoo Creek beach (**WA 1427**) located 14 km to the north. This is a 1.5 km long strip of low high tide beach, with the road running right behind the centre of the beach. It is bordered by mangroves to either end and fronted by 2 km wide tidal flats. The northern end of the beach is a popular basic camping area.

WA 1428-1431 CARNARVON BEACHES

No.	Beach	Rating	Type	Length
WA 1428	Pelican Pt Beach	2	R	4 km
WA 1429	Whitmore Is	2	R+ridged sand flats	500 m
WA 1430	Whitmore Pt	2	R+ridged sand flats	6.5 km
WA 1431	Miaboolya Beach	3	LTT→R	31.5 km
Spring & neap tidal range = 0.9 & 0.5 m				

Carnarvon is an well established town of 8000 on the southern banks of the Gascoyne River mouth (Figs. 4.322, 223), 900 km north of Perth. The rich soils of the surrounding Gascoyne River delta and floodplain, plus bore water from the river has lead to the development of intensive agriculture along the river, while its location at the river mouth resulted in the construction of the first jetty in 1904. Today is has a population of 7 000 and offers all range of accommodation, facilities and services. The Gascoyne River transports sand to the river mouth, which has resulted in progradation of the delta into Shark Bay. The river sand is also the start of a sediment cell with sand moved northward by the prevailing southerly waves and winds. The four town beaches (WA 1428-1431) reflect the nature of the river mouth and the northward transport system.

There are two Carnarvon swimming beaches on the shores of Shark Bay one at Pelican Point (WA 1428) at the southern end of Babbage Island (Fig. 4.323), and one 8 km north of the town at Miaboolya Beach (WA 1431) (Fig. 4.324). Both are protected from high ocean waves by Dorre and Bernier islands, which form the western side of Shark Bay, and by the more than 1 km wide sand flats which front both beaches.

Pelican Point Beach (WA 1428) occupies the 4 km long western shore of Babbage Island and is the main town beach (Fig. 4.325). The island has been built by low waves reworking the river sand into a 4 km long west to southwest-facing beach and island. It is composed of four sets of westward prograding low foredune ridges, grading to recurved spits towards the river channel in the north. The northern tip of the low sandy island ends at the river mouth, and the 2 km long Carnarvon jetty crosses the island, 1 km south of the tip. A small jetty to service an oil terminal is located in the middle of the island. The island is in a natural state south of the terminal and the southern 2 km and the southern tip called Pelican Point are the main recreation beach area. The main area on the southern rear of the island faces into the Fascine tidal creek, where there is a car park, amenities, several sun shelters and a boat ramp. The dominant wind waves generated within Shark Bay and occasional swell around Bernier Island average less than 0.5 m at the beach and tend to surge up a moderately steep beach, while at mid to low tide they spill across the wide low sand flats, which are exposed at low tide towards the northern end.

Region 5C CARNARVON: Carnarvon - Warroora

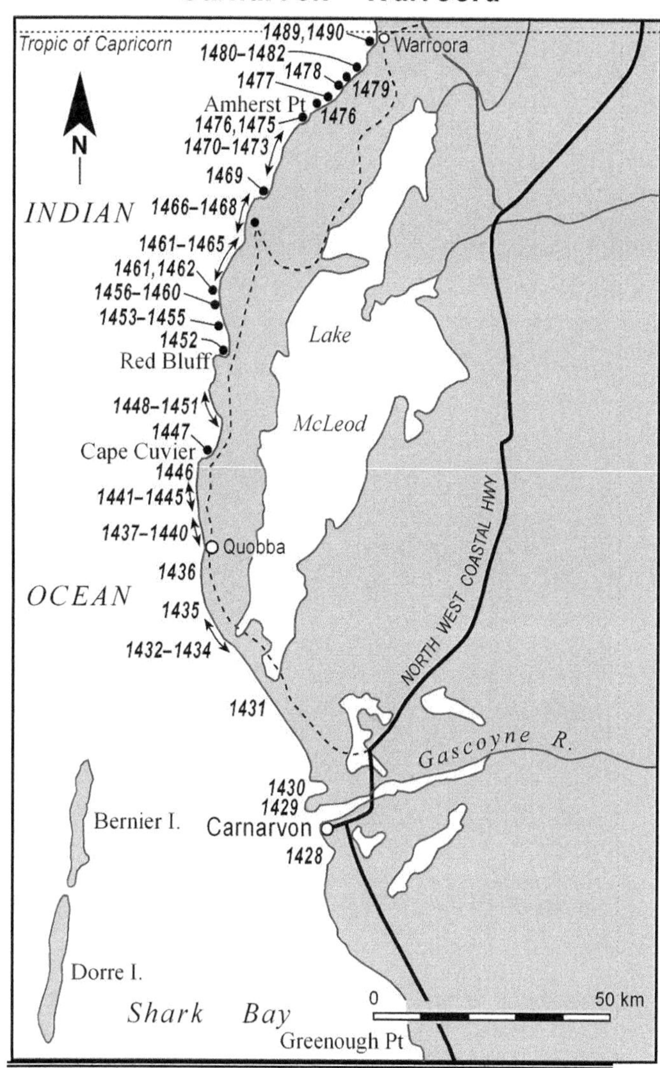

Figure 4.322 The Carnarvon to Warroora section of coast contains beaches WA 1428-1490.

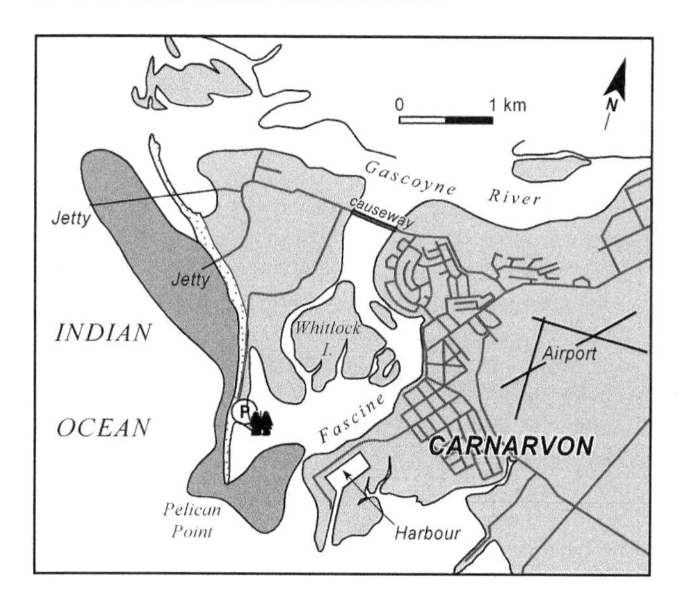

Figure 4.323 Carnarvon is located at on the deltaic mouth of the Gascoyne River. See Figure 4.300 for location.

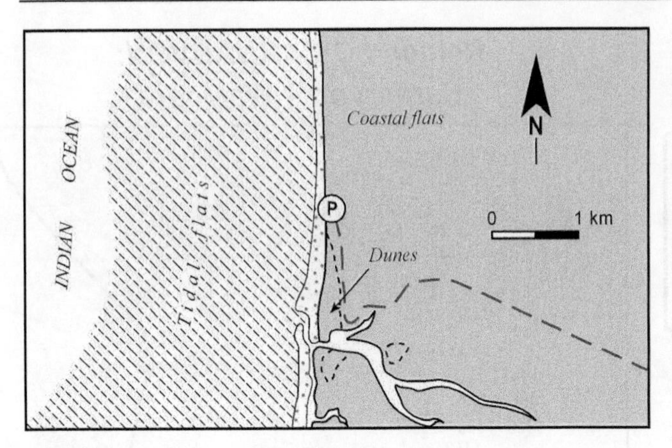

Figure 4.324 Miaboolya Beach is located 8 km north of Carnarvon. See Figure 4.306 for location.

Figure 4.325 Carnarvon (right), Babbage island and Pelican Point beach (left) at the mouth of the Gascoyne River.

Whitmore Island (**WA 1429**) is a 500 m long sand island located in the centre of the 1 km wide river mouth and can only be accessed by boat. It is fronted by 2 km wide ridged sand flats and backed by 1 km wide series of foredune ridges and spits, set in amongst mangroves. Other temporary sand islands and spits form off and to either side of the island/river mouth, depending on the nature of recent flood events.

Whitmore Point forms the northern entrance to the Gascoyne River and the beginning of beach **WA 1430**, which trends to the north for 6.5 km to Miaboolya Creek mouth. This is the first beach to receive sand moving northwards from the river mouth area and as a consequence varies considerable though time. The beach is backed by a 1 km wide series of beach-foredune ridges, which have evolved from episodic recurved spits moving northward and merging onto the shore. At anyone time one to three elongate spits up to 2 km long may be moving along the shore. The outer spits are fronted by 1 km wide ridged sand flats, while mangroves fill the inter-spit swales. The spits dominate the first 3 km of the beach and usually become welded to the shore along the northern 3 km where they manifest as northward migrating sand waves. Because of the backing mangrove wetlands there is no vehicle access to the shore.

Miaboolya Creek drains a 200 ha area of salt flats and mangroves and crosses the shore as a narrow, shallow tidal creek. The Miaboolya Road reaches the beach 500 m north of the creek mouth, which marks the beginning of the 31 km long **Miaboolya Beach (WA 1431).** The beach access is located 8 km north and 22 km by road from Carnarvon. The gravel road negotiates board salt flats and low sandy ridges and a causeway to reach to long, low beach (Fig. 4.326). There is a car park and vehicle access to the beach, but no facilities. The beach faces due west and receives low waves averaging about 0.5 m. These break over a wide shallow continuous bar and sand flats. Rips are usually absent, except during higher waves. The beach initially trends north and is backed by a 10 km wide beach-foredune ridge plain consisting of a 5 km wide outer, 3 km wide depression and up to 3 km wide inner plain, all part of the delta complex. Waves average about 0.5 m at the creek mouth and gradually increase up the beach as it becomes more exposed to ocean swell refracting around Bernier Island, located 45 km due west of the southern end of the beach. As the waves pick up the initial 2 km wide delta shoals, diminish 9 km up the beach, with normal beach conditions along the northern half of the beach. The beach has finer sand and low tide terrace along the southern half grading into a coarser, steep reflective beach along the northern half backed by the transgressive dunes. Possibly the finer fraction has been blown out into the dunes, leaving the coarser fraction behind.

Figure 4.326 Miaboolya beach is backed by a 10 km wide foredune ridge plain, part of the Gascoyne River delta deposits.

As the beach trends to the north then northwest a transformation takes place in the beach and arrangement of the sediments. The wide beach-foredune ridge plain narrows to 1 km, 15 km up the beach and is replaced by vegetated north-trending longwalled parabolics extending 2-5 m inland and rising to 20-30 m. The parabolics occupy the remaining 14 km of the beach, including the active 200 ha Bejaling Sandpatch (Fig. 4.327), and continue on into beach WA 1432. Parts of the central section of dunes are transgressing onto the southern shores of Lake MacLeod, the largest coastal salt lake in Australia. There are additional access tracks to the beach across the foredune plain and though the dunes from the southern end of Lake Macleod.

Figure 4.327 Wave energy and dune activity increases along the northern section of Miaboolya beach, shown here at Bejaling Sandpatch.

Swimming: The lower waves of Shark Bay and wide low beaches at both Babbage Island and Miaboolya produce relatively safe swimming beaches. Both are better at high tide when the sand flats are covered.

Surfing: None apart form very low beach breaks.

GASCOYNE RIVER DELTA extends from the mangrove-lined shore of Mangrove Creek, located 8 km south of the river mouth, for 20 km to the north. It protrudes up to 10 km west into Shark Bay adjacent to the town, and tapers to the north as the sandy sediment is transported along Miaboolya Beach. In total the delta covers about 15 000 ha. Between Mangrove Creek and Miaboolya Beach wave energy increases and nearshore gradient steepen, resulting in a pronounced increase in wave height at the shore and a transformation in the shoreline from mangroves to sand flats to low and finally higher energy beaches. At the northern end of the delta the sand is blown off the beach and into the dunes that now block the southern end of Lake MacLeod.

WA 1432-1436 **POINT QUOBBA (S)**

No.	Beach	Rating	Type	Length
WA 1432	Pt Quobba (S4)	5	R+reefs	2.3 km
WA 1433	Pt Quobba (S3)	5	R+reefs	2.9 km
WA 1434	Pt Quobba (S2)	5	R+reefs	200 m
WA 1435	Pt Quobba (S1)	5	R+reefs	6.8 km
WA 1436	Pt Quobba	2	R+platform	800 m
Spring & neap tidal range = 0.9 & 0.5 m				

Point Quobba is a low calcarenite bluff and reef that marks an abrupt change in the shoreline from the sand beaches to the south to the beginning of a section of calcarenite bluffs that extend due north (Fig. 4.328). The point is accessible off the Blowholes Road, with only 4 WD tracks through the backing dunes to the southern beaches. Beaches WA 1432-1436 extend for 13 km to the southeast of the point to the boundary with Miaboolya Beach (WA 1431). All the beaches face southwest to west into the prevailing winds and waves. However waves are reduced by Bernier Island, then by the Darwin and Fitzroy reefs, and finally by inshore and shore attached reefs along most of the beaches. As a result breaker waves average about 1 m and reflective conditions predominate.

Figure 4.328 Point Quobba and the backing beach and camping area.

Beach **WA 1432** is separated from the longer beach WA 1431 by the first small outcrop of calcarenite on the eastern shore of Shark Bay. The beach trends to the northwest for 2.3 km to a 100 m wide reef-fringed embayment. Scattered inshore reefs extending up to 100 m offshore dominate the beach causing wave breaking over the reefs and inducing a series of reef-controlled rips. Beach **WA 1433** commences on the northern side of the embayment and continues northwest for 2.9 km, past a series of low calcarenite bluffs, to a 10 m high section of bluffs. The beach is paralleled by irregular reefs extending up to 200 m offshore resulting in low waves at the shore. Both beaches are backed by largely vegetated longwalled parabolics extending 2-3 km inland, with vehicle tracks through the dunes to both beaches.

Beach **WA 1434** is a 200 m long high tide beach, located below 10 m high bluffs. It is a steep reflective beach fronted by deeper reefs, resulting in lower waves at the shore, with the Fitzroy Reefs located 2 km to the west. Beach **WA 1435** commences on the northern side of the bluffs in lee of a prominent attached reef and continues to the northwest then north for 6.8 km. It is a crenulate beach with reefs widening and shoaling to the north resulting in a shallow, up to 500 m wide, reef-dominated surf zone. A 2-3 km wide zone of north-tending vegetated parabolic dunes backs both beaches, with 4 WD tracks through the dunes to access to both beaches.

Point Quobba beach (**WA 1436**) is located immediately south of a 5 m high calcarenite point and attached low island that comprises the point. A shallow 100-200 m wide platform extends between the point and shore, resulting in low waves and a steep beach. The 800 m long beach has built out up to 100 m across the platform forming a small vegetated sandplain. The Blowholes

Road terminates at the point and there is a basic camping and shack area that extends the length of the beach and sand plain. The area is popular with fishers and surfers and small fishing boats are usually moored off the beach. There is a *righthand surf break* that wraps around the point straight off the beach.

WA 1437-1440 **QUOBBA**

No.	Beach	Rating	Type	Length
WA 1437	Quobba	7	R+platform/reef	1 km
WA 1438	Quobba (N1)	7	R+platform/reef	400m
WA 1439	Quobba (N2)	7	R+platform/reef	500 m
WA 1440	Two Mile	7	R+platform/reef	400 m
Spring & neap tidal range = 0.9 & 0.5 m				

Quobba is located 10 km due north of Point Quobba and is the site of Quobba homestead. The homestead provides basic accommodation and camping facilities at its beachfront location. Between the point and homestead is a straight section of 10-20 m high calcarenite cliffs containing blowholes and wave deposited clifftop boulders. This is a popular area of clifftop rock fishing. However it is also a dangerous section of shore and signs warn of the threat of large waves while some lifebelts are provided. One kilometre south of the station is a memorial to HMAS Sydney, which was sunk west of here in 1942. Beaches WA 1437-1440 are four adjoining beaches that extend for 4 km due north from Quobba. A gravel road parallels the coast north from Point Quobba providing access to the beaches or the bluffs above the beaches.

Beach **WA 1437** is a straight 1 km long west-facing strip of steep high tide sand bordered by 10 m high calcarenite points. It is fronted by a 100 wide band of calcarenite platform degrading into reefs along the outer edge, with waves averaging about 1.5 m breaking on the outer reefs and providing some good surf (Fig. 4.329). A well vegetated foredune then the homestead back the southern part of the beach. Beach **WA 1438** commences 100 m to the north and is a similar narrow 400 m long high tide beach crossed by 50-70 m wide platform-reef. There is vehicle access to the crest of the backing 10 m high foredune.

Figure 4.329 Quobba beach and homestead.

Beach **WA 1439** is located on the northern side of a 50 m long low calcarenite point. It is a 500 m long high tide beach, with a slight central protrusion, also fronted by 70 m wide platform-reefs, with a left-hand break off the southern reefs. Beach **WA 1440** is located 400 m to the north, adjacent to Two Mile Well, and consists of a 400 m long strip of high tide sand and backing 10 m high foredune, with a 150 m wide platform paralleling the beach. The northern 100 m of beach is located behind a raised platform. There is vehicle access to both beaches with the gravel road paralleling the rear of the foredunes.

WA 1440-1446 **THREE MILE - PT CULVER**

No.	Beach	Rating	Type	Length
WA 1441	Three Mile	8	R+reef	100 m
WA 1442	Four Mile	7	R+platform/reef	200 m
WA 1443	Five Mile	7	R+platform/reef	100 m
WA 1444	Six Mile	7	R+platform/reef	200 m
WA 1445	Seven Mile	7	R+platform/reef	300 m
WA 1446	Nine Mile	6	R+platform	3.1 km
WA 1447	Pt Culver(S)	7	R+platform	2.7 km
Spring & neap tidal range = 0.9 & 0.5 m				

To the north of Two Mile Well the shoreline continues relatively straight and north for 16 km to the prominent Point Culver. The coast is dominated by calcarenite bluffs and cliffs averaging 20-30 m, and rising to 60 m at the point. The gravel road parallels the coast, moving 1 km inland north of Three Mile, with a 4 WD track following the blufftops. The road turns inland at Nine Mile Well and cuts across the rear of Point Culver. Point Culver is the site of a salt ship loading facility including a 300 m long jetty, with salt piled high on the headland. A private road runs out to the facility, however public access is restricted. Beaches WA 1141-1145 are located 3-7 miles respectively north of Quobba, while beaches WA 1446 & 1447 occupy the 8 km of coast south of the point, down to Nine Mile Well.

Beach **WA 1441** is located 400 m along the 4 WD coastal track, with the track skirting the crest of the 20 m high bluffs. The 100 m long beach is sited at the base of the bluffs with rock debris littering the rear of the beach, and no foredune. The only access to the beach is down the steep bluffs. An intertidal platform fronts the southern half of the beach, with a permanent rip existing through a break in the reefs off the northern half, while waves break on reefs up to 100 m offshore. Beach **WA 1142** is located 500 m to the north and is a 200 m long beach, backed by a foredune that has climbed the backing 10 m high bluffs and spread about 50 m inland. The vehicle track skirts around the rear of the dune, while the dune provides access to the beach, which is fronted by a continuous 70 m wide intertidal platform, with waves breaking heavily on the outer edge.

Beach **WA 1443** lies at the end of a straight east-west trending fence line and track and adjacent to the Five Mile mark. Twenty metre high eroding bluffs border and back the narrow 100 m long high tide sandy beach with rock debris littering the rear of the beach. It is fronted by a continuous 70 m wide platform with waves breaking heavily across the rocks. Beach **WA 1444** is located 1 km to the north and consists of a 200 m long high tide beach located at the base of 20 m high bluffs. A central 15 m high foredune has partially climbed the backing bluffs, covering some of the rock debris. A continuous 100 m wide platform fronts the beach, with waves breaking heavily on the outer edge.

Beach **WA 1445** begins the near continuous strip of sand that extends north for 8 km to Point Culver. The beach is 3.1 km long and consists of a high tide beach backed by a climbing foredune and in places transgressive dunes that have climbed the backing 60-70 m high slopes and blown 1-2 km to the north. The beach is crenulate along shore, lying to the lee of a continuous 100 m wide platform, with waves breaking on the outer edge, and a calmer 'lagoon' between the break and the shore. In places the beach is fringed by a low calcarenite bluff at the shore. A 200 m long section of low bluffs separates it from beach **WA 1446** which continues north for 2.7 km, terminating 1.5 km south of the point (Fig. 4.330). This is a similar beach, backed by slopes rising to 60 m, and fronted by a 50 m wide platform, with waves breaking over the edge and reaching the shore. A vehicle track winds down the bluffs and reaches the southern end of the beach, then continues north along the rear of the beach.

WA 1448-1451 **PT CULVER - 17 MILE WELL**

No.	Beach	Rating	Type	Length
WA 1447	Pt Culver	5	R	100 m
WA 1448	17 Mile Well	7	R+platform/reef	2.8 km
WA 1449	17 Mile Well (N1)	7	R+platform/reef	200m
WA 1450	17 Mile Well (N2)	7	R+platform/reef	2.2km
WA 1451	17 Mile Well (N3)	8	TBR	2.2 km
Spring & neap tidal range = 0.9 & 0.5 m				

Figure 4.330 Point Culver with the rock-dominated beach WA 1446 to the south.

Point Culver is the southern boundary of an exposed slightly curving 15 km long west-facing bay, which is fringed by 60 m high bluffs and cliffs its entire length. The southern half is devoid of beaches apart from the small beach WA 1447, while the northern half contains four beaches (WA 1448-1451) all accessible by 4 WD tracks.

Beach **WA 1447** is located at the base of 90 m high cliffs 2 km east of Point Culver. It is inaccessible by foot and consists of a steep northwest-facing reflective beach with waves averaging about 1 m. Higher waves wash across the beach to the backing steep cliffs, while cliff debris and talus litters the back of the beach. One kilometre north of the beach are the remains of the wreck of a salt ship, the *Korean Star*, blown onto the rocks during a cyclone in 1988. A vehicle track runs down the backing slopes to the wreck.

Beach **WA 1448** is located immediately west of 17 Mile Well in the centre of the embayment. It is a west-facing 2.8 km long high tide beach, fronted by low calcarenite bluffs toward each end and a central sandy section. The entire beach is paralleled by a 50 m wide calcarenite platform/reef. It is backed initially by 60 m high cliffs, while to the north an active sand ramp has climbed steep slopes and reached the top of the bluffs. A vehicle track runs down the backing slopes to provide access to the shore. Beach **WA 1449** is located just past the northern end of the beach, in a 200 m long gap in the low calcarenite shoreline bluffs. The sandy high tide beach is fronted by a 70 m wide surf zone dominated by calcarenite reefs.

Beach **WA 1450** commences 500 m to the north in lee of an inflection in the calcarenite. It then trends north for 2.2 km initially to the lee of a 200 m wide reef dominated surf zone, which narrows to 100 m in the north. It is backed by vegetated sand ramps that climb part way up the backing slopes. There is a *lefthand reef break* off the southern end of the beach. Beach **WA 1451** lies another 1 km to the north and is a 2.2 km long west-southwest-facing beach, that is relatively free of reefs and has a 100 m wide rip-dominated sandy surf zone, with waves averaging over 1.5 m. It is backed by partially active sand ramp that has climbed the backing 80-90 m high bluffs and transgressed a few hundred metres to the north (Fig. 2.24d), with older vegetated dunes extending up to 4 km north towards Red Bluff.

WA 1452-1460 **RED BLUFF**

No	Beach	Rating	Type	Length
WA 1452	Red Bluff	4→7	R→R+reef	2.5 km
WA 1453	Red Bluff (N1)	7	R+reef	2.6 km
WA 1454	Red Bluff (N2)	7	R+reef	500 m
WA 1455	Red Bluff (N3)	7	R+reef	2.6 km
WA 1456	Red Bluff (N4)	7	R+reef	1 km
WA 1457	Red Bluff (N4)	7	R+platform	300 m
WA 1458	Red Bluff (N5)	7	R+platform	100 m
WA 1459	Red Bluff (N6)	7	R+platform	150 m
WA 1460	Red Bluff (N7)	7	R+platform	300 m
Spring & neap tidal range =1.2 & 0.1				

Red Bluff is a world renowned surfing location with a classic long left-hand break wrapping along the northern side of the 100 m high red bluffs (Fig. 4.331). The break is a 1 km long walk or paddle from the beach (WA 1452) located to the lee of the headland. A vehicle track from the Blowhole Road reaches the beach and continues north to Three Mile Camp and Gnarloo Homestead. There are basic camping facilities at the bluff, with surfers camped on and behind the beach. There is a more formal camping area and store at Three Mile Well 20 km to the north adjacent to beach WA 1462. In between Red Bluff and Three Mile is a rocky shoreline dominated by calcarenite platforms and reefs, grading in the north to fringing coral reef, the most southern mainland coral reef in Western Australia. Eleven beaches (WA 1452-1462) are located in lee of the reefs, with waves usually breaking heavily on the outer edge and lower waves at the shore. Beach WA 1452-1459 occupy parts of the 15 km of coast north of Red Bluff.

Figure 4.331 Red Bluff with waves wrapping round the point and the beach and camping area behind.

Red Bluff beach (**WA 1452**) commences at the eastern base of the 60-100 m high limestone bluffs, and curves to the north for 2.6 km as a moderately protected, steep, reflective beach. Wave height increases to the north, as does the presence of attached calcarenite platforms and reefs, with waves breaking across the reefs. The beach is backed by a 5-10 m high foredune, then vegetated slopes rising to the 60 m high rim of the backing bluffs. The informal Red Bluff camping area and surfing community is scattered along the back of the foredune and backing slopes.

Beach **WA 1453** continues to the north in lee of reefs which widen to 300 m, forming a 'lagoon' between the outer edge of the reef and the beach. The beach is dominated by calcarenite including an intertidal and elevated 2 m rock platform. The 2.6 km long beach varies between sloping beachrock and sandy patches, all fronted by the continuous reef. It is backed by vegetated slopes rising to 30 m.

Beach **WA 1454** is a 500 m long relatively straight west-facing, high tide beach bordered by 20 m high bluffs to the south and lower sand draped-bluffs to the north. Waves break heavily on the outer edge of the reef up to 200 m offshore. A relatively calmer lagoon is located between the reefs and shore. Active dunes and deflation surfaces back the beach and link with dunes to the north to form an extensive transgressive dune field that extends for 10 km to Three Mile Well.

Beach **WA 1455** commences on the northern side of the rocks and continues north for 2.5 km. It has a shore parallel reef which commences about 200 m offshore, narrowing into a central gap in the reef which is the location of *Turtle* surf break (Fig. 4.332). The reefs produce a long left and shorter righthand break off either side of the gap, with deep water in between. North of the gap the reef narrows to about 100 m, with the beach terminating at a small, broken 20 high bluff. Beach **WA 1456** continues on the northern side of the bluff for 1 km to the next 20 m high bluffs. It is fronted by a reef, which narrows from 100 m in the south to 50 m in the north, permitting the broken waves to reach the shore. Both beaches are backed by a mixture of vegetated and reactivated dunes and deflation surfaces, criss-crossed by 4 WD tracks.

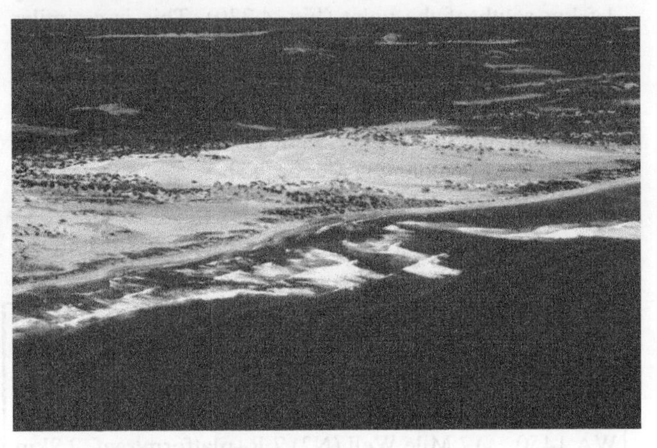

Figure 4.332 Beach WA 1455 is fringed by reefs, with the 'Turtle' surf break located in the gap in the reef (right).

Beach WA 1457-1460 are four pockets of sand located below and between 10-20 m high bluffs and spread along the next 3.5 km of shore. Beach **WA 1457** is a 300 m long strip of high tide sand located beneath 20 m high bluffs and backed by considerable rock debris. It is fronted by a 50-70 m wide platform, with waves breaking heavily on the outer edge, and some quieter tidal pools in the backing 'lagoon'. Beach **WA 1458** is a similar 100 m long strip of high tide sand backed by rocks and a 10 m high bluff, and fronted by a 70 m wide intertidal platform, with waves breaking on the outer edge, and some tidal pools close to shore. Beach **WA 1459** is located 200 m to the north and is a 150 m long high tide beach backed by rocks debris and a bluff. It has a 100 m wide platform off the southern end, narrowing towards the northern end of the beach. Beach **WA 1460** occupies the next small embayment. It faces west-southwest and is a 300 m long beach littered with rock debris and backed by an eroding 20 m high bluff. It is fronted by a continuous 100 m wide platform. A 4 WD track runs along the top of the buffs, with foot access to all four beaches down the steep bluffs.

THE CORAL COAST:
Three Mile Camp to North West Cape

Coast length 298 km (3970-4268 km)
Beaches 143 (WA 1460-1603)
Coral reefs: average 1.5 km offshore
(range 0.1-6 km)

Fringing and barrier coral reef systems begin to parallel the coast from about Three Mile Camp. They dominate much of the next 250 km of coast north to North West Cape, and include the well known Ningaloo Reef complex. The fringing reefs extends to the shore, while the barrier reefs have a distinct deeper lagoon between the reef and shoreline. However in reality variation in the distance of the reef off the shore (100 m to 6 km) results in a mixture of fringing and barrier reef sections along the coast, with an average distance offshore of 1.6 km.

The reefs have four major impacts on the backing shoreline and beaches:

1. It causes most ocean waves to break heavily over the reef edge resulting in a calmer lagoon to their lee and generally low ocean waves at the shore. The lagoon ranges up to 6 km in width.

2. Variation in the width, depth and morphology of the reef induces areas of greater and lesser wave breaking, resulting in some waves reaching the shore in lee of gaps in the reef.

3. The same variation in reef morphology induces wave refraction across and through the reef with the result that wave crests bend around the shoaler reefs. Over time this causes the shoreline to build more seaward in lee of the reefs and less so, or even erode in lee of the gaps, producing a crenulate shore, with salients in lee of reefs, and embayments in lee of gaps.

4. The reefs supply coral and algal carbonate detritus to the sandy lagoon floor and together with shell debris build the beaches and dunes.

A secondary impact of the salients, is to vary the orientation of the shore, which in turn impacts upon the receipt of onshore winds. The southern side of the salients tend to face more into the prevailing south winds and as a result are usually backed by both active and/or vegetated transgressive dunes.

WA 1461-1464 THREE MILE CAMP

No.	Beach	Rating	Type	Length
WA 1461	Three Mile Camp (S2)	3	R+coral reef	200 m
WA 1462	Three Mile Camp (S1)	4	R+coral reef	200 m
WA 1463	Three Mile Camp	3	R+coral reef	2 km
WA 1464	Three Mile Camp (N1)	7	R+coral reef	100m
Spring & neap tidal range = 1.2 & 0.1 m				

On the northern side of beach WA 1460, the shoreline turns and trends to the north-northeast for 3.5 km to Three Mile Camp. At the point the calcarenite bluffs are 20 m high, while to the north they decrease to 2 m at the camp. In between the shore is paralleled by fringing coral reefs extending from 100-200 m offshore. Waves break heavily along the reefs, with calmer lagoons in their lee and low waves at the shore. Just north of the point is a long lefthand break known as *The Point* and *Gnaraloo*. There is a car park on the backing bluffs with access down the bluff to the reef. Between the point and Three Mile Camp are three beaches (WA 1461-1463) with beach WA 1464 located north of the camp.

Beach **WA 1461** is located 1.5 km northeast of the point and consists of a 200 m long strip of high tide sand bordered and partly backed by a 2 m high bluffs, part of an elevated rock platform. The side of a partly vegetated parabolic dune backs the centre of the beach. The reef edge lies 150 m off the beach, with a lefthand break running along the reef (Fig. 4.333). A vehicle track reaches the bluffs above the southern end of the beach. Beach **WA 1462** lies 1 km further north along the low bluffed coast. It is a 200 m long strip of sand backed by a foredune that has blanketed the backing bluff. The reef edge lies 100 m off the beach, but is discontinuous to the north permitting waves averaging about 0.5 m to reach the shore. A vehicle track from the camp runs along the rear of the foredune.

Figure 4.333 View north along beaches WA 1461 to 1463 to Three Mile Camp. Note the reef and surf breaks.

Three Mile Camp is a camping reserve, which offers camping facilities and a store and fuel. The camp is located to the lee of a 2 m high calcarenite bluff, which dominates the shore to either side. The Three Mile Camp beach (**WA 1463**) commences immediately north of the camp, at the end of the bluffs. It trends north for 2 km as a slight protruding sandy beach located to the lee of a 100 m wide fringing coral reef which commences about 200 m up the beach, with a calmer lagoon between the reef edge and shore. There is a gap in the reef at the southern end of the beach, which permits low waves to reach the shore and is used to navigate small boats out through the reef. The beach is backed by an active sand sheet extending up to 300 m inland and reaching an inner

20 m high bluffline. The bluffline reaches the shore at the northern end of the beach and continues to the north. Beach **WA 1464** is located at the base of the bluff 1 km to the north. It is a narrow high tide beach, littered with cliff debris and partly fronted by a 2 m high bluff. The reef narrows to about 50 m along the bluffs and beach, with low waves reaching the shore.

WA 1465-1468 GNARALOO

No.	Beach	Rating	Type	Length
WA 1465	Gnaraloo	3	R+coral reef	3.4 km
WA 1466	Gnaraloo (N1)	3	R+coral reef	2 km
WA 1467	Gnaraloo (N2)	3	R+coral reef	2.6 km
WA 1468	Gnaraloo (N3)	3	R+coral reef	1.6 km
Spring & neap tidal range = 1.2 & 0.1 m				

Gnaraloo Homestead is a sheep station located 20 km north of Red Bluff. The homestead is sited on elevated ground 500 m east of the coast. The shoreline adjacent to the homestead trends north then northeast for 7 km to a low sandy foreland that forms the southern boundary of **Gnaraloo Bay**. In between are four sandy beaches (WA 1465-1468) all paralleled by 200 wide fringing coral reefs in the south widening to 500 m at the point. This is the southernmost and most poleward fringing reef in the world. Waves averaging about 1.5 m break heavily on the reef edge, while a relatively calm lagoon lies between the reef and shore with usually low waves to calm conditions at the shoreline (Fig. 4.334). All four beaches are accessible by 4 WD tracks.

Figure 4.334 Beaches WA 1465 and 1466 occupy the shore north of Gnaraloo.

Gnaraloo beach (**WA 1465**) commences 3 km south of the homestead at the northern end of 20 m high bluffs and trends to the north for 3.4 km to a small calcarenite-tied inflection in the shore. It is a slightly crenulate beach reflecting both irregularities in the reef, as well as outcrops of the low bluff along parts of the shore. Waves average about 0.5 m at the shore and surge up the moderately steep beach face. It is backed for the most part by a 10-15 m high foredune, then a mixture of vegetated and active dunes and deflation surfaces extending up to 300 m inland. The dunes widen and increase in activity to the north. The homestead backs the northern part of the beach and provides basic camping facilities.

Beach **WA 1466** commences in the inflection and trends due north for 2 km to the next inflection. A continuous 200 m wide reef system parallels the beach with a lagoon between the reef and shore, and low waves at the shore. The beach is backed by a foredune, with deflation surfaces in the south and two areas of active north-trending parabolics in the north.

Beach **WA 1467** is 2.6 km long and commences at the low sand inflection and trends more to the north-northeast as the reef moves up to 500 m offshore off the northern end. There is a wide lagoon between the reef and shore, with a central break in the reef producing a crenulation in the shore. It terminates at a curving sand foreland, with beach **WA 1468** continuing northeast for 1.6 km to the northern sandy foreland (Fig. 4.335). The reefs continue 400-500 m offshore, as does the wide calmer lagoon. There is a gap halfway along the reef mirrored by a curvature in the shore. Both beaches are backed by a 100-200 m wide vegetated foredune-dune system, and both have areas of stabilising parabolic dunes on their more southerly-facing shoreline. A landing ground is located to the east of beach WA 1467, on the flat plain between the foredune and inner bluffs.

Figure 4.335 View south of Granaloo Bay, with the world's southernmost fringing coral reefs paralleling the shore.

WA 1469-1474 GNARALOO BAY-CAPE FARQUHAR

No.	Beach	Rating	Type	Length
WA 1469	Gnaraloo Bay	3	R+fringing reef	7.2 km
WA 1470	Nine Mile Bore	3	R+fringing reef	3.5 km
WA 1471	Nine Mile Bore (N1)	3	R+fringing reef	2.8 km
WA 1472	Nine Mile Bore (N2)	3	R+fringing reef	1.8 km
WA 1473	Cape Farquhar (S)	3	R+fringing reef	3.5 km
WA 1474	Cape Farquhar (N)	3	R+fringing reef	3.5 km
WA 1475	Cape Farquhar (N1)	3	R+fringing reef	1.7 km
Spring & neap tidal range = 1.2 & 0.1 m				

Gnaraloo Bay commences at a 2 km long easterly inflection in the shore, 6 km north of Gnaraloo Homestead. The southern end of the bay marks the end of the gravel road, with 4 WD tracks to the north. Between the bay and Cape Farquhar, 17 km to the north, are seven reef-dominated beaches (WA 1469-1475), each paralleled by barrier reefs extending 500-1000 m offshore, with calmer lagoons to their lee and usually low waves at the shore. There are several *surfing breaks* located to either side of gaps in the reefs, with the southern side left-hand break usually providing the better break. The beaches are backed by a crenulate, 40 m high limestone bluffs, with the beaches usually building out from the bluffs, while in a few places they have eroded back exposing the bluffs to wave attack.

Gnaraloo Bay beach (**WA 1469**) commences in the southern arm of the bay, facing north. It soon curves round and trends north-northeast for 7.2 km to the next sandy foreland, 1 km southwest of Nine Mile Bore. The beach is paralleled by a near continuous barrier reef located 1.5 km off the southern corner of the bay, narrowing to several hundred metres along the central-northern section, with some breaks in the reef. Waves remain relatively low at the shore with steep, reflective conditions. A 200-300 m wide band of active dunes parallel the centre of the beach. There is vehicle access to the southern end of the beach, where there is a small car park and an area for launching boats off the beach.

Beach **WA 1470** commences on the northern side of the sandy foreland and protrudes to the north for 3.5 km trending northeast into the next embayment. The shape of the beach is controlled by a shore parallel reef located 400 m off the shore. The bulk of the beach protrudes to its lee, with a calm lagoon between the two, while the northern embayment lies to the lee of a 500 m wide gap in the reefs. An active sand sheet with several transverse dunes backs most of the beach, extending 200-300 m inland to the base of 40 m high bluffs. The embayment forms the boundary between beaches WA 1470 and 1471, with low refracted ocean waves reaching the shore. Beach **WA 1471** is a 2.8 km long beach that has prograded 200-300 m west as a series of crenulate foredune ridges, now experiencing some dune transgression. Behind the dune ridges, vegetated slopes rise to the crest of the 40 m high shore parallel bluffs. The reef lies 300-400 m off the beach, with parts extending to the shore. It terminates in the north where the backing bluffs reach the shore forming a 500 m long section of bluffed shoreline.

Beach **WA 1472** commences at the northern end of the bluffs and trends north, protruding in lee of the next section of reef. The 1.8 km long beach consists of the southern embayment and two protruding forelands and terminates at the tip of the northern foreland. The reef lies 400 m off the foreland, and up to 1.5 km off the northern embayment, which has formed in lee of a 700 m wide break in the reefs. Both forelands consist of a series of crenulate foredune ridges that have been subsequently transgressed, with some sand climbing up onto the backing 40 m high bluffs.

Beach **WA 1473** begins at the northern foreland and trends initially to the north-northeast into a 2 km wide embayment. The northern side of the embayment curves to the northwest, then north to terminate at the protruding, sandy Cape Farquhar, which has built out 1.2 km from the backing bluffline (Fig. 4.336). The southern half of the beach lies to the lee of a 700 m wide gap in the reef, with the beach reaching almost to the base of the bluffs. It widens to the cape as a foredune ridge plain, which has been transgressed by an active sand sheet, advancing northward across the rear of the cape. The reef narrows to 400 m off the tip of the cape. Beach **WA 1474** commences at the cape and curves to the northeast into the next embayment, which has eroded back to the bluffs, with the beach terminating at the beginning of a 300 m long section of bluffed shoreline. A low vegetated sand plain backs the northern side of the cape. There are a number of *surf breaks* on the reefs off the cape.

Figure 4.336 Cape Farquhar (left) is a prominent sandy foreland formed in lee of the fringing coral reef.

Beach **WA 1475** commences on the northern side of the eroding bluffs and forms the northern shore of the embayment. It is a 1.7 km long beach that terminates at a sandy foreland that has prograded up to 400 m out from the backing 30 m high bluffline, as a series of foredune ridges. These are being actively transgressed the length of the beach. The foreland lies to the lee of a reef system that extends 2 km offshore, and marks the end of the near continuous fringing reef that extends north from Gnaraloo Bay.

WA 1476-1482 **AMHERST & BULBARI POINTS**

No.	Beach	Rating	Type	Length
WA 1476	Lagoon Tank (S)	3	R+fringing reef	4.1 km
WA 1477	Lagoon Tank (N)	3	R+fringing reef	1.4 km
WA 1478	Amherst Pt (S)	3	R+fringing reef	2.3 km
WA 1479	Amherst Pt (N)	3	R+fringing reef	2.4 km
WA 1480	Bulbari Pt (N)	3	R+fringing reef	300 m
WA 1481	Bulbari Pt (N)	3	R+fringing reef	500 m
WA 1482	Bulbari Pt (N)	3	R+fringing reef	600 m
Spring & neap tidal range = 1.2 & 0.1 m				

At Cape Farquhar the shoreline turns and trends to the northeast for 20 km to Allison Point. Once past beach

WA 1475, the prominent fringing reefs cease and are replaced by deeper inshore reefs and in places shore parallel beachrock reefs close to shore. While the reef edge is absent, the shallow reef seafloor and northwest orientation of the coast results in lower ocean waves along the shore and reflective beach conditions. A vehicle track parallels the rear of the shore with access tracks off to most beaches. The limestone bluffline continues along the coast, in places fronted by sandy foreland, the bluffs exposed along a few shorter sections. Beaches WA 1476-1482 occupy the 13 km of northwest-facing shore either side of Amherst Point.

NINGALOO MARINE PARK

Established: 1987
Area: approx 400 000 ha
Coast length: 258 km (4018-4276 km)
Beaches: 138 (WA 1478-1606)
Coral reef: Fringing and barrier reefs located up to 6 km offshore

Ningaloo Marine Park encompasses a 258 km long section of coast from Amherst Point north to Point Murat. The park was gazetted in recognition of the outstanding fringing and barrier coral reef system that commences 40 km to the south at Three Mile Camp and continues all the way to the cape. The reef complex is a result of a combination of warm subtropical waters, an inshore platform to grow on, and warm-hot dry desert climate, which delivers negligible sediments to the shore. As a consequence the reef can grow right to the shore in places and one can literally drive right to the reef. In total it represents the world longest and most accessible fringing coral reef system.

The reef also exerts a dominating influence on the backing near continuous 138 beach systems. It causes breaking of the ocean wave over the reefs, producing in place excellent surf, but resulting in low wave to calm conditions at the shore. Variable wave refraction over and through the reef and gaps in the reef, induces shoreline crenulations, salients and embayments, which dominate the shoreline. Finally, it supplies carbonate sediments from the reef and lagoons to build the beaches and dunes, with the beach averaging 90 % carbonate (range 70-99 %).

Most of the shoreline is accessible by 4 WD, with a major tourist centre at Coral Bay, and accommodation at Warroora Homestead and 14 Mile campsite in the south, and a number of campsites between Yardi Creek and Neds Camp in the north, the latter within the Cape Range National Park.

Beach **WA 1476** commences at the tip of a low sandy foreland it shares with beach WA 1475. The tip is fringed by two arms of beachrock, which enclosed a small lagoon between the rocks and the shore, with a 50 m wide opening at the tip. The shoreline turns and begins it trend to the northeast, fringed by a reef dotted seafloor, without the prominent fringing reef. The beach is 4.1 km long and continues to the usually blocked Lagoon Tank creek mouth. It is backed by a series of up to 15 curving foredune ridges, 400 m wide in the south, narrowing to 100 m at the creek. The ridge lines indicate variation in the shoreline orientation through the Holocene, with some older lines extending seaward as beachrock reef midway along the beach. Shallow reefs and sandy seafloor extends 500 m offshore.

The creek impounds a small, narrow lagoon, which cuts through the backing Miocene limestone, the creek originating in the hinterland. Beach **WA 1477** commences on the northern side of the creek mouth and continues northeast for another 1.4 km. The backing bluffline approaches the coast to the north and forms a small point at the northern end of the beach, separating it from beach WA 1478. A shallow beachrock reef located about 200-300 m offshore parallels the entire beach, and together with the shallow reef and sand seafloor result in low waves at the shore.

Beach **WA 1478** continues northeast for 2.3 km to Amherst Point. The beachrock reef is attached or within 50 m of the shore, permitting waves averaging about 1 m to break over the reef and in places reach the shore, generating a few reef-controlled rips along the northern part of the beach. It is backed by a 50-100 m wide foredune ridge plain, then the 20 m high bluffs.

The bluffs emerge at the shore at Amherst Point, a 20 m high 100 m long bluffed section. Beach **WA 1479** commences on the northern side of the point and trends for 2.4 km to the north-northeast, then northeast as a convex sandy foreland formed in lee of scattered reefs extending up to 1 km offshore. The southern 1 km of beach is paralleled by a near continuous beachrock reef located 50 m offshore, with a lagoon between the reef and shore. Waves averaging about 1 m break over the reef and flow out as permanent rips though the gaps in the reef. Along the northern half the beachrock is attached to the shore, with wave breaking on the outer edge of the rocks. The beach is backed by a generally stable foredune ridge plain, which widens to 500 m in the centre, then the 15 m high bluffline. The beach terminates on the southern side of Bulbari Point, a 500 m long, 15 m high section of bluffs.

To the north of the point is a 1 km long curving embayment containing beaches WA 1480 and 1481, divided by a usually blocked creek mouth (Fig. 4.337). Beach **WA 1480** extends for 300 m between the southern point and the creek mouth. It is a steep reflective beach free of inshore rocks and reef, but with waves reduces by its northwest orientation and outer reefs. It is backed by a foredune, then the higher terrain. The creek meanders through the gentle backing slopes to the shore forming an elongate lagoon. Beach **WA 1481** continues on the northern side of the creek mouth for another 500 m to a 200 m long section of low eroding bluffs. It is backed by a foredune, then a deflated area with unstable dunes extending 200 m inland and up onto the backing bluffline.

Figure 4.337 Beaches WA 1480 and 1481 are located either side of a small creek just north of Bulbari Point (right).

Beach **WA 1482** is a 600 m long northwest-facing reflective beach bordered and backed by 10 m high calcarenite bluffs. Some calcarenite rocks and reef fringe part of the beach causing waves to break off the shore. It is backed by some dune instability that has blanketed the bluffs along the northern end of the beach.

WA 1483-1487 **BOOLBARLY WELL**

No.	Beach	Rating	Type	Length
WA 1483	Boolbarly Well	5	R+rocks	900 m
WA 1484	Boolbarly Well (N 1)	5	R+rocks	100 m
WA 1485	Boolbarly Well (N 2)	5	R+rocks	200 m
WA 1486	Boolbarly Well (N 3)	5	R+rocks	100 m
WA 1487	Boolbarly Well (N 4)	5	R+rocks	150 m
Spring & neap tidal range = 1.2 & 0.1 m				

Boolbarly Well is located at the mouth of a usually dry upland creek system, the creek blocked by beach **WA 1483**. The beach starts on the northern side of the sand draped-bluffs that separate it from beach WA 1482 and extends for 900 m to the north. The beach is backed and interrupted by an irregular outcrop of calcarenite, which forms bluffs at the northern end. It is backed by a reactivated older parabolic dune that parallels the beach. The southern end also blocks the Boolbarly Well drainage system, which only breaks out across the beach following rare heavy rain. A vehicle track terminates at the dry creek mouth.

The northern boundary bluffs continue for another 700 m. Located along the base of the bluffs are beaches WA 1484-1487. Beach WA **1484** is a 100 m long strip of sand backed by lower dune-draped bluffs, and fronted by shallow inshore reefs. Beach **WA 1485** is a similar 200 m long beach, with a 2 m platform outcropping along the northern end. Beaches WA 1486 and 1487 are backed by steep 15 m high bluffs. Beach **WA 1486** is a 100 m long strip of sand backed by the bluffs and fronted by continuous shallow inshore reefs. Beach **WA 1487** continues for 150 m to the northern end of the bluffs terminating at a small calcarenite point. Waves break over shallow reefs up to 100 m off the northern end of the beach.

WA 1488-1493 **ALISON PT-WARROORA-MAGIES REEF**

No.	Beach	Rating	Type	Length
WA 1488	Upper Boolbarly Well	3	R+fringing reef	1.8 km
WA 1489	Alison Pt	3	R+fringing reef	3.4 km
WA 1490	Warroora	3	R+fringing reef	3.3 km
WA 1491	Stevensons Well	3	R+fringing reef	3.8 km
WA 1492	Magies Reef	3	R+fringing reef	2.3 km
WA 1493	Magies Reef (N)	3	R+fringing reef	3.7 km
Spring & neap tidal range = 1.2 & 0.1 m				

Alison Point is a prominent sandy foreland that marks the southern boundary of a near continuous 45 km long section of fringing and barrier coral reefs, which extends north to Coral Bay and Point Maud. The reefs continue to induce wave attenuation and refraction leading to prominent shoreline salients to the lee of the reefs and embayments to the lee of gaps in the reefs, as well as supplying carbonate sand to the beaches and dunes. Because of the reefs most of the beaches receives low to no ocean waves at the shore. The first 13 km for shoreline north of the point contains six beaches (WA 1488-1493) and is backed by Warroora Homestead. All the beaches are accessible via 4 WD tracks that parallel the coast and cross the dunes where necessary to reach the shore.

Alison Point is bounded by the southern beach WA 1488 and the northern beach WA 1489. Beach **WA 1488** is backed by Upper Boolbarly Well. It commences at the end of the bluff-dominated beaches WA 1484-1487, and trends north as a sandy foreland terminating at a small gap-induced embayment 500 m south of the point. Fringing reefs parallel the beach 300 m offshore, with a lagoon between the reefs and shore. The southern half of the foreland is being actively transgressed by a series of transverse dunes, which along the eastern edge have climbed the boundary bluffline. They are overlapping an earlier transgressive system.

Beach **WA 1489** commences on the northern side of the gap and continues to the point where the foreland is 600 m wide and then curves to the northeast then east to the northern end of the foreland, where it narrows to 100 m. It has a total length of 3.4 km. On the northern side of the 200 m wide gap in the reefs is a long righthand *surf break*. The reef and lagoon continue around the point, terminating at the beginning of the boundary embayment. The beach is backed by a 100 m wide band of active dunes, then vegetated foredune ridges, which are being encroached on by the southern dunes.

Beach **WA 1490** is located 1.5 km due west of Warroora Homestead. It commences at the southern end of the embayment, formed by the break in the reef and continues north for 1.8 km as an open, more exposed beach, free of fringing reef. Wave height increases up the beach, breaking across some shallow inshore reefs and maintaining a 100 m wide surf zone, with two permanent rips along the northern few hundred metres of beach. It is backed by a 200 m wide stable foredune ridge plain in the south, grading into active dunes to the north.

Beach **WA 1491** commences at the northern end of the embayment and the beginning of a continuous fringing reef. The reef commences 200 m offshore, gradually widening to 300 m at the northern end of the 3.8 km long beach. A continuous lagoon runs between the reef and shore, with breaks in the reef towards either end inducing two small embayments, as well as *surfing breaks* either side of the northern gap. The beach is backed by a 100-300 m wide foredune ridge plan, which is being transgressed by dunes along the central Stevensons Well section. It terminates at a small pointed foreland, 200 m north of the **Tropic of Capricorn** (Fig. 4.338).

Region 5D CARNARVON: Warroora – North West Cape

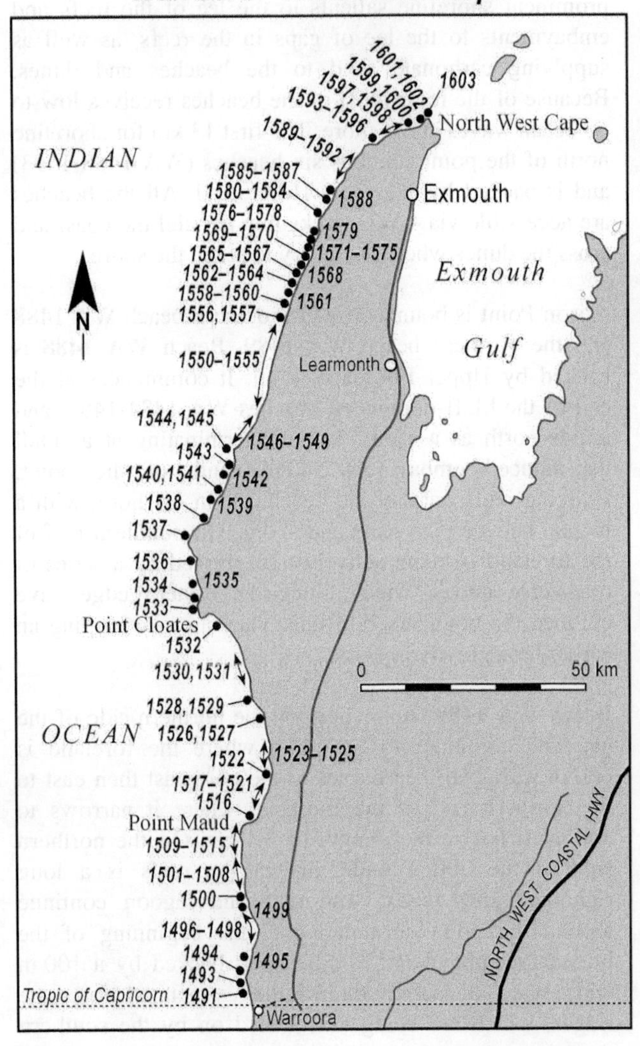

Figure 4.338 Map of region 5D extends from the Tropic of Capricorn to North West Cape; beaches WA 1491-1603.

Beach **WA 1492** commences on the northern side of the foreland and continues to trends north for 2.3 km to a 10 m high calcarenite bluff. The reef off the beach is degraded to the south allowing some waves to reach the shore. The reef becomes more continuous along the northern half with a lagoon extending to the bluffs. Generally vegetated parabolic dunes extending a few hundred metres to the north back the beach.

Beach **WA 1493** is a 3.7 km beach that commences on the northern side of the 200 m long bluffs, with a vehicle track reaching the southern corner of the beach. It lies to the lee of Magies Reef that widens from 500 m in the south to 1 km in the north, forming a broad lagoon system and low energy beach. Out on the widest reef section is a 100 m wide gap with a *righthand break* along the northern side. There are two gaps in the reef backed by a gently curving continuous embayment (Fig. 4.339). The beach is backed by vegetated nested parabolics extending up to 1.5 km inland and north to the lee of the northern beaches. A narrow series of foredunes has prograded out from the parabolics and back the present beach.

WA 1494-1500 PELICAN PT-14 MILE CAMP

No.	Beach	Rating	Type	Length
WA 1494	Pelican Pt	3	R+fringing reef	2 km
WA 1495	Pelican Pt (N)	3	R+fringing reef	6.5 km
WA 1496	Pelican Pt (N 1)	3	R+barrier reef	1.7 km
WA 1497	14 Mile Camp (S2)	3	R+barrier reef	100 m
WA 1498	14 Mile Camp (S1)	3	R+barrier reef	300 m
WA 1499	14 Mile Camp	3	R+barrier reef	2.7 km
WA 1500	14 Mile Camp (N)	3	R+barrier reef	1.6 km
Spring & neap tidal range = 1.2 & 0.1 m				

Figure 4.339 Maggies Reef extends up to 1 km offshore with the beach (WA 1493) curving behind.

Pelican Point is a prominent dune-capped sandy foreland that forms the southern boundary of an open 11 km wide embayment, all fringed by near continuous coral reefs located 1-1.5 km offshore. The embayment terminates in the north at 14 Mile Camp, with a 1 km wide break in the reef located off the centre of the embayment. Beaches WA 1494-1500 are located between the point and the camp, with 4 WD tracks providing access to all beaches.

Pelican Point beach (**WA 1494**) occupies the 2 km long west-facing outer side of the point, which is formed to the lee of a continuous reef located 500 m off the beach. Between the reef and shore is a shallow sand and reef

-filled lagoon. Vegetated parabolics that have originated from beach WA 1493 to the south, cover the backing foreland for up to 1.8 km to the east. A vehicle track runs along the shore to the northern tip of the point.

Beach **WA 1495** commences at the point and curves to the east then north for 6.5 km, terminating at a small sandy foreland 1 km south of the next gap in the reef. The entire beach is paralleled by a continuous barrier reef located 1.5 km offshore, with a wide shallow lagoon between the reef and shore (Fig. 2.21b). The beach is backed by a 200-300 m wide series of foredune ridges, which widen to 400 m at the northern foreland located to the lee of a wider section of reef and has been partly formed by wave refraction through a 1.5 km wide gap in the reef.

The wide gap permits refracted waves to reach the shore where they average up to 1 m in height. As a consequence the shoreline has been eroded back to a 1 km long central section of 10 m high calcarenite bluffs. Beach WA 1496 is located between the foreland and the bluffs, while beaches WA 1497 and 1498 occupy two small indentations in the bluffs. Beach **WA 1496** commences on the southern side of foreland and curves gently to the north for 1.7 km terminating at the buffs, the backing foredune ridges narrow to the bluffs. The beach faces out into South Passage and wave height increases from 0.5 m in the south to 1 m in the north, where they break over inshore calcarenite reefs.

Beach **WA 1497** is located 300 m along the bluffs, past a protruding section of calcarenite. It is a 100 m long strip of sand wedged between the base of the bluffs and inshore reefs, extending 100 m off the beach, with waves breaking over the reefs. Beach **WA 1498** lies 500 m to the north and is a similar 300 m long strip of sand , bordered and backed by the bluffs and fronted by reefs widening to 150 m. The bluffed section is backed by a vegetated and some active parabolic dunes extending a few hundred metres inland.

On the northern side of the bluffs the shore becomes increasingly protected by the continuation of the reefs located 1.5 km offshore. As a consequence a sandy foreland widens to the north. Beach **WA 1499** commences at the bluffs and continues north for 2.7 km to the tip of the 500 m wide foreland, known as 14 Mile Camp. The foreland consists of a well preserved sequence of foredune ridges, which splay to the north and are backed by a 20 m high line of bluffs, with older vegetated parabolics on top of the bluffs extending up to 1 km to the north.

Beach **WA 1500** begins at the foreland and trends essentially due north for 1.6 km to the next section of calcarenite bluffs. The beach forms the northern side of the foreland formed in lee of the reef located 1 km offshore. The northern end of the reef forms the southern boundary of Turtlet Bay, with refracted waves moving through the break to align the foreland. As a consequence wave height also increases towards the northern end, where they break over a 50 m wide band of inshore reefs.

WA 1501-1505 **TURTLET AT-PT ANDERSON**

No.	Beach	Rating	Type	Length
WA 1501	Turtlet Bay	6	R+barrier reef	500 m
WA 1502	Turtlet Bay (N1)	6	R+barrier reef	200 m
WA 1503	Turtlet Bay (N2)	6	R+barrier reef	700 m
WA 1504	Pt Anderson (S1)	3	R+barrier reef	1.2 km
WA 1505	Pt Anderson (S)	3	R+barrier reef	500 m
Spring & neap tidal range = 1.2 & 0.1 m				

Turtlet Bay occupies a 2 km wide embayment formed in lee of a 1.5 km break in the reefs, which are located 1.5 km offshore. The higher refracted waves moving through the gap and into the centre of the bay have eroded the shore back to 10-20 m high calcarenite bluffs. Beaches WA 1501-1503 are located in indentations along a 3 km long section of the bluffs. Active and vegetated parabolic dunes back the beaches and bluffs extending up to 1 km inland and rising to 64 m at Round Headed Hill.

Beach **WA 1501** is located directly east of the gap in the reefs permitting it to receive waves averaging about 1 m. These break across a 150 m wide area of inshore reefs which front the narrow 500 m long beach. Ten metre high bluffs border the beach, with a 2 m bluff outcropping along the shore, while predominately active parabolics back the beach.

Beach **WA 1502** is a 200 m long strip of sand backed by 5 m high bluffs, then a mixture of active and vegetated parabolics extending 1 km inland. It receives waves averaging less then 1 m, which break across a 50 m wide inner reef zone. Beach **WA 1503** commences 100 m to the north and is a 700 m long narrow beach, also backed by continuous 5 m high bluffs, then predominately vegetated parabolic dunes. Inshore reefs front the beach with waves maintaining a 50 m wide surf zone.

Beach **WA 1504** commences at the northern end of the bluffs. It lies to the lee of a reef system located 1.5 km off the beach and extends north for 3.5 km past Point Anderson. The reef shelters the beach from ocean waves with only low waves at the shore. It is a relatively straight 1.2 km long north-trending sandy beach, which terminates at a small outcrop of calcarenite. It is backed by a foredune, then semi-stable parabolics extending up to 3 km north to 50 m high Yalobia Hill. Beach **WA 1505** continues on the northern side of the outcrop for another 500 m to the tip of Point Anderson, a sandy foreland formed to the lee of the offshore reef system. Older beachrock reefs extend southwest of the point (Fig. 4.340), attesting to erosion of a formerly wider foredune-ridged point. The point is now covered with transgressive dunes.

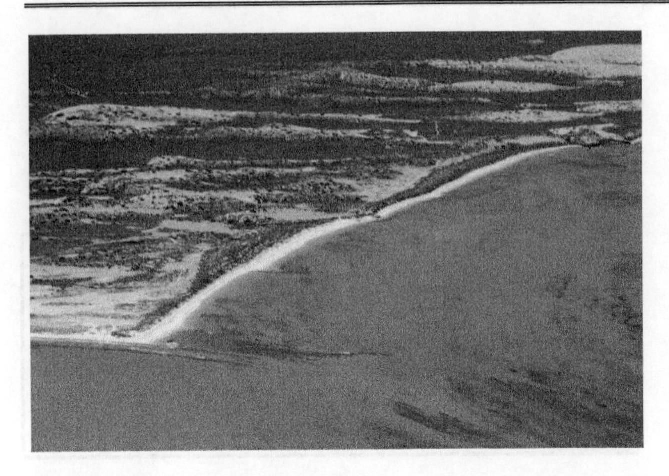

Figure 4.340 Anderson Point (left) is tied to beachrock reefs, with sandy beaches to either side.

WA 1506-1509 SOUTH PASSAGE-FIVE FINGER REEF

No.	Beach	Rating	Type	Length
WA 1506	Pt Anderson (N)	3	R+barrier reef	1.6 km
WA 1507	South Passage (S)	3	R	500 m
WA 1508	Bifocal Beach	3	R+barrier reef	1.7 km
WA 1509	Five Finger Reef	3	R+beachrock reef	300 m
Spring & neap tidal range = 1.2 & 0.1 m				

Point Anderson marks the southern boundary of a curving 4 km wide embayment that is bounded in the north by Five Finger Reef, with beaches WA 1506-1509 in between. The embayment is located to the lee of **South Passage**, a 700 m wide gap in the otherwise continuous reefs that lie 1 km off each point. The sandy forelands have built out to either end, with an eroding calcarenite bluff section in the centre. The entire shore is backed by predominately vegetated nested parabolic dunes rising to 55 m at Anchor Hill and extending 2 km in from Point Anderson and 1 km in from the rest of the embayment. Four wheel drive tracks wind through the dunes to reach all the beaches.

Beach **WA 1506** commences at the tip of Point Anderson and trends to the north-northeast for 1.6 km to the beginning of the 2 km long central bluffed section of the embayment. The beach receives low waves and is fronted by a wide sandy lagoon grading into reef flats towards the fringing reef. A north-trending active parabolic parallels the rear of the beach.

Beach **WA 1507** is a 500 m long beach wedged in along the base of the 10 m high bluffs, and fronted by inshore reefs, then the lagoon floor. It faces due west into the relatively narrow South Passage, as a result waves remain low at the shore. It is backed by largely vegetated parabolics. **Bifocal Beach (WA 1508)** commences on the northern side of the bluffs and curves to the north then northwest, terminating on the southern side of Five Finger Reef, which forms the tip of the northern foreland. The beach is backed by a foredune ridge plain, 200 m wide in the centre, then older vegetated parabolic dunes. The wide sand and patchy reef of the lagoon extends up to 1.5 km west to the main reef.

Five Finger Reef is formed by a series of truncated beachrock ridges, that one formed the northern side of the foreland. Erosion and reorientation of the shoreline has exposed the reef, which extend up to 100 m off the beach. They tend to break the 300 m long beach **(WA 1509)** into a series of small sandy compartments, each bordered by one of the fingers of beachrock. The beach is largely backed by an area of dune activity, extending 500 m to the north, then older vegetated parabolics, criss-crossed by 4 WD tracks.

WA 1510-1513 FALSE PASSAGE-BOGGY HILL

No.	Beach	Rating	Type	Length
WA 1510	False Passage (S)	3	R+barrier reef	1.6 km
WA 1511	False Passage (N)	3	R+barrier reef	200 m
WA 1512	View Rock	3	R+barrier reef	400 m
WA 1513	Boggy Hill	3	R+barrier reef	1 km
Spring & neap tidal range = 1.2 & 0.1 m				

False Passage is an irregular 200 m wide gap in the reef that leads to a shallow 2.5 km long embayment bordered by Five Finger Reef in the south and View Rock to the north. A sandy foreland is located to the lee of Five Finger Reef, with dune-capped calcarenite bluffs dominating the central-northern section of the embayment. Four beaches (WA 1510-1513) are located between Five Finger Reef and Boggy Hill 4 km to the north. All are backed by a 1 km wide band of vegetated, north-trending nested parabolics, and all are accessible by 4 WD tracks through the dunes.

Beach **WA 1510** trends north of Five Finger Reef for 1.6 km to the beginning of the 10 m high calcarenite bluffs. The coral reef lies 1 km off the southern end, widening to 1.5 km to the north, with a shallow lagoon in between and low waves at the shore. It is backed by the active dune that has emanated from Five Finger Reef beach (WA 1509).

Beach **WA 1511** is a 200 m long pocket of sand located between two 10-20 m high calcarenite bluffs and in line with the narrow False Passage, 2 km to the west. Waves remain low at the shore, with a sandy lagoon between the passage and beach. Beach **WA 1512** is a 400 m long strip of sand located towards the northern side of the bluffed section. It is backed by irregular 5-10 m high bluffs, capped by vegetated dunes rising to 31 m at View Hill. It is fronted by a shallow reef, then the sandy lagoon floor, with the fringing reef located 2 km offshore.

Beach **WA 1513** commences at the northern end of the bluffs and curves to the north-northwest, then northeast to the southern boundary of Coral Bay. The protruding, crenulate beach is fringed by a 50 m wide band of intertidal and then subtidal reef flats extending 1 km across the lagoon, with the main reef edge 1.5 km seaward. It is backed by vegetated dunes rising to 20 m

WA 1514-1515 **CORAL BAY**

No.	Beach	Rating	Type	Length
WA 1514	Coral Bay	3	R+barrier reef	1.4 km
WA 1515	Skeleton Bay	3	R+barrier reef	1.8 km
Spring & neap tidal range = 1.2 & 0.1 m				

Coral Bay, as the name implies is a shallow bay dominated by subtidal coral reefs. These are fringing reefs, which grow right to the shore. The small holiday settlement of about 200 people contains two caravan parks, a lodge, resort and store, as well as providing charter boats and diving equipment hire for viewing, diving and fishing the coral reefs. The reefs are part of the southern extension of the Ningaloo Reefs, the best and most extensive fringing coral reefs in Australia. Coral Bay lies 1020 km north of Perth and is 80 km off the highway, off the Exmouth road. However it is well worth a visit to see both the coral reef and many beautiful protected beaches that lie inside the reef.

Coral Bay itself is part of a 2.2 km wide west-facing embayment bordered by Boggy Hill to the south and Point Maud to the north (Fig. 4.341) and shares the embayment with the northern Skeleton Bay (Fig. 4.342). It has formed to the lee of 2 km wide gap in the reef, with a relatively shallow reef-filled lagoon maintaining low waves at the shore. Coral Bay beach (**WA 1514**) curves around the southern half of the embayment for 1.4 km. It faces north along the main public section, which is backed by the main accommodation areas and facilities, thereby providing shelter from the prevailing southerly winds. It then trends to the north along the main west-facing section, which is backed by a small cluster of houses, then generally vegetated nested-parabolic dunes. The southern corner of the beach has a small jetty and boat launching area and is the centre of boating activity in the area and usually busy with charter and recreational boating. However away from the boats, the beach is relatively steep and usually calm.

Skeleton Bay (**WA 1515**) occupies the northern half of the bay. It is a curving 1.8 km long beach, which initially tends north then curves to the west along the southern side of the prominent Point Maud. This beach is backed by a vegetated parabolic dune field and totally undeveloped.

Swimming: The near continuous coral reefs located 1 to 2 km offshore permit only low ocean waves reach the beach and calms are common. The beach at Coral Bay and others nearby tend to be relatively narrow and steep, with the water off the beaches sometimes containing small patches of coral.

Surfing: None at the shore, however if you have a boat you can explore the breaks over the outer coral reefs, particularly round the openings, or passages through the reefs.

Fishing: Coral Bay is a major centre for both viewing and fishing the reefs and a number of charter boats operate out of the bay. You can also fish from the steep beaches or the intervening rocky sections.

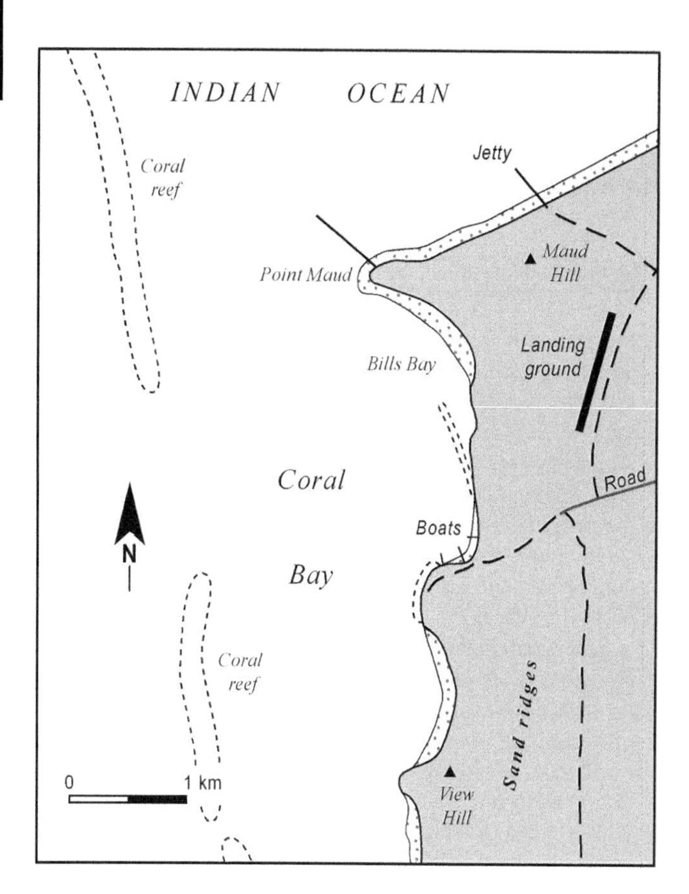

Figure 4.341 Coral Bay is located to the lee of fringing and barrier reefs, with the prominent Point Maud at its northern boundary.

Figure 4.342 Point Maud, Skeleton and Coral bays are sheltered by a coral barrier reef.

WA 1516-1523 MAUDS LANDING-DOUBLET HILL

No.	Beach	Rating	Type	Length
WA 1516	Mauds Landing	3	R+coral reef	8.5 km
WA 1517	Oyster Bridge	4	R+beachrock reef	150 m
WA 1581	False Teeth Reef	4	R+beachrock reef	1 km
WA 1519	The Lagoon	4	R+beachrock reef	600 m
WA 1520	Crayfish Rock	4	R+beachrock reef	300 m
WA 1521	Bolman Hill	4	R+beachrock reef	150 m
WA 1522	Gooseneck Hill	3	R+barrier reef	4.5 km
WA 1523	Doublet Hill	3	R+barrier reef	3 km
Spring & neap tidal range = 1.2 & 0.1 m				

Point Maud is a prominent sandy foreland formed to the lee of a fringing reef located 2 km off the tip of the point, and which continues north for 4 km. The northern end of the reef marks the beginning a 6 km wide gap in the reefs, which recommence 4 km offshore in line with Bolman Hill. Wave refraction through the gap has contributed to the progradation of Point Maud as a 1 km wide foredune ridge plain backed by a 1-2 km wide back barrier depression occupies by a 500 ha salt lake. The ridges narrow to the north with calcarenite bluffs and beachrock exposed between Oyster Bridge and Bolam Hill, where the shore faces more into the gap and waves are up to 1 m in height. North of Bolman Hill the shore is again protected by a continuous reef and trends north then northwest to the next smaller foreland at Bruboodjoo Point. Between Point Maud and Bruboodjoo Point is a 16 km wide west-facing embayment, bordered by the reefs to either end with the central 6 km wide gap. In between are eight near continuous beach systems (WA 1516-1523). All the beaches are accessible by 4 WD tracks that tend to follow the rear of the foredune or transgressive dunes, with tracks then crossing the dunes to the shore.

Point Maud beach (**WA 1516**) commences at the tip of the point and gently curves to the northeast then north for 8.5 km, terminating in the lee of a 200 m long beachrock reef at Oyster Bridge. For the most part the beach is sheltered by the offshore reef and its northwest orientation, with waves averaging less than 0.5 m and reflective conditions prevailing, together with sand waves along the southern lowest energy section. Wave height begins to pick up to over 0.5 m towards the northern end, with a 50 m wide lagoon formed between the northern reef and shore. The beach is backed by transgressive dunes originating from Coral Bay along the southern 1.5 km, then the foredune ridge plain and salt lake, with the old Mauds Landing (Fig. 4.343) located at the beginning of the ridges and Cardabia Homestead located on the ridges 1 km in from the centre of the beach. The foredunes continue to 1 km south of Oyster Bridge where active and vegetated parabolics dominate the next 3 km to Bolman Hill.

Oyster Bridge and False Teeth Reef are part of the shore-parallel beachrock reef that commences at the northern end of beach WA 1516 and continues north for 1.5 km to The Lagoon. **Oyster Bridge** beach (**WA 1517**) is a 150 m long beach located between a 50 wide gap in the reef, which forms a small open embayment. It is backed by a 20 m high foredune, then the transgressive dunes from the northern end of Point Maud beach. The beachrock reef continues north as **False Teeth Reef** and is backed by 1 km long beach **WA 1518**. The beach and reef alternate in dominance resulting in a beach of variable width and continuity, in places fronted by the beachrock while open to waves elsewhere. It is backed by vegetated transgressive dunes extending 500 m inland, with vehicle tracks across the dunes to the beach.

Figure 4.343 Mauds Landing is backed by vegetated transgressive dunes, with low a low energy beach grading into transverse sand waves.

The Lagoon is located at the northern end of the beachrock reef, which has produced a curving 600 m long beach (**WA 1519**) on its northern side. At the southern end it forms at very sheltered lagoon in lee of the reef, then as it trends north is fronted by discontinuous beachrock reef located 50-100 m offshore, with waves breaking over the reefs. It is backed by 20 m high vegetated dunes, with vehicle tracks to either end of the beach. **Crayfish Rock** (**WA 1520**) occupies a similar curving embayment 500 m to the north. It is sheltered by a section of beachrock reef on the south, with inter to subtidal reef filling much of the 300 m long embayment, resulting in wave breaking over the reefs and low waves at the shore. The beach is narrow, backed by 20 m high dunes and curves around to trend north, terminating against calcarenite bluffs. Beach **WA 1521** is located 200 m to the north and is a 150 m long strip of high tide sand backed by scarped 20 m high bluffs. It is fronted by shallow sand and reef seafloor with a discontinuous beachrock reef located 200 m offshore and patchy coral reefs beyond. It is backed by the 200 m wide northern end of the transgressive dunes, which rise to 31 m at Bolman Hill.

To the north of Bolman Hill wave height decreases owing to increasing protection from the offshore reefs. Beach **WA 1522** commences on the northern side of the hill and trends due north for 4.5 km to a small sandy foreland in front of Gooseneck Hill. This is a narrow eroding beach that is cutting into the backing 20-30 m high dunes. It is fronted by shallow sand and reef flats with the edge of the reefs up to 6 km offshore. It is backed by a 200-300 m wide band of vegetated north-trending parabolic dunes. The sandy foreland at Gooseneck Hill separates it from 3 km long beach **WA 1523** which continues to the north before curving to the northwest to terminate at Bruboodjoo Point, a sandy foreland that protrudes 1 km into the lagoon. The fronting sand and reef flats and backing dunes continue all the way to the point and include 21 m high Doublet Hill. The main vehicle track follows the rear of the dunes, with tracks across the dunes to the beaches.

WA 1524-1529 NINE MILE CAMP-TWIN HILL

No.	Beach	Rating	Type	Length
WA 1524	Nine Mile Camp	3	R+barrier reef	1.3 km
WA 1525	Nine Mile Camp (N)	3	R+barrier reef	400 m
WA 1516	Lund Hill	3	R+barrier reef	3.1 km
WA 1527	Twin Hill (S)	3	R+barrier reef	3 km
WA 1528	Twin Hill (S)	4	R+rocks/barrier reef	200 m
WA 1529	Twin Hill (S)	4	R+rocks/barrier reef	200 m
Spring & neap tidal range = 1.2 & 0.1 m				

Bruboodjoo Point is a 1 km wide sandy foreland that marks the southern boundary of a 32 km wide open west to west-southwest-facing embayment, bordered in the north by the large, prominent Point Cloates. Barrier reefs parallels the coast between 3-7 km offshore, with a northern 4 km wide gap called Black Rock Passage. The reefs lower ocean waves at the shore to usually 0.5 m or less resulting in reflective beach conditions. The 36 km of shoreline is occupied by nine beaches (WA 1524-1532), all fronted by a generally wide relatively shallow sand and reef filled lagoon. The coast track parallels the rear of the dunes behind the beaches, with numerous tracks across the dunes to the shore. The first 12 km of shore contains six beaches (WA 1524-1529).

Beach **WA 1524** is located on the northern side of Bruboodjoo Point and is known as **Nine Mile Camp** (Fig. 4.344). The 1.3 km long low energy beach initially faces north then curves round to face west. It is backed by a degraded low foredune, then 100-200 m wide salt flats, backed by 20 m high dunes from beach WA 1523. This is a reasonably popular campsite with campers usually nestled in lee of the dunes. They use the beach for launching small fishing boats. The beach terminates at the start of a 20 km long section of alternative calcarenite bluffs and beaches.

Beach **WA 1525** is located 200 m to the north at the start of the 15 m high bluffs, and is a 400 m long strip of high tide sand backed by dune-draped bluffs rising to 15-20 m and in two places protruding onto the beach. It is fronted by shallow reef flats, with sandy seafloor 300 m off the beach. The coast track parallels the rear of the dunes 200 m inland. Beach **WA 1526** commences 100 m to the north and is a 3.1 km long crenulate, north-trending low energy beach that terminates at the next set of bluffs. It widens from 100 m in the south to 300 m in the north and is paralleled by 300 m wide shallow reefs and backed by vegetated dunes that rise to 20 m and includes Lund Hill. Two vehicle tracks reach the beach. A 2.5 km long section of dune-capped calcarenite separates it from beach WA 1527.

Figure 4.344 Bruboodjoo Point, with Nine Mile Camp to the left.

Beach **WA 1527** commences on the northern side of 20 m high dune-capped bluffs. It initially trends north then turns and heads to the north-northwest, exposing the shore to the prevailing southerly winds. As a consequence, vegetated and some active parabolics back the beach and extend up to 1.5 km to the north, reaching heights of 29 m at Twin Hill. Two vehicle tracks cross the dunes to reach the central section of the 3 km long beach. It is fronted by shallow sand and reef lagoon floor that extends 4 km seaward to the reef system.

Beaches **WA 1528** and **1529** are neighbouring 200 m long beaches located 1 km north of Twin Hill, where the shoreline turns and again trends to the north. They are both bordered and backed by 10-20 m high dune-capped calcarenite bluffs and fronted by a mixture of calcarenite and coral reefs and patches of sandy seafloor. A winding vehicle track crosses the backing 500 m wide dunes to reach to top of the bluffs above both beaches.

WA 1530-1532 NORTH HILL-PT CLOATES

No.	Beach	Rating	Type	Length
WA 1530	North Hill	3	R+barrier reef	3.3 km
WA 1531	Carter Hill	3	R+barrier reef	2.3 km
WA 1532	PT Cloates	3	R+barrier reef	15 km
Spring & neap tidal range = 1.2 & 0.0 m				

North Hill is a 28 m high dune located approximately midway along the embayment, the embayment trending north-northwest and finally west for 22 km to terminate at Point Cloates. In between are three near continuous longer beaches, all backed by transgressive dunes which increase in size to the north as the coast swings to face into the prevailing southerly winds. The coast track follows the rear of the dunes with several access tracks though to the beaches.

Beach **WA 1530** commences at the northern end of a 2 km long section of calcarenite and 1 km south of North Hill. It trends north for 3.3 km to a 100 m long section of bluffed calcarenite called Carter Hill. The beach is paralleled by a 2 km wide channel, then 3 km of reef flats that extend west to the reef edge. It is backed by vegetated longwalled parabolics extending from 500 m in the south to 1 km inland in the north. The southern end of the beach is accessible via the 500 m long North Hill track.

Beach WA **1531** extends north-northwest from the Carter Hill bluffs for 2.3 km to the next section of bluffs at Daly Hill. It is backed by an active parabolic in the south, which impinges on the access track, and by three larger vegetated parabolics to the north, the largest 1.5 km in length. It is fronted by the channel and then the reef flats.

Beach **WA 1532** is one of the longer beaches along the Ningaloo coast. It starts on the northern side of the 500 m long Daly Hill bluffs and gently curves to the north-northwest, then northwest and finally west-southwest for the final 3 km to Point Cloates, where it terminates (Fig. 4.345). For much of its 15 km it is protected by the extensive barrier reefs located 5-6 km offshore, with waves at the shore averaging 0.3 m or less. The northernmost west-trending section of the beach, called Jane Bay, faces south into Black Rock Passage and does receive higher waves, which as a consequence have deepened the lagoon off this section. Most of the beach is however dominated by a relatively shallow sand and reef-filled lagoon and reflective beach conditions. The entire beach is backed by a mixture of active and vegetated parabolic dunes. They begin as a 1 km wide band in the south where they reach 58 m at Camp Hill, widening to as much as 3 km in the area around Entrance Hill where they reach 74 m, and extend 3.5 km across rear of Point Cloates where they reach 36 m. Because of the dunes access is limited to the northern Four Nile Tanks area, where the dunes narrow to 500 m.

Figure 4.345 Point Cloates is located to the lee of a continuous fringing reef.

WA 1533-1537 **PT CLOATES-PT BILLIE**

No.	Beach	Rating	Type	Length
WA 1533	Pr Cloates	3	R+fringing reef	1.5 km
WA 1534	Ningaloo	3	R+barrier reef	4.5 km
WA 1535	Beacon Pt (S)	3	R+barrier reef	5.7 km
WA 1536	Norwegian Bay	3	R+barrier reef	6.3 km
WA 1537	Pt Billie	3	R+fringing reef	4.2 km
Spring & neap tidal range = 1.2 & 0.0 m				

Point Cloates and Point Edgar-Point Billie are the two most prominent sandy forelands on the Ningaloo coast and protrude 4 km west to form the most western section of the shore and fringing reefs. Between the two points is a 15 km long crenulate west-facing embayment containing a central foreland, Beacon Point, formed to the lee of low sandy Frazer Island. Between the two points is 21 km of continuous low energy sandy shore, containing five reef-protected beaches. The reef lies 1 km off each point widening to 5 km offshore in the centre of the embayment off Beacon Point and Norwegian Bay. The shoreline is backed by Ningaloo Homestead in the south and the ruins of the Norwegian Bay whaling station in the north. Massive north-trending parabolic dunes cross Point Cloates in the south and Beacon Point and Point Edgar-Billie in the central-north. A gravel road connects the Exmouth Road with Ningaloo, while the coast track following the rear of the dunes, and occasional vehicle tracks cross the dunes to reach the shore.

Point Cloates beach (**WA 1533**) commences at the southern tip of the point and trends to the northwest, then north for 1.5 km, terminating in a small embayment 700 m west of the Point Cloates Light. Continuous fringing reefs lie 1 km off the point affording protection to the beach and backing lagoon, with low waves at the shore. The beach is backed by active and vegetated transgressive dunes that have partly buried the foredune ridges that built the foreland.

Ningaloo beach (**WA 1534**) commences in the southern corner of the embayment and trends north as a 4.5 km long convex beach, with Ningaloo Homestead located 200 m in from the western apex of the beach. It then continues north-northeast to the next small embayment at the Ningaloo shearing sheds. Continuous reefs front the beach and extend 5 km seaward of the northern end. The beach is backed by vegetated parabolic dunes, with the abandoned Point Cloates Lighthouse on the crest of a 40 m high dune. The homestead is located in a deflation basin between two parabolic dunes.

Beach **WA 1535** is a 5.7 km long gently curving north to north-northwest trending beach that starts at the shearing sheds and terminates at **Beacon Point**, the tip of the next foreland. The beach is very sheltered in the south where lighters used to moor to transport the wool. Near continuous barrier reefs lies 4-5 km off the beach, with one 500 m wide gap. Waves remain low at the shore, which is backed by active and vegetated parabolic dunes, which commence as soon as the shoreline turns to face slightly south. The north-trending dunes parallel the beach, extending up to 2 km in from the point and several kilometres to the north. The gravel road terminates at the shearing sheds, with the 4 WD coast track continuing to the north along the rear of the dunes.

Norwegian Bay is a curving 5 km wide bay bordered by Beacon Point and the northern Edgar Point. It contains a low energy 6.3 km long beach (**WA 1536**), with the ruins of the old whaling station located in the northern corner of the bay. Barrier reefs lie 4.5 km off Beacon Point narrowing to 1 km wide fringing reef off Point Edgar, with a 1 km wide gap in the reefs off the middle of the bay. Dunes from Beacon Point parallel the central-southern half of the beach, with dunes actively transgressing north behind the southwest-facing section of the beach (Fig. 4.346). These dunes have transgressed up to 1.5 km inland and are backed by older vegetated dunes that extend right across the 4.5 km wide foreland. The transgressive dunes are both exhuming and burying prior foredune ridges that built the foreland.

Figure 4.346 Edgar Point forms the northern boundary of Norwegian Bay, with active dunes transgressing north over the former foredune ridges.

At **Point Edgar** the shoreline turns and trends north-northeast to **Point Billie** and onto the next inflection in the shore, forming a 4.2 km long west-facing beach (**WA 1537**). The reef fringes the shore just 500 m off the beach, with low waves and reflective conditions prevailing at the shore. The beach is backed by the active dunes along its southern half extending 3 km to the east, with vegetated dunes to the north narrowing to 1.5 km. A 4 WD track crosses the northern dunes to reach the northern tip of the beach.

WA 1538-1539 **LEFROY BAY**

No.	Beach	Rating	Type	Length
WA 1538	Lefroy Bay (S)	3	R+barrier reef	2 km
WA 1539	Lefroy Bay	3	R+barrier reef	7.5 km
Spring & neap tidal range = 1.2 & 0.0 m				

Lefroy Bay is a 6 km wide northwest-facing sandy embayment bordered to the south by the northern side of the Billie Point foreland and to the north by the tapered V-shaped, Winderabandi Point foreland. Near continuous fringing reefs lies 1 km off each point and up to 3 km off the centre of the embayment. Two beaches (WA 1538 & 1539) occupy the 9.5 km of beach shore.

Beach **WA 1538** is a 2 km long northwest-facing slightly curving section of sandy shore that forms the northern shore of the 4 km long Point Edgar-Point Billie foreland. It is sheltered by barrier reefs, which extend 1 km off the southern end and 2 km off the north end, with a shallow sandy-reef filled lagoon in between. The beach is backed by a foredune ridge plain, which widens to 400 m in the centre and contains up to 8 ridges, then the rear of the vegetated transgressive dunes that have originated from Norwegian Bay to the south. It is separated from the main bay beach by a convex 300 m long low sandy point, fringed by reef and beachrock.

Lefroy Bay beach (**WA 1539**) commences on the northern side of the small point and trends to the northeast for 4 km before curving round to the northwest to terminate at the narrow tip of low sandy Winderabandi Point (Fig. 4.347). The low energy beach is backed by a narrow 50-100 wide crenulate strip of foredune, then the bedrock slopes that gently rise to the southern slopes of the Cape Range 500-1000 m to the east. The coast track parallels the rear of the beach, with several access tracks through to the shore. There is a camping reserve towards the southern end of the beach, which is usually occupied by a few campers, particularly during the winter months. A 2.5 km wide shallow lagoon fills most of the bay, with the reef beyond.

Figure 4.347 Winderabandi Point is a prominent sandy foreland that has built 1.5 km west in lee of continuous fringing reefs.

WA 1540-1545 WINDERBANDI-SANDY PTS

No.	Beach	Rating	Type	Length
WA 1540	Winderbandi Pt (N)	3	R+barrier reef	2.4 km
WA 1541	Winderbandi Pt (N1)	3	R+beachrock/barrier reef	700 m
WA 1542	Winderbandi Pt (N2)	3	R+beachrock/barrier reef	3.2 km
WA 1543	Sandy Pt (S 3)	3	R+beachrock/barrier reef	3 km
WA 1544	Sandy Pt (S 2)	3	R+fringing reef	200 m
WA 1545	Sandy Pt (S 1)	3	R+beachrock/fringing reef	300 m
Spring & neap tidal range = 1.2 & 0.0 m				

Winderbandi Point is a low sandy foreland, 1.5 km wide at its base, which protrudes 1.5 km to the west narrowing to a 100 m wide tip. It forms the northern boundary of Lefroy Bay and the southern boundary of the next open 1 km long north-northwest-facing embayment that terminates at Sandy Point in the north. The bay has 11.5 km of shoreline and contains six low energy beaches (WA 1540-1545), all fronted by a shallow sand and reef-filled lagoon that widens to 3 km along the central section of the bay. The beaches are backed by a narrow foredune plain and in places have eroded back to beachrock. The foot slopes of the Cape Range commence 200 m east of the southern half of the beach, widening to 2 km in the north. The 4 WD coast track parallels the foot slopes, with tracks off to most of the beaches.

Beach **WA 1540** occupies the northern, north-facing shoreline of Winderbandi Point. It is 2.4 km long, curving round to trend north where the point rejoins the base of the foothills. It is a low energy reflective beach fronted by shallow sand and patchy reef. It is backed by the foredune ridges of the point, which have been partly transgressed from the south by now vegetated parabolic dunes. The northern end of the beach lies close to the coastal track, with an informal camping area behind the southern corner of the beach. It terminates in the north at the beginning of a 400 m long section of beachrock.

Beach **WA 1541** occupies a slight 700 m long embayment located between two protruding sections of beachrock. It is fronted by both beachrock reefs, as well as shallow sandy and reef seafloor. A 50 m wide band of foredune, then the coast track parallels the rear of the beach. Beach **WA 1542** occupies the next 3.2 km long section of sandy shore, with beachrock forming the boundaries. It is a relatively straight, though crenulate beach, backed by a 50-100 m wide foredune, and fronted by shallow waters of the lagoon, with low waves breaking over patchy reefs that extend off the beach.

Beach **WA 1543** occupies the northern section of the bay shore. It is a relatively straight 3 km long beach backed by a 50-200 m wide foredune belt, with one central 1 km area of dune instability that parallels the beach. Shallow reefs dominate the 3 km wide lagoon, while the coast track and foothills increase in distance from the beach to the north.

Beaches WA 1544 and 1545 are located on the southern side of Sandy Point. Beach **WA 1544** is a narrow, 200 m long southwest-facing strip of high tide sand, backed by eroding 20 m high dunes, and fronted by the shallow lagoon floor. Beach **WA 1545** lies immediately to the west and is a similar more crenulate 300 m long beach backed by lower dunes with some beachrock outcropping along the shore. The dunes are part of vegetated dunes that have transgressed up to 1 km across Sandy Point. A vehicle track lies 200 m east of beach WA 1544, with no vehicle access through the dunes to either beach.

CAPE RANGE NATIONAL PARK

Established: 1964
Area: 50 581 ha
Coast length: 93 km (4183-4276 km)
Beaches: 56 (WA 1550-1606)

Cape Range National Park incorporates a substantial section of the Cape Range, which forms the backbone of the Exmouth Peninsula, together with the adjoining western shoreline. The range is composed of 100-300 m high Mesozoic limestone that has been deeply dissected by a series of gullies, some of which have been drowned at the coast forming winding cliff-lined creeks. The range commences as a gentle slopes to the lee of Point Cloates-Ningaloo, then to the east of Norwegian Bay forms a well defined north-south line of foothills rising to a dissected plateau which gradually increases in height to the north. The range parallels the remainder of the coast to Valming Head where it intersects the shore, a total distance of 105 km. The 93 km of coastline within the national park contains 56 beaches all fringed by the Ningaloo Reef system, and part of the Ningaloo Marine Park. The gravel Yardie Creek Road provides good access down to Yardie Creek, with several campsites located at some of the beaches. There is a visitor centre at Milyering.

WA 1546-1552 SANDY PT-BOAT HARBOUR

No.	Beach	Rating	Type	Length
WA 1546	Sandy Pt (N1)	3	R+beachrock/fringing reef	1.8 km
WA 1547	Sandy Pt (N2)	3	R+beachrock/fringing reef	500 m
WA 1548	Sandy Pt (N3)	3	R+fringing reef	800m
WA 1549	Sandy Pt (N4)	3	R+fringing reef	300 m
WA 1550	Sandy Pt (N5)	3	R+beachrock/barrier reef	3 km
WA 1551	Sandy Pt (N6)	3	R+beachrock/barrier reef	600 m
WA 1552	Boat Harbour	3	R+barrier reef	2 km
Spring & neap tidal range = 1.2 & 0.0 m				

The shoreline to the north of Sandy Point is paralleled by a 1.5 km wide lagoon then barrier reef, which widens to 2.5 km in the north. The reef terminates 6 km north of the point at a 4 km wide break in the reef, to the lee of which is an area of beachrock-dominated shoreline, called Boat Harbour. Between the point and the Boat Harbour area is a crenulate 9 km long section of northeast–trending coast containing seven low energy reflective beaches (WA 1546-1552). The coast is paralleled by the Ningaloo-Yardie track lying between 200-1500 m inland, with tracks off to all the beaches. The southern boundary of Cape Range National Park crosses the centre of beach WA 1550.

Beach **WA 1546** extends north of the tip of Sandy Point for 1.8 km to where the shoreline is bisected by an oblique beachrock reef (Fig. 4.348). The slightly crenulate beach is paralleled by shoreline beachrock reefs, then the 1.5 km wide shallow lagoon and outer reefs, resulting in low waves at the shore. It is backed by transgressive dunes from the southern side of the point, with a 4 WD track running straight across the dunes to the centre of the beach. Beach **WA 1547** commences in lee of the beachrock reef, which extends for 300 m along the beach. The rear of the beach is an informal camping area and the low dunes are crisscrossed by tracks and campsites. A second beachrock reef forms the northern boundary, the two reefs enclosing a lagoon, with a 100 m wide central opening. Beach **WA 1548** commences where the beachrock attaches to the shore and parallels the beachrock reef for 800 m, until the beachrock and low bluffs replace the beach for the next 600 m. It is backed by low vegetated dunes.

Figure 4.348 Sandy Point (foreground) with beaches WA 1546 & 1547 extending north in lee of fringing reefs.

Beach **WA 1549** forms the southern side of a small 300 m wide sandy foreland. The beach starts at the end of the beachrock and trends due north for 300 m to the tip of the foreland, with a shallow sand spit extending another 200 m into the 2.5 km wide lagoon. A degraded foredune ridge plain backs the beach. Beach **WA 1550** starts at the tip of the foreland and trends to the east, then northeast and finally north for 3 km, forming and open west-facing embayment, fronted by the 3 km wide lagoon and the reefs. While sandy forelands form either end of the beach, the central section has eroded back to beachrock, which outcrops along several hundred metres of the beach.

Beach **WA 1551** forms the northern side of the northern foreland. It occupies the truncated end of the foreland, with the original shoreline marked by two converging beachrock reefs, which continue out from the ends of the adjoining beaches (WA 1550 & 1552). The beach is sheltered by the southern reef and curves for 600 m between the two boundary reefs with beachrock also paralleling the shore. There is an open sandy section towards the northern end of the beach, behind which is an informal camping area.

Beach **WA 1552** starts where the beachrock reef joins the shore and trends to the north-northeast for 2 km to the beginning of a 1.5 km long section of beachrock located in the centre of the Boat Harbour area. The beach is backed by a 15-20 m high foredune, which continues north to Yardie Creek, with the Ningaloo-Yardie track paralleling the shore, then the foothills of the Cape Range 500 m to the east. The northern half of the beach faces out into the 2 km wide Boat Harbour channel. Wave height increases to the lee of the channel, with more hazardous conditions along the northern beachrock section.

WA 1553-1557 **YARDIE CREEK**

No.	Beach	Rating	Type	Length
WA 1553	Yardie Creek (S)	3	R+barrier reef	700 m
WA 1554	Yardie Creek (N)	3	R+barrier reef	3.7 km
WA 1555	Yardie Creek (N1)	3	R+barrier reef	3.7 km
WA 1556	Osprey Bay (S2)	3	R+beachrock/barrier reef	200 m
WA 1557	Osprey Bay (S1)	3	R+beachrock/barrier reef	1.7 km
Spring & neap tidal range = 1.2 & 0.0 m				

Yardie Creek flows through a drowned gully cut in the range limestone. The lower reaches of the creek are permanently flooded and usually impounded at the shore by a steep gravelly berm (Fig. 4.349). The creek is a popular tourist destination, which can be reached by 2 WD vehicles along the gravel Yardie Road.

Figure 4.349 Yardie Creek emerges from the Cape Range to reach the sea at a usually blocked mouth.

Beach **WA 1553** extends south of creek mouth for 700 m, to the northern end of the eroded beachrock section. It is paralleled by an intertidal beachrock reef, then the 2.5 km wide lagoon, which extends out to a broken section of the fringing reef. There are a range of *surfing breaks* out in the gaps in the reef, and which require a boat to reach. A degraded foredune area backs the beach, then the Ningaloo track and the lower slopes of the range. Formal campsites are located on either side of the creek mouth and 1 km south behind Boat Harbour.

Beach **WA 1554** extends north of the 100 m wide creek mouth for 3.7 km. It trends north-northeast as a crenulate sandy beach, with the crenulations induced by both patches of beachrock and fringing reef and a 500 m wide gap in the reefs. A foredune backs the southern end and widens to the north where it becomes unstable with a 1 km long section of shore-parallel active dunes extending north behind beach WA 1555. The gravel Yardie Creek road parallels the rear of the dunes.

Beach **WA 1555** commences at the sandy foreland and curves gently to the north for another 3.7 km to terminates at a small beachrock-fringed eroded foreland. It is backed by the active dunes in the south, then stable low foredunes along the central section, then another area of dune activity as the northern end of the beach trends slightly northwest to the next foreland. Reefs are located 2 km offshore off each end, with a central 500 m wide break in the reef off the narrow centre of the beach.

Beach **WA 1556** is a 200 m long pocket of sand located on the eroded tip of the foreland, with protruding beachrock reefs on either side enclosing a small 50 m wide lagoon. It is backed by a 300 m wide band of vegetated and active dunes originating on the southern side of the foreland, then the Yardie Creek road.

Beach **WA 1557** extends north of the foreland for 1.7 km to the beginning of a 500 m long section of low beachrock bluffs, with the Osprey Bay camp site located on the bluffs. The north-trending crenulate beach is backed by a 100 m wide zone of semi-stable foredune, then vegetated dunes, with the road 500 m inland. It has outcrops of beachrock along the beach, which induces crenulations, then sand flats, a shore parallel tidal channel, with the reef flats extending 1.3 km out to the reefs.

WA 1558-1560 **OPSREY & SANDY BAYS**

No.	Beach	Rating	Type	Length
WA 1558	Opsrey Bay (S)	3	R+beachrock/barrier reef	80 m
WA 1559	Opsrey Bay (N)	3	R+barrier reef	400 m
WA 1560	Sandy Bay	3	R+beachrock/fringing reef	2 km

Spring & neap tidal range = 1.0 & 0.4 m

Osprey and Sandy Bays are two adjoining embayments located either side of a small sandy foreland, with a central salt lake. The foreland has a 600 m long base and protrudes 500 m into the lagoon, with the fringing reef paralleling the shore 1 km offshore. Osprey Bay is located to the lee of a prominent 600 m long gap in the reefs (Fig. 4.350), with waves breaking heavily on the outer reef and two *surfing breaks* running along the reef either side of the gap. Waves however remain low at the shore.

The embayment to the lee of the gap has been eroded back to beachrock, with a 1 km long section exposed between beaches WA 1557 and 1558, with the Osprey Bay campsite located to the lee of the rocks. Beach **WA 1558** is an 80 m long pocket of sand with low beachrock bluffs to either side. Beach **WA 1559** lies 100 m to the north at the end of the beachrock and forms the southern side of the low sandy foreland. It trends to the northwest for 400 m to the tip of the foreland. Both beaches are fronted by shallow sand and reef flats, a narrow channel, and then the outer reef flats.

Figure 4.350 Osprey Bay is located in lee of a major gap in the fringing reef.

Sandy Bay beach (**WA 1560**) occupies the northern side of the foreland and curves to the northeast then trends north for 2 km to the next subdued foreland. The northern section is crenulate with some beachrock outcrops along the shore. The Sandy Bay picnic area is located behind the beach at the northern base of the foreland.

WA 1561-1564 **PILGONAMAN BAY**

No.	Beach	Rating	Type	Length
WA 1561	Pilgonaman Bay (S2)	3	R+beachrock/fringing reef	2 km
WA 1562	Pilgonaman Bay (S1)	3	R+beachrock/fringing reef	900 m
WA 1563	Pilgonaman Bay	3	R+beachrock/fringing reef	60 m
WA 1564	Pilgonaman Bay (N)	3	R+beachrock/fringing reef	300 m

Spring & neap tidal range = 1.0 & 0.4 m

Pilgonaman Bay is located at the mouth of the usually dry Pilgonaman Creek and located to the lee of a 200 m wide gap in the reefs, which lie 1 km offshore (Fig. 1.20). Beaches WA 1561-1564 are located either side of the creek mouth. There are two surfing breaks along either side of the gap. A 300 m long gravel track runs out to the creek mouth where there is a campsite and toilets, while the Yardie Creek road parallels the shore 200-300 m inland.

Beach **WA 1561** commences on the northern side of the subdued foreland it shares with the northern end of Sandy Bay beach (WA 1560) and continues north for another 2 km to the next pointed sandy foreland, which protrudes 200 m into the lagoon. Continues fringing reefs 1 km offshore parallel the beach, which has usually low waves and reflective conditions. There is no formal vehicle access to the beach.

Beach **WA 1562** commences at the tip of the foreland and curves to the north for 900 m terminating at an outcrop of beachrock on the southern side of the 100 m wide creek mouth. It is backed by the low foredunes of the foreland, which narrow to the north. Beach **WA 1563** occupies the creek mouth and consists of a 60 m long strip of coarse sand wedged between two beachrock outcrops and backed by the dry creek bed. The access track runs into the creek bed and the mouth is used to launch small boats.

Beach **WA 1564** extends north of the creek mouth for 300 m to the beginning of a section of low beachrock bluffs. The beach, which faces west-southwest, is backed by an unstable foredune, which has transgressed about 100 m to the north. A mixture of patchy reef and sand fill the lagoon off all the beaches.

Figure 4.351 Mandu Mandu Creek reaches the sea is lee of a gap in the reefs formed by the drowned creek valley.

WA 1565-1570 **BLOODWOOD CK-MANDU-TURQUOISE BAY**

No.	Beach	Rating	Type	Length
WA 1565	Bloodwood Ck (S)	3	R+fringing reef	150 m
WA 1566	Bloodwood Ck	3	R+fringing reef	200 m
WA 1567	Mandu Mandu	3	R+fringing reef	1.8 km
WA 1568	Oyster Stacks	3	R+fringing reef	900 m
WA 1569	Turquoise Bay (S)	3	R+fringing reef	1 km
WA 1570	Turquoise Bay	3	R+fringing reef	600 m
Spring & neap tidal range = 1.0 & 0.4 m				

To the north of Pilgonaman Creek the reef averages 1 km or less in width, with shallow reef flats extending right to the shore for the next 11 km to Turquoise Bay. While waves breaks over the reefs and remain low at the shore, much of the shoreline has been eroded back to beachrock and/low calcarenite bluffs. There are six beaches along this section (WA 1565-1570), which occupy only 4.6 km of the shore and include the boulder beaches at the mouths of Bloodwood and Mandu Mandu creeks. The Yardie Creek road parallels the rear of the beaches providing vehicle access to most of the beaches. There are campsites at Mandu Mandu, while Turquoise Bay is a popular picnic and snorkeling area.

Beach **WA 1565** is located 300 m south of Bloodwood Creek mouth and consists of a narrow 150 m long strip of high tide sand bordered by low calcarenite bluffs. Beach **WA 1566** is located across the mouth of the usually dry and blocked creek and consists of a 200 m long strip of coarse sand and rocks, backed by a low foredune, and bordered by low calcarenite bluffs. Shallow reefs flats extend 1 km off both beaches to the reef edge. The low bluffs extend 1 km to the north to the beginning of beach WA 1567.

Beach **WA 1567** continues north for 1.8 km to the 300 m wide blocked mouth of Mandu Mandu Creek (Fig. 4.351). The beach extends north as a gently curving sandy foreland backed by a 100-200 m wide series of low foredune ridges. At the creek mouth the sand is replaced by a steep cobble and boulder beach, the material delivered by the creek during infrequent rain events. The fringing reef deepens off the creek mouth with low waves reaching the shore. Camping areas are located either side of the creek mouth, which is also used for launching small boats out through the slightly deeper channel through the reef.

Beach **WA 1568** continues north of the creek mouth for 900 m to the next section of low calcarenite bluffs, which begins at the Oyster Stacks. The beach faces west across a 300 m wide shallow fringing reef. It can be accessed in the south at Mandu Mandu North and along the Oyster Stacks track, the latter a popular snorkeling area located where the reef flats narrow to 300 m.

Turquoise Bay lies to the lee of a prominent sandy foreland that has prograded 600 m seaward to the lee of 1 km wide section of fringing reef and its adjacent 200 m wide channel (Fig. 4.352). Turquoise Beach (**WA 1569**) occupies the southern side of the foreland. It is a west-facing 1 km long sandy beach fronted by a sandy bay floor grading into the shallow reef flats. It is backed by the low foredune ridges of the foreland.

Turquoise Bay beach (**WA 1570**) occupies a curving 600 m long section on the northern side of the foreland. It faces north across the shallow sandy inner bay floor, then the deeper water of the channel further out. The channel leads to a left-hand *surf break* on the southern side of the gap in the reef. Low surging ocean waves also increase in height along the northern section of the beach, which is bordered to the north by low dune-capped calcarenite bluffs. A gravel road runs out to the bay beach, with a sandy 4 WD track to the foreland beach.

Figure 4.352 Turquoise Bay is located to the lee of a sandy foreland and opens into a deep channel through the reef, with a left break on the reef.

WA 1571-1579 **TUKI BEACH-NEDS CAMP**

No.	Beach	Rating	Type	Length
WA 1571	Tuki Beach	3	R+fringing reef	800 m
WA 1572	Tuki Beach (N)	3	R+fringing reef	1.2 km
WA 1573	Channel	3	R+fringing reef	100 m
WA 1574	Camp 13	3	R+fringing reef	200 m
WA 1575	Camp 14	3	R+fringing reef	1.9 km
WA 1576	Sth T Bone Bay	3	R+fringing reef	1.2 km
WA 1577	T Bone Bay	3	R+barrier reef	2.1 km
WA 1578	Mesa Camp	3	R+barrier reef	600 m
WA 1579	Neds Camp	3	R+barrier reef	3 km
Spring & neap tidal range = 1.0 & 0.4 m				

Tuki Beach is located on the south side of the mouth of an unnamed dry creek, 2 km north of Turquoise Bay. The coast continues relatively straight to the north-northeast, past Tuki Beach, for 12 km to the prominent Low Point. The shoreline is initially fringed by coral reefs widening from 1 km off Tuki to 2 km off T Bone Bay, north of which a channel runs close to shore and they continue as barrier reefs located 2-2.5 km offshore. In between Tuki and Low Point are nine beaches (WA 1571-1579), which occupy most of the low shoreline. The Yardie Creek road parallels the shore usually running within a few hundred meters of the beaches, while the slopes of the Cape Range extend to within 1 km of the shoreline. Three creeks drain out of steep canyons from the 200-300 high range and form usually blocked dry creek mouth at the shore. The Cape Range National Park Milyaring Visitor Centre is located 1 km east of T Bone Bay (beach WA 1577).

Tuki Beach (**WA 1571**) is a curving 800 m long sandy beach that commences at the northern end of a 1.5 km long section of low calcarenite bluffs and terminates at the protruding usually blocked mouth of a dry creek. Low semi-stable foredune ridges back the beach extending up to 100 m inland. A subtidal cobble-boulder delta fronts the creek mouth, with beach **WA 1572** continuing to the north for another 1.2 km to the next section of low bluffs. It is backed by degraded foredune ridges extending up to 200 m inland. The Tuki Beach campsite is located behind the protruding southern end of the beach. Both beaches are paralleled by inshore shallow sand flats grading into the 1 km wide fringing reef.

Beach **WA 1573** is a 100 m long pocket of sand located at the mouth of usually blocked drainage channel, that drains a 50 ha shore parallel salt lake. Low bluffs lies to either side of the beaches which as no formal vehicle access.

Beach **WA 1574** extends 200 m south of the RAAF Creek mouth. The creek has deposited a 300 m long protruding cobble-boulder sub-tidal delta, with shallow sand flats beyond and the fringing reef 1.5 km offshore. A small 150 wide gap is located due west of the creek mouth, with a lefthand *surf break* along the southern side of the gap. The beach is bordered by low bluffs, including some between the northern end of the beach and the creek mouth. Camp 13 campsite is located at the northern end of the beach. Beach **WA 1575** extends north of the creek mouth for 1.9 km to the next creek mouth. Camp 14 is located at the northern end of the beach. Both beaches are backed by semi-stable 100-200 m wide foredune ridges and fronted by the fringing reef systems.

Lakeside beach (**WA 1576**) is named after the small elongate lagoon, which occupies the usually blocked mouth of the creek. The beach extends from the creek mouth for 1.2 km to the tip of the sandy foreland. Beachrock reefs parallel the southern end of the beach providing a sheltered spot, which is used for launching small boats off the beach in front of the Lakeside camping and picnic area. The beach then narrows before widening to the foreland where it is backed by a 200 m wide foreland covered with low semi-stable dunes. The fringing reefs widen to 2 km off the beach. The Milyering Visitor centre is located 1 km to the east.

T Bone Bay beach (**WA 1577**) is located on the northern side of the small foreland. It curves round initially in lee of the foreland, then trends north-northeast for 2.1 km to the next sandy foreland at Mesa Camp. T Bone Bay marks the southern limit of a lagoon channel that parallels the shore for 2 km to Neds Camp, before turning and linking with a 500 wide gap in the barrier reef off Neds Camp beach. The reef lies 2-2.5 km off beaches WA 1577-1579. T Bone Bay beach is backed initially by a narrow unstable foredune, which gradually widens to the north as the foreland develops, reaching a maximum of 200 m at the tip. Most of the dunes are unstable and migrating northward toward Mesa Camp.

Mesa Camp beach (**WA 1578**) is located on the northern side of the tip of the foreland. It consists of a 600 m long north-northwest-facing beach fronted by the deeper lagoon channel and terminating in the north to the lee of a 100 m long patch of shore-parallel beachrock. The beachrock is located just off the mouth of a dry sandy creek, which drains a 2 km long shore parallel salt lake. The Mesa Camp car park and camping area is located on the southern side of the creek mouth, which is also a popular spot for launching small boats, which can reach the deeper reef water via the Neds Camp channel.

Neds Camp beach (**WA 1579**) commences at the sandy creek mouth and curves to the northeast, them north for 3 km to the beginning of the low calcarenite bluffs that continue on as 1 km long Low Point. The beech is fronted by the deeper channel, with the centre of the beach facing due west into the gap in the reef, which permits low swell to reach the centre of the beach in the south. The beach is backed by a narrow foredune system where the Neds Camp camping area is located. The dunes widen to 300-400 in the north and become increasingly unstable, with an active blowout at the northern end of the beach extending 400 m to the north.

WA 1580-1583 MANGROVE BAY

No.	Beach	Rating	Type	Length
WA 1580	Low Pt (W)	3	R+barrier reef	300m
WA 1581	Low Pt (E)	3	R+barrier reef	200 m
WA 1582	Mangrove Bay (S)	2	R+sand flats	100 m
WA 1582	Mangrove Bay	2	R+fringing flats	600 m
Spring & neap tidal range = 1.0 & 0.4 m				

Mangrove Bay is a 700 m wide shallow bay partially protected by the sandy spit that extends north of Low Point (Fig. 4.353), as well as inshore fringing reefs and the barrier reef that extend 2.5 km offshore. As a consequence the first mangroves north of Carnarvon, 330 km to the south, are found in the small bay. Two beaches occupy Low Point (WA 1580 & 1581) and two the eastern bay shore (WA 1582 & 1583). There is a gravel road out to the rear of the northern beach WA 1583.

Figure 4.353 Mangrove Bay (left) is located to the lee of sandy Low Point.

Low Point is a 300 m long low sandy spit that has migrated about half way across the bay with beaches WA 1580 & 1581 located to either side of the spit. Beach **WA 1580** is a 300 m long west-facing sand beach fronted by sand flats, then a deeper channel and the barrier reef flats. It is attached to the low calcarenite bluffs of the point to the south and backed by a 300 m wide series of vegetated beach ridges that occupy the rear of the bay. Beach **WA 1581** occupies the eastern side of the spit. It is 200 m long and faces east into the small bay toward the mangrove-lined bay shoreline. It is fronted by sand flats and shallow bay floor.

Beach **WA 1582** is a 100 m long strip of high tide sand with mangroves to either side, as well as a central mangrove patch. It is also backed by mangroves then the bay beach ridges. Beach **WA 1583** occupies the northern west-facing shore of the bay and receive some low waves which maintain a slightly curving 600 m long beach fronted by fringing reefs which fill the northern half of the bay. It is backed by a low 100-200 m wide series of foredune ridges, with the access road running along the crest of the ridges to the beach. A small mangrove-lined calcarenite point forms its northern boundary and entrance to the bay. A bird observation hide is located on the point.

WA 1584-1586 MANGROVE BAY (N)

No.	Beach	Rating	Type	Length
WA 1584	Mangrove Bay (N1)	3	R+fringing reef	200 m
WA 1585	Mangrove Bay (N2)	3	R+fringing reef	750 m
WA 1586	Mangrove Bay (N3)	3	R+fringing reef	200 m
Spring & neap tidal range = 1.0 & 0.4 m				

On the northern side of the northern point of Mangrove Bay is an irregular 2 km of northeast trending shoreline containing three low energy beaches (WA 1584-1586). The beaches all lie to the lee of a combination of fringing reef and shore-parallel beachrock reef, then a 500 m wide lagoon channel followed by the barrier reef flats, which extend 3 km offshore.

Beach **WA 1584** extends for 200 m north of the northern bay point. It is a low, narrow low energy beach with patches of mangroves growing along and just off the beach, and with shallow fringing reef extending 500 m seaward. It terminates at the entrance to a small tidal creek which drains a backing 100 ha coastal lagoon occupied by mangroves, then salt flats. This beach can be readily accessed from the Mangrove Bay car park located 100 m to the south.

Beach WA **1585** commences on the northern side of the mangrove-lined creek mouth and curves to the north for 750 m to a low beachrock bluff. Three small patches of calcarenite outcrop along the low energy beach, which is in turn fronted by fringing reef, and narrows from 400 m in the south to 100 m the north. It is backed by a vegetated foredune and an older vegetated north-trending parabolic dune. Five hundred meters of low calcarenite bluffs and beachrock reef separate it from beach **WA 1586**. This is a curving 200 m long pocket of sand sheltered by the northern extension of the beachrock reef and shallow sand flats in between. More low bluffs border its northern end, while it is backed by the 10-20 m high vegetated transgressive dunes that originated from beach WA 1585. There is no formal access to either beach.

WA 1587-1592 TANTABIDDI WELL-JURABI PT

No.	Beach	Rating	Type	Length
WA 1587	Tantabiddi Well (S)	3	R+fringing/barrier reef	1.3 km
WA 1588	Tantabiddi Well	3	R+fringing/barrier reef	1.8 km
WA 1589	Tantabiddi Ck (S)	3	R+fringing/barrier reef	1.8 km
WA 1590	Tantabiddi Ck (S)	3	R+fringing/barrier reef	2.2 km
Spring & neap tidal range = 1.0 & 0.4 m				

The Tantabiddi Well area marks the beginning of a near continuous section of sandy beaches that extend 20 km northeast to North West Cape, the northern tip of the Ningaloo Reef complex. The first 7 km are occupied by four beaches (WA 1587-1590) centred on Tantabiddi Well and Tantabiddi Creek. The Yardie Creek road runs about 500 m in from the shore with vehicle access to all the beaches. The beaches are fronted by fringing reef, then the lagoon channel that commences off Low Point and widens to 1-1.5 km along the shore exiting via a 1.5 km wide passage west of beach WA 1590. The barrier reefs are located 3-4 km off the shore.

Beach **WA 1587** commences 300 m north of beach WA 1586 and trends to the northeast for 1.3 km as a protruding sandy foreland that terminates at the tip of the next foreland. It is fronted by 200-500 m wide fringing reef, then the channel and outer reefs, resulting in low waves at the shore. It is backed by a sandy foreland that widens to 200 m in the centre and has patches of dune instability. A 4 WD track winds out to the southern end of the beach.

Beach **WA 1588** commences at the tip of the foreland and curves to the northeast, then north for 1.8 km, past Tantabiddi Well, where the northern boundary of the Cape Range National Park bisects beach. It terminates at a section of low calcarenite bluffs. The ranger station is located at the park entrance 300 m east of the beach and a vehicle track winds out form the station to the centre of the beach. It is fronted by patchy fringing reef extending up to 500 m offshore, then the channel and outer reefs.

Beach **WA 1589** extends north of the low bluffs for 1.8 km as a curving protruding beach that terminates at the usually blocked mouth of Tantabiddi Creek (Fig. 4.354). Fringing reef parallels the beach and extends up to 500 m offshore. The beach is backed by some areas of unstable dunes in the south, then a 100 m wide foredune ridge plain, backed by a deflation basin. The Yardie Creek road parallels the rear of the beach 300-400 m inland, with a 4 WD track reaching the southern end of the beach.

Beach **WA 1590** commences at Tantabiddi Creek and trends to the north-northeast for 2.2 km to a low bluff-capped sandy foreland. Fringing reef parallels the beach, which faces west into the passage through the reef, located 3 km off the northern end of the beach. The passage permits low ocean swell to reach the beach. It is backed by the creek and a small lagoon in the south, adjacent to which a small research laboratory and public boat ramp are located. To the north dunes increase in instability with a 300 m wide area of active dunes backing the northern foreland.

Figure 4.354 Tantabiddi Creek separates beaches WA 1589 (right) and 1590.

WA 1591-1598 JURABI PT-JIMS BEACH

No.	Beach	Rating	Type	Length
WA 1591	Jurabi Pt (S)	3	R+fringing reef	2.5 km
WA 1592	Jurabi Pt (N)	3	R+fringing/barrier reef	4.2 km
WA 1593	False Is Pt	4	R+fringing reef	600 m
WA 1594	False Is Pt (N)	4	R+fringing reef	500 m
WA 1595	Torpedo Bay	4	R+fringing reef	900 m
WA 1596	Five Mile	4	R+fringing reef	1 km
WA 1597	Jensz	4	R+fringing reef	1.9 km
WA 1598	Jims Beach	4	R+fringing reef	4.2 km
Spring & neap tidal range = 1.0 & 0.4 m				

At **Jurabi Point** the shoreline turns and trends to the northeast for 20 km to North West Cape, the northern tip of the Exmouth Peninsula. The barrier reefs, 2 km wide off the point continue all the way to the cape, narrowing to fringing reef along much of the shore. Between Jurabi Point foreland and Vlaming Head are eight beaches (WA 1591-1598), which occupy most of the 17 km of shoreline. The Yardie Creek road runs a few hundred metres inland parallel to the shore, with numerous access points to most of the beaches. The Cape Range continues to parallel the shore 2-3 km inland, finally terminating where it intersects the coast at Vlaming Head.

Jurabi Point is a prominent sandy foreland that has prograded 1 km seaward, with beaches WA 1591 and 1592 located to either side of the foreland (Fig. 4.355). Beach **WA 1591** is a curving 2.5 km long west-facing beach located between two sandy forelands. It is fronted by a shallow lagoon that grades seaward into the fringing reef system. Waves average less than 0.5 m at the shore and reflective conditions prevail. The beach is backed by active dunes and deflation surfaces that widen to 1 km at the foreland. The dunes are transgressing over the foredune ridge systems that built the foreland. Two 4 WD tracks cross the dunes to reach the beach.

Figure 4.355 Jurabi Point is a prominent sandy foreland capped by active dunes and deflation surfaces.

Beach **WA 1592** commences at the tip of Jurabi Point and curves to the northeast for 4.2 km to False Island Point. The fringing reefs that contributed to the formation of the foreland lie 2 km off the point. They extend half way along the beach, beyond which the reef system narrows to a 500 m wide fringing reef system. The transgressive dunes from the southern side of the foreland continue right across the foreland and to the north gradually narrowing and finally terminating at False Island Point. Some of the dunes are actively supplying sand to the beach. Because of the dunes the only access is via a 4 WD track that reaches the tip of the foreland.

False Island Point has fringing reef growing to the shore with waves breaking across the 200-300 m wide reef system and larger waves reaching the shore. Low calcarenite bluffs fringe the point, with beach **WA 1593** located on its northern side. It is a 600 m long steep reflective beach receiving ocean swell averaging about 0.5 m at the shore. The beach is interrupted by outcrops of beachrock and low calcarenite bluffs, with fringing reef extending 400 m offshore. Four wheel drive tracks reach either end of the beach. Beach **WA 1594** lies 100 m to the north and is a similar 500 m long reflective beach bordered by small protruding low calcarenite bluffs, with low bluffs outcropping along much of the beach. The Baudin and Brooke access tracks lead to either end of the beach, which is a popular spot for rock fishing, with shallow fringing reefs extending 400 m offshore. Semi-stable foredunes back both beaches.

Torpedo Bay (**WA 1595**) occupies a 900 m long northwest-facing indentation in the shore, with beachrock bluffs bordering either end. The Trial access track leads to a car park behind the southern end of the beach, which is also a popular rock and beach fishing location. Wave averaging about 0.5 m surge up the steep reflective beach, with reefs extending 500-600 m offshore.

A 1 km long section of low bluffs separates Torpedo Bay from beach **WA 1596** that continues to the northeast for 1 km to the next section of low bluffs. The Five Mile access track leads to a car park at the southern end of the beach, with other tracks reaching the centre and northern end of the beach, which is backed by semi-stable foredunes. The fringing reef extends 600 m offshore, with low waves reaching the steep beach, and occasional outcrops of beachrock and bluffs along the shore.

Beach **WA 1597** occupies a 1.9 km long curving northwest-facing embayment bordered by the low bluffs to the south, and the subtle foreland at the southern end of Jims Beach to the north. Reefs extending 500 m offshore fringe either end of the beach, with a shallow gap in the centre resulting in the embayed shoreline. Waves break heavily on the boundary reefs, and most waves are attenuated through the gap with low waves at the shore, and reflective conditions the length of the beach. It is backed by foredune ridges to either end, which narrow to the centre, where the Wobiri and Jansez access tracks reach the shore.

Jims Beach (**WA 1598**) occupies a large sandy foreland composed of multiple foredune ridges, which has prograded up to 600 m seaward in the centre. The beach extends for 4.2 km to the northeast terminating at Vlaming Head. Waves break heavily on the outer end of the 500-600 m wide fringing reefs that parallel the shore (Fig. 4.356), resulting in low surging waves and reflective conditions the length of the beach. The Jacobzs, Mauritius and Hunter access tracks cross the foredunes to provide access along the northern half of the beach.

Figure 4.356 Jims Beach lies to the lee of a continuous fringing reef.

WA 1599-1603 **VLAMING HD-NORTH WEST CAPE**

No.	Beach	Rating	Type	Length
WA 1599	Vlaming Hd	5	R+fringing reef	500 m
WA 1600	Vlaming Hd (E)	5	R+fringing reef	3.3 km
WA 1601	Mildura Rd (1)	5	R+fringing reef	1.2 km
WA 1602	Mildura Rd (2)	5	R+fringing reef	1.1 km
WA 1603	North West Cape	5	R+fringing reef	500 m
Spring & neap tidal range = 1.0 & 0.4 m				

Vlaming Head marks the northern tip of the Cape Range, with 80 m high Lighthouse Hill located 500 m south of the cape. The slopes of the hill descend to the cape, which is fringed, by low beachrock bluffs and fringing reefs extending 200-300 m offshore. The shoreline turns at the head and trends east then northeast for 6 km to the cape (Fig. 4.357). In between are five near continuous sand beaches, all fringed by coral reefs extending 200-300 m offshore, the northernmost reef of the cape marking the northern end of the 260 km long Ningaloo coral reef system. A vehicle track from the Hunter access road terminates right at the head and provides a view along the beach.

Figure 4.357 Vlaming Head is located at the first bedrock outcrop on the Ningaloo shore. The shoreline turns to northeast where it is backed by beaches WA 1599 and 1600, with the caravan park behind.

Beach **WA 1599** commences at the head and trends east for 500 m to a 100 m long section of low beachrock bluffs. The beach is backed by a blowout that extends 100 m inland to the edge of the Yardie Creek road. Waves averaging about 0.5 m arrive at the beach after transgressing the fringing reefs. The *Vlaming Head* lefthand surfing break is located on the outer edge of the reef directly north of the head. The Lighthouse caravan park is located across the road opposite the eastern end of the beach.

Beach **WA 1600** commences on the eastern side of the low bluffs and trends east, then northeast for 3.3 km, backed by a 10-20 m high foredune and paralleled by the gravel Mildura Road along its northern half. The beach is fronted by reefs extending 100-200 m off the beach with low waves reaching the shore and a steep reflective beach. There are two vehicle access points in the south off the Yardie Creek road, and one in the north off the Mildura road. Good *surf breaks* are found out on the reef during moderate to higher swell.

Beach **WA 1601** is located on the north side of a slight beachrock-controlled inflection in the shore and continues northeast for 1.2 km, fringed all the way by 200-300 m wide reefs with waves and surf breaks on the outer reefs and low waves at the shore. There is vehicle access to either end of the beach, as well as a 4 WD track to the centre. Beach **WA 1602** continues northeast from the next slight inflection for another 1 km to the last inflection before the cape. It is a similar reflective beach with the reef widening to 300-400 m. It can be accessed from the northern track to beach WA 1601. Unstable foredunes rising to 15 m and extending 100-200 m inland back all the Mildura road beaches.

Beach **WA 1603** is located on the tip of **North West Cape**. It is a curving, convex-shaped 500 m long reflective beach, backed by a 100 m wide foredune system, that protrudes north as a sandy foreland between two low calcarenite bluffs (Fig. 4.358). Southwest waves break over reef off the western side where there is a lefthand *surf break*, while calmer conditions prevail to the north, with the reefs extending up to 500 m offshore. The wreck of the *SS Mildura* is clearly visible 200 m off the northern tip of the beach. The Mildura road terminates at a car park behind centre of the beach.

Figure 4.358 View south from North West Cape along beaches WA 1603 to 1600.

NORTH WEST CAPE

North West Cape marks the northern tip of Exmouth Peninsula and the Ningaloo Reef system. It also marks a dramatic change in the nature of the coast, with generally swell dominated, though reef-protected, coast to the south, while low nearshore gradients and wind waves increasingly dominate into Exmouth Gulf and to the northeast along the Pilbara coast. In addition as the wave height drops to the north the tide range, low (<1 m) along the entire Western Australian coast to the south, begins to increase in range, peaking at 10 m by Broome. As a consequence the wave dominated coast to the south is replaced by an increasingly tide-modified to tide-dominated coast to the north.

The cape also marks the terminus for sand moving north along both sides of the peninsula to converge in the low cape area north and east of Vlaming Head. In the west a 6 km long beach system (WA 1599-1603) has been build from the head northeast to the cape. In the east a 5 km long system has been built to Point Murat (WA 1604-1606), while beach 1604 and a section of calcarenite bluffs extends for 4 km to link Point Murat with the cape. Backing the three linked beach systems is a 1200 ha area of low salt flats, impounded by beaches and the Cape Range to the south, including a 3500 ha area of north-trending red desert longitudinal dunes. A 387 m high communication tower is located on the salt flats 2 km west of Point Murat.

EXMOUTH GULF

Coast length: 248 km (4272-4520 km)
Beaches: 50 (WA 1605-1654)
Sandy shoreline 93 km
Mangrove shoreline: 147 km
Figure 4.359

Exmouth Gulf is 45 km wide at the mouth bordered in the west by Point Murat at the northeast tip Exmouth Peninsula and Tubridgu Point on the low mangrove-lined mainland to the east. The U-shaped gulf extends for 85 km to the south terminating in the very low energy Gales and Giralia bays. Depths average over 10 m in the north and less than 10 m to the south. The western shore is exposed to the southerly winds and wind waves and is dominated by low coarse grained beaches, the sediment supplied from both fringing reefs and streams draining from the Cape Range. The eastern shore has a very low gradient and is dominated by intertidal mangroves and salt flats up to 20 km wide. As the waves decrease tide range increases into the gulf from a spring range of 2 m at Point Murat to 2.6 m at Learmouth. Apart from the perennial streams that drain off the range there are no major rivers or streams draining into the gulf.

WA 1604-1607 **PT MURAT**

No.	Beach	Rating	Type	Length
WA 1604	Pt Murat (N)	3	R+fringing reef	2.4 km
WA 1605	Pt Murat (S)	3	R+fringing reef	1.8 km
WA 1606	Bundegi	3	R+fringing reef	4.2 km
WA 1607	Bundegi Reef	3	R+sand flats/fringing reef	1.6 km

Spring & neap tidal range = 1.5 & 0.4 m

Point Murat is a prominent sandy foreland that forms the northeastern tip of Exmouth Peninsula (Fig. 4.360). Beach WA 1604 occupies the northern side of the foreland and beach WA 1605 the southern side.

Beach **WA 1604** commences at the eastern end of the 1.5 km long section of dune-capped calcarenite shore that tends southeast from North West Cape. The beach curves to the southeast for 2.4 km to the low sandy point. It faces northwest across a shallow 1 km wide fringing reef, which narrows to 100 m off the tip of the point. The beach is backed by an irregular foredune system and some vegetated transgressive dunes extending 100-200 m inland, then the restricted area around the high communication tower. There is public access to the eastern end of the beach near the point.

Beach **WA 1605** commences at the point and curves to the southwest for 1.8 km to a small sand foreland. There is a car park at the north end of the beach, which services a 200 m long jetty. The beach faces southeast across a fringing reef, which widens to 500 m in the south. It is backed by vegetated blowouts, and one active blowout, then the Murat Point road that terminates at the point and jetty.

Beach **WA 1606** extends from the southern side of the foreland for 4.2 km to a 500 m long section of low calcarenite bluffs. The Bundegi car park, small jetty and boat ramp is located just south of the foreland, with boats usually moored off the jetty in a small lagoon between the reef and shore. The fringing reef continues to the south paralleling the beach 500 m offshore in the north and widening to 1 km in the south. The Point Murat road parallels the beach 500 m inland, with low dunes and salt flats between the beach and road.

Beach **WA 1607** is a curving 1.6 km long, east-facing beach that is located between the northern low bluff section and a small calcarenite outcrop to the south. The beach is sheltered by ridged sand flats, then the fringing Bundegi Reef, which extends 1 km offshore. Scattered mangroves grow along the base of northern calcarenite bluffs and in the northern corner of the beach. A low foredune, then elongate salt lake backs the beach. There is a vehicle access to the southern end of the beach.

Region 5E CARNARVON: Exmouth Gulf – Cape Preston

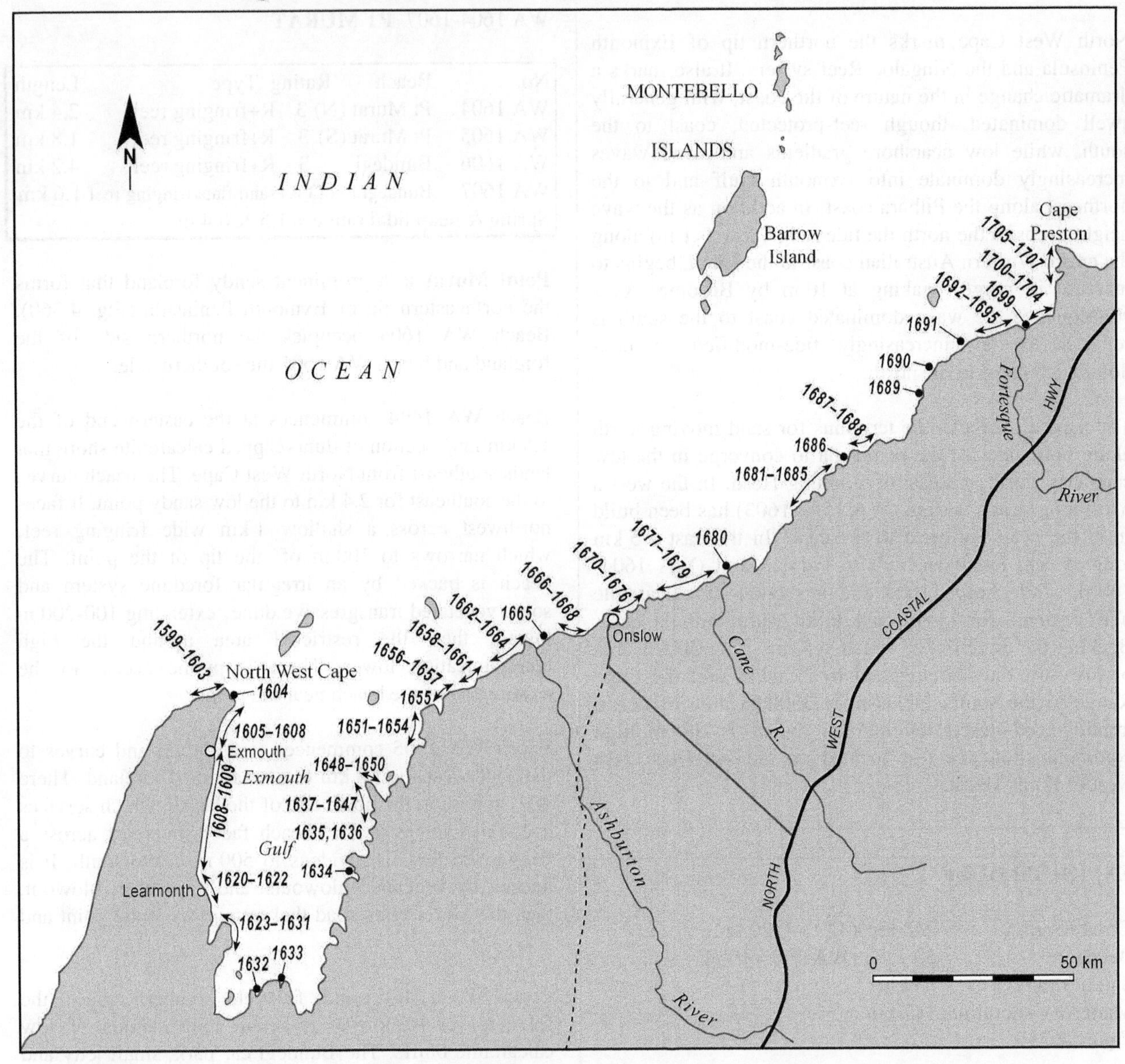

Figure 4.359 The northern section of the Carnarvon coast between Exmouth Gulf and Cape Preston contains beaches WA 1604-1707.

Figure 4.360 View from Point Murat across the top of the Exmouth Peninsula to North West Cape.

WA 1608-1611 EXMOUTH

No.	Beach	Rating	Type	Length
WA 1608	Exmouth (Town Beach)	3	R+sand flats/fringing reef	9 km
WA 1609	McLeod St Beach	3	R+sand flats/fringing reef	1.9 km
WA 1610	Mowbowra Well	3	R+sand flats/fringing reef	1.5 km
WA 1611	Town Common	3	R+sand flats/fringing reef	2.1 km

Spring & neap tidal range = 1.8 & 0.6 m

Exmouth is a planned town of 4000, which was established in 1963 when the joint US-Australian Naval Communication Station was commissioned. The well laid out town is located 15 km south of Point Murat and extends from the foothills of the Cape Range toward the shore of the gulf. The 0.5-1 km wide coastal fringe between the Exmouth Road and the shore is given over to a wide range of recreational facilities, including the boat harbour constructed during the mid 1990's and located on the southern side of the town. Exmouth is becoming increasingly popular as a tourist destination and base for exploring the Cape Range National Park and Ningaloo Reef Marine Park, as well as a range of boating and fishing activities in the gulf and over the reefs. Four near continuous beaches (WA 1608-1611) are located in front of and to either side of the town.

Beach **WA 1608** commences 6 km north of the town at the small calcarenite bluff that separates it from beach WA 1607. It then trends straight south for 9 km to the boat harbour, which has cut the beach into two (Fig. 4.361). For its entire length the beach is fronted by shallow sand flats extending 200-300 m into the gulf, with the fringing reef paralleling the beach approximately 1 km offshore. The beach is backed by a narrow band of foredunes rising 10-15 m in height. The communication station backs the northern half of the beach and access is restricted. Along the town section, called **Town Beach**, the best access and facilities are at Warne St, where there is a designated swimming area, foreshore reserve with picnic facilities, playground, and toilets. The Exmouth Yacht Club and a boat ramp are located at the southern end of the beach, while the 300 m long boat harbour breakwater is located 500 m to the south.

Figure 4.361 Exmouth's Town Beach and boat harbour.

Beach **WA 1609** commences on the southern side of the boat harbour and continues south for 1.9 km to the mouth of a usually dry creek. A 50-100 m wide foredune backs the beach, with a gravel road behind the dunes, then large acreage residential blocks. The racecourse backs the northern end of the beach adjacent to the boat harbour. The sand flats and reef continue along the front of the beach.

Beach **WA 1610** is a 1.5 km long beach located between the creek mouth and the protruding mouth of a creek adjacent to Mowbowra Well. The backing foredune narrows to 100 m, with a vehicle track paralleling the rear. Sand flats extend 200-300 m offshore, with the reef edge 1 km offshore.

Beach **WA 1611** extends south for 2.1 km past the mouth of the meandering Mowbowra Creek to a low calcarenite point, with the Town Common located behind the beach. The foot slopes of the Cape Range are located 500 m west of the beach. The low energy beach is paralleled by 300-400 m wide sand flats, then the reef extending up to 1 km offshore. A vehicle track reaches the southern end of the beach at the creek mouth.

WA 1612-1617 **QUALING POOL-KAILIS**

No.	Beach	Rating	Type	Length
WA 1612	Qualing Pool	1	R+sand flats/fringing reef	800 m
WA 1613	Pebble Beach	1	R+sand flats/fringing reef	4.7 km
WA 1614	Shothole Ck	1	R+sand flats/fringing reef	800 m
WA 1615	Badijirrajirra Ck	1	R+sand flats/fringing reef	5.2 km
WA 1616	Kailis	1	R+sand flats/fringing reef	3.5 km
WA 1617	Kailis (S)	1	R+sand flats/fringing reef	2.2 km
Spring & neap tidal range = 1.8 & 0.6 m				

The coast to the south of Exmouth trends roughly due south for 30 km to Learmouth where it bends to the southeast towards Heron Point. Beaches WA 1612-1617 occupy a 12 km long section of this shore beginning 7 km south of Exmouth and continuing past the Kailis Fisheries jetty. The seven beaches are relatively straight and continuous, with usually dry creek mouths draining out of the Cape Range forming the boundaries. The creeks have built a narrow coastal plain, with the beaches forming the eastern boundary. The Murat Point-Exmouth road parallels the coast usually within a few hundred metres of the shore, with vehicle tracks out to most of the beaches. Waves energy decreases down into the gulf, with waves dependent entirely on the prevailing winds and calms common. Tides gradually increase to the south from a spring range of 1.8 m at Exmouth to 2.1 m at Learmouth. All the beaches are fronted by 100-200 m wide sand flats, with coarse gravel and boulder beaches at the river mouths. Fringing reef lies within a few hundred metres of the shore, but degrade and become increasingly patchy to the south.

Beach **WA 1612** commences at the small calcarenite protrusion that separates it from beach WA 1611 and continues south for 800 m to the next creek mouth which contains a small elongate lagoon known at the Qualing Pool. Sand flats, then outer fringing reef parallel the beach, with the Exmouth Road located 600 m to the west and vehicle access tracks to either end of the beach, including the pool.

Pebble Beach (**WA 1613**) commences on the southern side of the pool and continues south for 4.7 km to the mouth of the creek that drains the Shothole Canyon. The creek delivers coarse gravel and pebbles to the beach, hence the name. A low foredune backs the northern half of the beach, with fringing reef growing to the shore along the northern part of the beach, and moving 200-400 m offshore to the south as 100 m wide sand flats parallel the beach. An airstrip parallels the rear of the beach, with a gravel road near the airstrip running straight out to the beach.

Beach **WA 1614** commences at the Shothole Creek mouth and trends to the south for another 800 m to the next creek mouth. Both creek mouths protrude slightly into the gulf, with the beach curving gently in between with vehicle tracks and car parks located beside each mouth. Sand flats and fringing reef continue to parallel the shore. Beach **WA 1615** commences at the creek mouth and extends 5.2 km south past three small blocked creeks to the larger blocked mouth of Badijirrajirra Creek, which drains out of canyons in the backing range. The road runs 200-300 m inland with a few vehicle tracks out to the beach. The beach is fronted by 100-200 m wide ridged sand flats, then the fringing reef.

Beach **WA 1616** commences at Badijirrajirra Creek mouth and trends south past the **Kailis Fisheries** and some small creek mouths for 3.5 km to the next prominent creek mouth. The Kailis complex includes a small jetty and breakwater, fish processing plant and caravan park, with fishing boats often moored off the jetty (Fig. 4.362) and dinghies pulled up on the beach. The beach contains coarse material from the creek. It is paralleled by the 100 m wide sand flats with patchy fringing reef up to 500 m offshore. Beach **WA 1517** curves to the south of the creek mouth for 2.2 km to the southernmost creek draining directly from the range to the shore. There are access tracks to both creek mouth and a central track off the Exmouth road, which runs just 200 m inland.

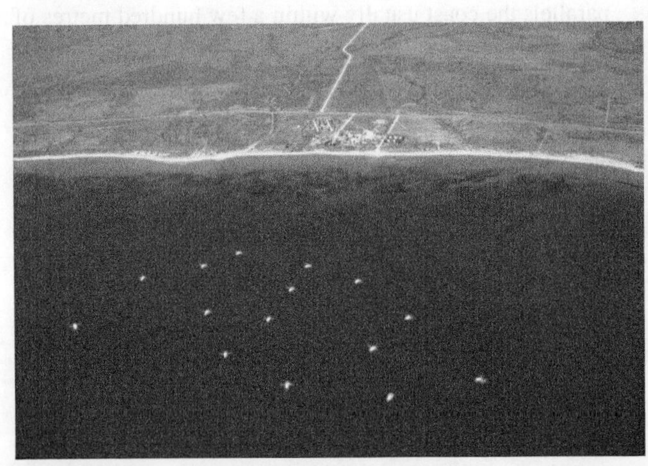

Figure 4.362 The Kailis fisheries complex, with boats moored off the beach.

WA 1618-1622 **LEARMOUTH-HERON PT**

No.	Beach	Rating	Type	Length
WA 1618	Potshot Beach	1	R+sand flats/fringing reef	6 km
WA 1619	Learmouth	1	R+sand flats/fringing reef	3.1 km
WA 1620	Heron Pt (N)	1	R+sand flats/fringing reef	4 km
WA 1621	Heron Pt (S)	1	R+sand flats	2.5 km
WA 1622	Bay of Rest	1	R+sand flats	1.5 km
Spring & neap tidal range = 2.1& 0.6 m				

South of Kailis the sediment movement is increasingly to the south, resulting in slight deflection of all the creek mouths to the south. In the vicinity of Potshot the southward transport begins to accumulate as part of a 17 km long barrier system containing three south-trending spits, with terminate in the south at the mangrove-lined Bay of Rest. Beaches WA 1618 and 1619 occupy the first spit, beach WA 1620, the second, and beaches WA 1621 and 1622 the third. The Learmouth airfield has been built on the low back barrier flats behind the two northern spits.

Potshot Beach (WA 1618) commences on the south side of the protruding creek mouth 4 km south of Kailis. The beach initially trends south, but after 1 km curves to the southeast as part of the 7 km long barrier spit. The spit is 200-300 wide and is paralleled for parts of its length by elongate salt flats. The beach has a total length of 6 km and terminates at the concrete Learmouth groyne and jetty, which partly blocks the southerly sand transport causing an offset in the beach. The beach is very low energy and often calm, with sand flats extending 100 m offshore and patchy fringing reef beyond. The Potshot memorial and a fish processing plant back the southern end of the beach.

Beach **WA 1619** commences on the southern side of the jetty and trends to the southeast, past a blocked creek mouth, and then south for 3.1 km to terminate at a tidal creek mouth which forms the end of the spit. The creek drains the backing mangrove-lined salt flats. The jetty is use to refuel the fishing boats which moor off the beach. The beach is capped by foredunes rising to 10 m, then the mangroves and salt flats, with the Exmouth road 1-2 km inland, and the Learmouth airfield on the western side of the road.

Beach **WA 1620** commences on the southern side of the 50 m wide creek mouth and curves to the south then southeast for 4 km to low sandy Heron Point (Fig. 4.363). One hundred metre wide sand flats parallel the beach, together with the southern most patches of the fringing reef in the gulf. There are vehicle tracks across the backing salt flats to both ends of the beach including the point area. Beach **WA 1621** continues to curve south of the point for 2.5 km. This is a slightly crenulate and increasingly low energy beach, which is part of a transition to the mangrove-lined shore to the south. It is backed by a low foredune, then the salt flats, with vehicle access to either end.

Figure 4.363 Heron Point with beaches WA 1620 (right) and 1621 (left).

Beach **WA 1622** forms the southernmost part of the spit system. It consists of a 1.5 km long series of irregular recurved spits that have been formed as sand has been episodically moved along the shore, possibly during tropical cyclone activity. The sand flats transform to tidal flats and widen to 1 km at the entrance to the Bay of Rest. The southern 1 km of the spit is embedded in mangroves. A 4 WD track runs down the spine on the low spit to the mangroves.

WA 1623-1631 **PT LEFROY-REST BAY**

No.	Beach	Rating	Type	Length
WA 1623	William Preston Pt	1	R+tidal flats	1 km
WA 1624	Pt Lefroy (W1)	1	R+tidal flats	2 km
WA 1625	Pt Lefroy (W)	1	R+tidal flats	700 m
WA 1626	Pt Lefroy	1	R+tidal flats	600 m
WA 1627	Rest Bay (1)	1	R+tidal flats	300 m
WA 1627	Rest Bay (2)	1	R+tidal flats	200 m
WA 1627	Rest Bay (3)	1	R+tidal flats	800 m
WA 1627	Rest Bay (4)	1	R+tidal flats	700 m
WA 1627	Lefroy Hill	1	R+tidal flats	500 m
Spring & neap tidal range = 2.1 & 0.6 m				

The Bay of Rest and Rest Bay occupy either side of a 3-4 km wide low peninsula, capped by vegetated longitudinal dunes in the southwest corner of the gulf. The mangrove-lined shoreline of the Bay of Rest dominates the western shore, with beaches WA 1623-1625 occupying the 3 km wide north shore, while patchy mangroves and very low energy beaches (WA 1626-1631) extend for 10 km down the eastern shore, finally giving way to the beginning of the 30 km long mangrove-lined shoreline of Gales Bay. All the beaches are very low energy, usually calm and fronted by tidal flats extending up to a few hundred metres into the gulf. They are also part of Exmouth Homestead with land access only via station vehicle tracks.

William Preston Point forms the northwestern tip of the peninsula, with beach **WA 1623** extending 1 km northeast from the point to the beginning of the northern shore of the peninsula. The southern half of the beach is backed by a low sandy foreland that narrows to the north. The beach is usually calm and fronted by 300 m wide tidal flats, which narrow to the south where the tidal channel for the bay lies 100 m off the beach. As a consequence a landing is located 500 m south of the beach, with a vehicle track to the southern end of the beach.

Beach **WA 1624** commences where the shoreline turns and trends east toward **Point Lefroy**. The beach is 2 km long, faces north up the gulf and is fronted by 500 m wide tidal flats, with patches of fringing reef increasing to the east (Fig. 4.364). It is backed by a low foredune and terminates at a small calcarenite inflection in the shore. Beach **WA 1625** continues east for 700 km to terminate at Point Lefroy, a low calcarenite-rimmed point that forms the northeastern tip of the peninsula. The tidal flats narrow to 200 m at the point while the patchy fringing reefs widen to 1 km. Straight tracks associated with oil exploration parallel the rear of the beaches and reach the eastern end.

Figure 4.364 Point Lefroy (left) and beach WA 1624.

Beaches WA 1626-1629 occupy the shore of a slightly curving 4.5 km long embayment called **Rest Bay**. Tidal flats extend 300-1000 m off the shore, with patchy reef up to 3 km offshore. Beach **WA 1626** commences on the southern side of Lefroy Point and trends to the south for 600 m. It consists of three slight embayments, with mangroves growing in each of the small bays. A solitary house is located behind the northern end of the beach, with a straight track paralleling the rear of the beach.

Beaches WA 1627-1629 are three patches of sand along an otherwise mangrove-lined shore of the bay. Beach **WA 1627** is 300 m long and located 1.5 km south of the point. Beach **WA 1628** is a 200 m long pocket of sand in the centre of the bay shore, while beach **WA 1629** occupies the southern 800 m terminating at the tip of a low sandy foreland, part of a south-trending barrier spit. Beach **WA 1630** occupies the southern half of the spit, extending from the eastern tip to curve to the south for 700 m, past a usually blocked creek mouth to terminate at the beginning of a 4 km long section of mangrove-lined shore. The low 200-300 m wide barrier backs beaches WA 1629 and 1630 then a mixture of salt flats and possibly older Holocene shorelines extending 1 km to the west.

Beach **WA 1631** is a very low energy 500 m long strip of beach that forms the southernmost exposure of sedimentary shoreline on the western gulf shore, with mangroves extending south for 17 km to the base of the Gales Bay and the gulf. The beach is located 4 km east of Exmouth Homestead, with a vehicle track leading to the beach and the shoreline to the north.

WA 1632-1633 **SANDLEWOOD PENINSULA**

No.	Beach	Rating	Type	Length
WA 1632	Pt Maxwell	1	R+tidal flats	800 m
WA 1633	Sandalwood	1	R+tidal flats	1.8 km
Spring & neap tidal range = 2.1 & 0.6 m				

The **Sandalwood Peninsula** occupies the centre of the southern shore of the gulf, with 0.5-1 km wide mangroves lining Gales Bay to the west and Giralia Bay to the east. A thinner band of mangroves line much of the shore of the peninsula, with only two areas relatively free of mangroves at beaches WA 1632 and 1633.

Beach WA **1632** is located at **Point Maxwell** on the northeastern tip of the peninsula. The 800 m long very low energy beach commences at Sandalwood Landing at the mouth of a small tidal creek. It then trends north protruding into the bay before curving round to terminate at the northern point. It is backed by a 10 m high foredune, and fronted by a 100 m wide tidal flats, then the deeper waters associated with the tidal channel.

From Point Maxwell mangroves dominate the shore for 7 km to the east across the top of the peninsula, to a slight 1.5 km wide embayment containing beach **WA 1633**. The beach is 1.8 km long, faces north up the gulf across 1 km wide tidal flats. It is backed by the gently rising terrain of the lightly wooded peninsula. Vehicle tracks parallel the rear of the beach and reach the eastern end of the beach.

This low energy peninsula marks the end of the Exmouth Peninsula and western gulf beaches. To the east of the beach is 6 km wide **Giralia Bay**, which typifies much of the shoreline to the east and north. The bay shoreline consists of 10 km wide supratidal salt flats, a 1 km wide band of mangroves and tidal flats extending up to 5 km into the bay, edged by deep tidal channels.

EASTERN EXMOUTH GULF

Coast length 120 km (4400-4520 km)
Beaches: 20 (WA 1634-1654)

The eastern coast of Exmouth Gulf extends for 120 km from mangrove-lined Giralia Bay in the south to the sandy shores of Tubridgi Point, the northeastern corner of the gulf. The entire shoreline is extremely low energy and low gradient. For the most part it consists of 2-5 km wide intertidal mud flats, grading into mangrove-lined tidal creeks, with mangroves up to 1 km wide, backed in turn by 10-15 km wide salt flats, the entire intertidal zone up to 20 km wide.

Mangroves and tidal creeks entirely dominate the first 40 km, with low energy beaches accumulating on the more exposed sections of the shore to the north. The beaches gradually increasing in size and extends, with Tubridgi Point backed by 1-2 km wide, low sand dunes. The shoreline is inaccessible by vehicle owing to the wide salt flats and dense mangroves.

WA 1634-1641 **HOPE-BURNSIDE ISLANDS**

No.	Beach	Rating	Type	Length
WA 1634	Hope Island	1	R+tidal flats	1.8 km
WA 1635	Simpson Is (W)	1	R+tidal flats	100 m
WA 1636	Simpson Is (N)	1	R+tidal flats	150 m
WA 1637	Burnside Is	1	R+tidal flats	2 km
WA 1538	Burnside Is (N1)	1	R+tidal flats	800 m
WA 1639	Burnside Is (N2)	1	R+tidal flats	500 m
WA 1640	Burnside Is (N3)	1	R+tidal flats	250 m
WA 1641	Burnside Is (N4)	1	R+tidal flats	600 m

Spring & neap tidal range = 2.1 & 0.6 m

Hope Island is low 700 ha island that has formed around an inner core of southwest-facing raised beach ridges. The southwest shore is now lined with mangroves and terminates at the low calcarenite bluffs of Hope Point. The bluffs extends 600 m north of the point to the beginning of 1.8 km long beach **WA 1634** which curves around the convex north-facing shoreline, grading into mangroves along its eastern end (Fig. 4.365). It is fronted by tidal flats, which widen from 200 m in the east to 1 km in the west.

Figure 4.365 Mangrove-fringed Hope Island and beach WA 1634 (left).

Simpson Island lies 2.5 km to the north, with a mangrove-lined tidal creek in between. The low island is 2 km long, 100-300 m wide and faces west across the gulf. Most of the eastern shore has been eroded to expose a low calcarenite bluff, with 100 m long beach **WA 1635** wedged in along the base of the bluff towards the southern end of the island. Beach **WA 1636** occupies the northern tip of the island. The 150 m long north-facing sandy beach is tied between low calcarenite outcrops, with tidal flats extending 100 m off the beach.

The southern tip of **Burnside Island** is located 1 km due east of the northern end of Simpson Island. The low, straight, narrow 2 km long island faces west-northwest into the gulf. It consists of the narrow high tide beach (**WA 1637**) backed by low scarped calcarenite bluffs, and a low grassy ridge widening to 200 m towards the northern end of the island. A 2 km wide mangrove-lined tidal creek backs the island.

One kilometer to the north is a more irregular, low 200 ha island consisting of a more exposed west-facing side, with a raised and active recurved spits forming the northern shoreline, the entire island backed by 1 km wide mangroves. Beach **WA 1638** occupies the northern 800 m of the western side of the island. It is fronted by 200-300 m wide sand flats, with low wind waves breaking on the narrow beach at high tide. The beach is backed by low dune-draped calcarenite, which outcrops to either end. Beach **WA 1639** commences at the western tip of the island and trends to the northeast for 500 m. This is a low energy high tide beach, with mangroves fringing either end. Beach **WA 1640** continues for another 250 m to the northeast and is a similar mangrove-fringed sandy beach, with continuous 300-400 m wide tidal flats linking the two beaches. Beach **WA 1641** consists of a raised recurved spit which extends in an irregular manner for 600 m to the east, where it grades into an active spit and mangrove-lined creek.

WA 1642-1640 **TENT ISLAND**

No.	Beach	Rating	Type	Length
WA 1642	Tent Is (S5)	1	R+tidal flats	2.1 km
WA 1643	Tent Is (S4)	1	R+tidal flats	700 m
WA 1644	Tent Is (S3)	1	R+sand flats	1.8 km
WA 1645	Tent Is (S2)	1	R+rock flats	80 m
WA 1646	Tent Is (S1)	1	R+rock flats	50 m
WA 1647	Tent Is (E)	1	R+sand flats	1.5 km
WA 1648	Tent Is (N1)	1	R+tidal flats	1.3 km
WA 1649	Tent Is (N2)	1	R+tidal flats	500 m
WA 1650	Tent Is (N3)	1	R+tidal flats	700 m
Spring & neap tidal range = 2.1 & 0.5 m				

Tent Island is a low irregular 8 km long, 2000 ha island that is composed of a series of low grassy beach ridges and recurved spits that have prograded southeast and east of two prominent western points, including the northern Tent Point. The ridges partly enclose a central-eastern 1000 ha area of salt flats and mangroves, with a 7 km long tidal creek separating the island from the mangrove-fringed mainland. The mainland shore is in turn backed by 15 km wide salt flats. Nine low energy beaches (WA 1642-1650) occupy most of the western and northern shore of the island.

Beaches 1642-1646 occupy much of the 7 km long southwestern side of the island and face into the prevailing southerly winds and low wind waves. The southern tip of the island consists of 1 km wide band of mangroves fronting an older beach ridge. As wave energy increases the mangroves give way to a small tidal creek on the northern side of which is beach **WA 1642** that continues northwest for 2.1 km. This is a low energy high tide beach fronted by irregular 1 km wide tidal flats, which narrow to 200 m at the northern end. An outcrop of mangrove-fringed calcarenite separates it from beach **WA 1643**, which continues northwest for another 700 m to the next small mangrove-fringe outcrop, with the narrower tidal flats continuing.

Beach **WA 1644** lies 200 m to the north on the northern side of a small sandy tidal inlet. This is the highest energy beach on the island with usually low wind waves breaking over the 200-300 m wide sand flats. The southern half of the 1.8 km long beach is an elongate spit backed by the mangrove-fringed tidal creek, with older beach ridges running perpendicular to the northern half of the beach. It terminates at the western point of the island, which contains a series of small calcarenite outcrops and bluffs, in amongst which are beaches WA 1645 and 1646.

Beach **WA 1645** is a curving 80 m long pocket of sand bordered by two small calcarenite points, with a few mangroves fringing the southern point. Low red bluffs separate it from beach **WA 1646,** 200 m to the north. It is a 50 m long pocket of sand, bordered and backed by 5 m high red bluffs. Irregular rock tidal flats extend a few hundred metres off the beach and point.

Beach **WA 1647** occupies the western side of the island and curves for 1.5 km to Tent Point (Fig. 4.366). The beach faces west and consists of a narrow high tide beach backed by a low foredune, then the older beach ridges. It is fronted by irregular rock and sand tidal flats extending up to 400 m off the centre of the beach, with a few scattered mangroves along the shore.

Figure 4.366 Tent Point (foreground) with beach WA 1647 (right) and 1648 (left).

At **Tent Point**, the northern tip of the island, the shoreline turns and trends to the east as three crenulate north-facing embayments containing beaches WA 1648-1650, all facing north across a 1 km wide tidal channel. Beach **WA 1648** occupies the first embayment and is a curving 1.3 km long, north-facing sandy beach, backed by a 50 m wide foredune, and fringed by a few mangroves. Mangroves surround the eastern point that separate it from beach **WA 1649**, a curving 500 m long sandy beach backed by a narrow foredune, and bordered in the east by a small calcarenite outcrop. Beach **WA 1650** continues east of the outcrop as a curving 700 m beach that terminates at the mouth of a small mangrove-lined tidal creek.

WA 1651-1656 **URALAU CK (S)-LOCKET ISLAND**

No.	Beach	Rating	Type	Length
WA 1651	Urala Ck (S1)	3	R+tidal channel	100 m
WA 1652	Urala Ck (S)	1	R+tidal flats	4 km
WA 1653	Urala Ck (N)	3	R+tidal channel	1.4 km
WA 1654	Locker Is (S)	2	R+sand/rock flats	6.5 km
WA 1655	Locker Is (N)	2	R+sand/rock flats	4 km
WA 1656	Locker Is (N1)	3	R+tidal channel	1 km
Spring & neap tidal range = 2.1 & 0.6 m				

Tubridgi Point is a low sandy point that forms the northeastern boundary of Exmouth Gulf. To the south of the point is 12 km of sandy shoreline that grades into the mangrove-dominated southeastern shoreline, while to the east sweeping sand beaches dominate to Onslow. The six beaches located either side of the point (WA 1651-1656) are located on two sandy islands, separated and backed by Urala Creek and it associated 10 km wide salt flats. The creek drains to either end of the island. There is no vehicle access to any of the beaches.

Beach **WA 1651** is a 100 m long strip of mangrove-fringed high tide sand located on the southern side of a 100 m wide tidal creek and fronted by the deep tidal channel. It is the northern part of a 1 km long beach ridge, the remainder of which is now encased in mangroves. Beach **WA 1652** commences on the northern side of the channel and continues to the northeast for 4 km to the southern entrance to Urala Creek. This beach is part of a barrier island which includes an inner 2 km wide splay of recurved spits, fronted by later more linear beach ridges, which splay to active recurved spits at either creek mouth. Tidal shoals lie off either end with narrower sand flats along the centre.

Locker Island commences on the northern side of the 200 m wide southern Urala Creek mouth. Beach **WA 1653** forms the first 1.4 km of the shore and is entirely located to the lee of the tidal channel and entrance shoals. As a consequence it is protected from waves, but paralleled by the deeper tidal channel, while it is backed by a 2 km wide series of older recurved spits. The remainder of the 11 km long island consists of three beaches (WA 1654-1656). Beach **WA 1654** occupies the eastern side of the island and extends due north of the creek mouth for 6.5 km to Tubridgi Point. The narrow high tide beach is part of an eroding shoreline. It is fronted, particularly to the north, by intertidal calcarenite reefs. The entire beach is backed by a 1-2 km wide active sand sheet consisting of transverse dunes that rise to several metres and are transgressing northward over a former beach ridge plain.

Beach **WA 1655** commences at Tubridgi Point and trends to the northeast for 4 km to the beginning of the northern entrance to Urala Creek. Calcarenite outcrops along much of the shoreline and as intertidal reefs, with the dune migrating from the south and spilling onto the beaches. Active dunes back the western half of the beach (Fig. 4.367), with low vegetated recurved spits behind the eastern half. Beach **WA 1656** occupies the southern shore of the northern Urala Creek mouth. The crenulate 1 km long beach faces north across the 100-200 m wide creek channel to beach WA 1657, and is for the most part fronted by the deep tidal channel.

Figure 4.367 Dune-covered Tubridgi Point (foreground) with beach WA 1655 extends 4 km to the north.

ONSLOW-CAPE PRESTON

Coast length: 250 km (4521-4770 km)
Beaches: 53 (WA 1655-1708)

Towns: Onslow, Dampier, Karratha

The 250 km long western Pilbara coast between Tubridgi Point and Cape Preston is a low energy northwest-facing low gradient shoreline. Most of the coast is dominated by broad intertidal salt flats with low energy beaches forming the shoreline and mangroves occupying both parts of the shoreline as well as dominating the tidal creeks. This is a low wave energy coast owing to four factors:

1. The southerly swell which dominates the open coast as far north as North West Cape, miss the east-trending coast west of Exmouth Gulf.
2. The southeast trade winds blow offshore.
3. The nearshore and inner shelf have low gradients.
4. Hundreds of islands and reefs extend from 20-50 km seaward of the shore.

All these combine to maintain calm to low wind waves at the shore. In addition tides begin to increase eastward along the coast from a spring range of 1.9 m at Onslow to 3.7 m at Dampier.

This is an arid sparsely populated coast, with only the small township Onslow (pop. 900) located 30 km northeast of Tubridgi Point. The North West Coast Highway is 100 km inland in the south reaching to within 15 km of the coast at Cape Preston with public access only possible at Onslow.

WA 1657-1658 **LOCKER POINT**

No.	Beach	Rating	Type	Length
WA 1657	Locker Pt (W2)	3	R+tidal channel	600 m
WA 1658	Locker Pt (W1)	3	R+sand/rock flats	4.9 km
Spring & neap tidal range = 1.9 & 0.6 m				

Locker Point is located 5 km northeast of the northern Urala Creek mouth (Fig. 3.368). Between the creek and the point are two beaches (WA 1657 & 1658). Beach **WA 1657** is a 600 m long low sandy, partly vegetated sand spit which forms the northern shore of the creek. It faces northwest and is fronted by tidal sand shoals then the deeper 100-200 m wide tidal channel. Beach **WA 1658** commences on the northern side of the beach and trends to the northeast for 4.9 km to the low sandy Locker Point. It consists of a high tide sandy beach fronted by intertidal beachrock, with an older beachrock reef paralleling the beach 1-1.5 km offshore. It is backed by a 2-3 km wide vegetated and partly active beach ridge plain, then 10-12 km wide salt flats. There is vehicle access to both beaches via the beaches to the north and east.

Figure 4.368 The northern Urala Creek mouth with beach WA 1657 on its northern side.

WA 1659-1661 **URALA-ROCKY PT**

No.	Beach	Rating	Type	Length
WA 1659	Whittackers Ck	2	R	3.3 km
WA 1660	Urala	2	R	8.7 km
WA 1661	Rocky Pt (S)	2	R	4 km
Spring & neap tidal range = 1.9 & 0.6 m				

The coast to the east of **Locker Point** sweeps in a series of six longer arcurate beaches for 25 km to the mouth of the Ashburton River. The first three beaches (WA 1659-1661) extend continuously for 16 km, with each separated by low sandy forelands. The Urala homestead backs beach WA 1660, with vehicle tracks from the homestead reaching all the beaches.

Beach **WA 1659** commences at Locker Point and curves gently to the east for 3.3 km to a subdued sand foreland. The low energy reflective beach receives waves averaging less than 0.5 m. It has a narrow beach face backed by generally vegetated beach to foredune ridges extending up to 2.5 km to the south. Whittackers Well is located behind the western end of the beach, with vehicle tracks paralleling the rear of the beach. Beach **WA 1660** is an 8.7 km long relatively straight northeast-trending beach, which terminates at the next slight sandy foreland. Urala homestead is located 1 km south of the northern end of the beach, with station tracks paralleling the rear of the beach. The reflective beach is backed by a moderately active foredune extending up to 300 m inland then vegetated foredune ridges for anther 1-2 km.

Beach **WA 1661** commences at the foreland and curves gently to the northeast for 4 km to Rocky Point, which marks the mouth of a 100 m wide rock-fringed tidal creek. The beach is backed by a 1-2 km wide series of beach-foredune ridges than are being actively transgressed by a dune sheet, which widens to 1 km against the northern creek.

ASHBURTON RIVER DELTA

The Ashburton River is one of the larger Pilbara river systems. The river extends 550 km to the southeast deep into the Pilbara, and drains a much of the central Pilbara region. It reaches the coast 20 km west of Onslow, the delta front extending for 21 between Rocky Point in the west and Entrance Point to the east and contains of beaches WA 1661-1666. The river only flows following occasional heavy rain, usually associated with tropical cyclones. It has built a 2-3 km wide deltaic plain out from the river mouth, consisting of splays of beach to foredune ridges and some intertidal swales. The site of Old Onslow (1883-1925) is located on an inner beach ridge splay 4 km southeast of the present river mouth. The plain extends for 10 km to the west and 9 km to the east with a total area of approximately 4500 ha.

WA 1662-1666 **AUHBURTON RIVER DELTA**

No.	Beach	Rating	Type	Length
WA 1662	Rocky Pt (N)	2	R	2.7 km
WA 1663	Baresand Pt (S)	2	R	2.6 km
WA 1664	Ashburton R (S)	2	R	2.3 km
WA 1665	Ashburton R (E)	2	R	6 km
WA 1666	Entrance Pt	2	R	2.6 km
Spring & neap tidal range = 1.9 & 0.6 m				

Rocky Point marks the westernmost distributary of the Ashburton River delta, with the creek flowing out in lee of the intertidal rock reefs that form the point. To the east of the creek beach-foredune ridges extend for 6 km to the Ashburton River channel. Beaches WA 1662-1666 form the sandy shoreline of the main delta front.

Beach **WA 1662** commences on the northern side of Rocky Point tidal creek and curves slightly to the northeast for 2.7 km to a low sandy foreland. The reflective beach is initially backed by vegetated beach ridges, which destablised to the north resulting in a 600 m wide band of active transverse dunes, which are overriding the outer ridges. Up to 30 ridges of the delta extend for 2 km to the southeast.

Beach **WA 1663** continues to the east of the foreland as a curving 2.6 km long beach to Baresand Point, which marks the mouth of a narrow tidal creek, a distributary of the main Ashburton River channel. The active dunes of its western neighbour narrow to a 50 m wide active foredune, backed by the beach-foredune ridge plain.

Beach **WA 1664** extends for 2.3 km from **Baresand Point** to the 300 m wide Ashburton River mouth (Fig. 4.369). The low energy reflective beach faces northwest. The central section of the beach is backed by a 1.5 km wide beach-foredune ridge plain, while the low bare ends of the beach are periodically eroded by major flood events. River mouth sand bars extend up to 1 km off the mouth.

Figure 4.369 The Ashburton River mouth.

Beach **WA 1665** commences at the **Ashburton River** mouth and sweeps to the east then northeast for 6 km to Entrance Point, the mouth of a second distributary channel. This is a low long beach that is backed by 200-300 m wide beach ridges in the west, which narrow to a low unvegetated beach to the east. The eastern beach-spit is formed by longshore transport of sand from the river mouth to the east, resulting in a dynamic beach that varies in shape and length.

A dynamic, migratory shallow creek mouth marks Entrance Point, with beach **WA 1666** extending for 2.6 km to the west then southwest to terminate at sandy Casuarina Point. This is a low, narrow dynamic beach, which continues the longshore transport of sand from the river mouth. The 5 km of beaches either side of Entrance Point are backed by a 2 km wide series of up to 10 discontinuous beach ridges separated by mangrove-filled swales, the entire system representing the northwards and eastward progradation of the eastern deltaic plain. The entire system rejoins the main shoreline at Casuarina Point.

WA 1667-1669 CASUARINA PT-BACK BEACH

No.	Beach	Rating	Type	Length
WA 1667	Casuarina Pt	2	R	3.4 km
WA 1668	Hooley Ck	2	R	2.8 km
WA 1669	Back Beach	2	R+LTT	7.8 km
Spring & neap tidal range = 1.9 & 0.6 m				

Casuarina Point marks the location where the dynamic migratory spits of the eastern Ashburton River delta link to the shore, with a near continuous shoreline curving round to the east then northeast for 14 km to Beadon Point. In between are three sandy beaches (WA 1667-1669).

Beach **WA 1667** commences in lee of the elongate spit that marks the eastern end of beach WA 1666. As this spit varies in length so too does the location of Casuarina Point and the length of the beach. Furthermore the eastern end of the beach is also migratory, sometimes deflecting the mouth of its eastern boundary Hooley Creek by more than 1 km. The beach averages 3.4 km in length and faces north. It consists of a reflective beach backed by a low foredune and older beach ridge-recurved spits. A vehicle track runs out straight across the old ridges to reach the centre of the beach.

Beach **WA 1668** commences on the eastern side of Hooley Creek and continues east for 2.8 km to Four Mile Creek. Depending on creek mouth conditions and the state of the migratory sand spits it may be continuous or cut in two. The low, narrow migratory spits move across the front of an inner core of older 1 km wide beach ridges, backed by 6 km wide salt flats.

Back Beach (**WA 1669**) extends from the 100 m wide Four Mile Creek for 7.8 km to the northeast to 20 m high Beadon Point. This is a slightly more exposed and higher energy beach with a moderately steep beach face fronted by a 50 m wide low tide bar. A sealed road from Onslow provides access to a car park and shelter at the northern end, which marks the main swimming area, while a gravel road runs down to the creek mouth at the southern end. The beach is backed by a 600 m wide outer and up to 3 km wide inner series of foredunes, then up to 9 km of salt flats. The Onslow Road crosses the rear foredune ridges.

WA 1670 ONSLOW

No.	Beach	Rating	Type	Length
WA 1670	Onslow	1	R+sand flats	3 km
Spring & neap tidal range = 1.9 & 0.6 m				

Onslow is a remote northwest town situated 1386 km north of Perth. The nearest towns are Exmouth 90 km across the gulf to the west, Carnarvon 400 km to the south and Dampier, 200 km to the northeast. Onslow itself lies 85 km off the North West Coast Highway. Originally a pearling centre, today Onslow is a quiet town of 800 people nestled along the usually quiet shores of Beadon Bay. The main street parallels Beadon Bay, with Onslow caravan park located at the western end of the bay overlooking the beach (Fig. 4.370).

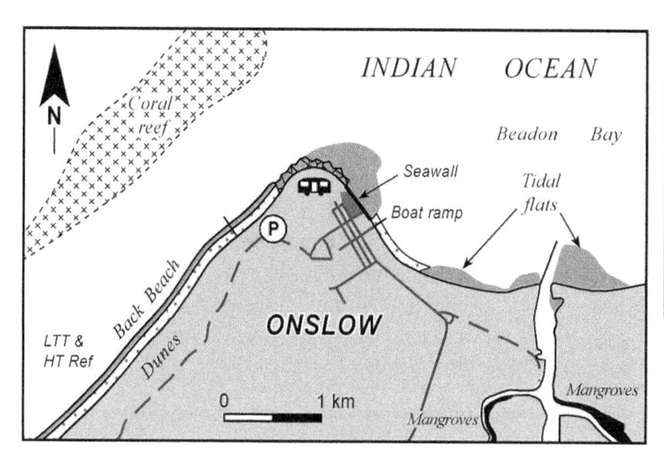

Figure 4.370 Onslow is located along the sheltered north-facing shore of Beadon Bay.

The Onslow coast is protected from most waves by its northerly orientation and scattered offshore reefs and shoals. At the same time the tide begins to increase in range, and for the first time in Western Australia reaching 2 m at Onslow and 3.7 km at Dampier. Onslow also prides itself on being the 'cyclone capital' of Australia, owing to the fact it was hit by at least 12 tropical cyclones during the 20th century, an Australian record.

Onslow beach (**WA 1670**) forms the western shore of Beadon Bay (Fig. 4.371). The 3 km long beach, curves to the southeast then east to the mouth of Beadon Creek. The creek houses the Onslow fishing fleet and its mouth is trained by a 500 m long rock groyne In addition to its northerly orientation it receives additional protection from Beadon Point. Waves are usually very low to calm at the shore. The beach itself is eroding into the backing pindan scarps and is backed by several seawalls and crossed by low rock groynes. At high tide the beach is narrow and awash, while at low tide it is fronted by tidal sand flats up to a few hundred meters wide. The township of Onslow, including the Bindi Bondi aboriginal community, back the beach, together with a road out to the creek. The eastern end terminates against the breakwater for the Beadon Creek marina.

Figure 4.371 Onslow with Back Beach (top) and the main town beach (foreground).

WA 1671-1676 BEADON BAY

No.	Beach	Rating	Type	Length
WA 1671	Beadon Ck (E)	1	R+sand flats	3.9 km
WA 1672	Beadon Ck (E1)	1	R+sand flats	2.3 km
WA 1673	Beadon Ck (E2)	1	R+sand flats	900 m
WA 1674	Beadon Ck (E3)	1	R+sand flats	3.1 km
WA 1675	Coolgra Pt (S2)	1	R+tidal flats	1.9 km
WA 1676	Coolgra Pt (S1)	1	R+tidal flats	150 m
Spring & neap tidal range = 1.9 & 0.6 m				

The southern and eastern shores of **Beadon Bay** consists of a series of six sandy beaches (WA 1671-1676) that trend to the northeast for 14 km gradually decreasing in energy to terminate amongst the mangroves of Coolgra Point. All the beaches and associated barriers are backed by 4-6 km wide salt flats and none of the beaches are accessible by vehicle.

Beach **WA 1671** extends east then northeast of Beadon Creek for a distance of 3.9 km to the next small tidal creek. Wave energy increases slightly up the beach, which is fronted by sand flats. It is backed by 200-300 m wide zone of active dunes climbing to 30 m high, then a 1.5 km wide beach ridge plain and then salt flats. Beach **WA 1672** continues on the eastern side of the tidal creek for 2.3 km to a small sandy foreland. It faces northwest and receives low wind waves, which break across intertidal sand flats. The beach is backed by a splay of several older beach ridges, which widen, from 1 km in the west to 2 km in the east. Beach **WA 1673** is a 900 m long easterly trending spit which forms the eastern end of the beach complex, as well as the southern shore of a 200 m wide tidal creek. This is a low energy north-facing beach, rimmed by 200 m wide mangrove-fringed tidal shoals, while a deeper tidal channel forms its western boundary.

Beach **WA 1674** is a 3.1 km long northwest-facing sandy high tide beach, fronted by 200 m wide sand flats, which widen to 500 m off the northern boundary against a 200 m wide tidal creek. A series of 10 beach ridges widen from 700 m in the west to 2 km in the east, indicative of past easterly longshore sand transport.

Beach **WA 1675** commences on the western side of the tidal creek and trends to the northeast for 1.9 km to terminate at a mangrove-fringed shore. It is fronted by 200-300 m wide tidal flats and backed by a 500 m wide series of several beach ridges, which are separated from the mainland by a 6 km long tidal creek. The mangroves at the northern end dominate the reminder of the shore for 4 km around to Coolgra Point. The only clear patch is occupied by 150 m long beach **WA 1676**. It occupies a slight indention in the sandy shoreline, with low waves reaching the shore at high tide, while 500 m wide tidal flats are exposed at low tide.

WA 1677-1680 **BARRIGOOMBAR CK**

No.	Beach	Rating	Type	Length
WA 1677	Barrigoombar Ck (W)	2	R	6.7 km
WA 1678	Barrigoombar Ck (E)	2	R	1.8 km
WA 1679	Yardie Landing (W)	1	R+tidal flats	100 m
WA 1680	Yardie Landing (E)	1	R+tidal flats	1.3 km
Spring & neap tidal range = 1.9 & 0.6 m				

To the east of Coolgar Point is a 120 km long section of northeast-trending coast that extends to the lee of the Mangroves and Passage Islands groups and associated coral reefs. The islands extend up to 40 km offshore and together with the northerly coastal orientation, shelter the shore from most ocean waves, resulting in usually low wind waves to calms at the shore. As a result 98 km (82 %) of the shore is lined with mangroves drained by 45 tidal creeks. There are four low energy beaches (WA 1677-1680) either side of Barrigoombar Creek and Cane River, followed by a 40 km long section of mangroves, then beaches WA 1681-1889, the latter also interspersed with mangroves.

Beach **WA 1677** is a 6.7 km long, north-facing curving sandy beach that commences 3 km southeast of Coolgra Point and terminates in the east at the protruding mouth of Barrigoombar Creek (Fig. 4.372). The beach receives low wind waves and is reflective, with tidal flats to either end together with the creek mouth shoals extending 500 m off the beach. In the west the beach is backed by a series of well developed grassy foredunes up to 1 km wide, which to the east have been destabilised into a 200 m wide band of reddish sand dunes, rising to 20 m towards the rear. Barrigoombar Well backs the centre of the beach, which can be accessed, via vehicle tracks.

Beach **WA 1678** commences on the eastern side of the creek mouth and trends to the northeast for 1.8 km before grading into a mangrove-fringed shoreline. There is some dune instability at the creek mouth, with a 500 m wide band of foredune ridges extending to the east and behind the mangroves. A vehicle track reaches the eastern end of the beach.

Figure 4.372 Barrigoombar Creek, with beach WA 1677 to the right and 1678 to the left.

Beach **WA 1679** is located 4 km to the northeast and consists of a 100 m wide gap in the otherwise continuous mangrove-fringed shoreline fronted by 200 m wide tidal flats. The mangroves fringe a former beach, backed by the foredune ridges, which extends continuously for 7 km from Barrigoombar Creek to Yardie Landing.

Beach **WA 1680** is a 1.3 km long series of eastward-trending recurved spits that commences 1 km northeast of Yardie Landing creek mouth. A few mangroves dot the low energy shoreline, which is fronted by 500 m wide tidal flats, while dense mangroves back the spit, together with a 2.5 km long series of former spits.

WA 1681-1688 **PORT WELD-PT ROBE**

No.	Beach	Rating	Type	Length
WA 1681	Port Weld (N1)	1	R+beachrock	1.4 km
WA 1682	Port Weld (N2)	1	R+beachrock	400 m
WA 1683	Port Weld (N3)	1	R+tidal flats	700m
WA 1684	Port Weld (N4)	1	R+tidal flats	200 m
WA 1685	Thringa Is (S)	1	R+tidal flats	200 m
WA 1686	Pt Robe (S2)	1	R+tidal flats	1.5 km
WA 1687	Pt Robe (S1)	1	R+tidal flats	1.3 km
WA 1688	Pt Robe (S)	1	R+tidal flats	400 m
Spring & neap tidal range = 1.9 & 0.6 m				

Port Weld is a 400 m wide tidal creek and part of a series of tidal creeks and six low energy mangrove-fringed barrier islands (WA 1681-1688) that extend for 21 km to the southwest of Point Robe. There is no vehicle access to any of the beaches. The islands are bordered and backed by mangroves-lined creeks, then 3-4 km of salt flats abutting the low gradient coastal plain. The Robe River reaches the coast to the lee of Robe Point and during floods its distributaries flow across the low gradient 5 km wide coastal plain to the rear of the islands. The usually dry distributaries extend for about 10 km east of the point, with Peter and Six Mile creeks reaching the salt flats further east.

Beach **WA 1681** is a 1.4 km long section of a 3 km long barrier island that is fringed by mangroves and 300-400 m wide tidal creeks to either end. The beach is dominated by exposed eroding beachrock, with small pockets of sand below the low scarped shoreline. It is backed by a low 200 m wide foredune system, then the mangroves and salt flats.

Beach **WA 1682** is a 400 m long strip of beachrock and small patches of sand that occupies the western end of the next island. The island consists of a series of two diverging recurved spits, now largely embedded in mangroves. Three hundred metre wide tidal creeks lie to either side of the beach, with tidal shoals extending 600 m offshore.

Beaches **WA 1683** and **1684** are two adjoining 700 m and 200 m long beaches. They occupy the centre of the next 3.3 km long barrier and consist of reflective high tide beaches fronted by 200 m wide tidal flats, with a patch of mangroves separating the two and bordering either end. Five hundred metre wide mangrove-lined tidal creeks border the island.

Beach **WA 1685** is a 200 m long patch of west-facing mangrove-dotted sand. It is located at the western end of 3.5 km long barrier island-spit, that is otherwise encased in mangroves. It is fronted by 200-300 m wide tidal flats then a 200 m wide tidal channel. Thringa Island is located 2 km northeast of the beach.

Beach **WA 1686** and **1687** occupy the southern arm of the northernmost of the barrier islands. The beaches are 1.5 km and 1.3 km long respectively, face northwest and are separated and fringed by mangroves. The low energy beaches are fronted by 1-1.5 km wide tidal flats, and backed by the low 300 m wide barrier, then 2 km wide salt flats. The sandy usually dry **Robe River** mouth reaches the shore to the lee of the island and links to the tidal creeks, which border the island. Beach **WA 1688** is located immediately west of Point Robe and occupies the northern part of the island. It consists of a 400 m long sandy reflective beach, bordered by mangrove-fringed beaches.

WA 1689-1697 **MT SALT-FORTESCUE R**

No.	Beach	Rating HT	Rating LT	Type	Length
WA 1689	Mt Salt (S2)	2	1	R+tidal flats	3 km
WA 1690	Mt Salt (S1)	2	1	R+tidal flats	400 m
WA 1691	Mt Salt	2	1	R+tidal flats	6.1 km
WA 1692	Mt Salt (N1)	2	1	R+tidal flats	3.4 km
WA 1693	Coonga Ck (S1)	2	1	R+tidal flats	2.1 km
WA 1694	Coonga Ck (S)	2	1	R+tidal flats	2.3 km
WA 1695	Diver Inlet (S)	2	1	R+tidal flats	3.7 km
WA 1696	Diver Inlet (N)	2	1	R+tidal flats	600 m
WA 1697	Diver Inlet	2	1	R+tidal channel	200 m

Spring & neap tidal range = 3.4 & 1.0 m

To the east of Point Robe is a 40 km long section of low gradient mangrove-fringed shoreline, backed by 6-10 km wide salt flats and drained by more than 20 tidal creeks. The area is backed by Mardie Station, with the homestead located 13 km inland and no vehicle access to the shore.

The next beach encountered is located 14 km northwest of the station and 5 km due west of the low Mount Salt, all of 16 m high. The 3 km long beach (**WA 1689**) emerges from a mangrove-fringed shore, and trends to the northeast finally curving east into a 200 m wide creek mouth (Fig. 4.373). Tidal flats widen to 300 m off the creek mouth. The beach is part of a 5 km long south-trending series of now largely inactive recurved spits. It is backed by a 300-400 m wide low barrier, then 5 km wide salt flats. Beach **WA 1690** is located on the eastern side of the creek mouth and is a curving 400 m long north-facing high tide beach, fringed by mangroves. It is fronted by 300 m wide tidal flats and backed by a 300 m wide low barrier then the salt flats. The barrier continues northeast for 15 km and contains beaches WA 1691-1694.

Figure 4.373 Beaches WA 1689 and 1690 are located either side of a mangrove-lined tidal creek.

Beach **WA 1691** commences 1.5 km to the north at the end of a mangrove-lined section of the barrier and on the northern side of a blocked creek mouth. This and other blocked creeks to the north may flow during major flooding events from the Robe and Fortescue rivers. The crenulate beach trends to the northeast for 6.1 km to the next blocked creek mouth. It consists of a high tide beach fronted by tidal flats. The backing barrier ranges from 200-400 m in width and up to 12 m in height and is in turn backed by 3-4 km wide salt flats, with Mount Salt located 3 km east of the southern dry creek. Beach **WA 1692** continues northeast of the creek mouth for another 3.4 km to the next protruding blocked dry creek. It is a crenulate beach with some outcrops of beachrock, backed by a 300 m wide barrier capped by a linear 11 m high foredune, then the salt flats. It terminates at the next blocked creek.

Beach **WA 1693** continues on the northern side of the dry boundary creek for another 2.1 km to the next protruding dry creek. It is a gently curing northwest-facing, low energy high tide sandy beach backed by a 0.5-1 km wide low barrier island, then 1-2 km wide salt flats. It is bordered in the northeast by an open 50 m wide, mangrove-lined tidal creek, with protruding creek mouth beachrock-shoals forming the boundary with beach **WA 1694**. This beach trends east-northeast for 2.3 km to the 100 m wide open mouth of **Coonga Creek**, a larger tidal creek draining the backing salt flats. It is backed by a 400 m wide barrier, capped by a 12 m high foredune called Delaney Hill, then the salt flats.

Beach **WA 1695** extends from the Coonga Creek for 3.7 km to the northeast, terminating at a beachrock protrusion that separates it from beach WA 1696. The beach is backed by a 400 m wide barrier containing a central foredune up to 15 m high. Beach **WA 1696** is a double crenulate 600 m long beach, fringed by a mangrove- dotted beachrock foreland, with a high tide sandy beach in between fronted by 500 m wide creek mouth shoals. It terminates at the mouth of Diver Inlet, which connects with the main channel of the **Fortescue River**, 5 km to the south. Beach **WA 1697** forms the northern end of the barrier and trends for 200 m into the inlet, as a sandy east-facing beach fronted by the 200 m wide tidal channel. During major floods much of the Fortescue River discharge flows out this channel. There is vehicle access across the backing 1 km wide salt flats and vehicle tracks running the length of the barrier providing access to all three beaches.

The **FORTESCUE RIVER** extends 550 m to the east and drains the northern side of the Hammersley Range. The river leaves the ranges at Mount Nicholson where it breaks into two distributaries. It crosses the highway 10 km to the north at Jilanjilan Pool beyond which it further bifracates into several usually dry distributaries, which form a broad, low gradient delta, fringed along its 21 km shoreline by beaches 1689-1697. Then distributaries reach the low gradient salt flats and mangroves that fringe the coast between Mount Salt and Driver Inlet, the latter the main river channel.

GREAT SANDY ISLAND NATURE RESERVE

Coast length:	103 km	(4677-4780 km)
Beaches	26	(WA 1684-1710)

WA 1698-1705 DIVER INLET-JAMES PT

No.	Beach	Rating HT	LT	Type	Length
WA 1698	Diver Inlet (E1)	2	1	R+tidal flats	200 m
WA 1699	Diver Inlet (E1)	2	1	R+tidal flats	1.3 km
WA 1670	James Pt (S3)	2	1	R+tidal flats	800 m
WA 1671	James Pt (S2)	2	1	R+tidal flats	900 m
WA 1672	James Pt (S1)	2	1	R+tidal flats	150 m
WA 1673	James Pt	2	1	R+tidal flats	100m
WA 1674	James Pt (E)	2	1	R+tidal flats	250 m
WA 1670	Mt Rough	2	1	R+tidal flats	350 m

Spring & neap tidal range = 3.4 & 1.0 m

To the northeast of Diver Inlet the shoreline and hinterland changes dramatically, as the low barrier island and extensive salt flats give way to more bedrock dominated shoreline, with bedrock reaching the shore at James Point and Cape Preston. This is the first bedrock at the shore since the Cape Range 240 km to the southwest. The spring tides increases to 3.4 m and the shoreline, which receives only low wind waves and calms, becomes increasingly tide-modified. Wide tidal flats front the limited beaches and mangroves fringe all but the Cape Preston beaches. The only vehicle access to the shore is around Mount Rough, though there are vehicle tracks on Cape Preston, which is separated from the mainland by 1 km wide mangroves and salt flats.

Beach **WA 1698** is a migratory multiple-recurved spit that occupies the eastern side of Diver Inlet (Fig. 4.374). It consists of a low, narrow 200 m long north-facing beach, with spits curving round either end, backed by 500 m wide band of mangroves then an older shoreline. Beach **WA 1699** is a continuation of the older shoreline and emerges from the mangroves for a distance of 1.3 km, trending east towards the mouth of the next 400 m wide inlet, which connects with the eastern distributary of the Fortescue River. Both beaches are fringed by mangroves and fronted by 1-1.5 km wide tidal flats.

Figure 4.374 Beach WA 1698 is a curving spit backed by mangroves (right), which grades into beach WA 1699 to the left.

Beach **WA 1700** is located on the eastern side of the inlet and is a very low energy high tide beach, partly fringed with mangroves. It trends to the north for 800 m to a mangrove-lined protrusion in the shore. Beach **WA 1701** commences on the northern side of the protrusion and continues for another 900 m until the mangroves again dominate the shore. Continuous 1 km wide tidal flats front both beaches. The mangroves continue for 2 km toward James Point

James Point is a 22 m high metasedimentary ridge with three small beaches located on the point (Fig. 4.375). Beaches **WA 1702** and **1703** are located on the western side of the point and are 150 m and 100 m long respectively. They consist of a high tide beach fronted by some outcrops of bedrock and backed by a partly active foredune that has blown over the backing grassy bedrock slopes, while the tidal flats narrow to 100 m at the point. Beach **WA 1704** is located on the eastern side of the point and is the exposed 250 m section of a 500 m long migratory spit that is moving south into a 2 km wide embayment dominated by mangroves. Yagobiddy Creek enters the southern part of the embayment 1.5 km east of the point.

Figure 4.375 James Point marks the boundary of the Carnarvon and Pilbara coasts.

The next 7 km of shore is dominated by mangroves and trends north in lee of the protecting Potter and Carey islands. Beach **WA 1705** is located 2 km due west of Mouth Rough and represents a 350 m long gap in the mangroves. Waves reach the beach at high tide, with 500 m wide tidal flats exposed at spring low tide. The beach is part of a 2 km long recurved spit, the northern arm of which form the southern entrance to a 1 km wide mangrove-lined tidal creek, which separated Cape Preston from the mainland. A vehicle track runs along the crest of the spit to the inlet.

Region 6 Pilbara Region: Cape Preston to Port Hedland

Length of Coast: 466 km (4770-5236 km)
Number of beaches: 188 (WA 1709-1896)
Fig. 4.376

The Pilbara coast is a predominately east-west trending section of low energy shoreline, located between the bedrock Cape Preston in the west and the low calcarenite shoreline of Port Hedland 250 km to the east. In between is 466 km of shoreline containing 188 beaches, which occupy only 132 km (28%) of the shore, the remainder occupied by mangroves, calcarenite barriers and the bedrock highs of the Dampier Peninsula and Cape Lambert. In addition six medium-sized rivers (Maitland, Harding, Sherlock, Peewah, Yule and Turner) reach the coast amidst mangroves, salt-flats and low gradient sand and mud flats. The major towns of Dampier, Karratha, Wickham-Point Samson and Port Hedland) have all been developed to export the Pilbara iron ore, which is delivered by rail to the ports. In addition Dampier services the North West shelf gas fields.

Coastal towns: Dampier, Karratha, Point Samson, (Cossack), Port Hedland
Other locations (km inland): Fortesque (25 km), Roebourne (10 km), Whim Creek (15 km)

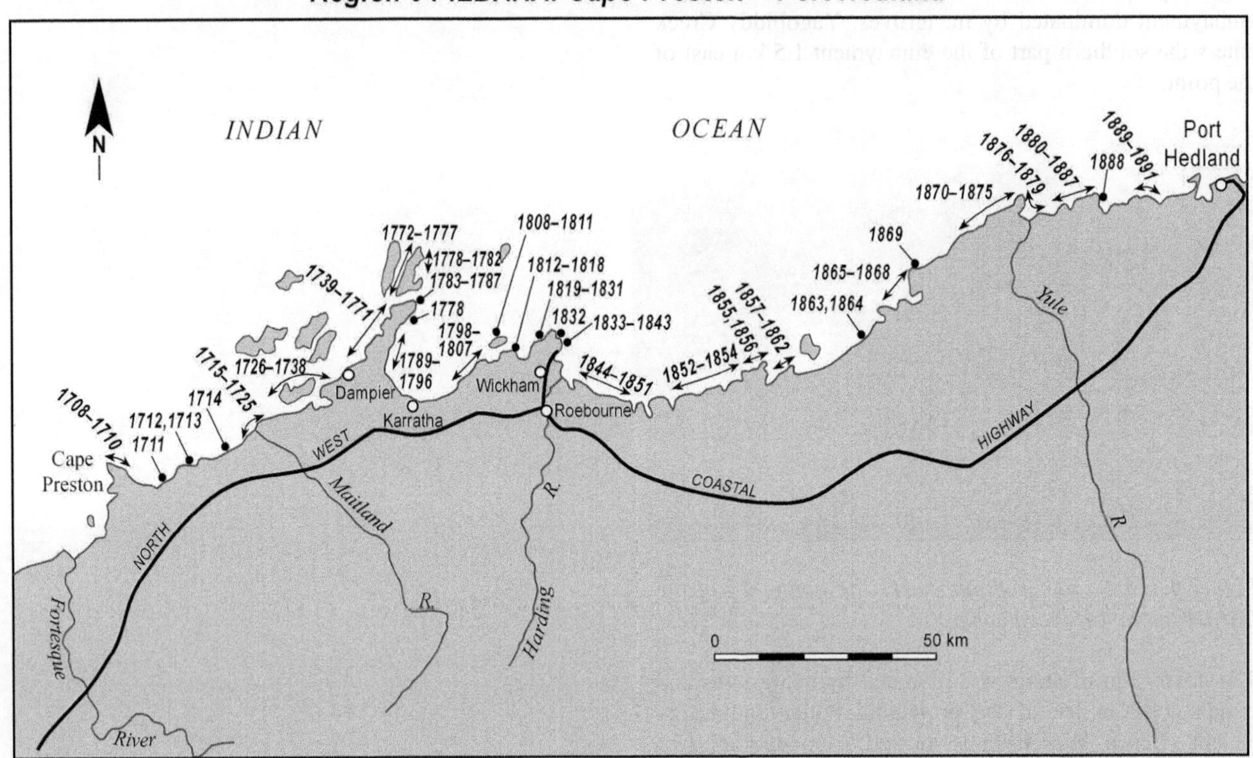

Figure 4.376 Region 6 includes the 466 km long Pilbara coast, which extends from Cape Preston to Port Hedland, and includes beaches WA 1708-1891.

EVOLUTION OF THE PILBARA SHORELINE

Sea level reached it present level along the Pilbara coast about 6 500 year ago. At that time the nearshore zone was deeper, and the coral reefs, which dot much of the shelf today, were non-existent or deeper. As a consequence of the deeper and less reef congested nearshore waters wave energy was higher along the shore. At this time many of the existing beaches, barriers and barrier islands, including those now fringed by mangroves were deposited. As sediment was delivered to the coast by the Ashburton, Cane, Robe, Fortesque, Yaanyare, Maitland, Harding, Sherlock, Peewah, Yule, Turner and De Grey rivers, the backing lagoons and nearshore gradually shoaled. At the same time the coral reefs grew to the surface, with both processes lowering waves at the shore and permitting the mangroves to spread. In addition a slight 1-2 m fall in sea level further shoaled the nearshore and backing lagoons. Today the lagoons have largely infilled to form high tide salt flats, while the shallow shoreline, provides extensive intertidal zones for mangroves development, and overall waves are lower along the shore.

WA 1706-1710 **CAPE PRESTON**

No.	Beach	Rating HT LT		Type	Length
WA 1706	Preston Spit (S)	2	1	R+sand flats	1.2 km
WA 1706	Preston Spit (N)	2	1	R+sand flats	4.5 km
WA 1706	Cape Preston (E1)	2	1	R+sand flats	400 m
WA 1706	Cape Preston (E2)	2	1	R+sand flats	1.7 km
WA 1706	Cape Preston spit	2	1	R+sand flats	5.3 km
Spring & neap tidal range = 3.4 & 1.0 m					

Cape Preston is an 800 ha vegetated bedrock outcrop, capped by 65 m high Mount Preston. It is the most prominent coastal landform in this region. Considerable sediment has accumulated to either side of the cape forming beaches WA 1706 and 1707 to the west, and beaches WA 1708-1710 to the east. The sediment has been sourced from the extensive sand flats that extend 1.5 km to the west and north of the cape. Beaches WA 1706 and 1707 lie either side of Preston Spit, a sandy cuspate foreland that has prograded west as a 1 km wide series of up to 30 foredune ridges, the highest rising to 15 m. Beach **WA 1706** is 1.2 km long and occupies the southern part of the salient. It faces west and curves from the mangrove-lined creek mouth to the foreland. It is backed by foredune ridges widening to 1 km in the north and fronted by 400 m wide tidal flats. Beach **WA 1707** forms the northern side of the salient, extending for 2.5 km as a curving beach backed by foredune ridges, then for another 2 km around a slightly protruding section of the backing cape, finally terminating against the final 500 m of bedrock shore that forms the cape (Fig. 4.377). Sand flats widen to 1.5 km off the northern end of the beach.

Figure 4.377 Cape Preston with beach WA 1707 to the right.

Beach **WA 1708** commences 500 m west of the tip of the cape and is a crenulate northeast-facing, 400 m long beach, bordered by low rocky points, with a central line of rocks almost cutting the beach in two. The vegetated slopes of 65 m high **Mount Preston** back the beach. Beach **WA 1709** continues on the eastern side of the rocks for 1.7 km to the next outcrop of rocks. The reflective beach has bedrock outcrops and is fronted by sand flats, which narrow from 500 m at the cape to 100 m at the eastern end. A 20 m high foredune backs the beach then the slopes of Mount Preston.

Beach **WA 1710** commences on the eastern side of the bedrock protrusion and curves to the southeast against gentle bedrock slopes, then east as a sand spit for a total length of 5.3 km. A continuous foredune up to 15 m high backs the beach, abutting the vegetated bedrock slopes to the west and forming the 3 km long Cape Preston spit to the east. Beachrock outcrops increasingly along the spit, and forms the eastern 500 m tip of the spit.

WA 1711-1713 **GNOOREA**

No.	Beach	Rating HT LT		Type	Length
WA 1711	Gnoorea (W)	2	1	R+tidal flats	1.7 km
WA 1712	Gnoorea (pipeline)	2	1	R+rock flats	300 m
WA 1713	Gnoorea (E)	2	1	R+tidal flats	5.5 km
Spring & neap tidal range = 3.4 & 1.0 m					

To the east of Cape Preston the shoreline trends east-northeast for 40 km to the high East Intercourse Island. The shoreline is for the most part low gradient, with 36 km (90%) fringed by mangroves and wide tidal flats, together with low energy barrier islands associated with the Yanyare and Maitland rivers (beaches WA 1714-1724), and one section of protruding beachrock shore (beaches WA 1711-1713). The only vehicle access is to the Goonrea beachrock section where a gas pipeline crosses shore at beach WA 1712.

Beach **WA 1711** emerges from the 7 km long, 1 km wide belt of mangroves that separate it from Cape Preston spit. The beach is clear of mangroves for 1.7 km before being fringed by a narrow band of mangroves for the next 2.5 km to the pipeline point. The beach faces northwest and is fronted by 2.5 km wide tidal flats and backed by a 100-200 m wide band of low foredunes, with a vehicle track reaching the centre of the beach. The pipeline parallels the rear of the beach.

Beach **WA 1712** occupies a 300 m long convex, north-facing sandy point, fringed by 50-100 m wide intertidal beachrock flats. It is backed by beachrock and a 20 m high foredune. The pipeline, constructed in 1997, crosses the eastern end of the beach, with a gravel road, built for the pipeline, also terminating at the point.

A band of beachrock extends for 300 m east of the point to the beginning of beach **WA 1713** that continues to the east for 5.5 km as a slightly crenulate beach to the next protruding beachrock point (Fig. 4.378). The beach is an older Holocene barrier that has been lithified into beachrock, which is exposed in places along the shore. It is backed by a 20 m high foredune, and a 200-300 m wide series of older lower foredunes, then 2 km wide salt flats. A vehicle track parallels the rear of the mangroves out to the point. Mangroves fringe the eastern end of the beach in lee of the point, which protrudes 600 m to the north.

Figure 4.378 Beach WA 1713 terminates at a low mangrove-fringed beachrock point.

WA 1714-1725 **YANYARE-MAITLAND R DELTA**

No.	Beach	Rating HT	LT	Type	Length
WA 1714	Yanyare R	2	1	R+tidal channel	300 m
WA 1715	Boodamurra Hill (S)	2	1	R+tidal flats	400 m
WA 1716	Boodamurra Hill (N)	2	1	R+tidal flats	700 m
WA 1717	Maitland R (W4)	1	1	R+tidal flats	200 m
WA 1718	Maitland R (W3)	1	1	R+tidal flats	600 m
WA 1719	Maitland R (W2)	2	2	R+tidal channel	300m
WA 1720	Maitland R (W1)	1	1	R+tidal flats	1 km
WA 1721	Maitland R (W)	2	2	R+tidal channel	300m
WA 1722	Maitland R (E)	2	2	R+tidal channel	400m
WA 1723	Maitland R (E1)	1	1	R+tidal flats	300 m
WA 1724	Maitland R (E2)	1	1	R+tidal flats	500 m
WA 1725	Maitland R (E3)	1	1	R+tidal flats	900 m

Spring & neap tidal range = 3.7 & 0.4 m

The **Yanyare** and larger **Maitland rivers** have deposited a 20 km long deltaic front consisting of low energy barrier islands, many fringed by mangroves, and backed by 1-2 km wide salts flats, which are drained by 10 tidal creeks, including those connected directly to the rivers. The eleven beaches along this section (WA 1214-1224) occupy only 6 km of the 20 km of shoreline, the remainder dominated by mangroves and creeks mouths. The delta front occupies an area of approximately 25 km^2. All the beaches are subject to changes induced by episodic flooding of the rivers, extension and retreat of the fringing mangroves, and in some places inlet mouth dynamics and channel migration. As a consequence the nature and length of some of the beaches changes over time. The delta front extends from the lee of the eastern Goorea point to the lee of East Intercourse Island. There is no vehicle access to the shore.

Beach **WA 1714** is located at the mouth a 200 m wide tidal creek, which links to the mouth of the usually dry Yanyare River, 3.5 km to the east. The beach is located 4.5 km east of the Goorea Point, with a 300 wide band of mangroves in between and older barrier islands behind. The 300 m long beach is the southern end of a recurved spit that has been built by sand moving into the western side of the creek mouth. It is free of mangroves because it is paralleled by the deeper inlet tidal channel.

Two mangrove-fronted barrier islands dominate the next 4 km of shore to the 2 km long convex-shaped Boodamurra Hill barrier island. Beach **WA 1715** occupies 400 m at the western end of the island where it is fronted by a deeper tidal channel, with mangroves to either side. Beach **WA 1716** occupies the eastern end of the island and extends for 700 m in part as an actively eastward growing spit, that is extending into the next creek mouth (Fig. 4.379). Both beaches are fronted by 1-2 km wide tidal flats.

Figure 4.379 Beach WA 1716 (right) terminates as a spit, which extends into the boundary tidal creek.

Beaches WA 1717-1719 occupy the eastern end of a 3.5 km long barrier island whose western half is fringed by mangroves, while beachrock outcrops to the east. Beach **WA 1717** occupies a 200 m long gap in the mangroves, with tidal flats extending 2-3 km offshore. It is part of a barrier spit, which continues east of the mangroves for another 600 m as beach **WA 1718**. The spit terminates as it attaches to a 100 m long outcrop of beachrock that forms the eastern boundary of the beach. The wide tidal flats and a tidal channel front the beach. Beach **WA 1719** commences at the beachrock and curves into the boundary creek mouth as a recurved spit for 300 m. The creek is a distributary of the Maitland River located 4 km to the east. All three beaches are backed by a 400 m wide series of older recurved spits that comprise the eastern end of the barrier, then 1-1.5 km wide salt flats.

Beaches 1720, 1721 and 1723 are located either side of the tidal creek that links with the 500 m wide Maitland River channel, 5 km to the southeast. Beach **WA 1720** is a 1 km long northwest-facing beach that occupies the northeastern shoreline of a 2 km long barrier island, the western end fringed with mangroves. The barrier island is the outermost of a 3.5 km wide sequence of barriers that have been deposited either side of the river mouth. The high tide beach is backed by a 500 m long section of unstable foredunes, while tidal flats extend 1 km eastward. Beach **WA 1721** continues to the northeast as a curving 300 m long active recurved spit, that forms the western shore of the river mouth a 100 m wide tidal creek. Beach **WA 1722** is a similar 400 long spit that occupies the eastern shore of the tidal creek. Both beaches parallel the deeper tidal channel, with channel and tidal flats extending 1 m to the north (Fig 3.1d).

Beach WA 1722 lies at the western end of a 2 km long north-facing, mangrove-lined barrier island, with beach **WA 1723** occupying a 300 m long gap in the mangroves towards the eastern end of the island. The island is one of a series of discontinuous spits that extend for up to 7 km east of the river mouth. It is backed by a 3.5 km wide series of inner barrier island-spits, mangroves and salt flats, that form the delta plain. Beach **WA 1724** occupies the eastern 400 m of the next 1 km long convex-shaped barrier island-spit. Mangroves are scattered off the shore, with tidal flats extending 1.5 km seaward.

Beach **WA 1725** is a 900 m long section of largely open sand and scattered mangroves, that occupies the western end of the eastern-most barrier spit. This is an irregular, 3 km long low narrow spit, largely encased in mangroves to either side. The spit terminates to the lee of the East Intercourse islands and is fronted by 2 km wide tidal flats.

DAMPIER COAST

The Dampier coast is dominated by a 50 km long, 3-5 km wide, 70-120 m high southwest-northeast trending ridge of sparsely vegetated dolerite that extends from West Intercourse Island, along the Burrup Peninsula to Dolphin Island. The bedrock high is connected to the mainland along the southern side of the peninsula by a 4-10 km wide area of salt flats. The peninsula is inturn ringed by the 42 high islands of the Dampier Archipelago. Much of the 120 km of rocky mainland shoreline is fringed by mangroves, with a series of 65 small beaches totalling only 18 km in length located on the west and north side of West Intercourse Island (WA 1726-1738), scattered along the northern shores of the peninsula (WA 1739-1771), around Dolphin Island (WA 1772-1783) and finally on the eastern side of the peninsula (WA 1784-1794). The port and town of Dampier is located on the peninsula in lee of East Intercourse Island, while the larger town of Karratha is located 15 km to the east across the salt flats on the mainland.

WA 1726-1729 **WEST INTERCOURSE IS (W)**

No.	Beach	Rating HT	Rating LT	Type	Length
WA 1726	W Intercourse Is (W1)	2	1	R+LTT	1.3 m
WA 1727	W Intercourse Is (W2)	2	1	R+sand flats	200 m
WA 1728	W Intercourse Is (W3)	2	1	R+sand flats	200 m
WA 1729	W Intercourse Is (W4)	2	1	R+sand flats	200 m
Spring & neap tidal range =3.7 & 1.0 m					

West Intercourse Island consists of three bedrock highs or 'islands' which reach a height of 70 m at False Summit, and are linked by mangroves and salt flats, and has a total area of 25 km^2. There is no vehicle access to the uninhabited island. Beaches WA 1726-1729 are located on the northwestern outcrop of the island and occupy its western and northern shore.

Beach **WA 1726** is the most exposed beach in the region, as it faces west down Mermaid Strait. It is a 1.3 km long sandy beach that receives wind waves averaging about 0.3 m. It is fronted by a 100 m wide low tide terrace and backed by a 200-300m wide series of low foredunes, then gently sloping bedrock that rises to 17 m. Mangrove-fringed low bedrock points border each end of the beach.

Beaches WA 1727-1729 are located along a 1.3 km long section of the northern shore of the island. Each is bordered by and backed by low bedrock and face north across the 500 m wide sand flats. Beach **WA 1727** is 300 m long, fringed by mangroves at its western end and a 100 m long point to the east. Beach **WA 1728** is a 100 m long pocket of sand between two low bedrock points, while beaches **WA 1729** is a more deeply embayed 150 m long pocket of sand, bordered by 300 m long mangrove-fringed points. Both beaches are backed by low splays of dune sand.

WA 1730-1738 **WEST INTERCOURSE IS (N)**

No.	Beach	Rating HT	Rating LT	Type	Length
WA 1730	W Intercourse Is (N1)	2	1	R+sand flats	200 m
WA 1731	W Intercourse Is (N2)	2	1	R+sand flats	300 m
WA 1732	W Intercourse Is (N3)	2	1	R+sand flats	300 m
WA 1733	W Intercourse Is (N4)	2	1	R+sand flats	100 m
WA 1734	W Intercourse Is (N5)	2	1	R+sand flats	200 m
WA 1735	W Intercourse Is (N6)	2	1	R+sand flats	500 m
WA 1736	W Intercourse Is (N7)	2	1	R+tidal flats	100 m
WA 1737	W Intercourse Is (N8)	2	1	R+tidal flats	50 m
WA 1738	W Intercourse Is (N9)	2	1	R+tidal flats	50 m
Spring & neap tidal range =3.7 & 1.0 m					

The northern and largest of the three interconnected West Intercourse 'islands' is fringed by mangroves along its east, south and western shores, with a series of small pocket, bedrock-bordered beaches (WA 1730-1735) along the central section of the 6 km long northern shore, together with three pockets on sand in amongst the rocks and mangroves on the northeast shore (WA 1736-1738). There is no vehicle access to the island or the beaches.

The northern six beaches all tend to face north into Mermaid Strait. They receive low wind waves generated in the strait and are fronted by sandy tidal flats averaging 400-500 m in width. They are all backed by sparsely vegetated bedrock slopes rising to 70 m at False Summit. Beach **WA 1730** is located at the northwestern tip of the 'island' and is a protruding 200 m long high tide beach, fronted by 200 m wide sand flats then the tidal channel that separates it from the western 'island', 1 km to the west.

Beaches **WA 1731, 1732, 1733** and **1734** are four adjoining pockets of high tide sand bordered and separated by low bedrock points and backed by dissected bedrock that rises to 40 m 500 m inland. They are 100-200 m long and occupy the mouth of usually dry creek beds. The are backed by low foredunes and face north across the 500 m wide sand flats.

Beach **WA 1735** is located 500 m to the east and is a 500 m long northwest-facing, slightly more exposed beach. It is fronted by the 500 m wide sand flats and backed by a 200 m wide band of low foredune ridges and wind blown sand then gently rising bedrock that reaches 30 m. The northern end of the beach is bordered by bedrock, which continues around the northern tip of the 'island' for 1.5 km to beach WA 1736 located on the northeast-facing shore of the island.

Beaches WA 1736-1738 are three pockets of sand that occupy gaps in the bedrock and mangroves along a 500 m long section of the 1.5 km long northeastern side of the 'island'. They are fronted by tidal flats that extend 3 km to the east and linked to three smaller islands (West Mid and East Mid Intercourse and Mistaken islands). Beach **WA 1736** is 100 m long and wedged between a northern mangrove-fringed low rocky point and a southern bedrock point, and backed by a 100 m wide zone of vegetated dunes. Beach **WA 1737** is a 50 m long pocket of sand bordered by low sloping bedrock, while beach **WA 1738** occupies the northern 50 m of a 200 long sandy plain that is elsewhere fronted by mangroves and rock outcrops.

WA 1739-1743 DAMPIER

No.	Beach	Rating HT	Rating LT	Type	Length
WA 1739	Dampier (Hampton)	2	3	R+tidal flats	900 m
WA 1740	Dampier (boat club)	2	3	R+tidal flats	50 m
WA 1741	Dampier (pool)	2	2	R+rock flats	800 m
WA 1742	Dampier (Harbour Mound)	2	3	R+rock flats	200 m
WA 1743	Dampier (power station)	2	2	R+rock flats	600 m

Spring & neap tidal range = 4.7 & 1.0 m

Dampier is located on the northern shores of Burrup Peninsula, with the major jetties and port facilities located on and in lee of East Intercourse Island (Fig. 4.380). The township of 2000 was constructed in the 1960's to provide a port facilities for the export of iron ore from the Hammersley Ranges. The long iron ore trains load ships at two harbour areas at Parker Point north of the town, and on East Intercourse Island west of the town. It is also used as a port facility for the Woodside oil and gas platforms, located up to 120 km off the coast. The township is located on slopes looking west across Hampton Harbour and the Hammersely Channel towards the Dampier Archipelago. The 3 km of public foreshore is given over to a range of recreational facilities, as well as the site of the towns five public beaches (WA 1739-1743). The two main recreational beaches are, Hampton and the swimming pool beach. Both beaches are well protected from ocean swell by the island. However the

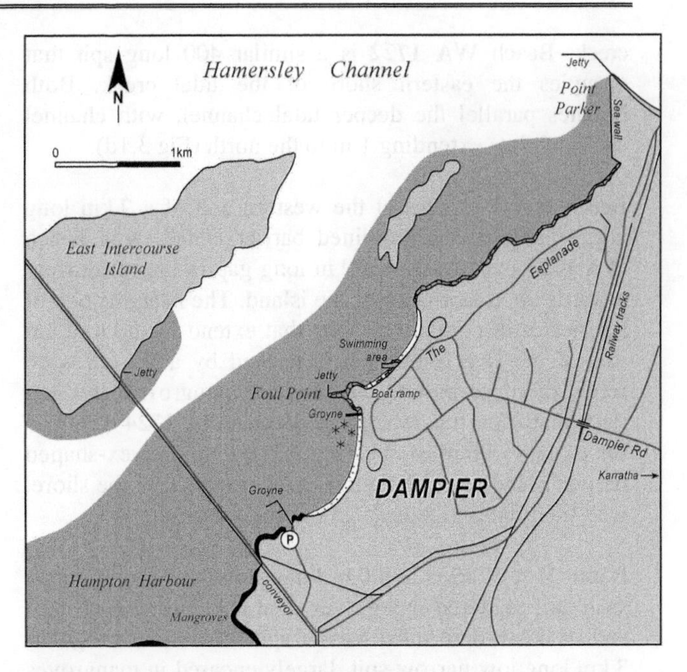

Figure 4.380 Dampier is located on rising ground that faces west across wide tidal flats to West Intercourse Island and the Hammersely Channel. Insert of Hearson Cove (beach WA 1794).

tide range in Dampier reaches 3.7 m resulting in a steep beach at high tide fronted by 1 km wide tidal flats at low tide.

Hampton Harbour beach (**WA 1739**) is the main recreational beach for Dampier (Fig. 4.381). It is a curving west to northwest-facing 900 m long strip of coarse high tide sand, with a mixture of sand and rock-flats exposed at low tide. The southern half of the beach is undeveloped, while the Esplanade runs along the rear of the northern half of the beach, which is also backed by Hampton sports oval and picnic and bar-b-que facilities located under a grove of coconut palms. Beach **WA 1740** is located at the northern end of the beach between two rock groynes that have been constructed either side of the main boat ramp. The beach is a 50 m long pocket of sand located next to the boat ramp and is used more for boats and landing passengers than for swimming. A large car park is located behind and to the west of the beach, occupying much of Foul Point.

Fig. 4.381 Back Beach is Dampier's main beach with the boat club and ramp at the northern end.

Beach **WA 1741** extends for 800 m to the north of 10 m high Foul Point. It has a curving-crenulate west-facing shoreline with several rock outcrops along the beach. A steep sandy beach dominates at high tide, with extensive intertidal rocks exposed a low tide. The Esplanade runs along the rear of the beach, with a netted swimming pool is located in a cleared area in the centre, parking at several locations and a sports oval located behind the northern end. Residential housing also backs the northern half of the beach. The northern end of the beach is bordered by a seawall and groyne that have been constructed to protect a boat ramp, with a large car park extending north behind the neighbouring the Harbour Mound beach (**WA 1742**) (Fig. 4.382). This beach curves around the protruding point for 200 m. Rocks outcrop along the beach, with some mangroves growing in the rocks, and extensive intertidal rocks exposed at low tide. Apart for a couple of clear sandy patches this beach is unsuitable for swimming.

Figure 4.382 Dampier's Harbour Mound beach and boat ramp.

Beach **WA 1743** commences on the northern side of Harbour Mound point and continues northeast for 600 m as a crenulate beach with numerous rock outcrops which dominate the intertidal zone. Housing backs the southern end of the beach, with the Esplanade continuing up to the power station at the northern end. A seawall built to protect the water intake for the power station forms the northern boundary.

WA 1744-1749 **KING BAY**

No.	Beach	Rating HT	Rating LT	Type	Length
WA 1744	Dampier (oil tanks)	2	2	R+rock flats	700 m
WA 1745	King Bay (1)	2	2	R+rock flats	150 m
WA 1746	King Bay (2)	2	2	R+rock flats	500 m
WA 1747	King Bay (3)	2	2	R+rock flats	500 m
WA 1748	King Bay (5)	2	2	R+tidal flats	80 m
Spring & neap tidal range = 3.7 & 1.0 m					

Parker Point separates the southern 3 km of Dampier public foreshore from the port facilities that extend 5 km north of the point into King Bay, with the Woodside port facilities located on the northern side of the bay. There are six low energy beaches (WA 1744-1749) located along this section, which because of their location in the port area they are not open to the public.

Beach **WA 1744** commences on the eastern side of the 350 m long rock seawall, that has a berthing facility at the end, and is backed by the 50 ha of red iron ore stockpile, then two large fuel tanks. A commercial boat slip is located between the seawall and the start of the beach. The beach then trends to the northeast for 700 m with beachrock and rock outcropping along the beach and a low rocky point forming the northern boundary. Rocks and rock flats dominate the intertidal beach. Beach **WA 1745** is located on the eastern side of the point and occupies the mouth of a small 100 m wide valley. The curving 150 m long beach plugs the mouth of the valley with a 5 ha wetland behind. It is a steep high tide beach, which is fronted by seagrass meadows.

Beach **WA 1746** commences 200 m to the north, and is a 500 m long high tide sand beach, with beachrock outcropping along the beach, together with red boundary rocks and hummocky outcrops. A vehicle track and a few low buildings back the beach. A small rock outcrop separates it from beach **WA 1747** that continues to the northeast for another 500 m grading at the eastern end into the mangrove-fringed rocks of inner King Bay. There is vehicle access and one building at the northern end of the beach.

Beach **WA 1748** is located 500 m to the north on the northern side of King Bay. It is a 50 m long pocket of sand wedged between two low mangroves-fringed rocky points. A road skirts the rear of the beach and facilities associated with the Woodside terminal surround the beach. A second beach and boulder spit once located 200-300 m to the west have been excavated to form the port berthing facilities, with a protective seawall and port loading area located on the boulder spit.

WA 1750-1753 **HOLDEN POINT**

No.	Beach	Rating HT	Rating LT	Type	Length
WA 1749	Phillip Pt (N)	2	2	R+rock flats	80 m
WA 1750	Holden Pt (S)	2	1	R+LTT	500 m
WA 1751	Holden Pt (N 1)	2	1	R+LTT	300 m
WA 1752	Holden Pt (N 2)	2	1	R+LTT	200 m
WA 1753	Woodside jetty	2	1	R+LTT	50 m
Spring & neap tidal range = 3.7 & 1.0 m					

The Woodside gas pipeline and treatment plant is located 10 km north of Dampier on the western shore of the Burrup Peninsula, between King and Withnell bays. The facility includes a supply deport, wharves and jetties in King Bay and around Phillip Point. Five kilometres to the north along the southern shores of Withnell Bay are two large jetties to supply gas tankers, with the gas treatment plant is located on the adjacent shore. In between the two bays is 7 km of rocky shoreline containing five pocket beaches (WA 1749-1753). The area is all located on Woodside land and off-limits to the public.

Beach **WA 1749** is a steep, 80 m long boulder beach located between the King Bay supply facilities and a jetty 500 m to the north. The beach has of a series of steep elevated boulder ridges and is bordered by jagged red dolerite points rising to 20 m.

Beach **WA 1750** occupies a 500 m wide west-facing embayment. The 500 m long beach curves around the shoreline with 10-20 m high red dolerite points to each end, as well as a central outcrop. Northwest winds have blown sand up to 200 m inland behind the southern end of the beach, with some minor dune activity also behind the northern end. There is vehicle access to the northern half of the beach.

Beaches WA 1751 and 1752 occupy part of the next small embayment located 500 m to the north to the lee of Holden Point (Fig. 4.383). Beach **WA 1751** curves around the southern corner of the bay for 250 m. It faces northwest and is backed by a low partly active foredune. It is bordered by 16 m high Holden Point to the west, and a small rock outcrop that separates it from beach WA 1752 to the north. Beach **WA 1752** continues to the north for another 200 m, terminating against some low rocks, with a small mangrove-fringed creek mouth 100 m to the north. Vehicle tracks reach the rear of both beaches.

Figure 4.383 Beaches WA 1751 and 1752 share a small embayment in lee of Holden Point (right).

Beach **WA 1753** is an artificial beach that has formed since the construction of the northern most rock jetty. It is a 50 m long pocket of high tide sand wedged in at the base of a V in the bordering seawalls. A 500 m long jetty is located immediately south of the beach, with the gas treatment plant behind.

WA 1754-1758 **WITHNELL BAY**

No.	Beach	Rating HT	Rating LT	Type	Length
WA 1754	Withnell Bay (S)	2	1	R+tidal flats	300 m
WA 1755	Withnell Bay (N1)	2	1	R+LTT	100m
WA 1756	Withnell Bay (N2)	2	1	R+LTT	60 m
WA 1757	Withnell Bay (N3)	2	1	R+LTT	150 m
WA 1758	Withnell Bay (W)	2	1	R+LTT/rocks	50 m

Spring & neap tidal range = 3.7 & 1.0 m

Withnell Bay is U-shaped, 200 ha bay located immediately north of the Woodside gas treatment plant with part of the plant bordering the southern shoreline. The bay has 4 km of predominately mangrove-fringed rocks shore, with one low energy beach on the southern side (WA 1754) and three small pockets of sand on the more exposed northern side (WA 1755-1757) (Fig. 4.384), and beach WA 1758 located 1 km north of the bay entrance.

Figure 4.384 The three pocket beaches WA 1755-1757 are located on the northern side of Withnell Bay.

Beach **WA 1754** is a crenulate 300 m long strip of high tide sand partly broken up by two areas of rocks and mangroves, with mangroves and rocky shore to either side. It faces northwest across the 1 km wide bay. A vehicle track runs along the back of the beach, which is fronted by tidal flats.

Beach WA 1755-1757 occupy three southwest-facing bedrock pockets inside the undeveloped northern entrance to the bay. They receive low wind waves and consist of a steep high tide beach with intertidal sand flats. Beach **WA 1755** is 100 m long with rocky points to either end and a small patch of grassy foredune behind. Beach **WA 1756** lies 300 m to the west and is a 60 m long pocket of sand wedged in between the rocky shore. Beach **WA 1757** is a 150 m long pocket of sand backed by a V-shaped grassy foredune area, with red rocks lining either side of the embayment. Its boundary western point has been smoothed by the deposition of a boulder spit, which curves into the bay mouth. There is no vehicle access to these beaches.

Beach **WA 1758** is located outside the bay, 1 km to the north and faces west across 10 km wide Maitland Sound. The 150 m long beach is surrounded by sloping red dolerite that rises to 40 m. Sand has blown up to 100 m up over the backing slopes partly blanketing the rocks with sand and dune grass.

WA 1759-1965 **CONZINC BAY**

No.	Beach	Rating HT	LT	Type	Length
WA 1759	Conzinc Bay (1)	2	1	R+LTT	100 m
WA 1760	Conzinc Bay (2)	2	1	R+LTT	800 m
WA 1761	Conzinc Bay (3)	2	1	R+LTT	250 m
WA 1762	Conzinc Bay (4)	2	1	R+LTT	350 m
WA 1763	Conzinc Bay (5)	2	1	R+LTT	500 m
WA 1763	Conzinc Bay (6)	2	1	R+tidal flats	100 m
WA 1765	Conzinc Bay (7)	2	1	R+tidal flats	100 m

Spring & neap tidal range = 3.7 & 1.0 m

Conzinc Bay an open 2 km long bedrock embayment located at the northwestern end of the Burrup Peninsula. Seven beaches (WA 1759-1765) are located along the 6 km of rocky bay shore. They tend to face west into Maitland Sound and receive low wind waves. The bay is undeveloped, however a 4 WD track follows the slope around the bay shore and provides vehicle access to the beaches.

Beaches WA 1759-1763 occupy the southern 2.5 km of the bay shore, all facing west to northwest out of the open bay and into the south. They are all separated by rocky points and backed by red dolerite slopes rising to 20-40 m. Each consists of a high tide sand beach fronted by a low tide terrace and in some places rock outcrops. Beach **WA 1759** is located at the southern entrance to the bay and is a 100 m long pocket of sand bordered by jagged dolerite points. Sand has blown from the beach up to 200 m to the south blanketing the backing rocky terrain. A 4 WD track runs along the rear of the dunes. A 200 m wide point separates it from beach **WA 1760**, an 800 m long northwest-facing sandy beach backed by 50 m wide partly unstable foredune, then grassy bedrock slopes. Some rock outcrops along the beach and a low rocky point forms the northern boundary. The 4 WD track continues along the rear of the dunes.

Beach **WA 1761** is located on the northern side of a low 100 m wide rocky point. It continues north for 250 m to a 200 m wide point. The beach is backed by an unstable foredune with sand blowing up to 150 m inland onto the backing gassy bedrock slopes with 4 WD tracks crisscrossing the backing dunes. Beach **WA 1762** extends north of the boundary point for 350 m, as a slightly curving sandy beach, backed by a 50 m wide low grassy foredune and bordered by low rocky points. Sand continues around the rear of the northern 100 m long point to beach **WA 1763**, which curves to the north for 500 m to terminate at a narrow sandy tidal creek mouth, with a rocky point on the northern side. The creek drains a narrow mangrove-lined tidal stream that runs behind the northern half of the beach. A continuous foredune extends along the spit and fronts bedrock along the southern half of the beach. Vehicle tracks reach the beach and creek mouth.

Beaches WA 1764 and 1765 are two low energy pockets of sand located in the mangrove-dominated northeastern corner of the bay. Beach **WA 1764** is a northwest-facing 100 m long pocket of sand, bordered by low rocky points with mangroves extending to the west for 1 km to the end of the bay. Beach **WA 1765** is located 600 m to the north on the northern side of the bay and is a southwest-facing 100 m long pocket of sand, partly fronted by mangroves and with mangrove-fringed rocky points to either end. Both beaches are fronted by sand flats and accessible via a 4 WD track.

WA 1766-1771 **SEARIPPLE PASSAGE**

No.	Beach	Rating HT	LT	Type	Length
WA 1766	Conzinc Bay (N1)	2	1	R+LTT	150 m
WA 1767	Conzinc Bay (N2)	2	1	R-boulder	100 m
WA 1768	Conzinc Bay (N 3)	2	1	R-boulder	300 m
WA 1769	Searipple Passage (W1)	2	1	R+LTT	100 m
WA 1770	Searipple Passage (W2)	2	1	R+LTT	80m
WA 1771	Searipple Passage (W3)	2	1	R+sand flats	400 m

Spring & neap tidal range =3.7 & 1.0 m

The northern end of the Burrup Peninsula consists of an east-west trending 6 km long rocky shoreline which is separated from Dolphin Island by a 0.5-1 km wide **Searipple Passage**, the ripple referring to the surface ripples caused by the strong tidal currents which flow through the passage. To the west the passage flows into Flying Foam Passage, another experiencing strong tidal currents and surface ripples and foam. While the centre of the passage's rocky shore is lined with mangroves, several small pocket sandy beaches are located at either entrance. Beaches WA 1769-1771 extend 2 km into the southwestern side of the entrance, while beaches WA 1766-1768 are located along the 1.5 km of rocky shore between the northern end of Conzinc Bay and the western entrance to the passage. 4 WD tracks across the backing bedrock slopes provide access to all six beaches.

Beach **WA 1766** is a 150 m long west-facing sandy beach, bordered by a 15 m high southern point that protrudes 250 m, with lower rocks at the north end. A partly unstable foredune extends 100 m inland, with a vehicle track crossing the foredune to reach the beach. Beaches WA 1767 and 1768 are two adjoining boulder beaches located 200 m to the north. Beach **WA 1767** is a 100 m long steep boulder beach wedged in between 10 m high red dolerite points. A vehicle track terminates at the rear of the beach. Beach **WA 1768** lies immediately to the north and is 300 m long with some patches of sand along its northern end and just off the beach. Rocks also outcrop along and just off the beach.

Beach **WA 1769** is a 100m long pocket of north-facing sand located immediately inside the western entrance to Searipple Passage. Partly vegetated sand has blown up to 200 inland over the backing gently rising bedrock slopes. A vehicle tracks runs down the western side of the beach to the shore. Beach **WA 1770** is located 600 m to the east, further into the passage. It is an 80 m long pocket of sand backed by a 100 m wide partly vegetated dune system. Moderately steep grassy valley sides border the beach.

Beach **WA 1771** is a curving, convex then concave, 400 m long beach that fills the outer 500 m of a V-shaped valley (Fig. 4.385). The beach-barrier consists of two parts. A narrow mangrove-lined creek runs along the eastern boundary against the valley sides, with overwash flats extending up to 300 m into the valley. Mangroves also border the eastern end of the beach at the creek mouth. The western half is covered by low vegetated dunes, which extends to the western valley side. A vehicle track enters the rear of the valley and runs along the side of the creek.

Figure 4.385 Beach WA 1771 fills a small bedrock valley.

WA 1772-1777 **FLYING FOAM PASSAGE**

No.	Beach	Rating HT	Rating LT	Type	Length
WA 1772	Flying Foam Passage (1)	2	1	R+sand flats	300 m
WA 1773	Flying Foam Passage (2)	2	1	R+sand flats	150 m
WA 1774	Flying Foam Passage (3)	2	1	R+sand flats	100 m
WA 1775	Flying Foam Passage (4)	2	1	R+sand flats	200 m
WA 1776	Flying Foam Passage (5)	2	1	R+sand flats	100 m
WA 1777	Flying Foam Passage (6)	2	1	R+sand flats	150m
Spring & neap tidal range = 3.7 & 1.0m					

Flying Foam Passage is a 1-2 km wide, 14 km long tidal passage between the bedrock Dolphin Island and Angel and Gidley islands to the west. Most of the Passage consists of steep rocky shoreline, with mangroves fringing in protected locations. Six small beaches (WA 1772-1777) are located along the southeastern entrance to the passage where they receive some of the southerly waves generated in Mermaid Sound. There is no vehicle access to or on Dolphin Island where the beaches are located.

Beach **WA 1772** is the highest energy of the six beaches, as it is located on the southwestern tip of Dolphin Island and faces southwest into the dominant wind and wind waves. The 300 m long beach is tied to the island bedrock at the northern end, while the southern end terminates in a wave and tide formed sand spit, with the deeper tidal channel of Searipple Passage to the south. It is backed by a 100-200 m wide grassy dune, then bedrock.

Beach **WA 1773** is located 400 m to the north in a narrow V-shaped valley that faces southwest into the passage. The 130 m long beach has partially filled in the valley and is ringed by the valley sides. Beach WA 1774 and 1775 lie 500 m to the north and are located either side of a rock tied point. Beach **WA 1774** is 100 m long and faces southwest into the southerly winds and waves, while its neighbour beach **WA 1775** is 200 m long and faces west into the passage.

Beach **WA 1776** is a 100 m long pocket of sand located 800 m to the north and 2 km into the passage. It is wedged into a small southwest-facing valley, permitting it to receive southerly waves that reach this far into the passage. Beach **WA 1777** lies immediately to the north and is a low energy 150 m long west-facing beach fronted by shallow sand flats, then the deeper waters of the passage. Deeper into the passage only rocky slopes and some fringing mangroves occupy the shore.

WA 1778-1782 **DOLPHIN ISLAND (NE)**

No.	Beach	Rating HT	Rating LT	Type	Length
WA 1778	Dolphin Is (NE1)	2	1	R+LTT	500 m
WA 1779	Dolphin Is (NE2)	2	1	R+LTT	400 m
WA 1780	Dolphin Is (NE)	2	1	R+LTT	400 m
WA 1781	Dolphin Is (NE4)	2	1	R+tidal flats	1 km
WA 1782	Dolphin Is (NE5)	2	1	R+tidal flats	200 m
Spring & neap tidal range = 4.1 & 0.7 m					

Dolphin Island is the northern extension of the Burrup Peninsula and continues north for 12 km to the lee of wave-formed, calcarenite Legendre and Hauy islands which trend east-west for 16 km. Much of the island shoreline is composed of steeply dipping dolerite slopes, in places forming cliffs rising to 130 m. The 20 km of eastern shore has six beaches, five located along the northeastern side of the island (WA 1778-1782) and one located on the southeast corner (WA 1783). The island is uninhabited with no development.

Beach **WA 1778** is located on the eastern side of the narrow northern tip of the island. It consists of a curving 500 m long east-facing beach bordered and backed by low rocky slopes. It is exposed to southerly waves generated in Nickol Bay.

Beaches WA 1779 and 1780 are two adjoining beaches located 2 km to the south at the base of steep slopes that rise to 50 m. Beach **WA 1779** is 400 m long, faces east and is fronted by narrow sand flats. A bulge in the backing slope separates it from 400 m long beach **WA 1780** which also faces east across narrow sand flats.

Beach **WA 1781** is located inside a 1 km wide circular-shaped bay, which has a 500 m wide opening to the east. The 1 km long beach occupies the western shore of the bay at the base of steep 90 m high slopes and faces east out the bay entrance. The beach is fronted by 300 m wide tidal flats, with mangroves occupying the southern corner of the bay.

Beach **WA 1782** lies just south of the southern entrance to the bay. It is a 200 m long pocket of east-facing sand, backed and bordered by 20-30 m high slopes, and fronted by 200 m wide sand flats.

WA 1783-1788 **SEARIPPLE PASSAGE (E)**

No.	Beach	Rating HT	Rating LT	Type	Length
WA 1783	Searipple Passage (E1)	3	3	R boulder	500 m
WA 1784	Searipple Passage (E2)	2	1	R+sand flats	250 m
WA 1785	Searipple Passage (E3)	2	1	R+sand flats	200 m
WA 1786	Searipple Passage (E4)	2	1	R+sand flats	100m
WA 1787	Searipple Passage (E5)	2	1	R+sand flats	80 m
WA 1788	Sloping Point (S)	3	3	R boulder	150m
Spring & neap tidal range =4.1 & 0.7 m					

The eastern end of Searipple Passage narrows to 500 m, with the entrance facing into Nickol Bay and receive waves generated across the bay by easterly and southerly winds. Beach WA 17983 is located long the northern entrance to the passage, while beaches WA 1784-1787 occupy small embayments along the southern 1 km of the passage shore, with beach WA 1788 located 2 km south of Sloping Point, which forms the southern entrance to the passage. There is no vehicle access to any of the beaches.

Beach **WA 1783** is part of a 1 km long recurved spit that has a boulder core and a sandy western tail, the latter fringed with mangroves. The beach occupies the 500 m long eastern section of the spit, which it is composed of boulders derived from the updrift dolerite slopes. Deep water and strong tidal flows lie off the beach.

Beaches WA 1784-1786 lie 500 m to the south and occupy a series of small embayments along the southern entrance to the passage and receive enough easterly waves to maintain the four small sandy northeast-facing beaches (Fig. 4.386). Beach **WA 1784** is a curving 250 m long beach and bordered by bare red dolerite boulders to the west and a sloping grassy point to the east. Vegetated dunes extend up to 150 m in from the beach. Beach **WA 1785** lies immediately to the east and is a similar 200 m long sandy beach occupying the mouth of a small grassy valley, with a usually dry creek running along the eastern side of the valley. A veneer of vegetated dunes has blown up to 50 m inland over the backing bedrock slopes. The eastern valley slopes rise to 100 m and a few mangroves fringe the eastern headland.

Figure 4.386 Beaches WA 1784 and 1785 occupy two small bedrock embayments and faces northeast across Searipple Passage.

Beach **WA 1786** is a 50 m long pocket of sand located 200 m to the east at the base of the next small indentation in the steep grassy slopes. The beach is fringed and partly fronted by rocks and mangroves. Beach **WA 1787** lies a further 300 m to the east and is wedged in at the base of small V-shaped valley. It is an 80 m long pocket of sand bordered by a large boulder point, with red boulders dominating the eastern slopes of the valley which rises to 100 m.

One hundred metre high red Sloping Point from the southern entrance to the passage, with beach **WA 1788** located at the mouth of a small valley 2 km south of the point. The 150 m long beach is composed of steeply sloping boulders and is backed by a usually dry creek, which drains a 2 km long V-shaped valley.

WA 1789-1794 **WATERING-COWRIE & HEARSON COVES**

No.	Beach	Rating HT	Rating LT	Type	Length
WA 1789	Watering Cove (N)	2	1	R+beachrock	800 m
WA 1790	Watering Cove (S)	2	1	R+tidal flats	50 m
WA 1791	Cowrie Cove (N1)	2	1	R+tidal flats	50 m
WA 1792	Cowrie Cove (N2)	2	1	R+tidal flats	200 m
WA 1793	Cowrie Cove (S1)	2	1	R+tidal flats	100 m
WA 1794	Cowrie Cove (S2)	2	1	R+tidal flats	100 m
WA 1794	Hearson Cove	2	1	R+tidal flats	800 m
Spring & neap tidal range = 4.1 & 0.7 m					

The eastern shore of **Burrup Peninsula** extends for 16 km from Sloping Point at the entrance to Searipple Passage down to the mangrove-lined shore that links the peninsula with the mainland at Karratha. Steep dolerite slopes rising 60-100 m dominate the first 8 km of shore south of the point. This is followed by three embayments each occupied by a cove – Watering Cove, Cowrie Cove and Hearson Cove. Seven beaches (WA 1789-1794) are associated with the three coves.

Beach **WA 1789** is an 800 m long northeast-facing sand beach overlying a beachrock core (Fig. 4.387). The northern half of the beach is sandy and backed by steep rocky slopes rising to over 100 m while the eastern half has prograded east to form the northern entrance to **Watering Cove**, whose mangroves back the rear of the beach. Beachrock is exposed increasingly along the shore towards the eastern end of the beach. Moderately active dunes back the beach extending up to 300 m inland to the base of the steep backing slopes. There is no vehicle access to the beach.

Figure 4.387 Beach WA 1789 is a straight northeast-facing beach backed by some active dunes.

Beach **WA 1790** is located on the southern entrance to Watering Cove. It consists of a 50 m long pocket of sand wedged in a low rocky point and facing northwest across the 500 m wide entrance. Mangroves fill the remainder of the cove, and there is 4 WD access across the backing salt flats to the southern point.

Cowrie Cove lies 2 km to the south and is linked to Watering Cove by continuous salt flats. The 500 m wide cove is also filled with mangroves and backed by salt flats then steep rocky slopes, with a vehicle track running along the base of the slopes. Beaches WA 1791 and 1792 are located on the northern entrance to the cove. Beach **WA 1791** is a 50 m long pocket of east-facing sand located at the tip of the northern point. It is fronted by tidal flats and scattered rock outcrops and receives easterly waves from Nickol Bay. Beach **WA 1792** is located 200 m to the west just inside the bay. It is a 200 m long south-facing sandy beach bordered and backed by 20 m high steep rocky slopes.

Beaches WA 1793 and 1794 are located on the southern entrance to the cove, with mangroves filling the shoreline in between. Beach **WA 1793** is a 100 m long strip of sand bordered by mangroves to the west, and mangroves-fringed rocks to the east. It faces north into the bay. Beach **WA 1794** lies 100 m to the east and is a 100 m long pocket of sand bordered and backed by 10 m high sloping rocks, and some mangroves. It lies just inside the southern entrance to the cove.

Two kilometres south of Cowrie Cove a rocky point extends 300 m out from the steep cliffed shoreline to impound beach **WA 1795**. It is a 200 m long east-facing sandy beach wedged between 30 m high bedrock slopes to the north and a point composed of large boulders to the south. It is backed by dunes that ramp up to 15 m high against the backing slopes and include some reddish early Holocene dunes.

Hearson Cove beach (**WA 1796**) occupies the western shore of the cove. The steep, shell-rich, 800 m long beach high tide faces east across 600 m wide low gradient tidal mud flats, with a few mangroves growing at the northern end of the beach (Figs. 2.20 and 4.388). It is backed by a low foredune, then a 200-300 m wide foredune plain, then salt flats that link to King Bay 3.5 km to the west. The beach and salt flats tie the Burrup Peninsula to the mainland. A gravel road runs out to the southern end of the beach, which is a popular tourist, picnic and swimming destination. It is however only suitable for swimming at mid to high tide as the wide tidal flats are exposed at low tide.

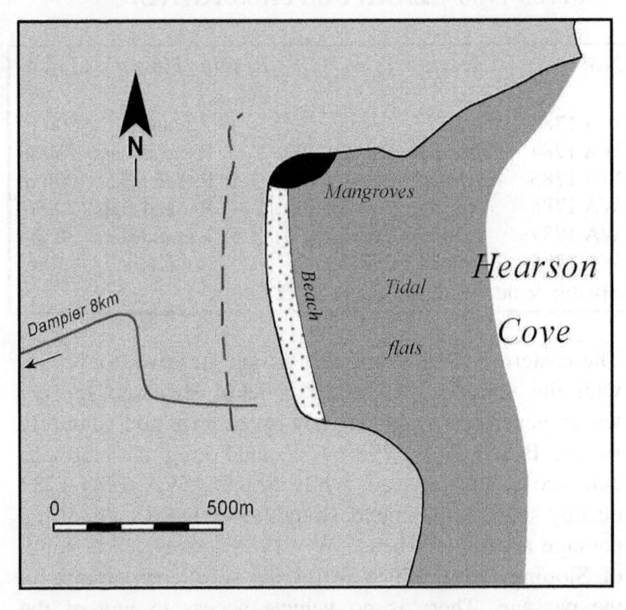

Figure 4.388 Hearson Cove is a popular recreational location just east of Dampier.

KARRATHA

Karratha was established in the late 1960's as a centre for the export of iron ore from the nearby port of Dampier. It has been boosted by the newer North West Gas Project, which has its landfall and gas treatment plant just north of Dampier. Both facilities are located on the more rugged Burrup Peninsula 15 km west of the town. Karratha is located on sloping terrain facing north in Nickol Bay, with extensive salt flats and mangroves fringing most of the shoreline. A 5 km long series of sandy spits that extend along much of the town shoreline, are now fronted by up to 2 km of salt flats and mangroves. The only open beach (WA 1797) on this section of shore is located 1.5 km east of the town.

The town of 12 000 is continuing to grow and offers a full range of services and accommodation.

WA 1797 **KARRATHA BEACH**

No.	Beach	Rating HT	LT	Type	Length
WA 1797	Karratha	2	1	R+tidal flats	500 m
Spring & neap tidal range = 4.1 & 0.7 m					

Karratha Beach (WA 1797) is the only section of open shoreline, partly free of mangroves along the entire southern shore of Nickol Bay. It is located at the northern most point of the Karratha shoreline and 4 km east of the shopping centre. It is a 200 m long strip of high tide sand bordered and backed by 20 m high bluffs of red Archaean ironstone. A road runs out to the top of the bluffs above the beach where there is a large car park. The beach consists of a steep high tide beach backed by a 50 m wide foredune that has blown up over the backing bluffs, with mangroves occupying the tidal flats along the eastern half of the beach. One kilometre wide tidal flats are exposed at low tide. The beach road also leads to a boat ramp located 100 m east of the eastern boundary bluff.

WA 1798-1801 **CLEAVERVILLE CK (W)**

No.	Beach	Rating HT	LT	Type	Length
WA 1798	Cleaverville Ck (W3)	2	2	R+rock flats	300 m
WA 1799	Cleaverville Ck (W2)	2	2	R+tidal/rock flats	100 m
WA 1800	Cleaverville Ck (W1)	2	2	R+tidal/rock flats	100 m
Spring & neap tidal range = 4.1 & 0.7 m					

Cleaverville Creek is a 2 km long tidal creek, located 12 km east-northeast of Karratha that drains out between two bedrock highs. Beaches WA 1798-1800 lie to the west of the creek mouth. Beach **WA 1798** is a 300 m long northwest-facing high tide sandy beach fronted by continuous ironstone rock flats, plus a few scattered mangroves. It is backed by a 50-100 m wide low foredune then 1 km wide salt flats. A vehicle track crosses the salt flats to reach the beach area.

Beaches WA 1799 and 1800 are located 1 km to the northeast on a 20 ha area of high ground that forms the western entrance to Cleaverville Creek. Both face north across the 100 m wide creek mouth. Beach **WA 1799** is a curving 100 m long strip of high tide sand bounded by low rocky points, with mangroves and rocks fringing either end. Beach **WA 1800** is a similar 100 m long beach located immediately to the east. Both beaches are backed by a low foredune and fronted by 100 m wide rock and tidal flats then the tidal channel. The high round is backed by 2 km wide band of mangroves and salt flats with no vehicle access to either beach.

WA 1801-1807 **CLEAVERVILLE**

No.	Beach	Rating HT	LT	Type	Length
WA 1801	Cleaverville (E1)	2	2	R+tidal/rock flats	300 m
WA 1802	Cleaverville (E2)	2	2	R+rock flats	100 m
WA 1803	Cleaverville (E3)	2	1	R+tidal flats	300 m
WA 1804	Cleaverville (E4)	2	1	R+sand flats	800 m
WA 1805	Cleaverville (E5)	2	1	R+sand flats	500 m
WA 1806	Cleaverville (E6)	2	1	R+sand flats	1.5 km
WA 1807	Cleaverville (E7)	2	2	R+sand/rock flats	100 m
Spring & neap tidal range = 4.1 & 0.7					

Cleaverville refers to a 7 km long section of north to northwest-facing shoreline that extends east of Cleaverville Creek to the entrance to Port Robinson. The shoreline is dominated by outcrops of red ironstones, with seven beaches located along the shore. A 13 km long gravel road runs out from the highway to the creek mouth, which is used for boat launching, then along the rear of the beaches to the port. The area is a popular camping and caravan area with numerous camps set-up along the shoreline during the winter months. Apart from the road there are however no facilities at the coast.

Beach **WA 1801** is a 300 m long, convex north-facing shelly beach that extends around the base of the 20 m high point that forms the eastern entrance to the creek mouth. It consists of a high tide beach fronted by intertidal rock flats. There is vehicle access to the centre of the beach, while the creek behind is used to launch boats.

Beaches WA 1802 and 1803 are located 500 m to the east and lie either side of a low 200 m long ironstone point. Beach **WA 1802** is a 100 m long pocket of high tide sand and rocks, bordered by rocks with a patch of mangroves at its eastern end and fronted by rock and boulder flats. Beach **WA 1803** faces north and is a slightly higher energy, high tide sandy beach fronted by intertidal boulder flats. Low partly vegetated dunes extend up to 100 m in from the beach, while there is vehicle access to either end.

Beach **WA 1804** occupies the next gentle embayment 1 km to the east. It is a curving 800 m long shelly high tide beach, with rocks outcropping along the eastern section, and boulders and rock flats exposed at low tide. There is vehicle access and several campsites at the western end of the beach.

Beach **WA 1805** is a further 1 km to the east. It is a 500 m long sandy high tide beach with beachrock outcrops, fronted by intertidal rocks flats, with a 100 m wide grassy dune behind. The main gravel road lies 200-300 m to the south, with sandy 4 WD tracks leading to the beach.

Beach **WA 1806** is the longest beach on this section of shore (Fig. 4.389). It extends from a small protruding 20 m high point for 1.5 km as a continuous straight north-facing beach, terminating at a shore parallel section of intertidal rocks. Beachrock also outcrops along the beach, while a 100 m wide grassy foredune backs the beach, with several campsites behind the western end of the beach. This is the most popular camping area on the Cleaverville shore. Beach **WA 1807** is located on the eastern side of the boundary rocks and consists of a curving 100 m long pocket of sand, that is almost encircled by rock flats to either side. A vehicle tracks leads to the sandy point that separates the two beaches.

Figure 4.389 The main Cleaverville camping area is located behind the 1.5 km long beach WA 1806.

WA 1808-1810 **DIXON ISLAND**

No.	Beach	Rating HT	LT	Type	Length
WA 1808	Dixon Is (W 1)	2	1	R+sand flats	900 m
WA 1809	Dixon Is (W 2)	2	2	R+rock flats	700 m
WA 1810	Dixon Is (N 1)	2	1	R	50 m
WA 1811	Dixon Is (N 2)	2	1	R	100 m
Spring & neap tidal range = 4.7 & 1.1 m					

Dixon Island is a 6 km long 0.5-1 km wide island composed of Archaean metasediments, which rise to a peak of 48 m, with a central capping of Pleistocene dunes. The exposed northern shore of the island is predominately rock, with mangroves fringing the southern shore, which borders Port Robinson and Bougner Entrance. There are only three beaches on the island (WA 1808-1811), which has no vehicle access.

Beaches WA 1808 and 1809 occupy the western end of the island and consist of a circular accumulation of sand around a 32 m high bedrock core. It forms the eastern entrance to 1.2 km wide Port Robinson, an undeveloped 500 ha tidal creek. Beach **WA 1808** is located on the southern side of the bedrock core and faces southeast into the tidal creek. It consists of a low energy sandy high tide beach and 500 m wide sand flats, with mangroves bordering the eastern end and rocks flats to the west. Beach **WA 1809** occupies the exposed northern side of the core and protrudes to the north in lee of 100 m wide intertidal rocks flats.

Beaches 1810 and 1811 occupy two small gaps in the otherwise rocky northern shore of the island. Beach **WA 1810** is located 1 km to the east and is a 50 m long pocket of sand, wedged in between rock flats and rocky shore to either side. Beach **WA 1811** is a further 2 km east and is a 100 m long pocket of sand in a slight indentation in the rocky shore. Both beaches are backed by small foredunes and fronted by relatively deep water that parallels the eroded northern shore of the island.

WA 1812-1818 **BOUGNER ENTRANCE**

No.	Beach	Rating HT	LT	Type	Length
WA 1812	Bougner Entrance (W4)	2	1	R+sand flats	300 m
WA 1813	Bougner Entrance (W3)	2	1	R+sand flats	300 m
WA 1814	Bougner Entrance (W2)	2	1	R+sand flats	250 m
WA 1815	Bougner Entrance (W1)	2	1	R+sand flats	150 m
WA 1816	Bougner Entrance (E1)	2	1	R+sand flats	200 m
WA 1817	Bougner Entrance (E2)	2	1	R+sand flats	150 m
WA 1818	Bougner Entrance (E3)	2	1	R+rocks	100 m
Spring & neap tidal range = 4.7 & 1.1 m					

Bougner Entrance is a 0.5-1 km wide tidal inlet that connects with Port Robinson and together separate 6 km long Dixon Island from the mainland. Beaches WA 1812-1815 are located along 1.5 km of the southern shore of the entrance, while beaches WA 1816-1818 extend for 1 km immediately east of the entrance. The beaches are located on a 300 ha area of high ground that is connected to the mainland by 500 m wide salt flats, with a vehicle track across the flat providing access to the beaches.

Beaches WA 1812-1815 all face northwest across the 500 m wide entrance channel. They are separated by low rocky outcrops and are fronted by 100-200 m wide sand flats then the deeper tidal channel. Beach **WA 1812** emerges from the mangroves that dominate the 6 km long southern shore of the port and entrance. It consists of a 300 m long high tide sandy beach, with mangroves fringing either end, and a small clump of rocks forming the northern boundary. The vehicle track across the tidal flats reaches the shore at this beach. It is backed by a 50-100 m wide low hummocky foredune.

Beach **WA 1813** is located immediately to the north and is a crenulate 300 m long high tide sandy beach, bordered by low mangrove-fringed rock outcrops, with two sets of rocks also outcropping along the beach. Low foredunes back the beach with the vehicle track behind. Beach **WA 1814** occupies the next 250 m long rock-bordered embayment. Mangroves fringe the rocks, with some rocks also outcropping on the beach. A vehicle track reaches the centre of the beach. Beach **WA 1815** is located immediately to the north at the southern inlet entrance. It is bordered by low mangrove-fringed rocks to the south and a 500 m long elongate low rocky point to the north, while it is backed by a 200 m wide band of low grassy foredunes (Fig. 4.390).

Figure 4.390 A long low rocky point separates beach WA 1815 (right) from beach WA 1816 (left).

Beaches WA 1816-1818 extends east of the rocky point and face northeast across 300 m wide sand flats. They are exposed to northerly waves and winds. A continuous 200 m wide band of low foredunes backs the beaches, with the vehicle track to the rear. Beach **WA 1816** consists of two pockets of sand totalling 200 m in length, with a central rock outcrop, bordered by the long point to the west and a rock outcrop to the east. Beach **WA 1817** is a straight 150 m long beach with the low rock boundary to the west and a 100 m long outcrop of beachrock to the east. Beach **WA 1818** continues on the eastern side of the beachrock for 100 m to a low rocky point, beyond which the shoreline trends to the south. Rocks outcrop along and off the beach and mangroves fringe the eastern point.

WA 1819-1826 **CAPE LAMBERT (S)**

No.	Beach	Rating HT	Rating LT	Type	Length
WA 1819	Cape Lambert (S8)	2	1	R+sand flats	800 m
WA 1820	Cape Lambert (S7)	2	1	R+sand flats	200 m
WA 1821	Cape Lambert (S6)	2	1	R+LTT	300 m
WA 1822	Cape Lambert (S5)	2	1	R+LTT	200 m
WA 1823	Cape Lambert (S4)	2	1	R+LTT	800 m
WA 1824	Cape Lambert (S3)	2	1	R+LTT	250 m
WA 1825	Cape Lambert (S2)	2	1	R+LTT	1.2 km
WA 1826	Cape Lambert (S1)	2	1	R+LTT/rocks	700 m

Spring & neap tidal range = 4.7 &1.1 km

Cape Lambert is located at the northern end of a 15 km long, 70-100 m high ridge of Proterozoic basalt. Because the ridge protrudes a few kilometers north of the general east-west trend of the coast it exposes the shore to higher waves as well as protrudes into deeper water. As a result of the former a number of beaches are located on either side of the cape, while the latter has resulted in it being developed as a port. The first port in the region was at Cossack established in 1863, which was replaced by Point Samson in the 1903, which has in turn be replaced in the 1970's by the 2.5 km long jetty at the tip of Cape Lambert that supplies iron ore to waiting ships. As the location of the ports has changed so too has the main coastal settlement, from Cossack to Port Samson, and now Wickham, gazetted in 1971, which houses the staff who work at Cape Lambert.

Beaches WA 1819-1830 are located along the western side of Cape Lambert. They emerge from the mangroves that dominate the 6 km long curving bay shoreline that separates the cape from Bougner Passage. They extend along the side of the cape range for 7 km to the tip of the cape. Beaches WA 1819-1824 are open to the public and accessible by car, while beaches WA 1825-1830 are located in the port area and closed to the public.

Beach WA **1819** is fringed by mangroves both along and off the southern end of beach. The 800 m long beach curves to the north to terminate in lee of a low rocky mangrove-fringed outcrop. Tidal flats extend 500 m off the beach to link with two low rocky outcrops. Beach **WA 1820** commences on the northern side of the boundary rocks and curves to the north for 200 m before a 200 m wide band of mangroves front the remainder of the sandy shoreline up to the next low rock outcrop 300 m to the north. The mangroves and adjacent tidal flats are sheltered by a low rocky outcrop located 400 m off the point and linked by the flats at low tide.

Beach **WA 1821** is the first of the more exposed beaches. It is a curving 300 m long beach bordered by low rocks to the south and low red bluffs grading into basalt to the north. It faces west towards Dixon Island and receives waves averaging about 0.3 m. It is backed by a 10 m high foredune with vehicle access to ether end of the beach. Beach **WA 1822** continues north of the low basalt for 200 m to the next basalt outcrop. A vehicle track crosses the foredune to the centre of the beach.

Beach **WA 1823** commences on the northern side of the 100 m long section of boundary basalt, and continues to the north for 800 m to a low red point. Basalt continues to outcrop along the southern 300 m of the beach, with the remainder consisting of a moderately steep high tide beach with sand and rocks exposed at low tide. A low foredune backs the beach with two small blowouts behind the northern end. The northern end of the beach is used to launching boats and small yachts and there is a cluster of buildings including the Port Walcott yacht club and large car park on the flat land behind the point. Beach **WA 1824** continues north of the 100 m wide point for another 250 m to a 300 m long section of low basalt and boulders. It is backed by a central blowout that extends 300 m inland, with the car park and buildings behind the southern end of the beach.

Beach **WA 1825** extends north of the basalt for 1.2 km to a 200 m long section of low rocks that separates it from beach WA 1826 (Fig. 2.16a). The beach initially consists of a mixture of sand and basalt boulders, before grading into the longer sandy section, with rocks again extending along the northern few hundred metres. A continuous 10 m high foredune backs the beach with vehicle assess in the south and centre. Beach **WA 1826** continues north of the rocky section for 700 m to a 20 m high section of deeply weathered basalt. The beach is backed by a continuous foredune, with rocks outcropping along most of the base of the beach. These two beaches are inside the port area and not accessible to the public.

WA 1827-1830 **CAPE LAMBERT (N)**

No.	Beach	Rating HT	LT	Type	Length
WA 1827	Cape Lambert (W4)	3	3	R/boulder	200 m
WA 1828	Cape Lambert (W3)	3	3	R+beachrock	150m
WA 1829	Cape Lambert (W2)	3	3	R+sand flats	50 m
WA 1830	Cape Lambert (W1)	2	3	R+rock flats	350 m
WA 1831	Cape Lambert	3	3	R/boulder+sand flats	150 m

Spring & neap tidal range = 4.7 & 1.1 m

The northern 2 km of **Cape Lambert** is dominated by iron ore stockpiles, conveyor belts and facilities for transporting the iron ore along two jetties at the tip of the cape to ore tankers. In addition along the western side of the cape are several oil storage tanks and a power station. Beaches WA 1827 to 1830 are located along the western side and tip of the cape.

Beach **WA 1827** is a curving, narrow 200 m long steep beach composed of basalt boulders. It commences at the northern end of the red bluffs and is backed and bounded in the north by similar bluffs. The oil tanks are located immediately south of the beach, with the power plant to the north. Beach **WA 1828** is a 150 m long strip of intertidal sand backed by raised beachrock bluffs and bordered by red bluffs. A groyne is located 50 m south of the beach, and marks the water intake for the power plant. Beach **WA 1829** lies 50 m to the north and is a 50 m long pocket of sand wedged in at the northwestern tip of the cape. It is bordered by low rugged rocky points, with rocks exposed off the beach at low tide.

Beach **WA 1830** connects the two tips of the cape. The beach is 350 m long faces north, with two 10 m high headlands extending 200 m seaward of each end (Fig. 4.391). It is backed by a 200 wide section of low dunes, with the open water outlet for the power station crossing the western end of the foredunes and draining out across the beach.

Beach **WA 1831** is located at the northern tip of the cape and has formed since large rocky seawalls and the two jetties where constructed. It is a 150 long and composed of cobbles and boulders. Today the 10 m high seawall backs the beach and the jetties border each end. Port facilities and plant are located immediately behind the beach and seawall.

Figure 4.391 Beach WA is located at the tip of Cape Lambert, with the iron ore loading facilities in the background.

WA 1832-1838 **POINT SAMSON**

No.	Beach	Rating HT	LT	Type	Length
WA 1832	Sams Creek	2	1	R+ridged sand flats	2.2 km
WA 1833	Point Samson (N)	2	3	R+beachrock	1.25 km
WA 1834	Point Samson	2	3	R+rock flats	700 m
WA 1835	Point Samson (E1)	2	3	R+rock flats	250 m
WA 1836	Point Samson (E2)	2	3	R+rock flats	150 m
WA 1837	Honeymoon Cove	2	1	R+LTT	100 m
WA 1838	Point Samson (E4)	2	3	R/boulder+rock flats	100 m

Spring & neap tidal range = 4.7 & 1.1 m

Point Samson is part of a 250 ha area of Pleistocene calcarenite tied to two low Archaean metasedimentary points. It is linked to Cape Lambert and the mainland by mangroves and salt flats, with a causeway across Pope Nose Creek connecting the town to the Roebourne Road. A tramway originally ran along the causeway from Roebourne to the port. The town, which was gazetted in 1909, is located on the eastern end of the outcrop and includes a 600 m long disused jetty, a marina at the mouth of the creek and a growing residential area (Fig. 4.392). It is a popular holiday location with a tavern, swimming beach, marina and good access to offshore fishing.

Beach **WA 1832** links Point Samson with Cape Lambert. The curving 2.2 km long beach commences in lee of the cape and curves to the south, then east to the 50 m wide mouth of Sams Creek. Eroding 10 m high calcarenite-capped bedrock of the point area forms the eastern side of the creek. The moderately steep high tide beach is fronted by 500 m wide ridged-sand flats and backed by a 400 m wide series of low foredunes, then the mangroves and salt flats of the creek. There is vehicle access to the western end of the beach. Just inside the creek mouth is a landing and a few buildings.

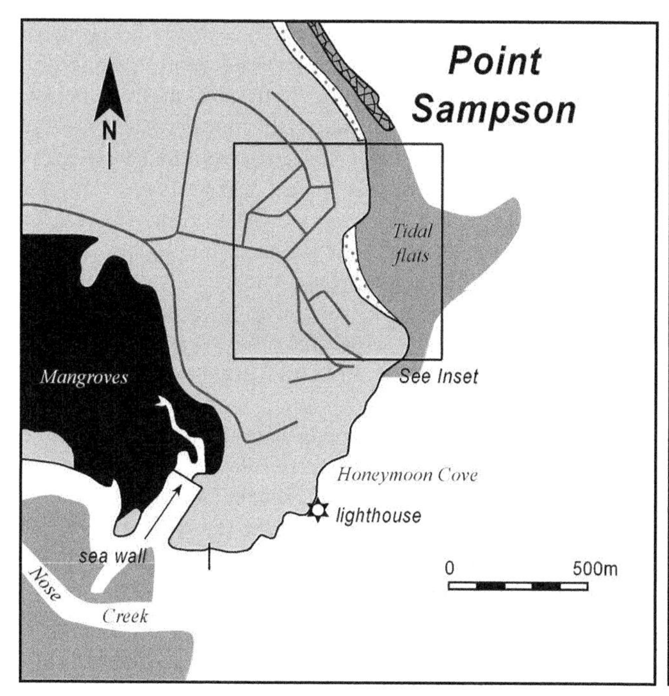

Figure 4.392 Point Sampson township.

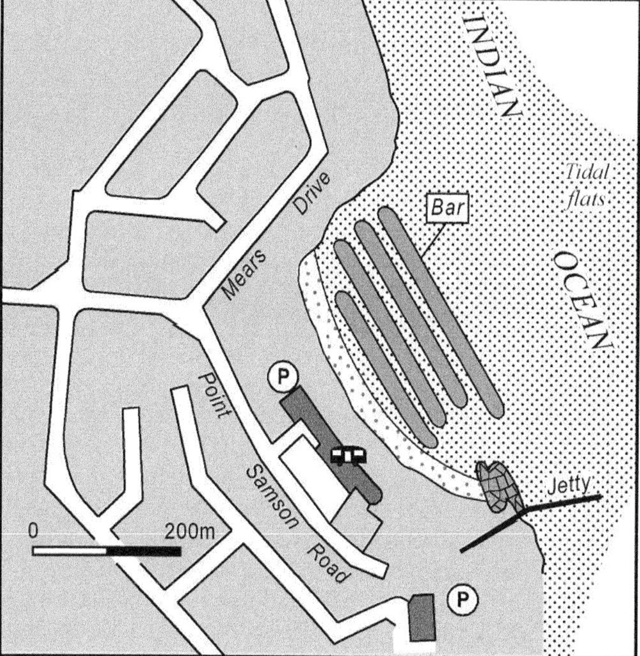

Figure 4.393 Point Sampson main beach.

Beach **WA 1833** commences 500 m east of the creek and trends to the southeast for 1.25 km terminating at the rounded 32 m high Point Samson. The beach is paralleled by continuous intertidal beachrock, which extends about 100-200 m offshore. It is backed by vegetated Holocene transgressive dunes including a parabolic dune that crossed the point to the creek mouth. There is a car park on the dune crest towards the southern end of the beach, with the houses of Point Samson commencing at the southern end.

Point Samson beach (**WA 1834**) fronts the original commercial area of the small settlement of the same name. The 700 m long beach begins at the western point then curves to the south to the edge of the bedrock, where the jetty is located (Figs. 4.393 & 394). It has a low gradient high tide beach with 500 m wide sand and rock flats exposed at low tide. The beach is also backed by a large car park, playground, picnic area, caravan park and tavern, with the growing residential area occupying the higher ground on the point.

Beaches WA 1835-1838 are four pocket beaches that occupy a 700 m long section of rocky shore than extends south of the jetty. There is car access to all four, with a walking track also to the first three. Beach **WA 1835** extends immediately south of the jetty for 250 m. Either end of the beach is fronted by intertidal rock flats, with only a central 100 m free of rocks at high tide. Beach **WA 1836** lies 100 m to the south and is a pocket 150 m long sand and cobble beach, almost enclosed by rocks and rock flats.

Figure 4.394 Point Sampson township and main beach at high tide (left).

Honeymoon Cove (**WA 1837**) is located 500 m south of the jetty and is a moderately steep 100 m long pocket of sand that is bordered and backed by 20 m high rugged metasedimentary rocks, with a dune partly blanketing the backing rocks. The rocks contain some interesting formations. The beach faces north and is free of rocks at high and low tide and a popular swimming area. There is a car park above the northern end of the beach with a ramped walkway to the beach.

Beach **WA 1838** is located another 150 m to the south and is a similar 100 m long beach except composed of cobble and boulders. It is backed by steep 10-20 m high bedrock bluffs, with no formal access to the beach. The Popes Nose Creek marina is located 200 m west of the beach.

WA 1839-1841 **PORT WALCOTT-COSSACK**

No.	Beach	Rating HT LT	Type	Length
WA 1839	Port Walcott	2 1	R+ridged sand flats	500 m
WA 1840	Mount Beach	2 1	R+ridged sand flats	700 m
WA 1841	Reader Head (W)	2 2	R+sand/rock flats	100 m
WA 1842	Reader Head (E)	2 2	R+sand/rocks flats	50 m
WA 1843	Cossack	2 1	R+tidal flats	300 m

Spring & neap tidal range = 4.7 & 1.1 m

Port Walcott is a 3 km wide east-facing embayment bordered by Point Samson to the north and Reader Point in the south. The port faces into the easterly winds and has filled with 1 km wide ridged sand flats. Popes Nose and two other tidal creeks drain into the port, with much of the inner shoreline fringed by 500 m wide mangroves. Three beaches are located in the port, one at the end of a central bedrock spur (WA 1839) and two below Mount Beach (WA 1840 & 1841). The historic town of Cossack is located 1 km south of Mount Beach.

Beach **WA 1839** is a low energy 500 m long easterly-facing beach that is backed by the narrow 30 m high ridge of metasedimentary rocks. The beach has prograded 100-200 m into the port, with mangroves fringing either end. A tidal creek parallels the centre of the beach, then 1.5 km wide ridged-sand flats extend into the bay. A vehicle track runs along the bedrock to reach the rear of the beach where a few small fishing shacks are located.

Mount Beach (WA 1840) is located 2 km south of the port beach and is attached to 34 m high Mount Beach at its southern end. A tidal creek and extensive flat and ridged sand flats separate the two beaches. The beach is 700 m long and faces northeast across 500 m wide ridged sand flats (Fig. 4.395). It is backed by a 300-400 m wide series of partly unstable, recurved foredunes, and bordered by the mount and Reader Point to the south. A gravel road from Cossack terminates at a car park at the southern end of the beach.

Figure 4.395 Mount Beach, shown here at high tide, is bordered by a tidal creek to the right and Reader Point to the left.

Beach **WA 1841** is located midway along the northern side of 500 m long Reader Point. The point is composed of metasedimentary rock, with the 100 m long pocket beach occupying a gap along the northern side of the point. It is composed of sand and rocks and fronted by sand and rock outcrops. A sealed road leads to a car park on the crest of the point above the western end of the beach. On the eastern side of the head is the smaller beach **WA 1842**, a 50 m long pocket of sand bordered and partly fronted by bedrock. The beach faces east into the prevailing winds, which have blown sand in the past part way up the backing 30 m high grassy slopes.

Beach **WA 1843** is a curving 300 m long strip of high tide sand fronted by 200 m wide tidal flats then the deep water of Butcher Inlet tidal channel. Mangroves fringe either end, together with a clump in the centre. The southern end of the beach is located 400 m north of the Cossack buildings, with the road passing 100 m west of the beach.

WA 1844-1847 **SHERLOCK BAY (W)**

No.	Beach	Rating HT LT	Type	Length
WA 1844	Butcher Inlet	2 1	R+tidal sand flats	1.7 km
WA 1845	Sherlock Bay	2 1	R+tidal channel	1.3 km
WA 1846	Sherlock Bay	2 1	R+tidal sand flats	3.5 km
WA 1847	Sherlock Bay	2 1	R+tidal sand flats	4.5 km

Spring & neap tidal range = 4.7 & 1.1 m

Sherlock Bay is an open north-facing 50 km wide shallow embayment that extends from Reader Head and Butcher Inlet in the west to Caroline Island to the east. Extensive salt flats, inner and outer beach ridges, and tidal sand flats associated with the mouths of the Harding, Jones, George and Sherlock rivers fill much of the central shoreline of the bay, most of which is only accessible by boat. The western distributary of the Harding River links to Butcher Inlet.

Beaches WA 1844-1847 are located on barrier islands along the westernmost 10 km of the bay. Beach **WA 1844** commences on the eastern side of Butcher Inlet directly opposite Cossack. This is a very low energy beach, with extensive mangroves and 1.5 km wide tidal flats. The mouth of a 400 m wide shallow inlet marks its eastern boundary where beach **WA 1845** curves round into the inlet mouth for 1.3 km. Both beaches are part of a 4.5 km long, 500 m wide barrier island that consists of a series of recurved spits to either end, with a central linear dune rising to 18 m. There is moderate dune activity along the beach and on the recurved spits. The island, which is aboriginal land, has vehicle access across the backing 5 km wide zone of salt flats and inner beach ridges.

Beach **WA 1846** is a westward migrating barrier island that has extended as a series of partly active recurved spits for 1.5 km across beach WA 1845 (Fig. 4.396). The beach extends east for 3.5 km to a 50 m wide creek mouth. The island is up to 400 m wide and backed by a 200 ha, 48 m bedrock high, then several kilometres of salt flats. Beach **WA 1847** continues east of the tidal creek for 4.5 km as a 300-400 m wide crenulate barrier island, to terminate at the next 100 m wide creek mouth. The island is backed by inner beach ridges and 3 km wide salt flats then the eastern distributary of the Harding River.

Figure 4.396 Beach WA 1846 extends into Butcher Inlet as a series of sandy recurved spits.

WA 1848-1854 **SHERLOCK BAY (CENTRE)**

No.	Beach	Rating HT	LT	Type	Length
WA 1848	Sherlock Bay (4)	2	1	R+tidal sand flats	800 m
WA 1849	Sherlock Bay (5)	2	1	R+tidal sand flats	700 m
WA 1850	Sherlock Bay (6)	2	1	R+tidal sand flats	2.7 km
WA 1851	Sherlock Bay (7)	2	1	R+tidal sand flats	1.7 km
WA 1852	Sherlock Bay (8)	2	1	R+tidal sand flats	800 m
WA 1853	Sherlock Bay (9)	2	1	R+tidal sand flats	1.3 km
WA 1854	Sherlock Bay (10)	2	1	R+tidal sand flats	3.5 km
WA 1855	Sherlock R (W)	2	1	R+tidal sand flats	3.8 km
WA 1856	Sherlock R (E)	2	1	R+tidal sand flats	1.2 km
Spring & neap tidal range = 4.7 & 1.1 m					

The central-eastern shoreline of **Sherlock Bay** extends for 34 km from beach WA 1848 to the mouth of the Sherlock River in lee of Caroline Island. This is a north-facing, very low gradient shoreline consisting of 1-2 km wide tidal flats, mangrove-fringed barrier islands, backed by 5-10 km of inner barriers and salt flats, all dissected by 10 mangrove-lined tidal creek, which link to the Harding, East Harding, Jones, George, Little Sherlock and Sherlock rivers. The spring tide range increase along the coast from 4.7 m in the west to 5 m in the east, resulting in some large tidal creeks and extensive intertidal areas. The entire deltaic complex covers 50 km of coast with an area of approximately 250 km^2. The only vehicle access to the coast is across the salt flats in the west to beach WA 1845, in the centre to the creek mouth between beaches WA 1847 and 1848, and in the east to beach WA 1856 at the mouth of the Sherlock River. Otherwise this is a low energy, low gradient difficult to access coast except by boat at high tide.

Beach **WA 1848** occupies the western 800 m of a low mangrove-fringed barrier island, with a 200 m wide tidal creek bordering the western end and a small 50 m wide creek separating it from beach **WA 1849**. This beach extends east for another 700 m to the next creek and makes up the remainder of the island. Both beaches are fronted by 1.5 km wide tidal sand flats, with extensive mangroves extending up to 300 m across the flats, and only a few patches of open sand. They are backed by 4 km of inner barriers and salt flats which grade into the mouth of the East Harding River.

Beaches WA 1850 and 1851 occupy part of the next 11 km long barrier island. Beach **WA 1850** is a relatively straight 2.7 km long beach fronted by near continuous 300 m wide mangroves, then 1.5 km wide tidal sand flats. Beach **WA 1851** continues east in a slight embayment for another 1.7 km, terminating at the mangrove-dominated western side of a 1 km wide tidal creek. The beaches are backed by a 1 km wide series of barrier island-beach ridges and recurved spits, then 4 km of inner barriers and salt flats.

Beach **WA 1852** occupies the outer north-facing 800 m long section of a Y-shaped barrier island that is bordered by 1 km wide tidal creeks to either side, with recurved spits extending up to 1.5 km into the creek mouth (Fig. 4.397). Dense mangroves border either end and the creek mouth, with only low scattered mangroves along the shoreline, with tidal sand flats then extending 3.5 km offshore. The two backing creeks are linked across 7 km wide salt flats to the mouth of the Jones River.

Figure 4.397 Beach WA 326 is located at the mouth of as large tidal creek.

Beach **WA 1853** occupies the western end of a 5 km long barrier complex. The 1.3 km long beach faces northwest across the mouth of a 1 km wide tidal creek, and consists of a series of unstable, amalgamated recurved spits, which are mobilised by the strong tidal currents. A few mangroves outcrop along the shore, which is backed by older beach ridges and salt flats. The remainder of the barrier is fronted and in the east, encased in dense mangroves.

Beach **WA 1854** occupies the central 3.5 km of a 6 km long barrier island, with the ends fronted by mangroves. The western 1 km of the beach appears to be eroding and is backed by active dunes extending up to 200 m inland. The centre and eastern section is composed of a 300-400 m wide series of low foredunes. The entire barrier is backed by a 3 km wide series of inner Holocene beach-foredune ridges systems, then 3 km wide salt flats which grade into the mouth of the Little Sherlock River.

Beaches WA 1855 and 1856 occupy the easternmost of the active barrier islands along this section of coast. Beach **WA 1855** is 3.8 km long and extends east from a small tidal creek to the mouth of the next creek, the mangrove-lined creek linking to one of the western distributaries of the Sherlock River. The beach is fronted by deeper sand flats, which permit low waves to reach the shore at high tide. It is backed by semi-stable foredunes extending 100-200 m inland, then a 2.5 km wide complex of older barriers. Beach **WA 1856** curves into the eastern creek mouth for 1.2 km and faces north across the tidal channel and shoals that extend 2 km into the bay. A vehicle track follows the Sherlock River and then crosses 3.5 km of sand flats to reach the eastern side of the creek, directly opposite the eastern end of beach WA 1856.

WA 1857-1862 **FORESTIER ISLAND (W)**

No.	Beach	Rating HT	Rating LT	Type	Length
WA 1857	Forestier Is (W1)	2	2	R+rock flats	80 m
WA 1858	Forestier Is (W2)	2	2	R+rock flats	250 m
WA 1858	Forestier Is (W3)	2	2	R+rock flats	2.5 km
WA 1860	Forestier Is (W4)	2	2	R+rock flats	200 m
WA 1861	Forestier Is (W5)	2	2	R+rock flats	100 m
WA 1862	Forestier Is (W6)	2	1	R+sand/rock flats	300 m
Spring & neap tidal range = 5.0 & 1.2 m					

The **Forestier Islands** are a linear group of eight uninhabited Pleistocene calcarenite islands that extend for 48 km between Caroline Island in the west and Cape Cossigny in the east and include the 155 m high dolerite Depot Island in the centre. They are backed by Forestier Bay which widens to 7 km in the centre, with the Sherlock River, Ball Creek and Peach River draining to the low gradient shoreline fringed by mangroves and backed by 5-10 km wide salt flats. The only vehicle access to the shore is at the mangrove-lined mouth of Balla Balla Creek, 20 km north of Whim Creek.

Beaches WA 1857-1862 occupy the unnamed 6 km long calcarenite island located immediately east of Caroline Island and north of the Sherlock River mouth. The island has a calcarenite core 200-500 wide and up to 20 m high, with a veneer of Holocene beaches and dunes occupying parts of the northern shore of the island. Because it is located 2 km north of the general east-west trend of the coast the ocean shoreline is exposed to wind waves and dominated by rock flats, with mangroves and flats dominating the tidal creeks that border either end.

Beach **WA 1857** is an 80 m long pocket of sand bordered by inter- to supra-tidal calcarenite rocks and backed by a veneer of sand over the calcarenite core. It has a steep reflective beach grading into the low tide rock flats.

Beach **WA 1858** commences 1 km to the east and is a 250 m long sand beach, grading into low calcarenite bluffs at either end, and backed by a 10 m high vegetated foredune. A 200 m long section of low bluffs separates it from beach **WA 1859**, the main island beach that continues east for 2.5 km. This is a relatively straight high tide reflective sandy beach, backed by a continuous foredune that rises to 20 m. Both beaches are fronted by the low tide rock flats.

Beach WA 1860 and 1861 are located in amongst the low calcarenite bluffs that extend 500 m east of the main beach (Fig. 4.398). Beach **WA 1860** occupies a 200 m wide embayment with the beach bordering the bluffs and some rocks outcropping in the centre. Beach **WA 1861** lies 200 m further east and is a 100 m long section of high tide sand located in amongst gaps in the low bluffs, with mangroves fringing the western end.

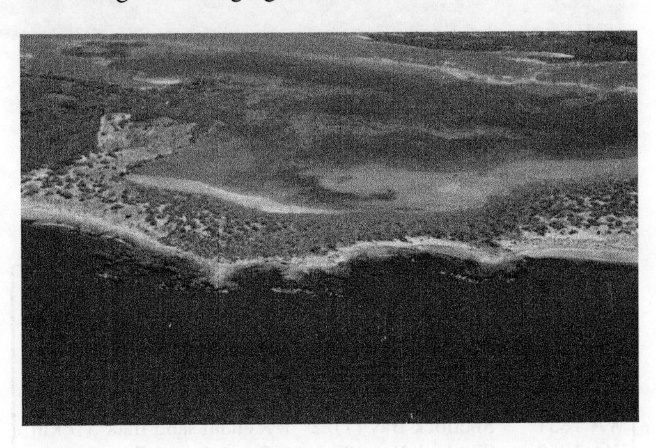

Figure 4.398 Beaches WA 1860 and 1861 are pockets of sand surrounded by Pleistocene calcarenite.

Beach **WA 1862** is located 1 km to the east on an inner part of the barrier system. It is a curving 300 m long low energy beach that grades into mangroves at either end, the mangroves extend 200-300 offshore and form a mangrove-lined embayment, while sand and rock flats extends 1 km north of the beach.

WA 1863-1864 **PEAWAH**

No.	Beach	Rating HT	Rating LT	Type	Length
WA 1863	Peawah (1)	2	1	R+tidal flats	1.8 km
WA 1864	Peawah (2)	2	1	R+tidal flats	200m
Spring & neap tidal range = 5.0 & 1.2 m					

The **Peawah River** reaches the coast 24 km north of Whim Creek in an area of 5-10 km wide salt flats and mangroves. Most of the shoreline is dominated by mangroves with a few scattered beach ridge-barriers. The best developed is located 4 km west of the terminal river channel and contains beaches WA 1863 and 1864.

Beach **WA 1863** is a 1.8 km long high tide reflective beach that occupies the western half of a 3 km long barrier island. It is bordered by mangroves to each end and backed by a moderately active foredune extending up to 200 m inland, with foredune ridges to the east. Beach **WA 1864** is located at the eastern end of the barrier and is a 200 m long open section of beach, with 50 m wide mangroves to either side. The entire barrier is backed by a 2 km wide series of four inner barriers, then 7 km of salt flats. There is no vehicle access to the island.

WA 1865-1868 **CAPE COSSIGNY (S)**

No.	Beach	Rating HT	Rating LT	Type	Length
WA 1865	Cape Cossigny (S3)	2	1	R+tidal flats	300 m
WA 1866	Cape Cossigny (S2)	2	2	R+tidal channel	350 m
WA 1867	Cape Cossigny (S1)	2	1	R+tidal flats	1.3 km
WA 1868	Cape Cossigny	2	1	R+tidal flats	500 m

Spring & neap tidal range = 5.0 & 1.2 m

Cape Cossigny is part of a cuspate foreland that has formed to the lee of Reef Island, the eastern most of the Forestier Islands. The cape is composed of low energy beaches and barriers, and is backed by up to 10 km wide salt flats. The 19 km between beach WA 1864 and the cape contains 30 km of mangrove-dominated shoreline including five tidal creeks, the central two more than a 1 km wide. Beach WA 1865 is located between the two main creeks, 6 km south of the cape, while beaches WA 1866-1868 are part of a beach-barrier system that terminates at the cape. There is no vehicle access to any of the beaches.

Beach **WA 1865** is a low energy 300 m high tide sand beach bordered by mangroves to either side and facing north across a 1.5 km wide tidal channel, with the deep tidal channel located 200 m off the beach.

Beach **WA 1866** is located at the mouth of a 50 m wide tidal creek. The creek meanders along the front of the 340 m long beach before turning and heading seaward. The beach is fringed by mangroves to either side and backed by a low partly vegetated sand spit. A 300 m long section of mangroves separate it from beach **WA 1867** that continues the beach and barrier north towards the cape. This is a 1.3 km long west-facing high tide beach, with mangroves to either end. It is backed by a mixture of partly active low dunes, vegetated dunes and a blocked creek mouth, while tidal flats extend 1.5 km to the west.

Beach **WA 1868** forms the western side of low sandy Cape Cossigny, the cape a curving 600 m long collection of recurved spits, bordered to each end by tidal creeks. The beach is 500 m long and faces west across 2 km wide tidal flats, which link to Reef Island located 2 km northwest of the cape.

WA 1869-1876 **CAPE COSSIGNY-CAPE THOUIN**

No.	Beach	Rating HT	Rating LT	Type	Length
WA 1869	Cape Cossigny (E)	2	2	R+LTT	12 km
WA 1870	Cowerie Well (1)	2	3	R+rock flats	600 m
WA 1871	Cowerie Well (2)	2	3	R+rock flats	3.6 km
WA 1872	Cowerie Well (3)	2	3	R+tidal channel	500 m
WA 1873	Cape Thouin (W2)	2	2	R+LTT	4.5 km
WA 1874	Cape Thouin (W1)	2	3	R+LTT	1.7 km
WA 1875	Cape Thouin	2	1	R+LTT	1.8 km
WA 1876	Cape Thouin (E)	2	1	R+LTT	3.5 km

Spring & neap tidal range = 5.0 & 1.2 m

Cape Cossigny and Cape Thouin are two low, protruding sandy capes located 30 km apart. Between the two capes in 36 km of near continuous sandy shoreline, backed by foredunes and some dune transgression, then the wide coastal plain of the Yule River, its western distributary and associated delta. The western-most distributary reaches the coast to the lee of Cape Cossigny, the Yule River West flows ultimately out past beach WA 1872, and the main channel past the end of beach WA 1876. There are 4 WD vehicle tracks to all the beaches between the capes, with the track crossing salt flats to reaches the Cape Thouin beaches.

Beach **WA 1869** is one of the longer beaches in the Pilbara region extending east of Cape Cossigny for 12 km to a low mangrove-fringed calcarenite foreland. It faces northwest and curves gently between the western boundary creek and the foreland. The western tip of the beach consists of a series of recurved spits that have migrated into the 300 m wide tidal creek that separates it from the cape (Fig. 2.16b). The beach then extends to the east as a steep high tide beach which receives waves up to 0.5 m at high tide. Active dunes rising to 20 m back much of the beach and extend 200-500 m inland terminating against a 500 m long section of mangroves that extends up to the tip of the foreland. They are backed by mangroves then salt flats for the first 5 km, then the low gradient deltaic plain. Vehicle tracks follow the rear of the dunes and reach the creek mouth. The eastern end of the beach.

Beach **WA 1870** commences on the eastern side of the foreland. It is a 600 m long north-facing, steep high tide beach, bordered by low calcarenite bluffs to either side. Sand and rock flats extends 300 m off the beach which is backed by three semi-stable foredunes, then vegetated dunes totalling 300 m in width, with the vehicle track behind reaching the foreland. A 1 km long section of low calcarenite bluffs separates it from beach **WA 1871** which continues to the northeast for 3.6 km to the mangrove-fringed mouth of the creek, that is linked by a distributary to the Yule River. A continuous dune-capped barrier links the beaches, with the beach backed by a semi stable 20 m high foredune, then 300-500 m wide vegetated dunes. The disused Cowerie Well is located behind the northern end of the beach, with two tall radio towers also backing the beach.

Beach **WA 1872** is a curving 500 m long beach located in the creek mouth. It faces northwest across the deep tidal channel, and is backed by a low foredune, then some active dune transgression extending up to 200 m inland. Patches of calcarenite extend across the eastern side of the creek mouth, then comprise the mainland for 2.5 km until the start of beach WA 1873.

Beaches WA 1873 and 1874 are two sections of sand in a 10 km long calcarenite barrier system that extends from the creek mouth to Cape Thouin. Beach **WA 1873** is a 1.8 km long sandy section located along a slight protrusion in the shoreline. It is followed by a second 3.5 km of low calcarenite bluffs, before 1.7 km long beach **WA 1874** continues to the point. The entire system faces northwest and receives waves averaging up 0.5 m which break against the low calcarenite bluffs at high tide and up the steep reflective beaches along the sandy sections. The beach and calcarenite sections are all backed by 10-20 m high semi-stable foredune, then vegetated older dunes increasing in width from 300 m in the south to 1 km at the cape. A vehicle track runs along the rear of the dunes to the cape area.

A strip of calcarenite forms the western side of the cape, with beach **WA 1875** curving to the east for 1.8 km (Fig. 4.399). It is a steep high tide reflective beach backed by low calcarenite bluffs to the east and fronted by 500 m wide sand flats. Sand moves eastward along the beach and forms recurved spits towards the eastern end. Beach **WA 1876** commences at the eastern tip and trends due south then southeast for 3.5 km, the finally 1 km a low narrow spit that forms the western side of the 1 km wide tidal creek at the mouth of the Yule River. The longshore movement of sand along this beach has formed a series of recurved spits that have deflected the backing creek 1.5 km to the south, with 1 km wide sand flats off the beach.

Figure 4.399 Cape Thouin with beach WA 1875 to the left.

WA 1877-1888 **YULE RIVER (E)**

No.	Beach	Rating HT	LT	Type	Length
WA 1877	Yule R mouth (1)	2	2	R+tidal channel	400 m
WA 1878	Yule R mouth (2)	2	1	R+tidal flats	250 m
WA 1879	Yule R mouth (3)	2	1	R+tidal flats	400 m
WA 1880	Yule R (E1)	2	1	R+sand flats	400 m
WA 1881	Yule R (E2)	2	1	R+sand flats	1.2 km
WA 1882	Yule R (E3)	2	1	R+sand flats	150 m
WA 1883	Yule R (E4)	2	1	R+sand flats	300 m
WA 1884	Yule R (E5)	2	1	R+sand flats	1.3 km
WA 1885	Yule R (E6)	2	1	R+sand flats	100 m
WA 1886	Yule R (E7)	2	1	R+sand flats	800 m
WA 1887	Yule R (E8)	2	1	R+sand flats	600 m
WA 1888	Yule R (E9)	2	1	R+sand flats	2.6 km

Spring & neap tidal range = 5.9 & 1.4 m

The **Yule River** is a medium sized Pilbara river system that extends 200 km to the south draining the northern side of the Chichester Range. It bifurcates 35 km from the coast into a West Branch and main Yule River channel reaching the coast via these usually dry sandy channels between Cape Coissigny and Cape Thouin, with the main channel flowing to the coast immediately east of Cape Thouin. Between the two entrances is 38 km of sandy shoreline, together with considerable beachrock outcrops, containing beaches 1865-1881. They are backed by a 250 km^2 low deltaic plain containing extensive areas of salt flats, fringed by the barriers and beachrock .

To the east of the river mouth is a 15 km long series of three barrier islands, the eastern two composed of core of Pleistocene calcarenite. The barriers contain 12 Holocene sandy beaches (WA 1877-1888) and terminate at a prominent sandy V-shaped point located immediately west of the Turner River mouth. The barriers are backed by 2-5 km wide band of mangroves and salt flats, then extensive low gradient coastal plain. There is vehicle access across the salt flats to the western two barrier islands.

Beaches WA 1877-1879 are located on the western 1.5 km of a low energy, partly mangrove-fringed discontinuous barrier island that extends for 5 km immediately east of the main river mouth (Fig. 4.400). The river mouth is a 1 km wide tidal creek, bordered on its eastern side by beach WA 1876. Beach **WA 1877** is a 400 m long recurved spit, that faces west across the river mouth. It is fronted by the deep tidal channel, with mangroves fringing either end of the spit. Beach **WA 1878** curves east of the mangroves for 250 m and is partly fringed by mangroves. At its eastern end the shoreline turns and trends due east for 400 m as beach **WA 1879**. This is a gently curving beach fringed by mangroves to either end, where it grades into recurved spits. The central open beach is fronted by 600 m wide tidal flats. The barrier which the beach occupy is backed by a 2 km wide series of inner beach ridge-barrier islands then 2-3 km wide salt flats. There is a vehicle track across

Figure 4,400 The Yule River mouth (right) and mangrove-fringed beaches WA 1877 to 1879 (left).

the salt flats to a tidal creek in the centre of the barrier, but not to any of the beaches.

Beaches WA 1880 and 1881 occupy part of the next barrier island, located 5 km east of the river mouth. The island faces north, is 2.5 km long and has recurved spits extending 1-1.5 km south into the boundary creek mouths. Low beachrock bluffs dominate either end of the island. Beach **WA 1880** is located in lee of the western tip of the island. It is a curving 400 m long west-facing, sandy high tide beach, fringed by mangroves to the south. It is fronted by 500 m wide tidal flats, then a 200 m wide tidal channel. A vehicle track crosses a 2 km wide section of salt flats to reach the island and terminate at the beach. Beach **WA 1881** is located along a 1.2 km long sandy central section of the island, with beachrock to either side. The steep reflective high tide beach is backed by a series of active blowouts extending up to 200 m inland and rising to 10-15 m, then the 300-500 m wide vegetated core of the island.

A 1 km wide tidal creek separates the central and eastern barrier islands. Beaches WA 1882-1888 occupy much of the shoreline of the easternmost and largest of the barrier islands. The island trends to the east then northeast for 6.5 km, where the shoreline is occupied by beaches WA 1882-1887, then turns and trends south into the Turner River mouth as a recurved spit containing beach WA 1888. There is a vehicle track across the 2 km wide backing salt flats that reaches the eastern end of the island. The entire northern shoreline is dominated by calcarenite outcrops, with the beaches occupying gaps in the calcarenite. Beach **WA 1882** is a 100 m long pocket of sand and rocks located in lee of the western tip of the island. It is bordered by beachrock, with a few scattered mangroves on the rocks and beach. It faces west across 400 m wide tidal flats to the 200 m wide channel. Beachrock bluffs extend 1 km east to beach **WA 1883**. This is a slightly curving 300 m long beach fronted by 500 m sand flats and bordered to either end by the beachrock. It is backed by a 200 m wide zone of moderately active dunes and blowouts, then 200 m of vegetated barrier.

Beach **WA 1884** lies 300 m to the east and commences in lee of a 100 m long shore-parallel extension of a beachrock reef. The beach initially curves round in lee of the reef before trending east for 1.3 km. It is backed by a low continuous foredune, then a 50 m wide zone of active dunes, then the vegetated core of the 300 m wide barrier. Beach **WA 1885** is located 200 m further east and consists of a curving, east-facing 100 m long pocket of sand bordered by beachrock outcrops. A small pond of water is located in lee of the western beachrock and drains across the northern end of the beach.

Beaches WA 1886 and 1887 occupy part of the 2.5 km long northwest-facing western end of the island. Beach **WA 1886** is a straight 800 m long high tide reflective beach bordered by beachrock to each end. An 800 m long section of beachrock separate it from beach **WA 1887**, which continues straight to the V-shaped rock-tipped point. Some beachrock also outcrops along the beach. Both beaches are backed by continuous active dunes, with blowouts, parabolics and some small areas of sand sheet extending up to 1 km to the east reach to the eastern side of the point.

At the point the shoreline abruptly turns and trends due south as beach **WA 1888**. This is a dynamics beach that terminates in an active spit that varies in shape and length. It can be up to 2.6 km long. Sand moving around and blown across the point supplies the beach and moves south into the western side of the Turner River mouth, where it forms a 1.5 km wide series of several eastward prograding spits, that have filled this western side of the river mouth. A vehicle track across the backing 2 km wide salt flats reaches the spit area, with a track through the dunes out to the point.

WA 1889-1891 **BOODARRIE**

No.	Beach	Rating HT	Rating LT	Type	Length
WA 1889	Boodarrie (W3)	2	1	R+tidal flats	200 m
WA 1890	Boodarrie (W2)	2	2	R+rock flats	250 m
WA 1891	Boodarrie (W1)	2	1	R+rock flats	50 m
Spring & neap tidal range = 5.9 & 1.4 m					

The **Turner River** is a medium sized Pilbara river that extends 150 m south to the Chichester Range. The river breaks into two main channels, the East and West Branch and a number of small distributaries to reach the coast over a 15 km section of shoreline between beach WA 1888 and Oyster Inlet, 10 km west of Port Headland. A series of Pleistocene calcarenite and Holocene beach ridges and recurved spits extend across the mouth as a series of five irregular barrier islands, all backed by extensive mangroves and salt flats between 2-6 km wide. Most of the barriers are fronted by 100-300 m wide belts of mangroves, with beaches only occupying parts of a 3 km long calcarenite island located 1.5 km west of Boodarrie landing. A vehicle track across the backing 2 km wide salt flats reaches the islands and provides access to the three beaches (WA 1889-1891).

Beach **WA 1889** is a curving, northwest-facing 200 m long strip of high tide sand located at the western tip of the island. It is bordered by low calcarenite bluffs, with a clump of mangroves fringing the northern end and tidal flats extending 300 m off the beach. Beach **WA 1890** is located 100 m to the east and consists of a curving, north-facing 250 m long high tide sand beach, fronted by a small embayment filled with mangrove-dotted calcarenite rock flats. Both beaches are backed by a veneer of Holocene dune sand over the Pleistocene calcarenite. Beach **WA 1891** is located towards the eastern end of the island in a curving 300 m long embayment that is dominated by low calcarenite bluffs (Fig. 4.401). The small 50 m long pocket of sand occupies a gap in the bluffs, with rock flats exposed at low tide. The entire embayment is backed by moderately active foredune extending up to 100 m inland.

Figure 4.401 Beach WA 1891 (centre) is a patch of sand on an otherwise Pleistocene calcarenite barrier.

WA 1892-1893 DOWNES ISLAND

No.	Beach	Rating HT LT	Type	Length
WA 1892	Downes Is (W)	2 1	R+tidal flats	200 m
WA 1893	Downes Is	2 2	R+rock flats	1.4 km
Spring & neap tidal range = 5.9 & 1.4 m				

Downes Island is a 3 km long calcarenite barrier island, with two sandy beaches occupying part of its northern shore and with no vehicle access The 0.5-1 km wide island is backed by 6 km of inner barriers encased in mangroves and inner salt flats. Beach **WA 1892** is located along the western shore of the island and faces west towards Weerdee Island and Oyster Inlet. The beach is 200 m long with a small calcarenite outcrop at the southern end, and a low narrow mangrove-fringed point at the northern end. Tidal flats extend 500 m off the beach and into the inlet. Beach **WA 1893** commences 500 m to the east and occupies 1.4 km of the northern shore of the island. It is a relatively straight, north-facing high tide sandy beach fronted by 300 m wide intertidal rock flats which narrow to the east. It is backed by a moderately active foredune up to 100 m wide then the vegetated core of the island.

WA 1894 FINUCARE

No.	Beach	Rating HT LT	Type	Length
WA 1894	Finucare	2 1	R+tidal flats	1 km
Spring & neap tidal range = 5.9 & 1.4 m				

Finucare Island is a 5 km long north-facing calcarenite island located immediately west of Port Hedland. The eastern end of the island is occupied by the small Finucare settlement and an iron ore jetty that extends into Port Hedland harbour. The entire exposed northern shore of the island is composed of a continuous calcarenite platform and bluffs, and includes an active Holocene and higher Pleistocene rock platform. A 2.5 km long recurved spit extends into the harbour, which is then linked to the mainland by a 4 km wide belt of mangroves and salt flats.

The only beach on the island is located along the first 1 km of the recurved spit. Finucare beach (**WA 1894**) is a 1 km long low energy beach that curves west from Hunt Point to finally faces south, with a 600 m long jetty extending into the harbour from the southern end. The beach is moderately steep and fronted 200-300 m wide tidal flats then the main harbour channel. The southern half of the beach is backed by iron ore stockpiles, with a foredune increasing in height towards the northern end of the beach. There is vehicle access to the beach from Finucare.

Region 7 Canning: Port Hedland - 80 Mile Beach-Roebuck Bay

Coast length: 618 km (5236-5834 km)
Beaches: 157 (WA 1895-2051)

Regional maps: 7A Fig. 4.402 Beaches WA 1894-1960 page 381
7B Fig. 4.414 Beaches WA 1961-2051 page 391

The Canning coast extends for 400 km between Port Hedland and Crab Creek on the northern side of Roebuck Bay. Much of the 618 km of shoreline is backed by the Great Sandy Desert. The Northern Highway, which parallels much of the coast, 10-20 km inland has only two roadhouses between Broome and Port Hedland, a road distance of 600 km. The coastline is dominated by the 122 km long shoreline of the De Grey River delta immediately east of Port Hedland, then the 222 km long Eighty Mile Beach, followed by a more irregular section of rocky, pindan and finally mangrove shoreline between Cape Bossut and Roebuck Bay.

Town: Port Hedland
Roadhouses: Pardoo, Sandfire
Camping: Cape Keraudren, 80 Mile Beach Caravan Park
Resort: Eco Beach
Homesteads: Anna Plains, Wallal Downs, Mandora, Frazier Downs (6 km), Yardoogarra,
Aboriginal community: Bidyadanga (La Grande), Yardoogrrra, Barn Hill

Region 7A CANNING: Port Hedland – Eighty Mile Beach

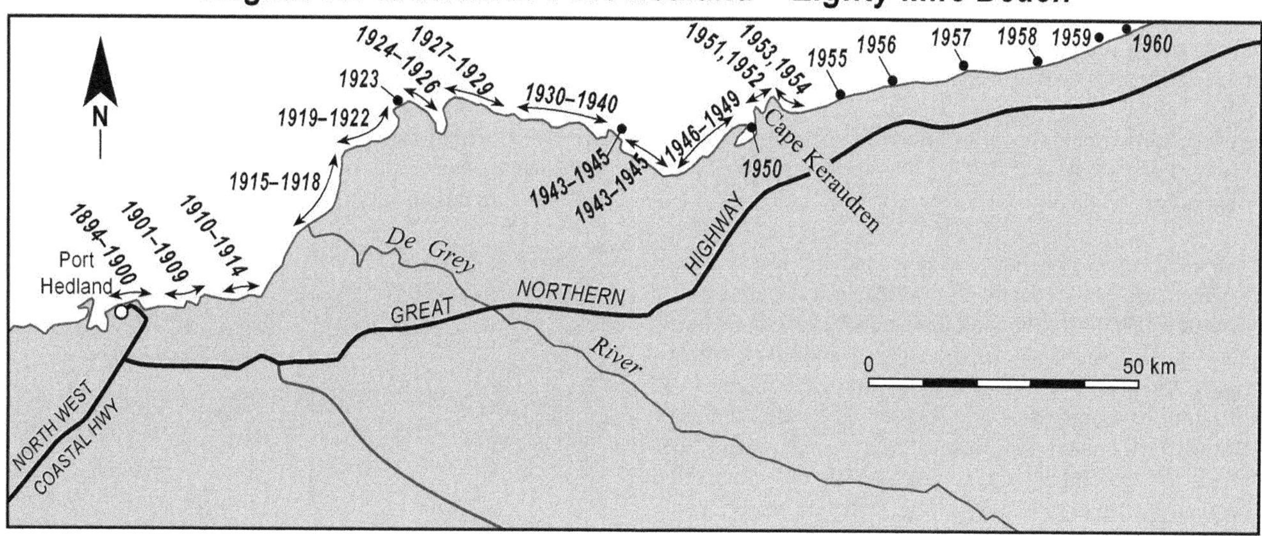

Figure 4.402 The western half of the Canning coastal region, beaches WA 1894-1960.

WA 1895-1902 **PORT HEDLAND**

No.	Beach	Rating HT	Rating LT	Type	Length
WA 1895	Airey Pt	2	1	R+tidal flats	200 m
WA 1896	Port Hedland groyne	2	2	R	100 m
WA 1897	Port Hedland marina	2	2	R	300 m
WA 1898	Port Hedland spoil bank (W)	2	2	R	2.7 km
WA 1899	Port Hedland spoil bank (E)	2	2	R+sandflats	2.9 km
WA 1900	Cemetery Beach	2	2	R+rock flats	1.3 km

Spring & neap tidal range = 5.9 & 1.4 m

Port Hedland is located on a 7 km long sand-capped calcarenite barrier island at the mouth of Stingray Creek (Fig. 4.403), a 500 m wide deep tidal creek. The creek provided a natural port, which has now been substantially upgraded to handle large bulk ore carriers. Railways from the Pilbara iron ore mines supply two ore loaders at the port and on adjacent Finucare Island. The old Port Hedland spreads along the low sand barrier for the full 7 km to the Cooke Point at the eastern end. Expansion of the port and the town resulted in a satellite suburb being built at the new Port Hedland 9 km to the south. Today the total population of Port Hedland is 15 000 the largest in the northwest.

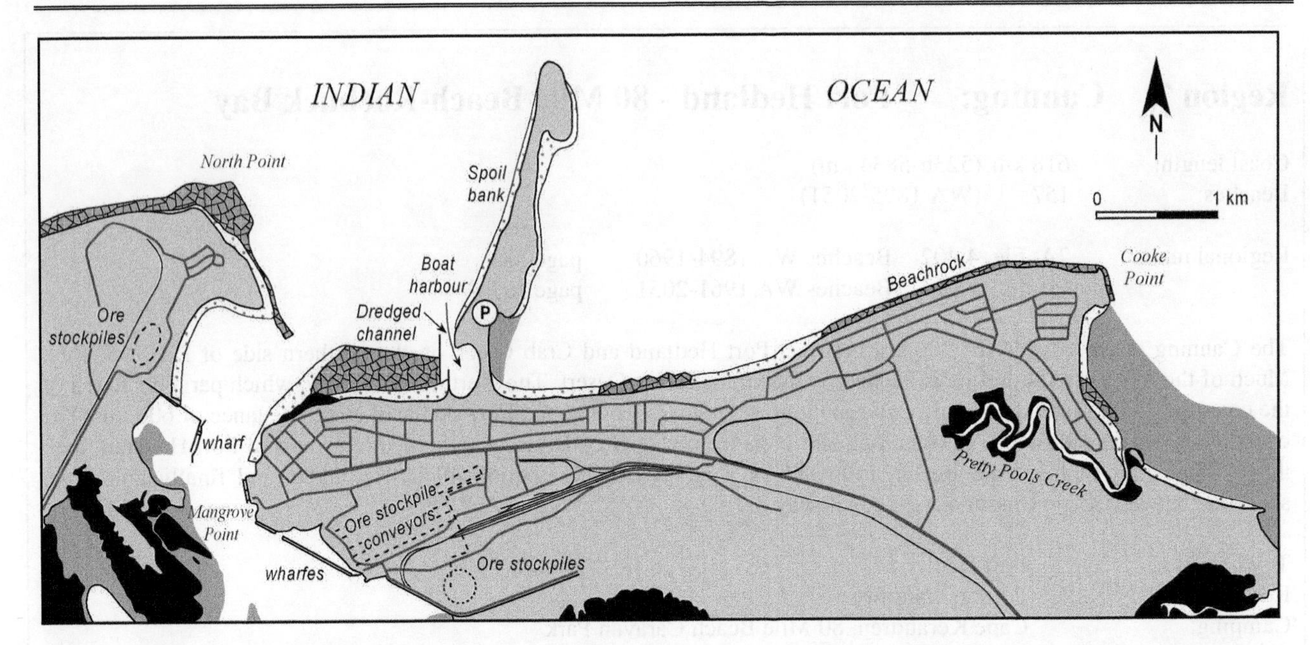

Figure 4.403 Port Hedland occupies a 7 km long Pleistocene barrier. The shoreline has been heavily modified for port facilities, with a few sandy beaches located on and to either side of the Spoil Bank.

The town is fronted by a continuous 7 km long, north-facing shoreline containing a mixture of sandy beach and low craggy beachrock bluffs. In addition a dredge spoil bank extends 2.5 km seaward of the beach. The result is four beach areas along the Port Hedland shoreline, the north-facing town beaches (WA 1895-1897), the east and west-facing newer beaches either side of the spoil bank (WA 1898-1899), Cemetery Beach (WA 1900), and the eastern more sheltered beach between Cooke Point and Pretty Pool (WA 1901-1902). All beaches have good road access and parking.

The once continuous northern Port Hedland beach is now broken into three sections by beachrock and coastal structures. Between the western Airey Point and the spoil bank is a 1.5 km long narrow strip of shoreline, fronted by rocky beachrock flats at low tide, and mangroves towards the point. Beach **WA 1895** is located amongst the western mangroves, immediately west of the boat ramp, groyne and large car park, which cuts across the original beach. It is a 200 m long high tide sandy beach, with a few mangroves along the front, then 200 m wide sand flats that narrow to the west into the main tidal channel and port area. The main port road and buildings back the beach.

Beachrock bluffs dominate the shore between the boat ramp and beach **WA 1896**, which is a 100 m long pocket of sand between the beachrock and the western groyne of the yacht club marina and second boat ramp. The beach is backed by a degraded foredune and paralleled by Sutherland St. Beach **WA 1897** occupies the shore of the marina, with the groyne forming the western boundary. The 300 m long beach trends east, then curves round to the north to terminate against a seawall that forms the northern shore of the marina, with a second groyne forming part of the entrance. The beach is backed by the yacht club, large car park and is used as a boat and yacht launching ramp.

The spoil bank is composed entirely of dredged shelly sand from the harbour channel. MacPherson Drive provides a circuit of the area and access to the 2 km long beaches on either side of the bank, together with car parks and shade areas at both beaches. The 2.7 km long west facing beach (**WA 1897**) is relatively steep and narrow and fronted by generally deeper water (Fig. 4.404), while the 2.9 km long east-facing beach (**WA 1898**) is fronted by low tide sand flats. Sand is accumulating naturally at the eastern base of the beach causing it to prograde 100-200 m along the northern beachrock shoreline.

Figure 4.404 The Port Hedland spoil bank, which has formed beach WA 1898 (foreground).

Cemetery Beach (WA 1899) extends for 1.3 km from in front of the cemetery east along the northern shoreline. It is a steep reflective high tide beach, fronted by inter to low tide calcarenite flats. A car park and picnic area is located at the western end, with a 50 m wide grassy foredune, then roads, houses, the council offices, swimming pool and water tower backing the beach, with access also at the western end.

WA 1901-1905 **COOKE PT-PRETTY POOL-FOUR MILE CK**

No.	Beach	Rating HT LT	Type	Length
WA 1901	Cooke Pt (1)	2 1	R+sand flats	500 m
WA 1902	Cooke Pt (2)	2 1	R+sand flats	700 m
WA 1903	Pretty Pool	2 1	R+sand flats	1.2 km
WA 1904	Four Mile Ck (W)	2 1	R+sand flats	800 m
WA 1905	Four Mile Ck (E)	2 1	R+sand flats	900 m
Spring & neap tidal range = 5.9 & 1.4 m				

Cooke Point is a low sand-capped calcarenite headland that forms the eastern end of the Port Hedland beach-barrier system (Fig. 4.405). At the point the shoreline turns and curves to the south then southeast for 6 km to the mouth of a 500 m wide tidal creek. Along the shore are five beaches (WA 1901-1905), separated by Pretty Pool Creek, Four Mile Creek and two unnamed creeks. The residential areas of Cooke Point and the newer Pretty Pool subdivision back beaches WA 1901-1904.

Figure 4.405 Cooke Point with Pretty Pool to the top.

Beach **WA 1901** is a curving 500 m long east-facing sandy high tide beach fronted by sand flats that widen from 300 m at the point to 1 km off Pretty Pool. It is backed by a 100 m wide grassy foredune then the residential area of Cooke Point. A 100 m long band of beachrock separates it from beach **WA 1902**, a 700 m long extension of the same system, that terminates at the mouth of Pretty Pool. A few mangroves are scattered across the flats as they widen along this section of beach. The foredune continues to the creek mouth, while the houses stop 200 m to the south The beach is a popular spit for watching the 'stairway to the moon' when the full moon rises at low tide over the ridged-sand flats.

Pretty Pool is a popular recreational spot for picnics and swimming in the tidal creek and adjacent beaches. It can be accessed by car through the Petty Pool subdivision, to a car park and picnic area by the creek, or on foot from the Cooke Point area. Beach **WA 1903** extends east from the mouth of the pool for 1.2 km to a low calcarenite point. The beach faces north and is fronted by 1 km wide sand flats and backed by a 200 m wide foredune with the Pretty Point subdivision backing the western end of the beach. The central-eastern section has a moderately stable foredune and backing vegetated dunes extending up to 500 m inland. Beach **WA 1904** continues southeast of the point for another 800 m into the mouth of Four Mile Creek. This is a very low energy high tide beach, with mangroves scattered across the inner 1 km wide sands flats. It terminates at a low calcarenite point. Vehicle tracks reach the creek mouth with numerous tracks through the backing degraded dunes.

Beach **WA 1905** is northeast-facing low energy beach that extends for 900 m east of Four Mile Creek. It is bordered by low calcarenite points, with numerous mangroves dotted across the 1 km wide tidal flats. There is vehicle access to both ends and the centre of the beach.

Swimming: All the Port Hedland beaches usually receive low waves and are often calm. However with a 6 m tide range the shoreline shifts dramatically each tide cycle and can induce some strong tidal currents. The best swimming is at mid to high tide, and its safest on a rising tide. If swimming in Pretty Pool check with locals for the presence of crocodiles.

Surfing: None.

Fishing: Port Hedland offers a wide range of fishing opportunities including beach and creek fishing, which is best as high tide, and the spoil bank, which has deeper water off the end.

Summary: Port Hedland is a mining port and its beaches have been modified to some degree, however they are very accessible and offer several kilometres of relatively safe swimming in usually calm conditions.

WA 1906-1909 **BEEBARINGARRA CREEK**

No.	Beach	Rating HT LT	Type	Length
WA 1906	Beebaringarra Ck (W3)	2 1	R+sand flats	2.1 km
WA 1907	Beebaringarra Ck (W2)	2 1	R+sand flats	1.4 km
WA 1908	Beebaringarra Ck (W1)	2 1	R+sand flats	1.9 km
WA 1909	Beebaringarra Ck (E)	2 1	R+sand flats	50 m
Spring & neap tidal range = 5.9 & 1.4 m				

Beebaringarra Creek is the tidal entrance to a medium sized upland river that reaches the coast 12 km east of Port Hedland. The mouth of the river diverges to flow out either side of a 6 km long Holocene-Pleistocene barrier island, with Beebaringarra Creek occupying the eastern channel. The mangrove-lined tidal creeks and 2-3 km wide salt flats separate the island from the low gradient coastal plain with vehicle access across the 2 km wide salt flats. Beaches WA 1906-1908 occupy the northern shoreline of the island.

Beach **WA 1906** occupies a 2.1 km long Holocene recurved spit that forms the western end of the island and the eastern boundary of a 500 m wide tidal creek, with beach WA 1905 located 1 km to the west. The beach faces northwest across 200-300 m wide sand flats. It is backed by 100-300 m wide, multiple recurved spits then a mangrove-lined tidal creek. Mangroves increase on the flats towards its eastern boundary, a low calcarenite point. Beach **WA 1907** commences on the eastern side of the point and trends east-northeast for 1.4 km to the next low point. This beach receives low waves, which break on the steep high tide beach. It is backed by a 50 m wide unstable foredune, then the 1 km wide low vegetated core of the island. A solitary shack is located behind the centre of the beach.

Beach **WA 1908** extends east of the boundary point for 1.9 km to the beginning of the eastern tip of the island, a 600 m long section of low calcarenite bluffs. The beach receives low wind waves and is backed by a foredune, then a 200 m wide zone of deflated and active dunes, then the inner vegetated core of the island. A 500 m wide tidal creek, which flows between low calcarenite bluffs, borders the eastern end of the island.

Beach **WA 1909**, is a 50 m long pocket of sand wedged in between mangrove-lined calcarenite shoreline at the eastern entrance to the tidal channel. It faces southwest across the creek mouth to beach WA 1908. A vehicle track across the sand flats reaches the beach, which is used for launching small fishing boats.

WA 1910-1914 **PETERMARER CREEK**

No.	Beach	Rating HT	Rating LT	Type	Length
WA 1910	Petermarer Ck (1)	2	1	R+tidal flats	1.5 km
WA 1911	Petermarer Ck (2)	2	1	R+rock flats	200 m
WA 1912	Petermarer Ck (3)	2	1	R+rock flats	300 m
WA 1913	Petermarer Ck (4)	2	1	R+rock/sand flats	500 m
WA 1914	Petermarer Ck (5)	2	1	R+tidal flats	1.8 km
Spring & neap tidal range = 5.9 & 1.4 m					

Petermarer Creek is located at the mouth of the medium-sized upland river of the same name. The river spreads into a 6 km wide arc of distributaries, which then drain across a 4 km wide zone of mangrove-fringed tidal creeks and salt flats, with 5 km long barrier island located off the western side of the river mouth area. Mangroves dominate the open shoreline to the east. Beaches WA 1910-1914 are located along the northern shore of the barrier island, which is composed of a series of eastern trending recurved spits. A vehicle track crosses the backing 1 km wide salt flats to reach the inner spits and provide access to the beaches.

Beach **WA 1910** is a curving 1.5 km long northwest-facing sandy high tide beach, located at the western end of the island and occupying the eastern side of a 300 m wide calcarenite-controlled tidal inlet (Fig. 4.406). Low calcarenite outcrops form the western end of the beach, with a low 100 m long projecting calcarenite point the eastern end. In between the curving beach parallels the

Figure 4.406 Beach WA 1910 occupies the curving inner side of this tidal inlet.

tidal channel. A vehicle track reaches the western tip of the point.

Beaches WA 1911 and 1912 are two adjoining beaches located at the northwestern tip of the island. Beach **WA 1911** is a 200 m long strip of high tide sand bordered by the calcarenite point to the west and a small outcrop to the east. Beach **WA 1912** continues east of the outcrop of another 300 m to the next intertidal calcarenite point. Both beaches faces northwest into the wind waves and are backed by a 50 m wide unstable foredune, then the vegetated inner core of the island which extends 3 km to the south. Vehicle tracks reach both beaches and points.

Beach **WA 1913** is located immediately east of the point and commences as a crenulate 500 m long high tide beach fronted by 100 m wide rock flats, that grade into 300 m wide mangrove-fringed sand flats, the latter part of a crenulate series of recurved pits, backed by a 500 m wide shallow lagoon dominated by salt flats, then older spits. Beach **WA 1914** continues as the eastern extension of the recurved spits for another 1.8 km to the end of the spit. Mangroves fringe both ends, with a more open 1 km long central section fronted by 200 m wide tidal flats. The entire spit is backed by a 2.5 km wide series of 4-5 inner spits, then salt flats.

DE GREY RIVER DELTA

Coast length 122 km (5268-5390 km)
Beaches 31 (WA 1915-1946)

The De Grey River is the northernmost and one of the largest rivers draining the Pilbara. It extends 450 km to the east, then south and together with its tributaries drains much of the northern Pilbara. The river reaches the coast across a 70 m wide front between its distributary, the Ridley River in the west, the main channel at Breaker Inlet in the centre, and Pardoo Creek in the west. The entire delta plain covers an area of 950 km^2 and includes extensive low gradient Pleistocene and Holocene deposits. The delta shoreline extends from 122 km and consists of both low energy beaches fronted by wide sand flats, which occupy 60% (70 km) of the shore, the remainder dominated by fringing mangroves, together with some Pleistocene calcarenite barriers in the east.

WA 1915-1920 **RIDLEY RIVER-SPIT POINT**

No.	Beach	Rating HT	LT	Type	Length
WA 1915	Ridley R mouth	2	1	R+tidal flats	500 m
WA 1916	Spit Pt (S 3)	2	1	R+sand flats	8.5 km
WA 1917	Spit Pt (S 2)	2	1	R+sand flats	7.8 km
WA 1918	Spit Pt (S 1)	2	1	R+sand flats	9.5 km
WA 1919	Spit Pt	2	1	R+sand flats	300 m
WA 1920	Spit Pt (E)	2	1	R+sand flats	3.5 km
Spring & neap tidal range = 5.9 & 1.4 m					

The Ridley River is the main western distributary of the large De Grey and smaller Strelly rivers. It marks the western boundary of the De Grey deltaic systems which covers 92 km of protruding shore and extends for 70 km east to Pardoo Creek, making it the largest delta in Pilbara. Between the Ridley River mouth and Spit Point is 26 km of continuous shoreline occupied by beaches (WA 1916-1918), with beach WA 1919 at the point, and beach WA 1920 extending east of the point. Vehicle tracks cross the backing low deltaic plain and salt flats to reach the coast and all the beaches.

Beach **WA 1915** is a low energy section of a 100 m wide curving sandy recurved spit located on the western side of the 400 m wide Ridley River mouth (Fig. 4.407). The beach extends west from the mangroves-fringed river channel for 500 m, with tidal flats widening to several hundred metres. A vehicle track crosses four inner spits that extend 5 km inland and then follows the salt flats along the river channel to reach the beach.

Figure 4.407 The Ridley River mouth with the low curving beach WA 1915 to the right.

The southern end of beach **WA 1916** occupies the eastern side of the river mouth. It is the terminal end of a 9 km long series of recurved spits and foredune ridges that have transported sand southward across the river mouth, deflecting it to the south in the process. The beach trends northeast for 8.5 km to a subtle sandy foreland. It is fronted by sand flats that widen to 1.5 km at the foreland and backed by a 100-200 m wide outer foredune area of variable stability, which include some blowouts. This in turn is backed by a 200 100 m wide outer barrier-spit, a 500 m wide inner barrier-spit, then several kilometres of salt flats including the tide dominated river channel. Vehicle tracks run along the inside of the inner and outer barriers to the river mouth.

Beach **WA 1917** commences at the tip of the foreland and curves to the northeast for 7.8 km to a small tidal creek. Sand flats, 300-400 m wide, parallel the beach, which is backed by a 2-4 km wide inner an outer series of foredune ridges, with up to 30 ridges backing the central section of the beach. The main access track via Midgee Midgee Well crosses the backing 15 km of floodplain to reach the rear of the barrier, with tracks then heading north and south along the coast.

Beach **WA 1918** extends from the tidal creek for 9.5 km northward to Spit Point, a low sand-calcarenite point located to the lee of a 6 km long northward trending shallow sand spit. The foredune ridges continue the length of this beach splaying to the north, where they separate into an outer barrier, then up to 3 km wide salt flats and two inner barriers, the total system up to 8 km wide. The outer barrier contains up to 12 ridges, which are stable, apart from some minor dune activity at the point. Beach **WA 1919** occupies a 300 m long northwest-facing curve in the shore that forms the point. It has a calcarenite core, which outcrops as low bluffs. The sandy beach is fronted by a large patch of mangroves then the sand flats that tie to the long spit.

Beach **WA 1920** extends immediately east of the point for 3.5 km as a 100-200 m wide spit that splays to the east widening to 1 km at its eastern end. Two hundred metre wide sand flats parallel the beach, which is backed by a mangrove-lined creek, then 2-4 km of salt flats, then two northwest-facing inner barriers.

WA 1921-1922 **SPIT POINT (E)**

No.	Beach	Rating HT	LT	Type	Length
WA 1921	Spit Pt (E 1)	2	1	R+tidal flats	300 m
WA 1922	Spit Pt (E 2)	2	1	R+tidal flats	300 m
Spring & neap tidal range = 5.9 & 1.4 m					

To the east of Spit Point is a 15 km wide, slightly curving, northwest facing embayment occupied by beach WA 1920 in the west, and the Larrey Point beaches (WA 1923 & 1924) in the east. In between is 9 km of mangrove-dominated shoreline containing two small cleared patches, which contain beaches WA 1921 and 1922.

Beach **WA 1921** is located on an eroding section of shore where the mangroves have been cut back to exposed the underlying peat. The 300 m long beach is a partly exposed chenier sand ridge sitting atop the high tide mud flats. A vehicle track crosses the backing high tide flats to reach a scarped eroding section of shore.

Beach **WA 1922** is located 1.5 km to the east in a 300 m wide gap in the 200 m wide belt of mangroves. The high tide beach is backed by small mangroves, with scattered mangroves dotted across the 500 m wide tidal flats. A vehicle track, which flooded by spring tides, crosses the backing flats to reach the beach.

WA 1923-1926 **LARREY POINT**

No.	Beach	Rating HT	LT	Type	Length
WA 1923	Larrey Pt (W)	2	1	R+tidal flats	3 km
WA 1924	Larrey Pt	2	1	R+tidal flats	600 m
WA 1925	Larrey Pt (E)	2	1	R+tidal flats	3 km
WA 1926	Navy Rise	2	1	R+tidal flats	1.8 km
Spring & neap tidal range = 5.9 & 1.4 m					

Larrey Point is a low sandy recurved spit that is part of the western side of the De Grey River mouth at Breaker Inlet. The spits curves for 8.4 km around the 6 km wide western boundary. The point is the outermost in a 4 km wide series of multiple beach-foredune ridges and recurved spits that have prograded northward during the Holocene, with the Pleistocene shoreline located 6 km to the south. Beaches WA 1923-1925 occupy the sandy component of the shoreline, with extensive mangroves to either end. All the beaches are low and prone to overwashing. The only vehicle access is across the backing flats to Navy Rise (beach WA 1926) in the east.

Beach **WA 1923** consists of a 3 km long series of multiple recurved spits that arc actively migrating south along the western side of the point (Fig. 4.408). As a consequence the nature and length of the spits and beach varies over a periods of years to decades. The spit is backed by a 1-2 km wide zone of mangroves and salt flats.

Figure 4.408 Larrey Point with the recurved spits of beach WA 1923 in the foreground.

Beach **WA 1924** is located at the eroding northern tip of the point. The erosion has exposed inter and high tide mud flats, as well as truncating some earlier spit complexes. A 600 m long high tide beach is wedged between the eroding high tide flats and the fronting intertidal tidal flats.

Beaches WA 1925 and 1926 are part of the eastern spit complex. Beach **WA 1925** extends east then southeast for 3 km to a small tidal creek. It is a low dynamic beach backed by a 300 m wide band of older ridges that narrows to the east. Beach **WA 1926** continues to the southeast of the creek mouth for another 1.8 km terminating at the end of the spit, which is migrating into the mangroves-lined western shore of Breaker Inlet. A 500 m wide series of three older spits back the end of the beach. These rise to a height of 12 m called Navy Rise. A vehicle track reaches the centre of the beach.

WA 1927-1929 **POISSONNIER PT**

No.	Beach	Rating HT	LT	Type	Length
WA 1927	Poissonnier Pt (S)	2	3	R+tidal channel	2.4 km
WA 1928	Poissonnier Pt	2	1	R+tidal flats	700 m
WA 1929	Poissonnier Pt (E)	2	3	R+tidal channel	200 m
Spring & neap tidal range = 5.9 & 1.4 m					

Poissonnier Point is a low sandy point located at the eastern head to **Breaker Inlet** and the **De Grey River**. Like its counterpart Larrey Point it consists of recurved spits attached to a central 7 km long mangrove-fringed chenier, backed by an 8 km wide series of up to 15 cheniers. Two beaches are located on the western side of the point (WA 1927 & 1928), with a 6 km long bank of mangroves separating these from a small solitary beach west of the point (WA 1929). A vehicle track reaches the eastern beach and crosses the interconnect cheniers to the western two beaches.

Beach **WA 1927** is a 3 km long active spit that extends south of Poissonnier Point and forms the eastern side of Breaker Creek entrance (Fig. 4.409). The main tidal channel runs along the side of the beach. To the east it is backed by 2 km wide salt flats then the inner chenier ridges. The vehicle track reaches the tip of the spit.

Figure 4.409 Poissonnier Point bordered by beaches WA 1927 (right) and 1928 (foreground).

Beach **WA 1928** commences at the point and trends east for 700 m before the mangroves take over and dominate the shoreline to the east. The nature and shape of the beach is affected by tidal and occasional river flows and changes from year to year. It is backed by a 500 m wide series of foredune ridges rising to 11 m at the point, which grade east into the cheniers.

Beach **WA 1929** is located at the eastern tip of the outer chenier plain, and is clear of mangroves because of its location adjacent to a 200 m wide tidal channel. The channel has scarped the backing inner ridges, with the high tide beach located between the scarp and channel. A vehicle track reaches the centre of the beach.

WA 1930-1932 **MANGROVE CK-SHELLBOROUGH**

No.	Beach	Rating HT	LT	Type	Length
WA 1930	Codini Landing	2	1	R+tidal sand flats	5.6 km
WA 1931	Mangrove Ck	2	1	R+tidal flats	1.8 km
WA 1932	Shellborough	2	1	R+rock flats	700m
Spring & neap tidal range = 5.9 & 1.4 m					

The Holocene shoreline trends to the east-southeast and merges the Pleistocene calcarenite barriers at Shellborough, 22 km east of the inlet. Mangroves dominate the first 11 km of shoreline east of the De Grey river mouth at Breaker Inlet. Three beaches (WA 1930-1932) are located past the mangroves either side of the junctions of the two barrier systems, while vehicle tracks from De Grey homestead reaching all the beaches.

Beach **WA 1930** commences at an abrupt change in the shoreline from mangrove-fronted broad cheniers, to a well-defined 400 m wide series of beach-foredune ridges, fronted by 800 m wide sand flats. The beach extends for 5.6 km to the east terminating at Codini Landing at the mouth of 50 m wide Mangrove Creek. The beach is backed by salt flats which narrow from 1 km in the west to merge with the backing Pleistocene barrier at the creek mouth. The 1-2 km wide Pleistocene barrier is in turn backed by up to 10 km of floodplain, then the low deltaic plain of the De Grey. Vehicle tracks run along the backing salt flats to each the beach.

Beach **WA 1931** extends east of the creek mouth for 1.8 km before begin taken over by a 300 m wide band of mangroves that continue east for another 3.5 km to the eastern end of the 6.5 km long **Shellborough** barrier. The open beach section is backed by degraded foredunes up to 20 m high that are actively blowing up to 500 m inland, the dune activity extending east behind the mangroves. The beach is backed by the disused Condon Well in the centre and the Shellborough landing ground on the backing salt flats. Beach **WA 1932** occupies the northern side of a 12 m high calcarenite point that forms the eastern end of the barrier system. The high tide sand beach protrudes north around the point for 700 m, fronted all the way by 100 m wide calcarenite flats, then a fringe of mangroves. Vehicle tracks reach the point.

WA 1933-1939 **CONDON CK-CARTAMINIA PT**

No.	Beach	Rating HT	LT	Type	Length
WA 1933	Condon Ck	2	2	R+tidal channel	700 m
WA 1934	Condon Ck (E)	2	2	R+tidal flats	500 m
WA 1935	Cartaminia Pt (1)	2	3	R+rock flats	400 m
WA 1936	Cartaminia Pt (2)	2	3	R+rock flats	300 m
WA 1937	Cartaminia Pt (3)	2	3	R+rock flats	800 m
WA 1938	Cartaminia Pt (4)	2	3	R+rock flats	1.6 km
Spring & neap tidal range = 5.9 & 1.4 m					

Condon Creek drains the rear of the Shellborough barrier, which grades into extensive floodplains, then the low deltaic plain of the De Grey. The creek marks a transition from the large De Grey system to the west and the smaller delta associated with Padroo Creek, which reaches the coast 20 km to the east. Two beaches (WA 1933 & 1934) are associated with the creek mouth, while beaches (WA 1935-1938) surround the 19 m high, dune-capped calcarenite of Cartaminia Point to the east. There are vehicle tracks to all the beaches.

Beach **WA 1933** is a curving 700 m long high tide beach located on the western side of Condon Creek mouth. It faces north across sand flats then the deep tidal channel to the calcarenite eastern end of the Shellborough barrier. A vehicle track terminates at the southern end of the beach (Fig. 4.410). Beach **WA 1934** commences past a clump of mangroves 100 m to the east. It trends east for 500 m before being taken over by a 200 m wide band of mangroves. The more open beach section is fronted by 800 m wide tidal flats, with scattered mangroves. It is backed by a low 300 m wide barrier plain, then up to 5 km of salt flats and floodplain, which links the western Condon Creek with an eastern unnamed tidal creek.

Figure 4.410 Beach WA 1933 is located inside to mouth of Condon Creek.

Beaches WA 1935 to 1938 occupy the northern and eastern side of the 4 km long calcarenite Cartaminia Point, which terminates at the eastern creek mouth. All are backed by a dune-capped calcarenite core, then the 2 km wide linked-slat flats of the two creeks. Beach **WA 1935** is a north-facing 400 m long high tide sand beach fronted by 50 m wide band of irregular intertidal calcarenite, that protrudes 100 m seaward at each end of the beach. It is backed by an unstable 50 m wide foredune, with vehicle tracks reaching each end of the beach. Beach **WA 1965** curves to the east for 300 m, with continuous intertidal rock flats and a 50 m wide degraded foredune to the rear.

Beach **WA 1937** extends to the southeast for another 800 m, with the rock flats narrowing to the east where there is a 300 m long section of intertidal sandy beach. It is backed by a 50-100 m wide deflated foredune area, the sand lightly blanketing the backing dunes. Beach **WA 1938** trends to the southeast for 1.6 km, with continuous crenulate 50 m wide rock flats, which are fronted by mangroves along the southern 500 m, the rocks terminating at the creek mouth. A 100 m wide foredune backs the beach, with some instability towards the northern end of the beach. A vehicle track follows the rear of the foredune to the creek mouth.

WA 1939-1943 **CARTAMINIA PT –MULLA MULLA DOWNS CK**

No.	Beach	Rating HT	Rating LT	Type	Length
WA 1939	Cartaminia Pt (4)	2	2	R+tidal channel	200 m
WA 1940	Cartaminia Pt (E)	2	2	R+tidal flats	2.7 km
WA 1941	Mulla Mulla Downs Ck (W 2)	2	2	R+rock flats	400 m
WA 1942	Mulla Mulla Downs Ck (W 1)	2	2	R+LTT	1.7 km
WA 1943	Mulla Mulla Downs Ck (W 2)	2	2	R+rock flats	300 m

Spring & neap tidal range = 5.9 & 1.4 m

To the east of Cartaminia Point is a curving 5 km wide north-facing embayment, bordered by calcarenite points to ether end, with a tidal creek and barrier occupying the centre. Beach WA 1939 is located at the eastern end of the point, beach WA 1940 occupies the barrier, and beaches WA 1941 and 1942 surround the eastern point. There is vehicle access across the backing floodplain and salt flats to all the beaches.

Beach **WA 1939** occupies the southeastern tip of the calcarenite recurved spit that extends from Cartaminia Point, with 500 m of mangroves separating it from beach WA 1938. It is a low energy mangrove-fringed crenulate beach that faces east across the tidal channel and associated sand shoals. Low vegetated dunes back the beach, with the vehicle track from the point terminating at the creek. Beach **WA 1940** is located on the opposite side of the 100 m wide creek mouth. It tends east for 2.7 km as a low energy high tide beach fronted by 1 km wide sand flats. It grades into mangroves in the east, which dominate the shoreline until the boundary calcarenite point. A 500 m wide low barrier with an inner 11 m high foredune ridge backs the beach. This is in turn backed by 2 km wide salt flats, which drains to tidal creeks to either side.

Beaches WA 1941-1943 occupy the shoreline of the 2 km long calcarenite point. Beach **WA 1941** is located at the western end of the point, and consists of a high tide sand beach fronted by 100 m wide intertidal calcarenite rock flats, with scattered mangroves The rocks extend as a narrow 500 m long reef off the western end of the beach. Unstable dunes back the beach, with a vehicle track to the western end of the beach.

Beach **WA 1942** is a straight 1.7 km long north-facing beach that receives low to moderate wind waves. It consists of the high tide beach and low tide terrace, with calcarenite bordering each end. Vegetated and active dunes extend 500-1000 m south of the beach to reach 27 m in elevation (Fig. 4.411). Beach **WA 1943** occupies the eastern end of the point. It is a 400 m long east-facing high tide and beach, fronted by 50-100 m wide calcarenite rock flats, which grade into mangroves at the southern end. **Mulla Mulla Downs Creek** is located 700 m south of the beach. A vehicle track runs round the top of the creek flats to reach the creek mouth and then up the shoreline to the eastern tip of the point. The point area is a used for fishing and camping.

Figure 4.411 Beach WA 1942 is a more exposed beach with a narrow surf zone and backing active dunes.

WA 1944-1949 **PARDOO**

No.	Beach	Rating HT	Rating LT	Type	Length
WA 1944	Padroo Well	2	1	R+sand/mud flats	6.3 km
WA 1945	Padroo Ck (W)	2	1	R+mud flats	2.7 km
WA 1946	Padroo Ck (E)	2	1	R+sand/mud flats	5.4 km
WA 1947	Red Pt	2	1	R+sand/mud flats	800 m
WA 1949	Red Pt (E 1)	2	1	R+mud flats	6.5 km
WA 1949	Red Pt (E 2)	2	1	R+sand/mud flats	1.5 km

Spring & neap tidal range = 5.9 & 1.4 m

Pardoo Creek is an eastern distributary of the De Grey River and drains into the centre of a 20 km wide north-facing open embayment, bordered by Cartaminia Point in the west and Mount Blaze in the east. Beaches WA 1944 and 1945 extend west from the creek to Mulla Mulla Downs Creek, while beaches WA 1946-1949 extends east to the lee of Mount Blaze. **Pardoo Homestead** is located 4 km southeast of the creek mouth.

Beach **WA 1944** commences 1 km southeast of Mulla Mulla Down Creek, with mangroves dominating the shoreline in between. The beach curves slightly to the southeast for 6.3 km to a clump of mangroves that mark an inflection in the shore. This is a low energy beach fronted by 1 km wide intertidal sand flats grading seaward into mud flats. It is backed by a 1 km wide series of up to eight foredune ridges, up to 20 m in height, then 2-3 km wide swamp. Two parabolic dunes have blown 1.5 km inland from the eastern end of the beach, the dunes activated by strong winter easterly winds. A straight vehicle track via Pardoo and Techelo wells crosses the foredunes to reach the centre of the beach.

Beach **WA 1945** commences on the southern side of the mangrove boundary. The 2.7 km long low energy beach initially trends south for 500 m before turning and trending east to the mangrove-lined mouth of Pardoo Creek. The high tide sand beach is fronted by 1-1.5 km wide mud flats and backed by a 0.5-1.5 km wide series of foredunes, which widen into the creek mouth. A vehicle track runs along the western side of the creek to the mouth and the eastern end of the beach.

A belt of mangroves up to 1 km wide dominate the creek mouth and extends for 1.5 km to the east. Beach **WA 1946** commences past the mangroves and curves gently to the east, then northeast for 6.5 m to a small tidal creek draining across the beach 500 m south of red Point. The beach is fronted by sand then mud flats that widen from 1 km in the west to 1.5 km near Red Point. It is backed by a 1 km wide series of 12 foredune ridges which reach to 22 m in height, with Pardoo Homestead located 3 km south of the western end of the beach, on the southern side of the intervening salt flats and swamp. Vehicle tracks and a central fence line cross the foredunes to reach the shore, with the fence extending out across the sand flats.

Beach **WA 1947** is a crenulate northwest-facing high tide sand beach that extends from the small tidal creek to the low calcarenite bluffs of Red Point. The sand flats narrow towards the point, with mud flats extending 1 km seaward of the point. A vehicle track terminates at the bluffs at the northern end of the beach.

Beach **WA 1948** commences 1 km east of the point at the end of the low bluffs, and continues to the northeast for 6.5 km to a small tidal creek. One kilometre wide mud flats parallel the beach all the way to the creek. It is backed by a 1-2 km wide barrier systems, with some minor foredune activity (Fig. 4.412), then a series of wide foredune-recurved spits, the spits all trending to the northeast. A vehicle track parallels the beach 500 m inland. Beach **WA 1949** continues northeast of the creek for 1.5 km terminates as a low vegetated, 100 m wide spit, that finally grades into an 800 m wide mangrove-filled tidal creek, with 11m high Mount Blaze on the northern side of the creek, 1 km north of the end of the beach. The entire barrier complex is backed by salt flats, which widen from 1 km behind Red Point to 5 km in lee of the terminal recurved spits.

Figure 4.412 Beach WA 1948 has a sandy beach and barrier fronted by wide intertidal mud flats.

WA 1950-1952 **BANNINGARRA CK**

No.	Beach	Rating HT	LT	Type	Length
WA 1950	Banningarra Ck	2	1	R+tidal flats	300
WA 1951	Pt Poolingerena (E1)	2	2	R+rock flats	50 m
WA 1952	Pt Poolingerena (E2)	2	2	R+tidal flats	50 m
Spring & neap tidal range = 5.9 & 1.4 m					

Mount Blaze-Point Poolingerena, an unnamed point and Cape Keraudren are three Pleistocene calcarenite barrier remnants that protrude up to 10 km north of the trend of the backing coastal plain and occupy 12 km of the coast. In between the three outcrops are two mangrove-lined embayments, with each of the outcrops connected to the coastal plain by salt flats, or in the case of Cape Keraudren by a beach system. The 25 km of shore between Mount Blaze and the cape is dominated by the three outcrops, which total 10 km together with 15 km of mangroves averaging 500 m in width. There is vehicle access to Mount Blaze, car access to Cape Keraudren, but none to the central barrier.

Beach **WA 1950** is located 3 km east of Cape Blaze in the western mangrove-lined embayment. It is a 300 m long strip of high tide sand in an opening in the otherwise mangrove-dominated embayment shoreline. It is backed by a low foredune, bordered by mangroves and fronted by 400 m wide tidal flats. Banningarra Creek enters the bay 1 km south of the beach.

Beach **WA 1951** is located on the central calcarenite barrier, 3 km east of Point Poolingerena and 500 m from the western tip of the barrier. It is a 50 m long pocket of northwest-facing sand, bordered by red intertidal calcarenite, and backed by the dune capping of the island. Beach **WA 1952** lies 1 km to the east in a circular, 200 m wide mangrove-lined embayment. The southern shore of the bay has a continuous sandy high tide shoreline, with the 50 m long beach located in lee of a central gap in the mangroves that dominate the remainder of the bay.

WA 1953-1955 **CAPE KERAUDREN**

No.	Beach	Rating HT	Rating LT	Type	Length
WA 1953	Cape Keraudren	2	2	R+rock/sand flats	1.5 km
WA 1954	Cape Keraudren	2	2	R+rock/sand flats	1.4 km

Spring & neap tidal range = 5.9 & 1.4 m

Cape Keraudren is a 14 km high dune-capped calcarenite point that forms the southern headland of the massive Eighty Mile Beach system. The cape is located 10 km due north of Pardoo Roadhouse and is a very popular area for camping and caravans and beach and boat fishing. While no facilities are provided, other than rubbish bins, many people camp around the cape beaches and creeks particularly during the winter months.

Beach **WA 1953** is located on the eastern side of the 14 m high cape. It curves to the south then southeast for 1.5 km and generally faces east across the sand flats (Fig. 4.413). The beach consists of a high tide strip of sand, a 1-2 m high beachrock bluff, then sand and some rock flats extending more than 500 m off the shore. It is backed by a vegetated foredune, now degraded by the many vehicles and campers, then a backing mangrove and calcarenite-lined bay when a boat ramp is located.

Beach **WA 1954** continues to the east as a curving, north-facing 1.4 km long beach, which terminates at the western side of Cootenbrand Creek. This is a similar beach with a high tide and beach, then a band of beachrock followed by the rock and sand flats, which widen to more than 1 km. Both beaches are popular camp sites, with many caravans and camps spread along the backing foredune and dune area and either side of Cootenbrand Creek during the winter.

Figure 4.413 Cape Keraudren (foreground) and the curving beach WA 1953 which is a popular camping site.

Australia's longest beaches

	Non-continuous km	Continuous km	Beach Type
Eighty Mile Beach	222	73	Ulradissipative
Ninety Mile Beach[1]	222	125	LBT-RBB-TBR-LTT
Coorong-Middleton[2]	212	194	R-LTT-Diss

[1] see Short (1996)
[2] see Short (2001)

WA 1955-1969 **EIGHTY MILE BEACH**

Eighty Mile Beach

Length 222 km (5430-5653 km)
Beaches 14 (WA 1955-1969)
Sand sections 14 (192 km)
Beachrock sections 10 km
Tidal creeks eight (Cootenbrand, Meetyou, Charleymia Well, unnamed, unnamed, Worro Well, unnamed, unnamed)

No.	Beach	Rating HT	Rating LT	Type	Length
WA 1955	Cootenbrand Ck	2	1	UD*	8 km
WA 1956	Meetyou Ck	2	1	UD	23.5
WA 1957	Planaires Bank	2	1	UD	4.6
				Beachrock	6.0
WA 1958	Ngooba Bore	2	1	UD	7.0
WA 1959	Louis Well	2	1	UD	3.6
WA 1960	Bumbiarra Well	2	1	UD	11.2
WA 1961	Chirrup Well	2	1	UD	8.7
WA 1962	Chirrup Bore	2	1	UD	5.6
WA 1963	Red Hill	2	1	UD	5.8
WA 1964	Waelburra Well	2	1	UD	4.1
WA 1965	Eighty Mile Beach	2	1	UD	30.0
WA 1966	Worroo Well	2	1	UD	10.0
WA 1967	Wyndhams Bore	2	1	UD	3.9
WA 1968	Maxwell Well	2	1	UD	16.0
WA 1969	Eighty Mile Beach (N)	2	1	UD	73.0
				Total	222.2 km

Spring & neap tidal range = 5.9 & 1.9 m increasing northward to 7.8 & 1.9 m
*UD = ultradissipative

Eighty Mile Beach is the longest beach in Western Australia extending for 222 km (138 miles) between the low calcarenite bluffs of Cape Keraudren in the southwest to Cape Missiessy in the north. The near continuous beach is broken into 14 section by 8 tidal creeks and two sections of beachrock totalling 10 km. It is however possible to drive the entire beach, with care, at low tide, particularly as the intertidal zone averages 1 km in width. Public access to the beach is only possible at Cape Keraudren and 93 km to the northeast at Eighty Mile Beach Caravan Park, just north of Wallal Downs Homestead. The old highway and vehicle tracks do however follow the rear of the entire beach with several tracks across the backing foredune ridges to the beach.

Region 7B CANNING: Eighty Mile Beach – Roebuck Bay

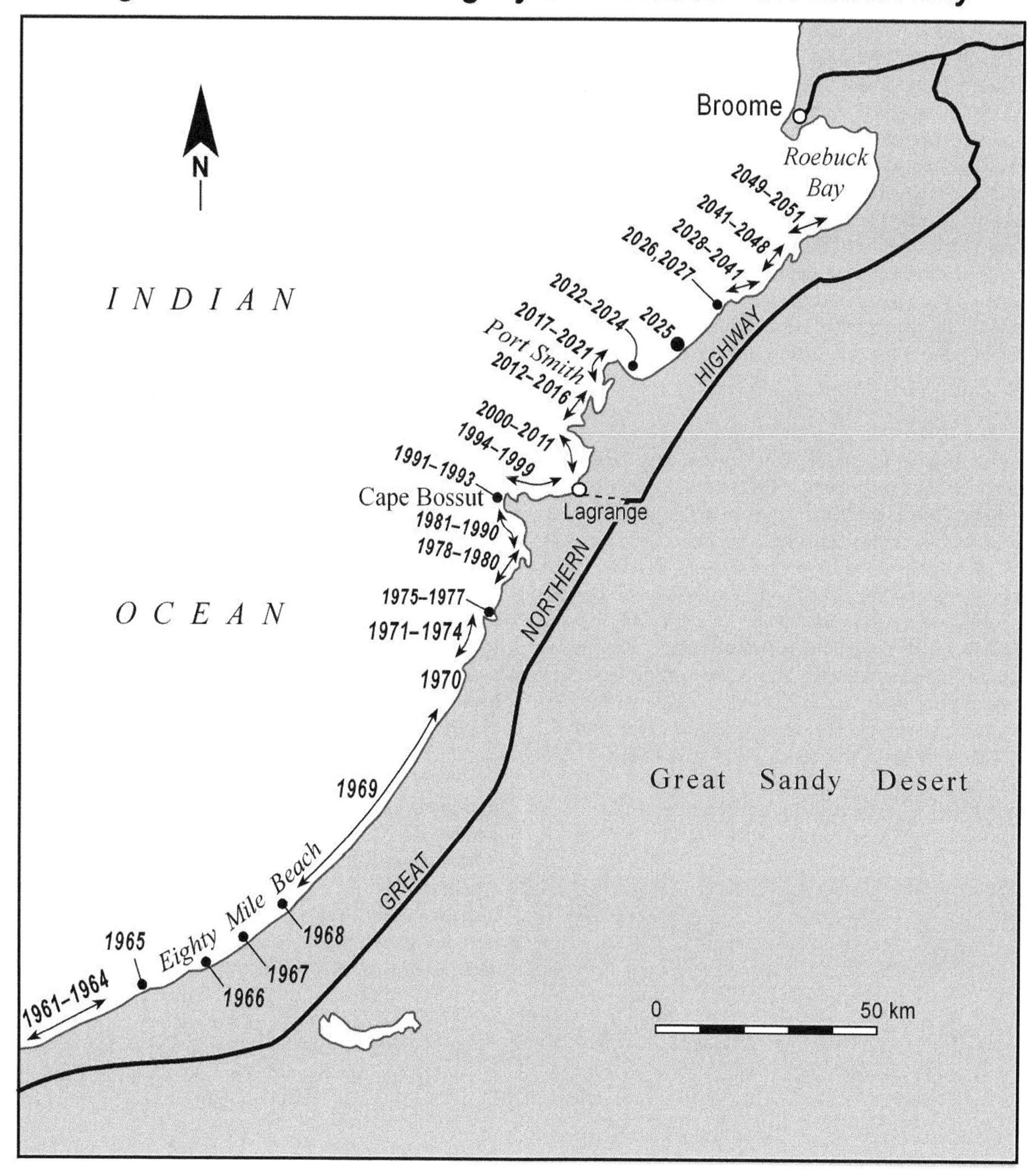

Figure 4.414   The northeastern section of the Canning coastal region includes most of Eighty Mile Beach (beaches WA 1961-2051).

Eighty Mile Beach at 222 km is the longest non-continuous beach system in Australia, on par with and Victoria's Ninety Mile beach (also 222 km) and just ahead of the South Australia's Coorong-Middleton beach (212 km). Non-continuous implies the beaches are broken into sections by bluffs, tidal creeks or inlets, though linked by a continuous low tide or surf zone. The longest continuous uninterrupted beaches is the Cape Jaffa-Murray Mouth section of the Coorong at 194 km, followed by a 125 km long section of Ninety Mile Beach (Lakes Entrance to Shoal Inlet), with the northernmost 73 km long section of Eighty Mile Beach the third longest.

The beach is backed by an equally long barrier system consisting of a southern 70 km long section from Cape Keraudren to Red Hill characterised by inner and outer discontinuous foredune ridges, some of which are now covered by sand sheets and transverse dunes. The 133 km long northern section begins north of Red Hill as a series of widely separated discontinuous foredune ridges and northward migrating recurved spits, set in a broad backbarrier depression. These converge at the beginning of the final section and form the longest foredune ridge system in Australia extending unbroken for 73 km to Cape Missiessy. This part of the system contains up to 20 ridges extending across a 2-2.5 km wide plain, backed in turn by a backbarrier depression up to 20 km wide.

Cootenbrand Creek marks the southwestern boundary of Eighty Mile Beach, with the dune-capped calcarenite bluffs of Cape Keraudren extending the final 3 km to the cape. Beach **WA 1955** commences at the shallow 300 m wide Cootenbrand Creek and trends to the east for 8 km to the 500 m wide shallow mouth of Meetyou Creek. Both creeks can be negotiable by vehicles at low tide. The beach is exposed to all waves out of the north which maintain a 1 km wide very low gradient ultradissipative beach (Fig. 2.19a). The beach is backed by a 2-3 km wide inner and outer barrier, the outer barrier composed of several well develop foredunes. The outer 500 m of the foredunes have been reactivated along a 6 km long section of the beach and now form a series of well developed bare transverse dunes, aligned to the dominant easterly winds (Fig. 2.24c).

Beach **WA 1956** is a 23.5 km long low gradient, 1 km wide ultradissipative beach that trends east from the 500 m wide mouth of Meetyou Creek to a similar shallow creek mouth near Charleymia Well (Fig. 4.415). The beach protrudes approximately 1 km seaward along its central section, with older recurved spits trending west towards Meetyou Creek and for 11 km east to the eastern creek mouth where they are fronted by a 100-200 m wide series of 3-4 well developed foredune ridges. The ridges have in turn been reactivated into a 300-400 m wide bare transverse dune sand sheets for 4 km east of Meetyou Creek. There is vehicle access to Meetyou Creek and via Fence Camp Well to the beach 5 km east of the creek. Easiest access is however along the shoreline at low tide. The outer barrier is backed by an irregular 3-5 km long swamp, then the low gradient coastal plain.

Figure 4.415 The northern end of the ultradissipative beach WA 1956 near Charleymia Well.

Beach **WA 1957** is a 4.6 km long north-facing section of ultradissipative beach that extends east of the creek mouth to the beginning of a 6 km long section of beachrock shoreline, and lies to the lee of Planaires Bank, a shallow section of the nearshore extending 2.5 km off the eastern end of the beach. The bank has induced a slight foreland bulge in the beach and adjoining beachrock section of shore. The foreland bulge and beach is backed by a 1-2 km wide series of Pleistocene barrier deposits, which also form a salient in lee of the bank. The beachrock section consists of a near continuous 5 m high bluff located along the inter to high tide shoreline, backed by a Pleistocene barrier that widens from 1 km at the foreland to 2.5 km in the east, and include 16 m high Shoonta Hill.

Beach **WA 1958** commences at the eastern end of the beachrock section and trends to the north-northeast for 7.7 km to the mouth of a small shallow creek. The low gradient beach is fronted by continuous 1 km wide ultradissipative beach, with some patches of beachrock outcropping towards the western end. It is backed by a continuous 200-500 m wide sand sheet capped by eastward migrating transverse dunes, then a 2-3 km wide swamp, which drains out the eastern creek mouth. Ngooba Bore is located 2.5 km south of the centre of the beach, with a track from the bore reaching the bare dunes. Beach **WA 1959** continues east of the small creek mouth for 3.6 km to the next small migratory creek, with the ultradissipative intertidal zone narrowing to 500 m at the creek mouth. The two creeks are linked by a meandering shallow tidal creek, which maintains a destabilised beach as it meanders into the foredune area. As a result the beach is backed by a low, sparsely vegetated foredune that is prone to overwashing, then the creek, which also meanders through a 1 m wide series of Holocene foredune. The barrier is in turn backed by 1-1.5 km wide swamp then the inner Pleistocene barrier. Louis Well is located on the Pleistocene barrier 2 km south of the eastern creek mouth.

Beach **WA 1960** commences on the eastern side of the migratory creek mouth and continues north-northeast for another 11.2 km to a slight beachrock-controlled easterly inflection in the shoreline, located 1.5 km north of Bimbiarra Well. To the east of the creek mouth the intertidal zone steepens and narrows to 200-300 m, while ultradissipative beach conditions continuing to dominate. The beach is backed by a low 300 m wide dune sheet in the west, which narrows and stabilizes to the west, where a 200-300 m wide band of foredunes form the barrier. These are all backed by a 1 km wide swamp, then the inner Pleistocene barrier. The old highway parallels the beach approximately 1 km inland.

Beach **WA 1961** commences at the beachrock inflection and continues to the northeast for 8.7 km to a more prominent beachrock controlled-inflection, the entire beach fronting a dune-capped Pleistocene barrier core. The dunes are stable in the west before destabilising into a series of deflation hollows and some parabolics towards the eastern end of the beach. The beach is initially backed by 3-5 m high beachrock bluffs and a narrow high tide beach, the spring tide reaching the bluffs. The bluffs decrease to the east, however patches of beachrock continue to outcrop across the intertidal. The ultradissipative intertidal zone widens from 200 m in the west to 300 m in the east. The old highway lies 1 km inland, with a second track running 200-300 m in from the eastern half of the beach with a track to the bluffs above the beach and the eastern inflection point. The old highway crosses a 3 km wide swampy interbarrier depression, with Chirrup Well located 3 km south of the centre of the beach.

Beach **WA 1962** commences at the inflection and trends northeast for 5.6 km to the next calcarenite induced easterly inflection in the shoreline. The beach fronts the core of a calcarenite Pleistocene barrier, with two additional beachrock outcrops in the intertidal, which also cause slight salients in the shoreline. The beach consists of a moderately steep shelly high tide beach, fronted by a low gradient 200 m wide ultradissipative upper intertidal beach, grading into a wider lower gradient lower intertidal. It is backed by a low dune-draped 800 m wide Pleistocene barrier, then an irregular usually dry backbarrier depression up to 5 km wide. The old highway runs along the inner edge of the barrier, with at least one track across to the shoreline. Chirrup Bore is located 2 km south of the centre of the each.

Beach **WA 1963** continues for 5.8 km to the northeast as an uninterrupted lower gradient ultradissipative beach (Fig. 2.15b). It terminates at the beginning of the next calcarenite section, which includes the low red bluffs of Red Hill. The upper intertidal is up to 300 m wide and is fronted by the wider lower gradient, lower intertidal. It is backed by a predominately unstable foredune up to 50 m wide and actively transgressing inland, then 200-300 m of vegetated foredunes. A backbarrier depression then extends 5 km to the south, with the old highway paralleling the rear of the beach 300-400 m inland.

Beach **WA 1964** commences at the Red Hill calcarenite bluffs and trends to the northeast for 4.1 km. The entire rear of the beach is dominated by outcropping 3-5 m high red calcarenite bluffs, together with calcarenite-beachrock in the upper intertidal. The calcarenite is fronted by the same low gradient upper and lower intertidal ultradissipative beach, with entire intertidal system up to several hundred metres wide.

The **Eighty Mile Beach** section (**WA 1965**) extends for 30 km from near Wallal Downs and the Eighty Mile Beach Caravan Park to the next tidal creek near Worroo Well. The southern end of this beach is the most publicly accessible and utilised section of the entire beach, and visited by thousands of people annually. The large beachfront caravan park (Fig. 4.416) is located 10 km due north of the highway. It provides accommodation, camp sites and basic facilities, as well as easy access to the beach for pedestrians, vehicles and small boats. The beach along this section is typical of the system, with a shell rich, moderately steep, 30 m wide high tide beach fronted by a 200 m wide upper intertidal, then a very low gradient lower intertidal extending several hundred metres seaward (Fig. 4.417). It is backed at the caravan park by a series of well developed foredunes extending 500 m inland. These widen and become discontinuous to the northeast, reaching 3 km in width at the eastern end of the beach section. As they widen a broad interdune depression dominate the back barrier area. At Mandora homestead towards the eastern end of the section the entire backbarrier depression is 6 km wide. This section of beach has also experienced easterly longshore sand transport, both to produce the splaying foredune ridges, as well as a 4 km long series of recurved spits at have deflected the eastern creek that distance longshore.

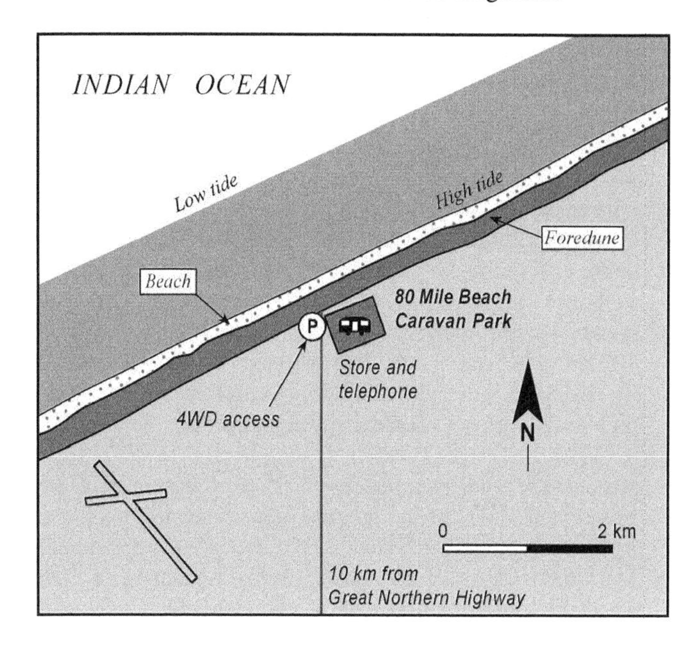

Figure 4.416 The Eighty Mile Beach caravan park is the best location to access this long beach.

Figure 4.417 Eighty Mile beach at low tide is up to several hundred metres wide.

Beaches WA 1966-1978 occupy a 30 km long central section of the beach, which is dominated by very low crossshore and backshore gradients, resulting in parts of the upper intertidal beach being covered with scattered stunted mangroves. In addition a series of migratory tidal creeks border the beaches and produce a relatively dynamic and unstable shoreline. The intertidal gradient also decreases as this zone widens to several hundred metres. Beach **WA 1966** commences on the eastern side of the shallow 500 m wide creek mouth that's separates it from beach WA 1965. The beach continues northeast for 10.5 km to the next shallow tidal creek. In between it is backed by a continuous 100-300 m wide foredunes, the rear reworked by the backing tidal creeks, which flow to either end, and are linked buy a continuous interbarrier depression. The depression totals 7 km in width with three to four sets of isolated foredune ridges extending across the first 3.5 km of the depression. **Mandora** homestead is located 7 km due south of the southwestern end of the beach. A vehicle track from the homestead reach the southern end of the beach where an outstation is located. It then runs along back of the beach and along the rear of the outer foredune.

Beach **WA 1967** is located between two eastward migrating creek mouths. The 3.9 km long beach itself is backed by an outer continuous series of recurved spits averaging 200-300 m in width, then two inner sets of spits, and finally an inner series of foredunes, the entire system totalling 3.5 km in width. Wyndhams Bore is located on the inner ridges. This is in turn backed by a 10 km wide back barrier depression. The beach is fronted by a 300 m wide upper intertidal ultradissipative zone, then a wider lower gradient, lower intertidal.

Beach **WA 1968** commences on the northeast side of the broad shallow creek mouth, and continues to the northeast for 16 km. This is the lowest energy upper intertidal section of the entire Eighty Mile Beach. The upper beach is covered with scattered mangroves, small tidal drainage creeks and an irregular beach-foredune boundary. It finally terminates where the gradient steepens and a well developed foredune replaces the scattered mangroves. The low foredune, which backs the beach, is backed by a 1.5 km wide salt flats, containing a few discontinuous scattered foredune ridges. Because of the swampy nature of the terrain there is no vehicle access from the rear, apart from a solitary fence line towards the centre of the beach. The salt flats are in turn bordered by an inner discontinuous foredune ridge, a second 1 km wide interbarrier depression, and then the inner foredune ridges, which are up to 1 km wide. The entire system is spread across 4 km of the backbarrier area, which are then backed by up to 20 km of backbarrier depression, extending all the way to the Great Northern Highway. The Sandfire Roadhouse is located 25 km southeast of the beach.

Beach **WA 1969** occupies the northern end of Eighty Mile Beach extending from the end of the mangrove section for 73 km. It curves to the northeast then north, to terminate at the low calcarenite of Cape Missiessy (Fig. 4.418). The cape is the first rocky point since Cape Keraudren 222 km to the southwest. In between is the great Eighty Mile Beach, which is actually 138 miles in length. The northern section is not only the longest continuous section of the beach, but also has the best developed and largest barrier system. A 1 km wide band of up to 20 foredune ridges extends northeast all the way to the cape, the width of the foredunes increasing to 2-2.5 km along much of the beach. The northern few kilometres of the barrier narrow to converge on the cape. They are predominately vegetated, apart from the northernmost 15 km which have or are experiencing moderate foredune activity in the form of blowouts and some parabolics, moving up to 300-400 m inland. The foredunes are backed initially by a 20 km wide backbarrier depression, which breaks down into a series of swampy embayments along the last 20 km. The continuous beach consists of the shell rich high tide beach, a low gradient upper intertidal, and then a very low gradient lower intertidal up to 1 km wide. At the northern end in lee of the cape, waves decrease and the outer tidal flats become covered in mud, as mangroves grow along the boundary with the cape. The old coast highway parallels the coast, running about 1 km inland in a foredune swale, to within 10 km of the cape where it turns and connects with the new Great Northern Highway. There is vehicle access to the shore at a few locations, the best developed 8 km west of Anna Plains homestead.

Figure 4.418 The northern end of Eight Mile Beach terminates at Cape Missiessy, where mangroves fringe the wide ultradissipative intertidal zone.

WA 1970-1972 **DESAULT BAY**

No.	Beach	Rating HT	LT	Type	Length
WA 1970	Desault Bay (1)	2	1	UD	4 km
WA 1971	Desault Bay (2)	2	1	UD	2.2 km
WA 1972	Desault Bay (3)	2	1	UD	3.7 km
Spring & neap tidal range = 7.8 & 1.9 m					

Desault Bay is an open 12 km long west-facing bay bordered by the low calcarenite points of Cape Missiessy in the south and Cape Jaubert to the north, which rise to 8 and 17 m respectively. The capes are connected by a 10 km long beach-barrier system, divided into three beaches (WA 1970-1972) by two shallow tidal creeks. There is vehicle access to both capes, with a track paralleling the rear of the barrier foredunes.

Beach **WA 1970** commences on the northern side of the low Cape Missiessy. The cape is surrounded by intertidal calcarenite flats dotted with mangroves. The beach trends north-northeast for 4 km to the 100 m wide mouth of Mangrove Creek. The creek drains a shallow lagoon located behind the outer series of foredunes-recurved spits, with up to 30 well developed ridges extending 2 km to the east of the creek mouth, where they are backed by a 2-3 km wide backbarrier depression. The recurved spits indicate the creek has migrated nearly 4 km northward from near the cape to its present location. At the same time up with 10 foredune ridges have prograded the shoreline up to 400 m seaward of the innermost spit. The southern end of the beach is fronted by low gradient sand flats with scattered mangroves. These grades into a several hundred metre wide ultradissipative beach, then the tidal shoals of the Mangrove Creek.

Beach **WA 1971** is located in the centre of the bay between Mangrove Creek and the 100 m wide northern creek. It is 2.2 km in length, faces west out of the centre of the bay, and is backed by the wide series of foredune ridges and backbarrier area. The northern creek meanders across the entire foredune ridge plain, and disconnects the northern and southern ridge systems. Between the two creeks the beach extends several hundred metres seaward as a low gradient ultradissipative system.

Beach **WA 1972** occupies the northern 3.7 km of shore from the northern creek to the rocky boundary with the cape. Wave energy increases slightly up the beach, due in part of the more moderate gradient of the intertidal. The beach consists of a well developed 500 m wide ultradissipative system, backed by more than 30 foredune ridges extending 1.8 km inland to the 1-1.5 km wide backbarrier depression. A vehicle track via Munganoo Well reaches the northern end of the beach and runs out along the cape.

The entire beach-barrier system has prograded up to 2 km west as a series of foredune ridges, truncated by the two creeks, and in the south complemented by the 4 km long series of outer recurved spits and foredune ridges.

WA 1973-1977 **CAPE JAUBERT**

No.	Beach	Rating HT	Rating LT	Type	Length
WA 1973	Cape Jaubert (S)	2	3	R+rock flats	2 km
WA 1974	Cape Jaubert	2	3	R+rock flats	700 m
WA 1975	Cape Jaubert (N1)	2	3	R+rock flats	200 m
WA 1976	Cape Jaubert (N2)	2	3	R+rock flats	400 m
WA 1977	Cape Jaubert (S)	2	3	R+rock flats	1.4 km
Spring & neap tidal range = 7.8 & 1.9 m					

Cape Jaubert is a calcarenite point that protrudes 2 km to the northwest, rising to 17 m at its tip. It forms the boundary between the southern Desault Bay and northern Geoffroy Bay. The cape is surrounded by five steep, narrow, high ride shell-rich beaches (WA 1973-1977), which are fronted by wide intertidal rocks flats on the southern and tip of the cape, and rocks, tidal flats and a creek on the northern side. A vehicle track runs right around the cape.

Beach **WA 1973** is separated from the northern end of beach WA 1972 by a 200 m long low red bluff. The bluff continues all the way to the point, with the 2 km long southwest facing crenulate beach consisting of discontinuous section of high tide sand fronted by upper intertidal rock flats grading into lower intertidal mixture of sand flats and scattered rocks. Beach **WA 1974** occupies the west-facing 700 m long tip of the cape. The high tide beach is up to 50 m wide and overlooks rock flats that extend up to 300 m seaward, where they grade into lower tide sand flats. Both beaches are backed by a low vegetated foredune over the backing calcarenite of the cape.

Beaches WA 1975-1977 are located along the 2.5 km long northern side of Cape Jaubert. Beach **WA 1975** is a 200 m long pocket of high tide sand wedged between low red calcarenite bluffs. The bluff and beach are backed by a 10 m high semi-stable foredune, while the beach is fronted by 500 m wide rock flats, with some veneers of sand, grading into a tidal channel that parallels the three beaches. Beach **WA 1976** is located 300 m to the east and is a more crenulate 400 m long high tide beach, also bordered by the low red bluffs and backed by the foredune that runs the length of the cape. There are several small blowouts in the foredune, with vegetated dune sand extending up to 200 m inland. It is fronted by a narrow band of rocks, 200 m wide tidal flats, then the tidal creek. Beach **WA 1977** commences 100 m to the east and continues for 1.4 km as a crenulate high tide beach into the entrance to Jaubert Creek. It is backed by low dune-draped calcarenite, and fronted by 100-200 m wide tidal flats and then the meandering creek channel.

WA 1978-1979 **GEOFFROY BAY**

No.	Beach	Rating HT	Rating LT	Type	Length
WA 1978	Geoffroy Bay (S)	2	1	UD-tidal flats	5.9 km
WA 1979	Geoffroy Bay (N)	2	1	UD-tidal flats	2.6 km
Spring & neap tidal range = 7.9 & 1.9 km					

Geoffroy Bay is an open, 8.5 km long, west-facing embayment bordered by Cape Jaubert to the south and 25 m high Cape Frezier to the north. In between are two beaches (WA 1978 & 1979), Jaubert Creek and a central tidal creek, which fill the entrance to the shallow 8 km deep embayment. The beaches are fronted by 1-2 km wide beach to tidal flat systems, including some low tide rock flats. There is vehicle access to the two cape and creek mouths.

Beach **WA 1978** commences on the northern side of 200 m wide Jaubert Creek that drains out along Cape Jaubert. The southern 1 km of the 5.9 km long beach is a recurved spit that has migrated part way across the creek mouth. The spit grades into a 1 km wide series of up to 16 foredune ridges, including inner recurved spits. This is in turn backed by an 8 km wide backbarrier depression drained by the two tidal creeks. The beach is fronted by an upper ultradissipative beach, which grades into extensive 1-2 km wide tidal flats associated with the southern tidal creek and sands accumulating in the southern half of the bay.

Beach **WA 1979** commences on the northern side of the 100 m wide northern tidal creek mouth. The beach continues north for 2.6 km to the inner boundary with the calcarenite rocks of Cape Frezier (Fig. 4.419). It is backed by a spit that widens from 200 to 500 m towards the creek mouth, then 1.5 km wide salt flats. A vehicle track runs around the perimeter of the salt flats to reach the creek mouth. It is fronted by a 1 km wide beach and tidal flats, consisting of an inner ultradissipative beach that narrows from 500 to 200 m in the north, then a 500 m wide veneer of sand and finally low gradient outer rock flats.

Figure 4.419 The northern end of Geoffrey Bay and Cape Frezier (foreground) and the wide tidal flats of beach WA 1979.

WA 1980-1981 **CAPE FREZIER**

No.	Beach	Rating HT	Rating LT	Type	Length
WA 1980	Cape Frezier (S)	2	3	R+rock/sand flats	400 m
WA 1981	Cape Frezier	2	3	R+rock/sand flats	1.1 km
Spring & neap tidal range = 7.8 & 1.9 m					

Cape Frezier is a 1 km long, up to 25 m high, dune-capped calcarenite point that protrudes 500 m to the west and separates Geoffroy Bay in the south from the larger Admiral Bay to the north. The generally low vegetated cape, has a band of red Pleistocene dunes along its western face, with high tide beaches located along the southern (WA 1980) and western shore (WA 1981).

Beach **WA 1980** extends for 400 m along the southern side of the cape. It is a southwest-facing high tide beach wedged between the backing 10 m high vegetated slope of the cape, and a 50-100 m wide zone of upper intertidal rocks, then the 1 km wide sand and tidal flats of its neighbour, beach WA 1979. A vehicle track reaches the eastern end of the beach. Beach **WA 1981** is located along the west-facing front of the cape. The beach extends for 1.1 km and consists of a moderately steep high tide beach backed by rock debris and the scarped red Pleistocene dunes, and fronted by low gradient sand flats, with scattered rocks particularly at each end. The northern 300 m trends to the northeast to the northern tip of the cape and is fronted by continuous 200-400 m wide irregular rock flats. A vehicle track reaches the northern tip.

WA 1982-1986 **CAPE DUHAMEL-MCKELSON CK**

No.	Beach	Rating HT	Rating LT	Type	Length
WA 1982	Cape Frezier (N)	2	1	UD+sand flats	3 km
WA 1983	Cape Duhamel (S)	2	3	R+rock/sand flats	500 m
WA 1984	Cape Duhamel	2	3	R+rock/sand flats	100 m
WA 1985	Cape Duhamel (N)	2	3	UD+sand flats	2 km
WA 1986	McKelson Ck (S)	2	1	R+tidal flats/channel	2.3 km
Spring & neap tidal range = 7.8 & 1.9 m					

To the north of Cape Frezier is an open 14 km wide embayment, the northern half called Admiral Bay, with low Tyron Point forming its northern boundary. The bay is dominated in the centre by McKelson Creek and its extensive tidal shoals. To either side of the creek are a series of beaches dominated by broad sand flats and/or calcarenite bluffs and outcrops. Beaches WA 1982-1986 are located along 8 km of shore between the cape and creek, with beaches WA 1987-1990 extending for 9 km from the creek mouth to the point. **Frazier Downs** homestead is located 7 km east of the creek mouth and there are numerous vehicle racks leading to most of the beaches.

Beach **WA 1982** commences at the northern base of Cape Frezier and curves slightly to the north-northeast for 3 km to the beginning of a section of calcarenite bluffs and rocks. The beach consists of a 50-100 m wide moderately steep high tide beach, fronted by extensive intertidal sand flats and some scattered rocks extending up to 500 m offshore. It is backed by a 100 m wide, scarped Holocene foredune, which grades north into a scarped red Pleistocene dune, both of which blanket the bluffs of the backing 15 m high coastal plain. Beach **WA 1983** commences at the bluffs and continues north for 500 m as a narrow high tide beach wedged in between the red 20-25 m high scarped bluffs, with intertidal rock then sand flats extending 600 m seaward. Beach **WA 1984** is located in a 50 m wide gap in the bluffs towards their northern end, which is marked by Cape Duhamel. It consists of a curving 100 m long pocket of sand, partly sheltered behind a red calcarenite band of rock. Inside the small shallow embayment the beach is backed by a 50 m wide foredune while outside the entrance are 500 m wide rock then sand flats.

Beach **WA 1985** commences on the northern side of Cape Duhamel, the northern tip of the bluffs, with a small tidal creek draining between the bluffs and the start of the beach. It tends north for 1.7 km to an easterly inflection in the shore, which marks the beginning of the wide McKelson Creek tidal shoals. The tidal shoals widen from 1 km off the cape to 2.5 km off the inflection. As a consequence waves are low at the shore and the beach is backed by a low foredune which is prone to overwashing (Fig. 4.420). This in turn has lead to dune activity forming a central deflation basin which floods on spring tide, backed by active dunes extending up to 1 km inland. The beach itself has a 100 m wide upper intertidal section fronted by the wide very low gradient sand flats and tidal shoals. Beach **WA 1986** commences at the inflection and tends to the northeast for 2.3 km as a crenulate recurved spit that forms the southern boundary of the 500 m wide creek mouth, with tidal shoals extending up to 4 km west of the creek mouth. The two beaches are part of a 4 km long, 0.5-1 km wide unstable barrier system, which is backed by the mangrove-lined creek, salt flats and a floodplain extending up to 9 km inland. Frazier Downs is located 6 km inland on the northern side of the floodplain.

Figure 4.420 Section of the low ultradissipative, overwash prone beach WA 1985.

WA 1987-1990 **ADMIRAL BAY**

No.	Beach	Rating HT	LT	Type	Length
WA 1987	McKelson Ck (N)	2	1	R+tidal channel/shoals	1.8 km
WA 1988	Admiral Bay (1)	2	1	UD+sand flats	3.5 km
WA 1989	Admiral Bay (2)	2	1	R+rock/sand flats	1.8 km
WA 1990	Admiral Bay (3)	2	1	R+rock/sand flats	1.4 km
Spring & neap tidal range = 7.8 & 1.9 m					

Admiral Bay occupies the northern part of the embayment from the mouth of McKelson Creek to Tyron Point. In between are four beaches (WA 1987-1990). Beach WA 1987 commences at the creek mouth and trends to the northwest to a northerly inflection which marks the beginning of the more open wave dominated shoreline. The beach is 1.8 km long and has slowly migrated northward eroding into the backing foredune ridges, truncating them at right angles. The erosion is caused by the northward migration of the southern spit (WA 1986) as it has moved into the creek mouth, causing the creek to move northward and erode into the ridges. The low energy beach faces southwest across sand shoals and the deep tidal channel, with tidal flats extending up to 4 km seaward.

Beach **WA 1988** begins at the inflection in the shore and curves gently to the north for 3.5 km to a small tidal creek. It is backed by a well developed 1.5-2 km wide series of up to 40 foredune ridges, the inner most ridges extending 4 km to the south in lee of the McKelson Creek. The barrier is one of the better formed and preserved foredune plains in the North West. The beach consists of a 300-400 m wide low gradient upper intertidal ultradissipative beach grading seaward into 2.5 km wide low tide sand flats.

Beach **WA 1989** commences on the northern side of the 100 m wide shallow creek mouth and extends north for 1.8 km to terminate at the next calcarenite-controlled tidal creek. This beach is dominated by calcarenite, which outcrops from the creek mouth the length of the beach. The calcarenite forms intertidal rock flats, backed by a narrow high tide beach, then a scarped 20 m high Holocene foredune that blankets the Pleistocene calcarenite core to a width of 100-500 m. The barrier is backed by 1 km wide salt flats, then the coastal plain, with vehicle tracks running across the salt flats to the beach..

Beach **WA 1990** commences on the northern side of the tidal creek and is a similar long narrow high tide beach that trends north for 1.4 km finally curving round to face south in lee of the protection of Tyron Point (Fig. 4.421). It is dominated by the near continuous calcarenite outcrops, with only the protected northern 200 m free of rocks. The calcarenite average 50-100 m in width grading into lower gradient rock and sand flats, the latter extending up to 2 km seaward. The beach is backed by a 100-300 m wide scarped foredune and dune-capped calcarenite barrier, which is backed by a mangrove-lined creek and salt flats, extending 1 km inland.

Figure 4.421 Tryon Point (left) and beach WA 1990 that is fronted by tidal flats up to 2 km wide.

WA 1991-1993 CAPE BOSSUT

No.	Beach	Rating HT	LT	Type	Length
WA 1991	Tryon Pt	2	3	R+rock flats	1.8 km
WA 1992	Cape Bossut	2	3	R+rock flats	2.3 km
WA 1993	Spit Pt	2	3	R+rock flats	1.2 km
Spring & neap tidal range = 7.8 & 1.9 m					

Tryon Point and Cape Bossut form either end of a straight 5 km long west-facing section of Pleistocene calcarenite that forms the western-most outcrops along this section of coast. It also forms the northern boundary of Admiral Bay and southern boundary of the larger Lagrange Bay to the north. Three beaches (WA 1991-1993) are located between the base of the bluffs and intertidal rock flats. All three are accessible by vehicle, with tracks crossing the backing coastal plain and salt flats to reach the 20 m high, 200 ha calcarenite outcrop.

Beach **WA 1991** commences at 14 m high Tryon Point and trends due north for 1.8 km to a slight protrusion in the shoreline. The beach consists of a steep, narrow high tide beach composed of sand, then rocks and beachrock, with a 500 m wide low gradient rock flats extending across the intertidal (Fig. 2.21a). It is backed by a scarped foredune, which drapes over the backing calcarenite core of the 300 m wide island. The island is separated from the mainland by 1 km wide salt flats.

Beach **WA 1992** commences at the protrusion and continues north for 2.3 km to 20 m high Cape Bossut. It is a similar steep sand and beachrock high tide beach with 600 m wide intertidal rock flats, which widen at the cape where they are criss-crossed by linear piles of rocks. The beach is backed by an unstable foredune then the barrier which widens to 1 km and which is backed by mangrove-lined Cape Bossut Creek.

Beach **WA 1993** extends northeast from Cape Bossut for 1.2 km to Spit Point, the beach backed by a Pleistocene calcarenite recurved spit. The beach is a steep narrow high tide beach, backed by a scarped 20 m high foredune, and fronted by rock flats, which narrow to 100 m at the northern end. The tip of the point forms the southern boundary of Cape Bossut Creek, with a 15 m high Pleistocene spit extending 2 km south along the western side of the creek.

WA 1994-1999 NECK-ISLET-ROCKY PTS

No.	Beach	Rating HT	LT	Type	Length
WA 1994	Ledge Pt	2	1	R+sand-rock flats	2.3 km
WA 1995	Middle Ck (W)	2	1	R+tidal flats	500 m
WA 1996	Middle Ck (E)	2	1	R+sand flats	3 km
WA 1997	Islet Pt (E)	2	1	R+rock/sand flats	500 m
WA 1998	Cliff Pt	2	1	R+tidal flats	600 m
WA 1999	Rocky Pt	2	1	R+tidal flats	2.2 km
Spring & neap tidal range = 7.8 & 1.9 m					

Lagrange Bay is a 17 km wide west-facing embayment containing 32 km of low energy shoreline, located between Cape Bossut in the south and False Cape Bossut to the north. The southern 12 km of shoreline between Cape Bossut and Cowan Creek at Rocky Point consists of an irregular north-facing shoreline containing six beaches (WA 1994-1999) bordered and separated by five points and five tidal creeks, all fronted by extensive tidal flats, with some mangroves along the shore (Fig. 4.422). There is vehicle access usually across backing salt flats, to all the beaches.

Figure 4.422 Cape Bossut Creek and Narrow Neck (left) and the western end of beach WA 1994, with its extensive tidal flats.

Beach **WA 1994** extends between the narrow Neck Point, the eastern border of Cape Bossut Creek, and Ledge Point, 2.3 km to the east. The beach faces north and consists of a western sandy section while the eastern half is backed by the calcarenite of Ledge Point. The western half is backed by a 500 m wide unstable barrier with areas of deflated and active dunes rising to 27 m, which in turn are backed by the 1 km wide mangroves and salt flats of a tributary of Cape Bossut Creek. The beach narrows to the west where it consists of a foredune blanketing the calcarenite, with beachrock outcropping along the base of the high tide beach. Sand flats extends several hundred metres off the western half with a mixture of rocks and sand flats off the eastern end.

Beach **WA 1995** is located 1.5 m southeast of Ledge Point, with calcarenite and two small tidal creeks between the point and the beach. It is a low energy, north-facing 500 m long beach with tidal creeks to either end and tidal channels and shoals extending 1.5 km seaward (Fig. 2.19a). It is backed by a 200 m wide series of low foredunes, which widen to 400 m at the eastern Middle Creek mouth, with an inner Pleistocene barrier behind.

Beach **WA 1996** commences on the eastern side of 200 m wide Middle Creek and curves to the northeast for 3 km to the low Islet Point, located to the lee of Red Islet, a solitary outcrop of red calcarenite. The beach is backed by a barrier, which has prograded southwest across the backing 2 km deep embayment, as a series of inner recurved spits. The spits are fronted by a 200 m, widening to 400 m in the west, band of foredune ridges, which are now being deflated and forming a series of blowouts. The relatively steep narrow high tide beach is fronted by 1 km wide sand flats, which merge with the Islet rocks to the east.

Beach **WA 1997** is located on the eastern side of Ledge Point and consists of a narrow 500 m long, north-facing high tide beach, backed by a 10-20 m high scarped foredune, and fronted by a 600 m wide mixture of rock and sand flats. Beach **WA 1998** lies 500 m to the south and is a straight 600 m long north-facing sandy beach backed by a 20-25 m high scarped white, then orange, sandy foredune. It is bordered by low calcarenite bluffs and rocks, including the eastern Cliff Point, and fronted by flat featureless 1 km wide sand flats.

Beach **WA 1998** extends from the lee of 25 m high Cliff Point for 2.2 km to Rocky Point at the mouth of Cowan Creek. This is a northwest-facing high tide beach, fronted by continuous high tide beachrock bluffs, then flat featureless 1 km wide tidal flats. The calcarenite core extends as a 200 m wide spit to the creek mouth, including a calcarenite recurved spit at the entrance. The spit is backed by 2 km wide mangroves and salt flats of the southern side of the creek. Vehicle tracks run around the side of the flats to reach the spit.

WA 2000-2004 **BLACK ROCK PT (BIDYADANGA)**

No.	Beach	Rating HT	Rating LT	Type	Length
WA 2000	Cowan Ck	2	1	UD-tidal sand flats	3.3 km
WA 2001	Black Rock Pt (1)	2	2	R+rock/sand flats	500 m
WA 2002	Black Rock Pt (2)	2	2	R+rock flats	600 m
WA 2003	Black Rock Pt (E1)	2	2	R+mud flats	2 km
WA 2004	Black Rock Pt (E2)	2	2	R+tidal flats	1.7 km
Spring & neap tidal range = 7.8 & 1.9 m					

Bidyadanga is a small aboriginal community (Lagrange) spread over about 100 ha of the low coastal plain to lee of Cowan Creek and Black Rock Point, and 10 km in from the highway. Beaches WA 2000-2004 are located either side of Black Rock Point and are all located within 2-3 km of the community, with several vehicle tracks running out to and along the beaches. The beaches occupy part of the southern shore of 14 km wide Lagrange Bay.

Beach **WA 2000** is a 3.3 km long spit that has extends south from Black Rock Point across the northern side of Cowan Creek, and helps enclose the 1 000 ha shallow lagoon filled with mangroves and extensive salt flats (Fig. 4.423). Since the original spit was formed the shoreline has prograded 500 m to the west, with the outer 100 m of foredune ridges reactivated and transgressing across the spit. A vehicle track runs the length of the rear of the barrier to reach the 200 m wide creek mouth. The beach has an upper ultradissipative section grading into 1 km wide low tide sand to mud flats.

Beaches WA 2001 and 2002 occupy the shore of **Black Rock Point**, a 15 m high dune-capped calcarenite point. Beach **WA 2001** is a curving 500 m long, north-facing, narrow high tide beach, fronted by irregular rocks flats grading into 800 m wide tidal flats. It is backed by a semi-stable, 50 m wide foredune. Beach **WA 2002** commences immediately to the east and continues for another 600 m to the next slight protrusion on the shore. It is a similar narrow high tide beach with rock then sand flats extending 700 m offshore. A gravel road from Bidyadanga reaches the southern boundary point and centre of the beach, with a shelter at the point. These two are the main recreational beaches for the community.

Figure 4.423 Black Point (foreground) and beach WA 2000 extending south to Cowan Creek.

Beach **WA 2003** curves to the northeast from the boundary protrusion for 2 km. It consists of a high tide beach backed by a low foredune then the coastal plain, while it is fronted by sand and mud tidal flats extending up to 1.5 km offshore. A few mangroves grow on the flats off the southern end of the beach, while vehicle tracks reach both ends of the beach. Beach **WA 2004** is a 1.7 km long low series of recurved spits, that extends north from the end of beach WA 2003. The spits have prograded across the southern end of an 8 km wide shallow inner section of Lagrange embayment, drained by Gururu Creek in the centre. The spit splays and widens to 300 m to the north. They are backed by 1 km wide salt flats, while tidal sand and mud flats extends up to 2 km into Lagrange Bay.

WA 2005-2007 **GURURU CREEK**

No.	Beach	Rating HT	Rating LT	Type	Length
WA 2005	Gururu Ck (S1)	2	1	R+tidal flats	900 m
WA 2006	Gururu Ck (S)	2	1	R+tidal flats	2.2 km
WA 2007	Gururu Ck (N)	2	1	R+tidal flats	900 m
Spring & neap tidal range = 7.8 & 1.9 m					

The eastern half of the bay consists of a 90 km^2 shallow embayment infilled with salt flats and a usually dry floodplain, drained by Gururu Creek. Beaches WA 2004-

2007 form the boundary shoreline between the inner embayment and the low gradient bay floor. The beaches are accessible by vehicle tracks cross the backing floodplain and salt flats.

Beach **WA 2005** is a 900 m long west-facing low beach ridge, that faces into the bay across 2 km wide sand and mud flats, together with tidal creeks draining out either end of the beach. It is backed by a 1.5-2 km wide series of inner beach ridge-barrier, of which beach WA 2004 forms the outer barrier of the system. Beach **WA 2006** is located 500 m to the north and is a 2.2 km long low narrow vegetated beach ridge, bordered by the tidal creek and fronted by high tide sand flats, then 1-1.5 km wide tidal sand and mud flats. It is backed by a 3 km wide series of five inner barriers and salt flats containing up to 20 beach ridges, then the 8 km wide inner floodplain.

Gururu Creek is a 200 m wide mangroves-lined creek located in the northeastern corner of Lagrange Bay. Beach WA 2006 lies 1 km south of the creek and beach WA 2007 1 km to the northwest. Beach **WA 2007** is a 900 m long low, narrow active sandy recurved spit that supplies sand to the northern side of the creek area. It is prone to overwashing and backed by a mangrove-filled intertidal depression, then two inner barrier and salt flats all extending 2.5 km inland, then the inner floodplain, which extends east for another 8 km.

WA 2008-2011 **FALSE CAPE BOSSUT**

No.	Beach	Rating HT	Rating LT	Type	Length
WA 2008	False Cape Bossut (S3)	2	2	R+beachrock-tidal flats	2.5 km
WA 2009	False Cape Bossut (S2)	2	2	R+beachrock-tidal flats	1.8 km
WA 2010	False Cape Bossut (S1)	2	2	R+rock flats	200 m
WA 2011	False Cape Bossut	2	2	R+rock flats	300 m

Spring & neap tidal range = 7.8 & 1.9 m

False Cape Bossut is located 18 km northeast of Cape Bossut, the two low calcarenite points form the respective headlands of Lagrange Bay. The False Cape lies to the lee of Outer Reef, with the southern shoreline extending into the bay as a 7 km long crenulate series of Holocene beaches (WA 2008 & 2009), backed by dune-capped Pleistocene barrier spits. Beaches WA 2010 and 2011 occupy the west-facing tip of the cape. There are vehicle tracks to the cape and all the beaches.

Beach **WA 2008** is a curving 2.5 km high tide and beach that is fronted by continuous band of intertidal beachrock, then 2 km wide tidal sand to mud flats. The beach is backed by a low vegetated foredune, draped over the backing and underlying Pleistocene barrier, which originally formed as a recurved spit. It terminates in the north at a rock-fronted protrusion. Beach **WA 2009** continues past the rocks and trends to northwest for 1.8 km to the cape. It is a crenulate high tide sandy beach fronted by upper intertidal rock flats, which widen from 200 m in the east to 500 m at the cape, with tidal sand and mud flats continuing 2 km into the bay. It is backed by a small vegetated foredune then the vegetated Pleistocene core of the cape.

Beach **WA 2010** occupies the southern half of the tip of the cape. It is a 200 m long west-facing strip of high tide sand, wedged between the red calcarenite rocky points, with rock flats extending several hundred metres west toward Outer Reef located 3 km off the cape. A vehicle track terminates at a small parking area behind the centre of the beach, and a foot track crosses the 50 m wide foredune to the beach. Beach **WA 2011** is located immediately to the north and is a similar 300 m long high tide beach, backed by a higher semi-stable foredune that rises to 25 m. The beach is fronted by rock flats that extend up to 1 km offshore towards the Outer Reef (Fig. 4.424). Between the beach and False Cape Creek 1 km to the west, is a 500 m wide zone of semi-stable parabolic dunes that originated from a former beach along this now rocky shoreline.

Figure 4.424 Beaches WA 2010 and 2011 (foreground) are fronted by rock flats that extend up to 1 km seaward to Outer Reef.

WA 2012-2016 **FALSE CAPE CK-PORT SMITH**

No.	Beach	Rating HT	Rating LT	Type	Length
WA 2012	False Cape Ck (S)	2	3	R+rock flats/channel	600 m
WA 2013	False Cape Ck (N)	2	2	R+rock flats	2.4 km
WA 2014	Port Smith (S 3)	2	2	R+rock flats	2.7 km
WA 2015	Port Smith (S 2)	2	1	R+rock/sand flats	3.2 km
WA 2016	Port Smith (S 1)	2	1	R+sand flats	700 m

Spring & neap tidal range = 7.8 & 1.9 m

At **False Cape Bossut** the shoreline turns and trends to the northeast for 18 km to Cape Latouche Treville. This is a calcarenite-dominated section of shore, tied to the two capes, with calcarenite also dominating most of the nine intervening beaches (WA 2012-2021). The shoreline is broken into three sections by False Cape Creek in the south, a second unnamed creek, and the larger Port Smith creek towards the centre. There is vehicle access to the southern beach across the backing salt flats, with Port Smith Caravan Park located 2 km east of the tidal entrance.

Beach **WA 2012** is located 1 km west of the cape at the western side of the calcarenite-controlled of the entrance to False Cape Creek (Fig. 4.425). The 600 m long beach is fronted by continuous 100-200 m wide intertidal rock flats, then the tidal channel along the eastern half of the beach. The calcarenite reefs on the northern side of the channel form a natural breakwater for the entrance. At low tide however the channel between the reefs narrows to less than 50 m. The beach is backed by a continuation of the semi-stable parabolic dunes that extend east of the cape to the creek, with three blowouts behind the eastern half of the beach.

Figure 4.425 Beachrock-controlled False Cape Creek with beaches WA 2012 (right) and 2013 (left).

Beach **WA 2013** commences immediately east of the rock-controlled creek mouth. It is a crenulate 2.4 km long northwest-facing strip of discontinuous high tide sand, set amongst calcarenite bluffs and fronted by continuous rock flats, which widen from 200 m at the creek to 500 m in the east. The beach is backed by a 400-500 m wide series of semi-stable blowouts and parabolics rising to 25 m, which overlap the Pleistocene calcarenite core of the barrier, then the 1-1.5 km wide mangroves and salt flats of the creek.

Beach **WA 2014** continues to the northeast as a relatively straight high tide sand beach, also backed by the semi-stable dunes and fronted by the continuous 500 m wide rock flats. It terminates at the mouth of a 50 m wide creek, which cuts through the barrier. It is also backed by 1-1.5 km band of mangroves to either end and a central section of salt flats, across which a vehicle track reaches the barrier. The mangroves drain both west to False Cape Creek and east to Port Smith.

Beach **WA 2015** commences on the northern side of the small creek mouth and continues the trend of the barrier to the northeast for another 3.2 km, to the southern side of Port Smith entrance. The high tide sand beach is initially fronted by the near continuous 500 m wide intertidal rock flats. Towards the port there are replaced by the entrance tidal sand shoals of Port Smith, which lowers wave energy at the shore causing the high tide beach to diminish in size. The tidal shoals extend 2 km south along the beach and together with a shore-parallel tidal channel extend up to 2 km offshore. The beach is backed by a 15-20 m high vegetated foredune, overlapping the Pleistocene calcarenite core. The Pleistocene barrier consists of a series of lithified north-trending recurved spits backed by the 1-2 km wide mangrove-filled southern arm of Port Smith.

Beach **WA 2016** is a narrow curving 700 m long low energy, high tide beach located at the eastern end of the calcarenite barrier adjacent to the port inlet. It is fronted by the extensive tidal shoals of the Port Smith inlet. It is backed by a Pleistocene recurved spit that trends east into the port.

WA 2017-2019 **PORT SMITH-CAPE LATOUCHE TREVILLE**

No.	Beach	Rating HT	Rating LT	Type	Length
WA 2017	Port Smith (N 1)	2	3	R+tidal channel	1.3 km
WA 2018	Port Smith (N 2)	2	1	R+tidal sand flats	1 km
WA 2019	Cape Latouche Treville (W2)	2	1	R+tidal channel	4.8 km

Sring & neap tidal range = 7.8 & 1.9 m

Port Smith is a 1200 ha tidal lagoon largely filled with mangroves and salt flats, with a deep 500 m wide tidal channel running out between two Pleistocene recurved spits. The channel continues 2 km seaward with the surrounding tidal sand shoals extending up to 3 km seaward (Fig. 2.19b). The Port Smith Caravan Park is located on the mainland directly opposite the 400 m wide creek mouth. It provides a range of facilities as well as good access to the eastern shore of the port and boat launching. Beaches WA 2107 and 2018 form the northern entrance to the port.

Beach **WA 2017** is a 1.3 km long narrow high tide beach, which is part of a Holocene recurved spit that has migrated into the northern side of the port. It faces southwest across the deep, 400 m wide tidal channel. A series of three older spits and intervening mangroves extend 2 km north of the modern spit, into the 250 ha northern arm of the bay. Beach **WA 2018** commences at a salient in the shore then turns and trends north away from the channel. The 1 km long beach is a transition from the tidal channel to the south and more wave dominated beach to the north. It is fronted by 2-3 km wide tidal sand shoals and the channel and backed by a low 2 km wide low series of beach ridge-recurved spits, then the older vegetated spits.

Beach **WA 2019** is a northwest-facing 4.8 km long beach and calcarenite system that extends from the northern entrance to Port Smith to the low rocks that form the western side of Cape Latouche Treville. The relatively exposed, straight beach is dominated by low calcarenite bluffs along most of the shoreline, with little or no beach at high tide. Intertidal rock flats dominate most of the beach, with a 1 km long sandy section at the northern end while scattered rock reefs lie off the beach. It is backed by a 1 km wide vegetated dune-draped Pleistocene barrier, that consists of an inner recurved spit and outer foredune ridges that enclosed the northern arm of Port Smith.

WA 2020-2024 **CAPE LATOUCHE TREVILLE**

No.	Beach	Rating HT	LT	Type	Length
WA 2020	Cape Latouche Treville (W1)	2	1	UD	700 m
WA 2021	Cape Latouche Treville	2	1	R+rock flats	1.6 km
WA 2022	Cape Latouche Treville (E1)	2	1	R+rock/sand flats	400 m
WA 2023	Cape Latouche Treville (E2)	2	1	R+sand flats-UD	800 m
WA 2024	Cape Latouche Treville (E3)	2	1	R+sand flats/beachrock	1.2 km

Spring & neap tidal range = 7.8 & 1.9 m

Cape Latouche Treville is a low calcarenite headland, capped by vegetated red Pleistocene dunes rising to 32 m. The dunes are part of a 500 ha of vegetated transgressive dunes, which cap the 3 km wide crest of the cape. There is vehicle access to the western side of the cape and to Gourdon Bay to the east, with no defined access to the cape area. Beaches WA 2020 and 2021 extend for 2.5 km east of the cape (Fig. 4.426), while beaches WA 2022-2024 occupying three rock-bounded embayments that extends for 3 km southeast of the cape and face northeast along Gourdon Bay.

Figure 4.426 Beaches WA 2019 to 2021 extend east of Cape Latouche Treville.

Beach **WA 2020** is a well developed 700 m long northwest-facing ultradissipative beach bordered at each end by intertidal rocky points extending 200-300 m seaward. In between is the medium gradient intertidal beach, backed by an incipient foredune at the base of vegetated 30 m high red Pleistocene dunes. Beach **WA 2021** extends immediately west for 1.6 km to the northern tip of the cape. It consists of a crenulate high tide sandy beach fronted by discontinuous 200-300 m wide intertidal rock flats and patches of sand. It is backed by a 10 m high foredune, then the Pleistocene dunes.

Beach **WA 2022** lies to the east of the protruding intertidal rocks that form the tip of the cape. It is a northeast-facing 400 m long high tide sand beach, backed by 10-20 m high scarped Pleistocene calcarenite, with rocks and debris littering the rear of the beach. It is fronted by a mixture of sand and rock flats, as well as some rock reefs. A 150 m long, scarped rocky point separates if from beach **WA 2023**. This beach curves to the southeast for 800 m. It is backed by a continuous 10-20 m high calcarenite scarp, with some rocks and sand flats along the northern half and more ultradissipative conditions along the more exposed southern half of the beach.

Beach **WA 2024** occupies the next rock-bordered embayment. It faces northeast and consists of a slightly curving 1.2 km long veneer of sand, backed by 10-20 m high scarped calcarenite, with rocks and beachrock dominating the intertidal zone. It is backed by the scarped Pleistocene dunes, then a 400 m wide former wetland.

WA 2025-2028 **GOURDON BAY**

No.	Beach	Rating HT	LT	Type	Length
WA 2025	Goudron Bay	2	1	UD	7.4 km
WA 2026	Cape Du Boulay (S)	2	1	UD	1.9 km
WA 2027	Cape Du Boulay (N)	2	1	UD	5.4 km
WA 2028	Cape Gourdon (S)	2	1	UD	800 m

Spring & neap tidal range = 7.8 & 1.9 km

Gourdon Bay is an open northwest-facing 15 km wide bay bordered by the calcarenite of Cape Latouche Treville to the south and the red Cretaceous sandstone of Cape Gourdon to the north. In between the bay shore curves for 19 km and contains seven near continuous beaches (WA 2022-2028). The only access and development in the bay region is a small cluster of buildings of the Yardoogarra community in the centre of the bay to the lee of beach WA 2025, with vehicle access only to beaches WA 2025-2027. Beaches WA 2025-2028 extend from the southern corner of the bay along the eastern shore past Cape Du Boulay to Cape Gourdon.

Beach **WA 2025** is a curving 7.4 km long ultradissipative beach that initially trends east of beach WA 2024, before curving to trend northeast. It is a continuous moderate gradient sandy beach, with a few rocks outcropping off the centre where they induce a slight salient. It terminates at the beginning of the deeply weathered red Cretaceous sandstone, south of Cape Du Boulay. It is for the most part backed by a semi-stable foredune that rises to 20 m, then the gently rising coastal plain. The small **Yardoogarra community** is located 500 m east of the centre of the beach with a vehicle track out to the beach.

Beach **WA 2026** commences on the northern side of the first sandstone-induced inflection and continues northeast for 1.9 km to the intertidal sandstone rocks that form Cape Du Boulay. The beach is a moderate gradient ultradissipative system, with rock flats outcropping on the beach towards the cape. It is backed by scarped, slumping and eroding red pindan, some of which protrude onto the northern end of the beach.

Beach **WA 2027** commences to the lee of the cape, initially curving to the east and then continues the trend of the shore to the northeast for 5.4 km terminating at a small usually blocked rock-bound creek mouth. In between is a continuous ultradissipative sandy beach, backed by scarped and gullied red pindan, with some rock flats in the lower intertidal. Rocks outcrop either side of the creek, with a large northern outcrop forming the boundary with beach **WA 2028**. This beach continues northeast for another 800 m to the red 20 m high eroding bluffs of Cape Gourdon. It is a moderate gradient ultradissipative beach, backed by gullies and in places protruding sandstone bluffs, with some limited foredune development in the gullies and against the bluffs.

WA 2029-2032 **CAPE GOURDON**

No.	Beach	Rating HT	Rating LT	Type	Length
WA 2029	Cape Gourdon	2	1	UD	900 m
WA 2030	Cape Gourdon (E1)	2	1	UD	300 m
WA 2031	Cape Gourdon (E2)	2	1	UD	350 m
WA 2032	Cape Gourdon (E3)	2	1	UD	1.2 km
WA 2033	Cape Gourdon (E4)	2	1	UD	250 m
Spring & neap tidal range = 8.3 & 1.9 m					

At **Cape Gourdon** the shoreline turns and trends east for 3 km before again trending to the northeast towards Cape Villaret. Along the easterly inflection are five rock-bounded beaches (WA 2029-2033) each bordered and backed by the deeply weathered red Cretaceous sandstone (Fig. 4.427), with numerous gullies cutting the sandstone and merging with the beaches. There is no vehicle access to this section of the shore.

Figure 4.427 Cape Gourdon (foreground) and beaches WA 2029 to 2032.

Beach **WA 2029** extends for 900 m east of Cape Gourdon as an ultradissipative beach, with the spring tide reaching to the backing scarped and gullied sandstone bluffs. Some of the bluffs protrude onto the beach and there is a central area of low tide rock flats. A small 20 m high cliffed point separates it from beach **WA 2030**, which continues east for another 300 m. This beach is backed by 10-20 m high sandstone scarps and extends seaward as a moderate gradient ultradissipative beach, with a line of rocks extending 500 m seaward of the western boundary point.

An eroding 200 m wide, 20 m high headland separates beaches WA 2030 and 2031. Beach WA **2031** is a 350 m long lower gradient ultradissipative beach backed by vegetated sloping sandstone, with patches of foredune in the gullies. A 100 m wide discontinuous headland forms the eastern boundary. Beach **WA 2032** commences on the eastern side of the headland and curves to the east then northeast for 1.2 km. A usually dammed creek is located at the southern end of the beach. Heavily eroded and gullied sandstone backs the beach rising to 60 m, 2 km inland. A foredune fills the southern gullies, while the sandstone is being eroded and scarped along the northern half of the beach. The eroding sandstone forms a number of scallops, with the spring tide reaching to the base of the bluffs.

A 200 m long section of protruding bluffs and scattered offshore rocks forms the boundary between beaches WA 2032 and 2033. Beach **WA 2033** continues north for another 250 m to a cliffed 30 m high red point. The beach is backed by continuous 20 m high cliffs, with rock debris at their base and some rocks outcropping across the ultradissipative beach.

WA 2034-2037 **BARN HILL**

No.	Beach	Rating HT	Rating LT	Type	Length
WA 2034	Cape Gourdon (E5)	2	1	UD	500 m
WA 2035	Cape Gourdon (E6)	2	1	UD	800 m
WA 2036	Cape Gourdon (E7)	2	1	UD	600 m
WA 2037	Barn Hill	2	1	UD	1.8 km
Spring & neap tidal range = 8.3 & 1.9 m					

The red Cretaceous sandstone continues to dominate the 12 km of shore between Cape Gourdon and Cape Villaret. The sandstone is heavily eroded and gullied and scarped along the shore, with protruding sections of the rocks forming the boundaries between the beaches. Beaches WA 2034-2037 occupy a 4 km long section of northeast trending shoreline between Cape Gourdon and Barn Hill. Two easterly inflections in the shore mark the beginning of beaches WA 2034 and 2037, with protruding sections of sandstone bluffs separating the other beaches. The only access to the shore is at Barn Hill.

Beach **WA 2034** is a curving north-northwest facing 500 m long beach that occupies the first easterly inflection in the shore. The ultradissipative beach is backed by scarped 10-20 m high red bluffs, with a 100 m long section of foredune filling the mouth of a gully towards the eastern end of the beach. A few rocks outcrop across the intertidal zone to either end of the beach. Beach **WA 2035** commences immediately past the eastern rocks and continues northeast for 800 m to the next 100 m long protrusion in the sandstone bluffs. Two creeks drain the backing gullied sandstone to reach the very southern end and centre of the beach, with vegetated dunes extending up to 100 m inland between the creeks. The beach continues as a moderate gradient ultradissipative system.

Beach **WA 2036** continues north of the bluffs for 600 m to the next inflection in the shore. It is backed by continuous 10-20m high red and white scarped bluffs, capped by vegetated, and in the north, eroded pindan (Fig. 4.428). Beach **WA 2037** commences in lee of a 200 m easterly inflection in the shore dominated by 20-30 m high white and red scarped sandstone. It trends northeast for 1.8 km to 30 m high Barn Hill. It is backed and bordered by scarped red 10 m high pindan bluffs with a central section of foredune. The small Barn Hill community is located behind the northern end of the beach, with a vehicle track reaching the centre of the beach. The beach is ultradissipative, with some rocks outcropping along the northern low tide zone.

Figure 4.428 Ultradissipative beach WA 2036, shown here at high tide, is backed by scarped red bluffs.

WA 2038-2042 **CAPE VILLARET (S)**

No.	Beach	Rating HT	LT	Type	Length
WA 2038	Barn Hill (N)	2	1	UD	3.2 km
WA 2039	Cape Villaret (S2)	2	1	UD	1 km
WA 2040	Cape Villaret (S1)	2	1	UD+rock flats	500 m
WA 2041	Cape Villaret (S	2	1	UD	250 m
Spring & neap tidal range = 8.3 & 1.9 m					

Between Barn Hill and Cape Villaret is a straight northeast-trending section of sandy shore backed by generally vegetated red pindan, which has slumped to the north of Barn Hill, and becomes increasingly scarped close to the cape. In between are four beaches, each separated by protruding section of scarped pindan-capped, red sandstone bluffs. A straight vehicle track from Barn Hill parallels the rear of the coast to the cape, with access tracks to some of the beaches.

Beach **WA 2038** commences on the northern side of Barn Hill in an easterly inflection in the shore. It trends straight northeast for 3.2 km to a 200 m long section of protruding bluffs. It is a moderate gradient ultradissipative beach backed by slumping 20-30 m high red pindan along the south-central section of the beach, with vegetated bluffs to the north, and a narrow strip of foredune along the central-northern section of the beach. There is a vehicle track down the bluffs at the very southern end of the beach. Beach **WA 2039** commences on the northern side of the headland and continues northeast for 1 km, to the beginning of the scarped red bluffs that dominate the remainder of the shoreline to the cape. The beach is backed by a central gully occupied by a usually dry creek bed, with 10-20 m high sand-draped bluffs to either side, the bluff becoming increasingly exposed towards the northern end of the beach.

Beach **WA 2040** is a 500 m long high tide ultradissipative beach backed by a narrow gully in the south then continuous pindan-capped sandstone bluffs. Rock debris littering the rear of the beach and intertidal tide rock flats dominated the central-northern section of the shore. Beach **WA 2041** is located on the western side of cape Villaret. It is a moderate gradient, 250 m long ultradissipative beach, bordered by a slight protrusion in the bluffs and rock flats to the south, and a 200 m long protrusion in the bluffs to the north that forms the 20 m high cape. It is backed by irregular bluffs, with a central dune-draped section.

WA 2043-2046 **CAPE VILLARET**

No.	Beach	Rating HT	LT	Type	Length
WA 2042	Cape Villaret	2	1	R+UD+rock flats	200 m
WA 2043	Cape Villaret (E1)	2	1	R+sand flats	200 m
WA 2044	Cape Villaret (E2)	2	1	R+sand flats/UD	300 m
WA 2045	Cape Villaret (E3)	2	1	R+sand flats	100 m
WA 2046	Cape Villaret (E4)	2	1	R+sand/rock flats	150 m
Spring & neap tidal range = 8.3 & 1.9 m					

Cape Villaret is a 30 m high dune-capped sandstone point, that consists of protruding arms of red sandstone bluffs, with sandy beaches in between. It marks a major easterly inflection in the shoreline and the beginning of the transition from the sandstone-dominated coast to the south, and the wide low gradient tidal flats of Roebuck Bay 30 km to the north. Five small beaches (WA 2042-2046) are located immediately east of the cape, in amongst the protruding sandstone.

Beach **WA 2042** is located at the tip of Cape Villaret. It is a 200 m long beach, with a 100 m wide high tide zone fronted by 200 m wide low tide rock flats. Twenty metre high bluffs border each end, with a ramped foredune against the backing bluffs.

Beach WA 2043 and 2044 are located 100 m to the east and share a 500 m wide curving embayment, bordered by 20 m high bluffs, with a narrow 100 m long arm of sandstone across the high tide beach separating the two which share a continuous low tide beach. Beach **WA 2043** is a slightly curving 200 m long high tide beach fronted by 300 m wide sand flats. The 100 m wide headland borders the western end, with the narrow arm of sandstone to the east. It is backed by steep, partly vegetated red Pleistocene foredune rising to 30 m, with a dirt road to the rear of the dune. Beach **WA 2044** continues east of the dividing arm for 200 m to a 20 m high bluffed headland. Wave energy increases slightly along the beach as the sand flats transform to a narrower ultradissipative beach by the headland. It is backed by 20 m high red bluffs, capped by another 20 m of truncated red Pleistocene transgressive sand dune.

Beach **WA 2045** is located immediately to the east and consists of a 100 m long pocket of high tide sand, bordered and backed by scarped red bluffs rising 20-30 m, with considerable rock debris along the rear of the beach. The inner and low tide beach is also littered with rock debris and flats. The flats link with beach **WA 2046** a 150 m long pocket of high tide sand bordered and backed by 10 m high bluffs capped by white Holocene dunes to the east and red Pleistocene dune to the west. It is fronted by 100 m wide high tide beach grading into low tide rock flats. **Eco Beach Resort** is located 500 m east of the beach (Fig. 4.429).

Figure 4.429 Eco Beach Resort is nestled in amongst the vegetation behind beach WA 2046.

WA 2047-2048 **ECO BEACH**

No.	Beach	Rating HT	LT	Type	Length
WA 2047	Eco Beach	2	1	UD	3.7 km
WA 2047	Yardoogarra	2	1	R+sand-tidal flats	3.7 km
Spring & neap tidal range = 8.3 & 1.9 km					

Eco Beach (**WA 2047**) is the name a sandy ultradissipative beach that sweeps from the lee of the rocks of Cape Villaret to the east, then northeast for 3.7 km to a sandy salient in lee of intertidal rocks. The once remote beach was developed as a wilderness retreat in the 1990's only to be devastated by Cyclone Rosita in 2000. It was still not reopened in 2005. The low impact beachfront retreat is nestled amongst the red Pleistocene dunes that back the southern corner of the beach, with a nature trail leading out to Cape Villaret and William Dampier Lookout.

The beach commences in lee of the rock immediately east of beach WA 2046 and trends east for 1 km before swinging to the northeast. It is backed by a white Holocene foredune then a 1 km wide zone of vegetated red Pleistocene transgressive dunes that extends east from the cape area and the beach. The beach has a moderate gradient high tide beach grading into a lower gradient intertidal zone. There are a few rocks in the intertidal just south of the retreat, with sand all the way to the salient, where rocks begin to dominate the intertidal.

Beach **WA 2048** continues east of the salient for 3 km where it is backed by the low coastal plain, before turning and trending northeast for 6 km, as a narrow spit grading into a dynamic recurved spit along its distal end. Yardoogarra Outstation is located 1 km inland, at the base of the spit, with a vehicle track out to the shore. Sand moving along the northern tip of the spit varies the beach length over period of years to decades. This section is backed by a tidal creek which has been deflected 3 km to the north and which periodically breaks out south of the tip, thereby shortening the beach. The beach is initially fronted by a mixture of 300 m wide sand and rock flats, that grade into 500 m wide tidal flats towards the creek mouth. The spit area is backed by a floodplain in the south grading into salt flats and then mangroves towards the creek.

WA 2049-2051 **SANDY POINT**

No.	Beach	Rating HT	LT	Type	Length
WA 2049	Sandy Pt	2	1	R+tidal flats	4 km
WA 2050	Bush Pt	1	1	R+tidal flats	300 m
WA 2051	Bush Pt	1	1	R+tidal flats	200 m
Spring & neap tidal range = 8.3 & 1.9 m					

Sandy Point (**WA 2049**) forms the distal end of the northerly sand transport from Cape Villaret. It consists of a 4 km long spiraling convex recurved spit (Fig. 4.430). The 0.5 to 1 km wide spit is composed of multiple recurves, which splay towards the curving northern end. The northward growth of the spit has deflected the backing Yardoogarra Creek 4 km to the north, with a 3 km wide zone of mangroves lining the creek. The mangroves link with the southern creek mouth and prevent vehicle access to the shore. The spit beach consists of a narrow high tide beach fronted by low gradient tidal flats, which extend 5-6 km offshore.

Figure 4.430 Sandy Point consists of a series of multiple recurved sandy spits, each separated by mangroves.

Bush Point is a low irregular 1 km long beach ridge that forms the northern side of the 1.5 km wide Yardoogarra Creek mouth. The point is fronted by 6 km wide tidal flats resulting in very low wave conditions at the shore. It consists of a low irregular beach ridge containing two beaches (WA 2050 & 2051) separated by a central 300 m long mangroves section. Beach **WA 2050** faces west across the creek mouth and consists of a convex-shaped 300 m long beach ridge which grades into an easterly trending recurved spit into the creek mouth. The deep tidal creek is located 300 m south of its eastern end. Beach **WA 2051** forms the northern end of the same ridge and consists of a 200 m long northerly-trending beach ridge-spit. Tidal flats extend 6 km west of the shoreline, with mangroves increasingly dominating the shoreline to the north. The beach ridge is the outermost of a 5 km wide series of six beach ridge-cheniers and wide intervening salt flats that are spread across the prograding coastal Racecourse Plain. There is no vehicle access to the shoreline.

ROEBUCK BAY

Roebuck Bay forms the boundary between the Pilbara coast to the south and the Kimberley coast to the north. The 20 km wide bay is bordered by Sandy and Bush points in the south and the prominent Entrance Point at Broome in the north. In between the shoreline curves for 70 km, extending 18 km to the east. The southern and eastern bay shore is dominated by 2 km wide mud flats, backed by a 500 m wide fringe of mangroves then the very low gradient Roebuck Plain (Fig. 4.431) which extends in places 25 km inland. The mangroves terminate in the northeast corner of the bay at Crab Creek, the beginning of the Kimberley beaches (K-1) and the 10 194 km long Kimberley coast. The 1360 beaches of the Kimberley coast are described in the companion book in this series *Beaches of the Northern Australian Coast: The Kimberley, Northern Territory and Cape York.*

Figure 4.431 The Roebuck Bay shoreline consists of wide intertidal mud flats, a distinct mangroves fringe, and the very wide salt and supratidal flats of the plain.

GLOSSARY

bar (sand bar) - an area of relatively shallow sand upon which waves break. It may be attached to or detached from the beach, and may be parallel (longshore bar) or perpendicular (transverse bar) to the beach.

barrier - a long term (1000s of years) shore-parallel accumulation of wave, tide and wind deposited sand, that includes the beach and backing sand dunes. It may be 100's to 1000's of metres wide and backed by a lagoon or estuary. The beach is the seaward boundary of all barriers.

beach - a wave deposited accumulation of sediment (sand, cobbles or boulders) lying between modal wave base and the upper limit of wave swash.

beach face - the seaward dipping portion of the beach over which the wave swash and backwash operate.

beach type - refers to the type of beach that occurs under wave dominated (6 types), tide-modified (3 types) and tide-dominated (3 types) beach systems. Each possesses a characteristic combination of hydrodynamic processes and morphological character, as discussed in chapter 2.

beach types
wave dominated (abbreviations, see Figures 2.4 & 2.5)
R - reflective
LTT - low tide terrace
TBR - transverse bar and rip
RBB - rhythmic bar and beach
LBT - longshore bar and trough
D - dissipative
tide-modified (abbreviations, see Figure 2.13)
R+LT – reflective + low tide terrace
R+BR – R + low tide bar and rip
UD – ultradissipative
tide-dominated (abbreviations, see Figure 2.16)
R+RSF – beach + ridged sand flats
R+SF – beach + sand flats
R+TSF –beach + tidal sand flats
R+TMF – beach +tidal mud flat

beach hazards - elements of the beach environment that expose the public to danger or harm. Specifically: water depth, breaking waves, variable surf zone topography, and surf zone currents, particularly rip currents. Also include local hazards such as rocks, reefs and inlets.

beach hazard rating - the scaling of a beach according to the hazards associated with its beach type as well as any local hazards.

beachrock – see calcarenite

berm - the nearly horizontal portion of the beach, deposited by wave action, lying immediately landward of the beach face. The rear of the berm marks the limit of spring high tide wave action.

blowout - a section of dune that has been destabilised and is now moving inland. Caused by strong onshore winds breaching and deflating the dune.

calcarenite – cemented calcareous sand grains. May form in coastal sand dunes due to pedogenesis, the formation of calcrete soils, commonly called dune calcarenite or dunerock. In beaches it forms from precipitation of calcium carbonate in the intertidal zone and is known as beach calcarenite or beachrock.

cusp - a regular undulation in the high tide swash zone (upper beach face), usually occurring in series with spacing of 10 to 40 m. Produced during beach accretion by the interaction of swash and sub-harmonic edge waves.

dunerock – see calcarenite

foredune - the first sand dune behind the beach. In Queensland it is usually vegetated by spinifex grass and ipomoea, then casuarina thickets.

gutter - a deeper part of the surf zone, close and parallel to shore. It may also be a rip feeder channel.

hole - a localised, deeper part of the surf zone, usually close to shore. It may also be part of a rip channel.

Holocene - the geological time period (or epoch) beginning 10 000 years ago (at the end of the last Glacial or Ice Age period) and extending to the present.

lifeguard - in Australia this refers to a professional person charged with maintaining public safety on the beaches and surf area that they patrol.

lifesaver - an Australian term referring to an active volunteer member of Surf Life Saving Australia, who patrol the beach to maintain public safety in the surf.

megacusp - a longshore undulation in the shoreline and swash zone, with regular spacing between 100 and 500 m, which match the adjacent rips and bars. Produced by wave scouring in lee of rips (megacusp or rip embayment) and shoreline accretion in lee of bars (megacusp horn).

megacusp embayment - see megacusp

megacusp horn - see megacusp

megarip – large scale topographically-controlled rip. Usually requires waves greater than 2-3 m to form.

parabolic dune - a blowout that has extended beyond the foredune and has a U shape when viewed from above; **longwalled** parabolic dunes, have long trailing arms, from a few hundred metres to kilometers long.

overwash – wave swash washing up and over a beach crest; deposits overwash sands as a landward sloping flat.

Pleistocene - the earlier of the two geological epochs comprising the Quaternary Period. It began 2 million years ago and extends to the beginning of the Holocene Epoch, 10 000 years ago.

reef – a submerged or partly submerged object; in Western Australia usually formed of bedrock or calcarenite in the central-south, with coral reefs in the north.

reef flats – intertidal area of coral reef extending seaward of the shoreline.

rip channel - an elongate area of relatively deep water (1 to 3 m), running seaward, either directly or at an angle, and occupied by a rip current.

rip current - a relatively narrow, concentrated seaward return flow of water. It consists of three parts: the *rip feeder current* flowing inside the breakers, usually close to shore; the *rip neck*, where the feeder currents converge and flow seaward through the breakers in a narrow 'rip'; and the *rip head*, where the current widens and slows as a series of vortices seaward of the breakers. see: megarip, topographic rip

rip embayment - see megacusp

rip feeder current - a current flowing along and close to shore, which converges with a feeder current arriving from the other direction, to form the basis of a rip current. The two currents converge in the rip (megacusp) embayment, then pulse seaward as a rip current.

rock flats – intertidal rocky area extending seaward of shoreline.

rock platform - a relatively horizontal area of rock, lying at the base of sea cliffs, usually lying above mean sea level and often awash at high tide and in storms. The platforms are commonly fronted by deep water (2 to 20 m)

rock pool - a wading or swimming pool constructed on a rock platform and containing sea water.

sea (waves) - ocean waves actively forming under the influence of wind. Usually relatively short, steep and variable in shape.

set-up - rise in the water level at the beach face resulting from low frequency accumulations of water in the inner surf zone. Seaward return flow results in a *set-down.* Frequency ranges from 30 to 200 seconds.

shore platform - as per rock platform.

swash - the broken part of a wave as it runs up the beach face or swash zone. The return flow is called *backwash.*

swell - ocean waves that have travelled outside the area of wave generation (sea). Compared to sea waves, they are lower, longer and more regular.

tidal pool - a naturally occurring hole, depression or channel in a shore platform, that may retain its water during low tide.

topographic rip – a rip current whose flow direction is controlled by a topographic feature (rocks, reef, headland, structure) located in the surf zone. The current is usually deflected seaward by the feature. Also known as headland rips.

transverse dune – a dune form perpendicular (i.e. transverse) to the dominant wind. Usually bare of vegetation; may reach 20-30 m in height, with a wave length of 100-200 m,

trough - an area of deeper water in the surf zone. May be parallel to shore or at an angle.

wave (ocean) - a regular undulation in the ocean surface produced by wind blowing over the surface. While being formed by the wind it is called a *sea* wave; once it leaves the area of formation or the wind stops blowing it is called a *swell* wave.

wave bore - the turbulent, broken part of a wave that advances shoreward across the surf zone. This is the part between the wave breaking and the wave swash and also that part caught by bodysurfers. Also called *whitewater*.

wave refraction - the process by which waves moving in shallow water at an angle to the seabed are changed. The part of the wave crest moving in shallower water moves more slowly than other parts moving in deeper water, causing the wave crest to bend toward the shallower seabed.

wave shoaling - the process by which waves moving into shallow water interact with the seabed causing the waves to refract, slow, shorten and increase in height.

REFERENCES AND ADDITIONAL READING:

Boreen, T, James, N, Wilson, C and Heggie, D, 1993, Surficial cool-water carbonate sediments on the Otway continental margin, southeastern Australia. *Marine Geology*, 112, 35-56.

Butt, & and Russell, P, 2002, Surf Science – an introduction to waves for surfing. University of Hawaii Press, Honolulu, 142 pp. An excellent introduction to the world of waves and surf.

Easton, A K, 1970, *Tides of the Continent of Australia.* Report 57, Horace Lamb Centre, Flinders University, Adelaide, 326 pp.

Jaggard, E, 1979, A Challenge Answered - A History of Surf Lifesaving in Western Australia. Surf Life Saving Association of Western Australia, Perth, 138 pp.

Laughlin, G, 1997, *The Users Guide to the Australian Coast.* Reed New Holland, Sydney, 213 pp. An excellent overview of the Australian coastal climate, winds, waves and weather.

Readers Digest, 1983, Guide to the Australian Coast. Readers Digest, Sydney, 479 pp. Excellent aerial photographs and information on the more popular spots along the Australian coast.

Ross, J (editor), 1995, *Fish Australia.* Viking, Melbourne, 498 pp. An excellent coverage of all South Australian coastal fishing spots.

Semeniuk, V, 1993, The Pilbara coast, a riverine coastal plain in a tropical arid setting, northwestern Australia. *Sedimentary Geology*, 83, 235-256.

Semeniuk, V, Keaneally, K F and Wilson, P G, 1978, *Mangroves of Western Australia.* Western Australian Naturalists Club, Perth, Handbook 12, 92 pp

Short, A D, 1993, *Beaches of the New South Wales Coast.* Sydney University Press, Sydney, 358 pp. The New South Wales version of this book.

Short, A D, 1996, *Beaches of the Victorian Coast and Port Phillip Bay.* Sydney University Press, Sydney, 298 pp. The Victorian version of this book.

Short, A D (editor), 1999, *Beach and Shoreface Morphodynamics.* John Wiley & Sons, Chichester, 379 pp. For those who are interested in the science of the surf.

Short, A D, 2000, *Beaches of the Queensland Coast: Cooktown to Coolangatta.* Sydney University Press, Sydney, 360 pp. The Queensland version of this book.

Short, A D, 2001, *Beaches of the South Australian Coast and Kangaroo Island.* Sydney University Press, Sydney 346 pp. The South Australian version of this book.

Surf Life Saving Australia, 1991, *Surf Survival; The Complete Guide to Ocean Safety.* Surf Life Saving Australia, Sydney, 88 pp. An excellent guide for anyone using the surf zone for swimming or surfing.

Warren, M, 1998, Mark Warren's Atlas of Australian Surfing. Angus & Robertson, Sydney, 232 pp. Covers main Australian surfing spots.

Williamson, J A, Fenner, P J, Burnett, J W and Rifkin, J F, (eds), 1996, *Venomous and Poisonous Marine Animals - a Medical and Biological Handbook,* University of New South Wales Press, Sydney, 504 pp. The definitive book on marine stingers.

Young, I R and Holland, G J, 1996, *Atlas of the Oceans, Wind and Wave Climate.* Elsevier, UK, 241 pp.

BEACH INDEX

Note: Patrolled beaches in **BOLD**

See also:
GENERAL INDEX p. 421
SURF INDEX p. 432

BEACH

GENERAL INDEX

A

B

C

D

SURF INDEX

Also see surfing locations by beach name in BEACH INDEX

BEACHES OF THE AUSTRALIAN COAST

Published by the Sydney University Press for the
Australian Beach Safety and Management Program
a joint project of

Coastal Studies Unit, University of Sydney and Surf Life Saving Australia

by

Andrew D Short
Coastal Studies Unit, University of Sydney

BEACHES OF THE NEW SOUTH WALES COAST
Publication: 1993 **ISBN:** 0-646-15055-3
358 pages, 167 original figures, including 18 photographs; glossary, general index, beach index, surf index.

BEACHES OF THE VICTORIAN COAST & PORT PHILLIP BAY
Publication: 1996 **ISBN:** 0-9586504-0-3
298 pages, 132 original figures, including 41 photographs; glossary, general index, beach index, surf index.

BEACHES OF THE QUEENSLAND COAST: COOKTOWN TO COOLANGATTA
Publication: 2000 **ISBN** 0-9586504-1-1
369 pages, 174 original figures, including 137 photographs, glossary, general index, beach index, surf index.

BEACHES OF THE SOUTH AUSTRALIAN COAST & KANGAROO ISLAND
Publication: 2001 **ISBN** 0-9586504-2-X
346 pages, 286 original figures, including 238 photographs, glossary, general index, beach index, surf index.

BEACHES OF THE WESTERN AUSTRALIAN COAST: EUCLA TO ROEBUCK BAY
Publication: 2005 **ISBN** 0-9586504-3-8
433 pages, 517 original figures, including 408 photographs, glossary, general index, beach index, surf index.

Order online from **Sydney University Press** at

http://www.sup.usyd.edu.au/marine

Forthcoming titles:

BEACHES OF THE TASMANIAN COAST AND ISLANDS (publication 2006) 1-920898-12-3

BEACHES OF NORTHERN AUSTRALIA: THE KIMBERLEY, NORTHERN TERRITORY AND CAPE YORK (publication 2007) 1-920898-16-6

BEACHES OF THE NEW SOUTH WALES COAST (2nd edition, 2007) 1-920898-15-8

SYDNEY UNIVERSITY PRESS